Birkhäuser

Nicolas Lerner
Institut de Mathématiques de Jussieu
Université Pierre et Marie Curie (Paris VI)
Paris, France

ISBN 978-3-0348-0693-0   ISBN 978-3-0348-0694-7 (eBook)
DOI 10.1007/978-3-0348-0694-7
Springer Basel Heidelberg New York Dordrecht London

Library of Congress Control Number: 2014933554

Printed on acid-free paper

Springer Basel is part of Springer Science+Business Media (www.birkhauser-science.com)

*This book is dedicated to the many students
who attended over the years my lectures on
integration theory; by their attention and their
work, they encouraged me to write this book.*

Contents

Preface

This volume is a textbook on Integration Theory, supplemented by 160 exercises provided with detailed answers. There are already many excellent texts on this topic and it is legitimate to ask whether it is worth while to add a new entry in an already long list of books on Measure Theory.

Nevertheless, the author's teaching experience has shown that many of these books were too difficult for a student exposed for the first time to integration theory. We have tried here to keep a rather elementary level, at least in the way of exposing our arguments and proofs, which are hopefully complete, detailed, sometimes at the cost of a lack of concision. Moreover, we hope that the many exercises (with answers) included at the end of each chapter will represent an asset for the present book.

A trend present in the contemporary textbook literature on integration theory is simply to omit the not-so-easy construction of Lebesgue measure. We are strongly opposed to this tendency, and we have made all efforts in our redaction to provide a complete construction of the mathematical objects used in the book, first and foremost for the construction of Lebesgue measure. Our point of view here is not exclusive of some compromises in the reading order which can be used by the reader trying to learn this material: the chapters of this book are of course ordered logically (chapter $n+1$ is using chapters $1, \ldots, n$ and never chapter $n+2, \ldots$), but some "construction" chapters, such as Chapter 2, parts of Chapters 4, 5, could be bypassed at first reading. We expect that a mathematically curious reader will feel the need of a construction after experiencing some of the most efficient (or more computational) parts of the theory and then will go back to these construction chapters.

Last but not least, we hope that this book could serve as a reasonable "entrance gate" to Integration Theory for scientists and mathematicians who are non-experts in measure theory. Another fact of mathematical life, say in the last thirty years, is that it is more and more difficult to learn some mathematics not directly connected with your professional area. Where is it possible for an Analyst to learn the algebraic properties of Theta functions? Where to find a text on Fourier Analysis accessible to an Algebraic Geometer? Although both questions above have (many) answers, it remains difficult to find a way to enter a domain with which you are not a priori conversant. It is the author's opinion that accessibility is now a rare commodity in the mathematical literature, and we hope that the present book will provide its share of that good.

Integration Theories

The initial goal of integration theory, founded more than two millennia ago[1] is to compute areas and volumes of various objects. A somewhat simplified mathematical version of this question is to consider a function $f : [0,1] \longrightarrow \mathbb{R}_+$ and try to evaluate the area A between the x-axis and the curve $y = f(x)$. The standard notation is

$$A = \int_0^1 f(x)dx.$$

Of course some assumptions should be made on the function f for this area to make sense.

Riemann's integral

Greek mathematicians of the third century B.C. were aware of volumes of spheres, cones, polyhedra, and of many classical geometric objects. Later, in the early eighteenth century, Gottfried Wilhelm LEIBNIZ (1646–1716) introduced the *Infinitesimal Calculus*, whereas Isaac NEWTON (1642–1727) defined the *Calculus of Fluxions*, both inventions (close to each other) closely linked with a notion of integral calculus. However the first systematic attempt to define the integral of a function is due to the German mathematician Bernhard RIEMANN (1826–1866): cutting the source space (here $[0,1]$) into tiny pieces,

$$[0 = x_0, x_1], \ldots [x_k, x_{k+1}], \ldots, [x_{N-1}, x_N = 1], \quad x_j \uparrow,$$

you expect A to be close to

$$S_N = \sum_{0 \leq k < N} (x_{k+1} - x_k)f(m_k), \quad \text{where } m_k \in [x_k, x_{k+1}],$$

since the area A should resemble the sum of the areas of the vertical rectangles with base (x_k, x_{k+1}) and height $f(m_k)$. In fact, assuming for instance f to be a uniform limit of step functions (a step function is a finite linear combination of characteristic functions of intervals), you obtain that S_N has a limit when

$$\sup_{0 \leq k < N} (x_{k+1} - x_k) \text{ goes to zero,}$$

and you define that limit as $\int_0^1 f(x)dx$. It is indeed a simple matter to show directly that this procedure works for a continuous function on $[0,1]$. That theory is quite elementary but has several downsides. The very first one is a terrible instability

---

[1]The Greek scientist Archimedes of Syracuse, who lived in the third century B.C., was able to provide a quadrature of the parabola.

with respect to small perturbations: in particular, if you modify the function f (say f continuous) on a rather small set such as the rational numbers $\mathbb{Q}$, you may ruin the integrability in the above sense. The rational numbers should be considered as "small" since it is a countable set $\{x_n\}_{n\in\mathbb{N}}$ and thus, for any $\epsilon > 0$,

$$\mathbb{Q} \subset \bigcup_{n\in\mathbb{N}} \left(x_n - \frac{\epsilon}{2^{n+2}}, x_n + \frac{\epsilon}{2^{n+2}} \right)$$

and thus the "length" ℓ of $\mathbb{Q}$ is such that for any $\epsilon > 0$,

$$\ell \leq \epsilon \sum_{n\in\mathbb{N}} 2^{-n-1} = \epsilon \Longrightarrow \ell = 0.$$

In particular, it is easy to show that the integral of $\mathbf{1}_{\mathbb{Q}\cap[0,1]}$ (a small perturbation of 0) cannot be defined by the procedure sketched above. Although the latter function may appear to be quite pathological, the effects of this instability are disturbing. Other difficulties occur with the Riemann integral, with complications in integrating unbounded functions and also in developing a comprehensive theory of multidimensional integrals.

The Lebesgue perspective

A key point in Lebesgue theory of integration (see, e.g., [8]) is that to calculate the integral of $f : X \longrightarrow \mathbb{R}$, one should not cut into small pieces the source space X (for instance in small subintervals if X is an interval of $\mathbb{R}$) but the *target space* should be cut into pieces depending on the function f itself. It is easy to illustrate this in the (very) simple case where

$$f : X = \{x_1, \ldots, x_m\} \longrightarrow \{y_1, \ldots, y_n\} = Y \subset \mathbb{R}.$$

We can evaluate $\sum_{x_j \in X} f(x_j)$ by sorting out the values taken by f:

$$\sum_{x_j \in X} f(x_j) = \sum_{y_k \in Y} y_k \, \mathrm{card}(\{x \in X, f(x) = y_k\}).$$

Also, playing around freely with the notation, say for f non-negative on $\mathbb{R}$, $H = \mathbf{1}_{\mathbb{R}_+}$,

$$\int_{\mathbb{R}} f(x)dx = \iint H(f(x) - y)H(y)dydx = \int \left(\int H(f(x) - y)dx \right) H(y)dy$$

$$= \int H(y) \, \mathtt{measure}(\{x \in \mathbb{R}, f(x) > y\})dy.$$

If we can "measure" the sets $\{x \in \mathbb{R}, f(x) > y\}$, it is thus quite natural to take as a definition for the integral of f the last expression. Note that this expression is very simple if f is taking a finite number of values $y_1, \ldots, y_N$: we have in that case

$$\int f(x)dx = \sum_{1 \leq k \leq N} y_k \, \mathtt{measure}(\{x \in \mathbb{R}, f(x) = y_k\}).$$

The set $\{x \in \mathbb{R}, f(x) = y_k\}$ could be quite complicated and we shall see that many functions could be well approximated by *simple functions*, i.e., finite linear combinations of characteristic functions. To overcome the difficulties linked to the integration of unbounded functions, we may consider $f(x) = \frac{1}{2}x^{-1/2}\mathbf{1}_{(0,1)}(x)$ (integral 1); we get according to the previous computation,

$$
\int_0^1 \frac{1}{2\sqrt{x}}dx = \int_0^{+\infty} \text{measure}\left(\left\{x \in (0,1), \frac{1}{2\sqrt{x}} > y\right\}\right)dy
$$

$$
= \int_0^{+\infty} \min\left(1, \frac{1}{4y^2}\right)dy = \int_0^{1/2} dy + \int_{1/2}^{+\infty} \frac{1}{4y^2}dy = \frac{1}{2} + \frac{1}{4\frac{1}{2}} = 1,
$$

and many other examples involving unbounded functions can be dealt with. If we go back to our stability problem, we may consider the function $q = \mathbf{1}_{\mathbb{Q}}, f : \mathbb{R} \to \mathbb{R}_+$, then the integral of f is equal to the integral of $f + q$:

$$
\int_{\mathbb{R}} (f + q)(x)dx = \int_0^{+\infty} \text{measure}(\{x \in \mathbb{R}, f(x) + q(x) > y\})dy
$$

$$
= \int_0^{+\infty} \text{measure}\left(\{x \in \mathbb{R}, f(x) > y\}\right)dy = \int f(x)dx,
$$

since the function q vanishes except on a set with measure 0. Since the reader may feel skeptical about the perturbation by this function q, let us give a finite version of it, illustrating the instability occurring with the Riemann approach, an instability which is not present with Lebesgue's simple method outlined above. We consider the interval $[0, 1]$ and for some large integer N the function

$$
f(x) = \sum_{0 \le k < N} \mathbf{1}_{[\frac{k}{N}, \frac{k+2-N}{N}]}(x).
$$

Applying Riemann's method, using the sequence $x_k = k/N, 0 \le k < N$, we deal with

$$
S = \sum_{0 \le k < N} \left(\frac{k+1}{N} - \frac{k}{N}\right)f(m_k), \quad m_k \in \left[\frac{k}{N}, \frac{k+1}{N}\right].
$$

We may for instance choose $m_k = x_k = k/N$, so that $f(m_k) = 1$ and $S = 1$. On the other hand, Lebesgue's method uses the fact that the function f is taking two values $0, 1$, and the evaluation of the integral by this method gives

$$
I = \text{measure}\{x \in [0, 1], f(x) = 1\} = \sum_{0 \le k < N} 2^{-N}/N = 2^{-N}.
$$

Nonetheless this value turns out to be the exact value of the integral, but also it goes to 0 when N goes to infinity whereas S is stuck at 1, very far from the true value I. It is of course a scaling problem, since choosing the sequence (x_k) such that $\sup_k |x_{k+1} - x_k| \le 2^{-N}$ will provide a more accurate value for S. Nevertheless this scaling phenomenon is a good illustration of the fact that a perturbation f with a small integral but with a large sup norm could trigger huge variations of S, although the Lebesgue calculation remains stable.

There is much more to say in favour of Lebesgue's point of view and in particular the fact that we can define a Banach space (complete normed vector space) of integrable functions, the space $L^1(\mathbb{R}^n)$, and also spaces $L^p(\mathbb{R}^n)$, $1 \le p \le +\infty$, other Banach spaces (L^2 is a Hilbert space), is of considerable interest and well tuned to the developments of functional analysis. Moreover, Lebesgue's theory provides its user with a remarkably simple convergence theorem, the so-called Lebesgue's dominated convergence theorem. The problem at hand is to decide whether $\int f_n(x)dx$ is converging with limit $\int f(x)dx$ when we have already a (weak) pointwise information, i.e., $\lim_n f_n(x) = f(x)$ for all x. A precise statement can be found in Chapter 1 (Theorem 1.6.8), but let just say here that a domination condition

$$\sup_n |f_n(x)| = g(x) \quad \text{is such that} \quad \int |g(x)|dx < +\infty,$$

will ensure nonetheless the sought convergence of integrals but also convergence of the sequence of functions $(f_n)_{n \in \mathbb{N}}$ in the functional space L^1.

Is there a downside to Lebesgue's integration theory[2]? Mathematically speaking, the answer is no, and that theory has been widely used, polished and sometimes generalized to many different situations. However, it is true that Lebesgue's theory of integration is not elementary and that its actual construction requires a significant effort. On the other hand the *Instruction Manual* for Lebesgue Integration is indeed quite simple and one should encourage the reader to enjoy the simplicity and efficiency of that theory before going back to the more austere construction aspects.

We may draw a comparison with the construction and use of the real numbers: the real line $\mathbb{R}$ is widely used in Calculus and elsewhere as a basic mathematical object, but few students actually go through a construction of $\mathbb{R}$. In fact, $\mathbb{R}$ is also a very complicated object, as could be seen through the many examples of the present book (cardinality questions, non-measurable subsets, Cantor ternary set, Cantor sets with positive measure, category and measure,...), but nobody (?) is suggesting that getting some familiarity with the real line should not be a part of a standard mathematical curriculum.

---

[2]An utterly pragmatic point of view was defended by Richard W. HAMMING (1915–1998), a computer scientist and mathematician: " *Does anyone believe that the difference between the Lebesgue and Riemann integrals can have physical significance, and that whether say, an airplane would or would not fly could depend on this difference? If such were claimed, I should not care to fly in that plane.*" In N. Rose Mathematical Maxims and Minims, Raleigh NC: Rome Press Inc., 1988. That criticism is surprising, since the norms of the functional spaces provided by Lebesgue theory are actually used in numerical approximations and their stability is expressed by inequalities involving those norms.

Description of the contents of the book

Chapter 1, entitled *General Theory of Integration*, presents the basic framework for integration theory, with the notion of measure space. We obtain rather easily the three classical convergence theorems (Beppo Levi, Fatou, Lebesgue's dominated convergence) and we can define the space of integrable functions $L^1(\mu)$. This abstract presentation of integration is not difficult to follow, but there is obviously a shortage of significant examples of measure spaces at this stage of the exposition.

The main examples are constructed in Chapter 2, *Actual Construction of Measure Spaces*; a first route is following the Riesz–Markov representation theorem via linear forms on continuous compactly supported functions. We present as well the more set-theoretic Carathéodory approach. At the end of this chapter, we introduce the notion of Hausdorff measure. Among the statements in the exercises, one may single out the construction of a non-measurable set, using the Axiom of Choice. The parts dealing with the construction of the Lebesgue measure are quite technical, and while using some earlier version of these notes for teaching a one-semester course, we always postponed the exposition of the details of the construction of Lebesgue measure to the very last week of class, after the students had acquired some familiarity with the scope and means of that integration theory.

Chapter 3 deals with *Spaces of Integrable Functions*. The important convexity inequalities (Jensen, Hölder, Minkowski) are studied and the definition of $L^p(\mu)$ spaces ($1 \leq p \leq \infty$) are given along with their main properties, most notably the fact that they are Banach spaces. We study as well integrals depending on a parameter, with continuity and differentiability properties; this part is of course related to many practical examples such as the Gamma function, Zeta function and many integrals or series depending on a parameter. The Riemann–Lebesgue Lemma, Egoroff's and Lusin's theorems are proven. The last section provides a survey of various notions of convergence encountered in the text. Some exercises are related to various explicit computations, others to more abstract questions, such as examples of non-separable spaces.

The fourth chapter, *Integration on a Product Space*, constructs integrals on product spaces, and contains statements and proofs of Tonelli's and Fubini's theorems. Some exercises are purely computational (e.g., computation of the volumes of the Euclidean balls in $\mathbb{R}^n$), others are more abstract, for instance with the study of the notion of monotone class.

Chapter 5 is entitled *Diffeomorphisms of Open Subsets of $\mathbb{R}^n$ and Integration*. We deal there with the change-of-variable formula and give some classical examples, such as polar coordinates. We also define integration on a smooth hypersurface of the Euclidean $\mathbb{R}^n$, using implicitly a distribution approach to construction of the simple layer. The last part of this chapter goes back to the notion of Hausdorff measures introduced in Chapter 2 and to the construction of Cantor sets. We give many details on construction of the classical Cantor ternary set, along with computation of its Hausdorff dimension and with study of the Cantor function (a.k.a. as "devil's staircase"). We study also Cantor sets with positive measure and compare the (unrelated) notions of category and measure. We calculate the cardinalities of the Borel and Lebesgue σ-algebras on $\mathbb{R}^n$: this requires some effort related to the introduction of cardinals and ordinals and we have devoted a lengthy appendix to these topics.

Convolution is the topic of Chapter 6, in which the Banach algebra $L^1(\mathbb{R}^n)$ is studied, as well as the classical Young's inequality. Weak L^p spaces are introduced and we give a proof of the Hardy–Littlewood–Sobolev inequality, following an explicit argument due to E. Lieb and M. Loss [43]. In the exercises, the reader will find various computations related to the heat equation and to the Laplace operator. We give also a study of Lorentz spaces and of the notion of decreasing rearrangement.

Chapter 7 is entitled *Complex Measures* and is essentially devoted to the proof of the classical Radon–Nikodym theorem, as well as to the expression of the dual of $L^p(\mu)$ for $1 \le p < \infty$. We give several examples with the spaces c_0, ℓ^p, and study various possible behaviors of weakly convergent sequences. The decomposition in absolutely continuous, pure point, singular continuous parts for a Borel measure on the real line is studied as well as the notion of polar decomposition of a vector-valued measure.

Basic Harmonic Analysis on $\mathbb{R}^n$ is the topic of Chapter 8. Here we have chosen to follow Laurent Schwartz' presentation of Fourier transformation, first via the space $\mathscr{S}(\mathbb{R}^n)$ of rapidly decreasing functions, for which it is truly easy to prove the Fourier inversion formula. Introducing the space $\mathscr{S}'(\mathbb{R}^n)$ of tempered distributions as the topological dual space of the Fréchet space $\mathscr{S}(\mathbb{R}^n)$ was impossible to resist, since the Fourier inversion formula follows almost immediately on the huge space $\mathscr{S}'(\mathbb{R}^n)$, by a trivial abstract nonsense argument. We took advantage of the fact that tempered distributions are much easier to understand than general distributions, essentially because the space $\mathscr{S}(\mathbb{R}^n)$ is simply a Fréchet space, whose topology is defined by a countable family of semi-norms. Understanding general distributions is complicated by the fact that the space of test functions is not metrizable. Anyhow, we recover easily the standard properties of the Fourier transformation as well as basic properties of periodic distributions. Along the way, we provide a proof of the Poisson summation formula using Gabor's wavelet method (Coherent States Method).

The last chapter is the ninth, *Classical Inequalities*, which begins with Hadamard's three-lines theorem and the Riesz–Thorin interpolation. Although this technique is useful to provide natural generalizations of Young's inequality, it falls short of dealing with natural operators such as the Hilbert transform: for that purpose, we give a proof of the Marcinkiewicz Theorem. We introduce the notion of maximal function, and prove the Lebesgue differentiation theorem. In order to study Sobolev spaces, we start with a classical inequality due to Gagliardo and Nirenberg. It turns out that this inequality is a perfect tool to handle Sobolev embedding theorems. We would have liked to expand that chapter to study Fourier multipliers and Hörmander–Mikhlin theorems as well as more general Sobolev spaces, including the homogeneous ones. The best way to do this would have been to introduce various tools of harmonic analysis, such as Calderón–Zygmund operators and pseudodifferential techniques: this would have been obviously too much and we refer the reader to [5] for these developments.

Let us go through our *Appendix*, essentially intended to reach a reasonable self-containedness for the present book. The first section is concerned with set theory, cardinals, ordinals: these notions are important for the understanding of many problems related to measure theory, and we have chosen a rather lengthy and elementary presentation of this topic. Section 2 deals with various topological questions, including the notion of filter, useful for the Tychonoff theorem. A proof of the Baire theorem is given and some classical consequences are recalled (Banach–Steinhaus, Open Mapping Theorem): these questions are important for the understanding of duality, which is also related to measure theory and L^p spaces. The last three sections of the appendix are concerned with basic formulas and classical computations related to integration. Although it might seem preposterous to provide again this widely available material in such a book, the author would like to point out in the first place that some of these formulas are not so easy to derive. But above all, it seems that the true absurdity would be to teach Lebesgue measure to people while ignoring basic formulas of integral calculus. These elementary computational aspects are here as a gentle reminder that Mathematics is also about computation, and that refined concepts and tools often find their motivations in intricate calculations.

Chapter 1

General Theory of Integration

1.1 Measurable spaces, σ-algebras

Definition 1.1.1. Let X be a set and $\mathcal{M} \subset \mathcal{P}(X)$ be a family of subsets of X. $\mathcal{M}$ is called a σ-algebra on X whenever

$$
\begin{aligned}
&(1) & A \in \mathcal{M} \quad &\Longrightarrow& A^c \in \mathcal{M}, \\
&(2) & (A_n \in \mathcal{M})_{n \in \mathbb{N}} \quad &\Longrightarrow& \cup_{n \in \mathbb{N}} A_n \in \mathcal{M} \\
&(3) & X \in \mathcal{M}. &&
\end{aligned}
$$

We shall say that $(X, \mathcal{M})$ is a measurable space.

Definition 1.1.2. Let $(X_1, \mathcal{M}_1), (X_2, \mathcal{M}_2)$ be two measurable spaces and $f : X_1 \to X_2$. The mapping f is said to be measurable if for all $A_2 \in \mathcal{M}_2$, $f^{-1}(A_2) \in \mathcal{M}_1$. That property will be symbolically denoted by $f^{-1}(\mathcal{M}_2) \subset \mathcal{M}_1$.

Properties (1), (2) in Definition 1.1.1 imply readily that a σ-algebra is stable by countable intersection. Moreover (3) follows from (1), (2) provided $\mathcal{M} \neq \emptyset$.

We call *countable* any set equipotent to a subset of $\mathbb{N}$, i.e., such that there exists an injection from D into $\mathbb{N}$. If D is a non-empty finite set, there exists a bijective mapping from D onto $\{1, \ldots, n\}$ where n is the cardinal of D. If D is infinite (i.e., not finite) countable, then it is equipotent to $\mathbb{N}$: we may in fact consider D as a subset of $\mathbb{N}$. We define

$$
d_1 = \min D, \ d_2 = \min D \backslash \{d_1\}, \ldots, d_{k+1} = \min D \backslash \{d_1, \ldots, d_k\}.
$$

Since D is infinite and $\mathbb{N}$ is well ordered (i.e., every non-empty subset of $\mathbb{N}$ has a smallest element) this definition makes sense for all $k \geq 1$. If $d \in D$, we may find $k \in \mathbb{N}$ such that $d_k \leq d < d_{k+1}$ since the sequence d_k is strictly increasing and we cannot have $d_k < d < d_{k+1}$ (that would contradict the very definition of d_{k+1}), so that we get $d = d_k$ and D is $\{d_k\}_{k \in \mathbb{N}}$, equipotent to $\mathbb{N}$. It is easy to show that $\mathbb{N}^* = \mathbb{N}\backslash\{0\}, 2\mathbb{N}, 2\mathbb{N}+1, \mathbb{Z}, \mathbb{N}\times\mathbb{N}$ are equipotent to $\mathbb{N}$. To get the latter, it is enough to note that

$$(p,q) \in \mathbb{N}\times\mathbb{N} \mapsto 2^p(2q+1) \in \mathbb{N}^*$$

is bijective[1]. Since the set $\mathbb{Q}$ of rational numbers can be injected in $\mathbb{Z}\times\mathbb{Z}$, we get from the preceding remark that $\mathbb{Q}$ is equipotent to $\mathbb{N}$ as well as $\mathbb{Q}^d$ (d integer ≥ 1). We shall see that the set $\mathbb{R}$ of real numbers is *not* countable since it is equipotent to $\mathcal{P}(\mathbb{N})$ (see Exercise 1.9.5). It is easy to show that for any set X, there is no surjection from X onto $\mathcal{P}(X)$ (see Exercise 1.9.2).

Let us give a couple of examples of σ-algebras. Let X be a set; $\{\emptyset, X\}$ is a σ-algebra on X as well as $\mathcal{P}(X)$. Moreover, if $\{A_k\}_{1\leq k\leq n}$ is a partition of X (each A_k is a non-empty subset of X, $A_k \cap A_l = \emptyset$ for $k \neq l$, $X = \cup_{1\leq k\leq n}A_k$), the set

$$\mathcal{M} = \{\cup_{k\in J}A_k\}_{J\subset\{1,\dots,n\}}$$

is a σ-algebra on X. In fact, defining $B(J) = \cup_{k\in J}A_k$, we get $B(J)^c = B(J^c)$, so that the stability by complement is fulfilled (stability by reunion is obvious). Let us note also that card $\mathcal{M}$ is 2^n since there is a bijection from $\mathcal{M}$ onto the subsets of $\{1,\dots,n\}$. Exercise 1.9.3 deals with a countable partition.

We can also note that for $(\mathcal{M}_i)_{i\in I}$ a family of σ-algebras on X, $\mathcal{M} = \cap_{i\in I}\mathcal{M}_i$ is also a σ-algebra on X: let $(A_n)_{n\in\mathbb{N}}$ be a sequence of $\mathcal{M}$, thus of $\mathcal{M}_i$ for each $i \in I$, then $\cup_{n\in\mathbb{N}}A_n$ belongs to $\mathcal{M}_i$ for each $i \in I$, thus to $\mathcal{M}$. Property (1) about the complement can be checked similarly (and $X \in \mathcal{M}$ since $X \in \mathcal{M}_i$ for all $i \in I$). Since a σ-algebra on X is included in $\mathcal{P}(X)$, we can give the following definition.

Definition 1.1.3. Let X be a set and $\mathcal{F} \subset \mathcal{P}(X)$. We define

$$\mathscr{M}(\mathcal{F}) = \bigcap_{\substack{\mathcal{M}\ \sigma\text{-algebra on } X \\ \mathcal{M}\supset\mathcal{F}}} \mathcal{M}.$$

We shall say that $\mathscr{M}(\mathcal{F})$ is the σ-algebra *generated* by $\mathcal{F}$ (or the smallest σ-algebra on X containing $\mathcal{F}$).

Lemma 1.1.4. *Let $(X_1,\mathcal{M}_1), (X_2,\mathcal{M}_2)$ be measurable spaces with $\mathcal{M}_2 = \mathscr{M}(\mathcal{F})$ and $f : X_1 \to X_2$ be a mapping. For f to be measurable, it is sufficient (and also necessary) that $f^{-1}(\mathcal{F}) \subset \mathcal{M}_1$, i.e., $\forall F \in \mathcal{F},\ f^{-1}(F) \in \mathcal{M}_1$.*

Proof. We set $\mathcal{N} = \{B \in \mathcal{M}_2, f^{-1}(B) \in \mathcal{M}_1\}$. This is a σ-algebra on X_2: if $B \in \mathcal{N}$, $f^{-1}(B^c) = \left(f^{-1}(B)\right)^c \in \mathcal{M}_1$. Moreover for a sequence $(B_n)_{n\in\mathbb{N}}$ of $\mathcal{N}$, we have $f^{-1}(\cup_{n\in\mathbb{N}}B_n) = \cup_{n\in\mathbb{N}}f^{-1}(B_n) \in \mathcal{M}_1$. Finally, $X_2 \in \mathcal{N}$, since $f^{-1}(X_2) = X_1 \in \mathcal{M}_1$. As a result, $\mathcal{N}$ is a σ-algebra containing $\mathcal{F}$ if $f^{-1}(\mathcal{F}) \subset \mathcal{M}_1$. This implies

$$\mathcal{M}_2 = \mathscr{M}(\mathcal{F}) \subset \mathcal{N} \subset \mathcal{M}_2 \Longrightarrow \mathcal{M}_2 = \mathcal{N}. \qquad \square$$

---

[1]see Exercise 1.9.1.

Lemma 1.1.5. *Let $(X, \mathcal{M})$ be a measurable space and $f : X \to Y$ be a mapping. Then the set $\mathcal{N} = \{B \subset Y, f^{-1}(B) \in \mathcal{M}\}$ is a σ-algebra on Y. It is the largest σ-algebra on Y making f measurable.*

Proof. For $B, B_n \in \mathcal{N}$, we have $f^{-1}(B^c) = (f^{-1}(B))^c \in \mathcal{M}$ and

$$f^{-1}(\cup_n B_n) = \cup_n f^{-1}(B_n) \in \mathcal{M}.$$

Since $Y \in \mathcal{N}$, we get the first result. $\mathcal{N}$ is the largest σ-algebra on Y such that f is measurable: if $(Y, \widetilde{\mathcal{N}})$ is a measurable space such that f is measurable, then for $B \in \widetilde{\mathcal{N}}$, the measurability of f implies $f^{-1}(B) \in \mathcal{M}$, so that $B \in \mathcal{N}$ and thus $\widetilde{\mathcal{N}} \subset \mathcal{N}$. $\qquad\square$

Lemma 1.1.6. *Let $(X, \mathcal{M})$, $(Y, \mathcal{N})$, $(Z, \mathcal{T})$ be measurable spaces and*

$$X \xrightarrow{f} Y \xrightarrow{g} Z$$

be measurable mappings. Then $g \circ f$ is measurable.

Proof. For $C \in \mathcal{T}$, we have $(g \circ f)^{-1}(C) = f^{-1}(g^{-1}(C)) \in \mathcal{M}$ since $g^{-1}(C) \in \mathcal{N}$ (g is measurable) and f measurable. $\qquad\square$

We have used above a simple property of the inverse image:

$$\begin{aligned}
(g \circ f)^{-1}(C) &= \{x \in X, g(f(x)) \in C\} \\
&= \{x \in X, f(x) \in g^{-1}(C)\} = f^{-1}(g^{-1}(C)).
\end{aligned} \tag{1.1.1}$$

Lemma 1.1.7. *Let $(X, \mathcal{M})$ be a measurable space and let $A \subset X$. The set*

$$\mathcal{M}_A = \{M \cap A\}_{M \in \mathcal{M}} \tag{1.1.2}$$

is a σ-algebra on A, the so-called σ-algebra trace on A of $\mathcal{M}$. It is the smallest σ-algebra on A such that the canonical injection ι_A of A into X is measurable. Moreover, if $A \in \mathcal{M}$, $\mathcal{M}_A = \{M \in \mathcal{M}, M \subset A\}$.

Proof. The properties in Definition 1.1.1 are obviously verified in both cases (A in or not in $\mathcal{M}$). We note also that a σ-algebra on A such that ι_A is measurable must contain $\iota_A^{-1}(M) = M \cap A$, for any $M \in \mathcal{M}$, proving the second statement. The last statement is obvious. $\qquad\square$

1.2   Measurable spaces and topological spaces

Definition 1.2.1. Let X be a set. A family $\mathcal{O}$ of subsets of X is a topology on X whenever the following conditions are satisfied,

(1) $O_i \in \mathcal{O}$ for $i \in I$ $\implies$ $\cup_{i \in I} O_i \in \mathcal{O}$,

(2) $O_1, O_2 \in \mathcal{O}$ $\implies$ $O_1 \cap O_2 \in \mathcal{O}$,

(3) $\emptyset, \, X \in \mathcal{O}$.

In other words, $\mathcal{O}$ is stable by union and by finite intersection[2]. We shall say that $(X, \mathcal{O})$ is a topological space.

Let $(X, \mathcal{O})$ be a topological space. A set $F \subset X$ is said to be *closed* whenever F^c is open. Of course, the intersection of a family of closed sets is closed as well as a finite union of closed sets. The *interior* of a set $A \subset X$ is defined as the union of the open sets included in A: the interior of A is open. The *closure* of a set $A \subset X$ is defined as the intersection of the closed sets containing A: the closure of A is closed. Denoting by $\bar{A}$ the closure of A and by $\mathring{A}$ its interior, we have

$$(\mathring{A})^c = \left[\bigcup_{\Omega \text{ open } \subset A} \Omega\right]^c = \bigcap_{\Omega \text{ open } \subset A} \Omega^c = \bigcap_{F \text{ closed } \supset A^c} F = \overline{A^c}, \qquad (1.2.1)$$

so that, defining the *boundary* of A as $\partial A = \bar{A} \backslash \mathring{A}$, we have

$$\partial A = \bar{A} \cap (\mathring{A})^c = \bar{A} \cap \overline{A^c} \qquad \text{(in particular, a closed set).} \qquad (1.2.2)$$

It is also easy to verify from the very definitions that

$$\text{interior}(A \cap B) = \mathring{A} \cap \mathring{B}, \qquad \text{closure}(A \cup B) = \bar{A} \cup \bar{B}. \qquad (1.2.3)$$

In fact $\mathring{A} \cap \mathring{B}$ is open included in $A \cap B$, thus included in the interior of $A \cap B$. Conversely the interior of $A \cap B$ is open included both in A and B so both in $\mathring{A}$ and $\mathring{B}$ and we get the first equality. To obtain the second one, we use the first and (1.2.1) with

$$\bigl(\text{closure}(A \cup B)\bigr)^c = \text{interior}(A^c \cap B^c) = \text{interior}(A^c) \cap \text{interior}(B^c) = (\bar{A})^c \cap (\bar{B})^c.$$

The following inclusions are satisfied whereas the equalities are not fulfilled in general[3],

$$\text{interior}(A \cup B) \supset \mathring{A} \cup \mathring{B}, \qquad \text{closure}(A \cap B) \subset \bar{A} \cap \bar{B}.$$

Let V be a subset of a topological space X and $x \in X$. We shall say that V is a neighborhood of x if $x \in \mathring{V}$, i.e., if V contains an open set containing x. The set of neighborhoods of a given point x will be denoted by $\mathcal{V}_x$. We can note that for $x \in X$,

$$V \subset W, \; V \in \mathcal{V}_x \Longrightarrow W \in \mathcal{V}_x, \qquad (1.2.4)$$

$$V_j \in \mathcal{V}_x, j = 1, 2 \Longrightarrow V_1 \cap V_2 \in \mathcal{V}_x, \qquad (1.2.5)$$

$$\emptyset \notin \mathcal{V}_x, \quad X \in \mathcal{V}_x. \qquad (1.2.6)$$

---

[2] We may note that stability by union implies for $I = \emptyset$ that $\emptyset \in \mathcal{O}$. Moreover stability by finite intersection implies for a set of empty indices that $X \in \mathcal{O}$. Condition (3) is somehow a consequence of (1) and (2).

[3] Taking in $\mathbb{C}$ the intersection of half-spaces $\pm \operatorname{Re} z > 0$, we find a counterexample to the second equality with $A \cap B = \emptyset$, $\bar{A} \cap \bar{B} = i\mathbb{R}$. To violate the first it is enough to use $\pm \operatorname{Re} z \geq 0$ with $A \cup B = \mathbb{C}$, $\mathring{A} \cup \mathring{B} = \{z, \operatorname{Re} z \neq 0\}$.

These properties define *filters*, a notion studied more extensively in Section 10.2 of our Appendix. *Metric spaces* are very important examples of topological spaces: a metric space is a set X equipped with a distance function d, i.e., $d : X \times X \to \mathbb{R}_+$ such that

$$d(x,y) = 0 \iff x = y \quad \text{(separation)}, \tag{1.2.7}$$
$$d(x,y) = d(y,x) \quad \text{(symmetry)}, \tag{1.2.8}$$
$$d(x,z) \le d(x,y) + d(y,z) \quad \text{(triangle inequality)}. \tag{1.2.9}$$

We define the topology $\mathcal{O}_d$ associated to the metric d as the family of sets which are unions of "open balls"

$$B(x,r) = \{y \in X, d(y,x) < r\} \quad (x \in X, r \ge 0 \text{ given}). \tag{1.2.10}$$

Stability by union follows from the definition and to show the stability by finite intersection, it is enough to note that for $x \in B(x_1,r_1) \cap B(x_2,r_2)$ we have $B(x,r) \subset B(x_1,r_1) \cap B(x_2,r_2)$ with $r = \min(r_1 - d(x,x_1), r_2 - d(x,x_2))$ since

$$d(y,x) < r \implies \begin{cases} d(y,x_1) \le d(y,x) + d(x,x_1) < r + d(x,x_1) \le r_1, \\ d(y,x_2) \le d(y,x) + d(x,x_2) < r + d(x,x_2) \le r_2. \end{cases}$$

As a result, a finite intersection of open balls is a union of open balls, implying that $\mathcal{O}_d$ is a topology. For $x \in X, r \ge 0$, the "closed ball" $B_c(x,r)$ is defined as

$$B_c(x,r) = \{y \in X, d(y,x) \le r\}, \tag{1.2.11}$$

and we note that $B(x,r) \subset \overline{B(x,r)} \subset B_c(x,r)$ since $B_c(x,r)$ is closed[4] and contains $B(x,r)$.

$\mathbb{R}^d$ is a metric space for the topology defined by the Euclidean distance. More generally, a vector space E on $\mathbb{C}$ or $\mathbb{R}$ equipped with a *norm*, i.e., a mapping

$$N : E \longrightarrow \mathbb{R}_+, \text{ so that } \begin{cases} N(x) = 0 \iff x = 0, \\ N(\alpha x) = |\alpha| N(x), \text{ for } \alpha \in \mathbb{C}, x \in E, \\ N(x+y) \le N(x) + N(y), \end{cases} \tag{1.2.12}$$

is a metric space for the distance $N(x - y)$. (E, N) is called a normed vector space. For instance, we may consider the space $C^0([0,1], \mathbb{R})$ of real-valued functions defined on $[0,1]$ equipped with the norm

$$\|f\|_\infty = \sup_{x \in [0,1]} |f(x)|.$$

---

[4] $\left(B_c(x,r)\right)^c = \cup_{y, d(x,y)>r} B(y, d(x,y) - r)$ since if $d(z,y) < d(x,y) - r$ and $d(x,y) > r$, we get

$$d(z,y) + r < d(x,y) \le d(x,z) + d(z,y) \implies r < d(x,z).$$

This implies that $(B_c(x,r))^c$ contains the above union and the inclusion is obvious.

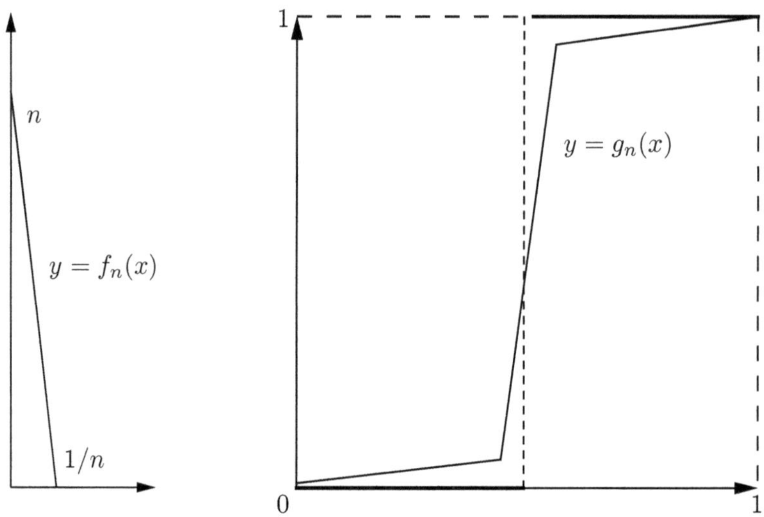

Figure 1.1: SEQUENCES f_n AND g_n.

Let us recall that on $\mathbb{R}^d$, all the norms are equivalent (see Exercise 1.9.8). The normed vector space $C^0([0,1], \mathbb{R})$ equipped with the norm $\|\cdot\|_\infty$ defined above is *complete*[5]. A complete normed vector space is called a Banach space.

We may notice that all the norms on $C^0([0,1], \mathbb{R})$ are *not* equivalent (see Ex. 1.9.8). We may consider the norm (Axioms (1.2.12) are easy to check)

$$\|f\|_1 = \int_0^1 |f(t)|\, dt.$$

The sequence f_n is bounded for the norm $\|\cdot\|_1$ and unbounded for $\|\cdot\|_\infty$. On the other hand the sequence of continuous functions g_n is a Cauchy sequence for $\|\cdot\|_1$ and converges for $\|\cdot\|_1$ towards the discontinuous function $\mathbf{1}_{[1/2,1]}$ (see Ex. 1.9.8). Of course, a topology fails in general to be stable by complement: on the Euclidean $\mathbb{R}$, the complement of the open set $]0, +\infty[$ is $]-\infty, 0]$ which is not open since it does not contain an open ball containing 0. In fact a topological space is said to be *connected* (intuitively made of a single piece) whenever the only sets which are both open and closed are the whole space and the empty set (see Appendix 10.2 on connectedness of topological spaces).

Lemma 1.2.2. *Let $(X, \mathcal{O})$ be a topological space and let A be a subset of X. The set*

$$\mathcal{O}_A = \{\Omega \cap A\}_{\Omega \in \mathcal{O}} \tag{1.2.13}$$

is a topology on A, the so-called induced topology on A by the topology of X, or the subspace topology. It is the smallest (coarsest, weakest) topology on A such that

---

[5]A normed vector space is said to be complete whenever all Cauchy sequences are convergent.

the canonical injection $\iota_A : A \to X$ is continuous. The closed sets of A for that topology are $C_A = \{F \cap A\}_{F\,closed\ in\ X}$, i.e., the "traces" of closed sets of X on A.

Proof. The properties of Definition 1.2.1 are obviously verified. The canonical injection is indeed continuous since $\iota_A^{-1}(\mathcal{O}) = \mathcal{O}_A$ and if $\widetilde{\mathcal{O}}$ is a topology on A making ι_A continuous, this implies $\iota_A^{-1}(\mathcal{O}) \subset \widetilde{\mathcal{O}}$. Let Φ be a closed set of A for this topology: we have, with complements in X, $\Phi^c \cap A = \Omega \cap A$ for some $\Omega \in \mathcal{O}$: as a consequence, since $\Phi \subset A$, we get $\Phi = A \cap (\Phi \cup A^c) = A \cap (\Omega^c \cup A^c) = A \cap \Omega^c$. $\square$

In a topological space, it is interesting to examine the σ-algebra generated by the topology.

Definition 1.2.3. Let $(X, \mathcal{O})$ be a topological space. The *Borel σ-algebra* on X is the σ-algebra generated by $\mathcal{O}$ (according to Definition 1.1.3).

Although the definition above is clear-cut, it does not give a very precise indication of what a Borel set is (an element of the Borel σ-algebra). For instance, a countable union of closed sets, called an F_σ, is a Borel set (the set $\mathbb{Q}$ of rational numbers, as a countable union of singletons is an F_σ). Its complement is a countable intersection of open sets (called a G_δ set): the set of irrational numbers on the real line is a G_δ set. Some subsets of $\mathbb{R}$ can be at the same time F_σ and G_δ such as $[0, 1]$, a closed set (thus F_σ) and G_δ since

$$[0, 1] = \cap_{n \geq 1} \left] -\frac{1}{n}, 1 + \frac{1}{n} \right[.$$

However $\mathbb{Q}^c$ (a G_δ set, according to the above argument) is not an F_σ. Otherwise, we could find a sequence of closed sets F_n such that $\mathbb{R} \backslash \mathbb{Q} = \cup_n F_n$; since $\mathbb{R} \backslash \mathbb{Q}$ does not contain any interval ($\mathbb{Q}$ is dense in $\mathbb{R}$) the interior of each F_n is empty. Finally, it would be possible to write $\mathbb{R}$ as a countable union of closed sets with empty interiors. The Baire theorem (see Section 10.2 in the Appendix) ensures that, in a complete metric space, a countable union of closed sets with empty interiors has also an empty interior. The previous equality describing $\mathbb{Q}^c$ as an F_σ set is thus absurd.

Lemma 1.2.4. *Let $(X_1, \mathcal{O}_1)$, $(X_2, \mathcal{O}_2)$ be topological spaces and let $f : X_1 \to X_2$ be a mapping. The following properties are equivalent.*

(i) The mapping f is continuous on X_1.
(ii) $f^{-1}(\mathcal{O}_2) \subset \mathcal{O}_1$, i.e., $\forall O_2 \in \mathcal{O}_2$, $f^{-1}(O_2) \in \mathcal{O}_1$.

Proof. Note that the continuity of f at a given point x_1 means that for all V_2 neighborhood of $f(x_1)$, there exists a neighborhood V_1 of x_1 such that $f(V_1) \subset V_2$. Using the notation introduced in Section 10.2, it means

$$\widetilde{f(\mathcal{V}_{x_1})} \supset \mathcal{V}_{f(x_1)}.$$

Since a neighborhood of a point contains an open set containing that point, we may replace in the previous definition the word *neighborhood* by *open neighborhood*. Continuity on the whole X_1 means continuity at each point of X_1. If f is continuous on X_1 and V_2 is an open subset of X_2, for $x_1 \in f^{-1}(V_2)$, there exists $V_1 \ni x_1$ such that $f(V_1) \subset V_2$, implying $V_1 \subset f^{-1}(f(V_1)) \subset f^{-1}(V_2)$. As a result, $f^{-1}(V_2)$ is open since it is a neighborhood of all its points. Conversely, assuming $f^{-1}(\mathcal{O}_2) \subset \mathcal{O}_1$, for $x_1 \in X_1$, $x_2 = f(x_1)$ and V_2 an open neighborhood of x_2, the set $V_1 = f^{-1}(V_2)$, is an open subset of X_1 containing x_1. We get $f(V_1) = f(f^{-1}(V_2)) \subset V_2$, providing the continuity of f. $\qquad\square$

Proposition 1.2.5. *Let $(X_1, \mathcal{O}_1)$, $(X_2, \mathcal{O}_2)$ be topological spaces and $\mathcal{B}_j, j = 1, 2$, their Borel σ-algebras. If $f : X_1 \to X_2$ is continuous, then it is measurable.*

Proof. Continuity means $f^{-1}(\mathcal{O}_2) \subset \mathcal{O}_1$. Since $\mathcal{B}_2$ is generated by $\mathcal{O}_2$ and $\mathcal{O}_1 \subset \mathcal{B}_1$, Lemma 1.1.4 proves that f is measurable. $\qquad\square$

Note that there exist functions which are continuous at only one point, such as

$$f : \mathbb{R} \to \mathbb{R}, \qquad f(x) = \begin{cases} x & \text{for } x \in \mathbb{Q}, \\ -x & \text{for } x \notin \mathbb{Q}, \end{cases} \qquad \text{continuous only at } 0.$$

One can show (Exercise 1.9.9) that the discontinuity set of a function f from $\mathbb{R}$ to itself is an F_σ, and that for any F_σ set A, there exists a function whose discontinuity set is A. In particular, there is no function from $\mathbb{R}$ into itself whose discontinuity set is $\mathbb{Q}^c$. On the contrary, the following function is continuous at $\mathbb{Q}^c$, discontinuous at $\mathbb{Q}$:

$$f(x) = \begin{cases} 1 & \text{for } x = 0, \\ 1/q & \text{for } x = p/q, p \in \mathbb{Z}^*, q \in \mathbb{N}^*, \text{ irreducible fraction}, \\ 0 & \text{for } x \notin \mathbb{Q}. \end{cases} \qquad (1.2.14)$$

On the other hand, an open subset of $\mathbb{R}$ is an F_σ set, as a countable union of closed intervals: let U be a non-empty open subset of $\mathbb{R}$ and let $x \in U$; there exists $\rho > 0$ so that $]x - \rho, x + \rho[\subset U$ and since $\mathbb{Q}$ is dense in $\mathbb{R}$, this implies the existence of $p, q \in \mathbb{Q}$ so that $x - \rho < p < x < q < x + \rho$, and thus $[p, q] \subset U$. The open set U is thus a union of compact intervals with rational endpoints. Now the mapping

$$[p, q] \mapsto (p, q)$$

is one-to-one from the set $\mathcal{Q}$ of compact intervals with rational endpoints into $\mathbb{Q} \times \mathbb{Q}$ which is equipotent to $\mathbb{N}$. As a result $\mathcal{Q}$ is (infinite) and equipotent to a subset of $\mathbb{N}$, thus equipotent to $\mathbb{N}$, proving the sought result. More generally, we have in any dimension the following result. We shall say that a *compact rectangle* of $\mathbb{R}^d$ is a set $\prod_{1 \le j \le d}[a_j, b_j]$ and that an *open rectangle* of $\mathbb{R}^d$ is a set $\prod_{1 \le j \le d}]a_j, b_j[$. Of course, compact rectangles are compact (even if for one j, $a_j > b_j$, since the empty set is compact) and open rectangles are open (even if for one j, $a_j \ge b_j$, since the empty set is open).

Lemma 1.2.6. *Let $d \geq 1$ be an integer. We define*

$$\mathcal{Q} = \left\{ \prod_{1 \leq j \leq d} [a_j, b_j] \right\}_{\substack{(a_j, b_j) \in \mathbb{Q}^2 \\ a_j < b_j}}.$$

$\mathcal{Q}$ is a countable family of compact rectangles of $\mathbb{R}^d$ such that any open set is a union (necessarily countable) of a subfamily of these compact rectangles.

Proof. First of all $\mathcal{Q}$ is infinite and can be injected into $\mathbb{Q}^{2d}$, and is thus equipotent to $\mathbb{N}$. As any neighborhood of $x = (x_1, \ldots, x_d) \in \mathbb{R}^d$ contains a cube

$$\left\{ y \in \mathbb{R}^d, \max_{1 \leq j \leq d} |y_j - x_j| < \rho \right\}$$

with $\rho > 0$, we may find $p_j, q_j \in \mathbb{Q}$ such that $x_j - \rho < p_j < x_j < q_j < x_j + \rho$. As a result, for any neighborhood U of x, there exists $P_{x,U} \in \mathcal{Q}$ with $x \in P_{x,U} \subset U$. Let Ω be an open subset of $\mathbb{R}^d$: for each $x \in \Omega$, there exists a neighborhood U_x of x, included in Ω. We have thus

$$\Omega = \cup_{x \in \Omega} U_x \supset \cup_{x \in \Omega} P_{x, U_x} \supset \cup_{x \in \Omega} \{x\} = \Omega \implies \Omega = \cup_{x \in \Omega} P_{x, U_x},$$

so that Ω is a union of a subfamily of elements of $\mathcal{Q}$ and since $\mathcal{Q}$ is countable, that union is necessarily countable. $\qquad\square$

Let $\mathcal{B}$ be the Borel σ-algebra of $\mathbb{R}^d$ and $\mathcal{R}_c, \mathcal{R}_o$ be the families of compact rectangles, open rectangles of $\mathbb{R}^d$. We have, following Definition 1.1.3 and the previous discussion

$$\mathcal{O} \subset \mathcal{M}(\mathcal{R}_c) \subset \mathcal{B} \implies \mathcal{B} = \mathcal{M}(\mathcal{R}_c).$$

Moreover, since $[p, q] = \cap_{n \geq 1}]p - 1/n, q + 1/n[$, any compact rectangle is a countable intersection of open rectangles and thus

$$\mathcal{R}_c \subset \mathcal{M}(\mathcal{R}_o) \implies \mathcal{B} = \mathcal{M}(\mathcal{O}) = \mathcal{M}(\mathcal{R}_c) \subset \mathcal{M}(\mathcal{R}_o) \subset \mathcal{B},$$

so that eventually

$$\mathcal{B} = \mathcal{M}(\mathcal{R}_c) = \mathcal{M}(\mathcal{R}_o). \tag{1.2.15}$$

We note that, for $p, q \in \mathbb{R}$,

$$[p, q] = [p, +\infty[\cap] - \infty, q] = [p, +\infty[\cap]q, +\infty[^c = \cap_{n \geq 1}]p - 1/n, +\infty[\cap]q, +\infty[^c,$$

so that the Borel σ-algebra on $\mathbb{R}$ is generated by the intervals $(]a, +\infty[)_{a \in \mathbb{R}}$ and thus also by the intervals $(] - \infty, a])_{a \in \mathbb{R}}$ or (since $]a, +\infty[= \cup_{n \geq 1}[a + 1/n, +\infty[)$ by the intervals $([a, +\infty[)_{a \in \mathbb{R}}$ and thus also by the intervals $(] - \infty, a[)_{a \in \mathbb{R}}$. Using Lemma 1.1.4 to check the measurability of $f : X \to \mathbb{R}$, it suffices to verify the measurability of $f^{-1}(]b, +\infty[)$. For instance if X is a Borel subset of $\mathbb{R}$ and f is monotonic on X, then f is Borel-measurable. In fact if f is increasing, $b \in \mathbb{R}$ with $A = f^{-1}(]b, +\infty[) \neq \emptyset$, we have

$$A = \cup_{a \in A}\big([a, +\infty[\cap X\big), \tag{1.2.16}$$

since if $X \ni x \geq a \in A$, we obtain $f(x) \geq f(a) > b$ and thus $x \in A$ (the other inclusion is trivial). As a result, we get

$$]\inf A, +\infty[\cap X \subset A \subset [\inf A, +\infty[\cap X, \tag{1.2.17}$$

since the second inclusion follows from (1.2.16) and the first is true if $\inf A = -\infty$, also from (1.2.16); if $\inf A > -\infty$, $\inf A \in \mathbb{R}$ since $A \neq \emptyset$ and for all $\epsilon > 0$, we can find $a_\epsilon \in A$ so that,

$$\inf A \leq a_\epsilon < \inf A + \epsilon \quad \Longrightarrow \quad]\inf A, +\infty[\cap X \subset \cup_{\epsilon > 0}[a_\epsilon, +\infty[\cap X \subset A,$$

where the last inclusion follows from (1.2.16). Whatever happens with $\inf A$, belonging or not to A, we find from (1.2.17) that

$$A = [\inf A, +\infty[\cap X \quad \text{or} \quad A =]\inf A, +\infty[\cap X,$$

Borel-measurable in both cases.

Theorem 1.2.7. *Let $(X, \mathcal{M})$, $(Y, \mathcal{N})$ be measurable spaces and $\mathbb{R}^d$ equipped with its Borel σ-algebra. Let $u_1, \ldots, u_d$ be measurable mappings from X in $\mathbb{R}$ and let $\Phi : \mathbb{R}^d \to Y$ be measurable. Then the mapping*

$$\begin{aligned} X &\to Y \\ x &\mapsto \Phi\big(u_1(x), \ldots, u_d(x)\big) \end{aligned}$$

is measurable. In particular, $f : X \to \mathbb{C}$ is measurable if (and only if) $\mathrm{Re}\, f, \mathrm{Im}\, f$ are measurable and then $|f|$ is also measurable. If $f, g : X \to \mathbb{C}$ are measurable, then $f + g, fg$ are also measurable. Moreover, if $A \in \mathcal{M}$, the indicator function of A is measurable.

Proof. From the composition Lemma 1.1.6, it is enough to check the measurability of $x \mapsto V(x) = \big(u_1(x), \ldots, u_d(x)\big)$ from X to $\mathbb{R}^d$. From Lemma 1.2.6 and Lemma 1.1.4, it suffices to verify that the inverse image by V of a compact rectangle of $\mathbb{R}^d$ belongs to $\mathcal{M}$. For that purpose, we note that

$$V^{-1}\left(\prod_{1 \leq j \leq d} [a_j, b_j]\right) = \bigcap_{1 \leq j \leq d} u_j^{-1}([a_j, b_j]) \in \mathcal{M},$$

since the u_j are measurable. The other statements in the theorem follow immediately (the very last assertion is obvious since $\mathbf{1}_A^{-1}(J) \in \{\emptyset, A, A^c, X\}$). $\qquad\square$

The following generalization of the previous theorem can be useful.

Theorem 1.2.8. *Let $(X, \mathcal{M})$, $(Y, \mathcal{N})$ be measurable spaces and let T be a separable metric space equipped with its Borel σ-algebra. Let $u_1, \ldots, u_d$ be measurable mappings from X into T and let $\Phi : T^d \to Y$ be a measurable mapping. Then the mapping*

$$\begin{aligned} X &\to Y \\ x &\mapsto \Phi\big(u_1(x), \ldots, u_d(x)\big) \end{aligned}$$

is measurable.

Proof. According to Lemma 1.1.6, it is enough to check the measurability of $x \mapsto V(x) = \big(u_1(x), \ldots, u_d(x)\big)$ from X in T^d. From Lemma 1.1.4 it suffices to check that the inverse image by V of an open set of T^d belongs to $\mathcal{M}$. Moreover for Ω an open subset of T^d and $x = (x_1, \ldots, x_d) \in \Omega$, there exist $r_1, \ldots, r_d$ positive numbers (that we may suppose rational numbers) so that the product of open balls

$$B(x_1, r_1) \times \cdots \times B(x_d, r_d) \ni x$$

is included in Ω. With $\mathbb{D}$ a countable dense subset of T, we may find $y_1, \ldots, y_d \in \mathbb{D}$ so that $\operatorname{dist}(x_j, y_j) < r_j/2$. Then the ball $B(y_j, r_j/2)$ is such that

$$x_j \in B(y_j, r_j/2) \subset B(x_j, r_j),$$

since $\operatorname{dist}(z, y_j) < r_j/2$ implies $\operatorname{dist}(z, x_j) \leq \operatorname{dist}(z, y_j) + \operatorname{dist}(y_j, x_j) < r_j/2 + r_j/2$ so that $z \in B(x_j, r_j)$. As a result, the open set Ω is a union of products

$$B(y_1, \rho_1) \times \cdots \times B(y_d, \rho_d), \qquad y_j \in \mathbb{D}, \rho_j \in \mathbb{Q}.$$

There is a surjection from $\mathbb{D}^d \times \mathbb{Q}^d$ (which is countable) onto the set $\mathcal{P}$ of these subsets and thus $\mathcal{P}$ is countable. We have

$$V^{-1}\left(\prod_{1 \leq j \leq d} B(y_j, \rho_j) \right) = \bigcap_{1 \leq j \leq d} u_j^{-1}(B(y_j, \rho_j)) \in \mathcal{M},$$

since the u_j are measurable. $\qquad\square$

Lemma 1.2.9. *Let $(X, \mathcal{O})$ be a topological space and $A \in \mathcal{B}_X$, the Borel σ-algebra on X. The Borel σ-algebra $\mathcal{B}_A$ on A is*

$$\mathcal{B}_A = \{M \in \mathcal{B}_X, M \subset A\} = \mathcal{M}(\mathcal{O}_A), \tag{1.2.18}$$

where $\mathcal{O}_A$ is the topology on A, given in Lemma 1.2.2.

Proof. From (1.2.13) and Definition 1.2.3, we have $\mathcal{B}_A = \mathcal{M}(\mathcal{O}_A)$. Since

$$\widetilde{\mathcal{B}} = \{M \in \mathcal{B}_X, M \subset A\}$$

is (obviously) a σ-algebra on A containing $\mathcal{O}_A$, it contains $\mathcal{M}(\mathcal{O}_A)$. Moreover $\widetilde{\mathcal{B}}$ makes the canonical injection ι_A measurable since $\iota_A^{-1}(\mathcal{B}_X) = \widetilde{\mathcal{B}}$. Also $\widetilde{\mathcal{B}}$ is the smallest σ-algebra on A making ι_A measurable since any σ-algebra making ι_A measurable must contain $\iota_A^{-1}(\mathcal{B}_X)$. We note now from Lemma 1.1.4 that the σ-algebra $\mathcal{M}(\mathcal{O}_A)$ on A is such that ι_A is measurable since $\iota_A^{-1}(\mathcal{O}) = \mathcal{O}_A$: as a result, we get that $\mathcal{M}(\mathcal{O}_A)$ contains $\widetilde{\mathcal{B}}$, proving the lemma. $\qquad\square$

Definition 1.2.10. The *extended real line* $\overline{\mathbb{R}}$ is the set obtained by adjoining two non-real distinct elements to the real line; it is the topological space $\mathbb{R}\cup\{-\infty, +\infty\}$, where the topology contains the open subsets of $\mathbb{R}$ and the sets

$$]a, +\infty[\cup\{+\infty\}, \qquad \{-\infty\}\cup]-\infty, a[$$

(denoted respectively by $]a, +\infty]$ and $[-\infty, a[$). The order relation on $\overline{\mathbb{R}}$ makes $-\infty$ the smallest element and $+\infty$ the largest. This order relation is compatible with the topology since the open sets are unions of intervals.

$\overline{\mathbb{R}}$ is easily shown to be homeomorphic to $[-1, 1]$ (i.e., there exists a bi-continuous bijective mapping ψ_0 from $\overline{\mathbb{R}}$ onto $[-1, 1]$), for instance by extending continuously

$$\mathbb{R} \ni x \mapsto \frac{x}{\sqrt{1+x^2}} = \psi_0(x) \in (-1, 1), \quad \psi_0(\pm\infty) = \pm 1,$$
$$(-1, 1) \ni y \mapsto \frac{y}{\sqrt{1-y^2}} = \psi_0^{-1}(y) \in \mathbb{R}, \qquad \psi_0^{-1}(\pm 1) = \pm\infty. \tag{1.2.19}$$

That homeomorphism is compatible with the order relation, i.e., is increasing. We note also that any monotone sequence (x_n) in $\overline{\mathbb{R}}$ converges since $\psi_0(x_n)$ is monotone in $[-1, 1]$ thus converging (since it is either increasing bounded from above or decreasing bounded from below) and since ψ_0^{-1} is continuous, we get the result. Since $\overline{\mathbb{R}}$ is compact, for any $A \subset \overline{\mathbb{R}}$, there exists a

$$\textit{least upper bound, or supremum, } \sup A = \inf\{M \in \overline{\mathbb{R}}, \forall a \in A, a \leq M\}, \tag{1.2.20}$$
$$\textit{greatest lower bound, or infimum, } \inf A = \sup\{m \in \overline{\mathbb{R}}, \forall a \in A, a \geq m\}. \tag{1.2.21}$$

If $A = \emptyset$, following the definition, we get $\sup A = -\infty, \inf A = +\infty$, the only case where the infimum is strictly larger than the supremum.

Definition 1.2.11. Let $(x_n)_{n\in\mathbb{N}}$ be a sequence in $\overline{\mathbb{R}}$. The sequences $(\inf_{k\geq n} x_k)_{n\in\mathbb{N}}$, $(\sup_{k\geq n} x_k)_{n\in\mathbb{N}}$, are monotone (the first is increasing, the next one decreasing). We define

$$\liminf x_n = \lim_{n\to+\infty}\left(\inf_{k\geq n} x_k\right) = \sup_{n\in\mathbb{N}}\left(\inf_{k\geq n} x_k\right),$$
$$\limsup x_n = \lim_{n\to+\infty}\left(\sup_{k\geq n} x_k\right) = \inf_{n\in\mathbb{N}}\left(\sup_{k\geq n} x_k\right).$$

Proposition 1.2.12. *Let $(x_n)_{n\in\mathbb{N}}$ be a sequence in $\overline{\mathbb{R}}$. Then $\liminf x_n$ is the smallest accumulation point of the sequence and $\limsup x_n$ the largest. We have*

$$\liminf x_n \leq \limsup x_n$$

and equality holds if and only if the sequence is converging to this value.

Proof. Using the homeomorphism ψ_0 defined above (cf. (1.2.19)) we can assume that $(x_n)_{n\in\mathbb{N}}$ is a sequence in $[-1,1]$. If y is an accumulation point of the sequence, i.e., a limit point of subsequence, $(x_{n_k})_{k\in\mathbb{N}}$, $(n_0 < n_1 < n_2 < \cdots < n_k < n_{k+1} < \cdots)$, then

$$y \underset{k\to+\infty}{\longleftarrow} x_{n_k} \leq \sup_{l\geq n_k} x_l \underset{k\to+\infty}{\longrightarrow} \limsup x_n,$$

where the second limit comes from a subsequence of a converging sequence. We get thus $y \leq \limsup x_n$ and similarly $y \geq \liminf x_n$. Moreover, $\limsup x_n$ is an accumulation point of the sequence since for all $\epsilon > 0, N \geq 1$, we may find $n_\epsilon \geq N$ with

$$\limsup x_n = \inf_n(\sup_{k\geq n} x_k) \leq \sup_{k\geq n_\epsilon} x_k < \limsup x_n + \epsilon,$$

so that, for $\eta > 0, \exists n(\varepsilon,\eta) \geq n_\varepsilon$ with

$$\limsup x_n - \eta = \inf_n(\sup_{k\geq n} x_k) - \eta \leq \sup_{k\geq n_\epsilon} x_k - \eta < x_{n(\epsilon,\eta)} \leq \sup_{k\geq n_\epsilon} x_k < \limsup x_n + \epsilon.$$

As a result, for all ϵ, η positive, for all $N \geq 1$, we can find $n(\epsilon,\eta) \geq n_\epsilon \geq N$ with

$$\limsup x_n - \eta < x_{n(\epsilon,\eta)} < \limsup x_n + \epsilon,$$

proving the result. If $\limsup x_n = \liminf x_n = l$, then

$$l \underset{n\to+\infty}{\longleftarrow} \inf_{k\geq n} x_k \leq x_n \leq \sup_{k\geq n} x_k \underset{n\to+\infty}{\longrightarrow} l,$$

implying $\lim x_n = l$. On the contrary, if $\liminf x_n < \limsup x_n$, the sequence has at least two different accumulation points and cannot converge. $\square$

Proposition 1.2.13. *Addition and multiplication of real numbers can be extended continuously respectively to[6]*

$$(\overline{\mathbb{R}} \times \overline{\mathbb{R}})\backslash\{(+\infty,-\infty),(-\infty,+\infty)\} \text{ and to } (\overline{\mathbb{R}} \times \overline{\mathbb{R}})\backslash\{(0,\pm\infty),(\pm\infty,0)\}.$$

---

[6]In other words, $x + y$ is meaningful for $x \in \overline{\mathbb{R}}, y \in \overline{\mathbb{R}}$, provided we avoid the "undetermined expression" $+\infty - \infty$. Same thing for the product and $0.\infty$. The adjective "undetermined" is justified by the fact that there is no continuous extension of the addition in $\mathbb{R}$ to $\overline{\mathbb{R}}$: if such an extension were existing, for $x_n = -n + l, y_n = n$, we would have for all values of the real parameter l, $l = \lim(x_n + y_n) = \lim x_n + \lim y_n = +\infty - \infty$. Somehow worse than this, with $x_n = -n + (-1)^n, y_n = n$, $+\infty - \infty$ would be the limit of the non-converging sequence $(-1)^n$.

Let $(x_n)_{n\in\mathbb{N}}, (y_n)_{n\in\mathbb{N}}$ be sequences of $\overline{\mathbb{R}}$ such that $x_n + y_n$, $\liminf x_n + \liminf y_n$ and $\limsup x_n + \limsup y_n$ are meaningful. Then, the following inequalities hold:[7]

$$\liminf x_n + \liminf y_n \leq \liminf (x_n + y_n)$$
$$\leq \limsup (x_n + y_n) \leq \limsup x_n + \limsup y_n.$$

Proof. Let us first assume that both sequences $(x_n), (y_n)$ are bounded in $\mathbb{R}$. For $k \geq n$, we have $x_k + y_k \leq \sup_{l \geq n} x_l + \sup_{l \geq n} y_l$ so that $\sup_{k \geq n}(x_k + y_k) \leq \sup_{l \geq n} x_l + \sup_{l \geq n} y_l$. As a result, taking limits for $n \to +\infty$, we get

$$\limsup (x_n + y_n) \leq \limsup x_n + \limsup y_n.$$

Noticing that

$$\liminf(-x_n) = \sup_n(\inf_{k \geq n} (-x_n)) = \sup_n(- \sup_{k \geq n} x_n) = - \inf_n(\sup_{k \geq n} x_n) = - \limsup x_n,$$

we get the result. We leave for the reader to check the remaining cases when at least one sequence is not bounded in $\mathbb{R}$. $\square$

The following result will be useful in the sequel.

Lemma 1.2.14. *Let $(a_{k,l})_{k\in\mathbb{N},l\in\mathbb{N}}$ be a double sequence of $\overline{\mathbb{R}}_+$. Then*

$$\sum_k \left(\sum_l a_{kl} \right) = \sum_l \left(\sum_k a_{kl} \right). \qquad \text{We shall write } \sum_{k,l} a_{kl} \text{ for that sum.}$$

Proof. We have seen above that series of elements of $\overline{\mathbb{R}}_+$ converge towards their supremum[8]. Thus, for all K, L, we have

$$\sigma = \sum_k \left(\sum_l a_{kl} \right) = \sum_k \sup_{L \geq 0} \left(\sum_{0 \leq l \leq L} a_{kl} \right) = \sup_K \sum_{0 \leq k \leq K} \left[\sup_{L \geq 0} \left(\sum_{0 \leq l \leq L} a_{kl} \right) \right]$$

$$\geq \sum_{0 \leq k \leq K} \left[\sup_{L \geq 0} \left(\sum_{0 \leq l \leq L} a_{kl} \right) \right] \geq \sum_{0 \leq k \leq K} \left[\sum_{0 \leq l \leq L} a_{kl} \right] = \sum_{0 \leq l \leq L} \left[\sum_{0 \leq k \leq K} a_{kl} \right],$$

and for all L, $\sigma \geq \sum_{0 \leq l \leq L}[\sum_k a_{kl}]$, which implies $\sigma \geq \sum_l (\sum_k a_{kl})$, and the result by exchanging k and l. $\square$

Remark 1.2.15. Addition of real numbers can be extended continuously to $\overline{\mathbb{R}}_+ \times \overline{\mathbb{R}}_+$; it is thus associative, commutative, with neutral element 0. Multiplication of real numbers cannot be extended continuously to $\overline{\mathbb{R}}_+ \times \overline{\mathbb{R}}_+$ but only to $\overline{\mathbb{R}}_+^* \times \overline{\mathbb{R}}_+^*$.

---

[7]Equalities are not true in general: check for instance $x_n = (-1)^n/2, y_n = (-1)^{n+1}$, for which $\liminf x_n + \liminf y_n = -1/2 - 1 < \liminf (x_n + y_n) = -1/2 < \limsup (x_n + y_n) = 1/2 < \limsup x_n + \limsup y_n = 1/2 + 1$.
[8]In particular, the infinite sums in the statement are meaningful.

We could use the (discontinuous[9]) convention $0 \cdot \infty = 0$ and[10] it is easy to verify that this new multiplication is associative, commutative, with neutral element 1, distributive with respect to the addition. The reader may also check Remark 1.3.4 below.

1.3 Structure of measurable functions

Proposition 1.3.1. *Let* $(X, \mathcal{M})$ *be a measurable space and let* $(f_n)_{n \in \mathbb{N}}$ *be a sequence of measurable functions from X into $\overline{\mathbb{R}}$. Then the functions* $\sup f_n$, $\inf f_n$, $\limsup f_n$, $\liminf f_n$ *are measurable. In particular, the pointwise limit of a sequence of measurable functions is measurable.*

Proof. Let us set $g = \sup f_n$, i.e., $g(x) = \sup_{n \in \mathbb{N}} f_n(x)$. For $a \in \mathbb{R}$, we have

$$g^{-1}(]a, +\infty]) = \cup_{n \in \mathbb{N}} f_n^{-1}(]a, +\infty]),$$

since $\sup_{n \in \mathbb{N}} f_n(x) = g(x) > a \iff \exists n_0 \in \mathbb{N}$, such that $f_{n_0}(x) > a$. Consequently, we get $g^{-1}(]a, +\infty]) \in \mathcal{M}$. According to Lemma 1.1.4, this proves the measurability of g since the Borel σ-algebra $\mathcal{B}_{\overline{\mathbb{R}}}$ of $\overline{\mathbb{R}}$ is equal to the σ-algebra generated by the intervals $]a, +\infty]$ (see (1.2.15), the discussion on page 9 and the increasing homeomorphism of $\overline{\mathbb{R}}$ with $[-1, 1]$ displayed in (1.2.19)). Thus $g = \sup f_n$ is measurable. Moreover the identities

$$\inf f_n = -\sup(-f_n), \qquad \limsup f_n = \inf_n (\sup_{k \geq n} f_k), \qquad \liminf f_n = \sup_n (\inf_{k \geq n} f_k)$$

give the other results. ⊔

Definition 1.3.2. Let $(X, \mathcal{M})$ be a measurable space. A measurable function $s : X \to [0, +\infty)$ is said to be *simple* if it takes only a finite number of values.

Let $\{\alpha_1, \ldots, \alpha_m\}$ be the image of s. Defining $A_k = s^{-1}(\{\alpha_k\})$, we get that $\{A_k\}_{1 \leq k \leq m}$ is a partition of X and

$$s(x) = \sum_{1 \leq k \leq m} \alpha_k \mathbf{1}_{A_k}(x),$$

where $\mathbf{1}_{A_k}$ is the indicator function of A_k.

---

[9] Considering the sequences in $(0, +\infty)$, $(x_n, y_n) \in \{(1/n, n^2), (1/n^2, n), (l/n, n), (\frac{2+(-1)^n}{n}, n)\}$, we see in each case $\lim x_n = 0$, $\lim y_n = +\infty$ and that the limit of $x_n y_n$ could be anything in $\mathbb{R}_+$ or that the sequence $x_n y_n$ is not converging. A somehow worse behaviour is given by the sequences

$$x_n = q_n/(n(1 + q_n)), \quad y_n = n(1 + q_n), \quad \text{where } \mathbb{Q}_+ = \{q_n\}_{n \geq 1}.$$

We have $\lim x_n = 0$, $\lim y_n = +\infty$ and the sequence $(x_n y_n)$ is dense in $\overline{\mathbb{R}}_+$.

[10] That commonly used convention refers to a "potential" vision of infinity: infinity is seen as something that can be reached by some limiting process. Looking at the product $0n = 0$ for all n, that convention looks natural. That potential vision is opposed to an "actual" viewpoint where infinity is there from the beginning. In measure theory, that convention is justified by the fact that integrating the zero function on any set, even of infinite measure, will give 0. Also integrating a function which is identically $+\infty$ on a set of measure 0 will give 0.

Theorem 1.3.3. *Let $(X, \mathcal{M})$ be a measurable space and let $f : X \longrightarrow \overline{\mathbb{R}}_+ = [0, +\infty]$ be a measurable mapping. There exists a sequence $(s_k)_{k \geq 1}$ of simple functions such that*

(1) $0 \leq s_1 \leq s_2 \leq \cdots \leq s_k \leq s_{k+1} \leq \cdots \leq f$,

(2) $\forall x \in X,\ \lim_k s_k(x) = f(x)$,

(3) *For f bounded, the limit is uniform:* $\lim_k \left(\sup_{x \in X} |f(x) - s_k(x)| \right) = 0$.

Proof. Let us first assume that $0 \leq f \leq 1$. We define[11]

$$s_k(x) = 2^{-k} E(2^k f(x)). \tag{1.3.1}$$

The function s_k takes finitely many values since $0 \leq 2^k f \leq 2^k$. We have also

$$2^k s_k \leq 2^k f < 2^k s_k + 1 \Longrightarrow 0 \leq f - s_k < 2^{-k}, \tag{1.3.2}$$

so that s_k converges uniformly towards f. Moreover, multiplying (1.3.2) by 2 and writing (1.3.2) for $k + 1$, we find

$$\mathbb{N} \ni 2^{k+1} s_k \leq 2^{k+1} f, \qquad 2^{k+1} s_{k+1} = E(2^{k+1} f).$$

Using the definition of the integer value, we obtain

$$2^{k+1} s_k \leq 2^{k+1} s_{k+1}, \text{ i.e., } s_k \leq s_{k+1},$$

proving that (s_k) is an increasing sequence. Every function s_k is measurable, as the composition of measurable functions[12]. If $0 \leq f \leq M$, for some positive real number M, we can apply the previous result to f/M. Let us go back to the case $0 \leq f \leq 1$ and set

$$\tilde{s}_k = s_k - 2^{-k} E(f).$$

If $f(x) < 1$, we have $s_k(x) = \tilde{s}_k(x)$. If $f(x) = 1$, we have $1 - 2^{-k} = \tilde{s}_k(x)$. In both cases, the sequences $(\tilde{s}_k(x))_{k \in \mathbb{N}}$ are increasing with limit $f(x)$ and $0 \leq \tilde{s}_k(x) < 1$. Using the homeomorphism ψ_0 defined in (1.2.19), which identifies $\overline{\mathbb{R}}_+$ to $[0, 1]$, we may consider

$$X \ \overset{f}{\longrightarrow} \ \overline{\mathbb{R}}_+ \ \overset{\psi_0}{\longrightarrow} \ [0, 1] \ \overset{\psi_0^{-1}}{\longrightarrow} \ \overline{\mathbb{R}}_+.$$

Using the previous arguments, we find a sequence of simple functions t_k valued in $[0, 1[$, increasing with limit $\psi_0 \circ f$. As a result, $\psi_0^{-1} \circ t_k$ is a simple function (in particular with finite values since t_k has values < 1) with limit f. The sequence $\psi_0^{-1} \circ t_k$ is increasing as t_k is and ψ_0^{-1} is increasing. The proof of the theorem is complete. $\qquad \square$

---

[11] $E(t)$ stands for the *integer value of* $t \in \mathbb{R}$, also called *floor function* or *greatest integer function*: $E(t)$ is the unique integer such that $E(t) \leq t < E(t) + 1$.

[12] The integer value is measurable since $E^{-1}([a, +\infty[) = [a, +\infty[$ if $a \in \mathbb{Z}$ and if $a \notin \mathbb{Z}$, $E^{-1}([a, +\infty[) = [E(a) + 1, +\infty[$.

Remark 1.3.4. Let f, g be measurable functions from X into $\overline{\mathbb{R}}_+$; then $f + g$ is well defined and measurable. It follows from Theorem 1.2.8 and the measurability of the (continuous) mapping

$$\begin{array}{ccc} \overline{\mathbb{R}}_+ \times \overline{\mathbb{R}}_+ & \longrightarrow & \overline{\mathbb{R}}_+ \\ (\alpha, \beta) & \mapsto & \alpha + \beta. \end{array}$$

Analogously, the symmetric (discontinuous) mapping M

$$\begin{array}{ccc} \overline{\mathbb{R}}_+ \times \overline{\mathbb{R}}_+ & \longrightarrow & \overline{\mathbb{R}}_+ \\ (\alpha, \beta) & \mapsto & \alpha \cdot \beta \end{array}$$

extending continuously to $\overline{\mathbb{R}}_+^* \times \overline{\mathbb{R}}_+^*$ the multiplication on $\mathbb{R}_+^* \times \mathbb{R}_+^*$ and defining $0.\infty = 0$ is Borel-measurable: for $a \in \overline{\mathbb{R}}_+$, the set

$$E_a = \{(x, y) \in \overline{\mathbb{R}}_+ \times \overline{\mathbb{R}}_+, M(x, y) > a\}$$

is included in $\overline{\mathbb{R}}_+^* \times \overline{\mathbb{R}}_+^*$ on which M is continuous. As a result, E_a is an open subset of $\overline{\mathbb{R}}_+^* \times \overline{\mathbb{R}}_+^*$, thus a Borel set of $\overline{\mathbb{R}}_+ \times \overline{\mathbb{R}}_+$. Using Theorem 1.2.8, we get that $f \cdot g$ is measurable.

1.4 Positive measures

Definition 1.4.1. Let $(X, \mathcal{M})$ be a measurable space. A *positive measure* on $(X, \mathcal{M})$ is a mapping $\mu : \mathcal{M} \to \overline{\mathbb{R}}_+$ satisfying $\mu(\emptyset) = 0$, and such that, for any sequence $(A_k)_{k \in \mathbb{N}}$ in $\mathcal{M}$ of pairwise disjoint sets ($k \neq l \Longrightarrow A_k \cap A_l = \emptyset$),

$$\mu(\cup_{k \in \mathbb{N}} A_k) = \sum_{k \in \mathbb{N}} \mu(A_k). \tag{1.4.1}$$

That property is called σ-additivity[13] and the triple $(X, \mathcal{M}, \mu)$ is called a *measure space* (where μ is a positive measure). When $\mu(X) = 1$, we shall say that μ is a *probability measure* and the triple $(X, \mathcal{M}, \mu)$ is called a *probability space*.

N.B. We shall define later in this text (Definition 7.1.1 in Chapter 7) the notion of *complex measure*.

Let us give a few simple examples.

(1) Let X be a finite set, equipped with the σ-algebra $\mathcal{P}(X)$, and let us define the *counting measure* μ_0 by $\mu_0(A) = \operatorname{card} A$.

(2) Let X be a non-empty finite set, (σ-algebra $\mathcal{P}(X)$), and let μ_1 be the probability measure μ_1 defined by

$$\mu_1(A) = \operatorname{card} A / \operatorname{card} X.$$

---

[13] or countable additivity.

(3) Let X be a set, (σ-algebra $\mathcal{P}(X)$). We define the *counting measure* on X by

$$\mu(A) = \begin{cases} \operatorname{card} A, & \text{when } A \text{ is finite,} \\ +\infty, & \text{when } A \text{ is infinite.} \end{cases}$$

To check that it is indeed a measure, we consider a sequence of pairwise disjoint subsets $(A_k)_{k\in\mathbb{N}}$: if one of them is infinite, (1.4.1) is obvious as well as when they are all finite with a finite union. If they are all finite with an infinite union, (1.4.1) follows from the inequalities

$$\operatorname{card}(\cup_{0\leq k\leq N} A_k) = \sum_{0\leq k\leq N} \operatorname{card} A_k \leq \sum_{0\leq k} \operatorname{card} A_k$$

and $\lim_{N\to+\infty} \operatorname{card}(\cup_{0\leq k\leq N} A_k) = +\infty$.

(4) Let X be a set (σ-algebra $\mathcal{P}(X)$). For $a \in X$ we define δ_a, the *Dirac measure at a* by

$$\delta_a(A) = \begin{cases} 1 & \text{if } a \in A, \\ 0 & \text{if } a \notin A. \end{cases}$$

(5) Series of positive measures on the same measurable space.

Lemma 1.4.2. *Let $(X, \mathcal{M})$ be a measurable space and let $(\mu_j)_{j\in\mathbb{N}}$ be a sequence of positive measures on $(X, \mathcal{M})$. For $A \in \mathcal{M}$, we define $\mu(A) = \sum_{j\in\mathbb{N}} \mu_j(A)$. Then μ is a positive measure on $(X, \mathcal{M})$.*

Proof. Let $(A_k)_{k\in\mathbb{N}}$ be a pairwise disjoint sequence in $\mathcal{M}$. We have

$$\mu(\cup_{k\in\mathbb{N}} A_k) = \sum_{j\in\mathbb{N}} \mu_j(\cup_{k\in\mathbb{N}} A_k) \underbrace{=}_{\substack{\sigma\text{-additivity} \\ \text{of each } \mu_j}} \sum_{j\in\mathbb{N}}\left(\sum_{k\in\mathbb{N}} \mu_j(A_k)\right)$$

$$\underbrace{=}_{\text{Lemma 1.2.14}} \sum_{k\in\mathbb{N}}\sum_{j\in\mathbb{N}} \mu_j(A_k) = \sum_{k\in\mathbb{N}} \mu(A_k). \qquad \square$$

(6) We want to construct a positive measure on $(\mathbb{R}, \mathcal{B}_{\mathbb{R}})$, where $\mathcal{B}_{\mathbb{R}}$ is the Borel σ-algebra on $\mathbb{R}$, such that, for $a \leq b$ real numbers,

$$\mu([a, b]) = b - a = \mu(]a, b]).$$

It is easy to construct μ on finite unions of pairwise disjoint intervals. **Although $\mathcal{B}_{\mathbb{R}}$ is generated by the intervals in the sense of Definition 1.1.3, extending μ to $\mathcal{B}_{\mathbb{R}}$ is a difficult task which is one of the main goals of this book.**

(7) Measure with density ν with respect to the Borel measure on $\mathbb{R}$. Let ν be a continuous non-negative function on $\mathbb{R}$; we want to construct a positive measure defined on $\mathcal{B}_{\mathbb{R}}$ such that for $a \leq b$ real numbers, we have

$$\mu_\nu([a, b]) = \int_a^b \nu(t)dt,$$

where the integral of ν is the Riemann integral. It is also easy to construct μ_ν on finite unions of pairwise disjoint intervals (this is the density version of the previous example in which $\nu \equiv 1$). *Also a difficult construction to be performed in the sequel.*

(8) Borel measure on $\mathbb{R}^d$. With $\mathcal{B}_{\mathbb{R}^d}$ standing for the Borel σ-algebra on $\mathbb{R}^d$, one of the goals of this book is to provide a construction of a positive measure defined on $\mathcal{B}_{\mathbb{R}^d}$, such that, for $a_j \leq b_j$ real numbers, we have

$$\mu\left(\prod_{1 \leq j \leq d} [a_j, b_j]\right) = \prod_{1 \leq j \leq d} (b_j - a_j).$$

It is the d-dimensional version of the example (6) on page 18.

(9) Cauchy probability on $\mathbb{R}$ with parameter $\alpha > 0$. It is the positive measure with density

$$\frac{1}{\pi}\frac{\alpha}{\alpha^2 + t^2}.$$

We note that $\int_{\mathbb{R}} \frac{1}{\pi}\frac{\alpha}{\alpha^2+t^2}\,dt = \left[\frac{1}{\pi}\arctan(t/\alpha)\right]_{-\infty}^{+\infty} = 1$. We define the *repartition function* of the probability μ on $\mathbb{R}$ as

$$F(t) = \mu\big((-\infty, t[\big).$$

The function F is increasing, tends to 0 (resp. 1) when t goes to $-\infty$ (resp. $+\infty$), and is left-continuous (see Exercise 1.9.25). In the specific case of the Cauchy probability, the repartition function is

$$F(t) = \frac{1}{\pi}\arctan\left(\frac{t}{\alpha}\right) + \frac{1}{2}.$$

(10) The Laplace–Gauss probability with *mean* (or *expectation*) m, *variance* σ^2 ($\sigma > 0$ is the *standard deviation*), has density

$$\frac{1}{\sigma\sqrt{2\pi}}\exp-\frac{(x-m)^2}{2\sigma^2}.$$

We note that $\int_{\mathbb{R}}\exp-\frac{(x-m)^2}{2\sigma^2}\,dx = \sigma\sqrt{2\pi}$ and

$$\int_{\mathbb{R}} x\exp-\frac{(x-m)^2}{2\sigma^2}\frac{dx}{\sigma\sqrt{2\pi}} = m, \qquad \int_{\mathbb{R}}(x-m)^2\exp-\frac{(x-m)^2}{2\sigma^2}\frac{dx}{\sigma\sqrt{2\pi}} = \sigma^2.$$

(11) Bernoulli probability with parameter $p \in [0,1]$: $p\delta_0 + (1-p)\delta_1$ on the set $X = \{0,1\}$.

(12) Binomial probability with parameters $n \in \mathbb{N}^*$ and $p \in [0,1]$,

$$\mu = \sum_{0 \leq k \leq n} C_n^k p^k (1-p)^{n-k}\delta_k,$$

where

$$C_n^k = \frac{n!}{(n-k)!k!} = \binom{n}{k}. \tag{1.4.2}$$

We can consider μ as a positive measure on $\{0, 1, \ldots, n-1, n\}$ so that $\mu(A) = \sum_{k\in A} C_n^k p^k (1-p)^{n-k}$.

(13) The Poisson probability with parameter $\lambda > 0$ is given by

$$e^{-\lambda} \sum_{k\in\mathbb{N}} \frac{\lambda^k}{k!} \delta_k,$$

which is meaningful, e.g., from Lemma 1.4.2. We may consider μ as defined on the subsets of $\mathbb{N}$ by $\mu(A) = e^{-\lambda} \sum_{k\in A} \frac{\lambda^k}{k!}$.

Lemma 1.4.3. *Let $(X, \mathcal{M}, \mu)$ be a measure space where μ is a positive measure and let $f : X \longrightarrow Y$ be a mapping. The set $\mathcal{N} = \{B \subset Y, f^{-1}(B) \in \mathcal{M}\}$ is a σ-algebra on Y: it is the largest σ-algebra on Y making f measurable. The so-called pushforward measure $f_*(\mu)$ is a positive measure defined on $\mathcal{N}$ by*

$$f_*(\mu)(B) = \mu\big(f^{-1}(B)\big).$$

If $g : Y \longrightarrow Z$ is another mapping, we have $(g \circ f)_* = g_* \circ f_*$.

Proof. The first statements follow from Lemma 1.1.5. To check that $f_*(\mu)$ is a positive measure defined on $\mathcal{N}$, we consider a sequence $(B_k)_{k\in\mathbb{N}}$ of pairwise disjoint elements of $\mathcal{N}$ and we note that $\big(f^{-1}(B_k)\big)_{k\in\mathbb{N}}$ is a pairwise disjoint sequence of $\mathcal{M}$ and thus

$$f_*(\mu)\big(\cup_{k\in\mathbb{N}} B_k\big) = \mu\big(f^{-1}(\cup_{k\in\mathbb{N}} B_k)\big) = \mu\big(\cup_{k\in\mathbb{N}} f^{-1}(B_k)\big)$$
$$= \sum_k \mu\big(f^{-1}(B_k)\big) = \sum_k f_*(\mu)(B_k).$$

Also we have trivially $f_*(\mu)(\emptyset) = \mu(f^{-1}(\emptyset)) = \mu(\emptyset) = 0$. The last "functorial" property[14] is obvious and follows from the other functorial property (see (1.1.1)) $(g \circ f)^{-1}(C) = f^{-1}\big(g^{-1}(C)\big)$: with $\mathcal{P} = \{C \subset Z, g^{-1}(C) \in \mathcal{N}\}$, we have for $C \in \mathcal{P}$,

$$\big((g \circ f)_*(\mu)\big)(C) = \mu\big((g \circ f)^{-1}(C)\big) = \mu\big(f^{-1}(g^{-1}(C))\big) = f_*(\mu)(g^{-1}(C))$$
$$= \big(g_*(f_*(\mu))\big)(C) = \big((g_* \circ f_*)(\mu)\big)(C). \quad \square$$

Proposition 1.4.4. *Let $(X, \mathcal{M}, \mu)$ be a measure space where μ is a positive measure.*

(1) *For $A, B \in \mathcal{M}$, $A \subset B \Longrightarrow \mu(A) \leq \mu(B)$.*

(2) *Let $(A_k)_{k\in\mathbb{N}}$ be an increasing sequence of $\mathcal{M}$ and $A = \cup_{k\in\mathbb{N}} A_k$; then $\mu(A_k) \uparrow \mu(A)$ in $\overline{\mathbb{R}}_+$.*

(3) *Let $(A_k)_{k\in\mathbb{N}}$ be a decreasing sequence in $\mathcal{M}$, such that $\mu(A_0) < +\infty$ and $A = \cap_{k\in\mathbb{N}} A_k$; then $\mu(A_k) \downarrow \mu(A)$ in $\mathbb{R}_+$.*

---

[14]This covariance property following from the contravariance property for inverse images explains also the notation with a $*$ at the bottom for the covariant pushforward and a -1 at the top for the contravariant inverse images.

Moreover the properties of Definition 1.4.1 are equivalent to $\mu(\emptyset) = 0$, (2) above and $\mu(A \cup B) = \mu(A) + \mu(B)$, for disjoint $A, B \in \mathcal{M}$.

Proof. The disjoint union of elements of $\mathcal{M}$, $B = (B \backslash A) \cup A$, implies $\mu(B) = \mu(B \backslash A) + \mu(A) \geq \mu(A)$ and thus (1). To get (2), we define $A_{-1} = \emptyset$, and prove inductively that[15]

$$A_k = \cup_{0 \leq l \leq k}(A_l \cap A_{l-1}^c),$$

so that $A = \cup_{k \geq 0} A_k = \cup_{k \geq 0}(A_k \cap A_{k-1}^c)$. For $k \neq l$ (say $k > l \geq 0$), since (A_j) is increasing, we have

$$(A_k \cap A_{k-1}^c) \cap (A_l \cap A_{l-1}^c) = A_k \cap A_l \cap A_{k-1}^c \cap A_{l-1}^c = A_l \cap A_{k-1}^c \subset A_{k-1}^c \cap A_{k-1} = \emptyset.$$

As a result, using (1.4.1), we obtain

$$\mu(A) = \sum_{k \geq 0} \mu(A_k \cap A_{k-1}^c) = \lim_{n \to \infty} \sum_{0 \leq k \leq n} \mu(A_k \cap A_{k-1}^c) = \lim_{n \to \infty} \mu(A_n), \quad \text{i.e., (2)}.$$

We check now (3). We have

$$A_0 \backslash A = A_0 \cap (\cup_{k \geq 0} A_k^c) = \cup_{k \geq 0} \underbrace{(A_0 \cap A_k^c)}_{\text{increasing of } k}.$$

Applying the already proven property (2), we get $\mu(A_0 \cap A_k^c) \uparrow \mu(A_0 \backslash A)$. For each k, we have

$$+\infty > \mu(A_0) = \mu(A_k) + \mu(A_k^c \cap A_0),$$

so that $\mu(A_0), \mu(A_k), \mu(A_k^c \cap A_0)$ are real numbers[16], and thus

$$\mu(A_k) = \mu(A_0) - \mu(A_k^c \cap A_0) \downarrow \mu(A_0) - \mu(A_0 \backslash A) = \mu(A),$$

proving (3). If μ is a positive measure, the properties mentioned in the last statement of Proposition 1.4.4 are fulfilled, as proven above. Conversely, we need to prove (1.4.1). Let $(A_k)_{k \in \mathbb{N}}$ be a pairwise disjoint sequence in $\mathcal{M}$: from property (2) in Proposition 1.4.4, using finite additivity[17], we get

$$\sum_{0 \leq k \leq n} \mu(A_k) = \mu(\cup_{0 \leq k \leq n} A_k) \uparrow \mu(\cup_{k \geq 0} A_k), \quad \text{i.e., } \sum_{k \geq 0} \mu(A_k) = \mu(\cup_{k \geq 0} A_k).$$

The proof of Proposition 1.4.4 is complete. $\qquad\square$

---

[15]True for $k = 0$; moreover $A_{k+1} = (A_{k+1} \cap A_k^c) \cup \overbrace{(A_{k+1} \cap A_k)}^{=A_k, \text{ since } A_j \uparrow} = (A_{k+1} \cap A_k^c) \cup A_k$.

[16]At this very point, we are using the assumption $\mu(A_0) < +\infty$, which is necessary as shown by the counting measure (Example (3) on page 18) on $\mathbb{N}$ with the decreasing sequence $A_k = [k, +\infty[\cap \mathbb{N}$: for each k, $\mu(A_k) = +\infty$ and $\mu(\cap_{k \geq 0} A_k) = \mu(\emptyset) = 0$.

[17]Trivial inductively from the additivity for two disjoint sets:

$$\mu(\cup_{0 \leq k \leq n+1} A_k) = \mu(\cup_{0 \leq k \leq n} A_k) + \mu(A_{n+1}) = \sum_{0 \leq k \leq n} \mu(A_k) + \mu(A_{n+1}).$$

Remark 1.4.5. Let $(X, \mathcal{M}, \mu)$ be a measure space where μ is a positive measure, let $(A_n)_{n\in\mathbb{N}}$ be a sequence in $\mathcal{M}$. Then

$$\mu(\cup_{n\in\mathbb{N}} A_n) \leq \sum_{n\in\mathbb{N}} \mu(A_n). \tag{1.4.3}$$

In fact, checking the increasing $B_n = \cup_{0\leq k\leq n} A_k$, we may apply property (2) in Proposition 1.4.4 so that

$$\mu(\cup_{n\in\mathbb{N}} A_n) = \mu(\cup_{n\in\mathbb{N}} B_n) = \sup_{n\in\mathbb{N}} \mu(B_n) \leq \sup_{n\in\mathbb{N}} \sum_{0\leq k\leq n} \mu(A_k) = \sum_{n\in\mathbb{N}} \mu(A_n),$$

since the inequality $\mu(B_n) \leq \sum_{0\leq k\leq n} \mu(A_k)$ holds trivially (inductively on n). See Exercise 1.9.19 for the Sieve Formula.

1.5 Integrating non-negative functions

We want now to define the "integral with respect to a measure μ" of simple functions as defined in Definition 1.3.2: let $s = \sum_{1\leq k\leq m} \alpha_k \mathbf{1}_{A_k}$, where the α_k are positive, distinct and each A_k belongs to $\mathcal{M}$. The integral will be defined as

$$\int_X s d\mu = \sum_{1\leq k\leq m} \alpha_k \mu(A_k),$$

which is a quite natural definition. We have to pay attention to the fact that since all $\alpha_k > 0$, although $\mu(A_k)$ could be $+\infty$, the product $\alpha_k \mu(A_k)$ is defined without ambiguity in $\overline{\mathbb{R}}_+$. We should also keep in mind that the elements of $\mathcal{M}$ could be awfully complicated: think for instance of the Borelian sets of type $F_\sigma, G_\delta, G_{\delta\sigma}, F_{\sigma\delta}, \dots$[18].

Lemma 1.5.1. *Let $(X, \mathcal{M}, \mu)$ be a measure space where μ is a positive measure and let s be a simple function, that is a measurable function $s : X \to [0, +\infty[$ taking a finite number of real non-negative distinct values $\alpha_1, \dots, \alpha_m$, in such a way that $s = \sum_{1\leq j\leq m} \alpha_j \mathbf{1}_{A_j}$, $A_j = s^{-1}(\{\alpha_j\})$. We define*[19]

$$I(s) = \sum_{\substack{1\leq j\leq m \\ \alpha_j > 0}} \alpha_j \mu(A_j), \tag{1.5.1}$$

---

[18] An F_σ is a countable union of closed sets, a G_δ is a countable intersection of open sets, a $G_{\delta\sigma}$ is a countable union of G_δ sets, a $F_{\sigma\delta}$ a countable intersection of F_σ sets, and so on. That terminology was introduced by the German mathematician Felix Hausdorff (1868–1942). The letter σ is a symbol for countable union (*Summe* in German) and δ is a symbol for countable intersection (*Durchschnitt*).

[19] We have only to handle products of positive real numbers α_j with elements of $\overline{\mathbb{R}}_+$. Moreover the consistency of our definition relies on the fact that the decomposition of s as such a sum is canonical since the α_j and thus the A_j are functions of s. The condition $I(0) = 0$ follows in fact from (1.5.1) since for $s = 0$, the summation takes place on an empty set of indices. We could have written $I(s) = \sum_{1\leq j\leq m} \alpha_j \mu(A_j)$ using the convention $0.\infty=0$. We have preferred to avoid that discontinuous convention, at a price of heavier notation.

and $I(0) = 0$. For s, t simple functions and $\lambda > 0$, we have

$$I(s) = \sup_{\substack{\sigma \text{ simple} \\ 0 \le \sigma \le s}} I(\sigma), \quad I(s+t) = I(s) + I(t), \quad I(\lambda s) = \lambda I(s). \tag{1.5.2}$$

Proof. [20] Let σ, s be simple functions such that $\sigma \le s$ (i.e., $\forall x \in X, \sigma(x) \le s(x)$). We have the canonical decomposition

$$\sigma = \sum_{1 \le k \le n} \beta_k \mathbf{1}_{B_k}, \quad s = \sum_{1 \le j \le m} \alpha_j \mathbf{1}_{A_j},$$

where $\{B_k\}_{1 \le k \le n}$ and $\{A_j\}_{1 \le j \le m}$ are partitions of X. The definition gives

$$I(\sigma) = \sum_{\substack{1 \le k \le n \\ \beta_k > 0}} \beta_k \mu(B_k) = \sum_{\substack{1 \le k \le n, 1 \le j \le m \\ \beta_k > 0, B_k \cap A_j \ne \emptyset}} \beta_k \mu(B_k \cap A_j).$$

Noticing that $B_k \cap A_j \ne \emptyset$ implies $\beta_k \le \alpha_j$ (since for $x \in B_k \cap A_j, \beta_k = \sigma(x) \le s(x) = \alpha_j$), and thus $\alpha_j > 0$ when $\beta_k > 0$, we get

$$I(\sigma) \le \sum_{\substack{1 \le k \le n, 1 \le j \le m \\ \alpha_j > 0}} \alpha_j \mu(B_k \cap A_j) = \sum_{\substack{1 \le j \le m \\ \alpha_j > 0}} \alpha_j \mu(A_j) = I(s),$$

proving the first result. To prove the next one, we note first that for s, t simple functions, the function $s + t$ is measurable as a sum of measurable functions and also simple since it takes only a finite number of non-negative real values. Using the canonical decomposition of s and t, we have

$$s = \sum_{1 \le j \le m} \alpha_j \mathbf{1}_{A_j}, \quad t - \sum_{1 \le k \le n} \beta_k \mathbf{1}_{B_k}, \quad \text{so that} \quad s + t - \sum_{\substack{1 \le j \le m \\ 1 \le k \le n}} (\alpha_j + \beta_k) \mathbf{1}_{A_j \cap B_k}.$$

The sets $A_j \cap B_k$ are measurable and pairwise disjoint ($A_j \cap B_k \cap A_{j'} \cap B_{k'} = \emptyset$ when $j \ne j'$ or $k \ne k'$), and since

$$X = \cup_{1 \le j \le m} A_j = \cup_{\substack{1 \le j \le m \\ 1 \le k \le n}} (A_j \cap B_k),$$

we get that $\{A_j \cap B_k\}_{\substack{1 \le j \le m, 1 \le k \le n \\ A_j \cap B_k \ne \emptyset}}$ makes a partition of X. Since $\mathbf{1}_\emptyset = 0$, we obtain

$$s + t = \sum_{\substack{1 \le j \le m, 1 \le k \le n \\ A_j \cap B_k \ne \emptyset}} (\alpha_j + \beta_k) \mathbf{1}_{A_j \cap B_k}. \tag{1.5.3}$$

---

[20]This proof is simple, but quite tedious, and could probably be omitted at first reading.

Whenever the $\alpha_j + \beta_k$ are distinct, Formula (1.5.3) provides the canonical decomposition of $s + t$ and we find

$$I(s+t) = \sum_{\substack{1 \le j \le m,\, 1 \le k \le n \\ A_j \cap B_k \ne \emptyset,\, \alpha_j + \beta_k > 0}} (\alpha_j + \beta_k)\mu(A_j \cap B_k). \tag{1.5.4}$$

When the $\alpha_j + \beta_k$ are not distinct and take the distinct positive values $\gamma_1, \ldots, \gamma_p$, we need to rewrite (1.5.3) as

$$s + t = \sum_{1 \le l \le p} \gamma_l \sum_{\substack{1 \le j \le m,\, 1 \le k \le n \\ A_j \cap B_k \ne \emptyset,\, \alpha_j + \beta_k = \gamma_l}} \mathbf{1}_{A_j \cap B_k}.$$

We get

$$I(s+t) = \sum_{1 \le l \le p} \gamma_l \mu\left(\bigcup_{\substack{1 \le j \le m,\, 1 \le k \le n \\ A_j \cap B_k \ne \emptyset,\, \alpha_j + \beta_k = \gamma_l}} (A_j \cap B_k) \right)$$

$$= \sum_{1 \le l \le p} \gamma_l \sum_{\substack{1 \le j \le m,\, 1 \le k \le n \\ A_j \cap B_k \ne \emptyset,\, \alpha_j + \beta_k = \gamma_l}} \mu(A_j \cap B_k) = \sum_{\substack{1 \le j \le m,\, 1 \le k \le n \\ A_j \cap B_k \ne \emptyset,\, \alpha_j + \beta_k > 0}} (\alpha_j + \beta_k)\mu(A_j \cap B_k),$$

so that (1.5.4) always hold. On the other hand, we have

$$I(s) + I(t) = \sum_{\substack{1 \le j \le m \\ \alpha_j > 0}} \alpha_j \mu(A_j) + \sum_{\substack{1 \le k \le n \\ \beta_k > 0}} \beta_k \mu(B_k)$$

$$= \sum_{\substack{1 \le j \le m,\, 1 \le k \le n \\ \alpha_j > 0}} \alpha_j \mu(A_j \cap B_k) + \sum_{\substack{1 \le j \le m,\, 1 \le k \le n \\ \beta_k > 0}} \beta_k \mu(A_j \cap B_k),$$

and using the notation $\mu_{jk} = \mu(A_j \cap B_k)$, we have

$$\sum_{\alpha_j > 0} \alpha_j \mu_{jk} + \sum_{\beta_k > 0} \beta_k \mu_{jk}$$

$$= \sum_{\alpha_j > 0, \beta_k > 0} \alpha_j \mu_{jk} + \sum_{\alpha_j > 0, \beta_k > 0} \beta_k \mu_{jk} + \sum_{\alpha_j > 0, \beta_k = 0} \alpha_j \mu_{jk} + \sum_{\alpha_j = 0, \beta_k > 0} \beta_k \mu_{jk}$$

$$= \sum_{\alpha_j > 0, \beta_k > 0} (\alpha_j + \beta_k)\mu_{jk} + \sum_{\alpha_j > 0, \beta_k = 0} (\alpha_j + \beta_k)\mu_{jk} + \sum_{\alpha_j = 0, \beta_k > 0} (\alpha_j + \beta_k)\mu_{jk}$$

$$= \sum_{\alpha_j + \beta_k > 0} (\alpha_j + \beta_k)\mu_{jk},$$

implying indeed

$$I(s) + I(t) = \sum_{\substack{1 \le j \le m,\, 1 \le k \le n \\ \alpha_j + \beta_k > 0,\ A_j \cap B_k \ne \emptyset}} (\alpha_j + \beta_k)\mu(A_j \cap B_k) = I(s+t).$$

Finally, with $\lambda > 0$ and s a simple function, we have

$$I(\lambda s) = I(\lambda \sum_{1 \le j \le m} \alpha_j \mathbf{1}_{A_j}) = \sum_{\substack{1 \le j \le m \\ \alpha_j > 0}} \lambda \alpha_j \mu(A_j) = \lambda I(s),$$

completing the proof of the lemma. $\qquad\square$

Thanks to this lemma, we can now define the integral of a measurable function $f : X \to [0, +\infty] = \overline{\mathbb{R}}_+$.

Definition 1.5.2. Let $(X, \mathcal{M}, \mu)$ be a measure space where μ is a positive measure and let $f : X \to [0, +\infty]$ be a measurable function. We define[21]

$$\int_X f d\mu = \sup_{\substack{s \text{ simple} \\ 0 \le s \le f}} I(s)$$

Note that from Lemma 1.5.1, for f simple, we have $\int_X f d\mu = I(f)$. Also $\int_X 0 d\mu = 0$ since $I(0) = 0$.

Remark 1.5.3. Going back to the list of examples starting on page 17, we can check how the integral of a non-negative measurable function is obtained from the measure of sets.

• Let $X = \{x_1, \ldots, x_n\}$ (σ-algebra $\mathcal{P}(X)$) with $\mu_0(A) = \operatorname{card} A$. We have

$$\int_X f d\mu_0 = \int_X \sum_{1 \le j \le n} f(x_j) \mathbf{1}_{\{x_j\}} d\mu_0 = f(x_1) + \cdots + f(x_n).$$

• Let $X = \{x_1, \ldots, x_n\}$ with the probability measure $\mu_1(A) = \operatorname{card} A / \operatorname{card} X$. We have

$$\int_X f d\mu_1 = \frac{f(x_1) + \cdots + f(x_n)}{n}.$$

• Let $X = \{x_1, \ldots, x_n\}$ and μ be the measure with density ν with respect to μ_0: we have

$$\int_X f d\mu = \sum_{1 \le j \le n} f(x_j) \nu_j.$$

In particular, if the non-negative real numbers ν_j are such that $\sum \nu_j = 1$, the measure μ is a probability measure on X.

• Let $(X = \{x_i\}_{i \in I}, \mathcal{P}(X))$ be equipped with the counting measure. We have

$$\int_X f d\mu = \sum_{i \in I} f(x_i) = \sup_{J \text{ finite} \subset I} \sum_{i \in J} f(x_i).$$

---

[21] The notation $\int_X f(x) d\mu(x)$, $\int_X f(x) \mu(dx)$ is also commonly used in the literature.

- Let X be a non-empty set and $a \in X$. With μ the Dirac mass at a, we have

$$\int_X f d\mu = f(a).$$

- For the Borel measure m on $\mathbb{R}^d$ (yet to be constructed), we shall use the same notation as for the Riemann integral $\int_{\mathbb{R}^d} f(x)dx$ and we shall see that this integral coincides with the Riemann integral for $f \in C_c^0(\mathbb{R}^d)$. We shall have also

$$\int_{\mathbb{R}} \mathbf{1}_{\mathbb{Q}}(x)dx = m(\mathbb{Q}) = 0.$$

- Let μ be a measure with density ν with respect to the Borel measure: we have

$$\int_{\mathbb{R}^d} f d\mu = \int_{\mathbb{R}^d} f(x)\nu(x)dx$$

so that $d\mu(x) = \nu(x)dx$ and we may consider symbolically that $\mu'(x) = \nu(x)$, explaining the notation $\int f(x)d\mu(x) = \int f(x)\nu(x)dx$. It is also tempting to use that notation, say for the Dirac mass at $0 \in \mathbb{R}$: awfully abusing the notation, making also a formal integration by parts, with $H = \mathbf{1}_{\mathbb{R}_+}$ (Heaviside function) we have, say for $f \in C_c^1(\mathbb{R})$,

$$f(0) = -\int_{\mathbb{R}} f'(x)H(x)dx = \int_{\mathbb{R}} f(x)H'(x)dx = \int_{\mathbb{R}} f(x)\delta(x)dx.$$

Distribution theory is necessary to handle properly these calculations, but the intuition given by the previous formula is not so bad: the Dirac mass at 0 appears as the "derivative" of the Heaviside function, is supported at 0, somehow $+\infty$ at 0 and 0 elsewhere.

- Let $(X, \mathcal{M}, \mu)$ be a measure space where μ is a positive measure and $\Phi : X \to Y$ be a mapping. We have seen in Lemma 1.4.3 the construction of a measure space $(Y, \mathcal{N}, \nu)$ where $\nu = \Phi_*(\mu)$ is the pushforward of μ. Let $g : Y \to \overline{\mathbb{R}}_+$ be a measurable function: then $g \circ \Phi$ is also measurable and

$$\int_Y g d\nu = \int_X (g \circ \Phi)d\mu$$

since for $g = \beta \mathbf{1}_B$,

$$\int_Y g d\nu = \beta\nu(B) = \beta\mu(\Phi^{-1}(B)) = \int_X \beta \mathbf{1}_{(\Phi^{-1}(B))}d\mu$$

$$= \int_X \beta(\mathbf{1}_B \circ \Phi)d\mu = \int_X (g \circ \Phi)d\mu,$$

and is the result by linearity for simple functions (see Exercise 1.9.23 for the general case).

Proposition 1.5.4. *Let $(X, \mathcal{M}, \mu)$ be a measure space where μ is a positive measure, let $f, g : X \to \overline{\mathbb{R}}_+$ be measurable functions, $A, B \in \mathcal{M}$ and $\alpha > 0$ a real number. We define*

$$\int_A f d\mu = \int_X \underbrace{f \cdot \mathbf{1}_A}_{f_A} d\mu, \quad \text{with} \quad f_A(x) = \begin{cases} f(x) & \text{when } x \in A, \\ 0 & \text{when } x \notin A. \end{cases} \tag{1.5.5}$$

The following properties hold.

(1) $0 \le f \le g \Longrightarrow \int_X f d\mu \le \int_X g d\mu, \qquad A \subset B \Longrightarrow \int_A f d\mu \le \int_B f d\mu.$

(2) $\int_X \alpha f d\mu = \alpha \int_X f d\mu.$

(3) $\mu(A) = 0 \Longrightarrow \int_A f d\mu = 0$, *even for* $f \equiv +\infty$.

(4) *Let s be a simple function, $E \in \mathcal{M}$, we define $\lambda_s(E) = \int_E s d\mu$. Then λ_s is a positive measure defined on $\mathcal{M}$.*

Proof. Property (1) follows from Definition 1.5.2 (the second part from $f_A \le f_B$) and (2) follows from Definition 1.5.2 and the last property in (1.5.2):

$$\int_X \alpha f d\mu = \sup_{s \text{ simple} \le \alpha f} I(s) = \sup_{s \text{ simple} \le \alpha f} I\left(\frac{\alpha s}{\alpha}\right) = \alpha \sup_{\frac{s}{\alpha} \text{ simple} \le f} I\left(\frac{s}{\alpha}\right) = \alpha \int_X f d\mu.$$

To get (3), we consider s simple $\le f_A$. We have $A^c \subset \{s = 0\}$, so that with

$$s = \sum_{\substack{1 \le j \le m \\ \alpha_j > 0}} \alpha_j \mathbf{1}_{A_j}$$

as the canonical decomposition of s and $\alpha_i \neq 0$, we have $A_i \subset A$ and thus $\mu(A_i) = 0$, implying $I(s) = 0$ and $\int_A f d\mu = 0$. Let us check (4): we note that $\lambda(\emptyset) = \int_\emptyset s d\mu = 0$ from the already proven (3) and $\mu(\emptyset) = 0$. Let $(E_j)_{j \ge 0}$ a sequence of pairwise disjoint sets in $\mathcal{M}$ and let $s = \sum_{1 \le k \le m} \alpha_k \mathbf{1}_{A_k}$ be a simple function. With $E = \cup_{j \ge 0} E_j$, from Lemma 1.5.1, Definition 1.5.2 and Lemma 1.2.14, we get

$$\lambda_s(E) = \int_X s_E d\mu = \sum_{1 \le k \le m} \alpha_k \mu(A_k \cap E) = \sum_{1 \le k \le m} \alpha_k \left(\sum_{j \ge 0} \mu(A_k \cap E_j)\right)$$

$$= \sum_{j \ge 0} \sum_{1 \le k \le m} \alpha_k \mu(A_k \cap E_j) = \sum_{j \ge 0} \lambda_s(E_j).$$

The proof of the proposition is complete. $\qquad \square$

1.6   Three basic convergence theorems

In the previous section, we were able to define $\int_X f d\mu$, the integral with respect
to a positive measure μ on X for a measurable function $f : X \to \overline{\mathbb{R}}_+$. We shall
soon see that for $f : X \to \mathbb{C}$ measurable such that $\int_X |f| d\mu < \infty$, it is easy to
define $\int_X f d\mu$.

We are now reaching the most interesting part of Integration Theory (essentially elaborated by Henri Lebesgue in his 1902 Ph.D. thesis defended at the
University of Nancy, under the directorship of Emile Borel, see, e.g., [8] for more
references and a historical perspective) and in particular, we shall state and prove
a couple of convergence theorems. Typically, it is our goal to prove that, under a
rather mild convergence assumption of a sequence f_n towards f, we obtain as well
the convergence of the sequence $\int_X f_n d\mu$ towards $\int_X f d\mu$ (at any rate, our convergence assumption on the f_n will be much weaker than uniform convergence). It
is also certainly a great achievement of Lebesgue theory of integration to provide
a vector space of integrable functions which is actually a Banach space. Our first
convergence theorem is due to Beppo Levi.

Theorem 1.6.1 (Monotone Convergence Theorem, a.k.a. Beppo Levi Theorem).
*Let $(X, \mathcal{M}, \mu)$ be a measure space where μ is a positive measure. Let $(f_n)_{n \geq 0}$ be a
sequence of measurable functions $X \to \overline{\mathbb{R}}_+$. Let us assume that*

$$\forall x \in X, \ f_n(x) \uparrow f(x), \quad i.e., \ f_n \ converges \ pointwise \ increasingly \ towards \ f.$$

Then the function f is measurable and

$$\lim_{n \to \infty} \int_X f_n d\mu = \sup_{n \geq 0} \int_X f_n d\mu = \int_X f d\mu.$$

We can note that the convergence assumption is reduced to pointwise convergence. Of course, without the additional hypothesis of monotonicity, the result
is not true in general[22].

Proof. From Proposition 1.3.1 we get that $\sup f_n$ is measurable and (1) in Proposition 1.5.4 implies that the sequence $(\int_X f_n d\mu)_{n \in \mathbb{N}}$ is increasing and bounded from
above by $\int_X f d\mu$. As a result, we have

$$\lim_{n \to \infty} \int_X f_n d\mu = \sup_{n \in \mathbb{N}} \int_X f_n d\mu \leq \int_X f d\mu. \tag{1.6.1}$$

We are left with the proof of the reverse inequality. Let $1 > \epsilon > 0$ and let s be a
simple function such that $0 \leq s \leq f$. We check the set

$$E_n = \{x \in X, (1 - \epsilon)s(x) \leq f_n(x)\},$$

---

[22]We may consider on $[0, 1]$, $f_n(x) = \begin{cases} xn^3, & \text{for } 0 \leq x \leq 1/n, \\ 2n^2 - xn^3, & \text{for } 1/n \leq x \leq 2/n, \\ 0 & \text{elsewhere.} \end{cases}$ The sequence of con-

tinuous functions (f_n) converges pointwise towards 0, nevertheless $\int_0^1 f_n(x) dx = n \to +\infty$.

which is measurable since s and f_n are both measurable and thus

$$f - (1 - \epsilon)s \quad \text{(meaningful since s takes finite values)}$$

is also measurable. We have $E_n = (f_n - (1 - \epsilon)s)^{-1}(\overline{\mathbb{R}}_+)$. Moreover, since the sequence (f_n) is increasing, we get $E_n \subset E_{n+1}$. Also we have $X = \cup_{n \in \mathbb{N}} E_n$ since if we could find $x_0 \in E_n^c$ for all $n \in \mathbb{N}$, we would have

$$+\infty > (1 - \epsilon)s(x_0) > f_n(x_0)(\geq 0)$$

so that $s(x_0) \in]0, +\infty[$ and

$$f(x_0) = \sup_n f_n(x_0) \leq (1 - \epsilon)s(x_0) < s(x_0) \leq f(x_0)$$

which is impossible. As a result, from (4) in Proposition 1.5.4 (λ_s is a measure), Proposition 1.4.4 (increasing convergence for the measure of sets) and (2) in Proposition 1.5.4 (homogeneity), we obtain

$$\int_{E_n} (1 - \epsilon)s d\mu = \lambda_{(1-\epsilon)s}(E_n) \uparrow \lambda_{(1-\epsilon)s}(X) = \int_X (1 - \epsilon)s d\mu = (1 - \epsilon)I(s). \quad (1.6.2)$$

But we have $(1 - \epsilon)s \cdot \mathbf{1}_{E_n} \leq f_n \cdot \mathbf{1}_{E_n} \leq f_n$, so that (1) in Proposition 1.5.4 implies

$$\int_{E_n} (1 - \epsilon)s d\mu \leq \int_{E_n} f_n d\mu \leq \int_X f_n d\mu. \quad (1.6.3)$$

We have thus $(1 - \epsilon)I(s) \underset{(1.6.2)}{=} \lim_n \int_{E_n} (1 - \epsilon)s d\mu \underset{(1.6.3)}{\leq} \sup_n \int_X f_n d\mu$, so that

$$(1 - \epsilon)\int_X f d\mu = (1 - \epsilon) \sup_{s \text{ simple } \leq f} I(s) \leq \lim_n \int_X f_n d\mu \leq \int_X f d\mu, \quad (1.6.4)$$

for all $\epsilon \in (0, 1)$. Taking the supremum on $\epsilon > 0$, yields the result[23]. $\qquad \square$

Corollary 1.6.2. *Let $(X, \mathcal{M}, \mu)$ be a measure space where μ is a positive measure. Let $(f_n)_{n \geq 0}$ be a sequence of measurable functions from $X \longrightarrow \overline{\mathbb{R}}_+$. We set $S(x) = \sum_{n \geq 0} f_n(x)$. Then S is non-negative measurable and*

$$\int_X S d\mu = \sum_{n \geq 0} \int_X f_n d\mu.$$

Proof. The measurability of S follows from Proposition 1.3.1 since in the first place

$$S_n(x) = \sum_{0 \leq k \leq n} f_k(x) \uparrow S(x),$$

---

[23]It is true even if $\int_X f d\mu = +\infty$ since, in that case, all the terms in inequality (1.6.4) are $+\infty$.

and the measurability of a finite sum of measurable functions valued in $\overline{\mathbb{R}}_+$ follows from the measurability (due to the continuity) of

$$
\begin{array}{ccc}
\overline{\mathbb{R}}_+ \times \overline{\mathbb{R}}_+ & \longrightarrow & \overline{\mathbb{R}}_+ \\
(\alpha, \beta) & \mapsto & \alpha + \beta
\end{array} .
$$

We can then apply Theorem 1.6.1 to get

$$
\int_X S d\mu = \sup_{n \geq 0} \int_X S_n d\mu. \tag{1.6.5}
$$

But we have

$$
\int_X S_n d\mu = \int_X \sum_{0 \leq k \leq n} f_k d\mu = \sum_{0 \leq k \leq n} \int_X f_k d\mu, \tag{1.6.6}
$$

where the second equality follows from Lemma 1.6.3 below. Assuming provisionally the results of this lemma, we see that (1.6.5)–(1.6.6) imply our corollary. $\quad\square$

Lemma 1.6.3. *Let $(X, \mathcal{M}, \mu)$ be a measure space where μ is a positive measure. Let $f_1, \ldots, f_N$ be measurable functions from $X \longrightarrow \overline{\mathbb{R}}_+$. Then $f_1 + \cdots + f_N$ is measurable and $\int_X (f_1 + \cdots + f_N) d\mu = \int_X f_1 d\mu + \cdots + \int_X f_N d\mu$.*

Proof. Using induction on N, it is enough to prove the lemma for $N = 2$. Let f_1, f_2 as in the lemma and, using Theorem 1.3.3, let $s_k^{(1)}, s_k^{(2)}$ be simple functions $0 \leq s_k^{(j)} \uparrow f_j, j = 1, 2$. From Theorem 1.6.1, we get

$$
\int_X s_k^{(j)} d\mu \uparrow \int_X f_j d\mu. \tag{1.6.7}
$$

As a result, from Lemma 1.5.1, Theorem 1.6.1 we obtain

$$
\int_X s_k^{(1)} d\mu + \int_X s_k^{(2)} d\mu = \int_X (s_k^{(1)} + s_k^{(2)}) d\mu \uparrow \int_X (f_1 + f_2) d\mu,
$$

providing along with (1.6.7) the result of the lemma. $\quad\square$

Lemma 1.6.4 (Fatou's Lemma). *Let $(X, \mathcal{M}, \mu)$ be a measure space where μ is a positive measure. Let $(f_n)_{n \geq 0}$ be a sequence of measurable functions from $X \to \overline{\mathbb{R}}_+$. The following inequality holds:*

$$
\int_X (\liminf_n f_n) d\mu \leq \liminf_n \left(\int_X f_n d\mu \right).
$$

Proof. We note first that the statement is meaningful since Proposition 1.3.1 implies the measurability of $\liminf f_n$ (valued in $\overline{\mathbb{R}}_+$). Recalling that $\liminf f_n = \sup_{n \in \mathbb{N}} (\inf_{k \geq n} f_k)$, we set $g_n = \inf_{k \geq n} f_k$, and find that g_n is measurable and such that $0 \leq g_n \uparrow \liminf f_n$. Applying then Beppo Levi's theorem 1.6.1, we get

$$
\int_X g_n d\mu \uparrow \int_X (\liminf f_n) d\mu. \tag{1.6.8}
$$

From Property (1) in Proposition 1.5.4, we obtain

$$\int_X g_n d\mu = \int_X (\inf_{k \geq n} f_k) d\mu \leq \int_X f_n d\mu,$$

implying[24] $\liminf \int_X g_n d\mu \leq \liminf \int_X f_n d\mu$ and from (1.6.8) the result of the lemma. $\qquad\square$

Proposition 1.6.5. *Let* $(X, \mathcal{M}, \mu)$ *be a measure space where* μ *is a positive measure. Let* $\nu : X \longrightarrow \overline{\mathbb{R}}_+$ *be a measurable mapping. For* $E \in \mathcal{M}$, *we define* $\lambda_\nu(E) = \int_E \nu d\mu$. *Then* λ_ν *is a positive measure defined on* $\mathcal{M}$. *For* $f : X \longrightarrow \overline{\mathbb{R}}_+$ *measurable, we have*

$$\int_X f d\lambda = \int_X f \cdot \nu \, d\mu$$

where $f \cdot \nu$ *is the measurable function[25] defined by the convention* $0.\infty = 0$. *We shall write* $d\lambda = \nu d\mu$ *and say that* λ *is the measure with density* ν *with respect to* μ.

Proof. We have trivially $\lambda_\nu(\emptyset) = \int_\emptyset \nu d\mu = 0$ from Property (3) in Proposition 1.5.4. Moreover, for $(A_j)_{j \in \mathbb{N}}$ a pairwise disjoint sequence of $\mathcal{M}$, Corollary 1.6.2 implies

$$\lambda_\nu(\cup_{j \geq 0} A_j) = \int_{\cup_{j \geq 0} A_j} \nu d\mu = \int_X \sum_{j \geq 0} \nu \cdot \mathbf{1}_{A_j} d\mu = \sum_{j \geq 0} \int_X \nu \cdot \mathbf{1}_{A_j} d\mu = \sum_{j \geq 0} \lambda_\nu(A_j),$$

proving the first statement in the proposition. For a simple function f, we have $f = \sum_{1 \leq j \leq m} \alpha_j \mathbf{1}_{A_j}$ and we may assume that the α_j are positive real numbers. We get then

$$\int_X f d\lambda_\nu = \sum_{1 \leq j \leq m} \alpha_j \lambda_\nu(A_j) = \sum_{1 \leq j \leq m} \alpha_j \int_X \nu \cdot \mathbf{1}_{A_j} d\mu,$$

and using Lemma 1.6.3, we obtain

$$\int_X f d\lambda_\nu = \int_X \sum_{1 \leq j \leq m} \alpha_j \mathbf{1}_{A_j} \cdot \nu d\mu = \int_X f \cdot \nu d\mu,$$

which is the sought result when f is a simple function. In the general case, we use the approximation Theorem 1.3.3 and Beppo Levi's theorem 1.6.1, providing with simple functions (s_k) converging pointwise increasingly to f,

$$\int_X f d\lambda_\nu \underset{\text{B. Levi}}{=} \sup_k \int_X s_k d\lambda_\nu \underset{s_k \text{ simple}}{=} \sup_k \int_X s_k \cdot \nu d\mu \underset{\text{B. Levi}}{=} \int_X f \cdot \nu d\mu.$$

---

[24] We are using here that for sequences $(x_n), (y_n)$ in $\overline{\mathbb{R}}$, the inequalities $\forall n, x_n \leq y_n$ imply $\liminf x_n \leq \liminf y_n$. This is obvious since for $l \geq n$, $\inf_{k \geq n} x_k \leq x_l \leq y_l$ so that $\inf_{k \geq n} x_k \leq \inf_{k \geq n} y_k$ and $\lim_n (\inf_{k \geq n} x_k) \leq \lim_n (\inf_{k \geq n} y_k)$.

[25] See Remark 1.3.4.

The reader may have noticed that we have used $\sup_k(s_k \cdot \nu) = (\sup_k s_k) \cdot \nu$, indeed obvious except if $\nu(x) = +\infty, \sup_k s_k(x) = 0$ or $\nu(x) = 0, \sup_k s_k(x) = +\infty$. In the latter case, we obtain 0 as well as in the first case since all the $s_k(x)$ are necessarily 0. $\qquad\square$

Definition 1.6.6. Let $(X, \mathcal{M}, \mu)$ be a measure space where μ is a positive measure. Let $f : X \to \mathbb{C}$ be a measurable mapping. We shall say that f belongs to $\mathcal{L}^1(\mu)$ if $\int_X |f| d\mu < +\infty$. We set then[26]

$$\int_X f d\mu = \int_X (\operatorname{Re} f)_+ d\mu - \int_X (\operatorname{Re} f)_- d\mu + i \int_X (\operatorname{Im} f)_+ d\mu - i \int_X (\operatorname{Im} f)_- d\mu,$$

which is meaningful since the integrals $\int_X (\operatorname{Re} f)_\pm d\mu$, $\int_X (\operatorname{Im} f)_\pm d\mu$ are bounded above (Proposition 1.5.4 (1)) by $\int_X |f| d\mu$, a finite quantity.

Proposition 1.6.7. *Let* $(X, \mathcal{M}, \mu)$ *be a measure space where* μ *is a positive measure. Then* $\mathcal{L}^1(\mu)$ *is a vector space on* $\mathbb{C}$ *and* $f \mapsto \int_X f d\mu$ *is a linear form on that space.*

Proof. Let f, g be in $\mathcal{L}^1(\mu)$ and α, β be complex numbers. Then $\alpha f + \beta g$ is a measurable function (Theorem 1.2.7) and since $|\alpha f + \beta g| \leq |\alpha||f| + |\beta||g|$, Proposition 1.5.4 (1)(2) and Lemma 1.6.3 imply $\alpha f + \beta g \in \mathcal{L}^1(\mu)$. If $f = f_1 + if_2, g = g_1 + ig_2$ is the decomposition in real and imaginary part, we have from Definition 1.6.6,

$$\operatorname{Re} \int_X (f + g) d\mu = \int_X (f_1 + g_1)_+ d\mu - \int_X (f_1 + g_1)_- d\mu. \tag{1.6.9}$$

But we have

$$\operatorname{Re}(f + g) = (f_1 + g_1)_+ - (f_1 + g_1)_- = f_1 + g_1 = (f_1)_+ - (f_1)_- + (g_1)_+ - (g_1)_-,$$

so that $(f_1 + g_1)_+ + (f_1)_- + (g_1)_- = (f_1)_+ + (g_1)_+ + (f_1 + g_1)_-$. Applying now Lemma 1.6.3, we get

$$\int_X (f_1 + g_1)_+ d\mu + \int_X (f_1)_- d\mu + \int_X (g_1)_- d\mu$$
$$= \int_X (f_1)_+ d\mu + \int_X (g_1)_+ d\mu + \int_X (f_1 + g_1)_- d\mu,$$

and using (1.6.9) (we manipulate here only real numbers and not $\pm\infty$),

$$\operatorname{Re} \int_X (f + g) d\mu = \int_X (f_1)_+ d\mu + \int_X (g_1)_+ d\mu - \int_X (f_1)_- d\mu - \int_X (g_1)_- d\mu$$
$$= \int_X \operatorname{Re} f d\mu + \int_X \operatorname{Re} g d\mu.$$

---

[26] For $x \in \mathbb{R}$, $x_+ = \max(x, 0)$, $x_- = \max(-x, 0)$ so that $x_\pm \geq 0$ and $x = x_+ - x_-, |x| = x_+ + x_-$.

Since we obtain analogously

$$\operatorname{Im}\int_X (f+g)d\mu = \int_X (f_2)_+ d\mu + \int_X (g_2)_+ d\mu - \int_X (f_2)_- d\mu - \int_X (g_2)_- d\mu$$
$$= \int_X \operatorname{Im} f d\mu + \int_X \operatorname{Im} g d\mu,$$

we get

$$\int_X (f+g)d\mu = \int_X \operatorname{Re} f d\mu + \int_X \operatorname{Re} g d\mu + i\int_X \operatorname{Im} f d\mu + i\int_X \operatorname{Im} g d\mu. \quad (1.6.10)$$

But from Definition 1.6.6, we have

$$\int_X f d\mu = \int_X \operatorname{Re} f d\mu + i\int_X \operatorname{Im} f d\mu,$$

so that (1.6.10) implies $\int_X (f+g)d\mu = \int_X f d\mu + \int_X g d\mu$. On the other hand if $\alpha = \alpha_1 + i\alpha_2$ is a complex number, we get from our reasoning above

$$\int_X \alpha f d\mu = \int_X \alpha_1 f_1 d\mu - \int_X \alpha_2 f_2 d\mu + \int_X i\alpha_1 f_2 d\mu + \int_X i\alpha_2 f_1 d\mu.$$

But for α_1, f_1 real-valued, Definition 1.6.6 and Proposition 1.5.4 (2) provide (with a discussion on the sign of α_1) $\int_X \alpha_1 f_1 d\mu = \alpha_1 \int f_1 d\mu$. We are left with the proof of $\int_X i f_1 d\mu = i\int f_1 d\mu$, which follows immediately from Definition 1.6.6. The proof of the proposition is complete. $\qquad\square$

Theorem 1.6.8 (Lebesgue dominated convergence theorem).[27] *Let $(X, \mathcal{M}, \mu)$ be a measure space where μ is a positive measure. Let $(f_n)_{n\in\mathbb{N}}$ be a sequence of measurable functions from X into $\mathbb{C}$ such that the following properties hold.*

(1) **Pointwise convergence:** $\forall x \in X$, $\lim_{n\to\infty} f_n(x) = f(x)$.
(2) **Domination:** $\exists g : X \to \overline{\mathbb{R}}_+$ *measurable, with* $\int_X g d\mu < +\infty$, *so that*

$$\forall n \in \mathbb{N}, \forall x \in X, \qquad |f_n(x)| \le g(x).$$

Then f is measurable and $\int_X |f|d\mu < +\infty$; moreover we have

$$\lim_{n\to\infty}\int_X |f - f_n|d\mu = 0, \quad \textit{implying} \quad \lim_{n\to\infty}\int_X f_n d\mu = \int_X f d\mu.$$

Proof. The measurability of f follows from Proposition 1.3.1. Moreover, Proposition 1.5.4 (1) implies

$$\int_X |f_n|d\mu \le \int_X g d\mu < +\infty.$$

---

[27] We shall give later a slightly more general version taking into account negligible sets.

Fatou's lemma 1.6.4 entails then

$$\int_X |f|\,d\mu = \int_X \liminf_n |f_n|\,d\mu \le \liminf_n \int_X |f_n|\,d\mu \le \int_X g\,d\mu < +\infty.$$

On the other hand, the inequality $|f_n - f| \le 2g$ and the Fatou's lemma imply

$$\int_X 2g\,d\mu = \int_X \liminf_n \left(2g - |f_n - f|\right)d\mu \le \liminf_n \int_X \left(2g - |f_n - f|\right)d\mu$$

$$\le \int_X 2g\,d\mu < +\infty.$$

From Lemma 1.6.3, we obtain thus

$$\int_X \left(2g - |f_n - f|\right)d\mu + \int_X |f_n - f|\,d\mu = \int_X 2g\,d\mu \le \liminf_n \int_X \left(2g - |f_n - f|\right)d\mu.$$

As a result, we get

$$\limsup_n \int_X |f_n - f|\,d\mu$$

$$\le \liminf_n \int_X \left(2g - |f_n - f|\right)d\mu + \limsup_n -\left[\int_X \left(2g - |f_n - f|\right)d\mu\right] = 0,$$

since the numerical sequence $\int_X \left(2g - |f_n - f|\right)d\mu$ is bounded. $\qquad\square$

1.7 Space $L^1(\mu)$ and negligible sets

The next proposition introduces the notion of a property true *almost everywhere* in a measure space $(X, \mathcal{M}, \mu)$. We shall write for short μ-a.e. for μ-almost everywhere.

Proposition 1.7.1. *Let $(X, \mathcal{M}, \mu)$ be a measure space where μ is a positive measure and let $f, g : X \to \overline{\mathbb{R}}_+$ be measurable mappings.*

(1) *$\int_X f\,d\mu = 0$ is equivalent to $f = 0, \mu$-a.e., i.e., $\mu(\{x \in X, f(x) \ne 0\}) = 0$.*
(2) *If $f \le g, \mu$-a.e., i.e., $\mu(\{x \in X, f(x) > g(x)\}) = 0$, then*

$$\int_X f\,d\mu \le \int_X g\,d\mu.$$

(3) *If $f = g, \mu$-a.e., i.e., $\mu(\{x \in X, f(x) \ne g(x)\}) = 0$, then $\int_X f\,d\mu = \int_X g\,d\mu$.*
(4) *If $\int_X f\,d\mu < +\infty$, then $f < +\infty, \mu$-a.e., i.e., $\mu(\{x \in X, f(x) = +\infty\}) = 0$.*

Proof. Let us prove (1): if $\int_X f\,d\mu = 0$, we define for any integer $k \ge 1$,

$$F_k = \{f \ge 1/k\}.$$

The sequence F_k is increasing measurable and $\cup_{k \geq 1} F_k = \{f > 0\}$. From Proposition 1.4.4, we obtain $\mu(F_k) \uparrow \mu(\{f > 0\})$ when $k \to +\infty$. But we have

$$\mu(F_k) = \int_X \mathbf{1}(f \geq 1/k)d\mu \overset{\text{Proposition 1.5.4(1)}}{\leq} \int_X k \cdot f d\mu$$

$$\underset{\text{Proposition 1.5.4(2)}}{=} k \int_X f d\mu = 0 \Longrightarrow \mu(\{f > 0\}) = 0.$$

Conversely, if $\mu(E) = 0$ with $E = \{f > 0\}$, since $f = f \cdot \mathbf{1}_E$, we obtain

$$\int_X f d\mu = \int_X f \cdot \mathbf{1}_E d\mu = \int_E f d\mu = 0, \quad \text{from Proposition 1.5.4(3).}$$

In particular, for $f \in \mathcal{L}^1(\mu)$, we have

$$\int_X |f| d\mu = 0 \Longrightarrow f = 0, \mu\text{-a.e.} \tag{1.7.1}$$

Let us prove (2). We consider the set E with measure 0 defined by $E = \{x \in X, f(x) > g(x)\}$. We have

$$f = f \cdot \mathbf{1}_E + f \cdot \mathbf{1}_{E^c}, \quad g = g \cdot \mathbf{1}_E + g \cdot \mathbf{1}_{E^c}, \tag{1.7.2}$$

and $f \cdot \mathbf{1}_{E^c} \leq g \cdot \mathbf{1}_{E^c}$. From Proposition 1.5.4 and Lemma 1.6.3, we see that it is enough to prove

$$\int_X f \cdot \mathbf{1}_E d\mu = 0 = \int_X g \cdot \mathbf{1}_E d\mu,$$

which is indeed fulfilled since $\int_X f \cdot \mathbf{1}_E d\mu = \int_E f d\mu = 0$, from Proposition 1.5.4(3). Using (1.7.2) for $E = \{x \in X, f(x) \neq g(x)\}$, Lemma 1.6.3 and Proposition 1.5.4, we obtain (3). To prove (4), we define $E = \{f = +\infty\}$, and we note that $\mu(E) > 0$ implies for all integers $n \geq 1$, that

$$\int_X f d\mu \geq \int_E f d\mu \geq n \int_E d\mu = n\mu(E),$$

entailing $\int_X f d\mu = +\infty$. $\qquad\qquad\qquad\qquad\qquad\qquad\qquad\qquad\square$

Definition 1.7.2. Let $(X, \mathcal{M}, \mu)$ be a measure space where μ is a positive measure. The space $L^1(\mu)$ is defined as the quotient of $\mathcal{L}^1(\mu)$ (cf. Definition 1.6.6) by the equivalence relation of *equality μ-a.e.* ($f \sim g$ means $\mu(\{x \in X, f(x) \neq g(x)\}) = 0$).

Remark 1.7.3. We note that $L^1(\mu)$ is a complex vector space as the quotient of the vector space $\mathcal{L}^1(\mu)$ by the subspace $\{f \in \mathcal{L}^1(\mu), f \sim 0\}$.[28] On the other hand, the

---

[28] For $f_1, f_2 \in \mathcal{L}^1(\mu)$ vanishing respectively on N_1^c, N_2^c with $\mu(N_j) = 0$, then for $\alpha_1, \alpha_2 \in \mathbb{C}$, we have $\alpha_1 f_1 + \alpha_2 f_2 = 0$ on $(N_1 \cup N_2)^c$ thus μ-a.e. since $\mu(N_1 \cup N_2) = 0$.

linear mapping $f \mapsto \int_X f d\mu$ defined on $\mathcal{L}^1(\mu)$ is compatible with the equivalence relation, i.e., depends only on the equivalence class of f: if $f \sim 0$, we have

$$\int_X f d\mu = \int_X (\operatorname{Re} f)_+ d\mu - \int_X (\operatorname{Re} f)_- d\mu + i \int_X (\operatorname{Im} f)_+ d\mu - i \int_X (\operatorname{Im} f)_+ d\mu = 0,$$

from Proposition 1.7.1(1). Similarly, for $f, g \in \mathcal{L}^1(\mu)$ real-valued and

$$f \leq g \ \mu\text{-a.e., then} \quad \int_X f d\mu \leq \int_X g d\mu. \tag{1.7.3}$$

This follows immediately from Proposition 1.6.7 and from

$$g - f \sim (g - f)\mathbf{1}_{N^c} \geq 0, \quad \text{with } \mu(N) = 0,$$

providing (1.7.3) using Proposition 1.7.1.

Theorem 1.7.4. *Let $(X, \mathcal{M}, \mu)$ be a measure space where μ is a positive measure.*

(1) *The mapping from $L^1(\mu)$ into $\mathbb{C}$ defined by $f \mapsto \int_X f d\mu$ is a linear form.*

(2) *The mapping from $L^1(\mu)$ into $\mathbb{R}_+$ defined by $f \mapsto \int_X |f| d\mu = \|f\|_{L^1(\mu)}$ is a norm and for $f \in L^1(\mu)$*

$$\left| \int_X f d\mu \right| \leq \int_X |f| d\mu. \tag{1.7.4}$$

N.B. We postpone to Section 3.2 in Chapter 2 the introduction of spaces $L^p(\mu)$ along with the proof that these spaces are complete.

Proof. Property (1) follows from Remark 1.7.3, and for the same reason, the mapping defined in (2) makes sense on the quotient space $L^1(\mu)$. If $f \in \mathcal{L}^1(\mu)$ is such that $\|f\|_{L^1(\mu)} = 0$, Proposition 1.7.1(1) implies $f \sim 0$, i.e., $f = 0$ in $L^1(\mu)$. Proposition 1.5.4(2) provides the homogeneity of this mapping, whereas the triangle inequality follows from

$$\|f + g\|_{L^1(\mu)} = \int_X |f + g| d\mu \leq \int_X (|f| + |g|) d\mu = \|f\|_{L^1(\mu)} + \|g\|_{L^1(\mu)}.$$

Finally, let us prove (1.7.4). We define the complex number

$$z = \int_X f d\mu = |z| e^{i\theta},$$

and using Proposition 1.6.7, Definition 1.6.6 and (1.7.3), we get

$$\left| \int_X f d\mu \right| = \operatorname{Re}\left(e^{-i\theta} \int_X f d\mu \right) = \operatorname{Re} \int_X e^{-i\theta} f d\mu$$

$$= \int_X \operatorname{Re}(e^{-i\theta} f) d\mu \leq \int_X |e^{-i\theta} f| d\mu = \int_X |f| d\mu. \qquad \square$$

Theorem 1.7.5 (Lebesgue dominated convergence theorem). *Let $(X, \mathcal{M}, \mu)$ be a measure space where μ is a positive measure. Let $(f_n)_{n \in \mathbb{N}}$ be a sequence of measurable functions from X into $\mathbb{C}$ such that the following properties hold.*

(1) **Pointwise convergence:** $\lim_{n \to \infty} f_n(x) = f(x), \mu\text{-}a.e.$[29]
(2) **Domination:** $\exists g : X \to \overline{\mathbb{R}}_+$ with $\int_X g\,d\mu < +\infty$, such that $\forall n \in \mathbb{N}, |f_n| \leq g$, $\mu\text{-}a.e.$[30] *Then the function f is*[31] *measurable,* $\int_X |f|\,d\mu < +\infty$ *and*

$$\lim_{n \to \infty} \int_X |f - f_n|\,d\mu = 0, \quad implying \quad \lim_{n \to \infty} \int_X f_n\,d\mu = \int_X f\,d\mu.$$

Proof. Taking into account our footnotes, we set

$$B = N \cup \bigcup_{n \in \mathbb{N}} M_n, \quad (\text{we have } B \in \mathcal{M} \text{ and } \mu(B) = 0), \quad \widetilde{f}(x) = \lim_n f_n(x) \mathbf{1}_{B^c}(x).$$

The sequence $\widetilde{f}_n = \mathbf{1}_{B^c} f_n$ satisfies the assumptions of Theorem 1.6.8, so that $\widetilde{f} \in \mathcal{L}^1(\mu)$ and

$$\lim_{n \to +\infty} \int_X |\widetilde{f}_n - \widetilde{f}|\,d\mu = 0.$$

Since $|f - f_n| = |\widetilde{f} - \widetilde{f}_n| + |f - f_n|\mathbf{1}_B$ and $f = \widetilde{f} + f\mathbf{1}_B$ with $\mu(B) = 0$, we get from Proposition 1.5.4(3) that $f \in \mathcal{L}^1(\mu)$ and the result

$$\lim_{n \to +\infty} \int_X |f_n - f|\,d\mu = 0. \qquad \square$$

Remark 1.7.6. We may reformulate this theorem in a more concise and elegant way by saying that whenever $(f_n)_{n \in \mathbb{N}}$ is a sequence of $L^1(\mu)$ converging pointwise to f with a domination condition $|f_n| \leq g \in L^1(\mu)$, then f_n converges towards f in the space $L^1(\mu)$. To sum-up, for a sequence (f_n) in $L^1(\mu)$,

$$\left. \begin{array}{c} f_n \xrightarrow[\text{convergence}]{\text{pointwise}} f \\ \text{and} \\ |f_n| \leq g \in L^1(\mu) \end{array} \right\} \implies f_n \xrightarrow[L^1(\mu)]{} f. \qquad (1.7.5)$$

The following lemma is taken from [16] (and has also an L^p version).

Lemma 1.7.7. *Let $(X, \mathcal{M}, \mu)$ be a measure space where μ is a positive measure. Let $(f_n)_{n \in \mathbb{N}}$ be a sequence of measurable functions from X into $\mathbb{C}$ such that the following properties hold.*

(1) *Pointwise convergence:* $\lim_{n \to \infty} f_n(x) = f(x), \mu\text{-}a.e.,$
(2) $\sup_n \int_X |f_n|\,d\mu < +\infty.$
Then $f \in \mathcal{L}^1(\mu)$ *and* $\|f_n - f\|_{L^1(\mu)} + \|f\|_{L^1(\mu)} - \|f_n\|_{L^1(\mu)} \longrightarrow 0.$

---

[29] $\exists N \in \mathcal{M}$, such that $\mu(N) = 0$ and $\forall x \in N^c$, $(f_n(x))_{n \in \mathbb{N}}$ is convergent with limit $f(x)$.
[30] $\forall n \in \mathbb{N}, \exists M_n \in \mathcal{M}$ with $\mu(M_n) = 0$ such that $\forall x \in M_n^c, |f_n(x)| \leq g(x)$.
[31] We define $f(x) = \mathbf{1}_{N^c}(x) \lim_n f_n(x)$.

Proof. Fatou's lemma implies

$$\int_X |f|\,d\mu = \int_X \liminf |f_n|\,d\mu \leq \liminf \int_X |f_n|\,d\mu \leq \sup_n \int_X |f_n|\,d\mu < +\infty.$$

On the other hand, we have $|f_n| \leq |f_n - f| + |f| \leq |f_n| + 2|f|$, so that

$$0 \leq |f_n - f| + |f| - |f_n| \leq 2|f|.$$

The Lebesgue dominated convergence theorem yields the result. $\qquad\square$

An important consequence is the following result.

Proposition 1.7.8. *Let $(X, \mathcal{M}, \mu)$ be a measure space where μ is a positive measure. Let f be in $L^1(\mu)$ and let $(f_n)_{n \in \mathbb{N}}$ be a sequence of functions in $L^1(\mu)$ such that the following properties hold.*

(1) *Pointwise convergence:* $\lim_{n \to \infty} f_n(x) = f(x), \mu\text{-a.e.,}$
(2) $\lim_n \|f_n\|_{L^1(\mu)} = \|f\|_{L^1(\mu)}.$

Then $\lim_n \|f_n - f\|_{L^1(\mu)} = 0$.

Remark 1.7.9. To sum-up, for a sequence (f_n) in $L^1(\mu)$, $f \in L^1(\mu)$,

$$\left.\begin{array}{c} f_n \xrightarrow[\text{convergence}]{\text{pointwise}} f \\ \text{and} \\ \lim_n \|f_n\|_{L^1(\mu)} = \|f\|_{L^1(\mu)} \end{array}\right\} \implies f_n \xrightarrow[L^1(\mu)]{} f. \qquad (1.7.6)$$

The following proposition is an important consequence of the Lebesgue dominated convergence theorem.

Proposition 1.7.10. *Let $(X, \mathcal{M}, \mu)$ be a measure space where μ is a positive measure. Let $f : X \longrightarrow \overline{\mathbb{R}}_+$ be a measurable mapping such that $\int_X f\,d\mu < \infty$.*

(1) *The set $N = \{x \in X, f(x) = +\infty\} \in \mathcal{M}$ and $\mu(N) = 0$.*
(2) *For any $\epsilon > 0$, there exists $\alpha > 0$ such that for all $E \in \mathcal{M}$ satisfying $\mu(E) \leq \alpha$, we have $\int_E f\,d\mu < \epsilon$. In other words, $\lim_{\substack{\mu(E) \to 0 \\ E \in \mathcal{M}}} \int_E f\,d\mu = 0$.*

In particular, for $u \in L^1(\mu)$, we have

$$\lim_{\substack{\mu(E) \to 0 \\ E \in \mathcal{M}}} \int_E |u|\,d\mu = 0. \qquad (1.7.7)$$

Proof. (1) The set $N = \{x \in X, f(x) = +\infty\}$ belongs to $\mathcal{M}$ as the inverse image of the closed set $\{+\infty\}$ by the measurable f. For all integers k, $k\mathbf{1}_N \leq f$, so that $k\mu(N) \leq \int_X f\,d\mu < +\infty$. The non-negative sequence $(k\mu(N))_{k \in \mathbb{N}}$ is bounded so that $\mu(N) = 0$.

(2) Let $E \in \mathcal{M}$ and $n \in \mathbb{N}$: since $\mu(N) = 0$, we have

$$(\natural) \qquad \int_E f d\mu = \int_{E \cap N^c} f d\mu = \int_{E \cap N^c \cap \{f \le n\}} f d\mu + \int_{E \cap N^c \cap \{f > n\}} f d\mu$$

$$\le n\mu(E) + \int f \mathbf{1}_{E \cap N^c \cap \{f > n\}} d\mu \le n\mu(E) + \int f \mathbf{1}_{n < f < +\infty} d\mu.$$

The sequence $g_n = f \mathbf{1}_{\{n < f < +\infty\}}$ is such that $g_n(x) = 0$ for $n \ge f(x)$, which is verified for $x \in N^c$ if n is large enough. Since $g_n(x) = 0$ for $x \in N$, we find

$$(\flat) \qquad\qquad \forall x \in X, \quad g_n(x) \to 0.$$

Moreover

$$(\sharp) \qquad\qquad 0 \le g_n \le f \mathbf{1}_{N^c} \quad \text{and} \quad f \mathbf{1}_{N^c} \in L^1(\mu).$$

The Lebesgue dominated convergence Theorem 1.7.5 shows that $(\flat)$ and $(\sharp)$ imply the convergence of g_n towards 0 in $L^1(\mu)$. From $(\natural)$, we get

$$0 \le \int_E f d\mu \le n\mu(E) + \theta_n, \quad \text{with} \quad \theta_n \xrightarrow[n \to +\infty]{} 0_+.$$

Let $\epsilon > 0$ be given: $\exists N \in \mathbb{N}$ such that $\theta_N < \epsilon/2$. Defining $\alpha = \frac{\epsilon}{2N+1}$ (we have indeed $\alpha > 0$), we get for $\mu(E) \le \alpha$,

$$0 \le \int_E f d\mu \le \frac{N\epsilon}{2N+1} + \theta_N < \epsilon/2 + \epsilon/2 = \epsilon, \quad \text{qed.}$$

A slightly shorter reasoning from $(\natural)$ would be

$$\forall n \in \mathbb{N}, \quad 0 \le \limsup_{\mu(E) \to 0} \int_E f d\mu \le \theta_n \implies 0 \le \limsup_{\mu(E) \to 0} \int_E f d\mu \le \lim_n \theta_n = 0. \qquad \square$$

1.8 Notes

Let us follow alphabetically the names of mathematicians encountered in our text above. Much more details can be obtained on the web and in particular at the very complete `http://www-groups.dcs.st-andrews.ac.uk/history/BiogIndex.html`

René BAIRE (1874–1932) was a French mathematician; the Baire category theorem is certainly the most basic and important theorem in Functional Analysis.

Stefan BANACH (1892–1945), a Polish mathematician who set the basis of Functional Analysis.

BERNOULLI (The reader will have certainly noted the spelling of the name with only a single "i".) The brothers Jacques (1654–1705) and Jean (1667–1748) Bernoulli as well as Daniel (1700–1782), son of Jean, lived in Basel and contributed to the development of Integral Calculus (Jacques), Mechanics (Jean), Kinetic Theory of Gas (Daniel). Jacques Bernoulli (quoted in the example page 19) contributed also to the calculus of probabilities with the *Law of large numbers* (see a simple version in Exercise 1.9.21(3)).

Emile BOREL (1871–1956), a French mathematician and politician, one of the creators of measure theory.

Augustin CAUCHY (1789–1857), a French mathematician, is one of the founders of Analysis.

Paul DIRAC (1902–1984) was a British physicist, one of the creators of Quantum Mechanics.

Pierre FATOU (1878–1929) was a French mathematician, author of the lemma bearing his name, a cornerstone of measure theory.

Carl-Friedrich GAUSS (1777–1855) was the most important German mathematician of his times.

Felix HAUSDORFF (1869–1942) was a German mathematician, founder of General Topology.

Pierre-Simon LAPLACE (1749–1827) was a French astronomer and mathematician.

Henri LEBESGUE (1875–1941) created modern measure theory in 1901, generalizing Riemann theory of integration.

Beppo LEVI (1875–1961) was an Italian mathematician, professor at the university of Genova, also an expert in algebraic geometry; he was forced into exile in 1938 by the antisemitic persecutions of the Mussolinian regime. There is now a Mathematics Research Institute named after Beppo Levi in the Argentinian town of Rosario, where he found refuge.

Denis POISSON (1781–1840) was a French mathematician.

Bernhard RIEMANN (1826–1866) was a German mathematician who contributed to many different areas of mathematics, ranging from Number Theory to Mathematical Analysis.

Lebesgue's dominated convergence theorem was first proven by Lebesgue on probability spaces, before B. Levi proved his monotone convergence theorem for non-negative functions. The latter result implies Fatou's lemma, from which follows easily the more general version of Lebesgue's dominated convergence.

1.9 Exercises

Elementary set theory

Exercise 1.9.1. *Show that the mapping $(p, q) \in \mathbb{N} \times \mathbb{N} \mapsto 2^p(2q+1) \in \mathbb{N}^*$ is bijective.*

Answer. Let $m \in \mathbb{N}^*$. Then m can be written as $2^p \times$ an odd integer with $p \in \mathbb{N}$, proving surjectivity. Moreover, if p_j, q_j are natural integers and $2^{p_1}(2q_1 + 1) = 2^{p_2}(2q_2 + 1)$, assuming as we may $p_1 \leq p_2$, we get that the odd number

$$2q_1 + 1 = 2^{p_2 - p_1}(2q_2 + 1),$$

implying that $p_2 = p_1$ and thus $q_2 = q_1$, proving injectivity.

Exercise 1.9.2. *Let X be a set and $\mathcal{P}(X)$ the set of its subsets. Let $f : X \to \mathcal{P}(X)$ be a mapping. Show that f cannot be onto.*

Answer. Let us consider $A = \{x \in X, x \notin f(x)\}$. Let us assume that there exists $a \in X$ such that $A = f(a)$. If $a \in f(a) = A$, then $a \notin A$, which is impossible. If $a \notin f(a) = A$, then $a \in A$, which is also impossible. As a result there does not exist $a \in X$ such that $A = f(a)$ and f is not onto.

Comment. We have proven more than what was actually required, since we produced an explicit construction. Let f be a mapping from X into $\mathcal{P}(X)$, then the set A is not in the image of f. This example is a version of the liar's paradox, already known in the ancient Greek civilization. Does somebody claiming "I lie" speak the truth? If yes, then he is indeed lying and thus does not speak the truth. If not, he is lying in saying that he lies and thus speaks the truth. . .

Back to mathematics, a very important consequence of this exercise is the so-called Russell's paradox[32] after which there is not a set of all sets. In fact, if such a "universe" $\mathcal{U}$ existed, it would contain its powerset and the inclusion $\mathcal{P}(\mathcal{U}) \subset \mathcal{U}$ would imply the existence of a surjection from $\mathcal{U}$ onto $\mathcal{P}(\mathcal{U})$. We could also consider

$$Y = \{x \in \mathcal{U}, x \notin x\},$$

and note that if $Y \in Y$, from the definition of Y we would have $Y \notin Y$. If $Y \notin Y$ then from the definition of Y, we would get $Y \in Y$, reaching a contradiction in both cases. Note that for finite sets, it is trivial to prove directly that $\forall n \in \mathbb{N}, n < 2^n$ (induction works with $n + 1 \leq 2^n$).

---

[32] Bertrand Russell (1872–1970) is a British logician, co-author of the monumental treatise *Principia Mathematica*, a joint work with A.N. Whitehead (1861–1947), elaborated between 1910 and 1913. In 1895, Georg Cantor (1845–1918) did create Set Theory, "a paradise from which we cannot be expelled" according to the words of David Hilbert. Seven years later, it was clear that serious difficulties occurred in Cantor theory, in particular with the very notion of set. Russell was an extraordinary character: Nobel prize winner for literature in 1950, he fought with great energy against the development of nuclear weapons and founded the very influential *Russell Tribunal*. For more on B. Russell: `http://www-history.mcs.st-and.ac.uk/history/Mathematicians/Russell.html` `http://www.nobel.se/literature/laureates/1950` and on liar's paradox: `http://www.utm.edu/research/iep/p/par-liar.htm`

Exercise 1.9.3.

(1) *Let X be a set and $A_1, \ldots, A_n$ be a finite partition of X. What is the σ-algebra generated by $A_1, \ldots, A_n$ and what is its cardinal?*

(2) *Let X be a set and $(A_k)_{k \in \mathbb{N}}$ be a partition of X. What is the σ-algebra generated by $(A_k)_{k \in \mathbb{N}}$? Show that it is equipotent to $\mathcal{P}(\mathbb{N})$.*

Answer. Question 1 is dealt with on page 2: the cardinal of that σ-algebra is 2^n.
(2) We define $\mathcal{T} = \{\cup_{j \in J} A_j\}_{J \subset \mathbb{N}}$. For all $j \in \mathbb{N}$, $A_j \in \mathcal{T}$ and every σ-algebra $\mathcal{A}$ such that all $A_j \in \mathcal{A}$ will contain $\mathcal{T}$. Moreover $\mathcal{T}$ is stable by reunion since

$$\cup_{i \in I} \cup_{j \in J_i} A_j = \cup_{j \in \cup_{i \in I} J_i} A_j, \quad \text{and } \cup_{i \in I} J_i \subset \mathbb{N}.$$

It is also stable by complement since $(A_k)_{k \in \mathbb{N}}$ is a partition: $\left(\cup_{j \in J} A_j\right)^c = \cup_{j \in J^c} A_j$. $\mathcal{T}$ contains also $X = \cup_{j \in \mathbb{N}} A_j$ and thus is the σ-algebra generated by the A_j. Let us now check the mapping

$$\mathcal{P}(\mathbb{N}) \ni J \mapsto \cup_{j \in J} A_j \in \mathcal{T},$$

which is obviously onto. This mapping is also one-to-one since, for J, K subsets of $\mathbb{N}$ such that

$$\cup_{j \in J} A_j = \cup_{k \in K} A_k,$$

we get for $j_0 \in J$, $A_{j_0} = \cup_{k \in K} \left(A_{j_0} \cap A_k\right) = \emptyset$ if $j_0 \notin K$. Since $A_{j_0} \neq \emptyset$, we obtain $J \subset K$ and similarly $K \subset J$, i.e., $J = K$ and a one-to-one mapping. We can write symbolically

$$\operatorname{card} \mathcal{T} = 2^{\aleph_0}$$

since we have proven that $\mathcal{T}$ is equipotent to $\mathcal{P}(\mathbb{N})$ and the cardinal of $\mathbb{N}$ is denoted by $\aleph_0$, pronounced aleph null (first letter in the 22-letters Hebrew alphabet).

This symbolic notation is justified by the general notation Y^X for the set of all mappings from a set X to a set Y and the fact that $\mathcal{P}(X)$ is equipotent to $\{0, 1\}^X$: the mapping

$$\Phi : \ \{0, 1\}^X \ni f \mapsto f^{-1}(\{1\}) \in \mathcal{P}(X)$$

is a bijection since it is one-to-one ($f^{-1}(\{1\}) = g^{-1}(\{1\})$ implies $f^{-1}(\{0\}) = \left(f^{-1}(\{1\})\right)^c = \left(g^{-1}(\{1\})\right)^c = g^{-1}(\{0\})$ and $f = g$) and onto since for $A \subset X$, $\mathbf{1}_A$ the indicator function of A (which is 1 on A, 0 elsewhere), we have

$$\Phi(\mathbf{1}_A) = A.$$

As a result, $\mathcal{P}(X)$ is equipotent to $\{0, 1\}^X$ and $\operatorname{card} \mathcal{P}(X) = 2^{\operatorname{card} X}$, as we have defined

$$(\operatorname{card} Y)^{\operatorname{card} X} = \operatorname{card}(Y^X).$$

The reader will find more on set theory and cardinals in Section 10.1 of our appendix.

Exercise 1.9.4. *Let X be a set and let $\mathcal{M}$ be a countable σ-algebra on X.*

(1) *Show that for $x \in X$, $A(x) = \bigcap_{\substack{M \in \mathcal{M} \\ x \in M}} M$ belongs to $\mathcal{M}$.*

(2) *Show that for $x, x' \in X$, we have either $A(x) \cap A(x') = \emptyset$ or $A(x) = A(x')$.*

(3) *Show that $\mathcal{M}$ is a σ-algebra generated by a countable partition. Show that $\mathcal{M}$ is finite (hint: use Exercise 1.9.3).*

Answer. (1) $A(x)$ is a countable intersection (since $\mathcal{M}$ is countable) of elements of $\mathcal{M}$, and thus belongs to $\mathcal{M}$.

(2) Let x, x' be elements of X. If $x \in A(x')$, we get $A(x) \subset A(x')$ and thus $A(x) = A(x') \cap A(x)$. Consequently, if $x \in A(x')$ and $x' \in A(x)$, we find

$$A(x) = A(x') \cap A(x) = A(x').$$

If $x \notin A(x')$ then $A(x')^c$ belongs to $\mathcal{M}$ and contains x so that $A(x) \subset A(x')^c$, entailing $A(x) \cap A(x') = \emptyset$ (same result if $x' \notin A(x)$).

(3) We define

$$\mathcal{N} = \{B \subset X, \exists x \in X, B = A(x)\}.$$

It is a subset of $\mathcal{M}$ and thus it is a countable set. Moreover, from (2) if $B \neq B' \in \mathcal{N}$, we have $B \cap B' = \emptyset$. With D countable, we note $\mathcal{N} = \{B_k\}_{k \in D}$ and find that $\mathcal{N}$ is a partition of X: if $X \neq \emptyset$ (if $X = \emptyset$, $\mathcal{M} = \{\emptyset\}$) no B_k is empty and $B_k \cap B_l = \emptyset$ for $k \neq l \in D$. We have also $\cup_{k \in D} B_k = X$ since for $x \in X$, there exists $k \in D$, such that $A(x) = B_k$. The σ-algebra $\mathcal{M}$ contains the σ-algebra generated by $\mathcal{N}$, which is uncountable when D is infinite from Exercise 1.9.3. This implies that D is finite as well as the σ-algebra generated by $\mathcal{N}$. Moreover, if $C \in \mathcal{M}$, we find

$$C = \cup_{x \in C} A(x)$$

since for $x \in C$, $C \supset A(x)$ and $x \in A(x)$; as a result C is a (countable) union of elements of $\mathcal{N}$. The σ-algebra $\mathcal{M}$ is thus the σ-algebra generated by $\mathcal{N}$, which is finite.

Exercise 1.9.5. *Show that $\mathbb{R}$ is equipotent to $\mathcal{P}(\mathbb{N})$ (hint: use dyadic expansions). Show that $\mathbb{R}$ is not countable.*

Answer. The last assertion follows from the first and Exercise 1.9.2. The mapping ψ_0 defined in (1.2.19) is bijective from $\mathbb{R}$ onto $(-1, 1)$, which is equipotent to $(0, 1)$ $(x \mapsto (x+1)/2)$. We have seen in the previous exercise that $\mathcal{P}(\mathbb{N})$ is equipotent to $\{0, 1\}^{\mathbb{N}}$, the set of mappings from $\mathbb{N}$ into $\{0, 1\}$. We have thus to prove that $\{0, 1\}^{\mathbb{N}}$ is equipotent to $(0, 1)$.

Let x be in $(0, 1)$. With E standing for the floor function (see the footnote on page 16), we define for any integer $k \geq 1$,

$$x_k = E(2^k x) - 2E(2^{k-1} x) = p_k(x).$$

Note that $E(t) = \max\{n \in \mathbb{Z}, n \leq t\} = \min\{n \in \mathbb{Z}, t < n + 1\}$. We have

$$E(2^k x) \leq 2^k x < E(2^k x) + 1,$$
$$E(2^{k-1} x) \leq 2^{k-1} x < E(2^{k-1} x) + 1,$$

and thus $2E(2^{k-1} x) \leq 2^k x < 2E(2^{k-1} x) + 2$, which implies

$$2E(2^{k-1} x) \leq E(2^k x) \leq 2^k x < E(2^k x) + 1 \leq 2E(2^{k-1} x) + 2.$$

This gives

$$0 \leq x_k = p_k(x) = E(2^k x) - 2E(2^{k-1} x) < E(2^k x) + 1 - 2E(2^{k-1} x) \leq 2.$$

Since x_k is an integer, we get $x_k \in \{0, 1\}$ and the series $\sum_{k \geq 1} \frac{x_k}{2^k}$ converges. We note that for any integer $n \geq 1$,

$$\sum_{1 \leq k \leq n} \frac{x_k}{2^k} = \sum_{1 \leq k \leq n} \frac{E(2^k x) - 2E(2^{k-1} x)}{2^k} = \sum_{1 \leq k \leq n} \frac{E(2^k x)}{2^k} - \sum_{1 \leq k \leq n} \frac{E(2^{k-1} x)}{2^{k-1}}$$

$$= \sum_{1 \leq k \leq n} \frac{E(2^k x)}{2^k} - \sum_{0 \leq k \leq n-1} \frac{E(2^k x)}{2^k} = \frac{E(2^n x)}{2^n} - E(x) = 2^{-n} E(2^n x).$$

Since $2^{-n} E(2^n x) \leq x < 2^{-n} E(2^n x) + 2^{-n}$, this implies $\lim_n 2^{-n} E(2^n x) = x$ and thus

$$x = \sum_{k \geq 1} \frac{x_k}{2^k}$$

with $x_k \in \{0, 1\}$. We have just constructed a mapping Ψ (dyadic expansion)

$$\begin{array}{rcl} \Psi : (0, 1) & \longrightarrow & \{0, 1\}^{\mathbb{N}^*} \\ x & \mapsto & \big(x_k = p_k(x)\big)_{k \geq 1}. \end{array}$$

This map is one-to-one since for $x, y \in (0, 1)$ such that for all $k \geq 1$, $x_k = y_k$, then $x = \sum_{k \geq 1} x_k 2^{-k} = \sum_{k \geq 1} y_k 2^{-k} = y$. The mapping Ψ is not onto (e.g., the zero sequence has no preimage), however we shall prove that the complement of the image of Ψ is countable. Let $(x_k)_{k \geq 1} \in \mathcal{D}^c$, with

$$\mathcal{D} = \{(x_k)_{k \geq 1} \in \{0, 1\}^{\mathbb{N}^*}, \exists N, \forall k \geq N, x_k = 1\} \cup \{0\}, \tag{1.9.1}$$

so that $(x_k)_{k \geq 1}$ is a sequence in $\{0, 1\}$ which is not the zero sequence nor identically 1 for k large enough. We note that $\mathcal{D}$ is countable since it can be injected into

$$\{0\} \cup_{N \geq 1} \{0, 1\}^{N-1}.$$

Let us set $X = \sum_{k \geq 1} x_k 2^{-k}$. We have $0 < X < \sum_{k \geq 1} 2^{-k} = 1$. Then

$$\frac{x_1}{2} \leq X \leq \frac{x_1}{2} + \sum_{k \geq 2} \frac{x_k}{2^k} < \frac{x_1}{2} + \sum_{k \geq 2} 2^{-k} = \frac{x_1}{2} + \frac{1}{2},$$

so that $x_1 \leq 2X < x_1 + 1$ and thus $E(2X) = x_1$ with $x_1 = p_1(X)$. We prove similarly

$$p_k(X) = E(2^k X) - 2E(2^{k-1} X) = x_k.$$

In fact assume that for an integer $n \geq 1$, we know that $\forall k \in \{1, \ldots, n\}, x_k = p_k(X)$; then

$$\sum_{1 \leq k \leq n+1} x_k 2^{-k} \leq X < \sum_{1 \leq k \leq n+1} x_k 2^{-k} + \sum_{n+2 \leq k} 2^{-k} = \sum_{1 \leq k \leq n+1} x_k 2^{-k} + 2^{-n-1},$$

entailing

$$\sum_{1 \leq k \leq n} p_k(X) 2^{-k} + x_{n+1} 2^{-n-1} \leq X < \sum_{1 \leq k \leq n} p_k(X) 2^{-k} + x_{n+1} 2^{-n-1} + 2^{-n-1},$$

i.e., $2^{-n} E(2^n X) + x_{n+1} 2^{-n-1} \leq X < 2^{-n} E(2^n X) + x_{n+1} 2^{-n-1} + 2^{-n-1}$, that is

$$2E(2^n X) + x_{n+1} \leq 2^{n+1} X < 2E(2^n X) + x_{n+1} + 1,$$

so that $x_{n+1} \leq 2^{n+1} X - 2E(2^n X) < x_{n+1} + 1$, implying

$$x_{n+1} = E\big(2^{n+1} X - 2E(2^n X)\big) = E(2^{n+1} X) - 2E(2^n X) = p_{n+1}(X), \qquad \text{qed.}$$

As a result Ψ is bijective from $(0,1)$ onto $\Psi((0,1))$ and $\Psi((0,1)) \supset \mathcal{D}^c$ where $\mathcal{D}$ is a countable set (thus as well as $\mathcal{D}_0 = \Psi\big((0,1)\big)^c$). It suffices now to prove that $\{0,1\}^{\mathbb{N}} \backslash \mathcal{D}_0$ is equipotent to $\{0,1\}^{\mathbb{N}}$. Let us consider $\mathcal{C}$ equipotent to $\mathbb{N}$ disjoint of $\mathcal{D}_0$ in $\{0,1\}^{\mathbb{N}}$ (such a $\mathcal{C}$ exists since $\{0,1\}^{\mathbb{N}}$ is not countable),

$$\{0,1\}^{\mathbb{N}} = \big(\{0,1\}^{\mathbb{N}} \backslash \mathcal{D}_0\big) \cup \mathcal{D}_0 = \big(\{0,1\}^{\mathbb{N}} \backslash (\mathcal{D}_0 \cup \mathcal{C})\big) \cup \big(\mathcal{D}_0 \cup \mathcal{C}\big).$$

But $\mathcal{D}_0 \cup \mathcal{C}$ is countable infinite, thus equipotent to $\mathbb{N}$ and thus to $\mathcal{C}$. Consequently, $\{0,1\}^{\mathbb{N}}$ is equipotent to $\big(\{0,1\}^{\mathbb{N}} \backslash (\mathcal{D}_0 \cup \mathcal{C})\big) \cup \mathcal{C} = \{0,1\}^{\mathbb{N}} \backslash \mathcal{D}_0$, qed.

Exercise 1.9.6. *Let $f : X \to Y$ be a mapping.*

(1) *Show that for a family $(B_i)_{i \in I}$ of subsets of Y,*

$$f^{-1}\big(\bigcup_{i \in I} B_i\big) = \bigcup_{i \in I} f^{-1}(B_i), \qquad f^{-1}\big(\bigcap_{i \in I} B_i\big) = \bigcap_{i \in I} f^{-1}(B_i).$$

(2) *Show that for a family $(A_i)_{i \in I}$ of subsets of X, $f(\bigcup_{i \in I} A_i) = \bigcup_{i \in I} f(A_i)$.*

(3) *Show that if f is one-to-one, $f(\bigcap_{i \in I} A_i) = \bigcap_{i \in I} f(A_i)$. Prove that the previous equality is not true in general (without the injectivity assumption).*

Answer. (1) $x \in f^{-1}(\bigcup_{i \in I} B_i)$ means $f(x) \in \bigcup_{i \in I} B_i$, equivalent to

$$\exists i \in I, \ f(x) \in B_i \iff \exists i \in I, \ x \in f^{-1}(B_i) \iff x \in \cup_{i \in I} f^{-1}(B_i).$$

Similarly, $x \in f^{-1}(\bigcap_{i \in I} B_i)$ means $f(x) \in \bigcap_{i \in I} B_i$, equivalent to

$$\forall i \in I, \ f(x) \in B_i \iff \forall i \in I, \ x \in f^{-1}(B_i) \iff x \in \cap_{i \in I} f^{-1}(B_i).$$

(2) $y \in f(\bigcup_{i \in I} A_i)$ means $\exists x \in \cup_{i \in I} A_i$ such that $y = f(x)$, that is

$$\exists i \in I, \exists x \in A_i, y = f(x) \iff \exists i \in I, y \in f(A_i) \iff y \in \cup_{i \in I} f(A_i).$$

(3) We note that $A \subset A' \subset X \implies f(A) \subset f(A')$. For all $j \in I$, we have thus $f(\bigcap_{i \in I} A_i) \subset f(A_j)$ so that $f(\bigcap_{i \in I} A_i) \subset \bigcap_{i \in I} f(A_i)$. If $y \in \bigcap_{i \in I} f(A_i)$,

$$\forall i \in I, \exists x_i \in A_i, \quad y = f(x_i),$$

which implies for $i, j \in I$, $f(x_i) = f(x_j)$. The injectivity of f implies thus for $i, j \in I$, $x_i = x_j$, so that $y = f(x)$ with $x \in \cap_{i \in I} A_i$, qed. We consider the mapping

$$f : \{0, 1\} \longrightarrow \{1\}, \quad f(0) = f(1) = 1,$$

and we set $A_i = \{i\}$. We have $f(A_0 \cap A_1) = f(\emptyset) = \emptyset \subsetneq f(A_0) \cap f(A_1) = \{1\}$. *Comment.* Let us note that, conversely, if that property holds then f is injective. In fact, if $x_1 \neq x_2$ belongs to X, since

$$\emptyset = f(\emptyset) = f(\{x_1\} \cap \{x_2\}) = f(\{x_1\}) \cap f(\{x_2\}) = \{f(x_1)\} \cap \{f(x_2)\}$$

we get $f(x_1) \neq f(x_2)$.

Exercise 1.9.7. *Let X be a set. A **partition** of X is a family $(A_i)_{i \in I}$ of non-empty subsets of X, pairwise disjoint ($i \neq j$ implies $A_i \cap A_j = \emptyset$), with union X.*

(1) *Let $(A_i)_{i \in I}$ be a partition of X. Show that the relation $x \mathcal{R} y$ defined by*

$$\exists i \in I \quad \text{such that } x \in A_i \text{ and } y \in A_i$$

    is an equivalence relation on X.

(2) *Show that every equivalence relation on X can be obtained as in Question (1).*

(3) *Describe the partition of $\mathbb{Z}$ associated to the equality modulo n.*

Answer. (1) $\mathcal{R}$ is reflexive since $X = \cup_{i \in I} A_i$: for $x \in X$, there exists $i \in I$ such that $x \in A_i$ and thus $x \mathcal{R} x$. Symmetry of $\mathcal{R}$ follows from the definition, itself symmetrical in x, y. Let x, y, z be in X such that $x \mathcal{R} y$ and $y \mathcal{R} z$. Then there exists $i, j \in I$ such that

$$x, y \in A_i, \quad y, z \in A_j.$$

Since the A_i are pairwise disjoint and $y \in A_i \cap A_j$, we find $A_i = A_j$ and $x \mathcal{R} z$ (transitivity).

(2) Let $\mathcal{R}$ be an equivalence relation on X. The quotient set $\mathcal{Q} = \{C_j\}_{j \in J}$ is the set of equivalence classes. No equivalence class is empty since C_j is defined as the equivalence class of an element of X. Moreover,

$$X = \cup_{j \in J} C_j$$

since for $x \in X$, the equivalence class of x is one of the C_j which thus contains x. Two distinct classes are disjoint since, if $C_j \cap C_k \neq \emptyset$, there exists $z \in C_j \cap C_k$, for $x_j \in C_j, x_k \in C_k$ and we have

$$x_j \mathcal{R} z \text{ and } z \mathcal{R} x_k \implies x_j \mathcal{R} x_k \implies C_j = C_k.$$

(3) Let n be an integer ≥ 2. The equality modulo n is the equivalence relation on $\mathbb{Z}$ given by

$$x \equiv y \quad (n) \iff x - y \in n\mathbb{Z} \iff n | (x - y), \quad \text{i.e., } n \text{ divides } x - y.$$

It is obviously an equivalence relation and the quotient set is denoted by $\mathbb{Z}/n\mathbb{Z}$. The related partition of $\mathbb{Z}$ is the family with n elements

$$A_r = r + n\mathbb{Z} = \{r + nq\}_{q \in \mathbb{Z}}, \quad 0 \leq r \leq n - 1.$$

This follows from Euclidean division: for $m \in \mathbb{N}$ there exists a unique couple (q, r) of integers such that $m = nq + r, \quad 0 \leq r \leq n - 1$. This equivalence relation is also compatible with the ring structure of $\mathbb{Z}$, i.e., with $p_n : \mathbb{Z} \to \mathbb{Z}/n\mathbb{Z}$ the canonical mapping sending an integer to its equivalence class modulo n, we may define addition and multiplication on $\mathbb{Z}/n\mathbb{Z}$ with

$$p_n(a) \oplus p_n(b) = p_n(a + b), \quad p_n(a) \otimes p_n(b) = p_n(ab)$$

and it is easily verified that for $a \equiv a' \quad (n), \ b \equiv b' \quad (n)$, the results are unchanged. A good exercise for the reader would be to write the multiplication table of $\mathbb{Z}/n\mathbb{Z}$ for $2 \leq n \leq 11$, and verify that $\mathbb{Z}/n\mathbb{Z}$ is a field iff n is a prime number. One may also look for the divisors of 0 in $\mathbb{Z}/n\mathbb{Z}$ for $n \in \{4, 6, 8, 9, 10\}$ and ... read an introduction to Arithmetic such as [4].

Topology

Exercise 1.9.8.

(1) *Show that all the norms on $\mathbb{R}^n$ are equivalent (two norms N_1 and N_2 on a real or complex vector space E are said to be equivalent whenever there exists $C > 0$ such that for all $x \in E$, $C^{-1} N_1(x) \leq N_2(x) \leq C N_1(x)$).*

(2) *Show that on $C^0([0, 1]; \mathbb{R})$, the norms*

$$\|f\|_1 = \int_0^1 |f(t)| dt, \quad \|f\|_\infty = \sup_{x \in [0,1]} |f(x)|,$$

are not equivalent.

(3) *Looking at Figure 1.1 on page 6, find a sequence g_n of continuous functions converging for $\|\cdot\|_1$ towards the discontinuous step function $\mathbf{1}_{[1/2,1]}$.*

Answer. **(1)** For $x = (x_j)_{1 \le j \le n} \in \mathbb{R}^n$, the Euclidean norm is

$$\|x\|_2 = \left(\sum_{1 \le j \le n} x_j^2 \right)^{1/2}.$$

Let N be another norm on $\mathbb{R}^n$. From the triangle inequality and the homogeneity, for $x, h \in \mathbb{R}^n$, we get

$$N(x+h) - N(x) \le N(h) \le \sum_{1 \le j \le n} |h_j| N(e_j) \le \|h\|_2 \left(\sum_{1 \le j \le n} N(e_j)^2 \right)^{1/2},$$

where $(e_j)_{1 \le j \le n}$ is the canonical basis of $\mathbb{R}^n$. We get the same estimate from above (by the same argument) for $N(x) - N(x+h)$ so that $|N(x+h) - N(x)| \le C \|h\|_2$ and (Lipschitz) continuity holds for N. As a result, we obtain on the compact set $\mathbb{S}^{n-1} = \{x \in \mathbb{R}^n, \|x\|_2 = 1\}$,

$$0 < c_1 = \inf_{x \in \mathbb{S}^{n-1}} N(x) \le c_2 = \sup_{x \in \mathbb{S}^{n-1}} N(x)$$

so that, by homogeneity, for all $x \in \mathbb{R}^n$, $c_1 \|x\|_2 \le N(x) \le c_2 \|x\|_2$, proving the equivalence of N with the Euclidean norm.

(2) We have of course $\|f\|_1 \le \|f\|_\infty$, but choosing as in Figure 1.1 for $n \ge 1$, the continuous function

$$f_n(x) = \begin{cases} n - n^2 x & \text{for } 0 \le x \le 1/n, \\ 0 & \text{for } 1/n < x \le 1, \end{cases}$$

we find $\|f_n\|_\infty = n$, $\|f_1\|_1 = 1/2$ so that there does not exist $C > 0$ such that for all $f \in C^0([0,1]; \mathbb{R})$, $\|f\|_\infty \le C \|f\|_1$.

(3) Let us define for $n \ge 1$, the continuous function

$$g_n(x) = \begin{cases} \dfrac{x}{n} & \text{for } 0 \le x \le \frac{1}{2} - \frac{1}{n}, \\ \left(\dfrac{n-1}{2} + \dfrac{1}{n}\right)\left(x - \dfrac{1}{2}\right) + \dfrac{1}{2} & \text{for } \frac{1}{2} - \frac{1}{n} \le x \le \frac{1}{2} + \frac{1}{n}, \\ \dfrac{x}{n} + 1 - \dfrac{1}{n} & \text{for } \frac{1}{2} + \frac{1}{n} \le x \le 1. \end{cases}$$

Noticing that g_n is valued in $[0,1]$, we have

$$\|g_n - \mathbf{1}_{[1/2,1]}\|_1 = \int_0^{\frac{1}{2} - \frac{1}{n}} g_n(x)\,dx + \int_{\frac{1}{2} - \frac{1}{n}}^{\frac{1}{2}} g_n(x)\,dx$$

$$+ \int_{\frac{1}{2}}^{\frac{1}{2} + \frac{1}{n}} |1 - g_n(x)|\,dx + \int_{\frac{1}{2} + \frac{1}{n}}^{1} |1 - g_n(x)|\,dx$$

$$\le \int_0^{\frac{1}{2} - \frac{1}{n}} g_n(x)\,dx + \frac{1}{n} + \frac{1}{n} + \int_0^{\frac{1}{2} - \frac{1}{n}} \left(1 - g_n\left(t + \frac{1}{2} + \frac{1}{n}\right)\right) dt$$

$$\le \frac{1}{2n} + \frac{2}{n} + \int_0^{\frac{1}{2} - \frac{1}{n}} \frac{1}{n}\left(\frac{1}{2} - \frac{1}{n} - t\right) dt \le \frac{3}{n}.$$

Exercise 1.9.9.

(1) *Let $f : \mathbb{R} \longrightarrow \mathbb{R}$ be a function. Show that the set of discontinuity of f is an F_σ set.*

(2) *Show that given an F_σ set A of $\mathbb{R}$, there exists $f : \mathbb{R} \longrightarrow \mathbb{R}$ whose discontinuity set is A.*

(3) *Show that there does not exist a function $f : \mathbb{R} \longrightarrow \mathbb{R}$ whose discontinuity set is $\mathbb{Q}^c$.*

(4) *Find $f : \mathbb{R} \longrightarrow \mathbb{R}$ whose discontinuity set is $\mathbb{Q}$.*

Answer. (1) We define the oscillation function of f by

$$\omega(x) = \limsup_{y \to x} |f(y) - f(x)|,$$

and note that $\omega : \mathbb{R} \to \overline{\mathbb{R}}_+$, and is such that the set S of points of discontinuities of f is

$$S = \{x \in \mathbb{R}, \omega(x) > 0\} = \cup_{k \geq 1} \underbrace{\left\{x \in \mathbb{R}, \omega(x) \geq \frac{1}{k}\right\}}_{S_k}.$$

Let $k_0 \geq 1$ and $(x_j)_{j \in \mathbb{N}}$ be a sequence in S_{k_0} converging to some point a. For each $j \in \mathbb{N}$, we can find a sequence $(y_{j,l})_{l \in \mathbb{N}}$ such that $\lim_l y_{j,l} = x_j$ and

$$|f(y_{j,l}) - f(x_j)| \geq \frac{1}{2k_0}.$$

The point a must belong to S: otherwise, if f were continuous at a,

$$|f(y_{j,l}) - f(a)| \geq |f(y_{j,l}) - f(x_j)| - |f(x_j) - f(a)| \geq \frac{1}{2k_0} - |f(x_j) - f(a)|.$$

Let $r > 0$ be given: for $j \geq j_r$, we have $|x_j - a| \leq r$ and for each j, we can find $l_{r,j}$ such that $|y_{j,l_{r,j}} - x_j| \leq r$. We obtain

$$\sup_{|y-a| \leq 2r} |f(y) - f(a)| + \sup_{|x-a| \leq r} |f(x) - f(a)| \geq \frac{1}{2k_0},$$

an inequality which is incompatible with the continuity of f at a. As a result, we have proven that

$$\overline{S_k} \subset S \implies \cup_{k \geq 1} \overline{S_k} \subset S = \cup_{k \geq 1} S_k \implies S = \cup_{k \geq 1} \overline{S_k}, \quad \text{indeed } F_\sigma.$$

(2) Let $(F_n)_{n \geq 1}$ be a sequence of closed subsets of $\mathbb{R}$ and let $S = \cup_{n \geq 1} F_n$ be an F_σ set. We may assume that the sequence $(F_n)_{n \geq 1}$ is increasing since we can consider the sequence of closed sets $(\cup_{1 \leq j \leq n} F_j)_{n \geq 1}$ which has the same union S. We define for $x \in S$,

$$n(x) = \min \underbrace{\{n \geq 1, x \in F_n\}}_{\substack{\text{non-empty} \\ \text{subset of } \mathbb{N}^*}}, \quad \text{and } f(x) = \begin{cases} \frac{1}{n(x)}, & \text{for } x \in S \cap \mathbb{Q}, \\ -\frac{1}{n(x)}, & \text{for } x \in S \cap \mathbb{Q}^c, \\ 0, & \text{for } x \notin S. \end{cases}$$

(2.1) We want first to show that f is continuous at S^c. Since $f = 0$ at S^c, the function f is continuous on the interior of S^c. Let a be in $S^c \backslash \text{interior}(S^c) = S^c \cap \overline{S}$: let $(x_j)_{j \geq 1}$ be a sequence of S with limit a. No subsequence of $\big(n(x_j)\big)_{j \geq 1}$ can be bounded, otherwise we could find some $N_0 \geq 1$ such that

$$\lim_l j_l = +\infty, n(x_{j_l}) \leq N_0 \Longrightarrow \forall l \geq 1, x_{j_l} \in F_{N_0}$$

$$\Longrightarrow a = \lim_l x_{j_l} \in \overline{F_{N_0}} = F_{N_0} \subset S,$$

which is impossible. As a result $\lim_j n(x_j) = +\infty$ and $\lim_j f(x_j) = 0 = f(a)$, proving continuity.

(2.2) Let us prove now that f is discontinuous at S.

(2.2.1) Let $a \in S \cap \mathbb{Q}$: we have in particular $a \in F_{n(a)}, a \notin F_{n(a)-1}$ (defining $F_0 = \emptyset$) and $f(a) = 1/n(a)$. If $a \in \text{interior}(F_{n(a)})$, there is a sequence $(x_j)_{j \geq 1}$ of $S \cap \mathbb{Q}^c$ converging to a and $f(x_j) < 0$, so that $\limsup_j f(x_j) \leq 0$ proving the discontinuity property at a. If $a \in \partial(F_{n(a)})$, then any open neighborhood V of a intersects $F_{n(a)}^c$. In the open set $V \cap F_{n(a)}^c$, an irrational number can be found: thus there is a sequence of irrational numbers $(x_j)_{j \geq 1}$ converging to a and $f(x_j) \leq 0$, entailing discontinuity at a.

(2.2.2) Let $a \in S \cap \mathbb{Q}^c$: we have in particular $a \in F_{n(a)}, a \notin F_{n(a)-1}$ (defining $F_0 = \emptyset$) and $f(a) = -1/n(a)$. If $a \in \text{interior}(F_{n(a)})$, there is a sequence $(x_j)_{j \geq 1}$ of $S \cap \mathbb{Q}$ converging to a and $f(x_j) > 0$, so that $\liminf_j f(x_j) \geq 0$ proving the discontinuity property at a. If $a \in \partial(F_{n(a)})$, then any open neighborhood V of a intersects $F_{n(a)}^c$. In the open set $V \cap F_{n(a)}^c$, a rational number can be found: thus there is a sequence of rational numbers $(x_j)_{j \geq 1}$ converging to a and $f(x_j) \geq 0$, entailing discontinuity at a.

(3) As proven on page 7, the Baire category theorem (see Section 10.2 in the Appendix) implies that $\mathbb{Q}^c$ is not an F_σ set, so that the already solved question 1 in this exercise answers that one as well.

(4) The function (1.2.14) does that job. In the first place, f is discontinuous at $\mathbb{Q}$, since in any neighborhood of a point $a \in \mathbb{Q}$, an irrational number can be found, so there is a sequence of irrational numbers (x_j) with limit a and $f(x_j) = 0$, $f(a) > 0$. Moreover f is continuous at $\mathbb{Q}^c$ since if $(x_j = p_j/q_j), p_j \in \mathbb{Z}^*, q_j \in \mathbb{N}^*$ is a sequence converging to $a \notin \mathbb{Q}$, we must have $\lim_j q_j = +\infty$: otherwise, we could find a bounded subsequence $(q_{j_l})_{l \geq 1}$ of $(q_j)_{j \geq 1}$ in $\mathbb{N}^*$, providing a constant subsequence $(q = q_{j_{l_m}})_{m \geq 1}$ in $\mathbb{N}^*$, and since $aq = \lim_m p_{j_{l_m}}$, we find that the sequence $(p_{j_{l_m}})_{m \geq 1}$ is constant for m large enough and $a \in \mathbb{Q}$, which contradicts the assumption.

Exercise 1.9.10. *Let (X, d) be a metric space, and let $f : X \to \mathbb{R}$ be a function. We define for $\varepsilon > 0$,*

$$C(f, \varepsilon) = \{x \in X, \exists \delta > 0, d(x, x'), d(x, x'') < \delta \Longrightarrow |f(x') - f(x'')| < \varepsilon\}.$$

(1) *Show that $C(f, \varepsilon)$ is open.*

(2) *We define $S = \{x \in X, f$ is not continuous at $x\}$. Show that S is a F_σ set (hint: prove that f is continuous at x iff $x \in \cap_{n \geq 1} C(f, 1/n)$).*

Answer. (1) Let $x \in C(f, \varepsilon)$: for some positive δ and F defined on $X \times X$ by $(x', x'') \mapsto F(x', x'') = f(x') - f(x'')$, we have $F(B(x, \delta) \times B(x, \delta)) \subset [0, \varepsilon)$. Let $y \in B(x, \delta/2)$: then we have

$$F(B(y, \delta/2) \times B(y, \delta/2)) \subset F(B(x, \delta) \times B(x, \delta)) \subset [0, \varepsilon),$$

entailing that $B(x, \delta/2) \in C(f, \varepsilon)$.

(2) Let $x \in S^c$: then for any $n \geq 1$, $x \in C(f, 1/n)$. Conversely, if the latter property holds and $\varepsilon > 0$ is given, we can take $n \geq 1/\varepsilon$ and find $\delta > 0$ such that $|f(B(x, \delta)) - f(x)| < 1/n \leq \varepsilon$, proving continuity at x. As a result

$$S^c = \cap_{n \geq 1} C(f, 1/n), \text{ which is a } G_\delta \text{ set, so that } S \text{ is a } F_\sigma \text{ set.}$$

See [36] for more on this topic: in particular for a (non-empty) metric space X without isolated points (a point x in a topological space is said to be isolated if the singleton $\{x\}$ is open) and a given F_σ set S, there exists a function $f : X \to \mathbb{R}$ such that the points of discontinuity of f are exactly S.

Measure theory

Exercise 1.9.11. *Let $(X, \mathcal{M})$ be a measurable space and let $f, g : X \to \overline{\mathbb{R}}$ be measurable mappings. Show that the following sets belong to $\mathcal{M}$.*

$$A = \{x \in X, f(x) \leq g(x)\},$$
$$B = \{x \in X, f(x) < g(x)\},$$
$$C = \{x \in X, f(x) = g(x)\}.$$

Answer. The mapping $X \ni x \mapsto \Phi(x) = (f(x), g(x)) \in \overline{\mathbb{R}} \times \overline{\mathbb{R}}$ is measurable from the proof of Theorem 1.2.8. We have then $A = \Phi^{-1}(L)$ with

$$L = \{(\alpha, \beta) \in \overline{\mathbb{R}} \times \overline{\mathbb{R}}, \alpha \leq \beta\}$$

which is a closed subset of $\overline{\mathbb{R}} \times \overline{\mathbb{R}}$. Similarly, we have

$$M = \{(\alpha, \beta) \in \overline{\mathbb{R}} \times \overline{\mathbb{R}}, \alpha < \beta\}, \quad B = \Phi^{-1}(M),$$
$$N = \{(\alpha, \beta) \in \overline{\mathbb{R}} \times \overline{\mathbb{R}}, \alpha = \beta\}, \quad C = \Phi^{-1}(N),$$

with M open, N closed.

Exercise 1.9.12. *Let $(X, \mathcal{M})$ be a measurable space and $f_n : X \to \mathbb{C}$ be a sequence of measurable functions. Show that the set*

$$A = \{x \in X, \text{ the sequence } (f_n(x))_{n \in \mathbb{N}} \text{ is convergent}\}$$

belongs to $\mathcal{M}$.

Answer. Using the Cauchy criterion, we find

$$A = \{x \in X, \forall \epsilon \in \mathbb{Q} \cap]0,1], \exists N, \forall n \geq N, \forall k \geq 0, |f_{n+k}(x) - f_n(x)| \leq \epsilon\},$$

so that

$$A = \bigcap_{\epsilon \in \mathbb{Q} \cap]0,1]} \left\{ \bigcup_{N \in \mathbb{N}} \left(\cap_{n \geq N, k \geq 0} \{x \in X, \ |f_{n+k}(x) - f_n(x)| \leq \epsilon\} \right) \right\}.$$

Since the f_n are measurable, the set $\{x \in X, \ |f_{n+k}(x) - f_n(x)| \leq \epsilon\}$ belongs to $\mathcal{M}$ (cf. Theorem 1.2.7). As a countable intersection of countable union of countable intersection of elements of $\mathcal{M}$, A belongs to $\mathcal{M}$.

Exercise 1.9.13. *Let $(X, \mathcal{M})$ be a measurable space and let $(u_n)_{n \in \mathbb{N}}$ be a sequence of measurable functions from X into $\mathbb{R}$. Show that the following sets are measurable:*

$$A = \{x \in X, \ \lim_{n \to +\infty} u_n(x) = +\infty\}, \quad B = \{x \in X, \ (u_n(x))_{n \in \mathbb{N}} \text{ is bounded}\}.$$

Answer. We have $A = \{x \in X, \forall m \in \mathbb{N}, \exists N \in \mathbb{N}, \forall n \geq N, u_n(x) \geq m\}$, so that defining

$$A_{n,m} = \{x \in X, u_n(x) \geq m\},$$

we find $A = \cap_{m \in \mathbb{N}} \left(\cup_{N \in \mathbb{N}} \left(\cap_{n \geq N} A_{n,m} \right) \right)$ which is measurable as every $A_{n,m}$ is. Similarly, we have

$$B = \{x \in X, \exists m \in \mathbb{N}, \forall n \in \mathbb{N}, |u_n(x)| \leq m\} = \cup_{m \in \mathbb{N}} \left(\cap_{n \in \mathbb{N}} B_{n,m} \right),$$

with $B_{n,m} = \{x \in X, |u_n(x)| \leq m\}$.

Exercise 1.9.14. *Let X, Y be topological spaces, with X a Hausdorff space, and let $f : X \to Y$ be continuous outside of a countable set D. Show that f is measurable $(X, Y$ are equipped with their Borel σ-algebra).*

Answer. The mapping $F : X \backslash D \to Y$ defined by $F(x) = f(x)$ is continuous: let $x \in X \backslash D$. Since f is continuous at x, for every neighborhood W of $f(x)$, there exists a neighborhood V of x, such that $f(V) \subset W$; thus $F(V \cap D^c) = f(V \cap D^c) \subset W$ and F is continuous at x (see Lemma 1.2.4). Let V be an open set of Y. We have

$$f^{-1}(V) = \{x \in X, f(x) \in V\} = \{x \in X \backslash D, f(x) \in V\} \cup \left(f^{-1}(V) \cap D \right)$$
$$= F^{-1}(V) \cup \left(f^{-1}(V) \cap D \right) = \left(U \cap (X \backslash D) \right) \cup \left(f^{-1}(V) \cap D \right),$$

where U is an open subset of X. Since X is a Hausdorff space, singletons $\{x\}$ are closed: the complement $\{x\}^c$ is open since if $x' \in X, x' \neq x$, there exist neighborhoods $V' \in \mathcal{V}_{x'}, V \in \mathcal{V}_x$ with $V \cap V' = \emptyset$ and thus $V' \subset \{x\}^c$ which is thus a neighborhood of x'. As a result, the set D is measurable as a countable union of points and $U \cap D^c$ is measurable. Moreover $f^{-1}(V) \cap D$ is countable thus measurable. Finally, $f^{-1}(V)$ is measurable and Lemma 1.1.4 proves that f is measurable.

Exercise 1.9.15. *Let X be a non-empty set and $\mathcal{M}$ be the σ-algebra generated by the singletons $\{x\}$ where $x \in X$.*

(1) *Show that $A \in \mathcal{M}$ iff A or A^c is countable.*

(2) *We assume that X is not countable and we define for $A \in \mathcal{M}$*

$$\mu(A) = \begin{cases} 0 & \text{when } A \text{ is countable,} \\ 1 & \text{when } A \text{ is not countable.} \end{cases}$$

Show that μ is a positive measure defined on $\mathcal{M}$.

Answer. (1) If A is a countable subset of X, A is a countable union of singletons and thus belongs to $\mathcal{M}$. Since $\mathcal{M}$ is also stable by complementation, we find as well that A^c countable implies $A \in \mathcal{M}$. We define

$$\mathcal{N} = \{A \subset X, A \text{ or } A^c \text{ is countable}\}.$$

We have proven $\mathcal{N} \subset \mathcal{M}$, and we see that $\mathcal{N}$ is stable by complementation, contains X and all singletons. Let $(A_n)_{n\in\mathbb{N}}$ be a sequence in $\mathcal{N}$. If all A_n are countable, then $\cup_{n\in\mathbb{N}}A_n$ is countable and thus belongs to $\mathcal{N}$. If there exists $k \in \mathbb{N}$ such that A_k is not countable, then A_k^c is countable and since

$$\left(\cup_{n\in\mathbb{N}}A_n\right)^c = \cap_{n\in\mathbb{N}}A_n^c \subset A_k^c,$$

we find that $\left(\cup_{n\in\mathbb{N}}A_n\right)^c$ is countable, entailing $\cup_{n\in\mathbb{N}}A_n \in \mathcal{N}$. The set $\mathcal{N}$ is thus a σ-algebra which contains all singletons, so that $\mathcal{M} \subset \mathcal{N}$ and eventually $\mathcal{M} = \mathcal{N}$, proving (1).

(2) We have $\mu(\emptyset) = 0$; let $(A_n)_{n\in\mathbb{N}}$ be a pairwise disjoint sequence of $\mathcal{M}$. If all A_n are countable, then $\cup_{n\in\mathbb{N}}A_n$ is countable and

$$\mu(\cup_{n\in\mathbb{N}}A_n) = 0 = \sum_{n\in\mathbb{N}} \mu(A_n).$$

If there exists $k \in \mathbb{N}$ such that A_k is not countable, then A_k^c is countable and $\cup_{n\in\mathbb{N}}A_n$ is not countable. Since

$$A_k^c \supset \cup_{n\neq k}A_n,$$

we get that for $n \neq k$, A_n is countable, thus $\mu(A_n) = 0$. As a result, we have

$$\mu(\cup_{n\in\mathbb{N}}A_n) = 1 = \mu(A_k) = \mu(A_k) + \sum_{n\in\mathbb{N}, n\neq k} \mu(A_n) = \sum_{n\in\mathbb{N}} \mu(A_n).$$

Exercise 1.9.16. *Let $(X, \mathcal{M})$ be a measurable space and $f : X \to \mathbb{C}$ be a measurable function. Prove that there exists a measurable function $\alpha : X \to \mathbb{C}$ satisfying $|\alpha| \equiv 1$, such that $f = \alpha|f|$.*

Answer. Since f is measurable, $E = f^{-1}(\{0\}) \in \mathcal{M}$ and $\mathbf{1}_E$ is measurable. Noticing that $f(x) + \mathbf{1}_E(x)$ is always different from 0, (1 for $x \in E$, $f(x) \neq 0$ otherwise), we set

$$\alpha(x) = \frac{f(x) + \mathbf{1}_E(x)}{|f(x) + \mathbf{1}_E(x)|},$$

so that α is measurable as a composition of measurable functions:

$$
\begin{array}{ccccc}
 & \text{measurable} & & \text{continuous} & \\
X & \to & \mathbb{C}^* & \to & \mathbb{S}^1 \\
x & \mapsto & f(x) + \mathbf{1}_E(x) = t & \mapsto & t/|t|
\end{array}
$$

and $f(x) + \mathbf{1}_E(x) = \alpha(x)|f(x) + \mathbf{1}_E(x)|$, so that for $x \notin E$, $f(x) = \alpha(x)|f(x)|$ and for $x \in E$, $f(x) = 0 = \alpha(x)|f(x)|$.

Exercise 1.9.17. *Let $(X, \mathcal{M}, \mu)$ be a probability space (measurable space where μ is a positive measure such that $\mu(X) = 1$). Defining $\mathcal{T} = \{A \in \mathcal{M}, \ \mu(A) = 0 \text{ or } \mu(A) = 1\}$, show that $\mathcal{T}$ is a σ-algebra on X.*

Answer. If $A \in \mathcal{T}$, then $A^c \in \mathcal{M}$, since $\mu(A^c) + \mu(A) = \mu(X) = 1$, so that $\mu(A^c) = 1 - \mu(A) \in \{0, 1\}$. If $A_n \in \mathcal{T}$, $n \in \mathbb{N}$, $A = \cup_{n \in \mathbb{N}} A_n \in \mathcal{M}$ and if for all n, $\mu(A_n) = 0$, then $\mu(A) = 0$. If there exists n_0 such that $\mu(A_{n_0}) = 1$, then $1 = \mu(A_{n_0}) \leq \mu(A) \leq \mu(X) = 1$, so that $\mu(A) = 1$. Moreover $X \in \mathcal{T}$ since $\mu(X) = 1$.

Exercise 1.9.18. *Let $(X, \mathcal{M})$ be a measurable space and let $(\mu_j)_{j \in \mathbb{N}}$ be a sequence of positive measures defined on $\mathcal{M}$ such that $\forall A \in \mathcal{M}, \forall j \in \mathbb{N}, \mu_j(A) \leq \mu_{j+1}(A)$. For $A \in \mathcal{M}$, we set $\mu(A) = \sup_{j \in \mathbb{N}} \mu_j(A)$.*

(1) *Show that μ is a positive measure defined on $\mathcal{M}$.*

(2) *Let $f : X \longrightarrow \overline{\mathbb{R}}_+$ be a measurable function. Show that*

$$\int_X f d\mu = \sup_{j \in \mathbb{N}} \int_X f d\mu_j \quad \text{(hint: start with simple functions).}$$

(3) *Let $j \in \mathbb{N}$ and let ν_j be defined on $\mathcal{P}(\mathbb{N})$ by*

$$\nu_j(A) = \text{card}(A \cap [j, +\infty[)$$

(card E as usual whenever E is finite, card $E = +\infty$ for E infinite). Show that for all $j \in \mathbb{N}$, $(\mathbb{N}, \mathcal{P}(\mathbb{N}), \nu_j)$ is a measure space. Show that

$$\forall A \subset \mathbb{N}, \quad \nu_j(A) \geq \nu_{j+1}(A).$$

Defining $\nu(A) = \inf_{j \in \mathbb{N}} \nu_j(A)$, show that $\nu(\mathbb{N}) = +\infty$ and for all $k \in \mathbb{N}$, $\nu(\{k\}) = 0$. Show that $(\mathbb{N}, \mathcal{P}(\mathbb{N}), \nu)$ is not a measure space.

Answer. (1) Let $(A_n)_{n\in\mathbb{N}}$ be a pairwise disjoint sequence of $\mathcal{M}$. We check

$$\mu(\cup_{n\in\mathbb{N}}A_n) = \sup_{j\in\mathbb{N}}\mu_j(\cup_{n\in\mathbb{N}}A_n) = \sup_{j\in\mathbb{N}}\left\{\sum_{n\in\mathbb{N}}\mu_j(A_n)\right\}.$$

We consider the measure space $(\mathbb{N},\mathcal{P}(\mathbb{N}),\lambda)$, where λ is the counting measure on $\mathbb{N}$ ($\lambda(A) = \mathrm{Card}A$ for A finite and $\lambda(A) = +\infty$ for A infinite). We find with $f_j(n) = \mu_j(A_n)$ that $0 \le f_j \le f_{j+1}$ (since $\mu_j(A) \le \mu_{j+1}(A)$), so that Beppo Levi's Theorem 1.6.1 implies

$$\sup_{j\in\mathbb{N}}\int_{\mathbb{N}}f_j d\lambda = \int_{\mathbb{N}}(\sup_{j\in\mathbb{N}}f_j)d\lambda,$$

i.e., $\sup_{j\in\mathbb{N}}\left\{\sum_{n\in\mathbb{N}}\mu_j(A_n)\right\} = \sum_{n\in\mathbb{N}}\sup_{j\in\mathbb{N}}\left\{\mu_j(A_n)\right\} = \sum_{n\in\mathbb{N}}\mu(A_n)$, providing σ-additivity for μ on $\mathcal{M}$. Moreover we have $\mu(\emptyset) = 0$.

(2) For a simple function $s = \sum_{1\le k\le m}\alpha_k\mathbf{1}_{A_k}$ with $A_k \in \mathcal{M}$ and $\alpha_k > 0$, using the fact that the sequences $(\mu_j(A_k))_{j\in\mathbb{N}}$ are increasing, we have

$$\int_X s\,d\mu = \sum_{1\le k\le m}\alpha_k\mu(A_k) = \sum_{1\le k\le m}\alpha_k\sup_{j\in\mathbb{N}}(\mu_j(A_k)) = \sum_{1\le k\le m}\alpha_k\lim_{j\to\infty}\mu_j(A_k)$$

$$= \lim_{j\to\infty}\left[\sum_{1\le k\le m}\alpha_k\mu_j(A_k)\right] = \sup_{j\in\mathbb{N}}\left[\sum_{1\le k\le m}\alpha_k\mu_j(A_k)\right] = \sup_{j\in\mathbb{N}}\int_X s\,d\mu_j.$$

Moreover, for $f : X \longrightarrow \overline{\mathbb{R}}_+$ a measurable function, we can find an increasing sequence (s_k) of simple functions converging pointwise to f. Theorem 1.6.1 and the previous result imply

$$(*) \qquad \int_X f d\mu = \sup_{k\in\mathbb{N}}\int_X s_k d\mu = \sup_{k\in\mathbb{N}}\left(\sup_{j\in\mathbb{N}}\int_X s_k d\mu_j\right).$$

Moreover, if $(a_{jk})_{j,k\in\mathbb{N}}$ is a double sequence in $\overline{\mathbb{R}}$, for all $l, m \in \mathbb{N}$, we have

$$\alpha = \sup_{j\in\mathbb{N}}(\sup_{k\in\mathbb{N}}a_{jk}) \ge \sup_{k\in\mathbb{N}}a_{lk} \ge a_{lm} \Longrightarrow \sup_{l\in\mathbb{N}}a_{lm} \le \alpha \Longrightarrow \sup_{m\in\mathbb{N}}(\sup_{l\in\mathbb{N}}a_{lm}) \le \alpha,$$

so that, exchanging the indices in the previous line,

$$(**) \qquad \sup_{j\in\mathbb{N}}(\sup_{k\in\mathbb{N}}a_{jk}) = \sup_{k\in\mathbb{N}}(\sup_{j\in\mathbb{N}}a_{jk}).$$

As a result, from $(*)$ and $(**)$, we get

$$\int_X f d\mu = \sup_{j\in\mathbb{N}}\left(\sup_{k\in\mathbb{N}}\int_X s_k d\mu_j\right) = \sup_{j\in\mathbb{N}}\left(\int_X f d\mu_j\right),$$

where the second equality follows from Theorem 1.6.1.

(3) With λ the counting measure on $\mathbb{N}$, ν_j is a the measure with density $\mathbf{1}_{[j,+\infty[}$ with respect to λ and we use the notation $\nu_j = \mathbf{1}_{[j,+\infty[}\lambda$. Since $[j,+\infty[\supset[j+1,+\infty[$, we have $\mathbf{1}_{[j,+\infty[} \geq \mathbf{1}_{[j+1,+\infty[}$ and thus $\nu_j(A) \geq \nu_{j+1}(A)$ for all $A \subset \mathbb{N}$. As $\nu(A) = \inf_{j\in\mathbb{N}}\nu_j(A)$, and $\nu_j(\mathbb{N}) = +\infty$, we obtain $\nu(\mathbb{N}) = +\infty$. Moreover for all $k \in \mathbb{N}$, $\nu(\{k\}) = \inf_{j\in\mathbb{N}}\nu_j(\{k\}) = 0$, since $\nu_j(\{k\}) = \lambda(\{k\}\cap[j,+\infty[) = 0$ if $j > k$. Thus ν is not a measure on $\mathbb{N}$ since $+\infty = \nu(\mathbb{N}) > \sum_{k\in\mathbb{N}}\nu(\{k\}) = 0$.

Exercise 1.9.19 (Inclusion-exclusion principle, sieve formula). *Let $(X,\mathcal{M},\mu)$ be a measure space where μ is a positive measure such that $\mu(X) < +\infty$. Let $\{A_j\}_{1\leq j\leq n}$ be a finite set of elements of $\mathcal{M}$. Prove that*

$$\mu\big(\cup_{1\leq j\leq n}A_j\big) = \sum_{1\leq k\leq n}(-1)^{k+1}\left\{\sum_{\substack{J\subset\{1,\ldots,n\}\\ \operatorname{card} J=k}}\mu\big(\cap_{j\in J}A_j\big)\right\}. \qquad (1.9.2)$$

(Hint: write and prove the formula for $n = 2,3$, then apply induction on n.)

Answer. For $n = 2$, $A_1 \cup A_2$ is equal to the pairwise disjoint union

$$\big(A_1\backslash(A_1 \cap A_2)\big) \cup \big(A_2\backslash(A_1 \cap A_2)\big) \cup (A_1 \cap A_2),$$

so that $\mu(A_1 \cup A_2) = \mu(A_1) - \mu(A_1 \cap A_2) + \mu(A_2) - \mu(A_1 \cap A_2) + \mu(A_1 \cap A_2) = \mu(A_1) + \mu(A_2) - \mu(A_1 \cap A_2)$, which is the sought formula. Let us assume that the formula is true for some $n \geq 2$ and let us prove it for $n + 1$. Applying the formula for $n = 2$, we find

$$\mu\big(\cup_{1\leq j\leq n+1}A_j\big) = \mu\big(\cup_{1\leq j\leq n}A_j\big) + \mu(A_{n+1}) - \mu\big(\cup_{1\leq j\leq n}(A_j \cap A_{n+1})\big),$$

so that applying twice the formula for n we get

$$\mu(\cup_{1\leq j\leq n+1}A_j) = \mu(A_{n+1}) + \sum_{1\leq k\leq n}(-1)^{k+1}\left\{\sum_{\substack{J\subset\{1,\ldots,n\}\\ \operatorname{card} J=k}}\mu\big(\cap_{j\in J}A_j\big)\right\}$$

$$+ \sum_{1\leq k\leq n}(-1)^{k}\left\{\sum_{\substack{J\subset\{1,\ldots,n\}\\ \operatorname{card} J=k}}\mu\big(A_{n+1}\cap_{j\in J}A_j\big)\right\}$$

$$= \sum_{1\leq l\leq n}(-1)^{l+1}\left\{\sum_{\substack{L\subset\{1,\ldots,n,n+1\}\\ \operatorname{card} L=l,\ n+1\notin L}}\mu\big(\cap_{j\in L}A_j\big)\right\}$$

$$+ \sum_{2\leq l\leq n+1}(-1)^{l+1}\left\{\sum_{\substack{L\subset\{1,\ldots,n,n+1\}\\ \operatorname{card} L=l,\ n+1\in L}}\mu\big(\cap_{j\in L}A_j\big)\right\} + \mu(A_{n+1})$$

$$= \sum_{1\leq l\leq n+1}(-1)^{l+1}\left\{\sum_{\substack{L\subset\{1,\ldots,n,n+1\}\\ \operatorname{card} L=l}}\mu\big(\cap_{j\in L}A_j\big)\right\}.$$

N.B. It is possible to avoid the assumption $\mu(X) < +\infty$ and write (1.9.2) as an equality between non-negative quantities, displaying the odd (resp. even) k on the rhs (resp. lhs).

Exercise 1.9.20. *Let X be a finite set with N elements.*

(1) *Find the number $d(N)$ of permutations σ of X (bijections from X onto X) without fixed points ($\forall x \in X, \sigma(x) \neq x$). Find an equivalent of $d(N)$ when N tends to infinity.*

(2) *Let Y be a finite set with p elements. Find the number of surjections from X onto Y.*

Answer. (1) Let $d(N)$ be the sought number. The total number of permutations of X is $N!$. The number of permutations of X with exactly $N - 2$ fixed points is $d(2)C_N^2 = C_N^2$. The number of permutations of X with exactly $N - 3$ fixed points is $d(3)C_N^3$, so that

$$N! = \sum_{0 \leq k \leq N} d(k)C_N^k, \quad \sum_{0 \leq k \leq N} \frac{d(k)}{k!}\frac{1}{(N-k)!} = 1, \quad \tilde{d} * f = u,$$

with $\tilde{d} = (d(k)/k!)_{k \in \mathbb{N}}$, $f = (1/k!)_{k \in \mathbb{N}}$, $u = (u_k = 1)_{k \in \mathbb{N}}$. With $g = (x^k/k!)_{k \in \mathbb{N}}$, we get

$$(f * g)(k) = \sum_{0 \leq j \leq k} \frac{1}{(k-j)!}\frac{x^j}{j!} = (1+x)^k/k!,$$

and thus $\tilde{d} * f * g = u * g$, i.e.,

$$\sum_{0 \leq j \leq N} \frac{d(j)}{j!}\frac{(1+x)^{N-j}}{(N-j)!} = \sum_{0 \leq j \leq N} \frac{x^j}{j!},$$

so that for $x = -1$ and $N \geq 1$,

$$d(N) = \sum_{0 \leq j \leq N} \frac{(-1)^j N!}{j!} = N!\left(\sum_{0 \leq j \leq N} \frac{(-1)^j}{j!}\right) \underset{N \to +\infty}{\sim} \frac{N!}{e}. \tag{1.9.3}$$

(2) Let $S(N, p)$ be the sought number. We have the following partition of Y^X (the set of all mappings from X into Y)

$$Y^X = \sqcup_{1 \leq k \leq p}\{\phi \in Y^X, \operatorname{card} \phi(X) = k\},$$

so that $p^N = \sum_{1 \leq k \leq p} S(N, k)C_p^k$, i.e., $\frac{p^N}{p!} = \sum_{1 \leq k \leq p} \frac{S(N,k)}{k!}\frac{1}{(p-k)!}$. Following the same calculations as above (with p replacing N and N fixed), we find

$$\left(\frac{p^N}{p!}\right)_{p \geq 1} = \left(\frac{S(N, p)}{p!}\right)_{p \geq 1} * (f_q)_{q \geq 1},$$

so that $\sum_{1\leq j\leq p}\frac{j^N}{j!}\frac{x^{p-j}}{(p-j)!} = \sum_{1\leq j\leq p}\frac{S(N,j)}{j!}\frac{(1+x)^{p-j}}{(p-j)!}$, and for $x=-1$,

$$S(N,p) = p! \sum_{1\leq j\leq p}\frac{j^N}{j!}\frac{(-1)^{p-j}}{(p-j)!} = \sum_{1\leq j\leq p}C_p^j j^N(-1)^{p-j}.$$

Notice that $S(N,1) = 1$, $S(N,2) = 2^N - 2$,

$$S(N,p) = p^N \sum_{1\leq j\leq p}C_p^j\left(\frac{j}{p}\right)^N (-1)^{p-j}. \tag{1.9.4}$$

It is a consequence of that formula that for $0 \leq N < p$, $S(N,p) = 0$: it is also a fact that can be verified directly as follows. We have in that case

$$\left(\frac{d}{dx}\right)^N\{(1+x)^p\} = \frac{(1+x)^{p-N}}{(p-N)!} = \sum_{N\leq j\leq p}C_p^j\frac{j!}{(j-N)!}x^{j-N}$$

and for $x = -1$, we get $0 = \sum_{N\leq j\leq p}(-1)^j C_p^j j(j-1)\ldots(j-N+1)$, i.e.,

$$\sum_{0\leq j\leq p}C_p^j(-1)^j = 0, \qquad \sum_{0\leq j\leq p}C_p^j(-1)^j j = 0, \qquad \sum_{0\leq j\leq p}C_p^j(-1)^j j(j-1) = 0,$$

implying that $\sum_{0\leq j\leq p}C_p^j(-1)^j j^2 = 0$, and the other equalities $S(N,p) = 0$ for $0 \leq N < p$ follow by limited induction on N.

We have used the standard definition for the binomial coefficient,

$$C_n^p = \mathrm{card}\{A \subset \{1,\ldots,n\}, \mathrm{card}\,A = p\}, \tag{1.9.5}$$

$$C_n^p = \frac{n!}{(n-p)!p!} \quad \text{for } 0 \leq p \leq n, \ C_n^p = 0 \quad \text{otherwise.} \tag{1.9.6}$$

We have the classical formulas, easily proven by induction on n,

$$(x_1 + x_2)^n = \sum_{0\leq p\leq n}C_n^p x_1^p x_2^{n-p}, \qquad \frac{(x_1+x_2)^n}{n!} = \sum_{p_1+p_2=n}\frac{x_1^{p_1}}{p_1!}\frac{x_2^{p_2}}{p_2!}, \tag{1.9.7}$$

$$\frac{1}{n!}(x_1 + \cdots + x_k)^n = \sum_{p_1+\cdots+p_k=n}\frac{x_1^{p_1}}{p_1!}\cdots\frac{x_k^{p_k}}{p_k!}, \tag{1.9.8}$$

$$\frac{1}{n!}(x_1 + \cdots + x_k)^n = \sum_{\substack{|\alpha|=n\\ \alpha\in\mathbb{N}^k}}\frac{x^\alpha}{\alpha!}, \qquad x^\alpha = x_1^{\alpha_1}\ldots x_k^{\alpha_k}, \tag{1.9.9}$$

where for $\mathbb{N}^k \ni \alpha = (\alpha_1,\ldots,\alpha_k)$, $\alpha! = \alpha_1!\ldots\alpha_k!$, $|\alpha| = \alpha_1 + \cdots + \alpha_k$.

Note also the immediate consequence of (1.9.5)

$$C_{n+1}^p = C_n^p + C_n^{p-1} \tag{1.9.10}$$

which implies $C_{n+1}^{q+1} = C_n^{q+1} + C_n^q = C_{n-1}^{q+1} + C_{n-1}^q + C_n^q = C_{n-2}^{q+1} + C_{n-2}^q + C_{n-1}^q + C_n^q$ and thus inductively on n, $\sum_{q \leq k \leq n} C_k^q = C_{n+1}^{q+1}$. Also we have for $n \geq p \geq 1$,

$$\text{card}\{(j_k)_{1 \leq k \leq p} \in \{1,\ldots,n\}^p, \text{ s.t. } j_1 \leq j_2 \leq \cdots \leq j_p\} = C_{n+p-1}^p, \qquad (1.9.11)$$

since an increasing sequence $(j_k)_{1 \leq k \leq p}$ of $g\{1,\ldots,n\}$ can be identified with the strictly increasing sequence $(j_k + k - 1)_{1 \leq k \leq p}$ of $\{1,\ldots,n+p-1\}$, that is to a subset with p elements of the latter. Moreover

$$\text{card}\{\alpha \in \mathbb{N}^d, |\alpha| = l\} = C_{l+d-1}^{d-1}, \qquad (1.9.12)$$

since defining $\beta_1 = \alpha_1 + 1, \beta_2 = \alpha_1 + \alpha_2 + 2, \ldots, \beta_{d-1} = \sum_{1 \leq j \leq d-1} \alpha_j + d - 1$, we identify $\{\alpha \in \mathbb{N}^d, |\alpha| = l\}$ with the set of strictly increasing sequences $(\beta_j)_{1 \leq j \leq d-1}$ valued in $\{1,\ldots,l+d-1\}$, whose cardinal is C_{l+d-1}^{d-1}.

Exercise 1.9.21.

(1) *Let $(X, \mathcal{M}, \mu)$ be a measure space where μ is a positive measure and let $f : X \to \overline{\mathbb{R}}$ be a measurable function. Prove the Chebyshev inequality:*

$$\forall t > 0, \quad \mu\big(\{x \in X, |f(x)| \geq t\}\big) \leq t^{-2} \int_X |f|^2 d\mu. \qquad (1.9.13)$$

(2) *Let $(\Omega, \mathcal{A}, \mathbb{P})$ be a probability space and $X : \Omega \longrightarrow \mathbb{R}$ be a random variable (i.e., a measurable mapping) such that $\int_\Omega |X|^2 d\mathbb{P} < +\infty$. Show that $\int_\Omega |X| d\mathbb{P} < +\infty$ and defining the expectation $E(X)$ and the variance $\sigma(X)^2$ of X as*

$$E(X) = \int_\Omega X d\mathbb{P}, \quad \sigma(X)^2 = \int_\Omega |X - E(X)|^2 d\mathbb{P}, \qquad (1.9.14)$$

prove the Bienaymé–Chebyshev inequality: for $a > 0$,

$$\mathbb{P}\big(|X - E(X)| \geq a\big) \leq \frac{\sigma(X)^2}{a^2}. \qquad (1.9.15)$$

(3) *Let $(\Omega, \mathcal{A}, \mathbb{P})$ be a probability space and $X_j : \Omega \longrightarrow \mathbb{R}$ be a sequence of random variables ($j \geq 1$ integer) such that for each $j, \int_\Omega |X_j|^2 d\mathbb{P} < +\infty$. Let us assume that there exist $m, s \in \mathbb{R}$ such that*

$$\forall j \geq 1, E(X_j) = m \quad \text{and} \quad \forall j, k \geq 1, \int_\Omega (X_j - m)(X_k - m) d\mathbb{P} = \delta_{j,k} s^2.$$

Defining $Y_n = \frac{1}{n} \sum_{1 \leq j \leq n} X_j$, prove that Y_n converges in probability to m, i.e.,

$$\forall \varepsilon > 0, \quad \mathbb{P}\big(|Y_n - m| \geq \varepsilon\big) = 0. \qquad (1.9.16)$$

Answer. (1) We have for $t > 0$,

$$\mu(\{x \in X, |f(x)| \geq t\}) = \int_X \mathbf{1}_{\{|f| \geq t\}} d\mu \leq \int_X \mathbf{1}_{\{|f| \geq t\}} t^{-2} |f|^2 d\mu \leq t^{-2} \int_X |f|^2 d\mu.$$

(2) We have $\int_\Omega |X| d\mathbb{P} = \int_{|X| \leq 1} |X| d\mathbb{P} + \int_{|X| > 1} |X| d\mathbb{P} \leq 1 + \int_{|X| > 1} |X|^2 d\mathbb{P} < +\infty$. Applying the first question to the function $X - E(X)$ with $t = b\sigma(X)$, assuming $\sigma(X) > 0$, $b > 0$, we get

$$\mathbb{P}(|X - E(X)| \geq b\sigma(X)) \leq \frac{1}{b^2 \sigma(X)^2} \int_\Omega |X - E(X)|^2 d\mathbb{P} = b^{-2}.$$

If $\sigma(X) = 0$, we find $X \equiv E(X)$ and for $a > 0$, the result (1.9.15) is obvious. We may note

$$\begin{aligned}
\int_\Omega |X - E(X)|^2 d\mathbb{P} &= \int_\Omega |X|^2 d\mathbb{P} + \int_\Omega |E(X)|^2 d\mathbb{P} - 2 \int_\Omega X E(X) d\mathbb{P} \\
&= \int_\Omega |X|^2 d\mathbb{P} - |E(X)|^2.
\end{aligned} \tag{1.9.17}$$

(3) We find $E(Y_n) = m$ and thus for $\varepsilon > 0$, from the Bienaymé–Chebyschev inequality, we get

$$\mathbb{P}(|Y_n - m| \geq \varepsilon) \leq \frac{\sigma(Y_n)^2}{\varepsilon^2}.$$

We calculate $\sigma(Y_n)^2 = \int_\Omega |Y_n - m|^2 d\mathbb{P} = n^{-2} \sum_{1 \leq j,k \leq n} \int_\Omega (X_j - m)(X_k - m) d\mathbb{P}$ and our assumption gives $\sigma(Y_n)^2 = n^{-2} n s^2$, so that $\mathbb{P}(|Y_n - m| \geq \varepsilon) \leq \frac{s^2}{n\varepsilon^2}$, proving the sought convergence.

Exercise 1.9.22. *For k, n positive integers, we define $a_{k,n} = \delta_{k,n}((-1)^n + 2)$. Show that, for each k, $\lim_n a_{k,n} = 0$, $a_{k,n} \geq 0$, $|\sum_k a_{k,n}| \leq 3$. Prove that the sequence $(\sum_{k \geq 1} a_{k,n})_{n \geq 1}$ does not have a limit. Check that the domination assumption in Lebesgue dominated convergence Theorem 1.7.5 is violated.*

Answer. The first limit is obvious and we have also for $n \geq 1$,

$$\sum_{k \geq 1} a_{k,n} = (-1)^n + 2, \quad \text{a divergent sequence.}$$

Of course the assumption of domination in the Lebesgue dominated convergence theorem is not satisfied since $\sup_n a_{k,n} = (-1)^k + 2$, a non-summable sequence. As a result for the measure space $(X, \mathcal{M}, \mu) = (\mathbb{N}^*, \mathcal{P}(\mathbb{N}^*),$ counting measure$)$, it is possible to find a sequence $(f_n)_{n \geq 1}$ of non-negative bounded functions with bounded integrals such that for all $x \in X$, $\lim_n f_n(x) = 0$, but so that the sequence $(\int_X f_n d\mu)_{n \geq 1}$ is divergent. This proves that the domination assumption cannot be dispensed with in general.

Exercise 1.9.23. *Let $(X, \mathcal{M}, \mu)$ be a measure space where μ is a positive measure.*

(1) *Let $f : X \longrightarrow Y$ be a mapping and let $\varphi : Y \to \overline{\mathbb{R}}_+$ be a measurable function. Prove that*

$$\int_Y \varphi \, d\big(f_*(\mu)\big) = \int_X (\varphi \circ f) d\mu. \tag{1.9.18}$$

(2) *Prove (1.9.18) for $\varphi \in \mathcal{L}^1(f_*(\mu))$.*

(3) *Let $f : X \longrightarrow Y$ and $g : Y \longrightarrow Z$ be two mappings. Show that*

$$(g \circ f)_*(\mu) = g_*\big(f_*(\mu)\big).$$

Answer. (1) That formula is satisfied for a simple function $\varphi = \sum_{1 \leq j \leq m} \alpha_j \mathbf{1}_{B_j}$:

$$\int_Y \varphi \, d\big(f_*(\mu)\big) = \sum_{1 \leq j \leq m} \alpha_j f_*(\mu)(B_j) = \sum_{1 \leq j \leq m} \alpha_j \mu(f^{-1}(B_j))$$

$$\text{(using } \mathbf{1}_{B_j} \circ f = \mathbf{1}_{f^{-1}(B_j)}) = \sum_{1 \leq j \leq m} \alpha_j \int (\mathbf{1}_{B_j} \circ f) d\mu = \int_X (\varphi \circ f) d\mu.$$

Beppo Levi's theorem 1.6.1 and the approximation Theorem 1.3.3 give the result.
(2) By linearity, that formula holds as well for $\varphi \in \mathcal{L}^1\big(f_*(\mu)\big)$.
(3) Following Lemma 1.4.3, the pushforward $f_*(\mu)$ is defined on the σ-algebra

$$\mathcal{N} = \{B \subset Y, f^{-1}(B) \in \mathcal{M}\}$$

by $f_*(\mu)(B) = \mu\big(f^{-1}(B)\big)$. The pushforward $g_*\big(f_*(\mu)\big)$ is defined on the σ-algebra

$$\mathcal{T} = \{C \subset Z, g^{-1}(C) \in \mathcal{N}\} = \{C \subset Z, f^{-1}(g^{-1}(C)) \in \mathcal{M}\}$$
$$= \{C \subset Z, (g \circ f)^{-1}(C) \in \mathcal{M}\}$$

by $g_*\big(f_*(\mu)\big)(C) = f_*(\mu)\big(g^{-1}(C)\big) = \mu\Big(f^{-1}\big(g^{-1}(B)\big)\Big) = \mu((g \circ f)^{-1}(C))$. As a result, the measures $g_*\big(f_*(\mu)\big)$ and $(g \circ f)_*(\mu)$ coincide on the σ-algebra $\mathcal{T}$.

Exercise 1.9.24. *Let X be a Hausdorff σ-compact topological space, let $\mathcal{B}$ be the Borel σ-algebra on X and let μ be a positive measure defined on $\mathcal{B}$ such that $\mu(K) < +\infty$ for K compact (μ is a Borel measure on X).*

(1) *Prove that the singletons are closed. We define*

$$D = \{a \in X, \mu(\{a\}) > 0\}.$$

(2) *Let n, l be integers ≥ 1. Assuming $X = \cup_{n \geq 1} K_n$, with K_n compact, we set*

$$D_{n,l} = \{a \in K_n \ \text{and} \ \mu(\{a\}) \geq 1/l\}.$$

Show that $D_{n,l}$ is finite and D is countable.

(3) *For $E \in \mathcal{B}$, we define $\lambda(E) = \mu(D \cap E)$. Show that it is meaningful and that λ is a Borel measure on X. Show that*

$$\lambda = \sum_{a \in D} \mu(\{a\})\delta_a,$$

where δ_a is the Dirac mass at a (i.e., $\delta_a(E) = \mathbf{1}_E(a)$).

(4) *Show that $\mu = \lambda + \nu$ where ν is a Borel measure on X such that for all $x \in X$, $\nu(\{x\}) = 0$.*

Answer. (1) For $x' \notin \{x\}$, we find $V \in \mathscr{V}_x, V' \in \mathscr{V}_{x'}$ such that $V \cap V' = \emptyset$, so that the complement of the singleton $\{x\}^c \supset V'$ and is a neighborhood of each of its points, thus an open set.

(2) Let $a_1, \ldots, a_m$ be distinct in $D_{n,l}$. We have

$$+\infty > \mu(K_n) \geq \mu(\{a_1, \ldots, a_m\}) = \sum_{1 \leq j \leq m} \mu(\{a_j\}) \geq m/l$$

so that $m \leq \mu(K_n)l < +\infty$, proving finiteness for $D_{n,l}$. For $a \in D$, we may find an integer $l \geq 1$ such that $\mu(\{a\}) \geq 1/l$. Since a belongs to some K_n, we find $a \in D_{n,l}$. This implies $D \subset \cup_{n \geq 1, l \geq 1} D_{n,l}$: Since $D_{n,l} \subset D$ we find that D is a countable union of finite sets, thus is countable.

(3) The set D is a Borel set as a countable union of singletons (closed sets), and with $E \in \mathcal{B}$, $D \cap E \in \mathcal{B}$. We may thus define $\lambda(E) = \mu(D \cap E)$. This defines a Borel measure since $\lambda(\emptyset) = \mu(\emptyset) = 0$, and for E_n a sequence of pairwise disjoint Borel sets, K a compact set, we have

$$\lambda(\cup_{n \in \mathbb{N}} E_n) = \mu(\cup_{n \in \mathbb{N}}(E_n \cap D)) = \sum_{n \in \mathbb{N}} \mu(E_n \cap D) = \sum_{n \in \mathbb{N}} \lambda(E_n),$$

$$\lambda(K) = \mu(K \cap D) \leq \mu(K) < +\infty.$$

With $D = \{a_n\}_{n \in \mathbb{N}}$, we have

$$\lambda(E) = \mu(D \cap E) = \mu(\{a_n, a_n \in E\}) = \sum_{n, a_n \in E} \mu(\{a_n\})$$

$$= \sum_{n \in \mathbb{N}} \mu(\{a_n\})\delta_{a_n}(E) = \sum_{a \in D} \mu(\{a\})\delta_a(E), \quad \text{q.e.d.}$$

(4) For $E \in \mathcal{B}$, we have $\mu(E) = \mu(E \cap D) + \mu(E \cap D^c) = \lambda(E) + \nu(E)$, with $\nu(E) = \mu(E \cap D^c)$. As in question (3), we find that ν is a Borel measure. For $x \in D$, we have $\nu(\{x\}) = \nu(\{x\} \cap D^c) = \nu(\emptyset) = 0$. For $x \notin D$, we find $0 = \mu(\{x\}) = \lambda(\{x\}) + \nu(\{x\}) = \nu(\{x\})$, so that for all x, we have $\nu(\{x\}) = 0$.

Exercise 1.9.25. *Let $\mathcal{B}$ be the Borel σ-algebra on $\mathbb{R}$ and μ be a positive measure defined on $\mathcal{B}$, finite on the compact sets.*

(1) *For $a \in \mathbb{R}$, we define*

$$F_a(t) = \begin{cases} \mu([a, t[) & \text{for } t > a, \\ -\mu([t, a[) & \text{for } t \leq a. \end{cases}$$

Show that F_a is increasing and left-continuous.

(2) *We assume that μ is a probability measure. We define the repartition function of the probability μ on $\mathbb{R}$ as*

$$F(t) = \mu\big((-\infty, t[\big).$$

Show that F is increasing, tends to 0 (resp. 1) when t goes to $-\infty$ (resp. $+\infty$), and is left-continuous.

Answer. (1) Let $s < t$ be real numbers. For $s > a$, we have $[a, s[\subset [a, t[$ and thus $F_a(s) = \mu([a, s[) \leq \mu([a, t[) = F_a(t)$. For $s \leq a < t$, we have $F_a(s) = -\mu([s, a[) \leq 0 \leq \mu([a, t[) = F_a(t)$. For $s < t \leq a$, we have $[t, a[\subset [s, a[$ and thus $F_a(s) = -\mu([s, a[) \leq -\mu([t, a[) = F_a(t)$. The function F_a is thus increasing.

Let $t_0 \in \mathbb{R}$ such that $t_0 > a$ and let $(t_n)_{n \geq 1}$ be an increasing sequence with limit t_0. We have

$$[a, t_0[= \cup_{n \geq 1} [a, t_n[$$

and using Proposition 1.4.4(2), we find

$$F_a(t_0) = \mu([a, t_0[) = \lim_{n \to \infty} \mu([a, t_n[) = \lim_{n \to \infty} F_a(t_n).$$

Let $t_0 \in \mathbb{R}$ such that $t_0 \leq a$ and let $(t_n)_{n \geq 1}$ be an increasing sequence with limit t_0. We have

$$[t_0, a[= \cap_{n \geq 1} [t_n, a[$$

using Proposition 1.4.4(3) along with $\mu([t_1, a[) \leq \mu([t_1, a]) < +\infty$, we find

$$F_a(t_0) = -\mu([t_0, a[) = -\lim_{n \to \infty} \mu([t_n, a[) = \lim_{n \to \infty} F_a(t_n).$$

(2) F is increasing since $t \mapsto (-\infty, t[$ is increasing, and tends to 1 when t goes to $+\infty$ from Proposition 1.4.4(2), tends to 0 when t goes to $-\infty$ from Proposition 1.4.4(3). The left-continuity is proven as in question (1) above.

Exercise 1.9.26. *Let $(X, \mathcal{M}, \mu)$ be a measure space where μ is a positive measure and let $f_n : X \to \overline{\mathbb{R}}_+$ be an increasing sequence of measurable functions such that $\sup_{n \in \mathbb{N}} \int_X f_n d\mu < +\infty$. Prove that $\sup_{n \in \mathbb{N}} f_n(x)$ is finite μ-a.e. Give an analogous statement for series of measurable functions valued in $\overline{\mathbb{R}}_+$.*

Answer. Thanks to Beppo Levi's theorem 1.6.1 we have, with $f = \sup_{n \in \mathbb{N}} f_n$,

$$\int_X f d\mu = \sup_{n \in \mathbb{N}} \int_X f_n d\mu,$$

so that f is a measurable function from X into $\overline{\mathbb{R}}_+$ such that $\int_X f d\mu < +\infty$. Proposition 1.7.1(4) shows that is finite μ-a.e. Similarly, for a sequence $(u_k)_{k\in\mathbb{N}}$ of measurable functions from X into $\overline{\mathbb{R}}_+$ such that

$$\sum_{k\geq 0}\int_X u_k d\mu < +\infty,$$

the series $\sum_{k\in\mathbb{N}} u_k(x)$ converges μ-a.e. towards a finite limit: in fact Corollary 1.6.2 implies

$$\int_X \Big(\sum_{k\in\mathbb{N}} u_k\Big) d\mu = \sum_{k\geq 0}\int_X u_k d\mu < +\infty$$

so that the function $\sum_{k\in\mathbb{N}} u_k(x)$ is finite μ-a.e.

Exercise 1.9.27. *Let $(X,\mathcal{M},\mu)$ be a measure space where μ is a positive measure and let $f : X \to \mathbb{C}$ be a function in $\mathcal{L}^1(\mu)$. We assume that for all $E \in \mathcal{M}$, $\int_E f d\mu = 0$. Show that f is vanishing μ-a.e.*

Answer. For $E = \{x \in X, \mathrm{Re}\, f(x) \geq 0\}$, we find

$$0 = \mathrm{Re}\Big(\int_E f d\mu\Big) = \int_E (\mathrm{Re}\, f) d\mu \implies \mathbf{1}_{\{\mathrm{Re}\, f\geq 0\}}\, \mathrm{Re}\, f = 0 \quad \mu\text{-a.e.,}$$

and since we have also $\mathbf{1}_{\{\mathrm{Re}\, f\leq 0\}}\, \mathrm{Re}\, f = 0$, μ-a.e., we get $\mathrm{Re}\, f = 0$, μ-a.e. We prove similarly that $\mathrm{Im}\, f = 0$, μ-a.e.

Exercise 1.9.28. *Let $(X,\mathcal{M},\mu)$ be a measure space where μ is a positive measure and let $(f_n)_{n\in\mathbb{N}}$ be a decreasing sequence of measurable functions from X into $\overline{\mathbb{R}}_+$ converging pointwise to a function f.*

(1) Prove that if there exists $N \in \mathbb{N}$ such that f_N belongs to $\mathcal{L}^1(\mu)$, then

$$\lim_n \int_X f_n d\mu = \inf_n \int_X f_n d\mu = \int_X f d\mu.$$

(2) Prove that this property does not hold if the assumption in (1) is removed.

Answer. (1) We can apply the Lebesgue dominated convergence Theorem 1.6.8, since for $n \geq N$, we have $0 \leq f_n \leq f_N \in \mathcal{L}^1(\mu)$.
(2) We note first that from Fatou's lemma 1.6.4, we have

$$\int_X f d\mu = \int_X \liminf f_n d\mu \leq \liminf_n \int_X f_n d\mu = \inf_n \int_X f_n d\mu.$$

Let us prove that we may have $0 \leq \int_X f d\mu < \inf_n \int_X f_n d\mu$. We consider $X = \mathbb{N}$ with the counting measure μ and the (decreasing) sequence $f_n = \mathbf{1}_{[n,+\infty)}$. We have $f = 0$ and $\int_X f_n d\mu = +\infty$ for all n.

Exercise 1.9.29. *Let $(X, \mathcal{M}, \mu)$ be a measure space where μ is a positive measure such that $\mu(X) < +\infty$. Let $(f_n)_{n \in \mathbb{N}}$ be a sequence of bounded complex-valued measurable functions converging uniformly towards a complex-valued function f on X:*

$$\lim_n \left(\sup_{x \in X} |f_n(x) - f(x)| \right) = 0.$$

(1) *Prove that each f_n and f belong to $\mathcal{L}^1(\mu)$ and $\lim_n \int_X |f_n - f| d\mu = 0$.*

(2) *Prove that this property does not hold if the assumption $\mu(X) < +\infty$ is removed.*

Answer. (1) As a pointwise limit of measurable functions, f is also measurable. We have also

$$\int_X |f_n(x) - f(x)| d\mu(x) \le \int_X \sup_{x \in X} |f_n(x) - f(x)| d\mu(x)$$

$$= \sup_{x \in X} |f_n(x) - f(x)| \mu(X),$$

proving the convergence. Also that inequality proves that, for each (large enough) n, $f_n - f$ belongs to $\mathcal{L}^1(\mu)$ and since each f_n is bounded, it belongs to $\mathcal{L}^1(\mu)$ (since $\mu(X) < +\infty$) as well as f.

(2) We consider $X = \mathbb{N}$ with the counting measure μ and the sequence $(n \ge 1)$

$$f_n(k) = \frac{1}{n} \mathbf{1}_{[n, 2n-1]}(k).$$

We have $\sup_{k \in \mathbb{N}} |f_n(k)| = 1/n$ which goes to zero when n goes to $+\infty$ but $\int_X f_n d\mu = 1$. Note that the sequence (f_n) as well as f belong to a bounded set of $\mathcal{L}^1(\mu)$. Of course the sequence (f_n) fails to be dominated by an L^1 function since for each $k \ge 1$,

$$\sup_{n \ge 1} f_n(k) \ge \frac{1}{k}, \quad \text{and} \quad \sum_{k \ge 1} \frac{1}{k} = +\infty.$$

Exercise 1.9.30. *Let $(X, \mathcal{M}, \mu)$ be a measure space where μ is a positive measure such that $\mu(X) < +\infty$ and let $f \in L^1(\mu)$ such that, for a given closed set T of $\mathbb{C}$,*

$$(\dagger) \qquad \forall E \in \mathcal{M} \text{ with } \mu(E) > 0, \quad \frac{1}{\mu(E)} \int_E f d\mu \in T.$$

Prove that $f(x) \in T$, μ-a.e.

Answer. For $z \in T^c, \exists \rho > 0$ with $\bar{B}(z, \rho) \subset T^c$. If we had $\mu\big(f^{-1}(\bar{B}(z, \rho))\big) > 0$, this would give, with $E = f^{-1}(\bar{B}(z, \rho))$, $\frac{1}{\mu(E)} \int_E f d\mu \in T$. However, we have

$$\frac{1}{\mu(E)} \int_{f^{-1}(\bar{B}(z,\rho))} f d\mu = \frac{1}{\mu(E)} \int_{f^{-1}(\bar{B}(z,\rho))} (f - z) d\mu + z,$$

and since

$$\left| \frac{1}{\mu(E)} \int_{f^{-1}(\bar{B}(z,\rho))} (f - z)d\mu \right| \leq \frac{\rho\mu(E)}{\mu(E)} = \rho,$$

this would imply $|z - T| \leq \rho$, which contradicts $\bar{B}(z,\rho) \subset T^c$. Consequently, $\mu\bigl(f^{-1}(\bar{B}(z,\rho))\bigr) = 0$. Since the open set T^c is a countable union of closed balls, this implies that $\mu\bigl(f^{-1}(T^c)\bigr) = 0$.

N.B. The assumption $\mu(X) < +\infty$ can be replaced by σ-finiteness: assuming ($\dagger$) for all E with positive finite measure, we get from the previous result that, for $X = \cup_{k\in\mathbb{N}}X_k, \mu(X_k) < +\infty$, since

$$\{x \in X, f(x) \in T^c\} = \cup_{k\in\mathbb{N}}\{x \in X_k, f(x) \in T^c\}$$

each $\{x \in X_k, f(x) \in T^c\}$ has 0 measure, as well as $f^{-1}(T^c)$.

Exercise 1.9.31. *Let $(X, \mathcal{M}, \mu)$ be a measure space where μ is a positive measure and let $(E_k)_{k\geq 1}$ be a sequence in $\mathcal{M}$ such that $\sum_{k\geq 1} \mu(E_k) < +\infty$. Prove that*

$$(\ddagger) \qquad\qquad \cup_{n\geq 0}\{x \in X, x \text{ belongs to } n \text{ subsets } E_k\}$$

has a complement with measure 0, i.e., almost all x lie in at most finitely many E_k.

Answer. The complement of the set ($\ddagger$) is $F = \cap_{n\geq 0}\bigl(\cup_{k>n} E_k\bigr)$: In fact if x belongs to infinitely many E_k, for each $n \geq 0$, there exists $k > n$ with $x \in E_k$; conversely, any $x \in F$ belongs to infinitely many E_k. We have

$$\mu(F) \leq \mu(\cup_{k>n} E_k) \leq \sum_{k>n} \mu(E_k) \xrightarrow[n\to+\infty]{} 0,$$

proving $\mu(F) = 0$.

Chapter 2

Actual Construction of Measure Spaces

In the previous chapter, we gave a presentation of integration theory along with convergence theorems and a functional space for integrable functions. All this seems to be very satisfactory, except for the fact that we do not have many examples: the counting measure is an example and its version on $\mathbb{N}$ is certainly a good way to present series and the space $\ell^1(\mathbb{N})$ of summable sequences of complex numbers $(a_n)_{n\in\mathbb{N}}$ (i.e., such that $\sum_{n\in\mathbb{N}} |a_n| < +\infty$).

However, our most important example is the construction of the Borel measure, defined on the Borel subsets of $\mathbb{R}$, such that $\mu([a,b]) = b - a$ for $a \le b$ real numbers. Everything remains to be done for this example: construction of such an object, proof of its uniqueness, various properties. The present chapter is essentially devoted to this construction.

2.1 Partitions of unity

Let X be a topological space and let $f : X \longrightarrow \mathbb{C}$ be a continuous function. We define the support of f as the set

$$\operatorname{supp} f = \{x \in X, \not\exists V \in \mathscr{V}_x \text{ such that } f_{|V} = 0 \}. \tag{2.1.1}$$

We note that $\operatorname{supp} f = \overline{\{x \in X, f(x) \ne 0\}}$: since $\left(\bar{A}\right)^c = \operatorname{int}(A^c)$, we have

$$x \notin \overline{\{x \in X, f(x) \ne 0\}} \iff x \in \operatorname{int}\{x \in X, f(x) = 0\} \iff \exists V \in \mathscr{V}_x, f_{|V} = 0,$$

which defines the complement of $\operatorname{supp} f$. As a result $\operatorname{supp} f$ is a closed subset of X since $(\operatorname{supp} f)^c$ is the union of open sets on which $f = 0$.

The vector space of continuous functions from X into $\mathbb{C}$ with compact support will be denoted by $C_c(X)$. For $f \in C_c(X)$, we have, if $\operatorname{supp} f \ne X$,

$$f(X) = f(\operatorname{supp} f) \cup \{0\}$$

and since the continuous image of the compact set $\operatorname{supp} f$ is compact, so is $f(X)$. If $f \in C_c(X)$ and $\operatorname{supp} f = X$, then X is compact and so is its image $f(X)$.

Lemma 2.1.1. *Let (X,d) be a metric space and let A be a non-empty subset of X. For $x \in X$, we set*

$$d(x, A) = \inf_{a \in A} d(x, a). \tag{2.1.2}$$

The function $d(\cdot, A)$ is Lipschitz continuous with Lipschitz constant ≤ 1, i.e., $|d(x_1, A) - d(x_2, A)| \leq d(x_1, x_2)$. That property implies uniform continuity for $d(\cdot, A)$. Moreover

$$\bar{A} = \{x \in X, d(x, A) = 0\}.$$

Proof of the lemma. For $x_1 \in X$ and $\epsilon > 0$, there exists $a \in A$ such that

$$d(x_1, A) \leq d(x_1, a) < d(x_1, A) + \epsilon.$$

Thus for $x_2 \in X$, we have $d(x_2, A) - d(x_1, A) \leq d(x_2, a) - d(x_1, a) + \epsilon \leq d(x_2, x_1) + \epsilon$, so that

$$d(x_2, A) - d(x_1, A) \leq d(x_2, x_1).$$

Switching x_1 with x_2, we get the sought $|d(x_2, A) - d(x_1, A)| \leq d(x_2, x_1)$. The set $\{x \in X, d(x, A) = 0\}$ is closed (since $d(\cdot, A)$ is continuous) and contains A, thus contains $\bar{A}$. Also, if $d(x, A) = 0$, there is a sequence $(a_k)_{k \in \mathbb{N}}$ in A such that $\lim_k d(x, a_k) = 0$, entailing $\lim_k a_k = x$ and $x \in \bar{A}$. $\qquad\square$

Proposition 2.1.2. *Let (X, d) be a locally compact metric space.*

(1) *Let A, B be disjoint non-empty closed subsets of X. Then, for all $x \in X$, $d(x, A) + d(x, B) > 0$ and the function $\psi_{A,B}$ defined on X by*

$$\psi_{A,B}(x) = \frac{d(x, B)}{d(x, A) + d(x, B)} = \begin{cases} 1, & \text{for } x \in A, \\ 0, & \text{for } x \in B, \end{cases} \tag{2.1.3}$$

belongs to $C(X; [0,1])$ and is supported in $\overline{B^c}$.

(2) *Let Ω be an open subset of X and let K be a compact subset of Ω. Then $0 < d(K, \Omega^c) = \inf_{x \in K, y \in \Omega^c} d(x, y)$. Moreover there exists a function $\varphi \in C_c(X)$ such that*

$$0 \leq \varphi \leq 1, \quad \varphi_{|K} = 1, \quad \operatorname{supp} \varphi \subset \Omega.$$

The function φ can be chosen to be identically 1 on a neighborhood of K.

Proof of Proposition 2.1.2. (1) From Lemma 2.1.1, we see that $d(x, A) + d(x, B) \geq 0$ and vanishes if and only if $x \in \bar{A} \cap \bar{B} = A \cap B = \emptyset$. If $\psi_{A,B}(x) \neq 0$, then $x \notin B$, thus $\operatorname{supp} \psi_{A,B} \subset \overline{B^c}$.

(2) Since K is a compact subset of Ω, we have

$$\epsilon_0 = \inf_{x \in K, \, y \notin \Omega} d(x, y) = d(K, \Omega^c) > 0, \tag{2.1.4}$$

otherwise, we could find sequences $x_k \in K$, and $y_k \in \Omega^c$ such that $\lim_k d(x_k, y_k) = 0$. Since K is a compact subset of X, we may find a subsequence $(x_{k_l})_{l \in \mathbb{N}}$ with limit $x \in K$. Since the sequence $d(y_{k_l}, x_{k_l})$ converges to 0, we get, using that Ω^c is closed,

$$\Omega^c \ni \lim_l y_{k_l} = x \in K,$$

which contradicts $K \subset \Omega$. Since X is locally compact, every point has a compact neighborhood: this implies that

$$\forall x \in \Omega, \exists r(x) > 0, \ \overline{B(x, r(x))} \text{ is compact} \subset \Omega.$$

Since K is compact and $K \subset \cup_{x \in K} B(x, r(x))$, we can find a finite set $(x_j)_{1 \leq j \leq N}$ with

$$K \subset \cup_{1 \leq j \leq N} B(x_j, r(x_j)) = \underbrace{U}_{\text{open}} \subset \underbrace{\cup_{1 \leq j \leq N} \overline{B(x_j, r(x_j))}}_{=L \text{ compact}} \subset \Omega. \tag{2.1.5}$$

Using the notation (2.1.3) we define $\varphi = \psi_{K, U^c}$: this is a continuous function, valued in $[0, 1]$, equal to 1 on K, supported in $\overline{U}$ which is a compact subset of Ω from (2.1.5). Note that applying this result to the compact set L, a subset of the open set Ω, we find a new function $\tilde{\varphi} \in C_c(X; [0, 1])$, $\operatorname{supp} \tilde{\varphi} \subset \Omega$, $\tilde{\varphi} = 1$ on L which is a neighborhood of K from the first inclusion in (2.1.5). $\qquad\square$

Theorem 2.1.3. *Let (X, d) be a locally compact metric space, let $\Omega_1, \ldots, \Omega_n$ be open subsets of X and let K be a compact set with $K \subset \Omega_1 \cup \cdots \cup \Omega_n$. Then for each $j \in \{1, \ldots, n\}$, there exists a function $\psi_j \in C_c(\Omega_j; [0, 1])$ such that $\sum_{1 \leq j \leq n} \psi_j \in C_c(\cup_{j=1}^n \Omega_j; [0, 1])$ and*

$$1 = \sum_{1 \leq j \leq n} \psi_{j|K}.$$

We shall say that $(\psi_j)_{1 \leq j \leq n}$ is a partition of unity on K, attached to $(\Omega_j)_{1 \leq j \leq n}$. In particular, for $\theta \in C_c(\cup_{1 \leq j \leq n} \Omega_j)$, using the previous result for $K = \operatorname{supp} \theta$, we get

$$\theta = \sum_{1 \leq j \leq n} \theta_j, \quad \text{with } \theta_j = \theta \psi_j \in C_c(\Omega_j).$$

Remark 2.1.4. The reader will see in Exercise 2.8.2 that this theorem can be extended to the case of a locally compact topological space. On the other hand, Exercise 2.8.8 deals with the $\mathbb{R}^m$ framework, and provides smooth partitions.

Proof. The case $n = 1$ is dealt with in Proposition 2.1.2. For all $x \in K$, there exists $r(x) > 0$ such that $K \subset \cup_{x \in K} B(x, r(x))$, where the closed ball $B_c(x, r(x))$ is included in one of the Ω_j. Applying the Borel–Lebesgue Lemma, we get

$$K \subset \cup_{1 \leq l \leq N} B(x_l, r(x_l)) \subset \cup_{1 \leq l \leq N} B_c(x_l, r(x_l)),$$

and defining

$$K_j = \bigcup_{\substack{1 \leq l \leq N, \\ B_c(x_l, r(x_l)) \subset \Omega_j}} \Big(B_c(x_l, r(x_l)) \cap K \Big),$$

we find $K \subset \cup_{1 \leq j \leq N} K_j$, with K_j compact $\subset \Omega_j$. Applying now Proposition 2.1.2, we find $\varphi_j \in C_c(\Omega_j; [0,1])$ such that $\varphi_{j|K_j} = 1$. We set then

$$\psi_1 = \varphi_1,$$
$$\psi_2 = (1 - \varphi_1)\varphi_2,$$
$$\cdots\cdots\cdots\cdots$$
$$\psi_n = (1 - \varphi_1)\ldots(1 - \varphi_{n-1})\varphi_n.$$

We have $\psi_j \in C_c(\Omega_j; [0,1])$ and inductively on n, the identity

$$\sum_{1 \leq j \leq n} \psi_j = 1 - \prod_{1 \leq j \leq n} (1 - \varphi_j). \tag{2.1.6}$$

In fact (2.1.6) holds for $n = 1$ and supposing it for some $n \geq 1$, we get

$$\sum_{1 \leq j \leq n+1} \psi_j = 1 - \prod_{1 \leq j \leq n} (1 - \varphi_j) + \varphi_{n+1} \prod_{1 \leq j \leq n} (1 - \varphi_j) = 1 - \prod_{1 \leq j \leq n+1} (1 - \varphi_j).$$

Equalities (2.1.6) and the previous one prove in particular that $\sum_{1 \leq j \leq n} \psi_j$ as well as each ψ_j are valued in [0,1] since it is the case for each φ_j. As a result, we have $K \subset \cup_{1 \leq j \leq n} K_j \subset \cup_{1 \leq j \leq n} \{\varphi_j = 1\} \subset \{\sum_{1 \leq j \leq n} \psi_j = 1\}$, concluding the proof. $\qquad\square$

2.2   The Riesz–Markov representation theorem

The results presented in this section concern a theorem proven by the Hungarian mathematician Frigyes RIESZ (1880–1956) and by Andreï MARKOV (1856–1922), a Russian mathematician; we follow the presentation of Walter RUDIN (1921–2010). The starting point is natural, although the proof has some technical aspects: it is not difficult to define the integral of compactly supported continuous functions, either directly or using the well-broomed Riemann theory of integration. In that case, using traditional notation, the mapping

$$C_c(\mathbb{R}^m) \ni f \mapsto \int_{\mathbb{R}^m} f(x)dx$$

is a linear form which is positive in the sense that the integral of a non-negative function is also non-negative. The theorem says that it is possible to construct a measure space $(\mathbb{R}^m, \mathcal{B}_m, \mu)$, where $\mathcal{B}_m$ is the Borel σ-algebra of $\mathbb{R}^m$, so that

$L^1(\mu) \supset C_c(\mathbb{R}^m)$ where $\int_{\mathbb{R}^m} f d\mu = \int_{\mathbb{R}^m} f(x) dx$ for $f \in C_c(\mathbb{R}^m)$. This is an extension of a Radon measure (continuous linear functional on $C_c(\mathbb{R}^m)$), which can be done also replacing $\mathbb{R}^m$ by a locally compact Hausdorff topological space; here we shall limit ourselves to locally compact metric spaces. A drawback of this point of view is that it uses heavily some topological structure on the base space. A purely set-theoretic extension could be implemented and we shall present later in this chapter that different approach.

Theorem 2.2.1. *Let (X, d) be a locally compact metric space. Let $L : C_c(X) \longrightarrow \mathbb{C}$ be a positive linear form (i.e., such that $f \geq 0 \Longrightarrow Lf \geq 0$; L is said to be a positive Radon measure[1] on X). Then there exists a σ-algebra $\mathcal{M}$ on X, containing the Borel σ-algebra $\mathcal{B}_X$, and a unique measure μ defined on $\mathcal{M}$ such that the following properties hold.*

(1) $\forall f \in C_c(X), \; Lf = \int_X f d\mu$.
(2) $\forall K$ *compact* $\subset X, \; \mu(K) < +\infty$.
(3) $\forall E \in \mathcal{M}, \; \mu(E) = \inf\{\mu(V), V \text{ open} \supset E\}$ *(outer regularity)*.
(4) $\forall E \in \mathcal{O}_X \cup \{E \in \mathcal{M}, \mu(E) < +\infty\}$,

$$\mu(E) = \sup\{\mu(K), K \text{ compact} \subset E\} \quad (\text{inner regularity}).$$

(5) $\forall E \in \mathcal{M}$ *such that* $\mu(E) = 0$, $A \subset E$ *implies* $A \in \mathcal{M}$ *(the σ-algebra $\mathcal{M}$ will be said μ-complete).*

N.B. Let us note that (1) is meaningful since a function f in $C_c(X)$ is Borel measurable, so that the inverse image of a Borelian of $\mathbb{C}$ belongs to $\mathcal{B}_X \subset \mathcal{M}$, proving the measurability of f. Moreover, since f is compactly supported, the inequality $|f| \leq \mathbf{1}_K \sup|f|$ and (2) imply $f \in \mathcal{L}^1(\mu)$.

Proof of the theorem, Uniqueness. Since μ satisfies (4) and open subsets are Borelian, we have for V open, $\mu(V) = \sup\{\mu(K), K \text{ compact} \subset V\}$. Property (3) shows then that μ is completely determined by its values on compact subsets of X. Let μ_1, μ_2 be two positive measures defined on a σ-algebra $\mathcal{M}$ containing $\mathcal{B}_X$ and satisfying (1-2-3-4). Let K be a compact subset of X. From (2), (3), we get that for all $\epsilon > 0$, there exists an open set $V_\epsilon \supset K$ such that

$$\mu_2(K) \leq \mu_2(V_\epsilon) < \mu_2(K) + \epsilon.$$

Let $\varphi \in C_c(V_\epsilon; [0,1])$ so that $\varphi_{|K} = 1$ (cf. Proposition 2.1.2). We have

$$\mu_1(K) = \int_X \mathbf{1}_K d\mu_1 \leq \int_X \varphi d\mu_1 = L\varphi$$

$$= \int_X \varphi d\mu_2 \leq \int_X \mathbf{1}_{V_\epsilon} d\mu_2 = \mu_2(V_\epsilon) < \mu_2(K) + \epsilon,$$

---

[1] cf. (2.8.7), (2.8.8) in Exercise 2.8.3.

which implies $\mu_1(K) \leq \mu_2(K)$. Switching μ_1 with μ_2, we get $\mu_2(K) = \mu_1(K)$, proving uniqueness.

Proof of the theorem, Existence. We shall now construct μ and $\mathcal{M}$:

$$\text{for } V \text{ an open set, we define} \quad \mu(V) = \sup\{L\varphi, \ \varphi \in C_c(V;[0,1])\}, \qquad (2.2.1)$$

$$\text{for any subset } E \subset X, \text{ we define } \mu^*(E) = \inf\{\mu(V), V \text{ open} \supset E\}. \qquad (2.2.2)$$

We define also

$$\mathcal{M}_F = \{E \subset X, \ \mu^*(E) < +\infty, \ \mu^*(E) = \sup_{K \text{ compact} \subset E} \mu^*(K)\}, \qquad (2.2.3)$$

$$\mathcal{M} = \{E \subset X, \forall K \text{ compact}, \ K \cap E \in \mathcal{M}_F\}. \qquad (2.2.4)$$

Lemma 2.2.2. *The mappings μ and μ^* are valued in $\overline{\mathbb{R}}_+$. If $V_1 \subset V_2$ are open sets, then $\mu(V_1) \leq \mu(V_2)$. If V is open, then $\mu(V) = \mu^*(V)$. Moreover $\mu(\emptyset) = 0$.*

Proof. Since $L\varphi \in \mathbb{R}_+$ for $\varphi \in C_c(X;[0,1])$, we have $\mu(V) \in \overline{\mathbb{R}}_+$ and thus the same for $\mu^*(E)$. If $V_1 \subset V_2$ are open, the inclusion $C_c(V_1;[0,1]) \subset C_c(V_2;[0,1])$ implies $\mu(V_1) \leq \mu(V_2)$. For V open, we have $\mu(V) = \mu^*(V)$ since whenever W open $\supset V$, we have $\mu(V) \leq \mu(W)$ so that $\mu^*(V) \leq \mu(V) \leq \mu^*(V)$. The last property follows from the very definition of

$$C_c(V;[0,1]) = \{\varphi : X \to [0,1], \text{continuous}, \text{supp}\,\varphi \text{ compact} \subset V\}.$$

When $V = \emptyset$, $\varphi \in C_c(V;[0,1])$ implies $\text{supp}\,\varphi = \emptyset$, so that $\varphi \equiv 0$ and thus $L\varphi = 0$, entailing $\mu(\emptyset) = 0$. $\qquad\square$

The σ-additivity of μ^* on $\mathcal{P}(X)$ does not hold in general[2], but we shall prove that it holds on a σ-algebra containing $\mathcal{B}_X$.

Lemma 2.2.3. *The mapping μ^* defined by (2.2.2) is increasing. Moreover, $\{E \subset X, \mu^*(E) = 0\} \subset \mathcal{M} \cap \mathcal{M}_F$. Also, $\mu^*(E) = 0$ implies $\mathcal{P}(E) \subset \mathcal{M}$.*

Proof. If $B \supset A$, we have

$$\{V \text{ open } \supset B\} \subset \{V \text{ open} \supset A\} \Longrightarrow$$

$$\mu^*(B) = \inf_{V \text{open} \supset B} \mu(V) \geq \inf_{V \text{open} \supset A} \mu(V) = \mu^*(A).$$

Moreover if $\mu^*(E) = 0$, then $E \in \mathcal{M} \cap \mathcal{M}_F$; in fact if $K \subset E$ is a compact subset of X, we have $\mu^*(K) = 0$ by monotonicity, so that $E \in \mathcal{M}_F$. Also $E \in \mathcal{M}$ since for K compact $\mu^*(K \cap E) \leq \mu^*(E) = 0$ so that $K \cap E \in \mathcal{M}_F$, from the above argument. Moreover, if $A \subset E$ and $\mu^*(E) = 0$, then $\mu^*(A) = 0$ and $A \in \mathcal{M}$. $\qquad\square$

---

[2]It is for instance possible to prove that there does not exist a positive measure defined on $\mathcal{P}(\mathbb{R}^m)$ which would coincide with the ordinary volume on compact rectangles $\prod_{1 \leq j \leq n}[a_j, b_j]$. As a matter of fact, this impossibility is the initial reason for the introduction of the notion of σ-algebra, to restrict the measure first to Borel sets, then to the completed σ-algebra, i.e., the σ-algebra generated by $\mathcal{B}_m$ and the subsets of sets with measure 0 (see Exercise 2.8.13).

Note also the monotonicity of L: for $f \leq g \in C_c(X; \mathbb{R})$ then
$$Lg = L(g - f + f) = L(g - f) + Lf \geq Lf.$$

Definition 2.2.4. Let X be a set and $\nu : \mathcal{P}(X) \to \overline{\mathbb{R}}_+$ be a mapping. We shall say that ν is an *outer measure* on X whenever

$$\nu(\emptyset) = 0, \tag{2.2.5}$$
$$A \subset B \subset X \Longrightarrow \nu(A) \leq \nu(B), \tag{2.2.6}$$
$$\text{for } (E_j)_{j \in \mathbb{N}} \text{ a sequence in } \mathcal{P}(X), \quad \nu(\cup_{j \in \mathbb{N}} E_j) \leq \sum_{j \in \mathbb{N}} \nu(E_j). \tag{2.2.7}$$

The last property is called *countable subadditivity*.

Lemma 2.2.5. *Let L be a positive Radon measure on X and μ, μ^* defined respectively on the open sets and on $\mathcal{P}(X)$ by (2.2.1), (2.2.2). Then μ^* is an outer measure on X.*

Proof. Property (2.2.5) follows from Lemma 2.2.2 and Property (2.2.6) from Lemma 2.2.3. Let us prove countable subadditivity for μ^*. Let V_1, V_2 be open subsets of X and $V = V_1 \cup V_2$. We have defined
$$\mu(V) = \sup_{\varphi \in C_c(V; [0,1])} L\varphi.$$

If $\varphi \in C_c(V; [0,1])$ and $K = \operatorname{supp} \varphi$, Theorem 2.1.3 implies that we can find $\theta_j \in C_c(V_j; [0,1]), j = 1, 2$, such that $\theta_1 + \theta_2 = 1$ on K. As a result, we get $\varphi = \theta_1 \varphi + \theta_2 \varphi$, so that with $\varphi_j = \theta_j \varphi$,
$$L\varphi = L\varphi_1 + L\varphi_2 \leq \sup_{\phi_1 \in C_c(V_1; [0,1])} L\phi_1 + \sup_{\phi_2 \in C_c(V_2; [0,1])} L\phi_2 = \mu(V_1) + \mu(V_2),$$

entailing $\mu(V_1 \cup V_2) \leq \mu(V_1) + \mu(V_2)$. Inductively on N, we get for $V_1, \ldots, V_N$ open,
$$\mu(\cup_{1 \leq k \leq N} V_k) \leq \sum_{1 \leq k \leq N} \mu(V_k). \tag{2.2.8}$$

To prove the lemma, we may assume that for all j, $\mu^*(E_j) < +\infty$ (otherwise the result is obvious). From (2.2.2), we obtain for all $\epsilon > 0$, for all $j \in \mathbb{N}$, the existence of an open set $V_{\epsilon,j} \supset E_j$ such that
$$\mu^*(E_j) \leq \mu(V_{\epsilon,j}) < \mu^*(E_j) + \epsilon 2^{-j-1}.$$

We set then $V_\epsilon = \cup_{j \in \mathbb{N}} V_{\epsilon,j}$ (an open set) and consider $\varphi \in C_c(V_\epsilon; [0,1])$. Since the support of φ is compact, there exists $N \in \mathbb{N}$ such that $\varphi \in C_c(\cup_{0 \leq j \leq N} V_{\epsilon,j}; [0,1])$. Consequently, from the definition (2.2.1) and (2.2.8) we get
$$L\varphi \leq \mu(\cup_{0 \leq j \leq N} V_{\epsilon,j}) \leq \sum_{0 \leq j \leq N} \mu(V_{\epsilon,j})$$
$$< \sum_{0 \leq j \leq N} \left(\mu^*(E_j) + \epsilon 2^{-j-1} \right) \leq \epsilon + \sum_{j \in \mathbb{N}} \mu^*(E_j).$$

As a result since μ^* is increasing and $\cup_{j\in\mathbb{N}}E_j \subset V_\epsilon$, we have for all $\epsilon > 0$,

$$\mu^*(\cup_{j\in\mathbb{N}}E_j) \leq \mu^*(V_\epsilon) = \mu(V_\epsilon) = \sup_{\varphi\in C_c(V_\epsilon;[0,1])} L\varphi \leq \epsilon + \sum_{j\in\mathbb{N}}\mu^*(E_j),$$

implying (2.2.7). $\qquad\square$

Lemma 2.2.6. *Let L be a positive Radon measure on X and μ, μ^* defined respectively on the open sets and on $\mathcal{P}(X)$ by (2.2.1), (2.2.2). Then, all compact subsets of X belong to $\mathcal{M}_F$; more precisely for K compact in X,*

$$\mu^*(K) = \inf\{L\varphi, \ \varphi \in C_c(X;[0,1]), \varphi_{|K} \equiv 1\}. \tag{2.2.9}$$

Proof. Let K be a compact subset of X, $\varphi \in C_c(X;[0,1])$, $\varphi_{|K} \equiv 1$ and $1 > \epsilon > 0$. The set $V_\epsilon = \{x \in X, \varphi(x) > 1 - \epsilon\}$ is open and contains K. For $\psi \in C_c(V_\epsilon;[0,1])$, we have

$$(1 - \epsilon)\psi \leq (1 - \epsilon)\mathbf{1}_{V_\epsilon} \leq \varphi,$$

so that from the monotonicity of L and the definition of μ^*, we get

$$\mu^*(K) \leq \mu^*(V_\epsilon) = \mu(V_\epsilon) = \sup_{\psi\in C_c(V_\epsilon,[0,1])} L\psi \leq (1 - \epsilon)^{-1}L\varphi. \tag{2.2.10}$$

This implies $\mu^*(K) \leq L\varphi < +\infty$ so that, since we have trivially by monotonicity

$$\mu^*(K) \leq \sup_{L \text{ compact} \subset K} \mu^*(L) \leq \mu^*(K), \quad \text{and thus equality,}$$

we get $K \in \mathcal{M}_F$. Moreover from (2.2.10), we get also

$$\mu^*(K) \leq \inf_{\varphi\in C_c(X,[0,1]), \ \varphi_{|K}\equiv 1} L\varphi. \tag{2.2.11}$$

To prove that (2.2.11) is an equality, we note, using $\mu^*(K) < +\infty$, that for all $\epsilon > 0$, there exists an open set W_ϵ containing K such that $\mu^*(K) \leq \mu(W_\epsilon) < \mu^*(K)+\epsilon$. Using Proposition 2.1.2, we find $\varphi \in C_c(W_\epsilon;[0,1]), \varphi_{|K} = 1$. Consequently, for all $\epsilon > 0$, we find $L\varphi \leq \mu(W_\epsilon) < \mu^*(K) + \epsilon$, entailing

$$\inf_{\varphi\in C_c(X,[0,1]), \ \varphi_{|K}\equiv 1} L\varphi < \mu^*(K) + \epsilon,$$

and the result of the lemma. $\qquad\square$

Lemma 2.2.7. *Let L be a positive Radon measure on X and μ, μ^* defined respectively on the open sets and on $\mathcal{P}(X)$ by (2.2.1), (2.2.2). Then any open set V is such that*

$$\mu(V) = \sup_{K \text{ compact} \subset V} \mu^*(K). \tag{2.2.12}$$

In particular $\mathcal{M}_F$ contains all the open sets V such that $\mu(V) < +\infty$.

Proof. We assume first $\mu(V) < +\infty$. For all $\epsilon > 0$, there exists $\varphi_\epsilon \in C_c(V; [0,1])$ such that

$$\mu(V) - \epsilon < L\varphi_\epsilon \le \mu(V).$$

Considering the compact set $K_\epsilon = \operatorname{supp} \varphi_\epsilon \subset V$ and W open containing K_ϵ, we have $\varphi_\epsilon \in C_c(W; [0,1])$ and thus $L\varphi_\epsilon \le \mu(W)$, which implies

$$L\varphi_\epsilon \le \inf_{W \text{ open} \supset K_\epsilon} \mu(W) = \mu^*(K_\epsilon).$$

Using monotonicity, this implies (2.2.12):

$$\mu(V) - \epsilon < \mu^*(K_\epsilon) \le \sup_{K \text{ compact} \subset V} \mu^*(K) \le \mu(V).$$

Moreover, for V open such that $\mu(V) < +\infty$, we have proven

$$\mu(V) = \sup_{K \text{ compact} \subset V} \mu^*(K), \quad \text{i.e., } V \in \mathcal{M}_F.$$

If $\mu(V) = +\infty$, we can find a sequence $\varphi_k \in C_c(V; [0,1])$ such that $L\varphi_k \ge k$. Considering $K_k = \operatorname{supp} \varphi_k \subset V$ and W open containing K_k, we have $\varphi_k \in C_c(W; [0,1])$ so that

$$L\varphi_k \le \mu(W) \implies L\varphi_k \le \inf_{W \text{ open} \supset K_k} \mu(W) = \mu^*(K_k).$$

This implies $\lim_k \mu^*(K_k) = +\infty$ and (2.2.12) in that case. $\qquad\square$

Lemma 2.2.8. *Let L be a positive Radon measure on X and μ, μ^* defined respectively on the open sets and on $\mathcal{P}(X)$ by (2.2.1), (2.2.2). Let $(E_j)_{j\in\mathbb{N}}$ be a pairwise disjoint sequence in $\mathcal{M}_F$: then,*

$$\mu^*(\cup_{j\in\mathbb{N}} E_j) = \sum_{j\in\mathbb{N}} \mu^*(E_j). \tag{2.2.13}$$

Whenever $\mu^*(\cup_{j\in\mathbb{N}} E_j) < +\infty$, we have $\cup_{j\in\mathbb{N}} E_j \in \mathcal{M}_F$.

Proof. We note first that for disjoint compact sets K_1, K_2, we have

$$\mu^*(K_1 \cup K_2) = \mu^*(K_1) + \mu^*(K_2). \tag{2.2.14}$$

In fact, we have $K_1 \subset K_2^c$ open and we may find $\varphi \in C_c(K_2^c; [0,1])$ such that $\varphi_{|K_1} = 1$. From Lemma 2.2.6, for all $\epsilon > 0$, there exists ψ_ϵ such that $\psi_{\epsilon|K_1 \cup K_2} = 1$, $\psi_\varepsilon \in C_c(X; [0,1]$ with

$$\mu^*(K_1 \cup K_2) \le L\psi_\epsilon < \mu^*(K_1 \cup K_2) + \epsilon.$$

Moreover, we have $\varphi\psi_{\epsilon|K_1} = 1$ and $(1 - \varphi)\psi_{\epsilon|K_2} = 1$. As a result for all $\epsilon > 0$,

$$\mu^*(K_1) + \mu^*(K_2) \overset{\overset{\text{Lemma 2.2.6}}{\le}}{} L(\varphi\psi_\epsilon) + L((1 - \varphi)\psi_\epsilon) = L(\psi_\epsilon)$$
$$< \mu^*(K_1 \cup K_2) + \epsilon \underset{\text{Lemma 2.2.5}}{\le} \mu^*(K_1) + \mu^*(K_2) + \epsilon,$$

providing (2.2.14). Let us return to the proof of the lemma. Since Lemma 2.2.5 provides an inequality when $\mu^*(\cup_{j\in\mathbb{N}}E_j) = +\infty$, we get the result in that case. Let us assume now that $\mu^*(\cup_{j\in\mathbb{N}}E_j) < +\infty$ and let $\epsilon > 0$. As $E_j \in \mathcal{M}_F$, we may find compact sets $K_{\epsilon,j} \subset E_j$ such that

$$\mu^*(E_j) - \epsilon 2^{-j-1} < \mu^*(K_{\epsilon,j}) \leq \mu^*(E_j).$$

As a result, for any $N \in \mathbb{N}$,

$$\mu^*(\cup_{j\in\mathbb{N}}E_j) \overset{\text{monotonicity}}{\geq} \mu^*(\underbrace{\cup_{0\leq j\leq N}K_{\epsilon,j}}_{\substack{\text{compact}\\\text{pairwise disjoint}}}) \overset{\substack{(2.2.14)\text{ and}\\\text{induction on }N}}{=} \sum_{0\leq j\leq N}\mu^*(K_{\epsilon,j})$$

$$\geq -\epsilon + \sum_{0\leq j\leq N}\mu^*(E_{,j}),$$

proving the first assertion in the lemma. Let us now show that $E = \cup_{j\in\mathbb{N}}E_j \in \mathcal{M}_F$. Since the series $\sum_{j\in\mathbb{N}}\mu^*(E_j) = \mu^*(E)$ converges, for all $\epsilon > 0$, there exists N_ϵ such that

$$\mu^*(E) - \epsilon \leq \sum_{0\leq j\leq N_\epsilon}\mu^*(E_j) \leq \epsilon + \sum_{0\leq j\leq N_\epsilon}\mu^*(K_{\epsilon,j}) = \epsilon + \mu^*(\overbrace{\cup_{0\leq j\leq N_\epsilon}K_{\epsilon,j}}^{\text{compact }\subset E}).$$

Consequently, we have

$$\mu^*(E) \leq 2\epsilon + \mu^*(\cup_{0\leq j\leq N_\epsilon}K_{\epsilon,j}) \leq 2\epsilon + \sup_{K\text{ compact }\subset E}\mu^*(K) \overset{\text{monotonicity}}{\leq} 2\epsilon + \mu^*(E),$$

concluding the proof of Lemma 2.2.8. $\square$

Lemma 2.2.9. *Let L be a positive Radon measure on X and μ, μ^* defined respectively on the open sets and on $\mathcal{P}(X)$ by (2.2.1), (2.2.2). Let $E, A_1, A_2 \in \mathcal{M}_F$. Then*

(1) *$\forall \epsilon > 0$, $\exists K_\epsilon$ compact, $\exists V_\epsilon$ open such that $K_\epsilon \subset E \subset V_\epsilon$, and $\mu(V_\epsilon \backslash K_\epsilon) < \epsilon$.*
(2) *$A_1 \backslash A_2$, $A_1 \cup A_2$, $A_1 \cap A_2 \in \mathcal{M}_F$.*

Proof. From the definition of $\mathcal{M}_F$, we have

$$\mu^*(E) < +\infty, \quad \inf_{V\text{ open }\supset E}\mu(V) = \mu^*(E) = \sup_{K\text{ compact }\subset E}\mu^*(K).$$

As a result for all $\epsilon > 0$, there exists a compact set $K_\epsilon \subset E$ and an open set $V_\epsilon \supset E$ such that

$$\mu^*(E) - \epsilon/3 < \mu^*(K_\epsilon) \leq \mu^*(E) \leq \mu(V_\epsilon) < \mu^*(E) + \epsilon/3.$$

Since $V_\epsilon \backslash K_\epsilon$ is an open set such that $\mu(V_\epsilon \backslash K_\epsilon) < +\infty$, we find using Lemma 2.2.7 that $V_\epsilon \backslash K_\epsilon \in \mathcal{M}_F$. Lemmas 2.2.8–2.2.6 provide now

$$\mu^*(V_\epsilon \backslash K_\epsilon) + \mu^*(K_\epsilon) = \mu(V_\epsilon) \le \mu^*(E) + \epsilon/3 \implies \mu^*(V_\epsilon \backslash K_\epsilon) \le 2\epsilon/3,$$

proving (1). Using that result, we find for $A_1, A_2 \in \mathcal{M}_F$

$$K_j \text{ compact} \subset A_j \subset V_j \text{ open}, \quad \mu(V_j \backslash K_j) < \epsilon.$$

Since $A_1 \backslash A_2 \subset V_1 \backslash K_2 \subset (V_1 \backslash K_1) \cup (K_1 \backslash V_2) \cup (V_2 \backslash K_2),$[3] Lemma 2.2.5 gives

$$\mu^*(A_1 \backslash A_2) \le 2\epsilon + \mu^*(K_1 \backslash V_2),$$

and since $K_1 \backslash V_2$ is a compact set $\subset A_1 \backslash A_2$, we find $A_1 \backslash A_2 \in \mathcal{M}_F$. Moreover the equality $A_1 \cup A_2 = (A_1 \backslash A_2) \cup A_2$ and Lemma 2.2.8 give $A_1 \cup A_2 \in \mathcal{M}_F$. Also the identity

$$A_1 \cap A_2 = \underbrace{A_1}_{\in \mathcal{M}_F} \backslash \underbrace{(A_1 \backslash A_2)}_{\in \mathcal{M}_F}$$

and the beginning of our proof shows that $A_1 \cap A_2 \in \mathcal{M}_F$. $\qquad\square$

Lemma 2.2.10. *Let L be a positive Radon measure on X and μ, μ^* defined respectively on the open sets and on $\mathcal{P}(X)$ by (2.2.1), (2.2.2). Then $\mathcal{M}$ defined in (2.2.4) is a σ-algebra on X containing the Borel σ-algebra $\mathcal{B}_X$.*

Proof. Let K be a compact subset of X and $A \in \mathcal{M}$. Then we have

$$A^c \cap K = K \backslash A = K \backslash (A \cap K),$$

and since $K \in \mathcal{M}_F$ (Lemma 2.2.6) and $A \cap K \in \mathcal{M}_F$ (assumption $A \in \mathcal{M}$), we find, from Lemma 2.2.9 that $A^c \cap K \in \mathcal{M}_F$, implying $A^c \in \mathcal{M}$. Moreover if $(A_j)_{j \ge 1}$ is a sequence of $\mathcal{M}$ and K is a compact set, we have,

$$(\cup_{j \ge 1} A_j) \cap K = \bigcup_{N \ge 1} \left\{ (A_N \cap K) \backslash [\cup_{1 \le j < N} (A_j \cap K)] \right\}.$$

Since our assumption implies $A_j \cap K \in \mathcal{M}_F$, we get from Lemma 2.2.9 that for all N,

$$(A_N \cap K) \backslash [\cup_{1 \le j < N} (A_j \cap K)] \in \mathcal{M}_F.$$

But these sets are pairwise disjoint with union $A \cap K$ $(A = \cup_{j \ge 1} A_j)$, with a finite outer measure since $\mu^*(A \cap K) \le \mu^*(K) < +\infty$. We may thus apply Lemma

---

[3] We have $X_1 \backslash X_4 \subset (X_1 \backslash X_2) \cup (X_2 \backslash X_3) \cup (X_3 \backslash X_4)$ since

$$X_1 \backslash X_4 = X_1 \cap X_4^c = (X_1 \cap X_4^c \cap X_2^c) \cup (X_1 \cap X_4^c \cap X_2)$$
$$\subset (X_1 \backslash X_2) \cup (X_1 \cap X_4^c \cap X_2 \cap X_3^c) \cup (X_1 \cap X_4^c \cap X_2 \cap X_3)$$
$$\subset (X_1 \backslash X_2) \cup (X_2 \backslash X_3) \cup (X_3 \backslash X_4).$$

2.2.8, proving $A \cap K \in \mathcal{M}_F$ and thus $A \in \mathcal{M}$. Moreover for F closed, $F \cap K$ is compact thus belonging to $\mathcal{M}_F$, implying $F \in \mathcal{M}$. In particular X belongs to $\mathcal{M}$. Finally, $\mathcal{M}$ is a σ-algebra on X containing the closed sets, thus the Borel σ-algebra $\mathcal{B}_X$. $\qquad\square$

Lemma 2.2.11. *Let L be a positive Radon measure on X and μ, μ^* defined respectively on the open sets and on $\mathcal{P}(X)$ by (2.2.1), (2.2.2). With $\mathcal{M}_F$ and $\mathcal{M}$ defined in (2.2.3), (2.2.4), we have*

$$\mathcal{M}_F = \{E \in \mathcal{M}, \ \mu^*(E) < +\infty\}.$$

Proof. Let $E \in \mathcal{M}_F$ and K compact. Lemmas 2.2.6–2.2.9 show that $K, E \cap K \in \mathcal{M}_F$, which implies $E \in \mathcal{M}$. Conversely, if $E \in \mathcal{M}$ and $\mu^*(E) < +\infty$, there exists V open $\supset E$ such that $\mu(V) < +\infty$ and from Lemma 2.2.7, $V \in \mathcal{M}_F$. Using Lemma 2.2.9, we find that for all $\epsilon > 0$, there exists K compact such that $K \subset V$ and $\mu(V \backslash K) < \epsilon$. Since we have assumed $E \cap K \in \mathcal{M}_F$, there exists a compact set $L \subset E \cap K$ such that

$$\mu^*(E \cap K) - \epsilon < \mu^*(L) \leq \mu^*(E \cap K).$$

Moreover we have $E \subset \underset{\in \mathcal{M}_F}{(E \cap K)} \cup \underset{\in \mathcal{M}_F}{(V \backslash K)}$, thus we find from Lemma 2.2.8,

$$\mu^*(E) \leq \mu^*(E \cap K) + \mu^*(V \backslash K) < \mu^*(L) + 2\epsilon \leq \mu^*(E \cap K) + 2\epsilon \leq \mu^*(E) + 2\epsilon,$$

entailing $E \in \mathcal{M}_F$. $\qquad\square$

Lemma 2.2.12. *Let L be a positive Radon measure on X and μ, μ^* defined respectively on the open sets and on $\mathcal{P}(X)$ by (2.2.1), (2.2.2). Then with $\mathcal{M}$ defined in (2.2.4), μ^* is a positive measure defined on the σ-algebra $\mathcal{M}$, and denoting the measure space $(X, \mathcal{M}, \mu^*)$ by $(X, \mathcal{M}, \mu)$, we find $\forall \varphi \in C_c(X)$, $L\varphi = \int_X \varphi d\mu$.*

Proof. We have proven in Lemma 2.2.2 that $\mu^*(\emptyset) = \mu(\emptyset) = 0$. Let $(E_j)_{j \geq 1}$ be a pairwise disjoint sequence in $\mathcal{M}$. If there exists $j_0 \geq 1$ such that $\mu^*(E_{j_0}) = +\infty$, we obtain the result for the σ-additivity since $\mu^*(E_{j_0}) \leq \mu^*(\cup_{j \geq 1} E_j)$. We may thus suppose that $\forall j \geq 1, \mu^*(E_j) < +\infty$. From Lemma 2.2.11, $\forall j \geq 1, E_j \in \mathcal{M}_F$ and Lemma 2.2.8 gives the result. To obtain the second property, we may assume that φ is real valued and we have only to prove that $L\varphi \leq \int_X \varphi d\mu$ since we shall deduce from this

$$-L(\varphi) = L(-\varphi) \leq \int_X -\varphi d\mu = -\int_X \varphi d\mu \implies L\varphi \geq \int_X \varphi d\mu.$$

We note also $C_c(X) \subset \mathcal{L}^1(\mu)$, since for $\varphi \in C_c(X)$, we have

$$|\varphi| \leq \sup |\varphi| \mathbf{1}_{\operatorname{supp} \varphi} \in \mathcal{L}^1(\mu),$$

because $\operatorname{supp} \varphi$ is compact, implying $\mu(\operatorname{supp} \varphi) < +\infty$, and moreover, φ is measurable since $\mathcal{M}$ contains the Borel σ-algebra. Let us then consider φ real-valued

$\in C_c(X)$ with compact support K such that $\varphi(X) \subset [a,b]$ and let $\epsilon > 0$ be given. We consider $(y_j)_{1 \leq j \leq N}$ real numbers such that

$$y_0 < a < y_1 < \cdots < y_N = b, \quad 0 < y_{j+1} - y_j < \epsilon.$$

We define $E_j = \{x \in K, y_{j-1} < \varphi(x) \leq y_j\}$, $1 \leq j \leq N$. The sets E_j are pairwise disjoint Borel sets with union K. Consequently, there exist some open sets $V_j \supset E_j$ such that

$$\mu(E_j) \leq \mu(V_j) < \mu(E_j) + \frac{\epsilon}{N}. \tag{2.2.15}$$

We consider the open sets $W_j = V_j \cap \{x \in X, \varphi(x) < y_j + \epsilon\} \supset E_j$. We have

$$\mu(W_j) \leq \mu(V_j) < \mu(E_j) + \frac{\epsilon}{N}, \quad K = \cup_{1 \leq j \leq N} E_j \subset \cup_{1 \leq j \leq N} W_j. \tag{2.2.16}$$

From Theorem 2.1.3 on partitions of unity, we find some functions ψ_j belonging to $C_c(W_j; [0,1])$ such that on K, $\sum_{1 \leq j \leq N} \psi_j = 1$, implying $\varphi = \sum_{1 \leq j \leq N} \psi_j \varphi$. From Lemma 2.2.6 we get

$$\mu(K) \leq L\Big(\sum_{1 \leq j \leq N} \psi_j \Big) = \sum_{1 \leq j \leq N} L\psi_j, \tag{2.2.17}$$

and since $\psi_j \varphi \leq (y_j + \epsilon)\psi_j$ with $y_j - \epsilon < \varphi(x)$ for $x \in E_j$, we get

$$L\varphi = L\Big(\sum_{1 \leq j \leq N} \psi_j \varphi \Big) \leq L\Big(\sum_{1 \leq j \leq N} (y_j + \epsilon)\psi_j \Big) = \sum_{1 \leq j \leq N} (y_j + \epsilon)L\psi_j$$

$$= \sum_{1 \leq j \leq N} \underbrace{(|a| + y_j + \epsilon)}_{=y_j - a + \epsilon + a + |a| \geq 0} L\psi_j - |a| \sum_{1 \leq j \leq N} L\psi_j$$

$$(\text{using } (2.2.1) \text{ and } (2.2.17)) \leq \sum_{1 \leq j \leq N} (|a| + y_j + \epsilon)\mu(W_j) - |a|\mu(K)$$

$$(\text{using } (2.2.16)) \leq \sum_{1 \leq j \leq N} (|a| + y_j + \epsilon)\Big(\mu(E_j) + \frac{\epsilon}{N}\Big) - |a|\mu(K).$$

Consequently, we obtain for all $\epsilon > 0$,

$$L\varphi \leq \sum_{1 \leq j \leq N} (|a| + y_j + \epsilon)\Big(\mu(E_j) + \frac{\epsilon}{N}\Big) - |a| \sum_{1 \leq j \leq N} \mu(E_j)$$

$$= \epsilon|a| + \sum_{1 \leq j \leq N} (y_j + \epsilon)\Big(\mu(E_j) + \frac{\epsilon}{N}\Big)$$

$$(\text{and since on } E_j, y_{j-1} < \varphi \implies y_j + \varepsilon \leq y_{j-1} + 2\varepsilon \leq \varphi + 2\varepsilon)$$

$$\leq \epsilon|a| + \sum_{1 \leq j \leq N} \int_{E_j} (\varphi + 2\epsilon)d\mu + \epsilon(b + \epsilon)$$

$$\leq \epsilon(|a| + b + \epsilon) + \int_X \varphi d\mu + 2\epsilon\mu(K),$$

so that $L\varphi \leq \int_X \varphi d\mu$. $\qquad \square$

We have thus proven that $(X, \mathcal{M}, \mu)$ is a measure space where μ is a positive measure so that $\mathcal{M} \supset \mathcal{B}_X$. Property (1) in Theorem 2.2.1 follows from Lemma 2.2.12, Property (2) from Lemma 2.2.6, Property (3) from (2.2.2), Property (4) for open sets from Lemma 2.2.7, Property (4) for sets $E \in \mathcal{M}$ with $\mu(E) < +\infty$, from Lemma 2.2.11 and Property (5) is proven in Lemma 2.2.3. The proof of Theorem 2.2.1 is complete.

Definition 2.2.13. Let X be a locally compact metric space, $\mathcal{B}_X$ its Borel σ-algebra, and let $(X, \mathcal{B}_X, \mu)$ be a measure space where μ is a positive measure. When the measure μ is finite on the compact sets, we shall say that μ is a positive Borel measure on X. When Property (3) (resp. (4)) in Theorem 2.2.1 is satisfied for all $E \in \mathcal{B}_X$, we shall say that μ is outer regular (resp. inner regular); μ will be said regular when both properties hold.

Theorem 2.2.14. *Let (X, d) be a locally compact metric space which is also σ-compact (i.e., countable union of compact sets), let $L : C_c(X) \longrightarrow \mathbb{C}$ be a positive linear form and let $(X, \mathcal{M}, \mu)$ be the measure space given by Theorem 2.2.1. The following additional properties hold.*

(1) μ is a regular Borel measure on X.

(2) For $E \in \mathcal{M}$ and $\varepsilon > 0$, there exists V, F such that

$$F \, closed \subset E \subset V \, open, \quad \mu(V \backslash F) < \epsilon.$$

(3) E belongs to $\mathcal{M}$ if and only if there exists an F_σ set (countable union of closed sets) A, and a G_δ set B (countable intersection of open sets), such that

$$A \subset E \subset B, \quad and \quad \mu(B \backslash A) = 0. \tag{2.2.18}$$

We start with the proof of (2). Let K_N be a sequence of compact sets with $X = \cup_{N \geq 1} K_N$. We have

$$\mu(K_N \cap E) \underset{\text{monotonicity}}{\leq} \mu(K_N) \underset{\text{(2) in Th. 2.2.1}}{<} +\infty.$$

From (2.2.2) there exists V_N open such that $V_N \supset K_N \cap E$ such that

$$\mu(K_N \cap E) \leq \mu(V_N) < \mu(K_N \cap E) + \epsilon 2^{-N-2}.$$

Since $E, V_N, K_N \in \mathcal{M}$, we have $\mu(V_N \backslash (K_N \cap E)) \leq \epsilon 2^{-N-2}$ and with the open set $V = \cup_{N \geq 1} V_N \supset E$,

$$\mu(V \backslash E) = \mu(\cup_{N \geq 1}(V_N \backslash E)) \leq \mu\left(\cup_{N \geq 1}\left(V_N \backslash (E \cap K_N)\right)\right)$$

$$\leq \sum_{N \geq 1} \mu(V_N \backslash (E \cap K_N)) \leq \epsilon/4.$$

Applying this to E^c, we find an open set $W \supset E^c$ such that $\mu(W \backslash E^c) \leq \epsilon/4$. Finally we get

$$F = W^c \text{ closed} \subset E \subset V, \quad \mu(V \backslash E) \leq \epsilon/4, \quad \mu(E \backslash F) = \mu(W \backslash E^c) \leq \epsilon/4,$$

implying the result.

Let us now prove (1). Outer regularity and finiteness on compact sets follow from Theorem 2.2.1. To get inner regularity, we have only to check Property (4) of Theorem 2.2.1 for Borel sets with infinite measure. Let E be a Borel set with infinite measure: from the already proven (2), there exists

$$F_1 \text{ closed} \subset E \subset V_1 \text{ open}, \quad \mu(V_1 \backslash F_1) < 1.$$

Since $\mu(E) = \mu(E \backslash F_1) + \mu(F_1) \leq 1 + \mu(F_1)$, we have $\mu(F_1) = +\infty$. We consider now the closed set $F_1 = \cup_{N \geq 1}(F_1 \cap K_N)$. Then from Proposition 1.4.4 (2), we find

$$\mu\underbrace{\left(F_1 \cap \left(\cup_{1 \leq j \leq N} K_j\right)\right)}_{L_N \text{ compact} \subset E} \underset{N \to +\infty}{\uparrow} \mu(F_1) = +\infty,$$

so that $\lim_N \mu(L_N) = +\infty$, providing Property (1) of Theorem 2.2.14.

We are left with the proof of (3). Let E be in $\mathcal{M}$. From the already proven (2) in this theorem, for all integers $j \geq 1$, there exists a closed set F_j and an open set V_j such that $F_j \subset E \subset V_j$ with $\mu(V_j \backslash F_j) \leq 1/j$. We get then

$$A = \cup_{j \geq 1} F_j \subset E \subset \cap_{j \geq 1} V_j = B,$$

and for all $j \geq 1$, $\mu(B \backslash A) \leq \mu(V_j \backslash F_j) \leq 1/j$, implying $\mu(B \backslash A) = 0$ and the first part of the statement. Conversely, if (2.2.18) holds, we have

$$E = (E \backslash A) \cup A, \quad E \backslash A \subset B \backslash A,$$

and since the σ-algebra $\mathcal{M}$ is complete, we have $E \backslash A \in \mathcal{M}$ as a subset of the negligible Borel set $B \backslash A \in \mathcal{B}_X \subset \mathcal{M}$, entailing finally $E \in \mathcal{M}$. $\qquad\square$

Remark 2.2.15.

(1) The Riesz–Markov representation Theorem 2.2.1 remains true when X is a locally compact Hausdorff topological space. Theorems on partition of unity must be proven in that framework and require some effort (see Exercise 2.8.2).

(2) Theorem 2.2.14 is true when X is a locally compact Hausdorff topological space which is σ-compact.

(3) Let us also note that a positive linear form on $C_c(X)$ is continuous (cf. Exercise 2.8.3).

2.3　Producing positive Radon measures

After the proof of the Riesz–Markov Theorem 2.2.1, we are in a good position
to produce some significant examples of measure spaces, in particular of Borel
measures on $\mathbb{R}^m$. However, we still need to provide a positive Radon measure to
apply the theorem. A standard way of doing this is to use another classical theory
of integration, due to Bernhard Riemann, but it is certainly overkilling since we
only need a Radon measure, that is integrating continuous functions with compact
support. We shall see here that for this sole purpose, it is not necessary to resort
to another integration theory.

Proposition 2.3.1. *Let $a \leq b$ be real numbers. For $f \in C([a,b])$ (real-valued continuous functions defined on $[a,b]$), there exists a unique differentiable function F
defined on $[a,b]$ such that*

$$F(a) = 0, \quad \forall x \in [a,b], \ F'(x) = f(x). \tag{2.3.1}$$

We shall note that unique solution as $F(x) = \int_a^x f(t)dt$. The mapping

$$C([a,b]) \ni f \mapsto \int_a^b f(t)dt \quad \text{is a positive linear form.}$$

*Moreover, defining for $f \in C([a,b])$, $\int_b^a f(t)dt = -\int_a^b f(t)dt$, we find Chasles'
identity,*

$$\int_a^b f(t)dt + \int_b^c f(t)dt = \int_a^c f(t)dt, \tag{2.3.2}$$

*for $f \in C(I)$, where I is an interval containing a,b,c. If $f \in C_c(\mathbb{R})$, with $\operatorname{supp} f \subset
[a,b]$ we define $\int_\mathbb{R} f(t)dt = \int_a^b f(t)dt$ and we have, for all $s \in \mathbb{R}$,*

$$\int_\mathbb{R} f(t - s)dt = \int_\mathbb{R} f(t)dt \tag{2.3.3}$$

Proof. We note first that the mean value theorem and (2.3.1) imply

$$\sup_{x \in [a,b]} |F(x)| \leq (b - a) \sup_{x \in [a,b]} |f(x)|. \tag{2.3.4}$$

Let us prove first uniqueness. If F, G are differentiable on $[a,b]$ and satisfy (2.3.1)
then $(F - G)' = 0$ on $[a,b]$ and the mean value theorem implies $\forall x \in [a,b]$,
$F(x) - G(x) = F(a) - G(a) = 0$. Moreover, if $(f_n)_{n \in \mathbb{N}}$ is a sequence of continuous
functions converging uniformly towards f on $[a,b]$, such that for all $n \in \mathbb{N}$, there
exist F_n so that (2.3.1) holds, then the sequences $(F_n), (F_n')$ converge uniformly
towards F, f, and F is differentiable on $[a,b]$ with $F' = f$: in fact, using (2.3.4),
we have

$$\sup_{x \in [a,b]} |F_{n+p}(x) - F_n(x)| \leq (b - a) \sup_{x \in [a,b]} |f_{n+p}(x) - f_n(x)|,$$

implying uniform convergence of F_n towards a function $F \in C([a, b])$ such that $F(a) = 0$. We have, for $x, x + h \in [a, b]$,

$$F_n(x + h) - F_n(x) = f_n(x)h + \left(f_n(x + \theta_n h) - f_n(x)\right)h,$$

for some $\theta_n \in (0, 1)$ and thus

$$\begin{aligned}
|F_n(x + h) &- F_n(x) - f_n(x)h| \\
&\leq |h| |f_n(x + \theta_n h) - f(x + \theta_n h) + f(x + \theta_n h) - f(x) + f(x) - f_n(x)| \\
&\leq |h| \left[2\|f_n - f\|_{C([a,b])} + \sup_{|t| \leq |h|} |f(x + t) - f(x)|\right],
\end{aligned} \tag{2.3.5}$$

so that

$$|F_n(x + h) - F_n(x) - f_n(x)h| \leq |h|\left[\epsilon_n + \omega(h)\right], \quad \text{with } \lim_n \epsilon_n = 0, \ \lim_{h \to 0} \omega(h) = 0.$$

We find $|F(x+h)-F(x)-f(x)h| \leq |h|\omega(h)$ so that F is differentiable with $F' = f$. We note that (2.3.1) holds trivially for continuous piecewise affine functions (see Exercise 2.8.9), and also that this type of functions can approximate uniformly continuous functions on $[a, b]$: with the previous remarks we get the existence. Using the notation $F(x) = \int_a^x f(t)dt$, we find that for $\alpha, \beta \in \mathbb{R}, f, g \in C([a, b])$,

$$\int_a^x \left(\alpha f(t) + \beta g(t)\right)dt = \alpha \int_a^x f(t)dt + \beta \int_a^x g(t)dt,$$

since if F, G satisfy (2.3.1) for f, g, then $\alpha F + \beta G$ satisfies (2.3.1) for $\alpha f + \beta g$. Moreover, if $f \geq 0$, then $F' = f \geq 0$ and $F(x) \geq F(a) = 0$ for $x \in [a, b]$. Let I be an interval of $\mathbb{R}$, $f \in C(I)$ and let $a, b, c \in I$. If $a \leq b \leq x \in I$, defining

$$F(x) = \int_a^x f(t)dt, \quad G(x) = \int_a^b f(t)dt + \int_b^x f(t)dt,$$

we find $F'(x) = f(x) = G'(x), F(b) = G(b)$, so that $F(x) = G(x)$, proving Chasles' identity (2.3.2) when $a \leq b \leq c$. Let us now consider I an interval of $\mathbb{R}$, $f \in C(I)$ and $x_0 \leq x_1 \leq x_2 \in I$. We have

$$\int_{x_1}^{x_0} f(t)dt \underbrace{=}_{\text{definition}} -\int_{x_0}^{x_1} f(t)dt \underbrace{=}_{\text{already proven}} \int_{x_1}^{x_2} f(t)dt - \int_{x_0}^{x_2} f(t)dt$$

$$= \int_{x_1}^{x_2} f(t)dt + \int_{x_2}^{x_0} f(t)dt,$$

proving Chasles' identity (2.3.2) in the general case. In particular for $f \in C_c(\mathbb{R})$, with supp $f \subset [a, b]$ we define $\int_{\mathbb{R}} f(t)dt = \int_a^b f(t)dt$, a consistent definition since if

$\operatorname{supp} f \subset [a', b']$, Chasles' identity induces $\int_a^b f(t)dt = \int_{a'}^{b'} f(t)dt$. Let us prove now (2.3.3): assuming $\operatorname{supp} f \subset [a, b]$, we have $\int f(t-s)dt = \int_{s+a}^{s+b} f(t-s)dt$ and with

$$F(x) = \int_a^x f(t)dt, \quad G(x) = \int_{s+a}^{s+x} f(t-s)dt,$$

we find that F, G are both differentiable with $G'(x) = f(s+x-s) = f(x) = F'(x)$ and since $F(a) = G(a) = 0$, we get $F = G$ and the result. $\qquad\square$

Proposition 2.3.2 (Fundamental theorem of calculus).

(1) *Let $a \leq b$ be real numbers and $f \in C([a, b])$. Defining for $x \in [a, b]$, $F(x) = \int_a^x f(t)dt$, the function $F \in C^1([a, b])$ and $F' = f$.*

(2) *Let $a \leq b$ be real numbers and $f, g \in C^1([a, b])$. Then for $x \in [a, b]$,*

$$\int_a^x f'(t)dt = f(x) - f(a)$$

and

$$\int_a^b f'(t)g(t)dt = \big[f(t)g(t)\big]_a^b - \int_a^b f'(t)g(t)dt.$$

(3) *Let I_1, I_2 be two intervals of $\mathbb{R}$, let $\kappa : I_1 \longrightarrow I_2$ be a C^1 mapping and let $f : I_2 \longrightarrow \mathbb{R}$ be continuous. Then for all $a_1, b_1 \in I_1$,*

$$\int_{\kappa(a_1)}^{\kappa(b_1)} f(t_2)dt_2 = \int_{a_1}^{b_1} f(\kappa(t_1))\kappa'(t_1)dt_1.$$

Proof. Property (1) is exactly Definition (2.3.1). To prove (2), we set for $x \in [a, b]$, $F(x) = \int_a^x f'(t)dt$. According to (2.3.1), we have $F(a) = 0$, $F' = f'$, implying $F(x) - f(x) = F(a) - f(a)$, which is the sought formula. Using Leibniz' $(fg)' = f'g + fg'$, the last part follows from the first. Let us prove (3): we set for $x_2 \in I_2$, $x_1 \in I_1$,

$$F(x_2) = \int_{\kappa(a_1)}^{x_2} f(t_2)dt_2, \quad G(x_1) = \int_{a_1}^{x_1} f(\kappa(t_1))\kappa'(t_1)dt_1.$$

We have $F(\kappa(a_1)) = 0 = G(a_1)$ and for $x_1 \in I_1$,

$$\frac{d}{dx_1}\big(F(\kappa(x_1))\big) = F'\big(\kappa(x_1)\big)\kappa'(x_1) = f\big(\kappa(x_1)\big)\kappa'(x_1) = G'(x_1),$$

so that $F(\kappa(x_1)) = G(x_1)$ and with $x_1 = b_1$, this is the result. $\qquad\square$

The previous propositions show that integrating continuous functions of one variable with compact support does not require any theoretical effort. For several variables, it is not much more complicated.

Proposition 2.3.3. *Let $m \geq 1$ be an integer and let $C_c(\mathbb{R}^m)$ be the vector space of complex-valued continuous functions with compact support. There exists a unique positive linear form on $C_c(\mathbb{R}^m)$ such that for $f(x) = \prod_{1 \leq j \leq m} f_j(x_j)$, $f_j \in C_c(\mathbb{R})$,*

$$Lf = \prod_{1 \leq j \leq m} \int_{\mathbb{R}} f_j(x_j)dx_j. \tag{2.3.6}$$

We shall note $Lf = \int_{\mathbb{R}^m} f(x)dx$. For all $t \in \mathbb{R}^m$, and all $f \in C_c(\mathbb{R}^m)$, we have

$$\int_{\mathbb{R}^m} f(x - t)dx = \int_{\mathbb{R}^m} f(x)dx. \tag{2.3.7}$$

Proof. Let us prove the existence for $m \geq 2$. We set

$$\int_{\mathbb{R}^m} f(x)dx = \int_{\mathbb{R}^{m-1}} \left(\int_{\mathbb{R}} f(x_1, x')dx_1 \right) dx',$$

which is meaningful if we know what is the integral of functions with compact support in $m - 1$ dimensions: in fact defining

$$g(x') = \int_{\mathbb{R}} f(x_1, x')dx_1,$$

we find that g is continuous with compact support since f is continuous with compact support and (2.3.4) implies

$$|g(x') - g(y')| \leq \sup_{x_1} |f(x_1, x') - f(x_1, y')| \operatorname{diam}(\operatorname{supp} f).$$

Moreover (2.3.6) as well as linearity and positivity are trivially satisfied. To prove uniqueness, we shall use the following lemma.

Lemma 2.3.4. *Let $m \geq 1$ be an integer. The vector space $\otimes_{1 \leq j \leq m} C_c(\mathbb{R})$ is dense in $C_c(\mathbb{R}^m)$.*

Proof of the lemma. We note first that $1 = \sum_{j \in \mathbb{Z}} (1 - |t - j|)_+$ since that function is 1-periodic and for $t \in [0, 1[$, the condition $|t - j| < 1$ implies

$$\max(0, j - 1) \leq t < \min(1, j + 1) \implies 0 \leq j \leq 1,$$

implying $\sum_{j \in \mathbb{Z}, |t-j| \leq 1} (1 - |t - j|)_+ = (1 - t) + \left(1 - (1 - t)\right) = 1$. Also, defining $\varphi(t) = (1 - |t|)_+$ and

$$\Phi(t_1, \ldots, t_m) = \prod_{1 \leq l \leq m} \varphi(t_l),$$

we find

$$1 = \prod_{1 \leq l \leq m} \sum_{j_l \in \mathbb{Z}} \varphi(t_l - j_l) = \sum_{j \in \mathbb{Z}^m} \Phi(T - j), \text{ with } T = (t_1, \ldots, t_m).$$

Consequently, for $\epsilon > 0$, $T \in \mathbb{R}^m$, $k = \epsilon j \in \epsilon \mathbb{Z}^m$ defining $\Phi_{k,\epsilon}(T) = \Phi\big(\epsilon^{-1}(T-\epsilon j)\big)$, we have,

$$1 = \sum_{j \in \mathbb{Z}^m} \Phi(\epsilon^{-1}T - j) = \sum_{j \in \mathbb{Z}^m} \Phi\big(\epsilon^{-1}(T - \epsilon j)\big) = \sum_{k \in \epsilon \mathbb{Z}^m} \Phi_{k,\epsilon}(T),$$

with $\Phi_{k,\epsilon} \in C_c(\mathbb{R}^m)$, $\operatorname{supp}\Phi_{k,\epsilon} = \{t, \|t - k\|_\infty \leq \epsilon\}$ (here for $t \in \mathbb{R}^m$, $\|t\|_\infty = \max_{1 \leq j \leq m} |t_j|$). Let $f \in C_c(\mathbb{R}^m)$; since $\operatorname{supp} f$ is compact, the following sums are finite and

$$f(t) = \sum_{k \in \epsilon \mathbb{Z}^m} \Phi_{k,\epsilon}(t)\big(f(t) - f(k)\big) + \sum_{k \in \epsilon \mathbb{Z}^m} \Phi_{k,\epsilon}(t)f(k).$$

Since we have

$$\sum_{k \in \epsilon \mathbb{Z}^m} \Phi_{k,\epsilon}(t)|f(t) - f(k)| \leq \sum_{k \in \epsilon \mathbb{Z}^m} \Phi_{k,\epsilon}(t) \sup_{\|t-s\| \leq \epsilon} |f(t) - f(s)| = \sup_{\|t-s\| \leq \epsilon} |f(t) - f(s)|,$$

the uniform continuity of f implies uniform convergence for $\sum_{k \in \epsilon \mathbb{Z}^m} \Phi_{k,\epsilon}(t)f(k)$ towards f. But $\Phi_{k,\epsilon}$ is a tensor product of continuous functions with compact support defined on $\mathbb{R}$, concluding the proof of the lemma. $\square$

Uniqueness in the proposition follows then from the linearity and continuity of L (which follows from positivity (see Exercise 2.8.3)): let L_1, L_2 be linear forms satisfying the assumptions of Proposition 2.3.3 and let $f \in C_c(\mathbb{R}^m)$. From Lemma 2.3.4, f is a uniform limit of a sequence f_n belonging to the vector space spanned by tensor products on which L_1 and L_2 coincide. We find

$$(L_1 - L_2)(f) = \lim_n (L_1 - L_2)(f_n) = \lim_n 0 = 0.$$

Property (2.3.7) is a consequence of uniqueness and of that property for $m = 1$, which is (2.3.3) in Proposition 2.3.1. $\square$

2.4   The Lebesgue measure on $\mathbb{R}^m$, properties and characterization

Definition 2.4.1. Let m be a positive integer. Let us consider the positive linear form defined on $C_c(\mathbb{R}^m)$ by Proposition 2.3.3: to $\varphi \in C_c(\mathbb{R}^m)$, we associate its "Riemann integral" $\int_{\mathbb{R}^m} \varphi(x)dx$. Applying the Riesz–Markov representation theorem 2.2.1 and Theorem 2.2.14, we find a measure space $(\mathbb{R}^m, \mathcal{L}_m, \lambda_m)$ where λ_m is a positive measure satisfying the properties of these theorems. We shall say that λ_m is the Lebesgue measure on $\mathbb{R}^m$ and $\mathcal{L}_m$ is the Lebesgue σ-algebra on $\mathbb{R}^m$.

N.B. Note in particular that $\mathcal{L}_m$ contains the Borel σ-algebra $\mathcal{B}_m$ on $\mathbb{R}^m$, and that λ_m is finite on compact sets as well as regular and complete. We shall note the space $L^1(\lambda_m)$ as $L^1(\mathbb{R}^m)$.

Theorem 2.4.2. *Let $m \geq 1$ be an integer and let $(\mathbb{R}^m, \mathcal{L}_m, \lambda_m)$ be the Lebesgue measure space $\mathbb{R}^m$ defined above. The σ-algebra $\mathcal{L}_m$ is stable by translation, contains the Borel σ-algebra $\mathcal{B}_m$, and is such that*

(1) $\lambda_m \left(\prod_{1 \leq j \leq d} [a_j, b_j] \right) = \prod_{1 \leq j \leq m} (b_j - a_j),$ *for* $a_j \leq b_j,$

(2) $\forall E \in \mathcal{L}_m, \forall x \in \mathbb{R}^m, \quad \lambda_m(E + x) = \lambda_m(E).$

(3) *If μ is a positive measure defined on $\mathcal{B}_m$, finite on the compact sets, invariant by translation (i.e., such that (2) holds) and such that $\mu([0,1]^m) = 1$, then $\mu = \lambda_m$ on $\mathcal{B}_m$.*

Proof. Let us prove (1), assuming first $a_j < b_j$ for all $1 \leq j \leq m$. Let $\epsilon > 0$ such that $\forall j \in \{1, \dots, m\}, a_j + \epsilon < b_j - \epsilon$ and $\varphi_j \in C_c(\mathbb{R}; [0,1])$ such that

$$
\varphi_j(x_j) = \begin{cases} 1 & \text{for } x_j \in [a_j + \epsilon, b_j - \epsilon], \\ \text{affine} & \text{for } x_j \in [a_j, b_j] \backslash [a_j + \epsilon, b_j - \epsilon], \\ 0 & \text{for } x_j \notin]a_j, b_j[. \end{cases}
$$

We consider the function $\varphi \in C_c(\mathbb{R}^m; [0,1])$ defined by $\varphi(x) = \varphi_1(x_1) \dots \varphi_m(x_m)$. We have

$$
\int_{\mathbb{R}^m} \varphi(x) dx = \prod_{1 \leq j \leq m} \int_{\mathbb{R}} \varphi_j(x_j) dx_j = \prod_{1 \leq j \leq m} (b_j - a_j - 2\epsilon + \epsilon).
$$

Defining $P = \prod_{1 \leq j \leq m} [a_j, b_j]$ and for $\mathbb{N} \ni k > k_0 = \dfrac{2}{\min_{1 \leq j \leq m}(b_j - a_j)},$

$$
P_k = \prod_{1 \leq j \leq m} [a_j + \frac{1}{k}, b_j - \frac{1}{k}],
$$

we get for $\epsilon = 1/k$,

$$
\lambda_m(P_k) = \int_{\mathbb{R}^m} \mathbf{1}_{P_k} d\lambda_m \leq \overbrace{\int_{\mathbb{R}^m} \varphi(x) dx}^{= \prod_{1 \leq j \leq m}(b_j - a_j - \epsilon)} = \int_{\mathbb{R}^m} \varphi d\lambda_m \leq \int_{\mathbb{R}^m} \mathbf{1}_P d\lambda_m = \lambda_m(P),
$$

so that, from Proposition 1.4.4(2) and $\mathring{P} = \cup_{k > k_0} P_k$ (increasing union),

$$
\lambda_m(\mathring{P}) = \lim_k \lambda_m(P_k) \leq \lim_k \prod_{1 \leq j \leq m} (b_j - a_j - \frac{1}{k}) = \prod_{1 \leq j \leq m} (b_j - a_j) \leq \lambda_m(P). \tag{2.4.1}
$$

This implies also that

$$
\lambda_m(\{x_1 = a_1\}) = 0, \tag{2.4.2}
$$

since for $\epsilon > 0$ and $M > 0$, we have

$$
\lambda_m \left(\{(x_1, x') \in \mathbb{R} \times \mathbb{R}^{m-1}, |x_1 - a_1| < \epsilon/2, \|x'\|_\infty < M/2\} \right) \leq \epsilon M^{m-1},
$$

so that $\lambda_m \left(\{ (x_1, x') \in \mathbb{R} \times \mathbb{R}^{m-1}, x_1 = a_1, \|x'\|_\infty < M \} \right) = 0$, entailing by countable union $\lambda_m(\{x_1 = a_1\}) = 0$. Since the difference $P \backslash \mathring{P}$ is included in a finite union of hyperplanes, Property (1) follows from (2.4.1), (2.4.2).

Let us prove now property (2) *in Theorem* 2.4.2. Let K be a compact subset of an open set V and let $\chi \in C_c(V; [0,1])$ such that $\chi_{|K} = 1$. We have from (2.2.9)

$$\lambda_m(K) \leq \int_{\mathbb{R}^m} \chi(x) dx = \int_{\mathbb{R}^m} \chi d\lambda_m \leq \int_{\mathbb{R}^m} \mathbb{1}_V d\lambda_m \leq \lambda_m(V),$$

and the inner regularity of λ_m ((4) in Theorem 2.2.1) implies

$$\lambda_m(V) = \sup_{K \text{compact} \subset V} \lambda_m(K) \leq \sup_{\chi \in C_c(V;[0,1])} \int_{\mathbb{R}^m} \chi(x) dx \leq \lambda_m(V). \qquad (2.4.3)$$

For $\theta \in \mathbb{R}^m$, we note τ_θ the translation of vector θ: we have $\tau_\theta(x) = x + \theta$, and $\tau_\theta = \tau_{-\theta}^{-1}$ is a homeomorphism, implying that $\tau_\theta(V)$ is open as the inverse image of an open set by a continuous map. We find then

$$\lambda_m(V + \theta) = \sup_{\chi \in C_c(V+\theta;[0,1])} \int_{\mathbb{R}^m} \chi(x) dx = \sup_{\psi \in C_c(V;[0,1])} \int_{\mathbb{R}^m} \psi(x + \theta) dx$$

$$= \sup_{\psi \in C_c(V;[0,1])} \int_{\mathbb{R}^m} \psi(x) dx = \lambda_m(V).$$

Since τ_θ is a homeomorphism, $\mathcal{B}_m$ is invariant by translation and using the outer regularity of Lebesgue's measure, we find for $E \in \mathcal{B}_m$ and $\theta \in \mathbb{R}^m$,

$$\lambda_m(E + \theta) = \inf_{W \text{open} \supset E+\theta} \lambda_m(W) = \inf_{V \text{open} \supset E} \lambda_m(V + \theta)$$

$$= \inf_{V \text{open} \supset E} \lambda_m(V) = \lambda_m(E). \qquad (2.4.4)$$

Let $E \in \mathcal{L}_m$. Using (3) in Theorem 2.2.14, we can find a F_σ set A, a G_δ set B such that $A \subset E \subset B$ and $\lambda_m(B \backslash A) = 0$. This implies for $\theta \in \mathbb{R}^m$,

$$A + \theta \subset E + \theta \subset B + \theta,$$

and moreover $A + \theta$ is still an F_σ set since τ_θ is a homeomorphism:

$$\tau_\theta(\cup_{n \in \mathbb{N}} F_n) = \cup_{n \in \mathbb{N}} \tau_\theta(F_n) = \cup_{n \in \mathbb{N}} \underbrace{\tau_{-\theta}^{-1}(F_n)}_{\text{closed}}.$$

We prove as well that $B + \theta$ is a G_δ set and using (2.4.4), we find

$$\lambda_m \left(\tau_\theta(B) \backslash \tau_\theta(A) \right) = \lambda_m \left(\tau_\theta(B \backslash A) \right) = \lambda_m(B \backslash A) = 0,$$

which implies from (3) in Theorem 2.2.14, that $E + \theta$ belongs to $\mathcal{L}_m$. We find moreover that

$$\lambda_m(E + \theta) = \lambda_m(A + \theta) = \lambda_m(A) = \lambda_m(E),$$

concluding the proof of (2).

Let us prove (3) *in Theorem* 2.4.2. We claim that

$$\mu\big(\{x_1 = 0\}\big) = 0. \tag{2.4.5}$$

In fact from Proposition 1.4.4(2) we have

$$\mu\big(\{x_1 = 0\}\big) = \sup_{M \in \mathbb{N}} \mu\big(\{x_1 = 0\} \cap \underbrace{\{\max_{2 \le j \le m} |x_j| \le M\}}_{K_M}\big),$$

and we note that for $M \in \mathbb{N}$,

$$\{\max_{1 \le j \le m} |x_j| \le M\} = \cup_{|\alpha| < M}(K_M + \alpha \vec{e_1}) \supset \cup_{\alpha \in \mathbb{Q}, |\alpha| \le M}(K_M + \alpha \vec{e_1}),$$

which implies

$$\sum_{\alpha \in \mathbb{Q}, |\alpha| \le M} \mu(K_M) = \sum_{\alpha \in \mathbb{Q}, |\alpha| \le M} \mu(K_M + \alpha \vec{e_1}) \le \mu\big(\{\max_{1 \le j \le m} |x_j| \le M\}\big) < +\infty,$$

so that $\mu(K_M) = 0 = \mu\big(\{x_1 = 0\}\big)$. From (2.4.5) and the invariance by translation of μ, we find that all affine hyperplanes parallel to the axes have measure 0.

Lemma 2.4.3. *Let* $(X, \mathcal{M}, \mu)$ *be a measure space where* μ *is a positive measure. Let* $(E_j)_{j \in \mathbb{N}}$ *be a sequence of* $\mathcal{M}$ *such that for* $j \ne k$, $\mu(E_j \cap E_k) = 0$. *Then we have*

$$\mu(\cup_{j \in \mathbb{N}} E_j) = \sum_{j \in \mathbb{N}} \mu(E_j).$$

Proof of the lemma. From Proposition 1.4.4(2), it is enough to prove that for all integers n, $\mu(\cup_{0 \le j \le n} E_j) = \sum_{0 \le j \le n} \mu(E_j)$. This is obvious inductively on n since

$$\mu(\cup_{0 \le j \le n+1} E_j) = \mu\big(\cup_{0 \le j \le n}(E_j \backslash E_{n+1})\big) + \mu(E_{n+1})$$
$$= \sum_{0 \le j \le n} \mu(E_j \backslash E_{n+1}) + \mu(E_{n+1})$$
$$= \sum_{0 \le j \le n} \big(\mu(E_j \backslash E_{n+1}) + \mu(E_j \cap E_{n+1})\big) + \mu(E_{n+1})$$
$$= \sum_{0 \le j \le n+1} \mu(E_j). \qquad \square$$

For $n \in \mathbb{N}^*$, we have

$$[0,1]^m = \cup_{0 \le k_j < n} \overbrace{\prod_{1 \le j \le m} \left[\frac{k_j}{n}, \frac{k_j + 1}{n}\right]}^{\text{rectangle } P_k}.$$

We note that we have n^m rectangles P_k which are all translated from the rectangle $P_0 = [0, 1/n]^m$ and such that $P_k \cap P_l$ is included in an affine hyperplane parallel to the axes for distinct multi-indices k, l. Using Lemma 2.4.3, consequences of (2.4.5) on the measure of hyperplanes parallel to the axes as well as translation invariance of μ, we find

$$1 = \mu([0,1]^m) = n^m \mu([0, 1/n]^m), \quad \text{i.e.,} \quad \mu([0, 1/n]^m) = n^{-m}.$$

Let us check now the compact rational rectangle,

$$P = \prod_{1 \le j \le m} [a_j, b_j], \quad a_j, b_j \in \mathbb{Q}, \quad [a_j, b_j] = \left[0, \frac{q_j}{n}\right] + \frac{c_j}{n}, \quad \frac{q_j}{n} = b_j - a_j, \quad q_j \in \mathbb{N}.$$

Since μ is translation-invariant, using again Lemma 2.4.3 and the previous arguments, we find

$$\mu(P) = \mu\left(\prod_{1 \le j \le m}\left[0, \frac{q_j}{n}\right]\right) = \mu\left(\cup_{0 \le k_j < q_j} \prod_{1 \le j \le m}\left[\frac{k_j}{n}, \frac{k_j+1}{n}\right]\right)$$

$$= q_1 \dots q_m n^{-m} = \prod_{1 \le j \le m} (b_j - a_j). \tag{2.4.6}$$

Lemma 2.4.4. *Let Ω be an open subset of $\mathbb{R}^m$. There exists a sequence of compact rational rectangles $(Q_n)_{n \in \mathbb{N}}$ such that for $n \ne m$, the intersection $Q_n \cap Q_m$ is included in an affine hyperplane parallel to the axes and*

$$\Omega = \cup_{n \in \mathbb{N}} Q_n.$$

Proof of the lemma. Lemma 1.2.6 provides a sequence $(P_n)_{n \in \mathbb{N}}$ of compact rational rectangles such that $\Omega = \cup_{n \in \mathbb{N}} P_n$. Consequently, defining

$$R_0 = P_0, R_1 = P_1 \backslash P_0, \dots, R_n = P_n \backslash (\cup_{0 \le j < n} P_j), \tag{2.4.7}$$

we get $\Omega = \cup_{n \in \mathbb{N}} R_n$, with R_n pairwise disjoint. Let us consider $(I_j)_{1 \le j \le m}$ and $(J_j)_{1 \le j \le m}$ bounded intervals of $\mathbb{R}$ with rational endpoints and the rational rectangles $S = \prod_{1 \le j \le m} I_j$, $T = \prod_{1 \le j \le m} J_j$. The set $S \backslash T$ is a finite union of pairwise disjoint rectangles and $S \cap T$ is a rational rectangle: it is true for $m = 1$ since $I \backslash J$ is a union of at most two disjoint intervals with rational endpoints and moreover for $m > 1$, with

$$S' = \prod_{1 \le j \le m-1} I_j, \quad T' = \prod_{1 \le j \le m-1} J_j,$$

we have

$$S \backslash T = (S' \times I_m) \backslash (T' \times J_m) = \overbrace{\left((S' \backslash T') \times I_m\right) \cup \left((S' \cap T') \times I_m \backslash J_m\right)}^{\text{disjoint union}}.$$

From the induction hypothesis $S'\backslash T'$ is a disjoint union of N_{m-1} rational rectangles and $S' \cap T'$ is a rational rectangle, we find that $S\backslash T$ is a union of N_m disjoint rational rectangles with

$$N_m \leq N_{m-1} + 2, \quad \text{so that } N_m \leq 2m.$$

Moreover, since

$$S \cap T = (S' \cap T') \times (I_m \cap J_m),$$

we find that $S \cap T$ is a rational rectangle. Going back to (2.4.7), we find that R_1 is a finite union of pairwise disjoint rational rectangles and inductively, it is also true for

$$R_{n+1} = P_{n+1}\backslash(\cup_{0 \leq j \leq n} P_j) = \left(P_{n+1}\backslash(\cup_{0 \leq j < n} P_j)\right)\backslash P_n.$$

We have proven that R_n is a finite disjoint union of rational rectangles, i.e.,

$$R_n = \cup_{1 \leq k \leq M_n} S_{k,n}, \quad S_{k,n} \text{ rational rectangle}, \ k \neq l \Longrightarrow S_{k,n} \cap S_{l,n} = \emptyset.$$

Moreover, since the R_n are pairwise disjoint, we have also

$$n \neq m \Longrightarrow S_{k,n} \cap S_{l,m} = \emptyset.$$

As a result we have

$$\begin{aligned}
\Omega &= \cup_{n \in \mathbb{N}} P_n = \cup_{n \in \mathbb{N}} R_n \\
&= \cup_{n \in \mathbb{N}} \cup_{1 \leq k \leq M_n} S_{k,n} \subset \cup_{n \in \mathbb{N}} \cup_{1 \leq k \leq M_n} \overline{S_{k,n}} \subset \cup_{n \in \mathbb{N}} P_n = \Omega,
\end{aligned} \tag{2.4.8}$$

and since the rational rectangles $S_{k,n}$ are pairwise disjoint, the intersection of their closure is included in an hyperplane parallel to the axes. The countable family $\left((\overline{S_{k,n}})_{1 \leq k \leq M_n}\right)_{n \in \mathbb{N}}$ of compact rational rectangle satisfies the properties asked for (Q_n) in Lemma 2.4.4, whose proof is now complete. $\qquad\square$

We obtain thus for an open set Ω, using Lemmas 2.4.3–2.4.4 and (2.4.6),

$$\mu(\Omega) = \sum_{n \in \mathbb{N}} \mu(Q_n) = \sum_{n \in \mathbb{N}} \lambda_m(Q_n) = \lambda_m(\Omega),$$

and this implies that λ_m coincide with μ on the open sets. Let $E \in \mathcal{B}_m$. Exterior regularity of λ_m (Theorem 2.2.1(3)), implies

$$\lambda_m(E) = \inf_{\Omega \text{ open} \supset E} \lambda_m(\Omega) = \inf_{\Omega \text{ open} \supset E} \mu(\Omega).$$

It suffices then that we prove outer regularity for μ. We consider the positive linear form

$$\Lambda(\varphi) = \int_{\mathbb{R}^m} \varphi \, d\mu,$$

defined on $C_c(\mathbb{R}^m)$: let us note that μ is finite on compact sets and since for $\varphi \in C_c(\mathbb{R}^m)$, $|\varphi| \leq \sup|\varphi| \mathbf{1}_{\text{supp}\,\varphi}$ (and φ measurable since continuous), Λ is indeed a positive linear form on $C_c(\mathbb{R}^m) \subset \mathcal{L}^1(\mu)$. Theorem 2.2.1 provides the existence of a regular measure ν, defined on $\mathcal{B}_m$ such that for $\varphi \in C_c(\mathbb{R}^m)$,

$$\int_{\mathbb{R}^m} \varphi d\nu = \int_{\mathbb{R}^m} \varphi d\mu. \tag{2.4.9}$$

Let Ω be an open subset of $\mathbb{R}^m$; from Lemma 1.2.6, there exists a sequence of compact sets $(K_j)_{j\geq 1}$ such that

$$\Omega = \cup_{j\geq 1} K_j. \tag{2.4.10}$$

We consider

$$\varphi_1 \in C_c(\Omega; [0,1]) \text{ such that } \varphi_{1|K_1} = 1$$
$$\varphi_2 \in C_c(\Omega; [0,1]) \text{ such that } \varphi_{2|K_1 \cup \text{supp}\,\varphi_1} = 1$$
$$\varphi_3 \in C_c(\Omega; [0,1]) \text{ such that } \varphi_{3|K_1 \cup K_2 \cup \text{supp}\,\varphi_1 \cup \text{supp}\,\varphi_2} = 1$$
$$\cdots\cdots$$
$$\varphi_{n+1} \in C_c(\Omega; [0,1]) \text{ such that } \varphi_{n+1|K_1 \cup \cdots \cup K_n \cup \text{supp}\,\varphi_1 \cup \cdots \cup \text{supp}\,\varphi_n} = 1.$$
$$\cdots\cdots$$

We have $0 \leq \varphi_n \leq \varphi_{n+1}, \quad \varphi_n(x) \uparrow \mathbf{1}_\Omega(x)$(from (2.4.10)). As a result, applying Beppo Levi's theorem for the measure ν, (2.4.9) and Beppo Levi's theorem for μ, we get

$$\nu(\Omega) = \lim_n \int_{\mathbb{R}^m} \varphi_n d\nu = \lim_n \int_{\mathbb{R}^m} \varphi_n d\mu = \mu(\Omega).$$

Thus ν is a regular measure coinciding with μ on the open sets. Using (3) in Theorem 2.2.14 for ν, we find for $E \in \mathcal{B}_m$ and for all $\epsilon > 0$,

$$F \text{ closed} \subset E \subset V \text{ open}, \quad \epsilon > \nu(\underbrace{V\backslash F}_{\text{open}}) = \mu(V\backslash F).$$

Consequently, we obtain

$$\mu(E) + \epsilon \geq \mu(E) + \mu(V\backslash F) \geq \mu(E) + \mu(V\backslash E) = \mu(V) \geq \mu(E)$$

so that $\mu(E) = \inf_{V \text{open} \supset E} \mu(V)$, concluding the proof of Theorem 2.4.2. $\qquad\square$

We shall prove in Chapter 5 a general theorem on changes of variables in integrals on $\mathbb{R}^m$, but the following lemma will be useful already in Chapter 2.

Lemma 2.4.5. *Let $m \in \mathbb{N}^*$ and let λ_m be the Lebesgue measure on $\mathbb{R}^m$. The space $L^1(\mathbb{R}^m)$ is invariant by translation and dilation, i.e., for $\theta > 0, T \in \mathbb{R}^m, f \in L^1(\mathbb{R}^m)$, the mappings $x \mapsto f(\theta x)$ and $x \mapsto f(x - T)$ belong to $L^1(\mathbb{R}^m)$ and*

$$\int_{\mathbb{R}^m} f(x)dx = \int_{\mathbb{R}^m} f(x - T)dx = \theta^m \int_{\mathbb{R}^m} f(x\theta)dx.$$

Proof. The first assertions are obvious using simple functions and Definition 1.5.2 since the mappings $x \mapsto \theta^{-1}x$ and $x \mapsto x + T$ are continuous thus measurable. Since the Lebesgue measure is invariant by translation, we get readily the first equality. For $\theta > 0$, we consider the positive measure μ_θ defined on $\mathcal{B}_m$ by

$$\mu_\theta(A) = \theta^m \lambda_m(\theta^{-1}A).$$

The measure μ_θ is finite on compact sets (for K compact, $\theta^{-1}K$ is compact), is invariant by translation (since λ_m is invariant by translation) and such that

$$\mu_\theta\big([0,1]^m\big) = \theta^m \lambda_m(\theta^{-1}[0,1]^m) = \theta^m \lambda_m([0,\theta^{-1}]^m) = 1.$$

Theorem 2.4.2 implies that $\mu_\theta - \lambda_m$, so that for $A \in \mathcal{B}_m$,

$$\int_{\mathbb{R}^m} \mathbf{1}_A(x)d\lambda_m(x) = \int_{\mathbb{R}^m} \mathbf{1}_A(x)d\mu_\theta(x) = \theta^m \lambda_m(\theta^{-1}A)$$

$$= \theta^m \int_{\mathbb{R}^m} \mathbf{1}_{\theta^{-1}A}(x)d\lambda_m(x) = \theta^m \int_{\mathbb{R}^m} \mathbf{1}_A(\theta x)d\lambda_m(x),$$

which implies the last equality for $f \in \mathcal{L}^1(\lambda_m)$. $\qquad\square$

2.5 Carathéodory theorem on outer measures

Definition 2.5.1. Let X be a set and let μ^* be an outer measure on X (see Definition 2.2.4). We define

$$\mathcal{M}_{\mu^*} = \{A \in \mathcal{P}(X), \forall Y \in \mathcal{P}(X), \quad \mu^*(Y) = \mu^*(Y \cap A) + \mu^*(Y \cap A^c)\}. \quad (2.5.1)$$

A subset E of X is said to be μ^*-negligible if $\mu^*(E) = 0$.

We note first that

$$X, \emptyset \in \mathcal{M}_{\mu^*}, \quad \big[A \in \mathcal{M}_{\mu^*} \Longleftrightarrow A^c \in \mathcal{M}_{\mu^*}\big], \qquad\qquad (2.5.2)$$
$$A \in \mathcal{M}_{\mu^*} \Longleftrightarrow \forall Y \in \mathcal{P}(X), \quad \mu^*(Y) \geq \mu^*(Y \cap A) + \mu^*(Y \cap A^c), \qquad (2.5.3)$$
$$\text{any negligible set belongs to } \mathcal{M}_{\mu^*}. \qquad\qquad\qquad\qquad\qquad (2.5.4)$$

In fact Property (2.5.1) is symmetrical in A, A^c and $\mu^*(\emptyset) = 0$, proving (2.5.2). Moreover, the subadditivity property (2.2.7) implies $\mu^*(Y \cap A) + \mu^*(Y \cap A^c) \geq \mu^*(Y)$, proving (2.5.3). Finally for E negligible and $Y \subset X$, from the monotonicity property (2.2.6), we obtain

$$\mu^*(Y \cap E) + \mu^*(Y \cap E^c) \leq \mu^*(E) + \mu^*(Y) = \mu^*(Y),$$

proving (2.5.4) from the already proven (2.5.3).

Lemma 2.5.2. *Let X be a set, μ^* be an outer measure on X and $\mathcal{M}_{\mu^*}$ be the subset of $\mathcal{P}(X)$ defined by (2.5.1). Then if A_1, A_2 belong to $\mathcal{M}_{\mu^*}$ so do $A_1 \cap A_2$ and $A_1 \cup A_2$. Moreover if $\{A_j\}_{j\geq 1}$ is a countable family of elements of $\mathcal{M}_{\mu^*}$, then $\cup_{j\geq 1}A_j$ belongs to $\mathcal{M}_{\mu^*}$.*

Proof. We have for $Y \subset X$, using $A_1 \in \mathcal{M}_{\mu^*}$,

$$\mu^*(Y \cap A_2^c) = \mu^*(Y \cap A_2^c \cap A_1^c) + \mu^*(Y \cap A_2^c \cap A_1),$$
$$\mu^*(Y \cap A_2) = \mu^*(Y \cap A_2 \cap A_1^c) + \mu^*(Y \cap A_2 \cap A_1),$$

so that, using $A_2 \in \mathcal{M}_{\mu^*}$,

$$\mu^*(Y) = \mu^*(Y \cap A_2) + \mu^*(Y \cap A_2^c)$$
$$= \underbrace{\mu^*(Y \cap A_2 \cap A_1^c)}_{[1]} + \underbrace{\mu^*(Y \cap A_2 \cap A_1)}_{[2]} + \underbrace{\mu^*(Y \cap A_2^c \cap A_1^c)}_{[3]} + \underbrace{\mu^*(Y \cap A_2^c \cap A_1)}_{[4]}.$$

Applying the previous equality to $Y \cap (A_1 \cup A_2)$, we find

$$\mu^*\big(Y \cap (A_1 \cup A_2)\big)$$
$$= \mu^*(Y \cap A_2 \cap A_1^c) + \mu^*(Y \cap A_2 \cap A_1) + \mu^*(\emptyset) + \mu^*(Y \cap A_1 \cap A_2^c), \qquad (2.5.5)$$

so that

$$\mu^*\big(Y \cap (A_1 \cup A_2)\big) + \mu^*\big(Y \cap (A_1 \cup A_2)^c\big) = [1] + [2] + [4] + [3] = \mu^*(Y),$$

proving that $A_1 \cup A_2$ belongs to $\mathcal{M}_{\mu^*}$ (as well as $A_1 \cap A_2$ by complement). Also we obtain inductively that $\cup_{1\leq j\leq n}A_j \in \mathcal{M}_{\mu^*}$ for $A_1, \ldots, A_n \in \mathcal{M}_{\mu^*}$. Let us consider now a countable family $(A_j)_{j\geq 1}$ of elements of $\mathcal{M}_{\mu^*}$. We may first consider

$$B_1 = A_1, \quad B_2 = A_2 \cap A_1^c, \ldots, \quad B_n = A_n \cap A_{n-1}^c \cap \cdots \cap A_1^c, \ldots$$

so that each $B_j \in \mathcal{M}_{\mu^*}$ (first part of the lemma), the family $(B_j)_{j\geq 1}$ is pairwise disjoint (since $B_n \subset A_n$ and $B_{n+m+1} \subset A_n^c$ for $m \geq 0$) and $\cup_{j\geq 1}B_j = \cup_{j\geq 1}A_j$ since $B_n \subset A_n$ and $A_n \subset \cup_{1\leq j\leq n}B_j$ (true for $n = 1$ and if true for some $n \geq 1$ $A_{n+1} = B_{n+1} \cup \big(A_{n+1} \cap (A_1 \cup \cdots \cup A_n)\big) \subset \cup_{1\leq j\leq n+1}B_j$). We have now for $Y \subset X$,

$$\mu^*\big(Y \cap (\cup_{1\leq j\leq n}B_j)\big) = \sum_{1\leq j\leq n} \mu^*(Y \cap B_j),$$

since that property is true for $n = 1$ and if true for some $n \geq 1$, we get since $\cup_{1\leq j\leq n}B_j, B_{n+1} \in \mathcal{M}_{\mu^*}$, applying (2.5.5) for $A_1 = \cup_{1\leq j\leq n}B_j$, $A_2 = B_{n+1}$, noting that $A_1 \cap A_2 = \emptyset$,

$$\mu^*\big(Y \cap (\cup_{1\leq j\leq n+1}B_j)\big) = \mu^*\big(Y \cap (\cup_{1\leq j\leq n}B_j)\big) + \mu^*(Y \cap B_{n+1}).$$

As a result, for $Y \subset X$,

$$\mu^*(Y) = \mu^*\big(Y \cap (\cup_{1 \le j \le n} B_j)\big) + \mu^*\big(Y \cap (\cap_{1 \le j \le n} B_j^c)\big)$$

$$= \sum_{1 \le j \le n} \mu^*(Y \cap B_j) + \mu^*\big(Y \cap (\cap_{1 \le j \le n} B_j^c)\big)$$

$$\ge \sum_{1 \le j \le n} \mu^*(Y \cap B_j) + \mu^*\big(Y \cap (\cap_{1 \le j} B_j^c)\big),$$

so that

$$\mu^*(Y) \ge \sum_{1 \le j} \mu^*(Y \cap B_j) + \mu^*\big(Y \cap (\cap_{1 \le j} B_j^c)\big), \tag{2.5.6}$$

and by subadditivity $\mu^*(Y) \ge \mu^*\big(Y \cap (\cup_{j \ge 1} B_j)\big) + \mu^*\big(Y \cap (\cup_{1 \le j} B_j)^c\big)$, proving via (2.5.3) that $\cup_{1 \le j} A_j = \cup_{1 \le j} B_j \in \mathcal{M}_{\mu^*}$, completing the proof of the lemma. $\qquad\square$

The following theorem, due to C. CARATHÉODORY (1873–1950) is a set-theoretic result allowing to construct a measure from an outer measure.

Theorem 2.5.3 (Carathéodory theorem on outer measures). *Let X be a set, μ^* be an outer measure on X and $\mathcal{M}_{\mu^*}$ be defined by (2.5.1). Then, with μ standing for the restriction of μ^* to $\mathcal{M}_{\mu^*}$, the triple $(X, \mathcal{M}_{\mu^*}, \mu)$ is a measure space where the σ-algebra $\mathcal{M}_{\mu^*}$ is μ-complete (contains all subsets of any $E \in \mathcal{M}_{\mu^*}$ such that $\mu^*(E) = 0$).*

Proof. Property (2.5.2) and Lemma 2.5.2 prove that $\mathcal{M}_{\mu^*}$ is a σ-algebra on X (see Definition 1.1.1). Moreover, we have $\mu^*(\emptyset) = 0$ (see Property (2.2.5) of an outer measure) and if $(B_j)_{j \ge 1}$ is a countable pairwise disjoint family of $\mathcal{M}_{\mu^*}$, applying (2.5.6) to $Y = \cup_{j \ge 1} B_j$, we find

$$\mu^*(\cup_{j \ge 1} B_j) \ge \sum_{j \ge 1} \mu^*(B_j) \underbrace{\ge}_{(2.2.7)} \mu^*(\cup_{j \ge 1} B_j),$$

concluding the proof (note that $\mathcal{M}_{\mu^*}$ is μ-complete from (2.5.4)). $\qquad\square$

The following result will be useful later on.

Theorem 2.5.4. *Let (X, d) be a metric space and μ^* be an outer measure on X such that for A, B subsets of X satisfying $d(A, B) > 0$, we have $\mu^*(A \cup B) = \mu^*(A) + \mu^*(B)$. Then the Borel σ-algebra $\mathcal{B}_X$ is included in $\mathcal{M}_{\mu^*}$.*

Proof. Since $\mathcal{M}_{\mu^*}$ is a σ-algebra, it is enough to prove that closed sets belong to $\mathcal{M}_{\mu^*}$. Let F be a closed subset of X: from (2.5.3), we need only to prove that for all $Y \subset X$ with $\mu^*(Y) < +\infty$, we have $\mu^*(Y) \ge \mu^*(Y \cap F) + \mu^*(Y \cap F^c)$. For $n \in \mathbb{N}^*$, we define

$$B_n = \{x \in Y \cap F^c, d(x, F) \ge 1/n\},$$

so that $B_n \subset B_{n+1}$ and $\cup_{n \geq 1} B_n = Y \cap F^c$: each B_n is included in $Y \cap F^c$ and conversely if $x \in Y \cap F^c$, we have $d(x, F) > 0$ since F is closed (see Lemma 2.1.1). As a result

$$d(Y \cap F, B_n) = \inf_{\substack{x' \in Y \cap F \\ x'' \in B_n}} d(x', x'') \geq \inf_{x'' \in B_n} d(x'', F) \geq 1/n > 0,$$

and thus $\mu^*(Y \cap F) + \mu^*(B_n) = \mu^*\big((Y \cap F) \cup B_n\big) \leq \mu^*(Y)$. To obtain the result we have only to prove $\lim_n \mu^*(B_n) = \mu^*(Y \cap F^c)$. We set for $n \geq 1$,

$$C_n = B_{n+1} \cap B_n^c = \{x \in Y \cap F^c, \frac{1}{n} > d(x, F) \geq \frac{1}{n+1}\},$$

and we note that for $|j - k| \geq 2$, say $j \geq k + 2$, $x_j \in C_j, x_k \in C_k$, we have

$$d(x_j, x_k) + \frac{1}{j} > d(x_j, x_k) + d(x_j, F) \geq d(x_k, F) \geq \frac{1}{k+1},$$

so that $d(C_j, C_k) \geq \frac{1}{k+1} - \frac{1}{j} > 0$, implying that[4]

$$\sum_{1 \leq j \leq N} \mu^*(C_{2j}) = \mu^*\big(\cup_{1 \leq j \leq N} C_{2j}\big) \leq \mu^*(Y) < +\infty,$$

$$\sum_{1 \leq j \leq N} \mu^*(C_{2j+1}) = \mu^*\big(\cup_{1 \leq j \leq N} C_{2j+1}\big) \leq \mu^*(Y) < +\infty.$$

As a result, $\sum_{j \geq 1} \mu^*(C_j) < +\infty$ and the subadditivity of μ^* implies

$$\mu^*(Y \cap F^c) \leq \mu^*(B_n) + \sum_{j \geq n} \mu^*(C_j),$$

so that $\mu^*(Y \cap F^c) \leq \liminf_n \mu^*(B_n) \leq \limsup_n \mu^*(B_n) \leq \mu^*(Y \cap F^c)$ proving the sought $\lim_n \mu^*(B_n) = \mu^*(Y \cap F^c)$. $\qquad\qquad\square$

2.6 Hausdorff measures, Hausdorff dimension

Definition, first properties

Let (X, d) be a separable metric space. Then, there exists a countable dense set $D = \{a_n\}_{n \in \mathbb{N}}$ in X so that for all $\varepsilon > 0$, $X = \cup_{n \in \mathbb{N}} B(a_n, \varepsilon)$ (any $x \in X$ is the limit of a sequence in D and thus for any $\varepsilon > 0$, there exists $a_n \in D$ with $d(x, a_n) < \varepsilon$). As a result, any subset E of X can be covered by a countable union of open sets with diameter $\leq 2\varepsilon$. We may thus give the following definition.

---

[4] If $(A_j)_{1 \leq j \leq N}$ are subsets of X such that $d(A_j, A_k) > 0$ for $j \neq k$, we have $\mu^*(\cup_{1 \leq j \leq N} A_j) = \sum_{1 \leq j \leq N} \mu^*(A_j)$: this is true for $N = 2$ and inductively for $N \geq 2$

$$\mu^*(\cup_{1 \leq j \leq N+1} A_j) = \mu^*(\cup_{1 \leq j \leq N} A_j) + \mu^*(A_{N+1})$$

since $d(A_{N+1}, \cup_{1 \leq j \leq N} A_j) \geq \min_{1 \leq j \leq N} d(A_{N+1}, A_j) > 0$, proving the property.

Definition 2.6.1. Let (X,d) be a separable metric space and let $\kappa \in \mathbb{R}_+$. For $\varepsilon > 0$, we define for $E \subset X$,

$$\mathfrak{h}^*_{\kappa,\varepsilon}(E) = \inf\left\{\sum_{n\in\mathbb{N}}(\operatorname{diam} U_n)^\kappa, \quad E \subset \cup_{n\in\mathbb{N}}U_n, \ U_n \text{ open}, \ \operatorname{diam} U_n \le \varepsilon\right\}.$$

Lemma 2.6.2. *With $X, d, \mathfrak{h}^*_{\kappa,\varepsilon}$ as above, for all $E \subset X$, the function $\mathbb{R}^*_+ \ni \varepsilon \mapsto \mathfrak{h}^*_{\kappa,\varepsilon}(E) \in \overline{\mathbb{R}}_+$ is decreasing. The function $\mathfrak{h}^*_\kappa$ defined on $\mathcal{P}(X)$ by*

$$\mathfrak{h}^*_\kappa(E) = \lim_{\varepsilon\to 0_+} \mathfrak{h}^*_{\kappa,\varepsilon}(E) = \sup_{\varepsilon>0} \mathfrak{h}^*_{\kappa,\varepsilon}(E), \tag{2.6.1}$$

is an outer measure on X (see Definition 2.2.4).

Proof. First of all we note that, say for subsets of $\overline{\mathbb{R}}$, the larger is the set, the smaller is the infimum and the larger is the supremum (let's call that the monotonicity principle). Let $\varepsilon_1 < \varepsilon_2$ be positive real numbers and let E be a subset of X. If $(U_n)_{n\in\mathbb{N}}$ is an open covering of E with $\operatorname{diam} U_n \le \varepsilon_1$, it is an open covering of E with $\operatorname{diam} U_n \le \varepsilon_2$, implying from the monotonicity principle that $\mathfrak{h}^*_{\kappa,\varepsilon_2}(E) \le \mathfrak{h}^*_{\kappa,\varepsilon_1}(E)$, which implies (2.6.1). We find also that $\mathfrak{h}^*_{\kappa,\varepsilon}(\emptyset) = 0$ and thus $\mathfrak{h}^*_\kappa(\emptyset) = 0$. Let $E_1 \subset E_2$ be subsets of X; then if $(U_n)_{n\in\mathbb{N}}$ is an open covering of E_2 with $\operatorname{diam} U_n \le \varepsilon$, it is also an open covering of E_1, implying from the monotonicity principle that

$$\mathfrak{h}^*_{\kappa,\varepsilon}(E_1) \le \mathfrak{h}^*_{\kappa,\varepsilon}(E_2) \implies \mathfrak{h}^*_\kappa(E_1) \le \mathfrak{h}^*_\kappa(E_2).$$

Let $(E_j)_{j\in\mathbb{N}}$ be a countable family of subsets of X such that $\mathfrak{h}^*_\kappa(E_j) < +\infty$ for all $j \in \mathbb{N}$ and let $\varepsilon > 0, \delta > 0$ be given; we have $\mathfrak{h}^*_{\kappa,\varepsilon}(E_j) \le \mathfrak{h}^*_\kappa(E_j) < +\infty$, so that there exists an open covering $(U_{n,j})_{n\in\mathbb{N}}$ of E_j with $\operatorname{diam} U_{n,j} \le \varepsilon$, and

$$\mathfrak{h}^*_{\kappa,\varepsilon}(E_j) \le \sum_n (\operatorname{diam} U_{n,j})^\kappa < \mathfrak{h}^*_{\kappa,\varepsilon}(E_j) + \delta 2^{-j-1}$$

and thus $\cup_{j\in\mathbb{N}}E_j \subset \cup_{j,n\in\mathbb{N}}U_{n,j}$, implying

$$\mathfrak{h}^*_{\kappa,\varepsilon}(\cup_{j\in\mathbb{N}}E_j) \le \sum_{j,n}(\operatorname{diam} U_{n,j})^\kappa \le \sum_j \mathfrak{h}^*_{\kappa,\varepsilon}(E_j) + \sum_j \delta 2^{-j-1} \le \sum_j \mathfrak{h}^*_\kappa(E_j) + \delta.$$

Since this inequality is true for any ε, δ positive, we get indeed

$$\mathfrak{h}^*_\kappa(\cup_{j\in\mathbb{N}}E_j) \le \sum_j \mathfrak{h}^*_\kappa(E_j). \tag{2.6.2}$$

Moreover that inequality is obviously satisfied when $\mathfrak{h}^*_\kappa(E_j) = +\infty$ for some j, completing the proof of the lemma. $\qquad\square$

Remark 2.6.3. For a subset E of a separable metric space, with

$$\mathcal{U}_\varepsilon(E) = \{\text{countable open covering } (U_n)_{n\in\mathbb{N}} \text{ of } E \text{ with diam}\, U_n \leq \varepsilon\},$$

and for $U = (U_n)_{n\in\mathbb{N}} \in \mathcal{U}_\varepsilon(E)$, $H(\kappa, U) = \sum_{n\in\mathbb{N}}(\text{diam}\, U_n)^\kappa$ we have

$$\mathfrak{h}^*_{\kappa,\varepsilon}(E) = \inf_{U\in\mathcal{U}_\varepsilon(E)} H(\kappa, U), \quad \mathfrak{h}^*_\kappa(E) = \sup_{\varepsilon>0}\Big\{\inf_{U\in\mathcal{U}_\varepsilon(E)} H(\kappa, U)\Big\}. \tag{2.6.3}$$

That formula implies readily

$$0 < \varepsilon_1 \leq \varepsilon_2 \Longrightarrow \mathcal{U}_{\varepsilon_1}(E) \subset \mathcal{U}_{\varepsilon_2}(E) \Longrightarrow \mathfrak{h}^*_{\kappa,\varepsilon_2} \leq \mathfrak{h}^*_{\kappa,\varepsilon_1}, \tag{2.6.4}$$

$$E_1 \subset E_2 \Longrightarrow \mathcal{U}_\varepsilon(E_2) \subset \mathcal{U}_\varepsilon(E_1) \Longrightarrow \mathfrak{h}^*_{\kappa,\varepsilon}(E_1) \leq \mathfrak{h}^*_{\kappa,\varepsilon}(E_2), \tag{2.6.5}$$

$$0 \leq \kappa_1 \leq \kappa_2, 0 < \varepsilon \leq 1 \Longrightarrow H(\kappa_2, U) \leq H(\kappa_1, U) \Longrightarrow \mathfrak{h}^*_{\kappa_2} \leq \mathfrak{h}^*_{\kappa_1}. \tag{2.6.6}$$

Lemma 2.6.4. *Let* $X, d, \mathfrak{h}^*_\kappa$ *be as above and let* A, B *be subsets of* X *such that*

$$0 < d(A, B) = \inf_{a\in A, b\in B} d(a, b).$$

Then we have $\mathfrak{h}^*_\kappa(A \cup B) = \mathfrak{h}^*_\kappa(A) + \mathfrak{h}^*_\kappa(B)$.

Proof. The subadditivity of $\mathfrak{h}^*_\kappa$ gives $\mathfrak{h}^*_\kappa(A \cup B) \leq \mathfrak{h}^*_\kappa(A) + \mathfrak{h}^*_\kappa(B)$. Let us prove the reverse inequality; we may of course assume that $\mathfrak{h}^*_\kappa(A \cup B) < +\infty$ and thus $h^*_\kappa(A), \mathfrak{h}^*_\kappa(B)$ are both finite. Then for ε, δ positive numbers with $\varepsilon \leq d(A, B)/2$, there exists an open covering $(U_n)_{n\in\mathbb{N}}$ of $A \cup B$ such that

$$\mathfrak{h}^*_{\kappa,\varepsilon}(A \cup B) \leq \sum_{n\in\mathbb{N}}(\text{diam}\, U_n)^\kappa < \mathfrak{h}^*_{\kappa,\varepsilon}(A \cup B) + \delta \leq \mathfrak{h}^*_\kappa(A \cup B) + \delta. \tag{2.6.7}$$

We define $N_A = \{n \in \mathbb{N}, U_n \cap A \neq \emptyset\}$ and we note that if $n \in N_A$, $U_n \cap B = \emptyset$: otherwise $\exists a \in U_n \cap A$, $\exists b \in U_n \cap B$ so that

$$d(A, B) \leq d(a, b) \leq \text{diam}\, U_n \leq \varepsilon = d(A, B)/2,$$

which is impossible since $d(A, B) > 0$. We get thus $N_A \cap N_B = \emptyset$; as a result since $A \cup B \subset \cup_{n\in\mathbb{N}} U_n$, we have from $A \cap B = \emptyset$,

$$A \subset \cup_{n\in N_A}(U_n \cap A) \subset \cup_{n\in N_A} U_n, \quad B \subset \cup_{n\in N_B}(U_n \cap B) \subset \cup_{n\in N_B} U_n,$$

so that $\mathfrak{h}^*_{\kappa,\varepsilon}(A) \leq \sum_{n\in N_A}(\text{diam}\, U_n)^\kappa$, $\mathfrak{h}^*_{\kappa,\varepsilon}(B) \leq \sum_{n\in N_B}(\text{diam}\, U_n)^\kappa$, and thus from (2.6.7) and $N_A \cap N_B = \emptyset$,

$$\mathfrak{h}^*_{\kappa,\varepsilon}(A) + \mathfrak{h}^*_{\kappa,\varepsilon}(B) \leq \mathfrak{h}^*_\kappa(A \cup B) + \delta$$

$$\Longrightarrow \lim_{\varepsilon\to 0_+}\big(\mathfrak{h}^*_{\kappa,\varepsilon}(A) + \mathfrak{h}^*_{\kappa,\varepsilon}(B)\big) \leq \mathfrak{h}^*_\kappa(A \cup B) + \delta,$$

implying $\mathfrak{h}^*_\kappa(A) + \mathfrak{h}^*_\kappa(B) \leq \mathfrak{h}^*_\kappa(A \cup B) + \delta$ for all $\delta > 0$, entailing the result. $\qquad\square$

Definition 2.6.5. Let (X, d) be a separable metric space and let $\kappa \geq 0$. The outer measure $\mathfrak{h}_\kappa^*$ on X is defined in (2.6.1). We define the Hausdorff measure $\mathfrak{h}_\kappa$ of dimension κ by using Theorem 2.5.3: $(X, \mathcal{M}_{\mathfrak{h}_\kappa^*}, \mathfrak{h}_\kappa)$ is a measure space where the complete σ-algebra $\mathcal{M}_{\mathfrak{h}_\kappa^*}$ is defined by (2.5.1) and $\mathfrak{h}_\kappa$ is the restriction of $\mathfrak{h}_\kappa^*$ to $\mathcal{M}_{\mathfrak{h}_\kappa^*}$. From Theorem 2.5.4 and Lemma 2.6.4, $\mathcal{M}_{\mathfrak{h}_\kappa^*}$ contains the Borel σ-algebra $\mathcal{B}_X$.

Lemma 2.6.6. *Let (X, d) be a separable metric space and let A be a subset of X such that $\mathfrak{h}_0^*(A) < +\infty$. Then A is a finite set and $\operatorname{card} A = \mathfrak{h}_0^*(A)$. The Hausdorff measure $\mathfrak{h}_0$ is the counting measure on X (see Example (3) on page 18).*

Proof. If $\mathfrak{h}_0^*(A) < +\infty$, we find that for all $\varepsilon > 0$, $\mathfrak{h}_{0,\varepsilon}^*(A) \leq \mathfrak{h}_0^*(A)$, so that

$$\exists N \geq 1, \quad \forall \varepsilon > 0, \quad \exists \text{ an open covering } (U_n)_{1 \leq n \leq N} \text{ of } A \text{ with } \operatorname{diam} U_n \leq \varepsilon.$$

Claim. This implies that the set A is finite with $\operatorname{card} A \leq N$. Assume that $a_1, \ldots, a_N, a_{N+1}$ are distinct elements of A. We set

$$\delta = \min_{1 \leq i \neq j \leq N+1} d(a_i, a_j).$$

It is not possible to find $(U_n)_{1 \leq n \leq N}$ covering A with $\operatorname{diam} U_n \leq \delta/2$: otherwise, we would have two points $a_i, a_j, i \neq j$ in the same U_n, so that

$$\delta/2 \geq \operatorname{diam} U_n \geq d(a_i, a_j) \geq \delta$$

which is not possible since $\delta > 0$, proving the claim. The claim implies as well $\operatorname{card} A \leq \mathfrak{h}_0^*(A)$. On the other hand, if A is a finite set, we can cover A with $\operatorname{card} A$ open balls with arbitrary small radius, which implies $\mathfrak{h}_{0,\varepsilon}^*(A) \leq \operatorname{card} A$ and eventually $\operatorname{card} A = \mathfrak{h}_0^*(A)$. For A infinite, we have proven $\mathfrak{h}_0^*(A) = +\infty$, proving the lemma. $\qquad \square$

Hausdorff dimension

Lemma 2.6.7. *Let (X, d) be a separable metric space, let $\kappa \geq 0$ be given and let A be a subset of X. Then if $\mathfrak{h}_\kappa^*(A) < +\infty$, we have $\mathfrak{h}_{\kappa'}^*(A) = 0$ for all $\kappa' > \kappa$ and if $\mathfrak{h}_\kappa^*(A) > 0$, we have $\mathfrak{h}_{\kappa''}^*(A) = +\infty$ for all $\kappa'' < \kappa$.*

Proof. If $\mathfrak{h}_\kappa^*(A) < +\infty$, we find that for all $\varepsilon > 0$, $\mathfrak{h}_{\kappa,\varepsilon}^*(A) \leq \mathfrak{h}_\kappa^*(A) < +\infty$. We can find a countable open covering $(U_n)_{n \in \mathbb{N}}$ of A such that $\operatorname{diam} U_n \leq \varepsilon$ and

$$\mathfrak{h}_{\kappa,\varepsilon}^*(A) \leq \sum_n (\operatorname{diam} U_n)^\kappa < \mathfrak{h}_{\kappa,\varepsilon}^*(A) + 1.$$

As a consequence, for $\kappa' > \kappa$, we have

$$\sum_n (\operatorname{diam} U_n)^{\kappa'} = \sum_n (\operatorname{diam} U_n)^{\kappa' - \kappa} (\operatorname{diam} U_n)^\kappa$$

$$\leq \varepsilon^{\kappa' - \kappa} \big(\mathfrak{h}_{\kappa,\varepsilon}^*(A) + 1\big) \leq \varepsilon^{\kappa' - \kappa} \big(\mathfrak{h}_\kappa^*(A) + 1\big).$$

As a result, we find $0 \le \mathfrak{h}^*_{\kappa',\varepsilon}(A) \le \sum_n (\operatorname{diam} U_n)^{\kappa'} \le \varepsilon^{\kappa'-\kappa}\big(\mathfrak{h}^*_\kappa(A)+1\big)$, so that

$$\mathfrak{h}^*_{\kappa'}(A) = \lim_{\varepsilon \to 0_+} \mathfrak{h}^*_{\kappa',\varepsilon}(A) = 0.$$

Let us assume now that $\mathfrak{h}^*_\kappa(A) > 0$ for some positive κ. For $\varepsilon > 0$, we can find a countable open covering $(U_n)_{n\in\mathbb{N}}$ of A such that $\operatorname{diam} U_n \le \varepsilon$. For $\kappa'' < \kappa$, we have

$$(\operatorname{diam} U_n)^{\kappa''} \varepsilon^{\kappa-\kappa''} \ge (\operatorname{diam} U_n)^{\kappa}$$

and thus $\sum_n (\operatorname{diam} U_n)^{\kappa''} \ge \sum_n \varepsilon^{\kappa''-\kappa}(\operatorname{diam} U_n)^{\kappa} \ge \varepsilon^{\kappa''-\kappa}\mathfrak{h}^*_{\kappa,\varepsilon}(A)$. As a result, we find

$$\mathfrak{h}^*_{\kappa'',\varepsilon}(A) \ge \varepsilon^{\kappa''-\kappa}\mathfrak{h}^*_{\kappa,\varepsilon}(A)$$

and since $\lim_{\varepsilon\to 0_+}\mathfrak{h}^*_{\kappa,\varepsilon}(A) = \mathfrak{h}^*_\kappa(A) > 0$, we get $\mathfrak{h}^*_{\kappa''}(A) = \lim_{\varepsilon\to 0_+}\mathfrak{h}^*_{\kappa'',\varepsilon}(A) = +\infty$. $\qquad\square$

Definition 2.6.8. Let (X,d) be a separable metric space and let A be a subset of X such that $\mathfrak{h}_0(A) = +\infty$. The Hausdorff dimension of A is defined as

$$D_\mathfrak{h}(A) = \sup\{\kappa \ge 0, \mathfrak{h}^*_\kappa(A) = +\infty\}. \tag{2.6.8}$$

A set such that $\mathfrak{h}_0(A) < +\infty$ is finite (Lemma 2.6.6): we define then $D_\mathfrak{h}(A) = 0$.

Note that we have also

$$D_\mathfrak{h}(A) = \kappa_+ = \inf\{\kappa \ge 0, \mathfrak{h}^*_\kappa(A) = 0\}. \tag{2.6.9}$$

In fact, if $\mathfrak{h}^*_\kappa(A) > 0$ for all $\kappa > 0$, Lemma 2.6.7 implies that $\mathfrak{h}^*_\kappa(A) = +\infty$ for all $\kappa \ge 0$ so that $D_\mathfrak{h}(A) = +\infty = \inf \emptyset$. If there exists $\kappa_0 > 0$ such that $\mathfrak{h}^*_{\kappa_0}(A) = 0$, then Lemma 2.6.7 implies $\mathfrak{h}^*_\kappa(A) = 0$ for $\kappa > \kappa_0$, proving that

$$\mathfrak{h}^*_\kappa(A) = 0 \quad \text{if } \kappa > \kappa_+ = \inf\{\kappa' \ge 0, \mathfrak{h}^*_{\kappa'}(A) = 0\}.$$

If $\kappa_+ = 0$, we get $\mathfrak{h}^*_\kappa(A) = 0$ on $(0,+\infty)$ and $\kappa_+ = 0 = D_\mathfrak{h}(A)$. If $\kappa_+ > 0$, we find $\mathfrak{h}^*_\kappa(A) = 0$ on $(\kappa_+,+\infty)$. Then for an increasing positive sequence with limit κ_+, $\kappa_n < \kappa_+$, we get

$$\mathfrak{h}^*_{\kappa_n}(A) > 0$$

so that $\mathfrak{h}^*_\kappa(A) = +\infty$ for $\kappa \in [0,\kappa_n)$ and thus on $[0,\kappa_+)$, proving $D_\mathfrak{h}(A) = \kappa_+$.

Hausdorff measures on $\mathbb{R}^m$

Lemma 2.6.9. *Let $\mathbb{R}^m$ be equipped with the distance d_∞ defined by*

$$d_\infty(x,y) = \max_{1\le j\le m} |x_j - y_j| \tag{2.6.10}$$

and let K be a compact subset of $\mathbb{R}^m$ with positive diameter δ for the distance d_∞. Then there exists $z'_j \le z''_j \le z'_j + \delta$ such that

$$K \subset \prod_{1\le j\le m} [z'_j, z''_j].$$

Proof. The continuous mapping $\mathbb{R}^m \ni x = (x_1, \ldots, x_m) \mapsto x_1 = \pi_1(x) \in \mathbb{R}$ is such that $\pi_1(K)$ is a compact subset of $\mathbb{R}$: $\pi_1(K) \subset [\inf \pi_1(K), \sup \pi_1(K)] = [z_1', z_1'']$ so that $z' = (z_1', \ldots, z_m'), z'' = (z_1'', \ldots, z_m'') \in K$ and thus

$$|z_1' - z_1''| \leq d_\infty(z', z'') \leq \delta,$$

proving the lemma. $\qquad\square$

Considering the separable metric space $(\mathbb{R}^m, d_\infty)$, Definition 2.6.5 provides a measure space $(\mathbb{R}^m, \mathfrak{h}_m, \mathcal{M})$ where $\mathcal{M}$ is $\mathfrak{h}_m$ complete and contains the Borel σ-algebra $\mathcal{B}_m$ on $\mathbb{R}^m$. Moreover, from its very definition, $\mathfrak{h}_m$ is translation invariant since $\mathfrak{h}_{m,\varepsilon}$ is translation invariant for any $\varepsilon > 0$; moreover $\mathfrak{h}_m$ is finite on compact sets since, for K bounded in $\mathbb{R}^m$, there exists $M > 0$ such that $K \subset [-M/2, M/2]^m$ and thus for $\varepsilon > 0, \delta > 0$, we have[5] with $a_k = -\frac{M}{2} + \varepsilon k$,

$$[-M/2, M/2]^m \subset \bigcup_{\substack{(k_1,\ldots,k_m) \\ 0 \leq k_j \leq [M/\varepsilon]}} \underbrace{\left(\prod_{1 \leq j \leq m} \,]a_{k_j} - \delta, a_{k_j} + \varepsilon + \delta[\right)}_{\text{open with } d_\infty \text{ diameter} = \varepsilon + 2\delta},$$

so that $\mathfrak{h}_{m,\varepsilon+2\delta}([-M/2, M/2]^m) \leq ([M/\varepsilon]+1)^m (\varepsilon+2\delta)^m \leq \left(M + \frac{M2\delta}{\varepsilon} + \varepsilon + 2\delta\right)^m$. With $\delta = \varepsilon^2/2$ we get

$$\mathfrak{h}_{m,\varepsilon+\varepsilon^2}([-M/2, M/2]^m) \leq (M + \varepsilon M + \varepsilon + \varepsilon^2)^m,$$

so that taking the limit of both sides when ε goes to 0, we obtain

$$\mathfrak{h}_m(K) \leq \mathfrak{h}_m([-M/2, M/2]^m) \leq M^m < +\infty. \tag{2.6.11}$$

Theorem 2.6.10. *Let $(\mathbb{R}^m, d_\infty)$ be as above. Definition 2.6.5 provides a measure space $(\mathbb{R}^m, \mathfrak{h}_m, \mathcal{M})$ where $\mathcal{M}$ is $\mathfrak{h}_m$ complete and contains the Borel σ-algebra $\mathcal{B}_m$ on $\mathbb{R}^m$. The Lebesgue measure space $(\mathbb{R}^m, \lambda_m, \mathcal{L}_m)$ given in Definition 2.4.1 is such that $\mathcal{L}_m \subset \mathcal{M}$ and λ_m coincides with $\mathfrak{h}_m$ on $\mathcal{L}_m$.*

Proof. Since $(\mathbb{R}^m, \lambda_m, \mathcal{L}_m)$ is given by Theorems 2.2.14–2.2.1, it is enough to prove that λ_m coincides with $\mathfrak{h}_m$ on the Borel σ-algebra $\mathcal{B}_m$: in fact the σ algebra $\mathcal{L}_m$ is generated by $\mathcal{B}_m$ and the subsets of λ_m-negligible Borel sets, so that, if we know that $\mathfrak{h}_m = \lambda_m$ on $\mathcal{B}_m$, the λ_m-negligible Borel sets will be also $\mathfrak{h}_m$-negligible and thus will belong to the $\mathfrak{h}_m$-complete $\mathcal{M}$.

On the other hand we already know that $\mathfrak{h}_m$ is a measure defined on the Borel σ-algebra $\mathcal{B}_m$, finite on compact sets, invariant by translation. To apply (3) in Theorem 2.4.2 and obtain our result, it is enough to prove that $\mathfrak{h}_m([0, 1]^m) = 1$

---

[5] Using the integer-valued floor function $[\cdot]$, defined in footnote page 16,

$$x \in [-M/2, M/2] \implies [(x + M/2)/\varepsilon] = k \leq [M/\varepsilon] \implies \varepsilon k \leq x + M/2 \leq \varepsilon(k+1)$$
$$\implies \varepsilon k - M/2 - \delta < \varepsilon k - M/2 \leq x \leq \varepsilon(k+1) - M/2 < \varepsilon(k+1) - M/2 + \delta.$$

and in fact, from (2.6.11) and translation invariance, we are reduced to the proof of $\mathfrak{h}_m([0,1]^m \geq 1$.

Let us assume that $\mathfrak{h}_m([0,1]^m) < 1$. Then for all $\varepsilon > 0$ we can find a (finite) collection of open bounded sets $(U_{j,\varepsilon})_{1 \leq j \leq N_\varepsilon}$ with diameter $\leq \varepsilon$, covering $[0,1]^m$ and such that $\sum_{1 \leq j \leq N_\varepsilon} (\operatorname{diam} U_{j,\varepsilon})^m \leq \mathfrak{h}_m([0,1]^m) < 1$. Since each $U_{j,\varepsilon}$ is relatively compact, we find from Lemma 2.6.9 and Theorem 2.4.2 that $\lambda_m(U_{j,\varepsilon}) \leq (\operatorname{diam} \overline{U_{j,\varepsilon}})^m = (\operatorname{diam} U_{j,\varepsilon})^m$ and this implies

$$1 = \lambda_m([0,1]^m) \leq \sum_{1 \leq j \leq N_\varepsilon} \lambda_m(U_{j,\varepsilon}) \leq \sum_{1 \leq j \leq N_\varepsilon} (\operatorname{diam} U_{j,\varepsilon})^m \leq \mathfrak{h}_m([0,1]^m) < 1,$$

which is impossible. The proof of Theorem 2.6.10 is complete. $\qquad\square$

It is important to note that we have found another way to construct the Lebesgue measure on $\mathbb{R}^m$, using the Carathéodory theorem on outer measures (Theorem 2.5.3), Theorem 2.5.4, and the definition and properties of the m-dimensional Hausdorff measure on $\mathbb{R}^m$. That construction is independent from the Riesz–Markov Theorem 2.2.1 and proceeds from a different perspective, a more set-theoretic approach without using a positive linear form as in the Riesz–Markov argument. It is however an interesting and important piece of information that the two measures constructed by these two different methods indeed coincide.

2.7 Notes

Let us follow the new names of mathematicians encountered along the text.

Constantin CARATHÉODORY (1873–1950) was a Greek mathematician.

Michel CHASLES (1793–1880) was a French mathematician.

Gottfried Wilhelm LEIBNIZ (1646–1716) was a German philosopher and mathematician, co-inventor with Isaac Newton of Infinitesimal Calculus.

Andrei MARKOV (1856–1922) was a Russian mathematician.

Frigyes (Frédéric) RIESZ (1880–1956) was a Hungarian mathematician who made fundamental contributions to functional analysis. His younger brother, Marcel RIESZ (1886–1969), was also a mathematician, author of basic contributions in Harmonic Analysis.

Johann RADON (1887–1956) was an Austrian mathematician.

2.8 Exercises

Topology

Exercise 2.8.1. *Let X be a topological space and let $f : X \to \mathbb{R}$. The function f is said to be lower semicontinuous at a point $a \in X$ when*

$$\forall \varepsilon > 0, \exists V_\varepsilon \in \mathscr{V}_a, \forall x \in V_\varepsilon, \quad f(a) - \varepsilon < f(x). \tag{2.8.1}$$

The function f is said to be upper semicontinuous at a point $a \in X$ when

$$\forall \varepsilon > 0, \exists V_\varepsilon \in \mathscr{V}_a, \forall x \in V_\varepsilon, \quad f(x) < f(a) + \varepsilon. \tag{2.8.2}$$

The function f is said to be lower (resp. upper) semicontinuous on X if it is lower (resp. upper) semicontinuous at every point of X.

(1) *Prove that f is continuous at $a \in X$ iff it is lower and upper semicontinuous at a.*

(2) *Prove that f is lower semicontinuous on X iff $\{x \in X, f(x) > \alpha\}$ is open for all $\alpha \in \mathbb{R}$. Prove that f is upper semicontinuous on X iff $\{x \in X, f(x) < \alpha\}$ is open for all $\alpha \in \mathbb{R}$.*

(3) *Let $A \subset X$. Prove that $\mathbf{1}_A$ is lower (resp. upper) semicontinuous iff A is open (resp. closed).*

(4) *Let $(f_i)_{i \in I}$ be a family of lower (resp. upper) semicontinuous functions on X. Then $\sup_{i \in I} f_i$ (resp. $\inf_{i \in I} f_i$) is lower (resp. upper) semicontinuous. Note that the former is valued in $(-\infty, +\infty]$ and the latter in $[-\infty, +\infty)$: our definitions of lower and upper semicontinuity are given by the conditions in (2).*

(5) *Let X be a non-empty compact topological space and let $f : X \to \mathbb{R}$ be a lower (resp. upper) semicontinuous function. Then there exists $a \in X$ such that $\inf_{x \in X} f(x) = f(a)$ (resp. $\sup_{x \in X} f(x) - f(a)$).*

(6) *Prove that a function $f : X \to \mathbb{R}$ is lower (resp. upper) semicontinuous at a point $a \in X$ iff $\liminf_{x \to a} f(x) = f(a)$ (resp. $\limsup_{x \to a} f(x) = f(a)$).*

We recall the following definitions, extending Definition 1.2.11: let X be a topological space, let f be a mapping from X into $\overline{\mathbb{R}}$ and let $a \in X$. We define

$$\liminf_{x \to a} f(x) = \sup_{V \in \mathscr{V}_a} \Big(\inf_{x \in V} f(x) \Big), \quad \limsup_{x \to a} f(x) = \inf_{V \in \mathscr{V}_a} \Big(\sup_{x \in V} f(x) \Big). \tag{2.8.3}$$

We have for $V_1, V_2 \in \mathscr{V}_a$,

$$\inf_{x \in V_2} f(x) \leq \inf_{x \in V_1 \cap V_2} f(x) \leq \sup_{x \in V_1 \cap V_2} f(x) \leq \sup_{x \in V_1} f(x),$$

so that $\inf_{x \in V_2} f(x) \leq \inf_{V_1 \in \mathscr{V}_a} \big(\sup_{x \in V_1} f(x) \big) = \limsup_{x \to a} f(x)$ which implies

$$\liminf_{x \to a} f(x) \leq \limsup_{x \to a} f(x). \tag{2.8.4}$$

Answer. (1) Continuity at $a \in X$ is expressed as:

$$\forall \varepsilon > 0, \exists V_\varepsilon \in \mathscr{V}_a, \quad f(V_\varepsilon) \subset (f(a) - \varepsilon, f(a) + \varepsilon),$$

thus is equivalent to the conjunction of upper and lower semicontinuity.

(2) We assume that f is lower semicontinuous on X: let $x_0 \in X$ and $\alpha < f(x_0)$. For $0 < \varepsilon = f(x_0) - \alpha$, we find $V \in \mathscr{V}_{x_0}$ such that

$$f(V) \subset (f(x_0) - \varepsilon, +\infty) = (\alpha, +\infty) \Longrightarrow V \subset f^{-1}(f(V)) \subset f^{-1}((\alpha, +\infty)),$$

implying that $f^{-1}((\alpha, +\infty))$ is open. Conversely, assuming $f^{-1}((\alpha, +\infty))$ open for all α, if we are given $x_0 \in X, \varepsilon > 0$, we know that

$$x_0 \in \{x \in X, f(x) > f(x_0) - \varepsilon\} \text{ is open,}$$

entailing (2.8.1). The result on upper semicontinuity can be obtained by remarking that f upper semi-continuous is equivalent to $-f$ lower semicontinuous.

(3) Let $A \subset X$; then we have

$$\{x \in X, \mathbf{1}_A(x) > \alpha\} = \begin{cases} \emptyset & \text{if } \alpha \geq 1 \\ A & \text{if } 1 > \alpha > 0 \,, \\ X & \text{if } \alpha \leq 0, \end{cases}$$

so that lower semicontinuity of $\mathbf{1}_A$ is equivalent to A open. Upper semicontinuity of $\mathbf{1}_A = 1 - \mathbf{1}_{A^c}$ is equivalent to lower semicontinuity of $\mathbf{1}_{A^c}$, which is equivalent to A^c open, i.e., to A closed.

(4) We have for $\alpha \in \mathbb{R}$ and $(f_i)_{i \in I}$ a family of lower semicontinuous functions

$$\{x \in X, \sup_{i \in I} f_i(x) > \alpha\} = \cup_{i \in I} \underbrace{\{x \in X, f_i(x) > \alpha\}}_{\text{open}},$$

so that $\sup_{i \in I} f_i$ is lower semicontinuous. Using $\inf_{i \in I} f_i = -\sup_{i \in I}(-f_i)$ gives that when $(f_i)_{i \in I}$ is a family of upper semicontinuous functions, so is $\inf_{i \in I} f_i$.

(5) Let f be a lower semicontinuous function on a non-empty compact set X. Then for $\alpha \in \mathbb{R}$, $K_\alpha = \{x \in X, f(x) \leq \alpha\}$ is a compact set. Let $\beta = \inf_{x \in X} f(x)$: we have $K_\alpha \subset K_{\alpha'}$ for $\alpha \leq \alpha'$ and for $\alpha > \beta$, $K_\alpha \neq \emptyset$

$$\cap_{\alpha > \beta}\{x \in X, f(x) \leq \alpha\} = \cap_{\alpha > \beta} K_\alpha \text{ is a non-empty compact set } L:$$

otherwise, we would have

$$X = \underbrace{\cup_{\alpha > \beta} K_\alpha^c}_{\substack{\text{compactness} \\ \text{of } X}} = \cup_{1 \leq j \leq N} K_{\alpha_j}^c = \left(\cap_{1 \leq j \leq N} K_{\alpha_j}\right)^c,$$

implying emptiness for $K_{\min_{1 \leq j \leq N} \alpha_j}$. Any $a \in L$ satisfies $f(a) = \beta$.

(6) Let us assume that f is upper semicontinuous at a: then for all $\varepsilon > 0, \exists V \in \mathcal{V}_a$ so that $f(V) \subset (-\infty, f(a) + \varepsilon)$ and thus $f(a) \le \sup_V f(x) \le f(a) + \varepsilon$, implying $\limsup_{x \to a} f(x) = f(a)$. Conversely, if the latter property holds, using the very definition of the infimum, we find that for all $\varepsilon > 0, \exists V \in \mathcal{V}_a$ so that,

$$f(a) \le \sup_V f(x) < f(a) + \varepsilon \Longrightarrow f(V) \subset (-\infty, f(a) + \varepsilon)$$

and upper semicontinuity at a.

Exercise 2.8.2 (Urysohn's Lemma). *Let Ω be an open subset of a locally compact Hausdorff topological space X and K be a compact subset of Ω. Show that there exists a function $\varphi \in C_c(X)$ such that*

$$0 \le \varphi \le 1, \quad \varphi_{|K} = 1, \quad \operatorname{supp} \varphi \subset \Omega.$$

Answer. Note that this result is proven in Proposition 2.1.2 for a metric space. Using the local compactness (see Proposition 10.2.36), we have

$$K \subset \cup_{x \in K} U_x, \quad x \in U_x \text{ open, relatively compact, } \overline{U_x} \subset \Omega,$$

and the compactness of K entails

$$K \subset \underbrace{\cup_{1 \le j \le N} U_{x_j}}_{=V_0, \text{ open}} \subset \overline{\cup_{1 \le j \le N} U_{x_j}} = \underbrace{\cup_{1 \le j \le N} \overline{U_{x_j}}}_{-\overline{V_0}, \text{ compact}} \subset \Omega.$$

Repeating the procedure, we can find V_1 open relatively compact such that

$$K \subset V_1 \subset \overline{V_1} \subset V_0 \subset \overline{V_0} \subset \Omega.$$

Let us assume that for $q_1 = 0, q_2 = 1, \ldots, q_n$ $(n \ge 2)$ distinct rational numbers in $[0, 1]$, we are able to find V_{q_j} open relatively compact such that

$$q_i < q_j \Longrightarrow K \subset \overline{V}_{q_j} \subset V_{q_i} \subset \overline{V}_{q_i} \subset \Omega.$$

Note that this property is proven for $n = 2$. Let $q_{n+1} \in \mathbb{Q} \cap (0, 1)$ in the complement of $E_n = \{q_1, \ldots, q_n\}$, and q_i the largest element of E_n such that $q_i < q_{n+1}$ and let q_j be the smallest element of E_n such that $q_j > q_{n+1}$. As above, we can find $V_{q_{n+1}}$ open relatively compact such that

$$\overline{V}_{q_j} \subset V_{q_{n+1}} \subset \overline{V}_{q_{n+1}} \subset V_{q_i}.$$

With $\{q_n\}_{n \ge 1} = \mathbb{Q} \cap [0, 1]$, we can construct $(V_{q_n})_{n \ge 1}$ open relatively compact such that

$$q, q' \in \mathbb{Q} \cap [0, 1], \quad q' < q \Longrightarrow K \subset \overline{V}_q \subset V_{q'} \subset \overline{V}_{q'} \subset \Omega.$$

We define now for $q \in \mathbb{Q} \cap [0, 1]$,

$$f_q = q \mathbf{1}_{V_q}, \qquad f = \sup_{\mathbb{Q} \cap [0,1]} f_q, \quad \text{valued in } [0, 1], \text{ lower s.c.,} \qquad (2.8.5)$$

$$g_q = (1 - q) \mathbf{1}_{\overline{V}_q} + q, \quad g = \inf_{\mathbb{Q} \cap [0,1]} g_q, \quad \text{valued in } [0, 1], \text{ upper s.c.} \qquad (2.8.6)$$

If $q' < q$, we have $V_q \subset V_{q'}$ and thus on V_q, $f_q = q \leq g_{q'} = (1 - q') + q' = 1$. If $q < q'$, we have $\overline{V}_{q'} \subset V_q$ and thus on V_q,

$$f_q = q \leq g_{q'} = \begin{cases} 1 & \text{on } \overline{V}_{q'}, \\ q' & \text{on } (\overline{V}_{q'})^c. \end{cases}$$

We have proven that for all q, q', $f_q \leq g_{q'}$ which implies $0 \leq f \leq g \leq 1$. On the other hand, $K \subset \cap_{q \in \mathbb{Q} \cap [0,1]} V_q$ so that for $x \in K$, $f_q(x) = q$ and thus $f(x) = 1$: $f_{|K} = 1$. We claim that for all x, $f(x) = g(x)$: otherwise, we could find x such that $f(x) < g(x)$ and thus $q, q' \in \mathbb{Q} \cap [0, 1]$ with

$$0 \leq f(x) < q < q' < g(x) \leq 1,$$

so that $x \notin V_q$ (since $f_q(x) < q$) and $x \in \overline{V}_{q'}$ (since $g_{q'}(x) > q'$) which is incompatible with $\overline{V}_{q'} \subset V_q$. Summing-up, the function f is 1 on K, valued in $[0, 1]$, lower s.c. by construction and upper s.c. since equal to g, so is eventually continuous. Since $V_1 \subset V_q \subset V_0 \subset \overline{V}_0$ for all $q \in \mathbb{Q} \cap [0, 1]$, the function $f = \sup_{q \in \mathbb{Q} \cap [0,1]} q \mathbf{1}_{V_q}$ is vanishing on the open set $(\overline{V}_0)^c$ so that $(\operatorname{supp} f)^c \supset (\overline{V}_0)^c$ and

$$\operatorname{supp} f \subset \overline{V}_0 \subset \Omega.$$

Exercise 2.8.3. *Let X be a topological space and let L be a positive linear form on $C_c(X)$. Show that L is continuous, in the sense that*

$$\forall K \, \text{compact} \subset X, \exists \gamma_K > 0, \ \forall f \in C_K(X), \ |Lf| \leq \gamma_K \sup |f|, \qquad (2.8.7)$$

$$\text{where } C_K(X) = \{f \in C_c(X), \operatorname{supp} f \subset K\}. \qquad (2.8.8)$$

Answer. For $f \in C_c(X), \operatorname{supp} f \subset K$ compact, we have with $\chi_K \in C_c(X; \mathbb{R}_+)$, $\chi_K = 1$ on K,

$$-\chi_K \sup |f| \leq f \leq \chi_K \sup |f| \implies |Lf| \leq L\chi_K \sup |f|.$$

Exercise 2.8.4. *Let (X, d) be a metric space such that all closed balls are compact: (X, d) is said to be proper and is locally compact (some locally compact metric spaces are not proper). Let Ω be an open subset of X and let K be a compact subset of Ω. Find a simpler proof of Proposition 2.1.2: there exists a function $\varphi \in C_c(X)$ such that*

$$0 \leq \varphi \leq 1, \quad \varphi_{|K} = 1, \quad \operatorname{supp} \varphi \subset \Omega.$$

The function φ can be chosen to be identically 1 on a neighborhood of K.

Answer. We have proven in (2.1.4) that $\epsilon_0 = \inf_{x \in K, \ y \notin \Omega} d(x, y) = d(K, \Omega^c) > 0$. As a result, we find

$$K_{\epsilon_0} := \cup_{x \in K} B(x, \epsilon_0) \subset \Omega \qquad (2.8.9)$$

since for $y \in B(x, \epsilon_0)$ and $x \in K$, we have $y \in \Omega$: otherwise, $y \in \Omega^c$ and

$$\epsilon_0 = d(K, \Omega^c) \leq d(x, y) < \epsilon_0,$$

which is impossible. Let us then define for some positive ϵ_1, ϵ_2 such that $\epsilon_1 + \epsilon_2 < \epsilon_0$,

$$\varphi(x) = \max\left(0, 1 - \frac{1}{\epsilon_2} d(x, K_{\epsilon_1})\right). \qquad (2.8.10)$$

That function is valued in $[0, 1]$ and is continuous as the maximum of two continuous functions. Moreover if $\varphi(x) \neq 0$, then $d(x, K_{\epsilon_1}) < \epsilon_2$ so that

$$\exists y \in K_{\epsilon_1} = \cup_{t \in K} B(t, \varepsilon_1), \quad d(x, y) < \epsilon_2, \quad \exists t \in K, \quad d(y, t) < \epsilon_1,$$

implying $d(t, x) < \epsilon_1 + \epsilon_2$ and $x \in K_{\epsilon_1 + \epsilon_2} \subset \{x, d(x, K) \leq \epsilon_1 + \epsilon_2\} = L$. The set L is closed (continuity of $d(\cdot, K)$) and included in Ω since $L \subset K_{\epsilon_0} \subset \Omega$, so that $\operatorname{supp} \varphi \subset L$. Moreover the set L is compact since if $(x_n)_{n \in \mathbb{N}}$ is a sequence in L, we find a sequence $(y_n)_{n \in \mathbb{N}}$ in K such that $d(x_n, y_n) \leq \epsilon_1 + \epsilon_2$. Extracting a convergent subsequence $(y_{n_k})_{k \in \mathbb{N}}$ with limit $y \in K$ of $(y_n)_{n \in \mathbb{N}}$, we get

$$d(x_{n_k}, y) \leq d(x_{n_k}, y_{n_k}) + d(y_{n_k}, y) \leq \epsilon_1 + \epsilon_2 + d(y_{n_k}, y)$$
$$\text{(for } k \text{ large enough)} \quad \leq \epsilon_1 + \epsilon_2 + \frac{1}{2}(\epsilon_0 - \epsilon_1 - \epsilon_2) = r < \epsilon_0.$$

The sequence $(x_{n_k})_{k \in \mathbb{N}}$ lies (for k large enough) in $B_c(y, r)$, which is assumed to be compact. We can extract another subsequence of $(x_{n_k})_{k \in \mathbb{N}}$, converging with limit $x \in B_c(y, r)$. The inequalities above ensure also that $d(x, y) \leq \epsilon_1 + \epsilon_2$ so that $x \in L$, proving the compactness of L.

Remark 2.8.5. Lemma 2.1.1 implies that φ is Lipschitz-continuous with a Lipschitz constant $1/\epsilon_2$. Since ϵ_2 can be chosen arbitrarily in $(0, d(K, \Omega^c))$, the function φ can be chosen Lipschitz continuous with a Lipschitz constant $> \frac{1}{d(K, \Omega^c)}$.

Exercise 2.8.6 (Partitions of unity on $\mathbb{R}^m$). *We define for* $x \in \mathbb{R}^m$,

$$\rho(x) = \begin{cases} \exp -(1 - |x|^2)^{-1} & \text{for } |x| < 1, \\ 0 & \text{for } |x| \geq 1, \end{cases}$$

where $|x| = (\sum_{1 \leq j \leq m} x_j^2)^{1/2}$ *stands for the Euclidean norm on* $\mathbb{R}^m$. *The function* ρ *is* C^∞ *with* $\operatorname{supp} f = \bar{B}(0, 1)$ *noted* $\mathbb{B}^m$.

Answer. Let us first consider the function ρ_0 defined by

$$\rho_0(t) = e^{-1/t} \text{ for } t > 0, \qquad \rho_0(t) = 0 \quad \text{for } t \leq 0.$$

Let us prove by induction on k that $\rho_0 \in C^k(\mathbb{R})$ is such that for $t \leq 0$, $\rho_0^{(k)}(t) = 0$ and for $t > 0$, $\rho_0^{(k)}(t) = p_k(1/t)e^{-1/t}$, where p_k is a polynomial. That property is true for $k = 0$ since $\lim_{t\to 0_+} e^{-1/t} = 0$. Assume that the property is true for some $k \geq 0$. Then, since $\rho_0 \in C^\infty(\mathbb{R}^*)$, we get for $t > 0$,

$$\rho_0^{(k+1)}(t) = e^{-1/t}\,\underbrace{t^{-2}\big(p_k(1/t) - p_k'(1/t)\big)}_{\text{polynomial } p_{k+1} \text{ in } t^{-1}}, \quad \text{and } \rho_0^{(k+1)}(t) = 0 \text{ for } t < 0.$$

We get also that $\lim_{t\to 0} t^{-1}\big(\rho_0^{(k)}(t) - \rho_0^{(k)}(0)\big) = \lim_{T\to +\infty} T p_k(T)e^{-T} = 0$, so that ρ_0 has a $(k+1)$th vanishing derivative at 0. The function $\rho_0^{(k+1)}$ is continuous since $\lim_{T\to +\infty} p_{k+1}(T)e^{-T} = 0$, completing the induction. As a result the function ρ_0 belongs to $C^\infty(\mathbb{R})$, with support $[0, +\infty)$ and is *flat* at the origin, i.e., $\forall k \in \mathbb{N}$, $\rho_0^{(k)}(0) = 0$. We have $\rho(x) = \rho_0(1 - |x|^2)$ so that $\rho \in C^\infty(\mathbb{R}^m; \mathbb{R})$, with support equal to the closed unit Euclidean ball.

N.B. The functions ρ, ρ_0 are paradigmatic examples of C^∞ functions which are not real-analytic: the function ρ_0 cannot be analytic at 0, since it is not identically 0 near the origin although its Taylor coefficients are all vanishing.

Exercise 2.8.7. *The vector space of C^∞ compactly supported functions from $\mathbb{R}^m$ into $\mathbb{C}$ will be noted $C_c^\infty(\mathbb{R}^m)$. Let Ω be an open subset of $\mathbb{R}^m$ and let K be a compact subset of Ω. Then there exists a function $\varphi \in C_c^\infty(\Omega; [0, 1])$ such that $\varphi = 1$ on a neighborhood of K.*

Answer. We recall that[6]

$$d(K, \Omega^c) = \inf_{x \in K, y \in \Omega^c} |x - y| > 0.$$

As a result, we have with $\mathbb{B}^m$ standing for the closed unit Euclidean ball of $\mathbb{R}^m$, $\epsilon_0 = d(K, \Omega^c)$, $K + \epsilon_0 \mathring{\mathbb{B}}^m \subset \Omega$: otherwise, we could find $|t| < 1, x \in K$ such that $x + \epsilon_0 t = y \in \Omega^c$, implying

$$0 < d(K, \Omega^c) \leq |x - y| < \epsilon_0 = d(K, \Omega^c),$$

which is impossible. With the function ρ defined in Exercise 2.8.6, we define with $0 < \epsilon \leq \frac{\epsilon_1}{2} < \frac{\epsilon_0}{4}$,

$$\varphi(x) = \int 1_{K + \epsilon_1 \mathbb{B}^m}(y)\rho\big((x - y)\epsilon^{-1}\big)\epsilon^{-n}dy\left(\int \rho(t)dt\right)^{-1}.$$

The function φ is C^∞ and such that

$$\text{supp}\,\varphi \subset K + \epsilon_1 \mathbb{B}^m + \epsilon \mathbb{B}^m \subset K + \frac{3\epsilon_0}{4}\mathbb{B}^m \subset K + \epsilon_0 \mathring{\mathbb{B}}^m \subset \Omega.$$

---

[6]We may assume that both K and Ω^c are non-empty, so that $d(K, \Omega^c)$ is a positive real number.

Moreover $\varphi = 1$ on $K + \frac{\epsilon_1}{2}\mathbb{B}^m$ (which is a neighborhood of K), since if $x \in K + \frac{\epsilon_1}{2}\mathbb{B}^m$, we have, for y satisfying $|x-y| \le \epsilon$, that $y \in K + \frac{\epsilon_1}{2}\mathbb{B}^m + \epsilon\mathbb{B}^m \subset K + \epsilon_1\mathbb{B}^m$. As a result, with $\tilde{\rho} = \rho\left(\int \rho(t)dt\right)^{-1}$, for $x \in K + \frac{\epsilon_1}{2}\mathbb{B}^m$, we have

$$1 = \int \tilde{\rho}((x-y)\epsilon^{-1})\epsilon^{-n}dy = \int \tilde{\rho}((x-y)\epsilon^{-1})\epsilon^{-n}\mathbf{1}_{K+\epsilon_1\mathbb{B}^m}(y)dy = \varphi(x).$$

We note also that, since $\tilde{\rho} \ge 0$ with integral 1, $\mathbf{1}_L(y) \in \{0,1\}$, we have, for all $x \in \mathbb{R}^m$, $0 \le \varphi(x) \le 1$.

Exercise 2.8.8. *Let $\Omega_1, \ldots, \Omega_n$ be open subsets of $\mathbb{R}^m$ and let K be a compact set with $K \subset \Omega_1 \cup \cdots \cup \Omega_n$. Then for each $j \in \{1, \ldots, n\}$, there exists a function $\psi_j \in C_c^\infty(\Omega_j; [0,1])$ such that $\sum_{1 \le j \le n} \psi_j \in C_c^\infty(\cup_{j=1}^n \Omega_j; [0,1])$ and*

$$1 = \sum_{1 \le j \le n} \psi_{j|K}.$$

We shall say that $(\psi_j)_{1 \le j \le n}$ is a partition of unity on K, attached to $(\Omega_j)_{1 \le j \le n}$. In particular, for $\theta \in C_c^\infty(\cup_{1 \le j \le n}\Omega_j)$, using the previous result for $K = \operatorname{supp}\theta$, we get

$$\theta = \sum_{1 \le j \le n} \theta_j, \quad \text{with } \theta_j = \theta\psi_j \in C_c^\infty(\Omega_j).$$

Answer. A simple inspection of the proof of Theorem 2.1.3 provides smooth functions.

Exercise 2.8.9 (Approximating continuous functions by piecewise affine functions). *A function $p : \mathbb{R} \to \mathbb{R}$ is said to be piecewise affine if there exists $x_1 < x_2 < \cdots < x_N$ real numbers such that the restriction of p to each interval (x_j, x_{j+1}) for $0 \le j \le N+1$ is affine $(x_0 = -\infty, x_{N+1} = +\infty)$. Prove that the vector space of compactly supported continuous piecewise affine functions is dense in $C_c(\mathbb{R}; \mathbb{R})$.*

Answer. Let ϕ be a function in $C_c(\mathbb{R}; \mathbb{R})$, supported in $[a,b]$ and let $\varepsilon \in (0,1)$ be given. We consider $N \in \mathbb{N}$ such that $N = [1 + (b-a)/\varepsilon] + 1$ and

$$x_1 = a < \cdots < x_j = a + (j-1)\varepsilon < \cdots < \underbrace{x_N = a + (N-1)\varepsilon}_{\ge b}.$$

We define

$$p(x) = \sum_{1 \le j < N} \mathbf{1}_{[x_j, x_{j+1})}(x)\left(\phi(x_j) + \frac{x - x_j}{x_{j+1} - x_j}(\phi(x_{j+1}) - \phi(x_j))\right).$$

The function p is piecewise affine, compactly supported in $[a, b+1]$, and verifies

$$p(x_j) = \phi(x_j) = p(x_{j+}), \quad \text{for } 1 \le j < N, \quad p(x_N) = 0 = \phi(x_N) = p(x_{N+}),$$
$$p(x_{j-}) = \phi(x_j), \quad\qquad \text{for } 2 \le j \le N, \quad p(x_1) = 0 = \phi(x_1) = p(x_{1-}),$$

thus continuous. We have

$$p(x) - \phi(x) = \sum_{1 \leq j < N} \mathbf{1}_{[x_j, x_{j+1})}(x)\left(\phi(x_j) - \phi(x) + \frac{x - x_j}{x_{j+1} - x_j}\left(\phi(x_{j+1}) - \phi(x_j)\right)\right),$$

and thus

$$p(x) - \phi(x)$$
$$= \sum_{1 \leq j < N} \mathbf{1}_{[x_j, x_{j+1})}(x)\left\{\left(\phi(x_j) - \phi(x)\right)\frac{x_{j+1} - x}{x_{j+1} - x_j} + \frac{x - x_j}{x_{j+1} - x_j}\left(\phi(x_{j+1}) - \phi(x)\right)\right\},$$

implying

$$\sup_{x \in \mathbb{R}} |p(x) - \phi(x)| \leq \sup_{|x' - x''| \leq \varepsilon} |\phi(x') - \phi(x'')| = \omega(\varepsilon).$$

Since ϕ is uniformly continuous, we get $\lim_{\varepsilon \to 0} \omega(\varepsilon) = 0$ and the result.

Exercise 2.8.10. *Let Ω be an open subset of $\mathbb{R}^n$. Prove that there exists a sequence $(K_j)_{j \geq 1}$ of compact subsets of Ω such that*

$$\Omega = \cup_{j \geq 1} K_j, \quad K_j \subset \mathring{K}_{j+1}. \tag{2.8.11}$$

Prove also that if K is a compact subset of Ω, there exists $j \in \mathbb{N}^*$ such that $K \subset K_j$.

Answer. Given an open set Ω of $\mathbb{R}^n$, we define for $j \geq 1$,

$$K_j = \{x \in \mathbb{R}^n, d(x, \Omega^c) \geq 1/j, |x| \leq j\}.$$

We note from the continuity of $d(\cdot, \Omega^c)$ and of the norm that K_j is a closed subset of $\mathbb{R}^n$; moreover it is also bounded and thus is a compact subset of $\mathbb{R}^n$, and in fact of Ω since $d(x, \Omega^c) > 0$ implies $x \notin \overline{\Omega^c} = \Omega^c$ (Ω is open). We have also for $j \geq 1$ that

$$K_j \subset \left\{x \in \mathbb{R}^n, d(x, \Omega^c) > \frac{1}{j+1}, |x| < j + 1\right\} \text{ which is open} \subset K_{j+1},$$

so that $K_j \subset \mathring{K}_{j+1}$. Finally, taking $x \in \Omega$, we have $d(x, \Omega^c) > 0$ (Ω^c is closed) and thus

$$j \geq \max\left(\frac{1}{d(x, \Omega^c)}, E(|x|) + 1\right) \implies x \in K_j,$$

proving $\Omega = \cup_{j \geq 1} K_j$ and the result, since the very last statement follows from

$$K \subset \Omega = \cup_{j \geq 1} \mathring{K}_{j+1},$$

which implies the result by the Borel–Lebesgue property and the fact that the sequence (K_j) is increasing.

Exercise 2.8.11 (Dini's Lemma).[7] *Let $a \leq b$ be real numbers and let $f_n : [a,b] \to \mathbb{R}$ be a sequence of continuous functions. We assume that for all $x \in [a,b]$, the sequence $(f_n(x))$ is decreasing with limit 0.*

(1) *Prove that (f_n) converges uniformly towards 0.*

(2) *Prove that the result of (1) does not hold without the assumption of decreasing monotonicity.*

Answer. (1) *Reductio ad absurdum*[8]: if the sequence (f_n) were not converging uniformly towards 0, the sequence $\omega_n = \sup_{x \in [a,b]} |f_n(x)|$ is such that there exists $\epsilon_0 > 0$ and a subsequence $(\omega_{n_k})_{k \in \mathbb{N}}$ such that for all k, $\omega_{n_k} > \epsilon_0$. As a result for all k, there exists $x_k \in [a,b]$ such that

$$f_{n_k}(x_k) > \epsilon_0.$$

Thanks to the compactness of $[a,b]$, we may find a subsequence of $(x_k)_{k \in \mathbb{N}}$ converging with limit $c \in [a,b]$. To simplify notation, let us assume that $(x_k)_{k \in \mathbb{N}}$ is converging towards c. For $l \geq 0$, we have $n_{k+l} \geq n_k$, and thus

$$f_{n_k}(x_{k+l}) \geq f_{n_{k+l}}(x_{k+l}) > \epsilon_0.$$

Since f_{n_k} is continuous, we find $f_{n_k}(c) \geq \epsilon_0 > 0$, contradicting the convergence of the sequence $(f_n(c))$ towards 0.

(2) Let us define φ_n piecewise affine on $[0,1]$,

$$\varphi_n(0) = 0, \quad \varphi_n(1/n) = 1, \quad \varphi_n(t) = 0 \text{ for } t \geq 2/n.$$

The sequence of continuous functions (φ_n) converges pointwise to 0, not uniformly since $\sup |\varphi_n| = 1$. Moreover the result is incorrect without the continuity assumption: defining ψ_n on $[0,1]$ by

$$\psi_n(0) = 0 = \psi_n(t) \text{ for } t \geq 1/n, \quad \psi_n(t) = 1 - nt \text{ for } 0 < t < 1/n,$$

we find that for all $t \in [0,1]$, the decreasing sequence $(\psi_n(t))_{n \in \mathbb{N}}$ goes to zero. However, the convergence is not uniform since $\sup_{[0,1]} |\psi_n| = 1$.

Exercise 2.8.12 (Support of an L^1 function). *Let $(X, \mathcal{M}, \mu)$ be a measure space where μ is a positive measure such that X is a topological space with $\mathcal{M} \supset \mathcal{B}_X$ and $\mu(\Omega) > 0$ for any non-empty open set Ω. Let $f \in \mathcal{L}^1(\mu)$.*

(1) *Defining*

$$\operatorname{supp} f = \{x \in X, \not\exists V \in \mathcal{V}_x, f_{|V} = 0, \mu\text{-a.e.}\}, \tag{2.8.12}$$

prove that $(\operatorname{supp} f)^c$ is open and is the largest open set on which $f = 0$ a.e.

---

[7] Ulisse DINI (1845–1918) is an Italian mathematician, who served as Director of *Scuola Normale Superiore* in Pisa. A bronze statue of Dini is located near the *Piazza dei Cavalieri*.

[8] About this method of proof, we may quote G.H. HARDY in *A Mathematician's Apology* [29]: Reductio ad absurdum, which Euclid loved so much, is one of a mathematician's finest weapons. It is a far finer gambit than any chess play: a chess player may offer the sacrifice of a pawn or even a piece, but a mathematician offers the game.

(2) *Prove that* supp f *depends only on the class of f modulo equality μ-a.e.*

(3) *Prove that* supp f *coincides with Definition (2.1.1) when f is a continuous function.*

(4) *Show by an example that it would be absurd to take (2.1.1) as a definition for non-continuous functions.*

Answer. (1) The complement of supp f is open, and every open set on which f vanishes a.e. is included in $(\text{supp } f)^c$.

(2) is obvious: if $f, \tilde{f}$ coincide a.e. $f_{|V} = 0$ a.e. is equivalent to $\tilde{f}_{|V} = 0$ a.e.

(3) It is enough to prove that for a continuous function f, and an open set V, $f_{|V} = 0$ a.e. implies $f_{|V} = 0$. If it were not the case, and $f(x_0) \neq 0$ for some $x_0 \in V$, the set

$$\{x \in V, |f(x)| > |f(x_0)|/2 > 0\}$$

would be open (thanks to the continuity of f) and non-empty (contains x_0), thus with a positive measure. As a consequence, f would not be 0 a.e. on V, contradicting the assumption.

(4) Taking $f = 1_{\mathbb{Q}}$, we see that $f = 0$, λ_1-a.e., so that supp $f = \emptyset$. Taking (2.1.1) as a definition for the support of f would imply supp $f = \overline{\mathbb{Q}} = \mathbb{R}$.

Measure theory

Exercise 2.8.13 (Completion of a measure). *Let $(X, \mathcal{M}, \mu)$ be a measure space where μ is a positive measure. Defining $\mathcal{N} = \bigcup_{E \in \mathcal{M}, \mu(E)=0} \mathcal{P}(E)$, prove that*

$$\mathcal{M}' = \{M \cup N\}_{M \in \mathcal{M}, N \in \mathcal{N}}$$

is the σ-algebra generated by $\mathcal{M} \cup \mathcal{N}$ and defining for $M \in \mathcal{M}, N \in \mathcal{N}, \mu'(M \cup N) = \mu(M)$, prove that this definition is consistent and $(X, \mathcal{M}', \mu')$ is a measure space such that $\mu'_{|\mathcal{M}} = \mu$.

Answer. Let $A = M \cup N \in \mathcal{M}', M \in \mathcal{M}, N \in \mathcal{N}, N \subset E \in \mathcal{M}, \mu(E) = 0$:

$$A^c = (M^c \cap N^c \cap E) \cup (M^c \cap N^c \cap E^c) = \underbrace{(M^c \cap N^c \cap E)}_{\in \mathcal{N}} \cup \underbrace{(M^c \cap E^c)}_{\in \mathcal{M}} \in \mathcal{M}'.$$

Let us consider sequences $A_n = M_n \cup N_n \in \mathcal{M}', M_n \in \mathcal{M}, N_n \in \mathcal{N}, N_n \subset E_n \in \mathcal{M}, \mu(E_n) = 0$:

$$\cup_{n \in \mathbb{N}} A_n = \underbrace{\left(\cup_{n \in \mathbb{N}} M_n\right)}_{\in \mathcal{M}} \cup \left(\cup_{n \in \mathbb{N}} N_n\right)$$

and since $\cup_{n \in \mathbb{N}} N_n \subset \cup_{n \in \mathbb{N}} E_n, \mu(\cup_{n \in \mathbb{N}} E_n) = 0$, we get $\cup_{n \in \mathbb{N}} A_n \in \mathcal{M}'$. As a result $\mathcal{M}'$ is a σ-algebra on X, containing $\mathcal{M} \cup \mathcal{N}$, so containing the σ-algebra generated by $\mathcal{M} \cup \mathcal{N}$. On the other hand $\mathcal{M}'$ is included in the σ-algebra generated by $\mathcal{M} \cup \mathcal{N}$.

Let $M' \in \mathcal{M}', M' = M_j \cup N_j, M_j \in \mathcal{M}, N_j \subset E_j \in \mathcal{M}, \mu(E_j) = 0, j =, 1, 2.$ Then since $M_1 \subset M_1 \cup N_1 = M_2 \cup N_2 \subset M_2 \cup E_2,$

$$\mu(M_1) \leq \mu(M_2 \cup E_2) = \mu(M_2) \implies \mu(M_1) \leq \mu(M_2),$$

(similarly $\mu(M_2) \leq \mu(M_1)$), so that $\mu'(M') = \mu(M_1)$ is defined without ambiguity. Let us consider a pairwise disjoint sequence in $\mathcal{M}'$: $A_n = M_n \cup N_n \in \mathcal{M}', M_n \in \mathcal{M}, N_n \in \mathcal{N}, N_n \subset E_n \in \mathcal{M}, \mu(E_n) = 0.$ We have

$$\mu'(\cup A_n) = \mu'\Big(\underbrace{(\cup_{n\in\mathbb{N}} M_n)}_{\in\mathcal{M}} \cup \underbrace{(\cup_{n\in\mathbb{N}} N_n)}_{\mathcal{N}}\Big) = \mu\big(\cup_{n\in\mathbb{N}} M_n\big)$$

and since the M_n are also pairwise disjoint $(M_n \subset A_n)$, we get

$$\mu'(\cup A_n) = \sum \mu(M_n) = \sum \mu'(A_n), \qquad \text{qed.}$$

Exercise 2.8.14. *Let $(X, \mathcal{M}, \mu)$ be a measure space where μ is a positive measure. Let $(f_n)_{n\in\mathbb{N}}$ be a sequence of complex-valued measurable functions on X. We shall say that $(f_n)_{n\in\mathbb{N}}$ converges locally in measure towards a measurable function f if*

$$\forall \alpha > 0, \forall Y \in \mathcal{M} \text{ with } \mu(Y) < \infty, \lim_n \mu\big(\{x \in Y, |f_n(x) - f(x)| > \alpha\}\big) = 0. \quad (2.8.13)$$

We shall say that $(f_n)_{n\in\mathbb{N}}$ converges globally in measure towards a measurable function f if

$$\forall \alpha > 0, \quad \lim_n \mu\big(\{x \in X, |f_n(x) - f(x)| > \alpha\}\big) = 0. \qquad (2.8.14)$$

Assume that μ is σ-finite, i.e., there exists a sequence $(X_k)_{k\in\mathbb{N}}$ in $\mathcal{M}$ such that $X = \cup_{k\in\mathbb{N}} X_k$ and for all $k \in \mathbb{N}$, $\mu(X_k) < +\infty$. Prove that the Lebesgue dominated convergence theorem holds with local convergence in measure replacing pointwise convergence in the assumptions.

Answer. Assuming $|f_n| \leq g \in \mathcal{L}^1(\mu)$, we have for $\alpha > 0$, $Y_k = \cup_{0 \leq l \leq k} X_l$,

$$\int_X |f_n - f| d\mu = \int_{Y_k \cap \{|f_n - f| \leq \alpha\}} |f_n - f| d\mu + \int_{Y_k^c \cap \{|f_n - f| \leq \alpha\}} |f_n - f| d\mu$$

$$+ \int_{Y_k \cap \{|f_n - f| > \alpha\}} |f_n - f| d\mu + \int_{Y_k^c \cap \{|f_n - f| > \alpha\}} |f_n - f| d\mu$$

$$\leq \alpha\mu(Y_k) + \int_{Y_k^c} 2g d\mu + \int_{Y_k \cap \{|f_n - f| > \alpha\}} 2g d\mu.$$

Using Proposition 1.7.10, we find that for all $\alpha > 0$ and all integers k,

$$\limsup_n \int_X |f_n - f| d\mu \leq \alpha\mu(Y_k) + \int_{Y_k^c} 2g d\mu, \qquad (2.8.15)$$

so that, for all k,

$$\limsup_n \int_X |f_n - f| d\mu \leq \int_X 2g 1_{Y_k^c} d\mu. \tag{2.8.16}$$

We have also $+\infty > \int_X 2g d\mu = \int_{Y_k} 2g d\mu + \int_{Y_k^c} 2g d\mu$, and by Beppo Levi's theorem $\lim_k \int_{Y_k} 2g d\mu = \int_X 2g d\mu$ which implies $\lim_k \int_{Y_k^c} 2g d\mu = 0$: the inequality (2.8.16) gives the result $\lim_n \int_X |f_n - f| = 0$.

To sum-up, for a sequence (f_n) in $L^1(\mu)$, f measurable,

$$\left. \begin{array}{c} f_n \xrightarrow[\text{in measure}]{\text{convergence}} f \\ \text{and} \\ |f_n| \leq g \in L^1(\mu) \end{array} \right\} \implies f_n \xrightarrow[L^1(\mu)]{} f. \tag{2.8.17}$$

Exercise 2.8.15. *Let $(X, \mathcal{M}, \mu)$ be a measure space where μ is a positive measure. Let $(f_n)_{n\in\mathbb{N}}$ be a sequence of measurable functions and f be a measurable function.*

(1) *Prove that if $(f_n)_{n\in\mathbb{N}}$ converges a.e. towards f, it implies that $(f_n)_{n\in\mathbb{N}}$ converges locally in measure towards f.*

(2) *Prove that the converse of the previous statement does not hold in general.*

Answer. (1) Let $\alpha > 0$ and Y a measurable set with finite measure:

$$\mu(\{x \in Y, |f_n(x) - f(x)| > \alpha\}) = \int_Y 1_{\{|f_n - f| > \alpha\}} d\mu.$$

The function $1_{\{|f_n - f| > \alpha\}}$ converges a.e. pointwise to 0 and is dominated by $1 \in \mathcal{L}^1(Y)$. As a result the Lebesgue dominated convergence gives the result.
(2) cf. Exercise 2.8.23.

Exercise 2.8.16 (Borel–Cantelli Lemma). *Let X be a set, μ^* be an outer measure on X and let $(A_n)_{n\in\mathbb{N}}$ be a countable family of subsets of X. Then*

$$\sum_{n\in\mathbb{N}} \mu^*(A_n) < +\infty \implies \mu^*(\limsup_n A_n) = 0,$$

where we have defined $\limsup_n A_n = \cap_{n\geq 0}(\cup_{k\geq n} A_k)$.

Answer. We have

$$0 \leq \mu^*(\limsup_n A_n) \leq \mu^*(\cup_{k\geq n} A_k) \leq \sum_{k\geq n} \mu^*(A_k) \xrightarrow[n\to+\infty]{} 0.$$

Exercise 2.8.17. *We define μ^* on $\mathcal{P}(\mathbb{R})$ by $\mu^*(A) = \inf\{\sum_{j\in\mathbb{N}}(b_j - a_j)\}$, where $\cup_{j\in\mathbb{N}}]a_j, b_j[$ runs among the coverings of A by open bounded intervals. Show that μ^* is an outer measure on $\mathbb{R}$ (see Definition 2.2.4).*

Answer. Since an open subset of $\mathbb{R}$ is a countable union of bounded open intervals, this appears as similar to Lemma 2.6.2 for $\kappa = 1$ and $X = \mathbb{R}$, but a simple direct proof may be useful. Properties (2.2.5), (2.2.6) are obvious. Let us prove (2.2.7). Let $(A_n)_{n\in\mathbb{N}}$ be a sequence of subsets of $\mathbb{R}$. We may assume that all $\mu^*(A_n)$ are finite, otherwise (2.2.7) is trivially satisfied. Let $\epsilon > 0$ be given. For each $n \in \mathbb{N}$, we consider a countable family of bounded open intervals $(I_k^n)_{k\in\mathbb{N}}$ such that

$$A_n \subset \cup_{k\in\mathbb{N}} I_k^n, \quad \mu^*(A_n) \leq \sum_{k\in\mathbb{N}} |I_k^n| < \mu^*(A_n) + \epsilon 2^{-n-1},$$

where $|I_k^n|$ is the length of I_k^n. We find $\cup_{n\in\mathbb{N}} A_n \subset \cup_{n,k\in\mathbb{N}} I_k^n$ and thus

$$\mu^*(\cup_{n\in\mathbb{N}} A_n) \leq \sum_{n,k\in\mathbb{N}} |I_k^n| = \sum_{n\in\mathbb{N}} \left(\sum_{k\in\mathbb{N}} |I_k^n| \right)$$

$$\leq \sum_{n\in\mathbb{N}} \left(\mu^*(A_n) + \epsilon 2^{-n-1} \right) = \epsilon + \sum_{n\in\mathbb{N}} \mu^*(A_n),$$

for any $\epsilon > 0$, proving the result.

Exercise 2.8.18. *Let $\varepsilon > 0$ be given. Construct a dense open subset Ω of $\mathbb{R}$ such that its Lebesgue measure $\lambda_1(\Omega) < \varepsilon$.*

Answer. We set $\mathbb{Q} = \{x_n\}_{n\geq 1}$ and we define

$$\Omega = \cup_{n\geq 1}]x_n - \epsilon 2^{-n-2}, x_n + \epsilon 2^{-n-2}[,$$

open as a union of open sets, dense since it contains $\mathbb{Q}$ and with Lebesgue measure

$$\lambda_1(\Omega) \leq \sum_{n\geq 1} \epsilon 2^{-n-1} = \epsilon/2 < \epsilon.$$

Exercise 2.8.19 (A non-measurable set). *We define on $[0,1]$ the equivalence relation $x \sim y$ means $x - y \in \mathbb{Q}$. Let us recall the statement of the Axiom of Choice: let I be a non-empty set and let $(X_i)_{i\in I}$ be a family of sets. Then*

$$\forall i \in I, X_i \neq \emptyset \implies \prod_{i\in I} X_i \neq \emptyset.$$

For $X \subset \mathbb{R}$ and $t \in \mathbb{R}$, we shall write $X + t = \{x + t\}_{x\in X}$.

(1) *Using the axiom of choice, show that there exists a subset A of $[0,1]$ defined by taking a single element in each equivalence class of $\sim$.*

(2) *Let $\varphi : \mathbb{N} \to \mathbb{Q} \cap [-1,1]$ be a bijective mapping. We define $A_n = A + \varphi(n)$. Show that*

$$[0,1] \subset \cup_{n\in\mathbb{N}} A_n \subset [-1,2].$$

(3) *Show that there is no positive measure μ defined on $\mathcal{P}(\mathbb{R})$, invariant by translation (i.e., such that $\mu(X) = \mu(X + t)$ for all subsets X of $\mathbb{R}$ and all real number t), and such that $\mu([a,b]) = b - a$ for $a \leq b$.*

(4) *Show that $A \notin \mathcal{L}$, where $\mathcal{L}$ is the Lebesgue σ-algebra on $\mathbb{R}$.*

Answer. (1) The quotient set $[0,1]/\sim$, is the set of equivalence classes $\{X_i\}_{i\in I}$. Each X_i is an equivalence class and is thus non-empty. Using the axiom of choice, we may find a family $(x_i)_{i\in I}$ of elements of $[0,1]$ such that X_i is the equivalence class of x_i. Let us define $A = \{x_i\}_{i\in I}$.

(2) The set $\mathbb{Q} \cap [-1,1]$ is infinite countable, thus equipotent to $\mathbb{N}$. Let $x \in [0,1]$. There exists $i \in I$ such that $x \sim x_i$, i.e., $x - x_i \in \mathbb{Q}$. As a consequence, $x = x_i + \rho$ with $\rho \in \mathbb{Q}$ and since both x and x_i belong to $[0,1]$, ρ belongs to $[-1,1]$ so that there exists $n \in \mathbb{N}$ such that $\rho = \varphi(n)$. This implies $x \in A_n = A + \varphi(n)$. Moreover, we have

$$A_n \subset [0,1] + [-1,1] \subset [-1,2].$$

(3) Let us assume that there exists such a measure. We note first that for $n \neq m$ integers, we have $A_n \cap A_m = \emptyset$: if $x \in A_n \cap A_m$, we get with $i, j \in I$,

$$x = x_i + \varphi(n) = x_j + \varphi(m)$$

and thus $x_i \sim x_j$, so that $x_i = x_j$, and $\varphi(n) = \varphi(m)$ entailing $m = n$ since φ is injective. We would have

$$1 = \mu([0,1]) \leq \mu(\cup_{n\in\mathbb{N}}A_n) = \sum_{n\in\mathbb{N}} \mu(A_n) = \sum_{n\in\mathbb{N}} \mu(A) \leq \mu([-1,2]) = 3,$$

which is impossible, since the first inequality implies $\mu(A) > 0$ whereas the next one gives $\mu(A) = 0$.

(4) The set A cannot belong to $\mathcal{L}$, since, if it were the case, the previous inequalities would hold for the Lebesgue measure λ_1 on $\mathbb{R}$, leading as above to a contradiction.

Calculations

Exercise 2.8.20.

(1) *Determine the values of the real parameter α for which $\int_0^1 \frac{dx}{x^\alpha}$ converges.*

(2) *Determine the values of the real parameter α for which $\int_1^{+\infty} \frac{dx}{x^\alpha}$ converges.*

(3) *Prove that the harmonic series (general term $1/n$) is divergent. Show that the sequence*

$$x_n = 1 + \frac{1}{2} + \frac{1}{3} + \cdots + \frac{1}{n-1} + \frac{1}{n} - \ln n$$

converges.

(4) *Show that $\lim_{A\to+\infty} \int_0^A \frac{\sin x}{x} dx$ exists.[9]*

(5) *Show that $\int_0^{+\infty} \left|\frac{\sin x}{x}\right| dx = +\infty$.*

Answer. Using $\ln x = \int_1^x \frac{dt}{t}$ for $x > 0$ and $\frac{d}{dx}\left(\frac{x^{-\alpha+1}}{-\alpha+1}\right) = x^{-\alpha}$ for $\alpha \neq 1$ we get
(1) for $\alpha < 1$ and
(2) for $\alpha > 1$.

---

[9]See Section 10.4 in the Appendix for the proof of $\int_0^{+\infty} \frac{\sin x}{x} dx = \pi/2$.

(3) We have

$$x_n = \sum_{1 \le k \le n} \left(\frac{1}{k} - \int_k^{k+1} \frac{dt}{t} \right) + \int_n^{n+1} \frac{dt}{t} = \sum_{1 \le k \le n} \int_k^{k+1} \frac{t-k}{kt} dt + \ln\left(1 + \frac{1}{n}\right).$$

For $1 \le k$, we have

$$0 \le \int_k^{k+1} \frac{t-k}{kt} dt \le \int_k^{k+1} \frac{1}{kt} dt \le \int_k^{k+1} \frac{1}{k^2} dt = \frac{1}{k^2},$$

so that the series with general term $\int_k^{k+1} \frac{t-k}{kt} dt$ converges. Since $\lim_{n \to \infty} \ln(1 + \frac{1}{n}) = 0$, we get convergence for the sequence (x_n). This implies that, with $\gamma = \lim x_n$,

$$\sum_{1 \le k \le n} \frac{1}{k} = \ln n + \gamma + \varepsilon_n, \quad \lim \varepsilon_n = 0.$$

(4) The function $\sin x / x$ is continuous on $\mathbb{R}$, takes the value 1 at $x = 0$. For $A \ge \pi/2$,

$$I(A) = \int_{\pi/2}^A \frac{\sin t}{t} dt = \left[\frac{-\cos t}{t} \right]_{\pi/2}^A - \int_{\pi/2}^A \frac{\cos t}{t^2} dt = -A^{-1} \cos A - \int_{\pi/2}^A \frac{\cos t}{t^2} dt.$$

Since $|\cos A| \le 1$ and $|t^{-2} \cos t| \le t^{-2}$, the rhs converges for $A \to +\infty$.
(5) For $A \ge 1$,

$$\ln A = \int_1^A \frac{dx}{x} = \int_1^A \frac{\cos(2x)}{x} dx + \int_1^A \frac{2\sin^2 x}{x} dx$$
$$\le \int_1^A \frac{\cos(2x)}{x} dx + 2 \int_1^A \frac{|\sin x|}{x} dx,$$

and the rhs goes to $+\infty$ with A. Since we can prove as in (3) that $\int_1^A x^{-1} \cos(2x) dx$ has a finite limit when $A \to +\infty$, we get the result.

N.B. The limit of the sequence (x_n) is the so-called *Euler constant*, denoted by the letter γ. An approximate value is

$$0.5772156649015328606065120900824024310421593359399235988057672348848677726777664$$

This important constant remains quite mysterious and it is not even known whether it is an algebraic number. For more mathematical details, see `http://mathworld.wolfram.com/Euler-MascheroniConstant.html`. To know the first 100 digits type with Mathematica `N[EulerGamma, 100]`.

Exercise 2.8.21.

(1) *Calculate* $\displaystyle\lim_{n\to\infty}\int_0^n \left(1-\frac{x}{n}\right)^n dx.$

(2) *Let $z\in\mathbb{C}$ such that $\operatorname{Re}z>0$. Show that*

$$\lim_{n\to\infty}\int_0^n x^{z-1}\left(1-\frac{x}{n}\right)^n dx = \int_0^{+\infty} x^{z-1}e^{-x}dx = \Gamma(z).$$

Answer. For $x\geq 0$, we have $\lim_{n\to+\infty}(1-\frac{x}{n})^n = e^{-x}$ (calculate the logarithm). Also for all $\theta>-1$, we have $\ln(1+\theta)\leq\theta$ ($\theta\mapsto\theta-\ln(1+\theta)$ is decreasing on $(-1,0]$, increasing on $[0,+\infty)$) so that for $0\leq x<n$, we have

$$\ln\left(1-\frac{x}{n}\right)\leq -\frac{x}{n}\quad\text{and thus}\quad 0\leq \mathbf{1}_{[0,n]}(x)\left(1-\frac{x}{n}\right)^n\leq e^{-x}.$$

We can use the Lebesgue dominated convergence theorem for both questions; the answer for (1) is $1=\Gamma(1)$.

Exercise 2.8.22. *Give an example of a sequence $(f_n)_{n\in\mathbb{N}}$ in $C_c^0([0,1],\mathbb{R}_+)$ converging pointwise to 0 such that*

$$\int_0^1 f_n(x)dx \to +\infty.$$

Answer. Piecewise affine f_n equal to n^2 at $1/n$, 0 at $0, 2/n$.

Exercise 2.8.23. *Find a sequence of step functions $f_n:[0,1]\longrightarrow\mathbb{R}_+$ such that $\lim_{n\to\infty}\int_0^1 f_n(x)dx=0$ and so that the sequence $(f_n(x))_{n\in\mathbb{N}}$ is divergent for any $x\in[0,1]$.*

Answer. For $0\leq k<m$ integers, we consider the function

$$F_{k,m}(x)=\mathbf{1}_{[\frac{k}{m},\frac{k+1}{m}[}(x)$$

and we set

$$f_0 = F_{0,1}$$
$$f_1 = F_{0,2},\quad f_2 = F_{1,2}$$
$$f_3 = F_{0,3},\quad f_4 = F_{1,3},\quad f_5 = F_{2,3}$$

$$\cdots$$

$$f_{\frac{m(m-1)}{2}} = F_{0,m},\ \cdots\ ,f_{\frac{m(m-1)}{2}+k}=F_{k,m},\ \cdots\ ,f_{\frac{m(m-1)}{2}+m-1}=F_{m-1,m}.$$

$$\cdots$$

A simple drawing will convince the reader that for a fixed x, the sequence $f_n(x)$ takes an infinite number of times the values 0 and 1, proving its divergence.

We prove this result formally below. We note that the sequence $\left(\frac{m(m-1)}{2}\right)_{m\geq 1}$ is strictly increasing, with value 0 for $m = 1$ and goes to $+\infty$. As a result, for all integer $n \geq 0$, there exists a unique integer $m_n \geq 1$ such that

$$\frac{m_n(m_n - 1)}{2} \leq n < \frac{m_n(m_n + 1)}{2}$$

so that

$$n = \frac{m_n(m_n - 1)}{2} + k_n, \quad \text{with } 0 \leq k_n < \frac{m_n{}^2}{2} = m_n.$$

We note $\lim_{n\to+\infty} m_n = +\infty$ since $m_n + 1 > \sqrt{2n}$. For $n \geq 0$, we set

$$f_n(x) - F_{k_n,m_n}(x).$$

We have

$$\int_0^1 f_n(x)dx = \int_0^1 F_{k_n,m_n}(x)dx = 1/m_n \longrightarrow 0 \quad \text{for } n \to +\infty.$$

Let $x \in [0, 1]$ be given. Let $n \geq 3$ be an integer such that $f_n(x) = 1$: then

$$\frac{k_n}{m_n} \leq x < \frac{1 + k_n}{m_n},$$

and if $k_n < m_n - 1$, we have

$$\frac{m_n(m_n - 1)}{2} \leq n < n+1 = \frac{m_n(m_n - 1)}{2} + k_n + 1$$

$$< \frac{m_n(m_n - 1)}{2} + m_n = \frac{m_n(m_n + 1)}{2},$$

so that $m_{n+1} = m_n$ and $f_{n+1}(x) = F_{1+k_n,m_n}(x) = 0$. If $f_n(x) = 1$ and $k_n = m_n-1$, we have $\frac{m_n-1}{m_n} \leq x < 1$ and

$$n+1 = \frac{m_n(m_n - 1)}{2} + m_n = \frac{m_n(m_n + 1)}{2}, \quad \text{so that} \quad m_{n+1} = 1+m_n, \ k_{n+1} = 0.$$

We get then

$$f_{n+1}(x) = F_{0,1+m_n}(x) = 0 \quad \text{since} \quad \frac{1}{1 + m_n} \leq \frac{m_n - 1}{m_n} \quad \text{since } m_n \geq 2.$$

As a result,

$$\text{for } n \geq 3, \quad f_n(x) = 1 \Longrightarrow f_{n+1}(x) = 0. \tag{2.8.18}$$

Moreover for $x \in [0, 1[$ and $n \geq 0$, we have $0 \leq m_n x < m_n$ so that

$$k = E(m_n x) \in \{0, \dots, m_n - 1\}.$$

We consider $n' = \frac{m_n(m_n-1)}{2} + k \geq \frac{m_n(m_n-1)}{2}$. We have

$$k \leq m_n x < 1 + k, \qquad \frac{k}{m_n} \leq x < \frac{1+k}{m_n}$$

so that

$$f_{n'}(x) = F_{k,m_n}(x) = 1, \tag{2.8.19}$$

implying that the sequence $f_n(x)$ takes the value 1 an infinite number of times. Since it takes also an infinite number of times the value 0 from (2.8.18), it cannot converge. We can also define $f_n(1) = (1 + (-1)^n)/2$. Using piecewise affine functions, it is possible to modify the above example so that the f_n are continuous.

Exercise 2.8.24. *Let $(X, \mathcal{M}, \mu)$ be a measure space where μ is a positive measure and let $f : X \to \mathbb{C}$ be a measurable function.*

(1) *Prove that if $f \in \mathcal{L}^1(\mu)$, then $\lim_n n\mu(\{|f| \geq n\}) = 0$. Is the converse true?*

(2) *Prove that if $f \in \mathcal{L}^1(\mu)$, then $\sum_{n \geq 1} \frac{1}{n^2} \int_{|f| \leq n} |f|^2 d\mu < +\infty$. Is the converse true?*

Answer. (1) We have

$$0 \leq n\mu(\{|f| \geq n\}) = \int_X n\mathbf{1}_{\{|f| \geq n\}} d\mu \leq \int_X |f|\mathbf{1}_{\{|f| \geq n\}} d\mu,$$

$$0 \leq g_n = |f|\mathbf{1}_{\{|f| \geq n\}} \leq |f| \in \mathcal{L}^1(\mu) \quad \text{and} \quad \lim_{n \to \infty} |f(x)|\mathbf{1}_{\{|f| \geq n\}}(x) = 0.$$

The Lebesgue dominated convergence theorem implies that

$$\lim_n \int_X |f|\mathbf{1}_{\{|f| \geq n\}} d\mu = 0$$

and the result. The converse is not true: the positive continuous function on $[0, e^{-1}]$ given by $g(x) = x \ln(x^{-1})$ has derivative $\ln(x^{-1}) - 1$ and is thus increasing on $[0, e^{-1}]$ from $g(0) = 0$ to $g(e^{-1}) = e^{-1}$. We have

$$\int_0^{e^{-1}} \frac{dx}{g(x)} = \int_0^{e^{-1}} \frac{dx}{x \ln(x^{-1})} = \int_e^{+\infty} \frac{du}{u \ln(u)} = \lim_{A \to +\infty} \ln(\ln A) = +\infty.$$

However for $n \geq 1$,

$$\left\{ x \in [0, e^{-1}], \frac{1}{g(x)} \geq n \right\} = \left\{ x \in [0, e^{-1}], g(x) \leq n^{-1} \right\} = [0, x_n]$$

where $x_n \in [0, e^{-1}]$ is characterized by $x_n \ln(x_n^{-1}) = g(x_n) = n^{-1}$, which implies

$$n\mu\left(\left\{ x \in [0, e^{-1}], \frac{1}{g(x)} \geq n \right\} \right) = nx_n = \frac{1}{|\ln x_n|} \longrightarrow 0,$$

since $x_n \longrightarrow 0_+$. Property in (1) can hold without f (here $1/g$) $\in \mathcal{L}^1$.

(2) For $f \in \mathcal{L}^1$, we have

$$\sum_{n \geq 1} \frac{1}{n^2} \int_{|f| \leq n} |f|^2 d\mu = \int \left(\sum_{n \geq 1} n^{-2} |f|^2 \mathbf{1}_{\{|f| \leq n\}} \right) d\mu.$$

With

$$F(x) = \sum_{n \geq 1} n^{-2} |f(x)|^2 \mathbf{1}_{\{|f| \leq n\}}(x) = \sum_{n \geq \max(|f(x)|, 1)} n^{-2} |f(x)|^2$$

$$= |f(x)|^2 \sum_{n \geq \max(|f(x)|, 1)} n^{-2}$$

using for $N \geq 1$, $\sum_{n \geq N} n^{-2} \leq \min\left(\frac{\pi^2}{6}, \frac{1}{N-1}\right)$, we get

$$0 \leq F(x) \leq \min\left(\frac{\pi^2}{6}, \frac{1}{\max(|f(x)|, 1) - 1}\right) |f(x)|^2 \leq \begin{cases} \frac{\pi^2}{6} |f(x)|^2 & \text{if } |f(x)| \leq 2, \\ \frac{|f(x)|^2}{|f(x)| - 1} & \text{if } |f(x)| > 2. \end{cases}$$

Since for $|f(x)| \leq 2$ we have $\frac{\pi^2}{6} |f(x)|^2 \leq \frac{\pi^2}{6} |f(x)||f(x)| \leq |f(x)| \frac{2\pi^2}{6} \leq 4|f(x)|$ and for $|f(x)| > 2$,

$$\frac{|f(x)|^2}{|f(x)| - 1} = \frac{|f(x)|}{|f(x)| - 1} |f(x)| \leq 2|f(x)|,$$

we get

$$0 \leq F(x) \leq 4|f(x)|,$$

proving the result. The converse is not true since with $f(x) - \frac{1}{x} \mathbf{1}_{[1, +\infty[}(x)$ (which is not in $\mathcal{L}^1$), we have nevertheless

$$\sum_{n \geq 1} \frac{1}{n^2} \int_{|f| \leq n} |f|^2 d\mu = \sum_{n \geq 1} \frac{1}{n^2} \int_{x \geq 1} \frac{1}{x^2} dx = \pi^2/6.$$

N.B. Looking at $F_f = \sum_{n \geq 1} \frac{1}{n^2} |f|^2 \mathbf{1}\{|f| \leq n\}$ we have

$$F_f = \mathbf{1}\{|f| \leq 1\} |f|^2 + \mathbf{1}\{|f| > 1\} |f|^2 \sum_{n \geq |f|} \frac{1}{n^2},$$

so that, with some positive constant C,

$$\mathbf{1}\{|f| \leq 1\} |f|^2 + C^{-1} \mathbf{1}\{|f| > 1\} |f| \leq F_f \leq \mathbf{1}\{|f| \leq 1\} |f|^2 + C \mathbf{1}\{|f| > 1\} |f|.$$

As a result, $F_f \in \mathcal{L}^1$ is equivalent to

$$\mathbf{1}\{|f| \leq 1\} |f|^2 \quad \text{and} \quad \mathbf{1}\{|f| > 1\} |f| \in \mathcal{L}^1.$$

When f belongs to $\mathcal{L}^1$, both conditions are satisfied. Conversely, we may have $F_f \in \mathcal{L}^1$ without $f \in \mathcal{L}^1$ since $\mathbf{1}\{|f| \leq 1\} |f|$ may fail to be integrable. However if $F_f \in \mathcal{L}^1$ and μ has finite total mass (i.e., $\mu(X) < +\infty$), we have $\mathbf{1}\{|f| \leq 1\} |f| \leq 1$ which is integrable, so that f is integrable.

Exercise 2.8.25. *What are the limits of the following sequences?*

$$u_n = \sum_{k \geq 1} \frac{n}{nk^2 + k + 1}, \quad v_n = \sum_{1 \leq k \leq 2n} \frac{n^2}{kn^2 + k^2}, \quad w_n = \sum_{1 \leq k \leq n^2} \frac{\sin k}{k^2} \left(\frac{k}{k+1} \right)^n.$$

Answer. For k integer ≥ 1, we have $\lim_{n \to +\infty} \frac{n}{nk^2 + k + 1} = \frac{1}{k^2}$ and moreover

$$\frac{n}{nk^2 + k + 1} = \frac{1}{k^2 + (k/n) + (1/n)} \leq \frac{1}{k^2} = F(k), \quad \sum_{k \geq 1} F(k) < \infty.$$

We apply the Lebesgue dominated convergence theorem on the measure space

$$(\mathbb{N}, \mu = \sum_{k \geq 1} \delta_k, \mathcal{P}(\mathbb{N}))$$

to the sequence $(f_n)_{n \geq 1}$ defined by $f_n(k) = \frac{n}{nk^2 + k + 1}$. This gives

$$u_n = \int_{\mathbb{N}} f_n d\mu \xrightarrow[n \to +\infty]{} \int_{\mathbb{N}} (\lim_n f_n) d\mu = \sum_{k \geq 1} \frac{1}{k^2} = \frac{\pi^2}{6}.$$

Using the same measure space and the sequence of positive functions $(g_n)_{n \geq 1}$ defined by

$$g_n(k) = \begin{cases} \frac{n^2}{kn^2 + k^2} & \text{for } 1 \leq k \leq 2n, \\ 0 & \text{otherwise}, \end{cases}$$

we get from Fatou's lemma,

$$+\infty = \sum_{k \geq 1} \frac{1}{k} = \int_{\mathbb{N}} \liminf_n g_n d\mu \leq \liminf_n \int_{\mathbb{N}} g_n d\mu = \liminf_n v_n,$$

so that $\lim_n v_n = +\infty$. With the same measure space and the sequence of functions $(h_n)_{n \geq 1}$ defined by

$$h_n(k) = \begin{cases} \frac{\sin k}{k^2} \left(\frac{k}{k+1} \right)^n & \text{if } 1 \leq k \leq n^2, \\ 0 & \text{otherwise}, \end{cases}$$

we note that $|h_n(k)| \leq F(k)$ where F is defined above. The Lebesgue dominated convergence theorem gives

$$\lim_n w_n = \int_{\mathbb{N}} (\lim_n h_n) d\mu = \sum_{k \geq 1} \lim_{n \to +\infty} \left(\frac{\sin k}{k^2} \left(\frac{k}{k+1} \right)^n \right) = \sum_{k \geq 1} 0 = 0.$$

The last point can be checked directly without using the Lebesgue dominated convergence theorem: for any integer $m \geq 1$, we have

$$|w_n| \leq \sum_{1 \leq k \leq m} \frac{1}{k^2} \left(\frac{m}{m+1} \right)^n + \sum_{k > m} \frac{1}{k^2} \leq \frac{\pi^2}{6} \left(\frac{m}{m+1} \right)^n + \sum_{k > m} \frac{1}{k^2},$$

and thus

$$\limsup_{n\to+\infty} |w_n| \le \sum_{k>m} \frac{1}{k^2}, \quad \text{so that} \quad \limsup_{n\to+\infty} |w_n| \le \inf_{m\ge 1} \sum_{k>m} \frac{1}{k^2} = 0$$

since $\sum_{k>m} \frac{1}{k^2}$ is the remainder of a converging series.

Exercise 2.8.26. *Determine the limits of the following sequences.*

$$I_n = \int_0^1 \frac{n}{1+x^2} \tanh\left(\frac{x}{n}\right) dx, \quad J_n = \int_0^1 \frac{ne^{-x}}{nx+1} dx.$$

Answer. Setting $f_n(x) = \frac{n}{1+x^2} \tanh\left(\frac{x}{n}\right)$, we find

$$\lim_{n\to+\infty} f_n(x) = \frac{x}{1+x^2} \quad \text{and} \quad |f_n(x)| \le \frac{x}{1+x^2} \sup_{\alpha>0}\left(\frac{\tanh\alpha}{\alpha}\right).$$

The Lebesgue dominated convergence theorem implies

$$\lim_{n\to+\infty} I_n = \int_0^1 \frac{x}{1+x^2} dx = \frac{1}{2}\left[\ln(1+x^2)\right]_0^1 = \frac{\ln 2}{2}.$$

We have $J_n = \int_0^1 e^{-x}\left(x + \frac{1}{n}\right)^{-1} dx$ and Fatou's lemma gives, with

$$g_n(x) = e^{-x}\left(x + \frac{1}{n}\right)^{-1},$$

$$+\infty = \int_0^1 e^{-x}x^{-1}dx = \int_0^1 \left(\liminf_n g_n(x)\right)dx \le \liminf_n \int_0^1 g_n(x)dx = \liminf_n J_n.$$

Chapter 3

Spaces of Integrable Functions

3.1 Convexity inequalities (Jensen, Hölder, Minkowski)

Definition 3.1.1 (Convex function of one real variable). Let I be an interval of $\mathbb{R}$. A function $\phi : I \to \mathbb{R}$ is said to be convex if for all $x_0, x_1 \in I$ and $\theta \in [0, 1]$, we have

$$\phi\big((1 - \theta)x_0 + \theta x_1\big) \leq (1 - \theta)\phi(x_0) + \theta\phi(x_1). \tag{3.1.1}$$

We note that $x_\theta = (1 - \theta)x_0 + \theta x_1$ ranges over the interval $[x_0, x_1]$ (or $[x_1, x_0]$) when θ ranges over $[0, 1]$ so that $x_\theta \in I$ and (3.1.1) makes sense. The function ϕ is said to be concave if $-\phi$ is convex.

The best explanation is encapsulated in Figure 3.1: a function is convex if the segments joining the points $(x_j, \phi(x_j)), j = 0, 1$ are above the curve of ϕ. In that picture, above the x-axis, we wrote only the y-coordinate of each point. Note also that on the vertical line $x = x_\theta$, the y-coordinate $(1 - \theta)\phi(x_0) + \theta\phi(x_1)$ can be calculated with the Thales theorem.

Proposition 3.1.2. *Let I be an interval of $\mathbb{R}$ and $\phi : I \to \mathbb{R}$ be a function.*

(1) *For ϕ differentiable, ϕ is convex iff ϕ' is increasing.*

(2) *For ϕ twice differentiable, ϕ is convex iff $\phi'' \geq 0$.*

(3) *If ϕ is convex, then ϕ is continuous on $\overset{\circ}{I}$.*

(4) *The function $x \mapsto e^x$ is convex on $\mathbb{R}$.*

Proof. Let us first give some equivalent properties to (3.1.1). A function $\phi : I \to \mathbb{R}$ is convex iff for all $x, y, z \in I$,

$$x_0 = x < y = x_\theta < x_1 = z \implies \phi(y) \leq \underbrace{\frac{z - y}{z - x}}_{1 - \theta} \phi(x) + \underbrace{\frac{y - x}{z - x}}_{\theta} \phi(z).$$

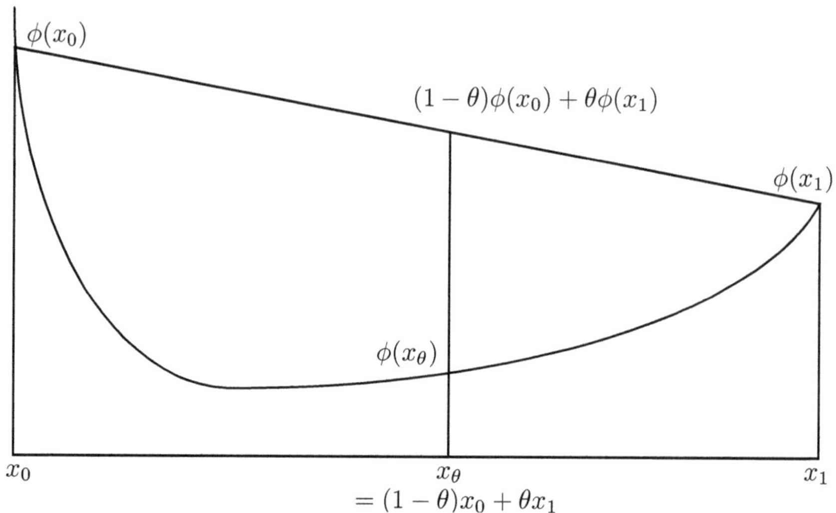

Figure 3.1: CONVEX FUNCTION

Property (3.1.1) is thus equivalent to the following: for all $x, y, z \in I$,

$$x < y < z \Longrightarrow \frac{\phi(y) - \phi(x)}{y - x} \le \frac{\phi(z) - \phi(x)}{z - x} \le \frac{\phi(z) - \phi(y)}{z - y}, \qquad (3.1.2)$$

since the first inequality is equivalent to $\phi(x_\theta) - \phi(x_0) \le \theta\big(\phi(x_1) - \phi(x_0)\big)$ and the second one to $(1 - \theta)\big(\phi(x_1) - \phi(x_0)\big) \le \phi(x_1) - \phi(x_\theta)$, both are equivalent to (3.1.1). Figure 3.2 describes (3.1.2). The lines XY, XZ, YZ through the points $X(x, \phi(x)), Y(y, \phi(y))$ and $Z(z, \phi(z))$ on the graph of ϕ have slopes increasing with lexicographic order: $XY \preccurlyeq XZ \preccurlyeq YZ$.

Let us prove (1). Let φ be a convex differentiable function on I and let $x_1 < x_2$ be points of I. For $0 < \epsilon < (x_2 - x_1)/2$, we have

$$x_1 < x_1 + \epsilon < x_2 - \epsilon < x_2.$$

Using inequalities (3.1.2) for the triples $x_1 < x_1 + \epsilon < x_2 - \epsilon$ and $x_1 + \epsilon < x_2 - \epsilon < x_2$, we get

$$\frac{\varphi(x_1 + \epsilon) - \varphi(x_1)}{\epsilon} \le \frac{\varphi(x_2 - \epsilon) - \varphi(x_1 + \epsilon)}{x_2 - x_1 - 2\epsilon} \le \frac{\varphi(x_2) - \varphi(x_2 - \epsilon)}{\epsilon},$$

so that, taking the limit when $\epsilon \to 0_+$, we obtain

$$\varphi'(x_1) \le \frac{\varphi(x_2) - \varphi(x_1)}{x_2 - x_1} \le \varphi'(x_2),$$

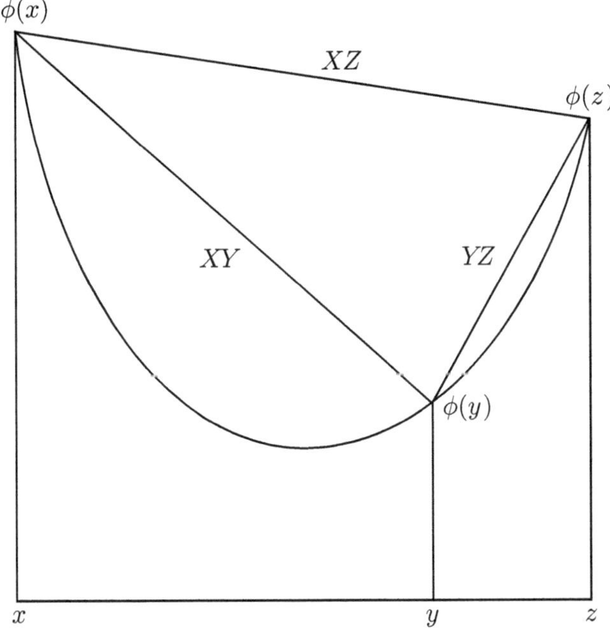

Figure 3.2: Description of (3.1.2)

proving that φ' is increasing. Conversely, let φ be a differentiable function on I, with an increasing derivative. For $x < y < z$ in I, there exists $\tilde{y} \in]x, y[$, $\tilde{z} \in]y, z[$ such that

$$\frac{\varphi(y) - \varphi(x)}{y - x} = \varphi'(\tilde{y}) \le \varphi'(\tilde{z}) = \frac{\varphi(z) - \varphi(y)}{z - y},$$

implying convexity for φ, completing the proof of (1). Property (2) follows from the equivalence, true for ψ differentiable on an interval I,

$$\psi \text{ increasing} \iff \psi' \ge 0.$$

Property (4) follows from (2). Let us prove (3). Let φ be a convex function defined on an interval I (with non-empty interior) and let $a < b$ be real numbers such that $[a, b] \subset \mathring{I}$. With $a < x_1 < x_2 < b$, applying (3.1.2), we find

$$\frac{\varphi(x_1) - \varphi(a)}{x_1 - a} \le \frac{\varphi(x_2) - \varphi(a)}{x_2 - a} \quad \text{and} \quad \frac{\varphi(b) - \varphi(x_1)}{b - x_1} \le \frac{\varphi(b) - \varphi(x_2)}{b - x_2}, \quad (3.1.3)$$

which implies

$$\frac{\varphi(x_1) - \varphi(a)}{x_1 - a}(x_2 - a) + \varphi(a) \le \varphi(x_2) \le \varphi(b) - (b - x_2)\frac{\varphi(b) - \varphi(x_1)}{b - x_1}.$$

Taking the limit when $x_2 \to x_{1+}$, we get

$$\varphi(x_1) = \varphi(x_1) - \varphi(a) + \varphi(a) \leq \liminf_{x_2 \to x_{1+}} \varphi(x_2) \leq \limsup_{x_2 \to x_{1+}} \varphi(x_2)$$

$$\leq \varphi(b) - \big(\varphi(b) - \varphi(x_1)\big) = \varphi(x_1),$$

implying

$$\lim_{x_2 \to x_{1+}} \varphi(x_2) = \varphi(x_1). \tag{3.1.4}$$

Similarly, from (3.1.3), we find

$$\varphi(b) - (b - x_1)\frac{\varphi(b) - \varphi(x_2)}{b - x_2} \leq \varphi(x_1) \leq (x_1 - a)\frac{\varphi(x_2) - \varphi(a)}{x_2 - a} + \varphi(a),$$

which implies

$$\varphi(x_2) = \varphi(b) - \big(\varphi(b) - \varphi(x_2)\big) \leq \liminf_{x_1 \to x_{2-}} \varphi(x_1) \leq \limsup_{x_1 \to x_{2-}} \varphi(x_1)$$

$$\leq \varphi(x_2) - \varphi(a) + \varphi(a) = \varphi(x_2),$$

so that $\lim_{x_1 \to x_{2-}} \varphi(x_1) = \varphi(x_2)$. The combination of left and right continuity ((3.1.4)) give the result. $\square$

Theorem 3.1.3 (Jensen inequality). *Let $(X, \mathcal{M}, \mu)$ be a probability space (measure space where μ is a positive measure such that $\mu(X) = 1$). Let I be a non-empty open interval of $\mathbb{R}$, $f : X \to I$ be a function in $\mathcal{L}^1(\mu)$ and let $\varphi : I \to \mathbb{R}$ be a convex function. Then $\varphi \circ f = \psi + g$, where $\psi \in \mathcal{L}^1(\mu)$ and g is measurable ≥ 0. Moreover $\int_X f d\mu \in I$ and*

$$\varphi\left(\int_X f d\mu\right) \leq \int_X (\varphi \circ f) d\mu,$$

with $\int_X (\varphi \circ f) \, d\mu = +\infty$ whenever $\int_X g d\mu = +\infty$.

Proof. We set $t_0 = \int_X f d\mu$ and $I = (a, b)$ where $-\infty \leq a < b \leq +\infty$. Let us prove first that $t_0 < b$: it is true whenever $b = +\infty$ since $f \in \mathcal{L}^1(\mu)$. If $b < +\infty$, since f is valued in I and μ is a probability measure, we have

$$t_0 = \int_X f d\mu \leq \int_X b d\mu = b\mu(X) = b.$$

If the equality $t_0 = b$ were satisfied, we would have $0 = \int_X (b - f)d\mu$, and since the function $b - f$ is non-negative and belongs to $\mathcal{L}^1(\mu)$, Proposition 1.7.1 implies $b = f$, μ-a.e., so at least in a point, which is not possible since f is valued in (a, b). We prove of course similarly that $t_0 > a$, so that $\int_X f d\mu \in I$. Using now the convexity of φ on I, we get

$$\beta = \sup_{\substack{s < t_0 \\ s \in I}} \frac{\varphi(t_0) - \varphi(s)}{t_0 - s} \leq \inf_{\substack{u > t_0 \\ u \in I}} \frac{\varphi(u) - \varphi(t_0)}{u - t_0} < +\infty. \tag{3.1.5}$$

As a consequence, we have

$$s \in I, s < t_0 \implies \varphi(t_0) - \varphi(s) \leq \beta(t_0 - s), \quad \text{i.e.,} \quad \varphi(s) \geq \varphi(t_0) - \beta(t_0 - s),$$

and moreover (3.1.5) implies

$$u \in I, u > t_0 \implies \varphi(u) - \varphi(t_0) \geq \beta(u - t_0), \quad \text{i.e.,} \quad \varphi(u) \geq \varphi(t_0) - \beta(t_0 - u),$$

so that $\forall \sigma \in I$, $\varphi(\sigma) \geq \varphi(t_0) - \beta(t_0 - \sigma)$. Since f is valued in I, we obtain

$$\forall x \in X, \quad \varphi(f(x)) \geq \varphi(t_0) - \beta(t_0 - f(x)),$$

entailing

$$\varphi \circ f = \underbrace{\varphi(t_0) - \beta(t_0 - f)}_{=\psi \in \mathcal{L}^1(\mu)} + \underbrace{\varphi \circ f - \varphi(t_0) + \beta(t_0 - f)}_{=g \text{ measurable} \geq 0},$$

since μ is a probability, $f \in \mathcal{L}^1(\mu)$ and $\varphi \circ f$ is measurable (φ is continuous from Proposition 3.1.2 and f is measurable). If g belongs to $\mathcal{L}^1(\mu)$, we find $\varphi \circ f \in \mathcal{L}^1(\mu)$ and

$$\int_X (\varphi \circ f) d\mu \geq \int_X (\varphi(t_0) - \beta(t_0 - f)) d\mu = \varphi(t_0) - \beta t_0 + \beta t_0 = \varphi\left(\int_X f d\mu\right).$$

If $\int_X g d\mu = +\infty$, with $0 \leq \psi_\pm \in \mathcal{L}^1(\mu)$, we have

$$\varphi \circ f + \psi_- = \psi_+ + g \geq 0 \implies \int_X (\varphi \circ f + \psi_-) d\mu = +\infty,$$

so that we may define $\int_X (\varphi \circ f) d\mu = +\infty$ in that case. $\qquad\square$

Remark 3.1.4. Let I be an interval of $\mathbb{R}$ and $\varphi : I \to \mathbb{R}$ be a convex function. Then for any integer $n \geq 1$ and any n-tuple $(\theta_1, \ldots, \theta_n)$ of non-negative real numbers such that $\sum_{1 \leq j \leq n} \theta_j = 1$, we have with $x_1, \ldots, x_n \in I$,

$$\varphi\left(\sum_{1 \leq j \leq n} \theta_j x_j\right) \leq \sum_{1 \leq j \leq n} \theta_j \varphi(x_j). \tag{3.1.6}$$

That property is equivalent to convexity (obviously stronger since (3.1.1) is (3.1.6) with $n = 2$): it follows from convexity as a consequence of Jensen's inequality applied to

$$X = \{1, \ldots, n\}, \quad \mu = \sum_{1 \leq j \leq n} \theta_j \delta_j, \quad \begin{array}{ccccc} X & \xrightarrow{f} & I & \xrightarrow{\varphi} & \mathbb{R} \\ j & \mapsto & x_j & \mapsto & \varphi(x_j) \end{array},$$

since Theorem 3.1.3 provides

$$\varphi\left(\sum_{1 \leq j \leq n} \theta_j x_j\right) = \varphi\left(\sum_{1 \leq j \leq n} \theta_j f(j)\right) = \varphi\left(\int_X f d\mu\right)$$

$$\leq \int_X (\varphi \circ f) d\mu = \sum_{1 \leq j \leq n} \theta_j (\varphi \circ f)(j) = \sum_{1 \leq j \leq n} \theta_j \varphi(x_j).$$

Note also that (3.1.6) is easily proven inductively on n for a convex function φ.

Lemma 3.1.5 (Geometric mean – Arithmetic mean inequality). *Let $a_1, \ldots, a_n$ be positive numbers and $\theta_1, \ldots, \theta_n$ be non-negative such that $\sum_{1 \leq j \leq n} \theta_j = 1$. Then*

$$\underbrace{\prod_{1 \leq j \leq n} a_j^{\theta_j}}_{\text{geometric mean of the } a_j} \quad \leq \quad \underbrace{\sum_{1 \leq j \leq n} \theta_j a_j}_{\text{arithmetic mean of the } a_j}, \qquad (3.1.7)$$

and equality holds iff $a_1 = \cdots = a_n$.

Proof. Using the previous remark along with the convexity of the exponential function, we find

$$\prod_{1 \leq j \leq n} a_j^{\theta_j} = \prod_{1 \leq j \leq n} e^{\theta_j \ln a_j} \leq \sum_{1 \leq j \leq n} \theta_j e^{\ln a_j} = \sum_{1 \leq j \leq n} \theta_j a_j.$$

Defining on $(\mathbb{R}_+^*)^n$ the function $\psi(a_1, \ldots, a_n) = \sum_{1 \leq j \leq n} \theta_j a_j - \prod_{1 \leq j \leq n} a_j^{\theta_j}$, we note that ψ is non-negative and we may assume that the numbers θ_j are all positive (if $\theta_j = 0$, the function ψ does not depend on a_j). If that smooth non-negative function is vanishing at some point of $(\mathbb{R}_+^*)^n$, then its differential should be 0. As a result, we have

$$0 = \frac{\partial \psi}{\partial a_j} = \theta_j - \theta_j a_j^{-1} \sum_{1 \leq k \leq n} a_k^{\theta_k} \implies a_j = \sum_{1 \leq k \leq n} a_k^{\theta_k},$$

since $\theta_j > 0$, proving the last result. $\qquad\qquad\square$

In the sequel to this book, we shall use the following notation: Let $1 < p < +\infty$ be a real number. We set $p' = \frac{p}{p-1}$ and we shall say that p' is the conjugate exponent of p, characterized by

$$\frac{1}{p} + \frac{1}{p'} = 1. \qquad (3.1.8)$$

When $p = 1$ (resp. $p = +\infty$) we define $p' = +\infty$ (resp. $p' = 1$).

Theorem 3.1.6 (Hölder & Minkowski inequalities). *Let $(X, \mathcal{M}, \mu)$ be a measure space where μ is a positive measure. Let $f, g : X \to \mathbb{C}$ be measurable functions, let $1 < p < +\infty$ and p' its conjugate exponent. Then,*

$$(1) \qquad \int_X |fg| d\mu \leq \left(\int_X |f|^p d\mu \right)^{1/p} \left(\int_X |g|^{p'} d\mu \right)^{1/p'} \qquad \text{(Hölder),}$$

$$(2) \qquad \left(\int_X |f + g|^p d\mu \right)^{1/p} \leq \left(\int_X |f|^p d\mu \right)^{1/p} + \left(\int_X |g|^p d\mu \right)^{1/p} \qquad \text{(Minkowski).}$$

Proof. We may assume that f, g are valued in $\mathbb{R}_+$. We can also suppose that $\int_X f^p d\mu > 0$ and $\int_X g^{p'} d\mu > 0$. Otherwise, from Proposition 1.7.1(1) we would have $f = 0$ μ-a.e. or $g = 0$ μ-a.e., so that $fg = 0$ μ-a.e., trivializing (1) since the lhs is 0. Also, we can assume that $\int_X f^p d\mu < +\infty$ and $\int_X g^{p'} d\mu < +\infty$: otherwise since these quantities are both positive, their product would be $+\infty$, trivializing (1) since the rhs is $+\infty$. Under these assumptions, we define

$$A = \left(\int_X f^p d\mu\right)^{1/p}, \quad B = \left(\int_X g^{p'} d\mu\right)^{1/p'} \quad \text{(we have } 0 < A, B < +\infty),$$

and

$$F = \frac{f}{A}, \quad G = \frac{g}{B} \quad \text{so that} \quad \int_X F^p d\mu = \int_X G^{p'} d\mu = 1.$$

From inequality (3.1.7), we get

$$FG = (F^p)^{1/p}(G^{p'})^{1/p'} \leq \frac{1}{p}F^p + \frac{1}{p'}G^{p'},$$

entailing

$$\int_X FG d\mu \leq \int_X \left(\frac{1}{p}F^p + \frac{1}{p'}G^{p'}\right) d\mu = 1, \quad \text{i.e.,} \quad \int_X fg d\mu \leq AB,$$

proving (1). Let us now prove (2), assuming as we may that f and g are non-negative such that $\int_X f^p d\mu$ and $\int_X g^p d\mu$ are finite. We have

$$(f + g)^p = f(f + g)^{p-1} + g(f + g)^{p-1},$$

and applying (1), we find

$$\int_X (f + g)^p d\mu \leq \left(\int_X f^p d\mu\right)^{1/p} \left(\int_X (f + g)^{(p-1)p'} d\mu\right)^{1/p'}$$
$$+ \left(\int_X g^p d\mu\right)^{1/p} \left(\int_X (f + g)^{(p-1)p'} d\mu\right)^{1/p'}.$$

Since $(p-1)p' = p$, we get

$$\int_X (f + g)^p d\mu \leq \left[\left(\int_X f^p d\mu\right)^{1/p} + \left(\int_X g^p d\mu\right)^{1/p}\right] \left(\int_X (f + g)^p d\mu\right)^{1/p'}.$$
$$(3.1.9)$$

The mapping $t \mapsto t^p$ from $\mathbb{R}_+$ into itself is convex since $p \geq 1$ (increasing derivative) and this implies $\left(\frac{f+g}{2}\right)^p \leq \frac{1}{2}f^p + \frac{1}{2}g^p$. As a result, the lhs of (3.1.9) is finite and we obtain the sought result

$$\left[\int_X (f + g)^p d\mu\right]^{1 - \frac{1}{p'} = \frac{1}{p}} \leq \left(\int_X f^p d\mu\right)^{1/p} + \left(\int_X g^p d\mu\right)^{1/p}. \qquad \square$$

3.2 L^p spaces

Definition 3.2.1. Let $(X, \mathcal{M}, \mu)$ be a measure space where μ is a positive measure and let $1 \leq p < +\infty$ be a real number. The space $\mathcal{L}^p(\mu)$ is the set of measurable functions $f : X \to \mathbb{C}$ such that

$$\int_X |f|^p d\mu < +\infty, \tag{3.2.1}$$

i.e., such that $|f|^p \in \mathcal{L}^1(\mu)$ (cf. Definition 1.6.6). As in Definition 1.7.2, we define $L^p(\mu) = \mathcal{L}^p(\mu)/\sim$ where $\sim$ stands for the equality μ-a.e. For $f \in \mathcal{L}^p(\mu)$, we set

$$\|f\|_{L^p(\mu)} = \left(\int_X |f|^p d\mu \right)^{1/p}. \tag{3.2.2}$$

Notation. We shall note $L^p(\mathbb{R}^d)$ the space $L^p(\lambda_d)$ where λ_d is the Lebesgue measure on $\mathbb{R}^d$ and $\ell^p(\mathbb{N})$ the space of complex-valued sequences $(a_k)_{k \in \mathbb{N}}$ such that $\sum_{k \in \mathbb{N}} |a_k|^p < +\infty$.

Lemma 3.2.2. *The quantity* (3.2.2) *depends only on the class of f in $\mathcal{L}^p(\mu)$ and $L^p(\mu)$ is a normed vector space for the norm* (3.2.2).

Proof of the lemma. We prove first that $\mathcal{L}^p(\mu)$ is a vector space on $\mathbb{C}$. Let $f, g : X \to \mathbb{C}$ be measurable functions and α, β be complex numbers. Minkowski's inequality implies for $f, g \in \mathcal{L}^p(\mu)$.

$$\|\alpha f + \beta g\|_{L^p(\mu)} \leq \|\alpha f\|_{L^p(\mu)} + \|\beta g\|_{L^p(\mu)} = |\alpha| \|f\|_{L^p(\mu)} + |\beta| \|g\|_{L^p(\mu)} < +\infty.$$

The space $L^p(\mu)$ is the quotient of $\mathcal{L}^p(\mu)$ by the subspace $\{f \in \mathcal{L}^p(\mu), f \sim 0\}$. Moreover (3.2.2) depends only on the class of f (cf. Proposition 1.7.1(1)) and is a norm on $L^p(\mu)$: The separation property follows from Proposition 1.7.1(1), homogeneity from Proposition 1.5.4(2) and triangle inequality from Theorem 3.1.6(2). $\qquad\square$

We want now to define the spaces $\mathcal{L}^\infty(\mu)$ and $L^\infty(\mu)$ of (essentially) bounded functions. Before we give such a definition, let us check the following example: we define

$$f : \mathbb{R} \to \mathbb{R}, \quad f(x) = \begin{cases} 1 & \text{for } x \notin \mathbb{Q}, \\ x & \text{for } x \in \mathbb{Q}, \end{cases}$$

easily seen to be measurable[1]. Although that function is not bounded, it is "essentially" bounded in the following sense: with λ_1 standing for the Lebesgue measure on $\mathbb{R}$, we have

$$\lambda_1(\{x \in \mathbb{R}, |f(x)| > 1\}) \leq \lambda_1(\mathbb{Q}) = 0.$$

---

[1] We have $f(x) = x \mathbf{1}_{\mathbb{Q}}(x) + \mathbf{1}_{\mathbb{Q}^c}(x)$, so that Theorem 1.2.7 implies the measurability of f.

Lemma 3.2.3. *Let $(X, \mathcal{M}, \mu)$ be a measure space where μ is a positive measure and let $f : X \to \mathbb{C}$ be a measurable mapping such that there exists $M \in \mathbb{R}_+$ such that*

$$\mu(\{x \in X, |f(x)| > M\}) = 0. \tag{3.2.3}$$

Then we shall say that f belongs to $\mathcal{L}^\infty(\mu)$. The set $\mathcal{L}^\infty(\mu)$ is a vector space on $\mathbb{C}$. The quantity

$$\|f\| = \inf\{M \in \mathbb{R}_+, \mu(\{|f| > M\}) = 0\}, \tag{3.2.4}$$

is a semi-norm on $\mathcal{L}^\infty(\mu)$ (i.e., satisfies homogeneity and triangle inequality). If f_1, f_2 belong to $\mathcal{L}^\infty(\mu)$ with $f_1 = f_2$ μ-a.e., then $\|f_1\| = \|f_2\|$.

Proof. We have

$$\forall k \geq 1, \ \mu\left(\left\{|f| > \frac{1}{k} + \|f\|\right\}\right) = 0,$$

and since $\{|f| > \|f\|\} = \cup_{k \geq 1}\{|f| > \frac{1}{k} + \|f\|\}$, we find (a countable union of negligible sets is negligible)

$$\mu(\{|f| > \|f\|\}) = 0. \tag{3.2.5}$$

Let f, g be in $\mathcal{L}^\infty(\mu)$. The inclusions

$$\{|f| \leq \|f\|\} \cap \{|g| \leq \|g\|\} \subset \{|f + g| \leq \|f\| + \|g\|\}$$

imply $\{|f| > \|f\|\} \cup \{|g| > \|g\|\} \supset \{|f + g| > \|f\| + \|g\|\}$, so that

$$\mu(\{|f + g| > \|f\| + \|g\|\}) = 0$$

and thus $f + g \in \mathcal{L}^\infty(\mu)$ along with $\|f + g\| \leq \|f\| + \|g\|$. Also for $\alpha \in \mathbb{C}$ and $f \in \mathcal{L}^\infty(\mu)$, we find readily $\alpha f \in \mathcal{L}^\infty(\mu)$ and $\|\alpha f\| = |\alpha| \|f\|$. To prove the last statement, we write with $N \in \mathcal{M}$, $\mu(N) = 0$, $|f_1|\mathbf{1}_{N^c} = |f_2|\mathbf{1}_{N^c}$ which implies for $M > 0$,

$$\mu(\{|f_1| > M\}) = \mu(\{|f_1|\mathbf{1}_{N^c} > M\}) = \mu(\{|f_2|\mathbf{1}_{N^c} > M\}) = \mu(\{|f_2| > M\}),$$

entailing $\|f_1\| = \|f_2\|$. $\qquad\qquad\square$

Definition 3.2.4. Let $(X, \mathcal{M}, \mu)$ be a measure space where μ is a positive measure. We define $L^\infty(\mu)$ as the quotient of $\mathcal{L}^\infty(\mu)$ by the relation of equality μ-a.e. For $f \in \mathcal{L}^\infty(\mu)$, we have

$$\|f\|_{L^\infty(\mu)} = \inf\{M \in \mathbb{R}_+, \mu(|f| > M) = 0\} := \operatorname{esssup} |f|. \tag{3.2.6}$$

This quantity depends only on the class of f in $\mathcal{L}^\infty(\mu)$ and $L^\infty(\mu)$ is a normed vector space for the norm (3.2.6). We shall denote by $L^\infty(\mathbb{R}^d)$ the space $L^\infty(\lambda_d)$ where λ_d is the Lebesgue measure on $\mathbb{R}^d$ and $\ell^\infty(\mathbb{N})$ the space of complex-valued sequences $(a_k)_{k \in \mathbb{N}}$ such that $\sup_{k \in \mathbb{N}} |a_k| < +\infty$.

Using the previous lemma, we have only to verify the separation axiom of the norm: if $\|f\| = 0$ for some $f \in \mathcal{L}^\infty(\mu)$, then for any $k \in \mathbb{N}^*$, we have

$$\mu(\{|f| > 1/k\}) = 0 \implies \mu(\{f \neq 0\}) = \mu(\cup_{k \geq 1}\{|f| > 1/k\}) = 0$$

so that (3.2.6) is a norm on the vector space $L^\infty(\mu)$.

Remark 3.2.5. Let f be in $L^\infty(\mu)$. We have

$$\|f\|_{L^\infty(\mu)} = \inf_{\substack{A \in \mathcal{M} \\ \mu(A^c) = 0}} \left(\sup_{x \in A} |f(x)| \right).$$

In fact if $f \in L^\infty(\mu), A \in \mathcal{M}, \mu(A^c) = 0$, we have $f \sim f\mathbf{1}_A$ and thus

$$\|f\|_{L^\infty(\mu)} = \|f\mathbf{1}_A\|_{L^\infty(\mu)} \leq \sup_{x \in A} |f(x)|.$$

Conversely if $f \in L^\infty(\mu)$, we saw that $\mu\left(\{|f| > \|f\|_{L^\infty(\mu)}\}\right) = 0$. Defining

$$A = \{|f| \leq \|f\|_{L^\infty(\mu)}\}$$

we find $\mu(A^c) = 0$ and $\|f\|_{L^\infty(\mu)} = \|f\mathbf{1}_A\|_{L^\infty(\mu)} \leq \sup_{x \in A} |f(x)| \leq \|f\|_{L^\infty(\mu)}$.

Proposition 3.2.6. *Let $(X, \mathcal{M}, \mu)$ be a measure space where μ is a positive measure. Let $1 \leq p, p' \leq +\infty$ be conjugate exponents (i.e., $\frac{1}{p} + \frac{1}{p'} = 1$), $f \in L^p(\mu)$ and $g \in L^{p'}(\mu)$. Then the product fg belongs to $L^1(\mu)$ and we have*

$$\|fg\|_{L^1(\mu)} \leq \|f\|_{L^p(\mu)}\|g\|_{L^{p'}(\mu)}.$$

Proof. For $1 < p < +\infty$, it is Hölder inequality (Theorem 3.1.6(1)). If $p = 1$, then $p' = +\infty$ and we have

$$|f(x)g(x)| \leq |f(x)|\|g\|_{L^\infty(\mu)} \quad \mu\text{-a.e.},$$

which gives the result by integration using Theorem 1.7.4. $\square$

Remark 3.2.7. Although the spaces $L^p(\mu)$ are quotients and its elements are classes of functions in $\mathcal{L}^p(\mu)$, we shall speak about *functions* of $L^p(\mu)$, keeping in mind that they could be modified on negligible sets.

Theorem 3.2.8. *Let $(X, \mathcal{M}, \mu)$ be a measure space where μ is a positive measure and let $p \in [1, +\infty]$. Then $L^p(\mu)$ is a Banach space (complete normed vector space) and $L^2(\mu)$ is a Hilbert space (complete pre-Hilbertian space).*

Proof. We assume first $1 \leq p < +\infty$ and consider a Cauchy sequence $(f_n)_{n \geq 1}$ in $L^p(\mu)$, i.e., such that

$$\forall \epsilon > 0, \ \exists N_\epsilon, \ \forall n, m \geq N_\epsilon, \ \|f_n - f_m\|_{L^p(\mu)} \leq \epsilon. \tag{3.2.7}$$

We claim that there exists a strictly increasing sequence of indices

$$1 \leq n_1 < n_2 < \cdots < n_k < n_{k+1} < \cdots \text{ such that } \|f_{n_{k+1}} - f_{n_k}\| \leq 2^{-k}. \quad (3.2.8)$$

In fact, using (3.2.7), we can find $n_1 \geq 1$ such that $\forall p \geq 0, \|f_{p+n_1} - f_{n_1}\| \leq 2^{-1}$. Let us assume that we have found $1 \leq n_1 < n_2 < \cdots < n_k$ such that

$$\forall p \geq 0, \forall j \in \{1, \ldots, k\}, \quad \|f_{p+n_j} - f_{n_j}\| \leq 2^{-j}. \quad (3.2.9)$$

From (3.2.7), we can find m_k such that $\forall p \geq 0, \forall m \geq m_k, \|f_{p+m} - f_m\| \leq 2^{-k-1}$. We define now

$$n_{k+1} = \max(1 + n_k, m_k),$$

and we check $n_k < n_{k+1}$ and $\forall p \geq 0, \|f_{p+n_{k+1}} - f_{n_{k+1}}\| \leq 2^{-k-1}$. This allows us to construct a strictly increasing sequence $(n_k)_{k \geq 1}$ satisfying (3.2.9) which implies Claim (3.2.8). For $k \geq 1$, we define now the non-negative measurable functions

$$g_k = \sum_{1 \leq j \leq k} |f_{n_{j+1}} - f_{n_j}|, \quad g = \sum_{j \geq 1} |f_{n_{j+1}} - f_{n_j}|. \quad (3.2.10)$$

Using (3.2.8) and the triangle inequality for the norm $L^p(\mu)$, we find

$$\|g_k\|_{L^p(\mu)} \leq \sum_{1 \leq j \leq k} \|f_{n_{j+1}} - f_{n_j}\|_{L^p(\mu)} \leq \sum_{1 \leq j \leq k} 2^{-j} \leq 1,$$

so that Fatou's lemma 1.6.4 implies

$$\int_X (|g|^p = \lim_k |g_k|^p = \liminf_k |g_k|^p) d\mu \leq \liminf_k \int_X |g_k|^p d\mu \leq 1,$$

proving $g \in L^p(\mu)$, $\|g\|_{L^p(\mu)} \leq 1$ and $0 \leq g < +\infty$ μ-a.e. (cf. Proposition 1.7.1 (4)). As a consequence, the telescopic series $\sum_{j \geq 1}(f_{n_{j+1}}(x) - f_{n_j}(x))$ is absolutely converging μ-a.e., i.e., on a measurable set A such that $\mu(A^c) = 0$. Let us define

$$f(x) = \left(f_{n_1}(x) + \sum_{j \geq 1}(f_{n_{j+1}}(x) - f_{n_j}(x))\right)\mathbf{1}_A(x).$$

Since $f_{n_1}(x) + \sum_{1 \leq j \leq k}(f_{n_{j+1}}(x) - f_{n_j}(x)) = f_{n_{k+1}}(x)$, we find

$$f(x) = \lim_k f_{n_k}(x), \quad \mu\text{-a.e.}$$

Let $\epsilon > 0$ be given and N_ϵ be an integer such that (3.2.7) is fulfilled. Fatou's lemma implies for $m \geq N_\epsilon$,

$$\int_X \left(|f - f_m|^p = \liminf_k |f_{n_k} - f_m|^p\right) d\mu \leq \liminf_k \int_X |f_{n_k} - f_m|^p d\mu \leq \epsilon^p.$$

As a result, $f - f_m$ belongs to $L^p(\mu)$ as well as $f = f - f_m + f_m$ and we have

$$\|f - f_m\|_{L^p(\mu)} \xrightarrow[m \to +\infty]{} 0, \quad \text{qed for } 1 \le p < +\infty.$$

In particular, $L^2(\mu)$ is complete for the norm

$$\|u\|_{L^2(\mu)} = \left(\int_X u\bar{u} d\mu \right)^{1/2}. \tag{3.2.11}$$

For $u, v \in L^2(\mu)$, Proposition 3.2.6 implies that $u\bar{v}$ belongs to $L^1(\mu)$ so that

$$L^2(\mu) \times L^2(\mu) \ni (u, v) \mapsto \int_X u\bar{v} d\mu = B(u, v) \tag{3.2.12}$$

is a sesquilinear Hermitian form, i.e., for $\lambda_1, \lambda_2 \in \mathbb{C}$, $u, v \in L^2(\mu)$,

$$B(\lambda_1 u_1 + \lambda_2 u_2, v) = \lambda_1 B(u_1, v) + \lambda_2 B(u_2, v), \quad \overline{B(v, u)} = B(u, v). \tag{3.2.13}$$

The vector space $L^2(\mu)$ equipped with the norm (3.2.11) is thus a Hilbert space. We need now to check the case $p = +\infty$. Let $(f_n)_{n \in \mathbb{N}}$ be a Cauchy sequence in $L^\infty(\mu)$. We define for $n, m \in \mathbb{N}$ the sets

$$A_n = \{x \in X, |f_n(x)| > \|f_n\|_{L^\infty(\mu)}\}, \tag{3.2.14}$$
$$B_{n,m} = \{x \in X, |f_n(x) - f_m(x)| > \|f_n - f_m\|_{L^\infty(\mu)}\}, \tag{3.2.15}$$

and we note that they are both negligible (from (3.2.5)). Let us define

$$E = \cup_{n \in \mathbb{N}} A_n \cup \cup_{k,l \in \mathbb{N}} B_{k,l}.$$

As a countable union of negligible sets, E is negligible and for $x \in E^c$, $n, m \in \mathbb{N}$,

$$|f_n(x) - f_m(x)| \le \|f_n - f_m\|_{L^\infty(\mu)}, \tag{3.2.16}$$
$$|f_n(x)| \le \|f_n\|_{L^\infty(\mu)} \le \sup_{\mathbb{N}} \|f_n\|_{L^\infty(\mu)} = M_0 < +\infty. \tag{3.2.17}$$

The very last inequality follows from the assumption (3.2.7) since the triangle inequality implies in a normed space

$$\|f_n\| \le \|f_n - f_m\| + \|f_m\|, \quad \|f_m\| \le \|f_n - f_m\| + \|f_n\|,$$

so that

$$\left| \|f_n\| - \|f_m\| \right| = \max\left(\|f_n\| - \|f_m\|, \|f_m\| - \|f_n\| \right) \le \|f_n - f_m\|, \tag{3.2.18}$$

proving that the sequence of real numbers $(\|f_n\|)_{n \in \mathbb{N}}$ is a Cauchy sequence, thus is bounded. For $x \in E^c$, the sequence of complex numbers $(f_n(x))_{n \in \mathbb{N}}$ is a Cauchy sequence, thus converging (with a limit $\le M_0$ in modulus). Let us now define

$$f(x) = \begin{cases} \lim_n f_n(x) & \text{for } x \in E^c, \\ 0 & \text{for } x \in E. \end{cases}$$

The function f belongs to $L^\infty(\mu)$ (note that f is measurable as a pointwise limit of the measurable $f_n \mathbf{1}_{E^c}$) and $\|f\|_{L^\infty(\mu)} \le M_0$. Moreover, using (3.2.7), for $\epsilon > 0$ and $n \ge N_\epsilon$ we have

$$|f_n(x) - f(x)|\mathbf{1}_{E^c}(x) = \lim_m |f_n(x) - f_m(x)|\mathbf{1}_{E^c}(x) \le \limsup_m \|f_n - f_m\|_{L^\infty(\mu)} \le \epsilon.$$

Since $\mu(E) = 0$, we find $\|f_n - f\|_{L^\infty(\mu)} \le \sup_{x \in E^c} |f_n(x) - f(x)|\mathbf{1}_{E^c}(x) \le \epsilon$, proving the convergence in $L^\infty(\mu)$ of the sequence $(f_n)_{n \in \mathbb{N}}$. The proof of Theorem 3.2.8 is complete. $\qquad\qquad\square$

Along the proof of the previous theorem, we have obtained the following result, which is of independent interest.

Lemma 3.2.9. *Let $(X, \mathcal{M}, \mu)$ be a measure space where μ is a positive measure, let $p \in [1, +\infty)$ and let $(f_n)_{n \in \mathbb{N}}$ be a convergent sequence in $L^p(\mu)$. Then there exists a subsequence $(f_{n_k})_{k \in \mathbb{N}}$ converging pointwise μ-a.e.*

We have seen in Exercise 2.8.23 that a sequence $(f_n)_{n \in \mathbb{N}}$ can be converging in $L^1(\mathbb{R})$ and nevertheless be such that for all $x \in \mathbb{R}$, the sequence $\big(f_n(x)\big)_{n \in \mathbb{N}}$ is divergent, proving that extracting a subsequence is necessary to get a.e. convergence from convergence in L^1.

The following theorem is an extension to L^p of Proposition 1.7.8.

Theorem 3.2.10. *Let $(X, \mathcal{M}, \mu)$ be a measure space where μ is a positive measure and let $p \in [1, +\infty)$. Let $f_n : X \to \mathbb{C}$ be a sequence of measurable functions converging μ-a.e. towards f.*

(1) *Let us assume that for all $n \in \mathbb{N}$, $f_n \in \mathcal{L}^p(\mu)$ and the numerical sequence $\|f_n\|_{L^p(\mu)}$ is bounded above. Then $f \in \mathcal{L}^p(\mu)$.*

(2) *We assume that $\lim_{n \to \infty} \|f_n\|_{L^p(\mu)} = \|f\|_{L^p(\mu)}$. Then the sequence $(f_n)_{n \in \mathbb{N}}$ converges in $L^p(\mu)$ towards f, i.e.,*

$$\lim_{n \to \infty} \int_X |f_n - f|^p d\mu = 0.$$

Proof. The function f is measurable as a simple limit of measurable functions. Moreover, Fatou's lemma implies

$$\int_X |f|^p d\mu = \int_X \liminf_n |f_n|^p d\mu \le \liminf_n \int_X |f_n|^p d\mu \le \sup_n \int_X |f_n|^p d\mu < +\infty,$$

which proves (1). Let us prove (2). We define

$$g_n = |f_n - f|^p - |f_n|^p + |f|^p, \tag{3.2.19}$$

and for a given $\lambda > 0$, we find

$$\int_X |g_n| d\mu = \underbrace{\int_X |g_n| \mathbf{1}_{\{|f_n| \le \lambda|f|\}} d\mu}_{\varepsilon_\lambda(n)} + \int_X |g_n| \mathbf{1}_{\{|f_n| > \lambda|f|\}} d\mu.$$

We note that $|g_n|\mathbf{1}_{\{|f_n|\le\lambda|f|\}}$ converges pointwise to 0 since g_n converges pointwise to 0. Moreover we have

$$\mathbf{1}_{\{|f_n|\le\lambda|f|\}}|g_n| \le \mathbf{1}_{\{|f_n|\le\lambda|f|\}}\left(|f_n|^p + |f|^p + \big||f_n| + |f|\big|^p\right)$$
$$\le |f|^p(\lambda^p + 1 + (\lambda+1)^p) \in \mathcal{L}^1(\mu).$$

Lebesgue's dominated convergence theorem implies then $\lim_n \varepsilon_\lambda(n) = 0$. Moreover, we have, noting that $f_n \ne 0$ on $\{|f_n| > \lambda|f|\}$,

$$|g_n|\mathbf{1}_{\{|f_n|>\lambda|f|\}} = \big||f_n - f|^p - |f_n|^p + |f|^p\big|\mathbf{1}_{\{|f_n|>\lambda|f|\}}$$
$$= |f_n|^p\big||1 - f/f_n|^p - 1 + |f/f_n|^p\big|\mathbf{1}_{\{|f_n|>\lambda|f|\}}.$$

For a complex number z such that $|z| < 1$, we have

$$\big||1 - z|^p - 1\big| \le p2^{p-1}\big||1 - z| - 1\big| \le p2^{p-1}|z| :$$

the first inequality comes from the mean value theorem for the function $t \mapsto t^p$ between 1 and $|1 - z|$, and the next one follows from the triangle inequalities

$$|1 - z| \le 1 + |z| \quad \text{and} \quad 1 \le |1 - z| + |z|.$$

As a result for $\lambda > 1$, we find

$$|g_n|\mathbf{1}_{\{|f_n|>\lambda|f|\}} \le |f_n|^p(1 + p2^{p-1})/\lambda,$$

which implies $\int_X |g_n|d\mu \le \varepsilon_n(\lambda) + \frac{1+p2^{p-1}}{\lambda}\int_X |f_n|^p d\mu$. Consequently, for all $\lambda > 1$, we get

$$\limsup_{n\to+\infty} \int_X |g_n|d\mu \le \frac{p2^{p-1}+1}{\lambda}\lim_{n\to+\infty}\int_X |f_n|^p d\mu,$$

implying $\lim_n \int_X |g_n|d\mu = 0$, and thus $\lim_n \int_X g_n d\mu = 0$. Going back to the definition of g_n in (3.2.19) we find now

$$0 = \lim_n \left(\int_X \left(|f_n - f|^p - |f_n|^p + |f|^p\right)d\mu\right)$$
$$= \lim_n \left(\int_X |f_n - f|^p d\mu - \int_X |f_n|^p d\mu + \int_X |f|^p d\mu\right).$$

Since we have assumed that $\lim_n \int_X |f_n|^p d\mu = \int_X |f|^p d\mu$, we obtain the result

$$\lim_n \int_X |f_n - f|^p d\mu = 0. \qquad \square$$

N.B. The statement of the previous theorem does not hold for $p = +\infty$: on the real line we may consider the L^∞ function $f = \mathbf{1}_{[-1,1]}$ which has norm 1. It is easy to find a sequence of continuous functions f_n with compact support in $[-2, 2]$

converging pointwise towards f with L^∞ norm equal to 1 with norm 1 (take f_n continuous piecewise affine, equal to 1 on $[-1, 1]$, equal to 0 on the complement of $(-1 - \frac{1}{n}, 1 + \frac{1}{n})$). However it is not possible that the sequence (f_n) converges in the L^∞ norm towards f, since the continuity of the (f_n) must be preserved by uniform limit, although f has discontinuity points.

Proposition 3.2.11. *Let $(X, \mathcal{M}, \mu)$ be a measure space where μ is a positive measure and let $1 \le p < +\infty$. We define*

$$S = \{s : X \to \mathbb{C}, measurable, \ s(X) \ finite \ with \ \mu(\{s \ne 0\}) < +\infty\}. \qquad (3.2.20)$$

The set S is dense in $L^p(\mu)$.

Proof. Let s be in S and let $\alpha_1, \ldots, \alpha_m$ be the distinct non-zero values taken by s. We have

$$s = \sum_{1 \le j \le m} \alpha_j \mathbf{1}_{A_j}, \quad A_j = s^{-1}(\{\alpha_j\}), \quad \mu(A_j) \le \mu(\{s \ne 0\}) < +\infty. \qquad (3.2.21)$$

Since the A_j are pairwise disjoint, we find

$$\int_X |s|^p d\mu = \sum_{1 \le j \le m} |\alpha_j|^p \mu(A_j) < +\infty, \qquad (3.2.22)$$

proving the inclusion $S \subset L^p(\mu)$. Let f be a positive function belonging to $L^p(\mu)$. Using the approximation Theorem 1.3.3, we find an increasing sequence of simple functions $(s_k)_{k \in \mathbb{N}}$ converging pointwise to f. Each s_k belongs to S since for s simple $\le f$, taking the distinct non-negative values $\alpha_1, \ldots, \alpha_m$ on the pairwise disjoint sets $A_1, \ldots A_m$, we have

$$s = \sum_{1 \le j \le m} \alpha_j \mathbf{1}_{A_j}, \quad \sum_{\substack{1 \le j \le m \\ \alpha_j > 0}} \alpha_j^p \mu(A_j) - \int_X s^p d\mu \le \int_X f^p d\mu < +\infty,$$

which implies $\mu(A_j) < +\infty$ whenever $\alpha_j > 0$ and thus

$$\mu(\{s \ne 0\}) = \sum_{\substack{1 \le j \le m \\ \alpha_j > 0}} \mu(A_j) < +\infty,$$

proving $s \in S$. Going back to the sequence $(s_k)_{k \in \mathbb{N}}$, we have

$$0 \le (f - s_k)^p = |f - s_k|^p \le f^p \in L^1(\mu) \text{ and } (f - s_k)^p \to 0 \text{ pointwise.}$$

Using Lebesgue's dominated convergence Theorem 1.6.8, this gives

$$\int_X |f - s_k|^p d\mu \to 0, \quad \text{i.e., } \lim_k \|f - s_k\|_{L^p(\mu)} = 0.$$

We conclude the proof by writing $f \in L^p(\mu)$ as

$$f = (\operatorname{Re} f)_+ - (\operatorname{Re} f)_- + i(\operatorname{Im} f)_+ - i(\operatorname{Im} f)_- . \tag{3.2.23}$$

$\square$

Remark 3.2.12. The previous proposition does not hold for $p = +\infty$ when $\mu(X) = +\infty$. For instance, the function 1 in $L^\infty(\mu)$ cannot be approximated in $L^\infty(\mu)$-norm by a function s which is 0 on the complement of a set A with finite measure: we have

$$\|1 - s\|_{L^\infty(\mu)} \ge \|(1 - s)\mathbf{1}_{A^c}\|_{L^\infty(\mu)} = \|\mathbf{1}_{A^c}\|_{L^\infty(\mu)} = 1,$$

since $\mu(A^c) = +\infty > 0$. However, when $p = +\infty$, we always have the following property.

Proposition 3.2.13. *Let $(X, \mathcal{M}, \mu)$ be a measure space where μ is a positive measure. We define*

$$S_\infty = \{s : X \to \mathbb{C}, \text{measurable}, s(X) \text{ finite}\}. \tag{3.2.24}$$

The set S_∞ is dense in $L^\infty(\mu)$. In particular, when $\mu(X) < +\infty$, we have $S_\infty = S$, where S is defined by (3.2.20), and Proposition 3.2.11 holds true in that case for $p = +\infty$.

Proof. Let $0 \le f \in L^\infty(\mu)$: we find N negligible such that $\tilde{f} = f\mathbf{1}_{N^c}$ is bounded non-negative. Theorem 1.3.3 implies that there exists an increasing sequence of simple functions $(s_k)_{k\in\mathbb{N}}$ converging uniformly towards $\tilde{f}$. Of course each s_k belongs to S_∞ and thus to $L^\infty(\mu)$ and we have

$$\|f - s_k\|_{L^\infty(\mu)} = \|\tilde{f} - s_k\|_{L^\infty(\mu)} \le \sup_{x\in X} |\tilde{f}(x) - s_k(x)| \xrightarrow[k\to+\infty]{} 0.$$

We conclude by decomposing f as in (3.2.23). $\square$

3.3   Integrals depending on a parameter

Continuity

Theorem 3.3.1. *Let $(X, \mathcal{M}, \mu)$ be a measure space where μ is a positive measure. Let Y be a metric space, let $y_0 \in Y$ and let $f : X \times Y \to \mathbb{C}$ be a mapping such that:*

(1) *For all $y \in Y$, the mapping $\begin{cases} X & \to & \mathbb{C} \\ x & \mapsto & f(x,y) \end{cases}$ belongs to $\mathcal{L}^1(\mu)$.*

(2) *The mapping $\begin{cases} Y & \to & \mathbb{C} \\ y & \mapsto & f(x,y) \end{cases}$ is continuous at y_0, μ-a.e. with respect to x.*

(3) *There exists a function $0 \le g \in \mathcal{L}^1(\mu)$ such that, μ-a.e. in $x \in X$, for all $y \in Y$, $|f(x,y)| \le g(x)$.*

Then the function F defined by

$$F(y) = \int_X f(x, y) d\mu(x) \tag{3.3.1}$$

is continuous at y_0. In particular, if the assumption (2) holds for all $y \in Y$, we find that F is continuous on Y.

Remark 3.3.2. Assumption (2) means that there exists $N \in \mathcal{M}$ such that $\mu(N) = 0$ so that for all $x \in N^c$, the mapping $y \mapsto f(x, y)$ is continuous at y_0. Assumption (3) means that there exists $N \in \mathcal{M}$ such that $\mu(N) = 0$ and

$$\sup_{y \in Y} |f(x, y)| \mathbf{1}_{N^c}(x) \in \mathcal{L}^1(\mu). \tag{3.3.2}$$

We note also that (1) allows us to define F by (3.3.1).

Proof. Let $(y_n)_{n \geq 1}$ be a sequence in Y converging towards y_0. We check

$$F(y_n) - F(y_0) = \int_X \underbrace{\left(f(x, y_n) - f(x, y_0) \right)}_{f_n(x)} d\mu(x).$$

Thanks to (2), the sequence $(f_n)_{n \geq 1}$ converges pointwise a.e. to 0; moreover (3) implies $|f_n| \leq 2g$, μ-a.e. We can apply Lebesgue's dominated convergence Theorem 1.7.5 entailing the sought result $\lim_{n \to +\infty} F(y_n) = F(y_0)$. $\qquad\square$

When the space Y is locally compact, the domination hypothesis (3) can be localized to any compact subset of Y.

Corollary 3.3.3. *Let $(X, \mathcal{M}, \mu)$ be a measure space where μ is a positive measure. Let Y be a locally compact metric space, and let $f : X \times Y \to \mathbb{C}$ such that (1) above is satisfied, as well as (2) for all y in Y. If for any compact subset K of Y, there exists a non-negative function $g_K \in \mathcal{L}^1(\mu)$ such that, μ-a.e. with respect to $x \in X$,*

$$\sup_{y \in K} |f(x, y)| \leq g_K(x), \tag{3.3.3}$$

then F defined by (3.3.1) is continuous on Y.

Proof. Since Y is locally compact, it is enough to check continuity for F restricted to any compact set, so we can apply the previous theorem. $\qquad\square$

Differentiability

Theorem 3.3.4. *Let $(X, \mathcal{M}, \mu)$ be a measure space where μ is a positive measure. Let Y be an open subset of $\mathbb{R}^m$, and $f : X \times Y \to \mathbb{C}$ be a mapping such that:*

(1) For all $y \in Y$, the mapping $\begin{cases} X & \to & \mathbb{C} \\ x & \mapsto & f(x, y) \end{cases}$ belongs to $\mathcal{L}^1(\mu)$.

(2) *For all $x \in X$, the mapping* $\begin{cases} Y & \to & \mathbb{C} \\ y & \mapsto & f(x,y) \end{cases}$ *is differentiable on Y.*

(3) *For any compact subset $K \subset Y$, there exists a non-negative function $g_K \in \mathcal{L}^1(\mu)$ such that, for all $x \in X$,*

$$\sup_{y \in K} \|d_y f(x,y)\| \le g_K(x). \tag{3.3.4}$$

Then the function F defined by (3.3.1) is differentiable on Y, $d_y f(\cdot, y) \in \mathcal{L}^1(\mu)$ and

$$dF(y) = \int_X d_y f(x,y)\, d\mu(x) \tag{3.3.5}$$

Remark 3.3.5. The differential $d_y f(x,y)$ is a vector in $\mathbb{C}^m$ (a complex-valued linear form on $\mathbb{R}^m$) whose Euclidean norm is taken in (3.3.4). For that vector, to belong to $\mathcal{L}^1(\mu)$ means that each component belongs to $\mathcal{L}^1(\mu)$. For all $T \in \mathbb{R}^m$, the mapping from X into $\mathbb{C}$, defined by $x \mapsto d_y f(x,y) \cdot T$ belongs to $\mathcal{L}^1(\mu)$: first of all,

$$d_y f(x,y) \cdot T = \lim_{k \to +\infty} k\big(f(x,y+T/k) - f(x,y)\big),$$

implying measurability, and also (3.3.4) gives

$$\int_X |d_y f(x,y) \cdot T| d\mu(x) < +\infty.$$

Proof. Let $y \in Y$ and $r > 0$ such that the closed ball $\bar{B}(y,r)$ is included in Y. For $h \in \mathbb{R}^m$ such that $\|h\| \le r$, we check

$$F(y+h) - F(y) = \int_X \big(f(x,y+h) - f(x,y)\big) d\mu(x)$$
$$= \int_X \Big[d_y f(x,y) \cdot h + \epsilon_{x,y}(h)\|h\|\Big] d\mu(x),$$

where we have

$$\lim_{h \to 0} \epsilon_{x,y}(h) = \epsilon_{x,y}(0) = 0. \tag{3.3.6}$$

Since the function $x \mapsto d_y f(x,y) \cdot h$ belongs to $\mathcal{L}^1(\mu)$, it is true also for $\epsilon_{x,y}(h)$ and we find

$$F(y+h) - F(y) = \int_X d_y f(x,y) \cdot h\, d\mu(x) + \int_X \epsilon_{x,y}(h) d\mu(x)\|h\|.$$

Using the mean value inequality, we get

$$|\epsilon_{x,y}(h)|\|h\| \le \sup_{\theta \in [0,1]} \|d_y f(x,y+\theta h)\|\,\|h\| + \|d_y f(x,y)\|\,\|h\|$$

so that from (3.3.4)

$$|\epsilon_{x,y}(h)| \leq 2 \sup_{z \in \bar{B}(y,r)} \|d_z f(x,z)\| \leq 2g_K(x) \in \mathcal{L}^1(\mu), \qquad (3.3.7)$$

with $K = \bar{B}(y,r)$. Inequalities (3.3.7) and (3.3.6) allow us to use Lebesgue's dominated convergence theorem to show that, for any sequence $(h_k)_{k \in \mathbb{N}}$ converging to 0 in $\mathbb{R}^m$, we have $\lim_{k \to +\infty} \int_X |\epsilon_{x,y}(h_k)| d\mu(x) = 0$. This implies

$$F(y+h) - F(y) = \int_X d_y f(x,y) \cdot h d\mu(x) + \eta_y(h)\|h\|,$$

with $\eta_y(h) = \int_X \epsilon_{x,y}(h) d\mu(x)$, and $\lim_{h \to 0} \eta_y(h) = 0$. We find thus that the mapping F is differentiable at any point $y \in Y$ with $dF(y) \cdot h = \int_X d_y f(x,y) \cdot h d\mu(x)$, concluding the proof. $\qquad \square$

Remark 3.3.6. It would be harmless of course to replace $\int_X f(x,y) d\mu(x)$ by $\int_{X \setminus N} f(x,y) d\mu(x)$ where N is negligible and thus to use a.e. assumptions. This is in fact a consequence of Theorem 3.3.4 where X could be replaced by $X \setminus N$. However, the situation here is slightly different from the a.e. assumption in Theorem 3.3.1: in the latter the hypothesis (3) is uniform with respect to $y \in Y$, as expressed by (3.3.2), whereas it is not the case for (1) when it is valid for all $y_0 \in Y$. In fact, in that case, (1) requires that for each $y_0 \in Y$, there exists a negligible set N (which could depend on y_0) so that, for all $x \in N^c$, the mapping $y \mapsto f(x,y)$ is continuous at y_0.

Holomorphy

Theorem 3.3.7. *Let $(X, \mathcal{M}, \mu)$ be a measure space where μ is a positive measure. Let U be an open subset of $\mathbb{C}$, and let $f : X \times U \to \mathbb{C}$ be a mapping satisfying the following properties.*

(1) *For all $z \in U$, the mapping $\begin{cases} X & \to & \mathbb{C} \\ x & \mapsto & f(x,z) \end{cases}$ belongs to $\mathcal{L}^1(\mu)$.*

(2) *For all $x \in X$, the mapping $\begin{cases} U & \to & \mathbb{C} \\ z & \mapsto & f(x,z) \end{cases}$ is holomorphic on U.*

(3) *For every compact subset K of U, there exists a non-negative function $g_K \in \mathcal{L}^1(\mu)$ such that for all $x \in X$,*

$$\sup_{z \in K} |f(x,z)| \leq g_K(x). \qquad (3.3.8)$$

Then the function F defined by (3.3.1) is holomorphic on U and for all $k \in \mathbb{N}$, the mapping

$$X \ni x \mapsto \frac{\partial^k f}{\partial z^k}(x,z) \in \mathbb{C}$$

belongs to $\mathcal{L}^1(\mu)$ and

$$F^{(k)}(z) = \int_X \frac{\partial^k f}{\partial z^k}(x, z)\,d\mu(x). \tag{3.3.9}$$

Remark 3.3.8. It is important to note that Assumption (3.3.8) is apparently very weak since we require only the local domination of the function itself, and not like in (3.3.4) a control of the derivative. In fact, the holomorphy assumption and Cauchy formula allow us to deduce from this some estimates for the derivatives. Generally speaking, the oscillations of holomorphic functions (e.g., the values of the derivatives) are controlled by the values of the functions. More precisely, the topology on $\mathcal{H}(U)$ (holomorphic functions on the open set U) is given by the countable family of semi-norms

$$\sup_{z \in K_j} |u(z)|, \quad K_j \text{ compact, such that } \cup_{j \in \mathbb{N}} K_j = U,$$

which makes $\mathcal{H}(U)$ a Fréchet space[2].

Proof. Let $z_0 \in U$ and $r_0 > 0$ such that the closed ball $K_0 = \bar{B}(z_0, r_0)$ is included in U. Let $(z_n)_{n \geq 1}$ be a sequence in $\bar{B}(z_0, r_0/2)\backslash\{z_0\}$ with limit z_0. With $z_n = z_0 + h_n$, let Γ_0 be the circle with center z_0 and radius r_0: we have, using Cauchy's formula

$$F(z_0 + h_n) - F(z_0) = \int_X \big[f(x, z_0 + h_n) - f(x, z_0)\big]\,d\mu(x)$$

$$= \int_X \frac{1}{2i\pi}\left[\oint_{\Gamma_0} \frac{f(x, \xi)}{\xi - z_0 - h_n}\,d\xi - \oint_{\Gamma_0} \frac{f(x, \xi)}{\xi - z_0}\,d\xi\right]\,d\mu(x)$$

$$= \int_X \frac{1}{2i\pi}\left[\oint_{\Gamma_0} \frac{f(x, \xi)}{\xi - z_0}\left(\frac{\xi - z_0}{\xi - z_0 - h_n} - 1\right)\,d\xi\right]\,d\mu(x)$$

$$= h_n \int_X \underbrace{\frac{1}{2i\pi}\left[\oint_{\Gamma_0} \frac{f(x, \xi)}{\xi - z_0}\frac{1}{\xi - z_0 - h_n}\,d\xi\right]}_{G_n(x)}\,d\mu(x).$$

We claim that for all $x \in X$,

$$\lim_{n \to +\infty} G_n(x) = \frac{1}{2i\pi}\oint_{\Gamma_0} \frac{f(x, \xi)}{(\xi - z_0)^2}\,d\xi = \frac{\partial f}{\partial z}(x, z_0). \tag{3.3.10}$$

---

[2]A Fréchet space is a complete metric vector space where the metric is given by a countable family of semi-norms $(p_j)_{j \in \mathbb{N}}$ (a semi-norm satisfies the properties of a norm – see (1.2.12) – except for the separation property); the family $(p_j)_{j \in \mathbb{N}}$ is assumed to be separating in the sense that $p_j(u) = 0$ for all $j \in \mathbb{N}$ implies that $u = 0$, and the metric is given by

$$d(u, v) = \sum_{j \geq 0} \frac{2^{-j} p_j(u - v)}{1 + p_j(u - v)}.$$

Indeed, for $\xi \in \Gamma_0$, we have $|\xi - z_0| = r_0$, $|\xi - z_0 - h_n| \geq |\xi - z_0| - |h_n| = r_0 - |h_n| \geq r_0/2$, which implies for all $x \in X$,

$$\frac{|f(x,\xi)|}{|\xi - z_0||\xi - z_0 - h_n|} \leq \frac{2|f(x,\xi)|}{r_0^2},$$

so that for $\xi = z_0 + r_0 e^{i\theta}$,

$$\frac{|ir_0 e^{i\theta}|}{|2i\pi|}\frac{|f(x, z_0 + r_0 e^{i\theta})|}{r_0|r_0 e^{i\theta} - h_n|} \leq \frac{1}{\pi}\frac{|f(x, z_0 + r_0 e^{i\theta})|}{r_0} = \Omega(\theta) \in L^1([0, 2\pi]). \quad (3.3.11)$$

Since for $\theta \in [0, 2\pi]$, we have

$$\lim_{n \to +\infty}\frac{ir_0 e^{i\theta}}{2i\pi}\frac{f(x, z_0 + r_0 e^{i\theta})}{r_0 e^{i\theta}(r_0 e^{i\theta} - h_n)} = \frac{ir_0 e^{i\theta}}{2i\pi}\frac{f(x, z_0 + r_0 e^{i\theta})}{r_0^2 e^{2i\theta}},$$

this implies from (3.3.11)

$$\begin{aligned}
\lim_{n \to +\infty} G_n(x) &= \lim_{n \to +\infty}\frac{1}{2i\pi}\oint_{\Gamma_0}\frac{f(x,\xi)}{(\xi - z_0)(\xi - z_0 - h_n)}d\xi \\
&= \lim_{n \to +\infty}\frac{1}{2i\pi}\int_0^{2\pi}\frac{f(x, z_0 + r_0 e^{i\theta})}{r_0 e^{i\theta}(r_0 e^{i\theta} - h_n)}ir_0 e^{i\theta}\,d\theta \\
&= \frac{1}{2i\pi}\int_0^{2\pi}\frac{f(x, z_0 + r_0 e^{i\theta})}{r_0^2 e^{2i\theta}}ir_0 e^{i\theta}\,d\theta \\
&= \frac{1}{2i\pi}\oint_{\Gamma_0}\frac{f(x,\xi)}{(\xi - z_0)^2}d\xi = \frac{\partial f}{\partial z}(x, z_0),
\end{aligned}$$

which proves Claim (3.3.10). Moreover, we have

$$|G_n(x)| \leq \frac{2\pi r_0}{2\pi}\frac{2}{r_0}\sup_{\xi \in \Gamma_0}|f(x,\xi)| \leq 2g_{K_0}(x) \in L^1(\mu).$$

Applying Lebesgue's dominated convergence to the sequence G_n, we find that the mapping $x \mapsto \frac{\partial f}{\partial z}(x, z_0)$ belongs to $L^1(\mu)$ and

$$\lim_{n \to +\infty} h_n^{-1}\big(F(z_0 + h_n) - F(z_0)\big) = \int_X \frac{\partial f}{\partial z}(x, z_0)d\mu(x),$$

for all $z_0 \in U$. We get then $(1), (2)$ for $\frac{\partial f}{\partial z}$ as well as (3) using Cauchy's formula. We conclude by a trivial induction argument. $\qquad\square$

Let us end this section with a couple of examples. In the first place, we consider the *Gamma function*, defined a priori on $H_0 = \{z \in \mathbb{C}, \operatorname{Re} z > 0\}$ by

$$\Gamma(z) = \int_0^{+\infty} t^{z-1}e^{-t}dt. \quad (3.3.12)$$

Thanks to Theorem 3.3.7, we prove that Γ is holomorphic on H_0, and is such that

$$\forall z \in H_0, \quad \Gamma(z+1) = z\Gamma(z), \tag{3.3.13}$$

a functional equation allowing us to extend Γ meromorphically to $\mathbb{C}$ with simple poles at $\{-k\}_{k\in\mathbb{N}}$ with residue $\frac{(-1)^k}{k!}$. We note that for $n \in \mathbb{N}$, we have $\Gamma(n+1) = n!$ as well as $\Gamma(1/2) = \sqrt{\pi}$.

The Zeta function is defined a priori on $H_1 = \{s \in \mathbb{C}, \operatorname{Re} s > 1\}$ by

$$\zeta(s) = \sum_{n \geq 1} \frac{1}{n^s}. \tag{3.3.14}$$

Theorem 3.3.7 implies that ζ is holomorphic on H_1. This function can be extended meromorphically to $\mathbb{C}$ with a single pole at 1 with residue 1. It can be proven also that for $\operatorname{Re} s > 1$,

$$\zeta(s) = \prod_{p\in\mathcal{P}} (1 - p^{-s})^{-1}, \tag{3.3.15}$$

where $\mathcal{P}$ stands for the sequence of prime numbers. Most notably, the distribution of prime numbers has an intimate connection with the location of the zeroes of the ζ function, as pointed out first by Riemann. In particular the Hadamard–de la Vallée-Poussin Theorem

$$\operatorname{card}\{p \in \mathcal{P}, p \leq x\} \overset{\text{def}}{=} \pi(x) \underset{x\to+\infty}{\sim} \frac{x}{\ln x}, \tag{3.3.16}$$

follows from the fact that the ζ function does not vanish on $\overline{H_1}$. The *Riemann hypothesis*, a most famous unsolved mathematical problem (November 2012 speaking) stated by Riemann in 1859, asserts that the non-real zeroes of the ζ function are located on the *critical line* $\{s \in \mathbb{C}, \operatorname{Re} s = \frac{1}{2}\}$. Another important function is the so-called function ξ, which is entire (i.e., holomorphic on $\mathbb{C}$), defined by

$$\xi(s) = \zeta(s)\Gamma(s/2)\pi^{-s/2}\frac{1}{2}s(s-1), \tag{3.3.17}$$

and which verifies the functional equation

$$\xi(s) = \xi(1-s). \tag{3.3.18}$$

The *Jacobi function* θ_J, is defined for $\operatorname{Re} z > 0$ by

$$\theta_J(z) = \sum_{n\in\mathbb{Z}} e^{-\pi n^2 z}. \tag{3.3.19}$$

Theorem 3.3.7 implies that θ_J is holomorphic on H_0. The Modular Property of θ_J is expressed as

$$\theta_J(1/z) = z^{1/2}\theta_J(z). \tag{3.3.20}$$

The *Beta function* is defined for $x, y \in H_0$ by

$$B(x, y) = \int_0^1 t^{x-1}(1 - t)^{y-1}dt, \tag{3.3.21}$$

and the following formula is easily proven:

$$B(x, y) = \frac{\Gamma(x)\Gamma(y)}{\Gamma(x + y)}. \tag{3.3.22}$$

Manifold other examples of applications of Theorem 3.3.7 occur in the mathematical literature and we refer the reader to the exercises sections as well as to our Appendix 10.5 for examples related to the Airy functions, Bessel functions, elliptic integrals, Fresnel integrals...

3.4 Continuous functions in L^p spaces

Theorem 3.4.1. *Let $1 \leq p < +\infty$ and let Ω be an open subset of $\mathbb{R}^m$. The space $C_c(\Omega)$ of complex-valued continuous compactly supported functions in Ω is dense in $L^p(\Omega)$.*

Proof. From Proposition 3.2.11, we know the density of S (see (3.2.20)) in $L^p(\Omega)$. Thus we need only to consider a Borel set $A \subset \Omega$ with finite measure and prove that we can approximate $\mathbf{1}_A$ in L^p-norm by a function of $C_c(\Omega)$.

Let $\epsilon > 0$ be given. From Theorem 2.2.14, we find a closed set F and an open set V of Ω such that

$$F \subset A \subset V, \quad \lambda_m(V \backslash F) < \epsilon^p/2^p, \tag{3.4.1}$$

which implies

$$\int_\Omega |\mathbf{1}_A - \mathbf{1}_V|^p d\lambda_m = \int_\Omega \mathbf{1}_{V\backslash A}^p d\lambda_m = \lambda_m(V \backslash A) < \epsilon^p/2^p. \tag{3.4.2}$$

Moreover we have

$$\lambda_m(V) = \lambda_m(A) + \lambda_m(V \backslash A) \leq \lambda_m(A) + \lambda_m(V \backslash F) \leq \lambda_m(A) + \epsilon^p/2^p < +\infty.$$

Using (2.4.3) in the proof of Theorem 2.4.2, we find $\chi \in C_c(V; [0, 1])$ such that

$$\lambda_m(V) - \epsilon^p/2^p < \int_\Omega \chi d\lambda_m \leq \lambda_m(V) = \sup_{\chi \in C_c(V; [0,1])} \int_\Omega \chi d\lambda_m < +\infty,$$

so that

$$\int_\Omega |\mathbf{1}_V - \chi|^p d\lambda_m = \int_V |1 - \chi|^p d\lambda_m \leq \int_V (1 - \chi)d\lambda_m = \lambda_m(V) - \int_V \chi d\lambda_m < \epsilon^p/2^p. \tag{3.4.3}$$

We get then from (3.4.2), (3.4.3) the inequality $\|\mathbf{1}_A - \chi\|_{L^p(\Omega)} < \epsilon$ and the result. $\qquad\square$

Remark 3.4.2. Of course Theorem 3.4.1 does not hold for $p = +\infty$ since for all $\chi \in C_c(\Omega)$, $\|\mathbf{1}_\Omega - \chi\|_{L^\infty(\Omega)} = 1$. On the other hand, thanks to Proposition 3.2.13, the space S_∞ is dense in $L^\infty(\Omega)$.

Theorem 3.4.3. *Let $1 \le p < +\infty$ and let Ω be an open set of $\mathbb{R}^m$. The space $C_c^\infty(\Omega)$ is dense in $L^p(\Omega)$.*

Proof. Let χ be in $C_c(\Omega)$ and $\rho_0 \in C_c^\infty(\mathbb{R}^d; \mathbb{R}_+)$, $\operatorname{supp} \rho_0 = \bar{B}(0,1)$, $\int_{\mathbb{R}^d} \rho_0(x)dx = 1$ (we may for instance consider the function ρ of Exercise 2.8.6 divided by its integral). For $\epsilon > 0$, we define

$$\chi_\epsilon(x) = \int_{\mathbb{R}^m} \rho_0\big((x - y)\epsilon^{-1}\big)\epsilon^{-m}\chi(y)dy. \tag{3.4.4}$$

Theorem 3.3.4 implies that χ_ϵ is a C^∞ function on $\mathbb{R}^m$. Moreover we have

$$\operatorname{supp} \chi_\epsilon \subset \operatorname{supp} \chi + \epsilon\bar{B}(0,1) \subset \Omega \quad \text{for } \epsilon \text{ small enough (cf. (2.1.4)).}$$

Using a dilation-translation change of coordinates in this integral of a compactly supported continuous function (see Lemma 2.4.5), we get

$$\chi_\epsilon(x) - \chi(x) = \int_{\mathbb{R}^m} \rho_0(z)\big(\chi(x + \epsilon z) - \chi(x)\big)dz$$

and since χ is uniformly continuous we find

$$|\chi_\epsilon(x) - \chi(x)| \le \sup_{|x_1 - x_2| \le \epsilon} |\chi(x_1) - \chi(x_2)| = \theta(\epsilon) \underset{\epsilon \to 0}{\to} 0,$$

so that $\int_{\mathbb{R}^d} |\chi_\epsilon(x) - \chi(x)|^p dx \le \theta(\epsilon)^p \lambda_m\big(\operatorname{supp} \chi + \epsilon\bar{B}(0,1)\big) \underset{\epsilon \to 0}{\to} 0.$ $\qquad\square$

Remark 3.4.4. For $1 \le p < +\infty$, the space $L^p(\Omega)$ is thus the completion of $C_c(\Omega)$ for the norm L^p. We could have defined $L^p(\Omega)$ using that completion argument, but we would have to manipulate classes of Cauchy sequences of continuous functions and this would be inelegant as well as complicated. Instead, we were able to realize L^p as a space of functions modulo the equality a.e. and it is much simpler this way. We shall see in Exercise 3.7.26 that the completion of $C_c(\mathbb{R}^m)$ for the L^∞ norm is not $L^\infty(\mathbb{R}^m)$ but $C_{(0)}(\mathbb{R}^m)$, the space of continuous functions going to 0 at infinity, i.e., continuous functions f on $\mathbb{R}^m$ such that

$$\lim_{R \to +\infty} \left\{ \sup_{|x| \ge R} |f(x)| \right\} = 0.$$

We shall end this chapter with an important consequence of Theorem 3.4.3.

Lemma 3.4.5 (Riemann–Lebesgue Lemma). *Let u be in $L^1(\mathbb{R}^m)$. We define*

$$\hat{u}(\xi) = \int_{\mathbb{R}^m} e^{-2i\pi x \cdot \xi} u(x)dx \qquad (\text{Fourier transform of } u). \tag{3.4.5}$$

Then we have $\hat{u}(\xi) \underset{|\xi| \to \infty}{\longrightarrow} 0$. *Moreover the function $\hat{u}$ is uniformly continuous on $\mathbb{R}^m$.*

Proof. We note first that (3.4.5) is meaningful as the integral of an L^1 function and we have also

$$\sup_{\xi \in \mathbb{R}^m} |\widehat{u}(\xi)| \leq \|u\|_{L^1(\mathbb{R}^m)}. \tag{3.4.6}$$

Let $\varphi \in C_c^\infty(\mathbb{R}^m)$. With $\alpha = (\alpha_1, \ldots, \alpha_m) \in \mathbb{N}^m$, we define

$$D^\alpha = D_1^{\alpha_1} \ldots D_m^{\alpha_m}, \quad D_j = \frac{1}{2i\pi} \frac{\partial}{\partial x_j}, \quad \xi^\alpha = \xi_1^{\alpha_1} \ldots \xi_m^{\alpha_m}. \tag{3.4.7}$$

Theorem 3.3.4 implies the identities

$$\xi_1 \widehat{\varphi}(\xi) = \widehat{D_1 \varphi}(\xi), \quad \widehat{D^\alpha \varphi}(\xi) = \xi^\alpha \widehat{\varphi}(\xi), \tag{3.4.8}$$

entailing $(1 + |\xi|^2) \widehat{\varphi}(\xi) = \text{Fourier}\left(\varphi + \sum_{1 \leq j \leq m} D_j^2 \varphi\right)$. We find thus

$$(1 + |\xi|^2) |\widehat{\varphi}(\xi)| \leq \|\varphi + \sum_{1 \leq j \leq m} D_j^2 \varphi\|_{L^1(\mathbb{R}^m)},$$

which implies $\lim_{|\xi| \to +\infty} \widehat{\varphi}(\xi) = 0$. For $u \in L^1(\mathbb{R}^m)$, we have

$$|\widehat{u}(\xi)| \leq |\widehat{(u - \varphi)}(\xi)| + |\widehat{\varphi}(\xi)| \leq \|u - \varphi\|_{L^1(\mathbb{R}^m)} + |\widehat{\varphi}(\xi)|,$$

so that for all $\varphi \in C_c^\infty(\mathbb{R}^m)$,

$$\limsup_{|\xi| \to \infty} |\widehat{u}(\xi)| \leq \|u - \varphi\|_{L^1(\mathbb{R}^m)} \implies \limsup_{|\xi| \to \infty} |\widehat{u}(\xi)| \leq \inf_{\varphi \in C_c^\infty(\mathbb{R}^m)} \|u - \varphi\|_{L^1(\mathbb{R}^m)} = 0.$$

We have also $\widehat{u}(\xi + \eta) - \widehat{u}(\xi) = \int_{\mathbb{R}^m} e^{-2i\pi x \cdot \xi}(e^{-2i\pi x \cdot \eta} - 1) u(x) dx$, so that

$$|\widehat{u}(\xi + \eta) - \widehat{u}(\xi)| \leq \int_{\mathbb{R}^m} |u(x)| \underbrace{|e^{-2i\pi x \cdot \eta} - 1|}_{\leq 2} dx,$$

and Lebesgue's dominated convergence theorem shows that, for all $\xi \in \mathbb{R}^m$,

$$\lim_{\eta \to 0} |\widehat{u}(\xi + \eta) - \widehat{u}(\xi)| = 0,$$

proving continuity, which is also a consequence of Theorem 3.3.1. We have also for $R > 1, |\eta| \leq 1$,

$$|\widehat{u}(\xi + \eta) - \widehat{u}(\xi)| \leq \sup_{|\xi| \leq R} |\widehat{u}(\xi + \eta) - \widehat{u}(\xi)| + 2 \sup_{|\xi| \geq R-1} |\widehat{u}(\xi)|$$

so that for $0 < \varepsilon < 1$, if ω_ρ is a modulus of continuity[3] of the continuous function $\widehat{u}$ on the compact set $\{|x| \leq \rho\}$

$$\sup_{|\eta| \leq \varepsilon, \xi \in \mathbb{R}^m} |\widehat{u}(\xi + \eta) - \widehat{u}(\xi)| \leq \omega_{R+1}(\varepsilon) + 2 \sup_{|\xi| \geq R-1} |\widehat{u}(\xi)|,$$

---

[3] For a continuous function v defined on a compact subset K of $\mathbb{R}^m$, the modulus of continuity ω is defined on $\mathbb{R}_+$ by $\omega(\rho) = \sup_{\substack{x,y \in K \\ |x-y| \leq \rho}} |v(x) - v(y)|$. We have $\lim_{\rho \to 0_+} \omega(\rho) = 0$.

proving that the lim sup of the lhs when ε goes to 0 is smaller than

$$2 \sup_{|\xi| \geq R-1} |\hat{u}(\xi)|, \quad \text{for all } R > 1.$$

Since that quantity is already proven to go to 0 when R goes to $+\infty$, we obtain the uniform continuity of $\hat{u}$. $\qquad\square$

The next result shows that, on a measure space X with finite measure, pointwise convergence of a sequence of (measurable) functions induces uniform convergence on a set with measure arbitrarily close to $\mu(X)$.

Theorem 3.4.6 (Egoroff's theorem). *Let $(X, \mathcal{M}, \mu)$ be a measure space where μ is a positive measure such that $\mu(X) < +\infty$. Let $f_n : X \to \mathbb{C}$ be a sequence of measurable functions converging pointwise towards a function f. Then for any $\epsilon > 0$, there exists $A_\epsilon \in \mathcal{M}$ with $\mu(A_\epsilon) < \epsilon$ and such that the sequence $(f_n)_{n \in \mathbb{N}}$ converges uniformly on $X \backslash A_\epsilon$.*

Proof. For $k \geq 1, n$, integers, we define

$$E_n^k = \cap_{p \geq n} \{ x \in X, |f_p(x) - f(x)| \leq 1/k \}.$$

Claim. For all $k \geq 1$, $X = \cup_{n \in \mathbb{N}} E_n^k$. In fact, for any $x \in X$, we have $\lim_m f_m(x) = f(x)$ so that for all $k \geq 1$, there exists an integer n such that for all $p \geq n$,

$$|f_p(x) - f(x)| \leq 1/k,$$

i.e., $x \in E_n^k$, proving the claim. We note also that $E_n^k \subset E_{n+1}^k$ and from Proposition 1.4.4(2), this gives $\lim_n \mu(E_n^k) = \mu(X)$. Since $\mu(X) < +\infty$, for all $\epsilon > 0$ and for all $k \geq 1$, there exists N_k such that

$$\forall n \geq N_k, \quad \mu(E_n^k) \geq \mu(X) - \epsilon 2^{-k}.$$

We may thus assume that there exists a sequence $(n_k)_{k \geq 1}$ strictly increasing such that

$$\mu(E_{n_k}^k) \geq \mu(X) - \epsilon 2^{-k}.$$

Indeed, we may define $n_k = k - 1 + \max_{1 \leq j \leq k} N_j$: we have then

$$N_k \leq n_k = k - 1 + \max_{1 \leq j \leq k} N_j \leq k - 1 + \max_{1 \leq j \leq k+1} N_j < k + \max_{1 \leq j \leq k+1} N_j = n_{k+1}.$$

Let $\epsilon > 0$ be given. We define $F = \cup_{k \geq 1} F_k$ with $F_k = \left(E_{n_k}^k \right)^c$. We have

$$\mu(F_k) = \mu(X) - \mu(E_{n_k}^k) \leq \epsilon 2^{-k}$$

and thus $\mu(F) \leq \sum_{k \geq 1} \mu(F_k) \leq \epsilon$. With $B = F^c$ and thus $\mu(B^c) \leq \epsilon$, we get

$$B = \cap_{k \geq 1} F_k^c = \cap_{k \geq 1} E_{n_k}^k,$$

providing $\sup_{x \in B} |f_n(x) - f(x)| \leq \sup_{x \in E_{n_k}^k} |f_n(x) - f(x)| \leq 1/k$ if $n \geq n_k$. The sequence $(\sup_{x \in B} |f_n(x) - f(x)|)_{n \in \mathbb{N}}$ is thus converging with limit 0. $\qquad\square$

Remark 3.4.7. The assumption $\mu(X) < +\infty$ is not dispensable. We consider the Lebesgue measure λ_1 on $\mathbb{R}$ and the sequence converging pointwise to 0 given by $f_n(x) = \mathbf{1}_{[0,1]}(x - n)$. If A is measurable with the Lebesgue measure $\leq 1/2$ and f_n converges uniformly on A^c, we must have

$$0 = \lim_n \big(\sup_{x \in A^c} \mathbf{1}_{[0,1]}(x - n)\big)$$

which implies $A^c \cap [n, n+1] = \emptyset$ for $n \geq N$, and thus

$$A \supset [n, n+1] \Longrightarrow \lambda_1(A) \geq 1, \quad \text{contradicting the assumption.}$$

Theorem 3.4.8 (Lusin's theorem)**.** *Let (X, d) be a locally compact metric space and let μ be a Borel measure on X such that Properties (1), (2), (3) in Theorem 2.2.14 are satisfied (this includes the case of the Lebesgue measure on $\mathbb{R}^m$).*

Let $f : X \to \mathbb{C}$ be a measurable function and let A be a measurable set such that $\mu(A) < +\infty$ and f vanishes on A^c. Let $\varepsilon > 0$ be given; then there exists $\phi \in C_c(X; \mathbb{C})$ such that

$$\mu(\{x \in X, f(x) \neq \phi(x)\}) < \varepsilon. \tag{3.4.9}$$

Proof. We assume first that $0 \leq f < 1$ and A is compact. We define s_k by (1.3.1) and we have $s_0 = 0$, $2E(2^{k-1}f) \leq E(2^k f) \leq 1 + 2E(2^{k-1}f)$ so that for $k \geq 1$,

$$2^k(s_k - s_{k-1}) = E(2^k f) - 2E(2^{k-1}f) \in \{0, 1\} = \mathbf{1}_{A_k},$$

and from Theorem 1.3.3, $f = \sum_{k \geq 1} 2^{-k}\mathbf{1}_{A_k}$. This implies in particular that $\cup_{k \geq 1} A_k \subset A$, since $x \in \cup_{k \geq 1} A_k \Longrightarrow f(x) > 0 \Longrightarrow x \notin A^c$. For each A_k and $\varepsilon_k > 0$ we can find

$$F_k \text{ closed} \subset A_k \subset V_k \text{ open}, \quad \mu(V_k \backslash F_k) < \varepsilon_k.$$

We may assume that A compact $\subset V_0$ open $\subset \overline{V_0}$ compact. Note that F_k is compact as a closed subset of the compact set A, so that we can find $\varphi_k \in C_c(V_k; [0, 1])$ with $\varphi_k = 1$ on F_k; we may also assume that $V_k \subset V_0$ since $A_k \subset A \subset V_0$. We set now $W = \cup_{k \geq 1}(V_k \backslash F_k)$, and choosing $\varepsilon_k = \varepsilon 2^{-k}$ for some positive number ε, we have

$$\mu(W) < \sum_{k \geq 1} \varepsilon_k \leq \varepsilon.$$

We define $\phi = \sum_{k \geq 1} 2^{-k}\varphi_k$, which belongs to $C_{\overline{V_0}}(X)$ and we have

$$\phi(x) - f(x) = \sum_{k \geq 1} 2^{-k} \underbrace{(\varphi_k(x) - \mathbf{1}_{A_k}(x))}_{=0 \text{ on } F_k \cup V_k^c} \Longrightarrow (\phi - f)\mathbf{1}_{W^c} = 0,$$

proving the result in that case and also in the case where f is bounded measurable and A compact.

Case f bounded and $\mu(A) < +\infty$. The inner regularity of μ implies that for any $\varepsilon > 0$, we can find K compact $\subset A$ such that

$$\mu(A) - \varepsilon < \mu(K) \le \mu(A) \implies \mu(A \backslash K) < \varepsilon.$$

We can find $\chi \in C_c(X)$ equal to 1 on K, supported in a neighborhood V of K such that $\mu(V \backslash K) < \varepsilon$ and $\overline{V}$ is compact. The function χf vanishes on $(\overline{V})^c$ and we may apply the previous result. Since $\chi = 1$ on K, we obtain the result in that case.

General case. We consider $B_n = \{x \in X, |f(x)| > n\} \subset A$ and we note that $\cap_n B_n = \emptyset$. This implies from Proposition 1.4.4(3) that $\lim_n \mu(B_n) = 0$. Since f coincides with the bounded function $\mathbf{1}_{B_n^c} f$, except on B_n, whose measure goes to 0, this gives the result. $\qquad\square$

3.5   On various notions of convergence

We collect in this section the various properties linked to the several convergence modes met in the text.

Definition 3.5.1. Let $(X, \mathcal{M}, \mu)$ be a measure space where μ is a positive measure and let $(f_n)_{n \in \mathbb{N}}$ be a sequence of measurable functions from X into $\mathbb{C}$.

(1) The sequence $(f_n)_{n \in \mathbb{N}}$ *converges almost everywhere* towards f if there exists $N \in \mathcal{M}$, such that $\mu(N) = 0$ and

$$\forall x \in N^c, \quad \lim_n f_n(x) = f(x).$$

(2) The sequence $(f_n)_{n \in \mathbb{N}}$ *converges in measure* towards f if

$$\forall \epsilon > 0, \quad \lim_n \mu\big(\{x \in X, |f_n(x) - f(x)| > \epsilon\}\big) = 0.$$

(3) The sequence $(f_n)_{n \in \mathbb{N}}$ *converges in the space* $L^1(\mu)$ towards $f \in L^1(\mu)$ if

$$\lim_n \|f_n - f\|_{L^1(\mu)} = 0.$$

(4) The sequence $(f_n)_{n \in \mathbb{N}}$ *satisfies the dominated convergence criterion* if **(1)** holds and if $g(x) = \sup_{n \in \mathbb{N}} |f_n(x)|$ is such that $g \in L^1(\mu)$.

Theorem 3.5.2. *Let $(X, \mathcal{M}, \mu)$ be a measure space where μ is a positive measure and let $(f_n)_{n \in \mathbb{N}}$ be a sequence of measurable functions from X into $\mathbb{C}$. With the notation of Definition 3.5.1, we have the following properties.*

(i) $(4) \implies (3) \cap (1)$.

(ii) $(3) \implies (2)$.

(iii) *(2) does not imply (1) in general, but it is true for a subsequence.*

(iv) *(3) does not imply (1) in general, but it is true for a subsequence.*

(v) *$(1) \implies (2)$ if $\mu(X) < +\infty$ and not in general without this condition.*

Proof. Assertion (i) is the Lebesgue dominated convergence Theorem 1.7.5. Statement (ii) follows from the inequality

$$\mu\big(\{x \in X, |f_n(x) - f(x)| > \epsilon\}\big) \le \int_X \frac{1}{\epsilon}|f_n - f|d\mu = \epsilon^{-1}\|f_n - f\|_{L^1(\mu)}.$$

The first part of (iii) follows from the example in Exercise 2.8.23 in which is displayed a sequence (f_n) of non-negative measurable functions converging in $L^1([0,1])$ towards 0 (thus in measure from the already proven (ii)) such that sequence $(f_n(x))_{n\in\mathbb{N}}$ diverges for every $x \in [0,1]$. Let us prove the second part of (iii): let (f_n) be a sequence converging to f in measure. This implies that

$$\forall k \ge 0, \exists N_k \in \mathbb{N}, \forall n \ge N_k, \quad \mu\big(\{|f_n - f| > 2^{-k}\}\big) < 2^{-k}.$$

Let us assume that we have found $N_0 < N_1 < \cdots < N_l$ such that the above property is true for $k = 0, \ldots, l$. Then using that

$$\lim_n \mu\big(\{|f_n - f| > 2^{-l-1}\}\big) = 0,$$

we may find $N_{l+1} > N_l$ such that $\mu\big(\{|f_n - f| > 2^{-l-1}\}\big) < 2^{-l-1}$ for $n \ge N_{l+1}$. Let us consider the subsequence $(f_{N_k})_{k\in\mathbb{N}}$. We define $E_k = \{x, |f_{N_k}(x) - f(x)| > 2^{-k}\}$. We know that $\mu(E_k) < 2^{-k}$ and $|f(x) - f_{N_k}(x)| \le 2^{-k}$ if $x \notin E_k$. Defining $F_m = \cup_{k > m} E_k$ we find that $\mu(F_m) \le 2^{-m}$ and moreover

$$\forall x \in F_m^c, \forall k > m, \quad |f_{N_k}(x) - f(x)| \le 2^{-k} \implies \forall x \in F_m^c, \lim_{k \to +\infty} f_{N_k}(x) = f(x).$$

The set $F = \cap_{m \ge 0} F_m$ has measure 0 and for each $x \in F^c = \cup_{m \ge 0} F_m^c$, we have $\lim_{k \to +\infty} f_{N_k}(x) = f(x)$, proving the sought result. The first part of statement (iv) follows from Exercise 2.8.23 and the second part from Lemma 3.2.9. The first part of statement (v) follows from (1) in Exercise 3.7.12 and the second part from Remark 3.7.13. $\qquad\square$

Theorem 3.5.3. *Let $(X, \mathcal{M}, \mu)$ be a measure space where μ is a positive measure and let $(f_n)_{n\in\mathbb{N}}$ be a sequence of measurable functions from X into $\mathbb{C}$. With the notation of Definition 3.5.1, we have the following properties.*

(j) *For $p \in [1, +\infty)$, (1) and $\lim_n \|f_n\|_{L^p(\mu)} = \|f\|_{L^p(\mu)}$ imply convergence in the space $L^p(\mu)$.*

(jj) *Local convergence in measure (see (2.8.14)) and domination ($\sup_n |f_n(x)| \in L^1(\mu)$) imply (3).*

Proof. Statement (j) is Theorem 3.2.10. Statement (jj) follows from Ex. 2.8.14. $\qquad\square$

3.6   Notes

Much more on the topic of convexity can be obtained from L. Hörmander's monograph, *Notions of Convexity* ([33]).

Let us follow alphabetically the names of mathematicians encountered in the text.

George AIRY (1801–1892) was an English mathematician and astronomer. The intensity of light near a caustic was the initial reason for his invention of the now called Airy function.

Friedrich BESSEL (1784–1846) was a German mathematician, astronomer.

Dmitri EGOROFF (1869–1931) was a Russian mathematician.

Augustin FRESNEL (1788–1827) was a French engineer who contributed significantly to the establishment of the theory of wave optics.

Jacques HADAMARD (1865–1963) was a French mathematician of extraordinary breadth and depth. He proved the Prime Number Theory at the same time as Charles DE LA VALLÉE-POUSSIN (1866–1962).

Otto HÖLDER (1859–1937), a German mathematician who proved his inequality in 1884.

Carl Gustav JACOBI (1804–1851) was a German mathematician, creator of the theory of elliptic functions.

Johan JENSEN (1859–1925) was a Danish mathematician, who proved in 1906 the fundamental inequality bearing his name.

Nikolai LUSIN (1883–1950) was a Russian mathematician, a Ph.D. student of Dmitri Egoroff.

Hermann MINKOWSKI (1864–1909) was a professor at the university of Göttingen. He also taught in Zürich where Albert Einstein attended his lectures.

THALES OF MILETUS lived from 624 BC to 547 BC. Miletus is a city in Asia Minor (now located in Turkey). Thales seems to be the first known Greek philosopher as well as a scientist, mathematician and a professional engineer. Thales' theorem is now in fact one of the axioms in the definition of vector spaces:

$$\lambda \cdot (x + y) = \lambda \cdot x + \lambda \cdot y$$

where λ is a scalar (e.g., a real number for real vector spaces) and x, y are vectors.

3.7 Exercises

Exercise 3.7.1. *Let $(a_j)_{1\leq j\leq n}$ and $(\theta_j)_{1\leq j\leq n}$ be as in Lemma 3.1.5. Prove the harmonic mean – geometric mean – arithmetic mean inequality*

$$\left(\sum_{1\leq j\leq n}\theta_j a_j^{-1}\right)^{-1}\leq\prod_{1\leq j\leq n}a_j^{\theta_j}\leq\sum_{1\leq j\leq n}\theta_j a_j,$$

and also that, if any of the inequalities above is an equality, we have $a_1=\cdots=a_n$.

Answer. The second inequality is proven in Lemma 3.1.5; also proven there is the fact that the equality holds iff all a_j are equal. With $b_j=a_j^{-1}$, the first inequality is equivalent to the second one, completing the answer.

N.B. The above inequality will be called **HGA** inequality and the second one **GA**.

Exercise 3.7.2 (Logarithmic convexity). *Let $f:I\longrightarrow\mathbb{R}_+^*$ be a function defined on an interval I of the real line. The function f is said to be log-convex when $\ln f$ is a convex function.*

(1) *Prove that a log-convex function is convex.*

(2) *Give an example of a convex function valued in $\mathbb{R}_+^*$ which is not log-convex.*

(3) *Prove that the Γ function is log-convex on $\mathbb{R}_+$.*

(4) *Prove that the Gamma function is the only positive valued function f defined on $\mathbb{R}_+^*$ such that*

- $f(1)=1$,
- $\forall x>0,\ f(x+1)=xf(x)$,
- f *is log-convex.*

Answer. (1) In the case where f is twice differentiable and log-convex, we have with ϕ convex twice differentiable

$$f=e^{\phi},\quad f'=e^{\phi}\phi',\quad f''=e^{\phi}\left(\phi'^2+\phi''\right)\geq 0,\tag{3.7.1}$$

implying convexity for f. Without the assumption of differentiability, we find with $x_0,x_1\in I,\theta\in(0,1),x_\theta=(1-\theta)x_0+\theta x_1$,

$$f(x_\theta)=e^{\phi(x_\theta)}\leq e^{(1-\theta)\phi(x_0)+\theta\phi(x_1)}\underbrace{\leq}_{\textsf{GA inequality}}(1-\theta)e^{\phi(x_0)}+\theta e^{\phi(x_1)},$$

proving convexity for f.

(2) The function $\mathbb{R} \ni x \mapsto x^2 + 1 \in \mathbb{R}_+^*$ is obviously convex but not log-convex. With $\phi(x) = \ln(1 + x^2)$, we have

$$\phi'(x) = \frac{2x}{1 + x^2}, \quad \phi''(x) = \frac{2(1 + x^2) - 2x2x}{(1 + x^2)^2} = \frac{2(1 - x^2)}{(1 + x^2)^2},$$

and since ϕ'' takes negative values, Proposition 3.1.2(2) implies that ϕ is not convex.

(3, 4) See Lemmas 10.5.4 and 10.5.5 in Section 10.5.

Exercise 3.7.3 (Hermite–Hadamard inequality). *Let $a < b \in \mathbb{R}$ and let $\phi : [a, b] \to \mathbb{R}$ be a convex function. Prove that*

$$\phi\left(\frac{a + b}{2}\right) \le \frac{1}{b - a} \int_a^b \phi(t)dt \le \frac{\phi(a) + \phi(b)}{2}. \tag{3.7.2}$$

Answer. Using an affine rescaling, we may assume that $[a, b] = [0, 1]$. We have for $\theta \in [0, 1/2]$,

$$\phi\left(\frac{1}{2}\right) = \phi\left(\frac{1}{2}\left(\frac{1}{2} - \theta\right) + \frac{1}{2}\left(\frac{1}{2} + \theta\right)\right) \le \frac{1}{2}\phi\left(\frac{1}{2} - \theta\right) + \frac{1}{2}\phi\left(\frac{1}{2} + \theta\right),$$

so that, integrating for $\theta \in [0, 1/2]$, we get

$$\frac{1}{2}\phi\left(\frac{1}{2}\right) \le \frac{1}{2}\int_0^{1/2} \phi\left(\frac{1}{2} - \theta\right) d\theta + \int_0^{1/2} \frac{1}{2}\phi\left(\frac{1}{2} + \theta\right) d\theta$$

$$= \frac{1}{2}\int_0^{1/2} \phi(t)dt + \frac{1}{2}\int_{1/2}^1 \phi(t)dt = \frac{1}{2}\int_0^1 \phi(t)dt,$$

which is the first inequality. On the other hand, for $t \in [0, 1]$, we have

$$\phi(t) = \phi((1 - t)0 + t1) \le (1 - t)\phi(0) + t\phi(1),$$

so that, integrating for $t \in [0, 1]$, we get

$$\int_0^1 \phi(t)dt \le \frac{\phi(0)}{2}[(1 - t)^2]_1^0 + \frac{\phi(1)}{2}[t^2]_0^1 = \frac{\phi(0) + \phi(1)}{2}.$$

Exercise 3.7.4 (Karamata's inequality). *Let $\phi : I \longrightarrow \mathbb{R}$ be a convex function defined on an interval of the real line. Prove that for $(x_j)_{1 \le j \le n}, (y_j)_{1 \le j \le n}$ decreasing finite sequences in I such that*

$$\text{for all } i \text{ with } 1 \le i < n, \quad \sum_{1 \le j \le i} y_j \le \sum_{1 \le j \le i} x_j, \quad \sum_{1 \le j \le n} y_j = \sum_{1 \le j \le n} x_j,$$

we have $\sum_{1 \le j \le n} \phi(y_j) \le \sum_{1 \le j \le n} \phi(x_j)$.

Answer. With $a_1 < a_2 < a_3 < a_4$ and $[ij] = \frac{\phi(a_j)-\phi(a_i)}{a_j-a_i}$, we have from the convexity of ϕ

$$[12] \leq [13] \leq [14] \leq [24] \leq [34].$$

This implies that for $x'' < x', y'' < y'$, we have, assuming all four points distinct,

$$\frac{\phi(y'') - \phi(x'')}{y'' - x''} \leq \frac{\phi(y') - \phi(x')}{y' - x'} \tag{3.7.3}$$

since one of the following situations occurs:

- $x'' < x' < y'' < y'$ so that (3.7.3) means $[13] \leq [24]$,
- $x'' < y'' < x' < y'$ so that (3.7.3) means $[12] \leq [34]$,
- $x'' < y'' < y' < x'$ so that (3.7.3) means $[12] \leq [34]$,
- $y'' < x'' < y' < x'$ so that (3.7.3) means $[12] \leq [34]$,
- $y'' < y' < x'' < x'$ so that (3.7.3) means $[13] \leq [24]$.

Assuming all the points are distinct, this proves that

$$\sigma_{i+1} = \frac{\phi(y_{i+1}) - \phi(x_{i+1})}{y_{i+1} - x_{i+1}} \leq \sigma_i = \frac{\phi(y_i) - \phi(x_i)}{y_i - x_i}$$

and thus, with $Y_i = \sum_{j \leq i} y_j$, $X_i = \sum_{j \leq i} x_j$

$$\sum_{1 \leq i \leq n} \left(\phi(x_i) - \phi(y_i) \right) = \sum_{1 \leq i \leq n} \sigma_i(x_i - y_i) = \sum_{1 \leq i \leq n} \sigma_i(X_i - X_{i-1} - Y_i + Y_{i-1})$$

$$= \sum_{1 \leq i \leq n} \sigma_i(X_i - Y_i) - \sum_{0 \leq i \leq n-1} \sigma_{i+1}(X_i - Y_i)$$

$$= \sum_{1 \leq i \leq n-1} (\sigma_i - \sigma_{i+1})(X_i - Y_i) + \sigma_n(X_n - Y_n)$$

$$= \sum_{1 \leq i \leq n-1} (\sigma_i - \sigma_{i+1})(X_i - Y_i) \geq 0.$$

We can get rid of the assumption that all points are distinct since we have only used the expression $\sigma_i(x_i - y_i) = \phi(x_i) - \phi(y_i)$, which is 0 whenever $x_i = y_i$.

Exercise 3.7.5. *Let $\varphi : I \to \mathbb{R}$ be a convex function defined on an interval I of $\mathbb{R}$. Let $[a, b] \subset \mathring{I}$ and $a < x_1 < x_2 < b$. Show that*

$$\frac{\varphi(x_1) - \varphi(a)}{x_1 - a}(x_2 - a) + \varphi(a) \leq \varphi(x_2) \leq \varphi(b) - (b - x_2)\frac{\varphi(b) - \varphi(x_1)}{b - x_1}.$$

Prove that φ is continuous on $\mathring{I}$. Give an example of a convex function defined on $[0, 1]$ and continuous only on $(0, 1)$.

Answer. Continuity of φ is proven in Proposition 3.1.2. On the other hand the function

$$\varphi(x) = \begin{cases} 1 & \text{if } x \in \{0, 1\}, \\ 0 & \text{if } x \in]0, 1[, \end{cases}$$

is convex on $[0, 1]$: Property (3.1.1) is verified for $\theta \in]0, 1[$ if $0 \le x_0 < x_1 \le 1$ since $x_\theta \in]0, 1[$; also (3.1.1) holds for $\theta \in \{0, 1\}$ and for $x_0 = x_1$.

Exercise 3.7.6. *Let u, v be positive log-convex functions defined on some interval I of the real line. Prove that $u + v$ is log-convex.*

Answer. We calculate for $\theta \in [0, 1], x_0, x_1 \in I$, setting $u = e^\phi, v = e^\psi$, with ϕ, ψ convex on I,

$$\ln\big(u((1-\theta)x_0 + \theta x_1) + v((1-\theta)x_0 + \theta x_1)\big)$$
$$= \ln\big(e^{\phi((1-\theta)x_0+\theta x_1)} + e^{\psi((1-\theta)x_0+\theta x_1)}\big)$$
$$\le \ln\big(e^{(1-\theta)\phi(x_0)+\theta\phi(x_1)} + e^{(1-\theta)\psi(x_0)+\theta\psi(x_1)}\big).$$

With $a_0 = u(x_0)^{1-\theta}, a_1 = u(x_1)^\theta, b_0 = v(x_0)^{1-\theta}, b_1 = v(x_1)^\theta$ we have from Hölder's inequality,

$$a_0 a_1 + b_0 b_1 \le (a_0^{1/(1-\theta)} + b_0^{1/(1-\theta)})^{1-\theta}(a_1^{1/\theta} + b_1^{1/\theta})^\theta,$$

so that

$$\ln\big(e^{(1-\theta)\phi(x_0)+\theta\phi(x_1)} + e^{(1-\theta)\psi(x_0)+\theta\psi(x_1)}\big)$$
$$\le (1-\theta)\ln\big(a_0^{1/(1-\theta)} + b_0^{1/(1-\theta)}\big) + \theta\ln\big(a_1^{1/\theta} + b_1^{1/\theta}\big)$$
$$= (1-\theta)\ln\big(u(x_0) + v(x_0)\big) + \theta\ln\big(u(x_1) + v(x_1)\big).$$

We have thus proven

$$\ln\big(u((1-\theta)x_0 + \theta x_1) + v((1-\theta)x_0 + \theta x_1)\big)$$
$$\le (1-\theta)\ln\big(u(x_0) + v(x_0)\big) + \theta\ln\big(u(x_1) + v(x_1)\big),$$

which is the log-convexity of $u + v$.

Exercise 3.7.7. *Determine the set of real numbers α, β, γ such that*

$$u_\alpha(t) = \frac{t^\alpha e^{-t}}{(1+t^{1/2})} \in L^1(\mathbb{R}_+), \quad v_\beta(t) = \frac{\sin t}{t^\beta e^t} \in L^1(\mathbb{R}_+), \quad w_\gamma(t) = \frac{\ln|t|}{|t|^\gamma} \in L^1([-1, 1]).$$

Answer. $\alpha > -1$: if that condition is fulfilled, u_α belongs to $L^1(\mathbb{R}_+)$ and conversely if $u_\alpha \in L^1(\mathbb{R}_+)$, then $u_\alpha \in L^1_{\text{loc}}(\mathbb{R}_+)$ and thus $t^\alpha \in L^1_{\text{loc}}(\mathbb{R}_+)$, implying $\alpha > -1$.

$\quad \beta < 2$: if that condition is fulfilled, v_β belongs to $L^1(\mathbb{R}_+)$ since $v_\beta \in L^1([r, +\infty))$ for all $\beta \in \mathbb{R}$, all $r > 0$ and $v_\beta(t) \sim t^{1-\beta}$ in a neighborhood of

0. Conversely, if $v_\beta \in L^1(\mathbb{R}_+)$, then $v_\beta \in L^1_{\text{loc}}(\mathbb{R}_+)$ and thus $t^{1-\beta} \in L^1_{\text{loc}}(\mathbb{R}_+)$, implying $1 - \beta > -1$, i.e., $\beta < 2$.

$\gamma < 1$: using the parity of w_γ setting $t = 1/x$, we find that $w_\gamma \in L^1([-1,1])$ is equivalent to $x^{\gamma-2} \ln x \in L^1([1,+\infty))$ which is equivalent to $\gamma - 2 < -1$, i.e., $\gamma < 1$.

Exercise 3.7.8.

(1) *Let E be a normed vector space. Prove that E is complete iff the normally convergent series are convergent (a series $\sum u_n$ is normally convergent whenever $\sum \|u_n\| < +\infty$).*

(2) *Let $(X, \mathcal{M}, \mu)$ be a measure space where μ is a positive measure and let $\sum u_n$ be a normally convergent series in $L^1(\mu)$. Prove that $\sum u_n(x)$ converges μ-a.e.*

(3) *Let $(f_n)_{n\geq 1}$ be a sequence in $L^1(\mu)$ such that $\sum_{n\geq 1} \|f_{n+1} - f_n\|_{L^1(\mu)} < +\infty$. Prove that the sequence (f_n) converges in $L^1(\mu)$ and also μ-a.e. Compare this with Exercise 2.8.22.*

Answer. (1) Let us assume first that E is complete; let $\sum u_n$ be a normally convergent series. We define $S_n = \sum_{0\leq k\leq n} u_k$, and we have for $p \geq 0$,

$$\|S_{n+p} - S_n\| = \Big\| \sum_{n<k\leq n+p} u_k \Big\| \leq \sum_{n<k\leq n+p} \|u_k\| \leq \sum_{n<k} \|u_k\| = \epsilon_n.$$

Since the numerical series $\sum \|u_k\|$ converges, we have $\lim_n \epsilon_n = 0$ and (S_n) is a Cauchy sequence, thus converges. Conversely, let E be a normed vector space in which normally convergent series are convergent. Let $(u_n)_{n\in\mathbb{N}}$ be a Cauchy sequence. For all $\epsilon > 0$, there exists N_ϵ such that, for $n \geq N_\epsilon, m \geq N_\epsilon$,

$$\|u_n - u_m\| \leq \epsilon.$$

Using that property, we may find $n_1 \in \mathbb{N}$ such that, for all $p \geq 0$,

$$\|u_{n_1+p} - u_{n_1}\| \leq 1/2.$$

Also, we may find $n_2 > n_1 \in \mathbb{N}$ such that for all $p \geq 0$,

$$\|u_{n_2+p} - u_{n_2}\| \leq 1/2^2,$$

and more generally, we may construct a strictly increasing sequence of integers $n_1 < n_2 < \cdots < n_j$ such that for all $p \geq 0$,

$$\|u_{n_j+p} - u_{n_j}\| \leq 2^{-j}.$$

The series $\sum_{j\geq 1}(u_{n_{j+1}} - u_{n_j})$ is normally convergent, thus converges. Since

$$\sum_{1\leq j<l} (u_{n_{j+1}} - u_{n_j}) = u_{n_l} - u_{n_1},$$

the sequence $(u_{n_l})_{l\in\mathbb{N}}$ is convergent. As a subsequence of a Cauchy sequence, this implies that $(u_n)_{n\in\mathbb{N}}$ is indeed convergent: let w be the limit of $(u_{n_l})_{l\in\mathbb{N}}$. We have

$$\|u_n - w\| \le \|u_n - u_{n_l}\| + \|u_{n_l} - w\|.$$

Let $\epsilon > 0$ be given and $n \ge N_\epsilon$. Since n_l goes to infinity with l, we get

$$\|u_n - w\| \le \limsup_{l\to+\infty} \|u_n - u_{n_l}\| + \limsup_{l\to+\infty} \|u_{n_l} - w\| \le \epsilon + 0 = \epsilon,$$

entailing convergence for the sequence (u_n).

(2) Since $L^1(\mu)$ is complete, the series $\sum u_n$ converges in $L^1(\mu)$. Moreover, since

$$\sum_{n\in\mathbb{N}} \int_X |u_n| d\mu < +\infty,$$

Corollary 1.6.2 implies $\int_X \left(\sum_{n\in\mathbb{N}} |u_n|\right) d\mu = \sum_{n\in\mathbb{N}} \int_X |u_n| d\mu < +\infty$, proving that $\sum_{n\in\mathbb{N}} |u_n|$ belongs to $L^1(\mu)$. As a result, that function is μ-a.e. finite, i.e., for $N \in \mathcal{M}$, with $\mu(N) = 0$,

$$\forall x \in N^c, \quad \sum_{n\in\mathbb{N}} |u_n(x)| < +\infty,$$

so that for all $x \in N^c$, the series $\sum_{n\in\mathbb{N}} u_n(x)$ converges.

N.B. Let $(f_n)_{n\in\mathbb{N}}$ be a convergent sequence in $L^1(\mu)$; we may find a subsequence converging μ-a.e. (Lemma 3.2.9). Extracting a subsequence cannot be dispensed with, as shown by Exercise 2.8.23. Moreover if $\lim_n f_n = f$ in L^1 and (f_n) converges μ-a.e. towards g, then $g = f$ μ-a.e.: for $\epsilon > 0, n \in \mathbb{N}$,

$$\mu(\{x, |f(x) - g(x)| \ge \epsilon\})$$
$$\le \mu(\{x, |f(x) - f_n(x)| \ge \epsilon/2\}) + \mu(\{x, |g(x) - f_n(x)| \ge \epsilon/2\}),$$

so that

$$\mu(\{x, |f(x) - g(x)| \ge \epsilon\}) \le 2\epsilon^{-1} \int_X |f - f_n| d\mu + \mu(\{x, |g(x) - f_n(x)| \ge \epsilon/2\}),$$

proving

$$\mu(\{x, |f(x) - g(x)| \ge \epsilon\}) \le \limsup_n \mu(\{x, |g(x) - f_n(x)| \ge \epsilon/2\}) = 0, \qquad \text{qed.}$$

(3) The series $\sum_k (f_k - f_{k-1})$ is normally convergent thus convergent in L^1 from (1). Since

$$S_n = \sum_{1\le k\le n} (f_k - f_{k-1}) = f_n - f_0,$$

the sequence (f_n) converges in L^1. Moreover from (2), $\sum_k (f_k(x) - f_{k-1}(x))$ converges μ-a.e., so that $(f_k(x))$ converges μ-a.e.

N.B. Convergence μ-a.e. does not imply convergence in L^1 as shown by Exercise 2.8.22. See however Exercises 2.8.15 and 2.8.14 for the weak notion of *convergence in measure*, weaker than μ-a.e. convergence, which along with a domination assumption, implies convergence in L^1.

Exercise 3.7.9. *Let $(X, \mathcal{M}, \mu)$ be a measure space where μ is a positive measure. $(X, \mathcal{M}, \mu)$ is said to be σ-finite whenever there exists a sequence $(X_n)_{n \in \mathbb{N}}$ of elements of $\mathcal{M}$ such that for all n, $\mu(X_n) < +\infty$ and $X = \cup_{n \in \mathbb{N}} X_n$ (see Exercise 2.8.14). Show that $(X, \mathcal{M}, \mu)$ is σ-finite iff there exists $f \in \mathcal{L}^1(\mu)$ such that for all $x \in X$, $f(x) > 0$.*

Answer. We suppose first that $(X, \mathcal{M}, \mu)$ is σ-finite. We consider

$$f(x) = \sum_{n \in \mathbb{N}} \frac{1_{X_n}(x)}{2^n \big(\mu(X_n) + 1\big)}. \tag{3.7.4}$$

For all $x \in X$, we have $f(x) > 0$ (since x belongs to one X_n) and

$$\int_X |f| d\mu \leq \sum_{n \in \mathbb{N}} \frac{\mu(X_n)}{2^n \big(\mu(X_n) + 1\big)} \leq 2.$$

Conversely, if there exists $f \in \mathcal{L}^1(\mu)$ such that for all $x \in X$, $f(x) > 0$, we define for $n \in \mathbb{N}$,

$$X_n = \{x \in X, f(x) > 1/(n + 1)\}.$$

We have $X = \cup_{n \in \mathbb{N}} X_n$ since for $x \in X$, $f(x) > 0$, so that $f(x) > 1/(n + 1)$ for $n \geq E(1/f(x))$. On the other hand since f is positive and belongs to $\mathcal{L}^1(\mu)$,

$$\mu(X_n) < \int_X (n + 1) f d\mu = (n + 1) \int_X f d\mu < +\infty.$$

Exercise 3.7.10. *Let $(X, \mathcal{M}, \mu)$ be a measure space where μ is a positive measure. Let $f : X \to \mathbb{C}$ be a measurable function such that $\mu(\{x \in X, f(x) \neq 0\}) > 0$. For $p \in [1, +\infty)$, we define*

$$\varphi(p) = \int_X |f|^p d\mu \quad \text{and} \quad J = \{p \in [1, +\infty), \varphi(p) < +\infty\}.$$

(1) *Let $p_0 \leq p_1 \in J$. With $\theta \in [0, 1]$ and $p_\theta = (1 - \theta)p_0 + \theta p_1$, show that $p_\theta \in J$ (hint: use Hölder's inequality).*

(2) *Prove that φ is positive on J and $\ln \varphi$ is convex on J.*

(3) *We assume that there exists $r_0 \in [1, +\infty)$ such that $f \in L^{r_0}(\mu) \cap L^\infty(\mu)$. Prove that $f \in L^p(\mu)$ for $p \in [r_0, +\infty]$. Show that*

$$\lim_{p \to +\infty} \|f\|_{L^p(\mu)} = \|f\|_{L^\infty(\mu)}.$$

(4) *We assume that there exists $r_0 \in [1, +\infty)$ such that $f \in L^p(\mu)$ for $p \in [r_0, +\infty)$. Show that if $f \notin L^\infty(\mu)$ we have*

$$\lim_{p \to +\infty} \|f\|_{L^p(\mu)} = +\infty.$$

Answer. The assumption $\mu(\{|f| > 0\}) > 0$ implies $\varphi(p) > 0$ for all $p \geq 1$ ($\varphi(p) = 0$ would imply $f = 0$, μ-a.e.). For $\theta \in (0, 1)$, using Hölder's inequality, we have

$$0 < \varphi(p_\theta) = \int_X \overbrace{|f|^{p_0(1-\theta)}}^{\in L^{\frac{1}{1-\theta}}} \overbrace{|f|^{p_1 \theta}}^{\in L^{\frac{1}{\theta}}} d\mu$$

$$\leq \left(\int_X |f|^{p_0} d\mu \right)^{1-\theta} \left(\int_X |f|^{p_1} d\mu \right)^{\theta} = \varphi(p_0)^{1-\theta} \varphi(p_1)^{\theta},$$

proving (1), (2).

(3) We have $|f| \leq \|f\|_{L^\infty}$ μ-a.e., so that $\int_X |f|^p d\mu \leq \int_X |f|^{r_0} d\mu \|f\|_{L^\infty}^{p-r_0} < +\infty$ for $p \geq r_0$ and thus

$$0 < \varphi(p)^{\frac{1}{p}} \leq \varphi(r_0)^{\frac{1}{p}} \|f\|_{L^\infty}^{1 - \frac{r_0}{p}} \xrightarrow[p \to +\infty]{} \|f\|_{L^\infty}.$$

We have also $\|f\|_{L^\infty} > 0$ (otherwise $f = 0$ μ-a.e.). Let ϵ such that $0 < \epsilon < \|f\|_{L^\infty}$; we note that

$$+\infty > \int_X |f|^p d\mu \geq \int_{|f| \geq \|f\|_{L^\infty} - \epsilon} |f|^p d\mu \geq (\|f\|_{L^\infty} - \epsilon)^p \mu(\{|f| > \|f\|_{L^\infty} - \epsilon\}),$$

entailing

$$\varphi(p)^{1/p} \geq \overbrace{\mu(\{|f| > \|f\|_{L^\infty} - \epsilon\})}^{>0}{}^{1/p} (\|f\|_{L^\infty} - \epsilon) \xrightarrow[p \to +\infty]{} \|f\|_{L^\infty} - \epsilon.$$

Finally, we obtain $\lim_{p \to +\infty} \|f\|_{L^p} = \|f\|_{L^\infty}$ since

$$\forall \epsilon > 0, \ \|f\|_{L^\infty} - \epsilon \leq \liminf_p \|f\|_{L^p} \leq \limsup_p \|f\|_{L^p} \leq \|f\|_{L^\infty}.$$

(4) Since $f \notin L^\infty$, for all $n \in \mathbb{N}$, $\mu(\{|f| > n\}) > 0$ and thus

$$\varphi(p) \geq \int_{|f| > n} |f|^p d\mu \geq n^p \mu(\{|f| > n\}) \Longrightarrow \|f\|_{L^p} \geq \mu(\{|f| > n\})^{1/p} n,$$

which implies $\forall n \in \mathbb{N}, \quad \liminf_{p \to +\infty} \|f\|_{L^p} \geq n$, and thus $\lim_{p \to +\infty} \|f\|_{L^p} = +\infty$.

Exercise 3.7.11. *Let $(X, \mathcal{M}, \mu)$ be a probability space. Let f, g be measurable functions from X into $]0, +\infty)$ such that for all $x \in X$, $f(x)g(x) \geq 1$. Show that $\int_X f d\mu \int_X g d\mu \geq 1$.*

Answer. We have $1 = \mu(X) \leq \int_X f^{1/2} g^{1/2} d\mu \leq \left(\int_X f d\mu \right)^{1/2} \left(\int_X g d\mu \right)^{1/2}$.

Exercise 3.7.12. *Let $(X, \mathcal{M}, \mu)$ be a probability space. Let $(f_n)_{n \in \mathbb{N}}$ be a sequence of measurable functions from X into $\mathbb{R}$. Let $f : X \to \mathbb{R}$ be a measurable function. The sequence $(f_n)_{n \in \mathbb{N}}$ is said to converge in measure[4] towards f if for all $\epsilon > 0$,*

$$\lim_n \mu(\{|f_n - f| > \epsilon\}) = 0.$$

(1) *Show that, if f_n converges towards f μ-a.e., then f_n converges towards f in measure.*

(2) *Let $p \in [1, +\infty]$ and $f_n, f \in L^p(\mu)$ such that f_n converges towards f in $L^p(\mu)$. Show that f_n converges towards f in measure.*

Answer. (1) If (f_n) converges towards f μ-a.e., there exists $N \in \mathcal{M}$ such that $\mu(N) = 0$ and $\forall x \in N^c$, $\lim_{n \to +\infty} |f_n(x) - f(x)| = 0$. As a result for $\epsilon > 0$, Lebesgue's dominated convergence implies

$$\lim_n \mu(\{|f_n - f| > \epsilon\}) = \lim_n \int_X \mathbf{1}_{\{|f_n - f| > \epsilon\}} d\mu = 0,$$

since $\mathbf{1}_{\{|f_n - f| > \epsilon\}}(x) = 0$ when $|f_n(x) - f(x)| \leq \epsilon$ and thus the sequence $\mathbf{1}_{\{|f_n - f| > \epsilon\}}$ converges towards 0 μ-a.e. and is bounded above by 1, which is in L^1 since μ is a probability.

(2) If $p < +\infty$ and $\epsilon > 0$, we have

$$\mu(\{|f_n - f| > \epsilon\}) = \int_X \mathbf{1}_{\{|f_n - f| > \epsilon\}} d\mu$$

$$\leq \epsilon^{-p} \int_X |f_n - f|^p d\mu = \epsilon^{-p} \|f_n - f\|_{L^p}^p \xrightarrow[n \to +\infty]{} 0.$$

If $p = +\infty$, we note that for $\alpha > 0$,

$$\|g\|_{L^\infty(\mu)} \leq \alpha \implies \mu(\{|g| > \alpha\}) = 0.$$

As a result if $\lim_n \|f_n - f\|_{L^\infty(\mu)} = 0$ and $\epsilon > 0$, we have

$$\text{for } n \geq N_\epsilon, \ \|f_n - f\|_{L^\infty(\mu)} \leq \epsilon$$

and thus $\mu(\{|f_n - f| > \epsilon\}) = 0$. The sequence $\left(\mu(\{|f_n - f| > \epsilon\})\right)_{n \in \mathbb{N}}$ is stationary equal to 0 for $n \geq N_\epsilon$.

N.B. Let $(X, \mathcal{M}, \mu)$ be a measure space where μ is a positive measure. Let $1 \leq p < +\infty$, and let $(f_n)_{n \in \mathbb{N}}$ be a sequence converging towards f in $L^p(\mu)$: then it converges as well in measure, as proven by the previous inequalities and there is no need here to assume $\mu(X) < +\infty$.

---

[4]See also Exercises 2.8.15 and 2.8.14.

Remark 3.7.13. On the contrary, the assumption $\mu(X) < +\infty$ cannot be dispensed with for (1) since for instance the sequence f_n defined on $\mathbb{R}$ by $f_n(x) = \frac{x}{n}\mathbf{1}_{[0,n^2]}(x)$ goes to 0 pointwise although

$$\mu(\{|f_n(x)| > \epsilon\}) = \mu(\{n^2 \geq x > n\epsilon\}) = n^2 - n\epsilon \xrightarrow[n\to+\infty]{} +\infty.$$

We may note that this sequence belongs to $\cap_{p\geq 1}L^p(\mathbb{R})$, without converging in any L^p since it would contradict (2).

Exercise 3.7.14. *Let $(X, \mathcal{M}, \mu)$ be a probability space and let $f \in L^\infty(\mu)$ be different from the zero function. We set $\alpha_n = \|f\|_{L^n(\mu)}^n$. Prove that α_{n+1}/α_n tends towards $\|f\|_{L^\infty(\mu)}$ (hint: use Exercise 3.7.10).*

Answer. We note first that $0 < \alpha_n < +\infty$ since on the one hand

$$\alpha_n \leq \|f\|_{L^\infty(\mu)}^n \mu(X) = \|f\|_{L^\infty(\mu)}^n < +\infty,$$

and on the other hand $\alpha_n = 0$ would imply $f = 0$ μ-a.e. and thus $f = 0$ in $L^\infty(\mu)$. For $n \in \mathbb{N}$, we have

$$\alpha_{n+1} = \int_X |f|^{n+1}d\mu \leq \|f\|_{L^\infty(\mu)} \int_X |f|^n d\mu = \|f\|_{L^\infty(\mu)}\alpha_n$$

and thus using Jensen's inequality (Theorem 3.1.3), we get

$$\alpha_n^{1+\frac{1}{n}} = \left(\int_X |f|^n d\mu\right)^{\frac{n+1}{n}} \leq \int_X (|f|^n)^{\frac{n+1}{n}} d\mu = \alpha_{n+1} \leq \|f\|_{L^\infty(\mu)}\alpha_n,$$

so that

$$\|f\|_{L^n(\mu)} = \alpha_n^{\frac{1}{n}} \leq \frac{\alpha_{n+1}}{\alpha_n} \leq \|f\|_{L^\infty(\mu)}.$$

Using Exercise 3.7.10 (3), we get $\lim_{n\to+\infty} \|f\|_{L^n(\mu)} = \|f\|_{L^\infty(\mu)}$, and the previous inequalities imply the result.

N.B. The same statement is true for a measure space $(X, \mathcal{M}, \mu)$ where μ is a positive measure and f such that

$$0 \neq f \in \cap_{p\geq 1}L^p(\mu).$$

In fact, we have as above $\alpha_{n+1} \leq \alpha_n\|f\|_{L^\infty(\mu)}$ and

$$\alpha_n = \int_X |f|^n d\mu = \|f\|_{L^1(\mu)} \int_X |f|^{n-1}\frac{|f|d\mu}{\|f\|_{L^1(\mu)}}.$$

Using Jensen's inequality, we obtain with the probability measure $d\nu = \frac{|f|d\mu}{\|f\|_{L^1(\mu)}}$,

$$\alpha_n^{1+\frac{1}{n-1}} = \|f\|_{L^1(\mu)}^{1+\frac{1}{n-1}}\left(\int_X |f|^{n-1}d\nu\right)^{\frac{n}{n-1}} \leq \|f\|_{L^1(\mu)}^{\frac{n}{n-1}} \int_X |f|^n d\nu$$

$$= \|f\|_{L^1(\mu)}^{\frac{n}{n-1}-1}\int_X |f|^{n+1}d\mu = \alpha_{n+1}\|f\|_{L^1(\mu)}^{\frac{1}{n-1}},$$

so that

$$\left(\alpha^{\frac{1}{n}}\right)^{-\frac{n}{n-1}}\|f\|_{L^1(\mu)}^{-\frac{1}{n-1}} = \alpha^{\frac{1}{n-1}}\|f\|_{L^1(\mu)}^{-\frac{1}{n-1}} \leq \frac{\alpha_{n+1}}{\alpha_n} \leq \|f\|_{L^\infty(\mu)}.$$

Using Exercise 3.7.10 (3), we get $\lim_{n\to+\infty}\|f\|_{L^n(\mu)} = \|f\|_{L^\infty(\mu)}$, and the previous inequalities imply the result.

Exercise 3.7.15. *Let $p \in [1,+\infty[$ and $h \in \mathbb{R}^d$. For $u \in L^p(\mathbb{R}^d)$, we define $(\tau_h u)(x) = u(x - h)$. Show that $\|\tau_h u\|_{L^p} = \|u\|_{L^p}$ and*

$$\lim_{h\to 0}\|\tau_h u - u\|_{L^p} = 0.$$

Answer. The equality of L^p norms is due to the translation invariance of Lebesgue's measure. Let $\varphi \in C_c^0(\mathbb{R}^d)$. Considering the compact set $K = \{x + t\}_{x\in\mathrm{supp}\,\varphi,|t|\leq 1}$, and $|h| \leq 1$, we have

$$\|\tau_h\varphi - \varphi\|_{L^p}^p = \int_{\mathbb{R}^d}|\varphi(x - h) - \varphi(x)|^p dx \leq \lambda_d(K)\sup_{x\in K}|\varphi(x - h) - \varphi(x)|^p \xrightarrow[h\to 0]{} 0,$$

from the uniform continuity of φ. This gives

$$\|\tau_h u - u\|_{L^p} \leq \|\tau_h u - \tau_h\varphi\|_{L^p} + \|\tau_h\varphi - \varphi\|_{L^p} + \|\varphi - u\|_{L^p} = \|\tau_h\varphi - \varphi\|_{L^p} + 2\|\varphi - u\|_{L^p},$$

so that for all functions $\varphi \in C_c^0(\mathbb{R}^d)$,

$$\limsup_{h\to 0}\|\tau_h u - u\|_{L^p} \leq 2\|\varphi - u\|_{L^p}.$$

We get $\limsup_{h\to 0}\|\tau_h u - u\|_{L^p} \leq 2\inf_{\varphi\in C_c^0(\mathbb{R}^d)}\|\varphi - u\|_{L^p} = 0$, since $C_c^0(\mathbb{R}^d)$ is dense in $L^p(\mathbb{R}^d)$ for all $p \in [1,+\infty[$.

Exercise 3.7.16. *Find the values of $p \in [1,+\infty]$ for which the following functions are in $L^p(\mathbb{R}_+)$: $f_1(t) = 1/(1+t)$, $f_2(t) = 1/(\sqrt{t}(1+t))$, $f_3(t) = 1/(\sqrt{t}(\ln t)^2 + 1)$, $f_4(t) = t^{-1/2}\sin(t^{-1})$.*

Answer. We have the following equivalences, justified below:

$$\int_0^{+\infty}|f_1(t)|^p dt = \int_0^{+\infty}\frac{dt}{(1+t)^p} < +\infty \qquad\Longleftrightarrow\qquad 1 < p,$$

$$\int_0^{+\infty}|f_2(t)|^p dt = \int_0^{+\infty}\frac{dt}{t^{p/2}(1+t)^p} < +\infty \qquad\Longleftrightarrow\qquad \frac{2}{3} < p < 2,$$

$$\int_0^{+\infty}|f_3(t)|^p dt = \int_0^{+\infty}\frac{dt}{\left(1 + \sqrt{t}(\ln t)^2\right)^p} < +\infty \qquad\Longleftrightarrow\qquad 2 \leq p,$$

$$\int_0^{+\infty}|f_4(t)|^p dt = \int_0^{+\infty}\frac{|\sin(t^{-1})|^p}{t^{p/2}} < +\infty \qquad\Longleftrightarrow\qquad \frac{2}{3} < p < 2.$$

We note that for f_3, the square of the L^2 norm is bounded above by

$$e + \int_e^{+\infty} \frac{dt}{t(\ln t)^4} = e + \int_1^{+\infty} s^{-4}ds < +\infty.$$

Since $t^{p/2}(\ln t)^{2p} \geq 1$ for $t \geq e$, the pth power of the L^p norm of f_3 for $1 \leq p < 2$ is bounded below by

$$2^{-p} \int_e^{+\infty} \frac{dt}{t^{p/2}(\ln t)^{2p}} = 2^{-p} \int_1^{+\infty} e^{\overset{>0}{\overbrace{s\left(1-\frac{p}{2}\right)}}} s^{-2p}ds = +\infty.$$

Moreover we have

$$\int_0^{+\infty} \frac{|\sin(t^{-1})|^p}{t^{p/2}}dt = \int_0^{+\infty} \frac{|\sin s|^p}{s^{2-\frac{p}{2}}}ds < +\infty$$

if $2 - \frac{p}{2} > 1$, $\frac{3p}{2} - 2 > -1$, i.e., if $\frac{2}{3} < p < 2$. Moreover if $p = 2$, the same computation gives for $\epsilon > 0$,

$$\int_\epsilon^{\epsilon^{-1}} \frac{|\sin(t^{-1})|^p}{t^{p/2}}dt = \int_\epsilon^{\epsilon^{-1}} \frac{\sin^2 s}{s}ds \geq \int_1^{\epsilon^{-1}} \frac{\sin^2 s}{s}ds \xrightarrow[\epsilon \to 0_+]{} +\infty,$$

by an argument similar to Exercise 2.8.20 (4): we note that $\frac{\sin^2 s}{s} = \frac{1-\cos(2s)}{2s}$ and the integral $\int_1^{+\infty} \frac{\cos(2s)}{s}ds$ converges. On the other hand, if $p > 2$, $p = 2 + 2\theta$, $\theta > 0$,

$$\int_\epsilon^{\epsilon^{-1}} \frac{|\sin(t^{-1})|^p}{t^{p/2}}dt \geq \int_1^{\epsilon^{-1}} \frac{(\sin s)^{2+2\theta}}{s^{1-\theta}}ds$$

$$\geq 2^{-2-2\theta} \int_{\{1 \leq s \leq \epsilon^{-1}, |\sin s| \geq 1/2\}} s^{\theta-1}ds \xrightarrow[\epsilon \to 0_+]{} +\infty,$$

since $\sin s \geq 1/2$ on $\cup_{k \in \mathbb{Z}}[\frac{\pi}{6} + 2k\pi, \frac{5\pi}{6} + 2k\pi]$ and consequently

$$\int_{\{1 \leq s \leq \epsilon^{-1}, |\sin s| \geq 1/2\}} s^{\theta-1}ds \geq \frac{1}{\theta} \sum_{\substack{k \geq 1 \\ \frac{5\pi}{6}+2k\pi \leq \epsilon^{-1}}} \left[\left(\frac{5\pi}{6} + 2k\pi\right)^\theta - \left(\frac{\pi}{6} + 2k\pi\right)^\theta\right]$$

$$\geq \sum_{\substack{k \geq 1 \\ \frac{5\pi}{6}+2k\pi \leq \epsilon^{-1}}} \frac{2\pi}{3}\left(\frac{5\pi}{6} + 2k\pi\right)^{\theta-1} \xrightarrow[\epsilon \to 0_+]{} +\infty.$$

For $p \leq 2/3$, the integrand is equivalent near 0_+ to $s^{\frac{3p}{2}-2}$ and $\frac{3p}{2} - 2 \leq -1$, so that the integral diverges.

Exercise 3.7.17. *Let $n \geq 1$ be an integer and f_n defined on $\mathbb{R}$ by*

$$f_n(x) = \frac{n^\alpha}{(|x| + n)^\beta}, \quad \text{with } \beta > 1.$$

(1) *For $1 \leq p \leq +\infty$, show that $f_n \in L^p(\mathbb{R})$ and calculate $\|f_n\|_p$.*

(2) *Prove that g_n defined by $g_n(x) = n^\gamma e^{-n|x|}$ belongs to $L^p(\mathbb{R})$ for all $p \geq 1$.*

(3) *Deduce from the previous questions that for $1 \leq p < q \leq +\infty$ the topologies on $L^p \cap L^q$ induced by L^p and L^q cannot be compared.*

Answer. (1) For $p \geq 1$, $\beta > 1$,

$$\|f_n\|_p^p = 2 \int_0^{+\infty} \frac{n^{\alpha p}}{(x+n)^{\beta p}} dx = 2 \int_0^{+\infty} \frac{n^{(\alpha-\beta)p+1}}{(y+1)^{\beta p}} dy$$

$$= 2n^{(\alpha-\beta)p+1} \left[\frac{(y+1)^{-\beta p+1}}{-\beta p + 1} \right]_0^{+\infty} = \frac{2n^{(\alpha-\beta)p+1}}{\beta p - 1},$$

so that $\|f_n\|_p = 2^{\frac{1}{p}} (\beta p - 1)^{-\frac{1}{p}} n^{\alpha - \beta + \frac{1}{p}}$. Moreover we have $\|f_n\|_\infty = n^{\alpha - \beta}$.

(2) We have $\|g_n\|_\infty = n^\gamma$ and for $p \geq 1$,

$$\|g_n\|_p^p = n^{\gamma p} 2 \int_0^{+\infty} e^{-npx} dx = \frac{n^{\gamma p} 2}{np}, \quad \text{i.e., } \|g_n\|_p = n^{\gamma - \frac{1}{p}} 2^{\frac{1}{p}} p^{-\frac{1}{p}}.$$

(3) We calculate for $1 \leq p < q \leq +\infty$,

$$\frac{\|f_n\|_p}{\|f_n\|_q} = n^{\frac{1}{p} - \frac{1}{q}} \overbrace{C_1(p, q, \beta)}^{\substack{\text{depends only} \\ \text{on } p, q, \beta}} \xrightarrow[n \to +\infty]{} +\infty, \qquad \frac{\|g_n\|_p}{\|g_n\|_q} = n^{\frac{1}{q} - \frac{1}{p}} \overbrace{C_2(p, q)}^{\substack{\text{depends only} \\ \text{on } p, q}} \xrightarrow[n \to +\infty]{} 0.$$

If the topologies on $L^p \cap L^q$ induced respectively by L^p and L^q were comparable, we would have for instance for a sequence (φ_n) of $L^p \cap L^q$,

$$\lim_{L^p} \varphi_n = 0 \implies \lim_{L^q} \varphi_n = 0.$$

This is contradicted by the choice $\varphi_n = n^{-\gamma + \frac{1}{q}} g_n$ since

$$\lim_n \|\varphi_n\|_p = \lim_n n^{\frac{1}{q} - \frac{1}{p}} 2^{1/p} p^{-1/p} = 0,$$

whereas $\|\varphi_n\|_q = 2^{1/q} q^{-1/q} > 0$ which is independent of n (true also for $q = +\infty$). It is not possible either to have for a sequence (φ_n) in $L^p \cap L^q$,

$$\lim_{L^q} \varphi_n = 0 \implies \lim_{L^p} \varphi_n = 0.$$

Choosing $\varphi_n = n^{-\alpha + \beta - \frac{1}{p}} f_n$ gives

$$\lim_n \|\varphi_n\|_q = \lim_n n^{-\alpha + \beta - \frac{1}{p} + \alpha - \beta + \frac{1}{q}} 2^{1/q} (\beta q - 1)^{-1/q} = 0$$

whereas $\|\varphi_n\|_p = n^{-\alpha + \beta - \frac{1}{p} + \alpha - \beta + \frac{1}{p}} 2^{1/p} (\beta p - 1)^{-1/p} = 2^{1/p} (\beta p - 1)^{-1/p} > 0$, is independent of n.

Exercise 3.7.18. *Let $(X, \mathcal{M}, \mu)$ be a measure space where μ is a positive measure. Let $p, p' \in\,]1, +\infty[$ such that $1/p + 1/p' = 1$. A sequence $(f_n)_{n \in \mathbb{N}}$ in $L^p(\mu)$ will be said to converge weakly towards $f \in L^p(\mu)$ if for all $g \in L^{p'}(\mu)$,*

$$\lim_n \int_X f_n g\, d\mu = \int_X f g\, d\mu.$$

(1) Show that convergence in L^p implies weak convergence.

(2) Show that the converse is not true.

Answer. (1) Let (f_n) be a sequence in L^p converging towards f in L^p. Then for all $g \in L^{p'}$, using Hölder's inequality

$$\left| \int_X (f_n - f) g\, d\mu \right| \leq \| f_n - f \|_{L^p} \| g \|_{L^{p'}} \xrightarrow[n \to +\infty]{} 0.$$

(2) The converse is untrue since $f_n(x) = \mathbf{1}_{[0,1]}(x) e^{inx}$ has norm 1 in $L^p(\mathbb{R})$ and converges weakly in L^p since for $g \in L^{p'}(\mathbb{R})$, and $\varphi \in C_c^\infty(\mathbb{R}), \operatorname{supp} \varphi \subset [0,1]$

$$\int_0^1 g(x) e^{inx}\, dx = \int_0^1 \big(g(x) - \varphi(x) \big) e^{inx}\, dx + \int_{\mathbb{R}} \varphi(x) e^{inx}\, dx.$$

Since we have $\int \varphi(x) e^{inx}\, dx = (in)^{-1} \int \varphi(x) \frac{d}{dx}(e^{inx})\, dx = (-in)^{-1} \int \varphi'(x) e^{inx}\, dx$, we get

$$\limsup_{n \to +\infty} \left| \int_0^1 g(x) e^{inx}\, dx \right| \leq \int_0^1 |g(x) - \varphi(x)|\, dx$$

$$\leq \left(\int_0^1 |g(x) - \varphi(x)|^{p'}\, dx \right)^{1/p'} = \| g \mathbf{1}_{[0,1]} - \varphi \|_{L^{p'}},$$

for all $\varphi \in C_c^\infty(\mathbb{R}), \operatorname{supp} \varphi \subset [0,1]$. Since these functions are dense in $L^{p'}([0,1])$ according to Theorem 3.4.3 (here $p > 1$ and thus $p' < +\infty$), we get

$$\inf_{\varphi \in C_c^\infty(\mathbb{R}), \operatorname{supp} \varphi \subset [0,1]} \| g \mathbf{1}_{[0,1]} - \varphi \|_{L^{p'}} = 0$$

and $\lim_n \int_0^1 g(x) e^{inx}\, dx = 0$.

N.B. 1. We note that (e^{inx}) goes to 0 in L^∞-weak*, which means that for all functions g in $L^1(\mathbb{R})$, $\lim_n \int_{\mathbb{R}} g(x) e^{inx}\, dx = 0$: this is the Riemann–Lebesgue lemma (Lemma 3.4.5).

N.B. 2. Let us give another counterexample. Let $f \in C_c(\mathbb{R})$ with norm 1 in $L^p(\mathbb{R})$; we consider the sequence (f_n) with norm 1 in $L^p(\mathbb{R})$ defined by $f_n(x) = n^{1/p} f(nx)$. That sequence goes to 0 weakly in $L^p(\mathbb{R})$, since for $g \in L^{p'}(\mathbb{R})$, we have for $\varphi \in C_c(\mathbb{R})$,

$$\int_{\mathbb{R}} f_n(x) g(x)\, dx = \int_{\mathbb{R}} f_n(x) \big(g(x) - \varphi(x) \big)\, dx + \int_{\mathbb{R}} f(y) \varphi(y/n)\, dy\, n^{\frac{1}{p}-1}$$

which implies, since $|f(y)\varphi(y/n)dyn^{\frac{1}{p}-1}| \le |f(y)|(\sup|\varphi|)n^{-1/p'}$,

$$\limsup_n \left| \int_{\mathbb{R}} f_n(x)g(x)dx \right| \le \|g - \varphi\|_{L^{p'}} \implies \lim_n \int_{\mathbb{R}} f_n(x)g(x)dx = 0.$$

Note that if $f \in L^p(\mathbb{R})$ has norm 1, the result remains the same since for $\psi \in C_c(\mathbb{R})$,

$$f_n(x) = n^{1/p}f(nx) = n^{1/p}\psi(nx) + n^{1/p}f(nx) - n^{1/p}\psi(nx).$$

For $g \in L^{p'}(\mathbb{R})$, we have thus

$$\limsup_n \left| \int_{\mathbb{R}} f_n(x)g(x)dx \right|$$

$$\le \limsup_n \left| \int_{\mathbb{R}} \psi_n(x)g(x)dx \right| + \limsup_n \int_{\mathbb{R}} n^{-1/p'}|f(y) - \psi(y)||g(y/n)|dy$$

$$\le \|g\|_{L^{p'}}\|f - \psi\|_{L^p},$$

which implies $\lim_n \int_{\mathbb{R}} f_n(x)g(x)dx = 0$. If $p = 1$ and f is a function in L^1 with integral 1, the sequence $f_n(x) = nf(nx)$ does not go to 0 weakly: in particular if $g \in C(\mathbb{R}) \cap L^\infty(\mathbb{R})$, we have

$$(\ddagger) \qquad\qquad \lim_n \int_{\mathbb{R}} f_n(x)g(x)dx = g(0).$$

In fact the function f_n is also in L^1 with integral 1 and

$$\int_{\mathbb{R}} f_n(x)g(x)dx - g(0) = \int_{\mathbb{R}} f_n(x)\big(g(x) - g(0)\big)dx = \int_{\mathbb{R}} f(y)\big(g(y/n) - g(0)\big)dy.$$

Since $|f(y)\big(g(y/n) - g(0)\big)| \le |f(y)|2\sup|g|$, and by continuity of g at 0,

$$\lim_n f(y)\big(g(y/n) - g(0)\big) = 0,$$

Lebesgue's dominated convergence gives the result $(\dagger)$.

N.B. 3. Another counterexample is given by $f_n(x) = f(x-n)$ where $f \in L^p(\mathbb{R})$ has norm 1 in L^p Each f_n has norm 1 in L^p and nevertheless for $g \in L^{p'}$, $\varphi \in C_c(\mathbb{R})$, we have for a fixed $A > 0$,

$$\int_{\mathbb{R}} f_n(x)g(x)dx = \int_{\mathbb{R}} f(y)\big(g(y+n) - \varphi(y+n)\big)dy$$

$$+ \int_{\{|y|\le A\}} f(y)\varphi(y+n)dy + \int_{\{|y|>A\}} f(y)\varphi(y+n)dy,$$

which implies

$$\limsup_n \left| \int_{\mathbb{R}} f_n(x)g(x)dx \right| \le \|g - \varphi\|_{L^{p'}} + \limsup_n \int_{\{|y|>A\}} |f(y)||\varphi(y+n)|dy$$

$$\le \|g - \varphi\|_{L^{p'}} + \left(\int_{\{|y|>A\}} |f(y)|^p dy \right)^{1/p} \|\varphi\|_{L^{p'}}.$$

Taking the infimum with respect to A in the rhs, we get

$$\limsup_n \left| \int_{\mathbb{R}} f_n(x)g(x)dx \right| \leq \|g - \varphi\|_{L^{p'}}, \quad \text{for all } \varphi \in C_c(\mathbb{R}),$$

so that $\lim_n \int_{\mathbb{R}} f_n(x)g(x)dx = 0$.

Exercise 3.7.19. *Let μ be a positive measure defined on the Borel σ-algebra of $\mathbb{R}$ such that $\mu(\mathbb{R}) < +\infty$. We define $f(x) = \int_{\mathbb{R}} e^{itx} d\mu(t)$. Show that f is continuous on $\mathbb{R}$. Show that if*

$$\frac{1}{h^2}\Big(2f(0) - f(h) - f(-h)\Big)$$

has a limit when h goes to 0, then $\int_{\mathbb{R}} t^2 d\mu(t) < +\infty$ and f is of class C^2.

Answer. Let (x_k) be a convergent sequence of real numbers with limit x. Using $\mu(\mathbb{R}) < +\infty$, we have

$$\left| e^{itx_k} - e^{itx} \right| \leq 2 \in L^1(\mu), \quad \text{and} \quad \lim_k e^{itx_k} = e^{itx},$$

and Lebesgue's dominated convergence theorem gives $\lim_k f(x_k) = f(x)$. We note that

$$\frac{1}{h^2}\Big(2f(0) - f(h) - f(-h)\Big) = h^{-2} \int_{\mathbb{R}} (2 - 2\cos th)d\mu(t) \xrightarrow[h \to 0]{} L.$$

From Fatou's lemma, we obtain

$$\int_{\mathbb{R}} \liminf_{h \to 0}\left(h^{-2}|2 - 2\cos th|\right)d\mu(t)$$

$$\leq \liminf_{h \to 0} \int_{\mathbb{R}} \left(h^{-2}|\underbrace{2 - 2\cos th}_{\geq 0}|\right)d\mu(t) = \liminf_{h \to 0} h^{-2} \int_{\mathbb{R}} (2 - 2\cos th)d\mu(t) = L.$$

Since

$$\lim_{h \to 0} h^{-2}(2 - 2\cos th) = \lim_{h \to 0} h^{-2}\big(2 - 2[1 - 2\sin^2(th/2)]\big) = \lim_{h \to 0} \frac{4\sin^2(th/2)}{h^2} = t^2,$$

we get

$$(\Upsilon) \qquad \int_{\mathbb{R}} t^2 d\mu(t) \leq L < +\infty.$$

We note incidentally that

$$\int_{\mathbb{R}} |t|d\mu(t) \leq \left(\int_{\mathbb{R}} t^2 d\mu(t)\right)^{1/2} \mu(\mathbb{R})^{1/2} \leq \left(L\mu(\mathbb{R})\right)^{1/2} < +\infty.$$

Using Theorem 3.3.4, we find that f is twice differentiable and

$$f''(x) = - \int_{\mathbb{R}} e^{itx} t^2 d\mu(t).$$

That formula and Condition (Υ) ensure continuity for f'', using Theorem 3.3.1.

Exercise 3.7.20. *Show that $\ell^\infty(\mathbb{N})$ and $L^\infty(\mathbb{R})$ are not separable. (Hint: reductio ad absurdum.)*

Answer. Assume that $\ell^\infty(\mathbb{N})$ contains a countable dense subset $\{x_n\}_{n\in\mathbb{N}}$. Each element x_n is a bounded sequence $(x_{n,k})_{k\in\mathbb{N}}$, i.e., such that

$$\sup_{k\geq 0} |x_{n,k}| = \|x_n\|_{l^\infty(\mathbb{N})} < +\infty.$$

The triangle inequality implies

$$2 \leq |1+x_{0,0}| + |1-x_{0,0}| \leq 2\max\big(|1+x_{0,0}|, |1-x_{0,0}|\big) = 2\max\big(|-1-x_{0,0}|, |1-x_{0,0}|\big)$$

and thus $\max\big(|-1-x_{0,0}|, |1-x_{0,0}|\big) \geq 1$. We may thus find $y_0 \in \{-1,1\}$ such that $|y_0 - x_{0,0}| \geq 1$. Let us assume that we have found $y_0, \ldots, y_k \in \{-1,1\}$ such that

$$\forall l \in \{0, \ldots, k\}, \quad |y_l - x_{l,l}| \geq 1.$$

As above, we may find $y_{k+1} \in \{-1,1\}$ such that

$$|y_{k+1} - x_{k+1,k+1}| \geq 1.$$

We have thus constructed a sequence $y = (y_k)_{k\in\mathbb{N}}$ such that $\forall k \in \mathbb{N}, |y_k| = 1$ (and thus this sequence belongs to $\ell^\infty(\mathbb{N})$) such that

$$\|y - x_n\|_{l^\infty(\mathbb{N})} = \sup_{k\in\mathbb{N}} |y_k - x_{n,k}| \geq |y_n - x_{n,n}| \geq 1.$$

This contradicts the density of $\{x_n\}_{n\in\mathbb{N}}$.

Let $\{\varphi_n\}_{n\in\mathbb{Z}}$ be a countable subset of $L^\infty(\mathbb{R})$. We have

$$\mathbb{R} = \cup_{m\in\mathbb{Z}} I_m, \quad I_m = [m, m+1[.$$

The triangle inequality implies

$$2 \leq \|1+\varphi_n\|_{L^\infty(I_n)} + \|1-\varphi_n\|_{L^\infty(I_n)} \leq 2\max\big(\|-1-\varphi_n\|_{L^\infty(I_n)}, \|1-\varphi_n\|_{L^\infty(I_n)}\big).$$

For all $n \in \mathbb{Z}$, we may thus find $\theta_n \in \{-1,1\}$, such that

$$\|\theta_n - \varphi_n\|_{L^\infty(I_n)} \geq 1.$$

The function

$$\psi(x) = \sum_{n \in \mathbb{Z}} \theta_n \mathbf{1}_{I_n}(x) = \sum_{n \in \mathbb{N}, \theta_n = 1} \mathbf{1}_{I_n}(x) - \sum_{n \in \mathbb{N}, \theta_n = -1} \mathbf{1}_{I_n}(x)$$

belongs to $L^\infty(\mathbb{R})$ and has norm 1 (ψ is measurable since it takes two values $-1, 1$ and $\psi^{-1}(\{1\})$ and $\psi^{-1}(\{-1\})$ are countable unions of intervals). Moreover for $n \in \mathbb{Z}$,

$$\|\psi - \varphi_n\|_{L^\infty(\mathbb{R})} \geq \|\psi - \varphi_n\|_{L^\infty(I_n)} = \|\theta_n - \varphi_n\|_{L^\infty(I_n)} \geq 1,$$

making impossible the density of $\{\varphi_n\}_{n \in \mathbb{Z}}$.

Exercise 3.7.21. *Here, L^p stands for the space $L^p(\mu)$ where μ is the Lebesgue measure on $]0, +\infty[$ and $\|u\|_p$ is the L^p norm of u.*

(1) *Let $f :]0, +\infty[\to \mathbb{R}$, be a continuous function with compact support in $]0, +\infty[$. For $x > 0$, we set*

$$(Hf)(x) = \frac{1}{x} \int_0^x f(t)\,dt.$$

For $p > 1$, show that Hf belongs to L^p.

(2) *For f as in (1), taking non-negative values, show that*

$$(\sharp) \qquad\qquad \|Hf\|_p \leq \frac{p}{p-1} \|f\|_p,$$

(hint: $F = Hf$ is also a non-negative function, integrate by parts in

$$\int_0^{+\infty} F(x)^p \frac{d}{dx}(x)\,dx. \quad)$$

(3) *For f as in (1), show $(\sharp)$.*

(4) *Show that the mapping $H : C_c(]0, +\infty[) \longrightarrow L^p$ is uniquely extendable to L^p and verifies $(\sharp)$ for all $f \in L^p$.*

(5) *Show that the constant $\frac{p}{p-1}$ in $(\sharp)$ cannot be replaced by a smaller constant (hint: take $f(x) = x^{-1/p}$ on $[1, \lambda]$, 0 elsewhere and let λ go to $+\infty$).*

Answer. (1) Since f is supported in $[a, b]$ with $0 < a \leq b < +\infty$, Hf vanishes on $]0, a]$ and is bounded above by $\frac{1}{x} \int_a^b |f(t)|\,dt$ elsewhere. As a result,

$$\int_0^{+\infty} |Hf(x)|^p\,dx \leq \int_a^{+\infty} x^{-p}\,dx \left(\int_a^b |f(t)|\,dt\right)^p < +\infty, \qquad \text{since } p > 1.$$

(2) With f as in (1), taking non-negative values, the function $F = Hf$ is also non-negative. Thus

$$\|F\|_p^p = \int_a^{+\infty} F(x)^p\,dx.$$

Since $xF(x) = \int_0^x f(t)dt$, the function $x \mapsto xF(x)$ is differentiable on $]0, +\infty[$ with a derivative f. Thus on $]0, +\infty[$, F is differentiable and $xF'(x) + F(x) = f(x)$. For $N \geq a$, we have

$$\int_a^N F(x)^p dx = [xF(x)^p]_a^N - \int_a^N xpF(x)^{p-1}F'(x)dx$$

$$= NF(N)^p - \int_a^N pF(x)^{p-1}\big(f(x) - F(x)\big)dx,$$

so that for $N \geq b$

$$p\int_a^b F(x)^{p-1}f(x)dx - p\int_a^N F(x)^{p-1}f(x)dx - (p-1)\int_a^N F(x)^p dx + NF(N)^p.$$

From (1), we know that $0 \leq F(N) \leq N^{-1}\int_a^b f(t)dt$, and taking the limit when N goes to $+\infty$ in the above equality, we get

$$p\int_a^b F(x)^{p-1}f(x)dx = (p-1)\int_a^{+\infty} F(x)^p dx,$$

i.e.,

$$\|F\|_p^p = \frac{p}{p-1}\int_0^{+\infty} F(x)^{p-1}f(x)dx.$$

With $1/p + 1/q = 1$, i.e., $q = p/(p-1)$, Hölder's inequality implies

$$\|F\|_p^p = \frac{p}{p-1}\int_0^{+\infty} F(x)^{p-1}f(x)dx$$

$$\leq \frac{p}{p-1}\left(\int_0^{+\infty} F(x)^{(p-1)q}dx\right)^{1/q}\left(\int_0^{+\infty} f(x)^p dx\right)^{1/p} = \frac{p}{p-1}\|F\|_p^{p-1}\|f\|_p,$$

which is ($\sharp$).

(3) With f as in (1), we set

$$f_+(x) = \max(f(x), 0) = \frac{1}{2}(|f(x)| + f(x)), \quad f_-(x) = \frac{1}{2}(|f(x)| - f(x))$$

so that the functions $f_\pm$ are non-negative continuous with compact support and $f = f_+ - f_-$ as well as $f_- f_+ = 0$, so that $Hf = Hf_+ - Hf_-$. Since the functions $Hf_\pm$ are non-negative, we have

$$\|Hf\|_p^p = \int_0^{+\infty} |(Hf_+)(x) - (Hf_-)(x)|^p dx$$

$$\leq \int_0^{+\infty} \max\Big([(Hf_+)(x)]^p, [(Hf_-)(x)]^p\Big)dx$$

$$\leq \int_0^{+\infty} \left([(Hf_+)(x)]^p + [(Hf_-)(x)]^p \right) dx = \|Hf_+\|_p^p + \|Hf_-\|_p^p$$

$$\leq \left(\frac{p}{p-1} \right)^p \left(\|f_+\|_p^p + \|f_-\|_p^p \right) = \left(\frac{p}{p-1} \right)^p \int_0^{+\infty} (f_+(x)^p + f_-(x)^p) dx$$

$$= \left(\frac{p}{p-1} \right)^p \int_0^{+\infty} |f(x)|^p dx = \left(\frac{p}{p-1} \right)^p \|f\|_p^p,$$

providing ($\sharp$) for continuous functions with compact support in $]0, +\infty[$.

(4) Let $1 < p < +\infty$ and let $f \in L^p$. Then in L^p, $f = \lim f_k$ where f_k is continuous with compact support. Since

$$\|Hf_k - Hf_l\|_p \leq \frac{p}{p-1} \|f_k - f_l\|_p,$$

the sequence (Hf_k) is a Cauchy sequence, thus is converging. On the other hand if $(\tilde{f}_k)$ is another sequence with limit f in L^p, we have

$$\|Hf_k - H\tilde{f}_k\|_p \leq \frac{p}{p-1} \|f_k - \tilde{f}_k\|_p,$$

and thus Hf_k and $H\tilde{f}_k$ are converging towards the same limit. We can thus define without ambiguity, $Hf = \lim Hf_k$. Moreover, if $\tilde{H}$ extends H on L^p and is continuous, we shall have

$$\tilde{H}f = \lim \tilde{H}f_k = \lim Hf_k = Hf.$$

Moreover since $\|u\|_p \leq \|v\|_p + \|u - v\|_p$ and $\|v\|_p \leq \|u\|_p + \|v - u\|_p$ and thus $| \|Hf\|_p - \|Hf_k\|_p | \leq \|Hf - Hf_k\|_p$, we get

$$\|Hf\|_p = \lim \|Hf_k\|_p \leq \limsup \frac{p}{p-1} \|f_k\|_p = \frac{p}{p-1} \|f\|_p.$$

(5) For $x > 0$, the function $t \mapsto \mathbf{1}_{(0,x)}(t) f(t)$ belongs to L^1 since it is the product of an L^p function with an L^q function ($\mathbf{1}_{(0,x)} \in L^q$ since $\int |\mathbf{1}_{(0,x)}(t)|^q dt = x$). Let $f \in L^p$. We may set for $x > 0$,

$$Kf(x) = \frac{1}{x} \int_0^x f(t) dt = \frac{1}{x} \int_0^{+\infty} \underbrace{\mathbf{1}_{(0,x)}(t) f(t)}_{\in L^1} dt.$$

Moreover if (f_k) is a sequence in $C_c(]0, +\infty[)$ such that $\lim f_k = f$ in L^p, then for all $x > 0$, $\lim f_k = f$ in $L^1(0, x) \supset L^p(0, x)$ since using Hölder's inequality $\|f\|_{L^1(0,x)} \leq x^{1/q} \|f\|_{L^p(0,x)}$. As a result for $x > 0$,

$$(Kf)(x) = \frac{1}{x} \lim_{\mathbb{R}} \int_0^x f_k(t) dt = \lim_{\mathbb{R}} (Hf_k)(x).$$

But the sequence (Hf_k) converges in L^p with limit Hf. We may thus extract a sequence converging almost everywhere towards Hf. Thus for almost all $x > 0$, $(Kf)(x) = (Hf)(x)$ and thus the functions Hf and Kf coincide in L^p.

For $\lambda \geq 1$, we consider the function f_λ defined in (5). Since $\frac{1}{q} = \frac{p-1}{p}$, we have

for $0 < x \leq 1$, $(Hf_\lambda)(x) = 0$,

for $1 \leq x \leq \lambda$, $(Hf_\lambda)(x) = x^{-1} \int_1^x t^{-1/p} dt = qx^{-1}(x^{1/q} - 1) = q(x^{-1/p} - x^{-1})$,

and for $\lambda \leq x$, $(Hf_\lambda)(x) = x^{-1} \int_1^\lambda t^{-1/p} dt = qx^{-1}(\lambda^{1/q} - 1)$.

Consequently, we get

$$\|Hf_\lambda\|_p = \left(\int_0^{+\infty} |(Hf_\lambda)(x)|^p dx \right)^{1/p}$$

$$= q \left(\int_1^\lambda (x^{-1/p} - x^{-1})^p dx + \int_\lambda^{+\infty} x^{-p}(\lambda^{1/q} - 1)^p dx \right)^{1/p}$$

$$= q \left(\int_1^\lambda x^{-1}(1 - x^{-1/q})^p dx + \frac{\lambda^{1-p}}{p-1}(\lambda^{1/q} - 1)^p \right)^{1/p}$$

$$= q \left(\ln \lambda + \int_1^\lambda x^{-1}\big((1 - x^{-1/q})^p - 1\big) dx + O(1) \right)^{1/p}$$

$$= q\big(\ln \lambda + O(1)\big)^{1/p}.$$

We have also $\|f_\lambda\|_p = \left(\int_1^\lambda x^{-1} dx \right)^{1/p} = (\ln \lambda)^{1/p}$ so that, defining

$$\mu = \sup_{f \in L^p, f \neq 0} \frac{\|Hf\|_p}{\|f\|_p},$$

we get $\frac{p}{p-1} = \lim_{\lambda \to +\infty} \frac{\|Hf_\lambda\|_p}{\|f_\lambda\|_p} \leq \mu \leq \frac{p}{p-1}$, proving $\mu = \frac{p}{p-1}$.

Exercise 3.7.22. *Let u be a function in $L^1(\mathbb{R})$. We set for $\xi \in \mathbb{R}$,*

$$\widetilde{u}(\xi) = \int_\mathbb{R} u(x) \cos(x\xi)\, dx.$$

(1) *Show that $\widetilde{u}$ belongs to L^∞. Show that the function $\widetilde{u}$ is uniformly continuous on $\mathbb{R}$.*

(2) *Show that for $\varphi \in C_c^1$, $\lim_{|\xi| \to +\infty} \widetilde{\varphi}(\xi) = 0$.*

(3) *Show that $\lim_{|\xi| \to +\infty} \widetilde{u}(\xi) = 0$.*

Answer. An immediate consequence of the Riemann–Lebesgue Lemma 3.4.5.

Exercise 3.7.23. *For $n \in \mathbb{N}$ and $x \geq 0$, we define $f_n(x) = \frac{ne^{-x}}{nx^{1/2}+1} \cos x$.*

(1) *Show that f_n belongs to $L^1(\mathbb{R}_+)$.*

(2) *Show that the sequence $a_n = \int_{\mathbb{R}_+} f_n(x)dx$ converges towards $\int_{\mathbb{R}_+} f(x)dx$ for some $f \in L^1(\mathbb{R}_+)$.*

Answer. (1) For $x \geq 0$, we have $|f_n(x)| \leq e^{-x}x^{-1/2} = g(x)$. The functions f_n and g are continuous on $\mathbb{R}_+^*$ and belong to $L^1(\mathbb{R}_+)$ since bounded from above by $\mathbf{1}_{[0,1]}(x)x^{-1/2} + \mathbf{1}_{[1,+\infty]}(x)e^{-x}$.

(2) For $x > 0$, we have $\lim_n f_n(x) = f(x) = e^{-x}x^{-1/2}\cos x$. Since $|f_n| \leq g \in L^1(\mathbb{R}_+)$, Lebesgue's dominated convergence theorem implies $f \in L^1(\mathbb{R}_+)$ and $\lim_n a_n = \int_{\mathbb{R}_+} f(x)dx$. Note that

$$\int_0^{+\infty} f(x)dx = \operatorname{Re} \int_0^{+\infty} e^{-(1+i)x}x^{-1/2}dx.$$

We have for $z > 0$, using Section 10.5,

$$\int_0^{+\infty} e^{-zx}x^{-1/2}dx = z^{-1/2}\int_0^{+\infty} e^{-t}t^{-1/2}dt = z^{-1/2}\Gamma(1/2) = \pi^{1/2}z^{-1/2}.$$

So with the results of Section 10.5, we obtain by analytic continuation of holomorphic functions on $\operatorname{Re} z > 0$,

$$\pi^{1/2}e^{-\frac{1}{2}\operatorname{Log} z} = \int_0^{+\infty} e^{-zx}x^{-1/2}dx,$$

implying $\int_0^{+\infty} e^{-(1+i)x}x^{-1/2}dx = \pi^{1/2}e^{-\frac{1}{2}\operatorname{Log}(1+i)} = \pi^{1/2}2^{-1/4}e^{-i\pi/8}$ and thus

$$\lim_n a_n = \pi^{1/2}2^{-1/4}\cos(\pi/8) = \pi^{1/2}\frac{\sqrt{1+\sqrt{2}}}{2}.$$

Exercise 3.7.24. *Let $(X, \mathcal{M}, \mu)$ be a measure space where μ is a positive measure.*

(1) *Let $(A_n)_{n\in\mathbb{N}}$ be a sequence of elements of $\mathcal{M}$ such that $\sum_{n\in\mathbb{N}} \mu(A_n) < +\infty$. For $n \in \mathbb{N}$, we set $B_n = \cup_{k\geq n} A_k$. Show that $\mu(\cap_{n\in\mathbb{N}} B_n) = 0$.*

(2) *Let ν be a positive measure on $(X, \mathcal{M})$. We shall say that ν is dominated by μ whenever*

$$\forall A \in \mathcal{M}, \ \mu(A) = 0 \implies \nu(A) = 0.$$

Assuming $\nu(X) < +\infty$, show that if ν is dominated by μ,

$$\forall \epsilon > 0, \ \exists \delta > 0, \ \forall A \in \mathcal{M}, \ \mu(A) < \delta \implies \nu(A) < \epsilon.$$

Answer. (1) We have $\mu(B_n) \leq \sum_{k\geq n} \mu(A_k)$ which goes to 0 when n goes to infinity as the remainder of a converging series. For all $n \in \mathbb{N}$, we have

$$0 \leq \mu(\cap_{k\in\mathbb{N}} B_k) \leq \mu(B_n) \implies \mu(\cap_{k\in\mathbb{N}} B_k) = 0.$$

(2) *Reductio ad absurdum:* if the required property is not satisfied, $\exists \epsilon_0 > 0$ such that for all $n \in \mathbb{N}$, $\exists A_n \in \mathcal{M}$ such that

$$\mu(A_n) < 2^{-n} \quad \text{and} \quad \nu(A_n) \geq \epsilon_0.$$

Since the series $\sum_n \mu(A_n)$ converges, we find from (1) that

$$0 = \mu\big(\cap_{n \in \mathbb{N}}(\cup_{k \geq n} A_k)\big) \quad \big(\Longrightarrow \nu\big(\cap_{n \in \mathbb{N}}(\cup_{k \geq n} A_k)\big) = 0\big).$$

Using $\nu(X) < +\infty$, we have $B_n = \cup_{k \geq n} A_k$, $B_n \supset B_{n+1}$ and $\lim \nu(B_n) = \nu(\cap_n B_n) = 0$ and thus

$$0 = \nu\big(\cap_{n \in \mathbb{N}}(\cup_{k \geq n} A_k)\big) = \lim_{n \to \infty} \nu\big((\cup_{k \geq n} A_k)\big) \geq \limsup \nu(A_n) \geq \epsilon_0 > 0,$$

which is a contradiction.

Exercise 3.7.25. *Let $(X, \mathcal{M}, \mu)$ be a measure space where μ is a positive measure such that $\mu(X) < +\infty$. A family of measurable functions $(u_i)_{i \in I}$ is said to be equi-integrable whenever*

$$\lim_{t \to +\infty} \left(\sup_{i \in I} \int_{E_i(t)} |u_i| d\mu \right) = 0, \quad \text{with} \quad E_i(t) = \{x \in X, |u_i(x)| > t\}.$$

(1) *Let $(u_i)_{i \in I}$ be a family of measurable functions from X into $\mathbb{C}$. Show that if $(u_i)_{i \in I}$ is equi-integrable, then*

$$\forall \epsilon > 0, \ \exists \delta > 0, \ \forall A \in \mathcal{M}, \ \mu(A) < \delta \Longrightarrow \sup_{i \in I} \int_A |u_i| d\mu < \epsilon.$$

(2) *Let $(u_n)_{n \in \mathbb{N}}$ be a sequence of measurable equi-integrable functions from X into $\mathbb{C}$, μ-a.e. converging towards a function u. Show that for $\epsilon > 0$, we have*

$$\lim_{n \to +\infty} \mu\big(\{|u_n - u| > \epsilon\}\big) = 0.$$

Show that the sequence $(u_n)_{n \in \mathbb{N}}$ converges in $L^1(\mu)$.

Answer. (1) Let us assume that the required condition does not hold. There exists $\epsilon_0 > 0$ such that for all $n \in \mathbb{N}$, there exists $A_n \in \mathcal{M}$ with $\mu(A_n) < 1/n$ and

$$\sup_{i \in I} \int_{A_n} |u_i| d\mu \geq \epsilon_0.$$

Consequently for $t \geq 0$

$$t\mu(A_n) + \sup_{i \in I} \int_{A_n \cap \{|u_i| > t\}} |u_i| d\mu$$

$$\geq \sup_{i \in I} \int_{A_n \cap \{|u_i| > t\}} |u_i| d\mu + \sup_{i \in I} \int_{A_n \cap \{|u_i| \leq t\}} |u_i| d\mu \geq \epsilon_0,$$

which implies for $t_n = \frac{n\epsilon_0}{2} \to +\infty$ with n,

$$\sup_{i \in I} \int_{|u_i| > t_n} |u_i| d\mu \geq \sup_{i \in I} \int_{A_n \cap \{|u_i| > t_n\}} |u_i| d\mu \geq \epsilon_0 - t_n \mu(A_n) \geq \frac{\epsilon_0}{2},$$

contradicting the assumption of equi-integrability.

(2) We check for $M > \epsilon > 0$;

$$\int_X |u_n - u| d\mu = \int_{|u_n - u| \leq \epsilon} |u_n - u| d\mu + \int_{|u_n - u| > \epsilon} |u_n - u| d\mu$$

$$\leq \epsilon \mu(X) + \int_{|u_n - u| > \epsilon, |u_n| \leq M} |u_n| d\mu + \int_{|u_n - u| > \epsilon, |u_n| > M} |u_n| d\mu + \int_{|u_n - u| > \epsilon} |u| d\mu$$

$$\leq \epsilon \mu(X) + M \mu(\{|u_n - u| > \epsilon\}) + \int_{|u_n| > M} |u_n| d\mu + \int_{|u_n - u| > \epsilon} |u| d\mu.$$

Consequently, we have

$$\limsup_{n \to \infty} \int_X |u_n - u| d\mu \leq \epsilon \mu(X) + M \limsup_{n \to \infty} \left(\mu(\{|u_n - u| > \epsilon\}) \right)$$

$$+ \sup_{n \in \mathbb{N}} \int_{|u_n| > M} |u_n| d\mu + \limsup_{n \to \infty} \int_{|u_n - u| > \epsilon} |u| d\mu.$$

But we know that for $\epsilon > 0$,

$$\lim_{n \to \infty} \left(\mu(\{|u_n - u| > \epsilon\}) \right) = 0.$$

In fact, we have $A_n = \{|u_n - u| > \epsilon\} \subset B_n = \cup_{k \geq n} \{|u_k - u| > \epsilon\}$ and B_n is decreasing (and $\mu(X) < +\infty$), so that with

$$B = \cap_{n \in \mathbb{N}} B_n, \qquad \mu(B) = \lim_n \mu(B_n).$$

Since $B = \cap_{n \in \mathbb{N}} B_n = \cap_{n \in \mathbb{N}} (\cup_{k \geq n} \{|u_k - u| > \epsilon\})$ for $x \in B$, for all $n \in \mathbb{N}$, there exists $k \geq n$ such that $|u_k(x) - u(x)| > \epsilon$, so that the sequence $u_l(x)$ does not converge towards $u(x)$. Since we have assumed that the convergence μ-a.e. holds, we get that B has zero measure and $\mu(B_n)$ converges to 0. As a result for all $M \geq \epsilon > 0$,

$$\limsup_{n \to \infty} \int_X |u_n - u| d\mu \leq \epsilon \mu(X) + \sup_{n \in \mathbb{N}} \int_{|u_n| > M} |u_n| d\mu + \limsup_{n \to \infty} \int_{|u_n - u| > \epsilon} |u| d\mu.$$

Taking the limit when $M \to +\infty$, we find, by using equi-integrability, that

$$\limsup_{n \to \infty} \int_X |u_n - u| d\mu \leq \epsilon \mu(X) + \limsup_{n \to \infty} \int_{|u_n - u| > \epsilon} |u| d\mu.$$

But we have proven that $\mu(A_n)$ goes to 0. From (1), we find that

$$\sup_{j \in \mathbb{N}} \int_{A_n} |u_j| d\mu \to 0, \qquad \text{for } n \to \infty.$$

Fatou's lemma implies

$$0 \le \int_{A_n} |u| d\mu = \int_{A_n} \liminf_j |u_j| d\mu \le \liminf_j \int_{A_n} |u_j| d\mu$$

$$\le \sup_{j \in \mathbb{N}} \int_{A_n} |u_j| d\mu \to 0 \quad (n \to +\infty),$$

and thus $\lim_n \int_{|u_n - u| > \epsilon} |u| d\mu = 0$. Finally for all $\epsilon > 0$,

$$\limsup_{n \to \infty} \int_X |u_n - u| d\mu \le \epsilon \mu(X),$$

providing the result $\lim_{n \to \infty} \int_X |u_n - u| d\mu = 0$.

Exercise 3.7.26. *Let X be a locally compact Hausdorff topological space. We define*

$$C_{(0)}(X) = \{f \in C(X; \mathbb{R}), \forall \varepsilon > 0, \exists K_\varepsilon \text{ compact}, \ \sup_{x \notin K_\varepsilon} |f(x)| \le \varepsilon\}. \qquad (3.7.5)$$

(1) *Prove that the functions of $C_{(0)}(X)$ are also bounded on X. Prove that $C_{(0)}(X) = C_c(X)$ whenever X is compact.*

(2) *Prove that $C_{(0)}(X)$ is a Banach space for the norm $\|f\| = \sup_{x \in X} |f(x)|$.*

(3) *Prove that $C_c(X)$ is dense in $C_{(0)}(X)$.*

N.B. This exercise proves in particular that the completion of $C_c(\mathbb{R}^m)$ for the L^∞ norm is $C_{(0)}(\mathbb{R}^m)$, a proper subset of $L^\infty(\mathbb{R}^m)$. We have seen in Theorem 3.4.1 that for $1 \le p < +\infty$, the completion of $C_c(\mathbb{R}^m)$ for the L^p norm is $L^p(\mathbb{R}^m)$.

Answer. (1) If f belongs to $C_{(0)}(X)$, there exists a compact set K_1 such that $\sup_{x \notin K_1} |f(x)| \le 1$: as a result,

$$\sup_{x \in X} |f(x)| \le \max\left(\sup_{x \notin K_1} |f(x)|, \sup_{x \in K_1} |f(x)|\right) < +\infty.$$

The last statement of the first question is obvious by taking $K_\varepsilon = X$.

(2) The mapping $C_{(0)}(X) \ni f \mapsto \|f\|$ obviously satisfies the axioms of a norm (see, e.g., (1.2.12)). Let us now consider a Cauchy sequence $(f_j)_{j \in \mathbb{N}}$ in $C_{(0)}(X)$: this implies that for every $x \in X$, the sequence of real numbers $(f_j(x))_{j \in \mathbb{N}}$ is a Cauchy sequence, thus converges. Let us define $f(x) = \lim_j f_j(x)$. Since X is locally compact, each point $x_0 \in X$ has a compact neighborhood K_0. Defining $g_j = f_{j|K_0}, g = f_{|K_0}$, we see that $(g_j)_{j \in \mathbb{N}}$ is a Cauchy sequence in $C(K_0; \mathbb{R})$

converging uniformly towards g: this implies that g is continuous on K_0 since, for $x, x' \in K$, the inequality

$$|g(x') - g(x)| \leq |g(x') - g_j(x')| + |g_j(x') - g_j(x)| + |g_j(x) - g(x)|$$

and the continuity of g_j implies

$$\limsup_{x' \to x} |g(x') - g(x)| \leq 2 \sup_{y \in K_0} |g_j(y) - g(y)| \leq 2 \limsup_{k} \|f_j - f_k\| = 2\varepsilon_j.$$

Since (f_j) is a Cauchy sequence, $\lim_j \varepsilon_j = 0$, and thus g is continuous on K_0, which is a neighborhood of x_0: this implies continuity for f on a neighborhood of any point, thus continuity of f on X.

Let $\delta > 0$ be given. We have, for $x \in X, j \in \mathbb{N}$,

$$|f(x)| \leq |f(x) - f_j(x)| + |f_j(x)| = \lim_{k} |f_k(x) - f_j(x)| + |f_j(x)|$$

$$\leq \limsup_{k} \|f_k - f_j\| + |f_j(x)| = \varepsilon_j + |f_j(x)|.$$

Let j be such that $\varepsilon_j \leq \delta/2$ (possible since $\lim_j \varepsilon_j = 0$) and let $K_{j,\delta}$ be a compact subset such that $\sup_{K_{j,\delta}^c} |f_j| \leq \delta/2$ (possible since $f_j \in C_{(0)}(X)$). We obtain $\sup_{K_{j,\delta}^c} |f| \leq \delta$ and f belongs to $C_{(0)}(X)$. Moreover the inequality

$$|f(x) - f_j(x)| = \lim_{k} |f_k(x) - f_j(x)| \leq \limsup_{k} \|f_k - f_j\| = \varepsilon_j$$

implies $\lim_j \|f - f_j\| = 0$, that is the convergence of the sequence (f_j) towards f in $C_{(0)}(X)$.

(3) Let $\varepsilon > 0$ be given and let $f \in C_{(0)}(X)$. There exists a compact set K such that $\sup_{K^c} |f| \leq \varepsilon$. On the other hand, using Urysohn's Lemma (cf. Exercise 2.8.2), we may find a function $\varphi \in C_c(X; [0,1])$ such that $\varphi_{|K} = 1$. The function $g = f\varphi$ belongs to $C_c(X)$ and we have

$$|g(x) - f(x)| = \mathbf{1}_{K^c}(x)|f(x)|(1 - \varphi(x)) \leq \varepsilon,$$

so that $\|g - f\| \leq \varepsilon$, proving the density of $C_c(X)$ in $C_{(0)}(X)$.

Exercise 3.7.27.

(1) *Let $(X, \mathcal{M}, \mu)$ be a measure space where μ is a positive measure. Show that if $\mu(X) < +\infty$, the assumptions $1 \leq q \leq p \leq +\infty$ imply $L^p(\mu) \subset L^q(\mu)$ continuously. Show that the conditions $1 \leq q \leq p \leq +\infty$ imply $L^p_{\mathrm{loc}}(\mathbb{R}^n) \subset L^q_{\mathrm{loc}}(\mathbb{R}^n)$.*

(2) *Show that the conditions $1 < q < p < +\infty$ imply $\ell^1(\mathbb{N}) \subset \ell^q(\mathbb{N}) \subset \ell^p(\mathbb{N}) \subset \ell^\infty(\mathbb{N})$ with continuous injections and strict inclusions. Show that the inclusion*

$$\ell^1(\mathbb{N}) \subset \cap_{q > 1} \ell^q(\mathbb{N}) \quad \text{is strict.}$$

(3) *Let $p, q \in [1, +\infty]$ be two distinct indices. Show that $L^p(\mathbb{R}^n)$ is not included in $L^q(\mathbb{R}^n)$.*

Answer. Exercise 3.7.10 gives several details related to the present exercise.

(1) Using Hölder's inequality, we get

$$\|u\|_{L^q}^q = \int_X |u|^q d\mu \leq \||u|^q\|_{L^{p/q}} \|1\|_{L^r}, \qquad \frac{q}{p} + \frac{1}{r} = 1,$$

so that $\|u\|_{L^q} \leq \|u\|_{L^p} \mu(X)^{\frac{1}{q}-\frac{1}{p}}$. The same proof gives the inclusion of local spaces since we integrate on compact sets. Note that for L^p_{loc} spaces, the exponent p is an index of regularity.

(2) Let $1 < q < p < +\infty$ and let $x = (x_n)_{n \in \mathbb{N}}$ an element ℓ^q. We have

$$\|x\|_{\ell^p}^p = \sum_{n \geq 0} |x_n|^p \leq \sup_{n \subset \mathbb{N}} |x_n|^{p-q} \sum_{n \geq 0} |x_n|^q \leq \left(\sum_{n \in \mathbb{N}} |x_n|^q\right)^{\frac{p}{q}-1+1}$$

so that $\|x\|_{\ell^p} \leq \|x\|_{\ell^q}$ and this works as well for $q = 1$ and $p = +\infty$, proving the continuous injections. The inclusions are strict since for $\alpha > 0, 1 \leq p < +\infty$, we have

$$(n^{-\alpha})_{n \geq 1} \in \ell^p \iff \alpha p > 1,$$

so that for $1 < r_1 < q < r_2 < p < r_3 < +\infty$,

$$(n^{-1/r_1})_{n \geq 1} \in \ell^q \backslash \ell^1, \quad (n^{-1/r_2})_{n \geq 1} \in \ell^p \backslash \ell^q, \quad (n^{-1/r_3})_{n \geq 1} \in \ell^\infty \backslash \ell^p.$$

Moreover the sequence

$$\left(\frac{1}{n \ln n}\right)_{n \geq 2} \in \cap_{q > 1} \ell^q(\mathbb{N}) \backslash \ell^1(\mathbb{N}),$$

proving the last assertion of question 2. Similarly the inclusion

$$L^\infty_{\text{loc}}(\mathbb{R}^n) \subset \cap_{1 \leq p < +\infty} L^p_{\text{loc}}(\mathbb{R}^n)$$

is strict since $\ln|x| \in \cap_{1 \leq p < +\infty} L^p_{\text{loc}}(\mathbb{R}^n) \backslash L^\infty_{\text{loc}}(\mathbb{R}^n)$. Also for $1 < q < +\infty$, the inclusion

$$L^q_{\text{loc}}(\mathbb{R}^n) \subset \cap_{1 \leq p < q} L^p_{\text{loc}}(\mathbb{R}^n)$$

is strict since $|x|^{-\frac{n}{q}} \in \cap_{1 \leq p < q} L^p_{\text{loc}}(\mathbb{R}^n) \backslash L^q_{\text{loc}}(\mathbb{R}^n)$.

(3) See Exercises 3.7.16, 3.7.17. We note also that for $1 \leq p < q \leq +\infty$ and $\chi \in C^0_c(\mathbb{R}^n), \chi(0) = 1$,

$$\chi(x)|x|^{-\frac{n}{p}+\epsilon} \in L^p, \quad \chi(x)|x|^{-\frac{n}{p}+\epsilon} \notin L^q,$$

provided $\epsilon > 0$, $-\frac{qn}{p} + q\epsilon < -n$, i.e., $0 < \epsilon < \frac{n}{q}\left(\frac{q}{p} - 1\right)$. Moreover, we have

$$(1+|x|)^{-\frac{n}{q}-\sigma} \in L^q, \quad (1+|x|)^{-\frac{n}{q}-\sigma} \notin L^p,$$

provided $0 < \sigma$, $\frac{np}{q} + \sigma p < n$ i.e., $0 < \sigma < \frac{n}{p}\left(1 - \frac{p}{q}\right).$

Exercise 3.7.28. *Let m be an integer ≥ 1. We denote by $\langle , \rangle$ the standard dot-product on $\mathbb{R}^m$. Let A be a real $m \times m$, positive definite symmetric matrix (i.e., $\langle Ax, x \rangle > 0$ for $x \neq 0$).*

(1) *Show that the function f defined by $f(x) = \exp\{-\langle Ax, x \rangle\}$ belongs to $L^1(\mathbb{R}^m)$.*

(2) *Show that*
$$\int_{\mathbb{R}^m} \exp\{-\langle Ax, x \rangle\} dx = \pi^{m/2} (\det A)^{-1/2}.$$

(3) *Let B be an $m \times m$ matrix. Show that*
$$\int_{\mathbb{R}^m} \langle Bx, x \rangle \exp -\{\langle Ax, x \rangle\} dx = \frac{1}{2} \pi^{m/2} (\det A)^{-1/2} \operatorname{trace}(BA^{-1}).$$

(4) *Let F be the function from $\mathbb{R}$ into $\mathbb{C}$ defined by $F(t) = \int_{\mathbb{R}} e^{itx} e^{-x^2} dx$. Show that F is of class C^1 on $\mathbb{R}$ and verifies $2F'(t) + tF(t) = 0$. Give an explicit expression for F.*

(5) *For $y \in \mathbb{R}^m$, calculate $\displaystyle\int_{\mathbb{R}^m} \exp\{i\langle y, x \rangle - \langle Ax, x \rangle\} dx$.*

Answer. (1) The function f is continuous on $\mathbb{R}^m$. There exists $\Omega \in O(m)$ such that $A = \Omega D {}^t\Omega$ where D is the diagonal matrix with the (positive) eigenvalues of A, denoted by α_i. The function f satisfies
$$\exp\{-\langle Ax, x \rangle\} \leq \exp\{-\alpha_{min} \|x\|^2\}, \quad \alpha_{min} = \min_{1 \leq i \leq m} \alpha_i > 0,$$

which implies integrability.

(2) We have
$$\int_{\mathbb{R}^m} \exp\{-\langle Ax, x \rangle\} dx = \int_{\mathbb{R}^m} \exp\{-\langle \Omega D {}^t\Omega x, x \rangle\} dx$$
$$= \int_{\mathbb{R}^m} \exp\{-\langle D {}^t\Omega x, {}^t\Omega x \rangle\} dx,$$

and with the change of variables $y = {}^t\Omega x$, we get, since $|\det \Omega| = 1$:
$$\int_{\mathbb{R}^m} \exp\{-\langle Ax, x \rangle\} dx = \int_{\mathbb{R}^m} \exp\{-\langle Dy, y \rangle\} |\det \Omega| \, dy$$
$$= \int_{\mathbb{R}^m} \exp\Big\{-\sum_{i=1}^{m} \alpha_i y_i^2\Big\} dy.$$

Since $\det A = \prod_{i=1}^{m} \alpha_i$, we get
$$\int_{\mathbb{R}^m} \exp\{-\langle Ax, x \rangle\} dx = \prod_{i=1}^{m} \int_{\mathbb{R}} \exp\{-\alpha_i y_i^2\} dy_i$$
$$= \prod_{i=1}^{m} \sqrt{\frac{1}{\alpha_i}} \int_{\mathbb{R}} \exp\{-t_i^2\} dt_i = \sqrt{\frac{\pi^m}{\det A}}.$$

(3) The same calculation as in the previous question gives

$$I_{A,B} = \int_{\mathbb{R}^m} e^{-\langle Ax,x\rangle}\langle Bx,x\rangle dx = \int_{\mathbb{R}^m} e^{-\langle Dy,y\rangle}\langle {}^t\Omega B\Omega y, y\rangle dy$$

$$= \int_{\mathbb{R}^m} \sum_{1\le j,k\le m} c_{j,k} y_j y_k \exp\left\{-\pi \sum_{1\le j\le m} \alpha_j y_j^2\right\} dy$$

$$= \int_{\mathbb{R}^m} \sum_{1\le j\le m} c_{j,j} y_j^2 \exp\left\{-\sum_{1\le j\le m} \alpha_j y_j^2\right\} dy,$$

with $(c_{j,k})_{1\le j,k\le m} = {}^t\Omega B\Omega$. We note that for $a > 0$,

$$\int_{\mathbb{R}} e^{-at^2} t^2 dt = -\frac{d}{da}\left(\int_{\mathbb{R}} e^{-at^2} dt\right) = -\frac{d}{da}(\pi^{1/2} a^{-1/2}) = \frac{1}{2}\pi^{1/2} a^{-3/2},$$

so that

$$I_{A,B} = \sum_{1\le j\le m} \frac{1}{2} c_{j,j} \alpha_j^{-1} \pi^{m/2} \prod_{1\le k\le m} \alpha_k^{-1/2} = \frac{1}{2}\pi^{m/2}(\det A)^{-1/2} \sum_{1\le j\le m} c_{j,j}\alpha_j^{-1}.$$

Since $\operatorname{trace} MN = \operatorname{trace} NM$, we have

$$\sum_{1\le j\le m} c_{j,j}\alpha_j^{-1} = \operatorname{trace}\left({}^t\Omega B\Omega D^{-1}\right) = \operatorname{trace}\left({}^t\Omega B\Omega ({}^t\Omega A\Omega)^{-1}\right)$$

$$= \operatorname{trace}\left({}^t\Omega B A^{-1}\Omega\right) = \operatorname{trace} B A^{-1},$$

which is the sought result.

(4) We may apply Theorem 3.3.4:

(i) For all t, the mapping $x \longmapsto e^{itx}e^{-x^2}$ is continuous and $\sup_{t\in\mathbb{R}} |e^{itx}e^{-x^2}| = e^{-x^2}$ which is integrable on $\mathbb{R}$. F is thus well defined on $\mathbb{R}$.

(ii) For all x in $\mathbb{R}$, the mapping $t \longmapsto e^{itx}e^{-x^2}$ is of class C^1 on $\mathbb{R}$ with derivative $ixe^{itx}e^{-x^2}$.

(iii) Moreover $\sup_{t\in\mathbb{R}} |ixe^{itx}e^{-x^2}| = |x|e^{-x^2}$ which is integrable on $\mathbb{R}$.

As a result F is of class C^1 on $\mathbb{R}$ and

$$\forall t \in \mathbb{R}, \qquad F'(t) = \int_{\mathbb{R}} ixe^{itx}e^{-x^2}\, dx.$$

Integrating by parts gives

$$F'(t) = \left[-\frac{i}{2}e^{itx}e^{-x^2}\right] + \frac{i}{2}\int_{\mathbb{R}} it e^{itx}e^{-x^2}\, dx = -\frac{t}{2}F(t).$$

Since $F(0) = \sqrt{\pi}$ we obtain $\forall t \in \mathbb{R}, \quad F(t) = \sqrt{\pi}\exp\{-t^2/4\}.$

(5) As in the first question

$$\int_{\mathbb{R}^m} \exp i\langle y, x\rangle \exp -\langle Ax, x\rangle dx = \int_{\mathbb{R}^m} \exp i\langle y, x\rangle \exp -\langle \Omega D^t \Omega x, x\rangle dx$$

$$= \int_{\mathbb{R}^m} \exp i\langle y, x\rangle \exp -\langle D^t \Omega x, {}^t\Omega x\rangle dx.$$

The change of variables $z = {}^t\Omega x$ gives

$$\int_{\mathbb{R}^m} \exp i\langle y, x\rangle \exp -\langle Ax, x\rangle dx = \int_{\mathbb{R}^m} \exp i\langle y, \Omega z\rangle \exp -\langle Dz, z\rangle |\det \Omega| dz$$

$$= \int_{\mathbb{R}^m} \exp i\langle {}^t\Omega y, z\rangle \exp -\langle Dz, z\rangle dz,$$

so that

$$\int_{\mathbb{R}^m} \exp i\langle y, x\rangle \exp -\langle Ax, x\rangle dx = \int_{\mathbb{R}^m} \prod_{j=1}^m \exp i({}^t\Omega y)_j z_j \exp -\alpha_j z_j^2 dz$$

$$= \prod_{j=1}^m \int_{\mathbb{R}} \exp i({}^t\Omega y)_j z_j \exp -\alpha_j z_j^2 dz_j.$$

Using the change of variable $x_j = \sqrt{\alpha_j} z_j$ in each integral we get

$$\int_{\mathbb{R}^m} \exp i\langle y, x\rangle \exp -\langle Ax, x\rangle dx = \prod_{j=1}^m \frac{1}{\sqrt{\alpha_j}} \int_{\mathbb{R}} \left(\exp i({}^t\Omega y)_j x_j / \sqrt{\alpha_j} \right) \exp -x_j^2 dx_j$$

$$= \prod_{j=1}^m \frac{1}{\sqrt{\alpha_j}} F(({}^t\Omega y)_j / \sqrt{\alpha_j}),$$

and the previous question gives

$$\int_{\mathbb{R}^m} \exp i\langle y, x\rangle \exp -\langle Ax, x\rangle dx = \sqrt{\frac{\pi^m}{\det A}} \exp -\frac{1}{4} \sum_{j=1}^m \frac{1}{\alpha_j} ({}^t\Omega y)_j^2$$

$$= \sqrt{\frac{\pi^m}{\det A}} \exp -\frac{1}{4} \langle D^{-1}\,{}^t\Omega y, {}^t\Omega y\rangle$$

so that, since $A^{-1} = \Omega D^{-1}\,{}^t\Omega$,

$$\int_{\mathbb{R}^m} \exp i\langle y, x\rangle \exp -\langle Ax, x\rangle dx = \sqrt{\frac{\pi^m}{\det A}} \exp -\frac{1}{4} \langle A^{-1} y, y\rangle.$$

Exercise 3.7.29. *We define* c_0 *as the space of sequences of complex numbers converging to* 0.

(1) *Show that the space* c_0 *is a closed subspace of* ℓ^∞.

(2) *Show that the spaces* c_0, ℓ^p, *for* $1 \le p < +\infty$ *are separable.*

Answer. (1) Let $(u_n)_{n\in\mathbb{N}}$ be a sequence in c_0 converging towards u in ℓ^∞. Each u_n is a sequence $(a_{k,n})_{k\in\mathbb{N}}$ such that $\lim_k a_{k,n} = 0$ and $u = (b_k)_{k\in\mathbb{N}} \in \ell^\infty$. We have

$$|b_k| \le |b_k - a_{k,n}| + |a_{k,n}| \le \|u - u_n\| + |a_{k,n}|,$$

so that $\forall n \in \mathbb{N}$, $\limsup_k |b_k| \le \|u - u_n\|$, and taking the infimum on n of the rhs implies $\limsup_k |b_k| = 0$, and $u \in c_0$.

(2) Let us define the countable set

$$D = \cup_{N\in\mathbb{N}}\{(y_k)_{k\in\mathbb{N}}, y_k \in \mathbb{Q} + i\mathbb{Q},\ y_k = 0 \text{ for } k > N\}.$$

Then D is dense in c_0: let $u = (x_k)_{k\in\mathbb{N}}$ be in c_0 and let $\epsilon > 0$ be given. Then there exists N_ϵ such that $\sup_{k \ge N_\epsilon} |x_k| < \epsilon/2$. Moreover, by density of $\mathbb{Q}$ in $\mathbb{R}$, there exists $(y_k)_{0\le k\le N_\epsilon}$ such that each $y_k \in \mathbb{Q} + i\mathbb{Q}$ and $\max_{0\le k\le N_\epsilon} |x_k - y_k| < \epsilon/2$. With $v = (y_k)_{k\in\mathbb{N}}$ ($y_k = 0$ for $k > N_\epsilon$), we have $v \in D$ and

$$\|u - v\|_\infty \le \max_{0\le k\le N_\epsilon} |x_k - y_k| + \sup_{k\ge N_\epsilon} |x_k| < \epsilon,$$

proving the sought property.

The set D is also dense in ℓ^p for $1 \le p < +\infty$: let $u = (x_k)_{k\in\mathbb{N}}$ be in ℓ^p and let $\epsilon > 0$ be given. Then there exists N_ϵ such that

$$\sum_{k\ge N_\epsilon} |x_k|^p < \epsilon^p/2.$$

Moreover, by density of $\mathbb{Q}$ in $\mathbb{R}$, there exists $(y_k)_{0\le k\le N_\epsilon}$ such that each $y_k \in \mathbb{Q} + i\mathbb{Q}$ and

$$\max_{0\le k\le N_\epsilon} |x_k - y_k|^p < \frac{\epsilon^p}{2N_\epsilon + 1}.$$

With $v = (y_k)_{k\in\mathbb{N}}$ ($y_k = 0$ for $k > N_\epsilon$), we have $v \in D$ and

$$\|u - v\|_p^p = \sum_{0\le k\le N_\epsilon} |x_k - y_k|^p + \sum_{k\ge N_\epsilon} |x_k|^p < \epsilon^p,$$

proving the sought property.

Exercise 3.7.30. *Let* $(X, \mathcal{M}, \mu)$ *be a measure space where* μ *is a positive measure. Prove that* $L^1(\mu) \subset L^\infty(\mu)$ *if and only if*

$$\inf_{\substack{E\in\mathcal{M} \\ \mu(E)>0}} \mu(E) > 0. \tag{3.7.6}$$

Prove that, when this condition is satisfied, we have for $1 \le p \le q \le \infty$, $L^p(\mu) \subset L^q(\mu)$. *Give an example of such a measured space.*

Answer. Let us assume first that (3.7.6) holds with an infimum equal to $\alpha > 0$ and let $f \in L^1(\mu)$ be different from 0. If f were not in $L^\infty(\mu)$, for every $k \in \mathbb{N}$, we would have

$$\mu\big(\underbrace{\{x \in X, |f(x)| > k\}}_{E_k}\big) > 0,$$

so that

$$+\infty > \|f\|_{L^1(\mu)} \geq \int_{E_k} |f|\,d\mu \geq k\mu(E_k) \geq k\alpha \xrightarrow[k \to +\infty]{} +\infty,$$

which is impossible. We obtain thus $f \in L^\infty(\mu)$. With $\epsilon > 0$, assuming that f is not 0 and $\epsilon \in \big(0, \|f\|_{L^\infty(\mu)}\big)$, we define

$$F_\epsilon = \{x \in X, |f| > \|f\|_{L^\infty(\mu)} - \epsilon\}.$$

We find $\mu(F_\epsilon) > 0$ and thus $\mu(F_\epsilon) \geq \alpha > 0$. As a result for every $\epsilon \in \big(0, \|f\|_{L^\infty(\mu)}\big)$, we get

$$\|f\|_{L^1(\mu)} \geq \int_{F_\epsilon} |f|\,d\mu \geq \big(\|f\|_{L^\infty(\mu)} - \epsilon\big)\mu(F_\epsilon) \geq \alpha\big(\|f\|_{L^\infty(\mu)} - \epsilon\big),$$

implying $\|f\|_{L^\infty(\mu)} \leq \alpha^{-1}\|f\|_{L^1(\mu)}$. We remark that if $1 \leq p < +\infty$, we find also under (3.7.6),

$$f \in L^p(\mu) \Longrightarrow |f|^p \in L^1(\mu) \Longrightarrow |f|^p \in L^\infty(\mu).$$

We note also that, assuming (3.7.6) and $1 \leq p < q < +\infty$, we find from the previous argument that if $f \in L^p(\mu)$, we obtain that $|f|^p$ belongs to $L^\infty(\mu)$ with

$$\||f|^p\|_{L^\infty} \leq \alpha^{-1}\||f|^p\|_{L^1} = \alpha^{-1}\|f\|_{L^p}^p$$

$$\Longrightarrow \int |f|^q\,d\mu \leq \int |f|^p d\mu \, \|f\|_{L^\infty}^{q-p} \leq (\alpha^{-1})^{\frac{q-p}{p}}\|f\|_{L^p}^{p+q-p}$$

$$\Longrightarrow \|f\|_{L^q(\mu)} \leq (\alpha^{-1})^{\frac{q-p}{pq}}\|f\|_{L^p},$$

proving that $f \in L^q(\mu)$ (with a continuous injection).

Conversely, let us assume that $L^1(\mu) \subset L^\infty(\mu)$. If for any $k \in \mathbb{N}^*$, we could find $E_k \in \mathcal{M}$ such that $0 < \mu(E_k) < 2^{-k}$, then

$$\|f = \sum_{k \geq 1} k\mathbf{1}_{E_k}\|_{L^1(\mu)} \leq \sum_{k \geq 1} k\mu(E_k) < +\infty \Longrightarrow f \in L^1(\mu) \Longrightarrow f \in L^\infty(\mu),$$

but since $\mu(E_k) > 0$, we have $\|f\|_{L^\infty(\mu)} \geq k$ for all $k \in \mathbb{N}$, which is impossible.

The most typical example is given by the ℓ^p spaces $(1 \leq p \leq +\infty)$ which are the L^p spaces for the measured space

$$(\mathbb{N}, \mathcal{P}(\mathbb{N}), \mu), \quad \mu = \sum_{k \in \mathbb{N}} \delta_k.$$

Here μ is the counting measure on $\mathbb{N}$ so that $\mu(E) \geq 1$ if E is not empty.

Exercise 3.7.31. *Let $(X, \mathcal{M}, \mu)$ be a measure space where μ is a positive measure. Let $f_1, \ldots, f_N$ be non-negative measurable functions and let $p_1, \ldots, p_N \in [1, +\infty]$ such that*

$$\sum_{1 \leq j \leq N} \frac{1}{p_j} = 1.$$

Prove that $\int_X f_1 \ldots f_N \, d\mu \leq \prod_{1 \leq j \leq N} \|f_j\|_{L^{p_j}(\mu)}.$

Answer. When $N = 2$, this is Hölder's inequality. We may assume that all f_j are not vanishing μ-a.e. (otherwise the lhs is 0) and that each f_j belongs to $L^{p_j}(\mu)$ (otherwise the rhs is $+\infty$ as the product of positive quantities in $\overline{\mathbb{R}}_+$ with one of them $+\infty$). Induction on N: let $N \geq 2$ and $p_1, \ldots, p_{N+1} \in [1, +\infty]$ with $\sum_{1 \leq j \leq N+1} \frac{1}{p_j} = 1$. Applying Hölder's inequality we find

$$(\sharp) \qquad \int_X f_1 \ldots f_N f_{N+1} \, d\mu \leq \left\| \prod_{1 \leq j \leq N} f_j \right\|_{L^{p'_{N+1}}(\mu)} \|f_{N+1}\|_{L^{p_{N+1}}(\mu)}.$$

Since $\sum_{1 \leq j \leq N} \frac{p'_{N+1}}{p_j} = 1$ (ensuring that $p_j / p'_{N+1} \geq 1$) and

$$\left\| \prod_{1 \leq j \leq N} f_j \right\|_{L^{p'_{N+1}}(\mu)} = \left\| \prod_{1 \leq j \leq N} f_j^{p'_{N+1}} \right\|_{L^1(\mu)}^{\frac{1}{p'_{N+1}}},$$

we may use the induction hypothesis to obtain

$$\left\| \prod_{1 \leq j \leq N} f_j \right\|_{L^{p'_{N+1}}(\mu)} \leq \left(\prod_{1 \leq j \leq N} \|f_j^{p'_{N+1}}\|_{L^{p_j/p'_{N+1}}} \right)^{\frac{1}{p'_{N+1}}}.$$

The rhs of that inequality equals $\prod_{1 \leq j \leq N} \|f_j\|_{L^{p_j}}$, and with $(\sharp)$ this provides the answer.

Chapter 4

Integration on a Product Space

4.1 Product of measurable spaces

Definition 4.1.1 (σ-algebra on a product space). Let $(X_1, \mathcal{M}_1), (X_2, \mathcal{M}_2)$ be measurable spaces. We define the *product σ-algebra* of $\mathcal{M}_1$ and $\mathcal{M}_2$ as the σ-algebra on $X_1 \times X_2$ generated by the sets $A_1 \times A_2$, where $A_j \in \mathcal{M}_j, j = 1, 2$ (such a set $A_1 \times A_2$ will be called a *Cartesian rectangle*, CAR for short). That σ-algebra will be denoted by $\mathcal{M}_1 \otimes \mathcal{M}_2$.

We note that $\mathcal{M}_1 \otimes \mathcal{M}_2$ is the smallest σ-algebra (i.e., the intersection of σ-algebras) on $X_1 \times X_2$ such that the canonical projections $\pi_j : X_1 \times X_2 \to X_j, \pi_j((x_1, x_2)) = x_j, j = 1, 2$ are measurable. First of all π_1 is measurable since for $A_1 \in \mathcal{M}_1$, we have $\pi_1^{-1}(A_1) = A_1 \times X_2$ which is a CAR, thus belongs to $\mathcal{M}_1 \otimes \mathcal{M}_2$ (same for π_2). Moreover if $\mathcal{T}$ is a σ-algebra on $X_1 \times X_2$ such that π_j are measurable, then for $A_j \in \mathcal{M}_j$, $\mathcal{T}$ contains $\pi_1^{-1}(A_1) = A_1 \times X_2$ and $\pi_2^{-1}(A_2) = X_1 \times A_2$, thus their intersection

$$(A_1 \times X_2) \cap (X_1 \times A_2) = A_1 \times A_2.$$

The σ-algebra $\mathcal{T}$ contains the CAR and thus the σ-algebra generated by the CAR, i.e., $\mathcal{M}_1 \otimes \mathcal{M}_2$.

Remark 4.1.2. Let $f_j : X_j \to \mathbb{C}$ $(j = 1, 2)$ be measurable mappings. We define the tensor product $f_1 \otimes f_2$ by

$$
\begin{aligned}
f_1 \otimes f_2 : \quad X_1 \times X_2 &\to \qquad \mathbb{C} \\
(x_1, x_2) &\mapsto \quad f_1(x_1) f_2(x_2).
\end{aligned}
$$

The mapping $f_1 \otimes f_2$ is the product $(f_1 \circ \pi_1)(f_2 \circ \pi_2)$; since each $f_j \circ \pi_j$ is measurable (cf. Lemma 1.1.6), Theorem 1.2.7 shows that their product is also measurable.

Proposition 4.1.3. *Let $(X_1, \mathcal{M}_1), (X_2, \mathcal{M}_2), (Y, \mathcal{T})$ be measurable spaces and let $f : X_1 \times X_2 \to Y$ be a measurable mapping. Then*

(1) $\forall x_1 \in X_1$, *the mapping* $f(x_1, \cdot) : x_2 \in X_2 \mapsto f(x_1, x_2) \in Y$ *is measurable,*

   $\forall x_2 \in X_2$, *the mapping* $f(\cdot, x_2) : x_1 \in X_1 \mapsto f(x_1, x_2) \in Y$ *is measurable.*

(2) *For* $A \in \mathcal{M}_1 \otimes \mathcal{M}_2$, *and* $(x_1, x_2) \in X_1 \times X_2$, *we define*

$$A(x_1, \cdot) = \{x_2 \in X_2, (x_1, x_2) \in A\},$$
$$A(\cdot, x_2) = \{x_1 \in X_1, (x_1, x_2) \in A\}. \tag{4.1.1}$$

The set $A(x_1, \cdot)$ *belongs to* $\mathcal{M}_2$ *and* $A(\cdot, x_2)$ *belongs to* $\mathcal{M}_1$.

Let us check first Figure 4.1 with the "vertical slice" $A(x_1, \cdot)$. Of course drawing an horizontal slice would be easy, but the picture would not gain much.

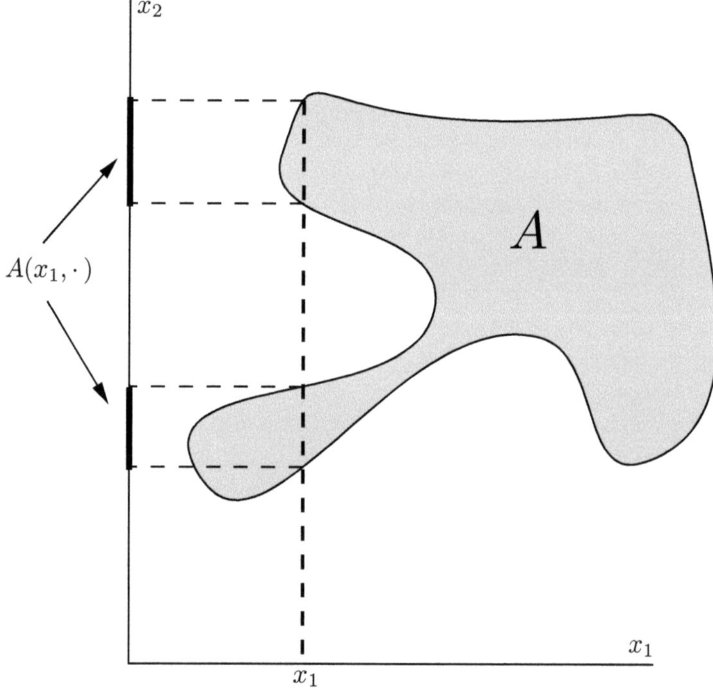

Figure 4.1: VERTICAL SLICE

$$\mathbf{A} \in \mathcal{M}_1 \otimes \mathcal{M}_2, \quad A(x_1, \cdot) = \{x_2 \in X_2, (x_1, x_2) \in A\}$$

Proof of the proposition. Let B be in $\mathcal{T}$. For $x_1 \in X_1$, we have

$$f(x_1, \cdot)^{-1}(B) = \{x_2 \in X_2, f(x_1, x_2) \in B\}$$
$$= \{x_2 \in X_2, (x_1, x_2) \in f^{-1}(B)\} = \left(f^{-1}(B)\right)(x_1, \cdot).$$

Since f is measurable, the set $f^{-1}(B)$ belongs to $\mathcal{M}_1 \otimes \mathcal{M}_2$; it is thus enough to prove (2) to obtain (1). We define

$$\mathcal{M} = \{E \subset X_1 \times X_2, \forall x_1 \in X_1, \forall x_2 \in X_2, E(x_1, \cdot) \in \mathcal{M}_2, E(\cdot, x_2) \in \mathcal{M}_1\}.$$

We note that $E \in \mathcal{M}$ implies $E^c \in \mathcal{M}$: for $x_1 \in X_1$, we have

$$(E^c)(x_1, \cdot) = \{x_2 \in X_2, (x_1, x_2) \in E^c\}$$
$$= \{x_2 \in X_2, (x_1, x_2) \notin E\} = \big(\underbrace{E(x_1, \cdot)}_{\substack{\in \mathcal{M}_2 \\ \text{since } E \in \mathcal{M}}} \big)^c$$

and thus $(E^c)(x_1, \cdot) \in \mathcal{M}_2$ since $\mathcal{M}_2$ is stable by complement as a σ-algebra. We prove as well that, for $x_2 \in X_2$, we have $(E^c)(\cdot, x_2) \in \mathcal{M}_1$, so that $E^c \in \mathcal{M}$. Moreover if $(E_k)_{k \in \mathbb{N}}$ is a sequence of $\mathcal{M}$, then $\cup_{k \in \mathbb{N}} E_k \in \mathcal{M}$: for $x_1 \in X_1$, we have

$$\big(\cup_{k \in \mathbb{N}} E_k\big)(x_1, \cdot) = \{x_2 \in X_2, (x_1, x_2) \in \cup_{k \in \mathbb{N}} E_k\} = \cup_{k \in \mathbb{N}} \big(\underbrace{E_k(x_1, \cdot)}_{\substack{\in \mathcal{M}_2 \\ \text{since } E_k \in \mathcal{M}}} \big)$$

which belongs to $\mathcal{M}_2$ since $\mathcal{M}_2$ is stable by countable union, as a σ-algebra. Since we can get by the same proof, mutatis mutandis, that for $x_2 \in X_2$, $\big(\cup_{k \in \mathbb{N}} E_k\big)(\cdot, x_2)$ belongs to $\mathcal{M}_1$, we have indeed proven that $\cup_{k \in \mathbb{N}} E_k \in \mathcal{M}$. We note also that the CAR belongs to $\mathcal{M}$: let A_j be in $\mathcal{M}_j$, $j = 1, 2$. For $x_1 \in X_1$, we have

$$(A_1 \times A_2)(x_1, \cdot) = \{x_2 \in X_2, (x_1, x_2) \in A_1 \times A_2\} = \left\{\begin{matrix} \emptyset, & \text{if } x_1 \notin A_1 \\ A_2, & \text{if } x_1 \in A_1 \end{matrix}\right] \in \mathcal{M}_2.$$

We prove as well that for $x_2 \in X_2$, we have $(A_1 \times A_2)(\cdot, x_2) \in \mathcal{M}_1$. As a result, $\mathcal{M}$ is a σ-algebra on $X_1 \times X_2$ containing the CAR, and thus the σ-algebra $\mathcal{M}_1 \otimes \mathcal{M}_2$, which is generated by the CAR. This completes the proof of (2) and of the Proposition. $\qquad\square$

Remark 4.1.4. Let $d \in \mathbb{N}$ and let $\mathcal{B}_d$ be the Borel σ-algebra on $\mathbb{R}^d$. Then if $d_1, d_2 \in \mathbb{N}$, we have

$$\mathcal{B}_{d_1+d_2} = \mathcal{B}_{d_1} \otimes \mathcal{B}_{d_2}. \tag{4.1.2}$$

We prove first $\mathcal{B}_{d_1+d_2} \supset \mathcal{B}_{d_1} \otimes \mathcal{B}_{d_2}$: $\mathcal{B}_{d_1+d_2}$ is a σ-algebra such that the projections are measurable (since they are continuous), thus contains the smallest σ-algebra $\mathcal{B}_{d_1} \otimes \mathcal{B}_{d_2}$ making these projections measurable. Moreover, from Lemma 1.2.6, the σ-algebra $\mathcal{B}_{d_1+d_2}$ is generated by the compact rectangles $\prod_{1 \leq j \leq d_1+d_2} [a_j, b_j]$ which are also CARs (equal to $\prod_{1 \leq j \leq d_1} [a_j, b_j] \prod_{d_1+1 \leq j \leq d_1+d_2} [a_j, b_j]$). Consequently, using the notation in Definition 1.1.3, we have

$$\mathcal{B}_{d_1} \otimes \mathcal{B}_{d_2} \subset \mathcal{B}_{d_1+d_2} = \mathcal{M}(\text{compact rectangles}) \subset \mathcal{M}(\text{CAR}) = \mathcal{B}_{d_1} \otimes \mathcal{B}_{d_2}. \qquad\square$$

4.2   Tensor product of sigma-finite measures

Lemma 4.2.1. *Let $(X_1, \mathcal{M}_1, \mu_1)$, $(X_2, \mathcal{M}_2, \mu_2)$ be measure spaces where the μ_j are positive σ-finite measures (i.e., $X_j = \cup_{k\in\mathbb{N}} E_k^j$, with $\mu_j(E_k^j) < +\infty$). Let A be in $\mathcal{M}_1 \otimes \mathcal{M}_2$. Defining $\varphi_1(x_1) = \mu_2(A(x_1, \cdot))$, $\varphi_2(x_2) = \mu_1(A(\cdot, x_2))$, the functions φ_j are $\mathcal{M}_j$ measurable ($j = 1, 2$) and*

(b)
$$\int_{X_1} \varphi_1 d\mu_1 = \int_{X_2} \varphi_2 d\mu_2.$$

Proof. Let us first assume that $A = A_1 \times A_2$ with $A_j \in \mathcal{M}_j$. We have

$$(A_1 \times A_2)(x_1, \cdot) = \{x_2 \in X_2, (x_1, x_2) \in A_1 \times A_2\} = \begin{cases} \emptyset & \text{if} \quad x_1 \notin A_1, \\ A_2 & \text{if} \quad x_1 \in A_1, \end{cases}$$

and this implies

$$\varphi_1(x_1) = \begin{cases} 0 & \text{if} \quad x_1 \notin A_1, \\ \mu_2(A_2) & \text{if} \quad x_1 \in A_1, \end{cases}$$

i.e., $\varphi_1 = \mu_2(A_2) \cdot \mathbf{1}_{A_1}$, $\varphi_2 = \mu_1(A_1) \cdot \mathbf{1}_{A_2}$, so that if $\mu_1(A_1)$ and $\mu_2(A_2)$ are both finite,

$$\int_{X_1} \varphi_1 d\mu_1 = \mu_2(A_2)\mu_1(A_1) = \int_{X_2} \varphi_2 d\mu_2.$$

Moreover if $\mu_2(A_2) = +\infty$ and $\mu_1(A_1) = 0$, we have $\varphi_1 = 0$, μ_1-a.e. and $\varphi_2 = 0$, proving the result in that case as well. If $\mu_2(A_2) = +\infty$ and $\mu_1(A_1) > 0$, we find $\int_{X_1} \varphi_1 d\mu_1 = +\infty = \int_{X_2} \varphi_2 d\mu_2$, so that the sought property is proven when A is a CAR. Let us now define

$$\mathcal{R} = \{A \in \mathcal{M}_1 \otimes \mathcal{M}_2, \text{(b) holds true}\}. \tag{4.2.1}$$

We have already proven that

$$\mathcal{R} \supset \text{CAR}. \tag{4.2.2}$$

Moreover, we claim that if $(A_j)_{j\in\mathbb{N}}$ is an increasing sequence of $\mathcal{R}$, then

$$\cup_{j\in\mathbb{N}} A_j \in \mathcal{R}. \tag{4.2.3}$$

Indeed, defining $\varphi_{1,j}(x_1) = \mu_2(A_j(x_1, \cdot))$, $\quad \varphi_{2,j}(x_2) = \mu_2(A_j(\cdot, x_2))$, the sequence $A_j(x_1, \cdot) = \{x_2 \in X_2, (x_1, x_2) \in A_j\}$ is increasing with union $A(x_1, \cdot)$. As a result, we have $0 \le \varphi_{1,j}(x_1) \uparrow \varphi_1(x_1)$, $0 \le \varphi_{2,j}(x_2) \uparrow \varphi_2(x_2)$, and Beppo Levi's theorem implies

$$\int_{X_1} \varphi_{1,j} d\mu_1 \uparrow \int_{X_1} \varphi_1 d\mu_1 \quad \text{and} \quad \int_{X_2} \varphi_{2,j} d\mu_2 \uparrow \int_{X_2} \varphi_2 d\mu_2.$$

Since each A_j belongs to $\mathcal{R}$, we have $\int_{X_1} \varphi_{1,j} d\mu_1 = \int_{X_2} \varphi_{2,j} d\mu_2$, proving Claim (4.2.3). Moreover, we claim that if $(A_j)_{j\in\mathbb{N}}$ is a sequence of pairwise disjoint elements of $\mathcal{R}$, we have

$$\cup_{j\in\mathbb{N}} A_j \in \mathcal{R}. \tag{4.2.4}$$

In fact, considering the increasing sequence $B_n = \cup_{0 \le j \le n} A_k$, and using (4.2.3), we see that it is enough to check that if A_1, A_2 are disjoint elements of $\mathcal{R}$, then $A_1 \cup A_2 \in \mathcal{R}$. We have indeed

$$(A_1 \cup A_2)(x_1, \cdot) = \{x_2 \in X_2, (x_1, x_2) \in A_1 \cup A_2\} = \underbrace{A_1(x_1, \cdot) \cup A_2(x_1, \cdot)}_{\text{disjoint union}},$$

so that $\mu_2(A_1 \cup A_2)(x_1, \cdot) = \mu_2(A_1(x_1, \cdot)) + \mu_2(A_2(x_1, \cdot))$ and

$$\int_{X_1} \overbrace{\mu_2(A_1 \cup A_2)(x_1, \cdot)}^{\varphi_1(x_1)} d\mu_1(x_1)$$
$$= \int_{X_1} \mu_2(A_1(x_1, \cdot)) d\mu_1(x_1) + \int_{X_1} \mu_2(A_2(x_1, \cdot)) d\mu_1(x_1).$$

Since both A_1, A_2 belong to $\mathcal{R}$, we have proven Claim (4.2.4). Moreover, for $A_1 \in \mathcal{M}_1, A_2 \in \mathcal{M}_2$ with $\mu_j(A_j) < \infty, j = 1, 2$, and for (Q_j) a decreasing sequence in $\mathcal{R}$ such that $A_1 \times A_2 \supset Q_j$, we claim that

$$Q = \cap_j Q_j \in \mathcal{R}. \tag{4.2.5}$$

Indeed, let us define

$$\varphi_{1,j}(x_1) = \mu_2(Q_j(x_1, \cdot)) = \mu_2(\{x_2 \in X_2, (x_1, x_2) \in Q_j\}) \le \mu_2(A_2) < +\infty.$$

Using Proposition 1.4.4(3), we get

$$\varphi_{1,j}(x_1) \to \varphi_1(x_1) = \mu_2(\{x_2 \in X_2, (x_1, x_2) \in Q\}),$$
$$\mu_1(Q_j(\cdot, x_2)) = \varphi_{2,j}(x_2) \to \varphi_2(x_2) = \mu_1(\{x_1 \in X_1, (x_1, x_2) \in Q\}).$$

We have also

$$0 \le \varphi_{1,j}(x_1) \le \mu_2(\{x_2 \in X_2, (x_1, x_2) \in A_1 \times A_2\}) = \psi_1(x_1).$$

But we have already seen in (4.2.1) that $\int_{X_1} \psi_1 d\mu_1 = \mu_1(A_1)\mu_2(A_2)$ (a finite quantity here). We may thus apply the Lebesgue dominated convergence theorem and get

$$\int_{X_1} \varphi_{1,j} d\mu_1 \to \int_{X_1} \varphi_1 d\mu_1 \quad \text{and} \quad \int_{X_2} \varphi_{2,j} d\mu_2 \to \int_{X_2} \varphi_2 d\mu_2.$$

Since Q_j belongs to $\mathcal{R}$, we find $\int_{X_1} \varphi_{1,j} d\mu_1 = \int_{X_2} \varphi_{2,j} d\mu_2$ proving Claim (4.2.5). We need a definition.

Definition 4.2.2. Let X be a set and $\mathcal{S}$ be a subset of the powerset $\mathcal{P}(X)$. The set $\mathcal{S}$ is said to be a *Monotone Class* on X when for $(A_j)_{j \in \mathbb{N}}$ increasing sequence of $\mathcal{S}$, $(B_j)_{j \in \mathbb{N}}$ decreasing sequence of $\mathcal{S}$, $\cup_{j \in \mathbb{N}} A_j \in \mathcal{S}$, $\cap_{j \in \mathbb{N}} B_j \in \mathcal{S}$. Note that if $(\mathcal{T}_i)_{i \in I}$ is a family of monotone classes on X, then $\cap_{i \in I} \mathcal{T}_i$ is also a monotone class on X.

Since μ_1 is σ-finite, we can find a sequence $(X_{1,k})$ of elements of $\mathcal{M}_1$ such that

$$X_1 = \cup_{k\in\mathbb{N}} X_{1,k}, \qquad \mu_1(X_{1,k}) < +\infty.$$

We may as well assume that the $X_{1,k}$ are pairwise disjoint. Let $(X_{2,l})$ be a sequence with the same properties with respect to $(X_2, \mathcal{M}_2, \mu_2)$. We define the set

$$\mathcal{S} = \{A \in \mathcal{M}_1 \otimes \mathcal{M}_2, \ \forall (k,l), \ A \cap (X_{1,k} \times X_{2,l}) \in \mathcal{R}\}. \tag{4.2.6}$$

Then we claim that $\mathcal{S}$ is a monotone class. Indeed, let $A_j \in \mathcal{M}_1 \otimes \mathcal{M}_2$ be an increasing sequence such that $A_j \cap (X_{1,k} \times X_{2,l}) \in \mathcal{R}$. From (4.2.3), we find that $\cup_j A_j$ belongs to $\mathcal{S}$. Similarly (4.2.5) and the fact that $\mu_1(X_{1,k}), \mu_2(X_{2,l})$ are both finite imply the property on decreasing sequences, proving the claim. As a result, $\mathcal{S}$ is a monotone class included in $\mathcal{M}_1 \otimes \mathcal{M}_2$, containing the CAR ((4.2.2)) and countable pairwise disjoint unions of CARs ((4.2.4)).

Lemma 4.2.3. $\mathcal{M}_1 \otimes \mathcal{M}_2$ *is the smallest monotone class on* $X_1 \times X_2$ *which contains finite unions of* CARs.

Let us take provisionally that lemma for granted. We get then $\mathcal{S} = \mathcal{M}_1 \otimes \mathcal{M}_2$. As a consequence, if $A \in \mathcal{M}_1 \otimes \mathcal{M}_2$, then $A \cap (X_{1,k} \times X_{2,l})$ satisfy the property of Lemma 4.2.1, so that using

$$A = \cup_{k,l}\Big\{A \cap (X_{1,k} \times X_{2,l})\Big\} \quad \text{(disjoint union)},$$

we find from (4.2.4) that $A \in \mathcal{R}$, concluding the proof of Lemma 4.2.1. $\qquad\square$

Proof of Lemma 4.2.3. $\mathcal{M}_1 \otimes \mathcal{M}_2$ is a σ-algebra, thus a monotone class. We may thus consider the monotone class $\mathcal{T}$ defined as

$$\mathcal{T} = \text{intersection of monotone classes containing the finite unions of CARs.}$$

Since $\mathcal{M}_1 \otimes \mathcal{M}_2$ is a monotone class containing the finite unions of CARs, we get that $\mathcal{M}_1 \otimes \mathcal{M}_2 \supset \mathcal{T}$. We need to prove the other inclusion. Note that it is enough to prove that $\mathcal{T}$ is a σ-algebra: if that it is so, $\mathcal{T}$ will contain the CAR, thus the σ-algebra generated by the CAR, that is $\mathcal{M}_1 \otimes \mathcal{M}_2$. We note that

$$(A_1 \times A_2) \cap (B_1 \times B_2) = (A_1 \cap B_1) \times (A_2 \cap B_2), \tag{4.2.7}$$
$$(A_1 \times A_2)\backslash(B_1 \times B_2) = \big[(A_1\backslash B_1) \times A_2\big] \cup \big[(A_1 \cap B_1) \times (A_2\backslash B_2)\big]. \tag{4.2.8}$$

We see that the difference of two CARs is a disjoint union of two CARs. Then the symmetric difference of two CARs is as disjoint union of four CARs, the union of two CARs is a disjoint union of five CARs. We find that the set

$$\mathcal{E} = \text{finite disjoint unions of CARs,} \tag{4.2.9}$$

is stable by union, intersection, and symmetric difference. For $P \subset X_1 \times X_2$, we set

$$\Omega(P) = \{Q \subset X_1 \times X_2, \ P\backslash Q, \ Q\backslash P, \ P \cup Q \in \mathcal{T}\}.$$

We see at once that

$$Q \in \Omega(P) \Longleftrightarrow P \in \Omega(Q). \tag{4.2.10}$$

Moreover, if $(Q_j)_{j\in\mathbb{N}}$ is an increasing sequence of $\Omega(P)$ and $Q = \cup_j Q_j$, we have

$$P\backslash Q = P \cap Q^c = P \cap \cap_j Q_j^c = \cap_j (P \cap Q_j^c),$$

and since $P \cap Q_j^c$ is decreasing and in $\mathcal{T}$ (which is a monotone class), we find that $P\backslash Q \in \mathcal{T}$. We prove similarly that $Q\backslash P, P \cup Q \in \mathcal{T}$. As a result, $\Omega(P)$ is a monotone class. Let $P \in \mathcal{E}$: if $Q \in \mathcal{E}$, we have $Q \in \Omega(P)$ since we have already seen that $\mathcal{E}$ is stable by union, intersection and symmetric difference. We find

$$\mathcal{E} \subset \Omega(P) \ \text{ for } P \in \mathcal{E}.$$

Since $\Omega(P)$ is a monotone class, using the very definition of $\mathcal{T}$, we find

(♯) $$\mathcal{T} \subset \Omega(P) \ \text{ for } P \in \mathcal{E}.$$

Consequently, if $Q \in \mathcal{T}$, we have

$$P \in \mathcal{E} \underset{(♯)}{\Longrightarrow} \mathcal{T} \subset \Omega(P) \underset{Q\in\mathcal{T}}{\Longrightarrow} Q \in \Omega(P) \underset{(4.2.10)}{\Longrightarrow} P \in \Omega(Q),$$

so that $\mathcal{E} \subset \Omega(Q)$. Since $\Omega(Q)$ is a monotone class, we find

$$\mathcal{T} \subset \Omega(Q) \ \text{ for } Q \in \mathcal{T}.$$

Finally for $P, Q \in \mathcal{T}$, we have $\mathcal{T} \subset \Omega(Q)$ which implies $P \in \Omega(Q)$ and thus $P\backslash Q, Q\backslash P, P \cup Q \in \mathcal{T}$. We get then

$$X_1 \times X_2 \in \mathcal{E} \subset \mathcal{T},$$

$$\text{if } Q \in \mathcal{T}, Q^c = \Big(\underbrace{X_1 \times X_2}_{\in\mathcal{T}} \backslash \overbrace{Q}^{\in\mathcal{T}} \Big) \in \mathcal{T},$$

$(Q_j \in \mathcal{T})_{j\in\mathbb{N}}, P_n = \cup_{1\le j\le n} Q_j \in \mathcal{T}$, monotone class, thus $\cup_n P_n \in \mathcal{T}$,

proving that $\mathcal{T}$ is a σ-algebra, completing the proof of Lemma 4.2.3. $\qquad\square$

Definition 4.2.4 (Tensor product of σ-finite measures). Let $(X_1, \mathcal{M}_1, \mu_1)$ and $(X_2, \mathcal{M}_2, \mu_2)$ be measure spaces where each μ_j is a σ-finite positive measure. For $A \in \mathcal{M}_1 \otimes \mathcal{M}_2$, using the notation (4.1.1) and Lemma 4.2.1 we set

$$(\mu_1 \otimes \mu_2)(A) = \int_{X_1} \mu_2(A(x_1, \cdot)) d\mu_1(x_1) = \int_{X_2} \mu_1(A(\cdot, x_2)) d\mu_2(x_2).$$

Then $\mu_1 \otimes \mu_2$ is a σ-finite positive measure. From the proof of Lemma 4.2.1 we find that for $A_j \in \mathcal{M}_j$, $j = 1, 2$, $(\mu_1 \otimes \mu_2)(A_1 \times A_2) = \mu_1(A_1) \cdot \mu_2(A_2)$ (with the convention $0 \cdot \infty = 0$).

Indeed, if $(A_k)_{k \in \mathbb{N}}$ is a pairwise disjoint sequence of $\mathcal{M}_1 \otimes \mathcal{M}_2$, if $x_1 \in X_1$, then $A_k(x_1, \cdot)$ is measurable (Proposition 4.1.3 (2)) and, using (4.1.1), we find $\left(\cup_{k \in \mathbb{N}} A_k\right)(x_1, \cdot) = \cup_{k \in \mathbb{N}} \underbrace{A_k(x_1, \cdot)}_{\text{pairwise disjoint}}$, so that

$$\mu_2\left(\left(\cup_{k \in \mathbb{N}} A_k\right)(x_1, \cdot)\right) = \sum_{k \in \mathbb{N}} \mu_2\left(A_k(x_1, \cdot)\right). \tag{4.2.11}$$

Lemma 4.2.1 implies that the mappings $x_1 \mapsto \mu_2\left(A_k(x_1, \cdot)\right)$ are measurable and Corollary 1.6.2 gives

$$(\mu_1 \otimes \mu_2)(\cup_{k \in \mathbb{N}} A_k) \overset{\text{déf.}}{=} \int_{X_1} \mu_2\left(\left(\cup_{k \in \mathbb{N}} A_k\right)(x_1, \cdot)\right) d\mu_1(x_1)$$

$$\overset{(4.2.2)}{=} \int_{X_1} \left(\sum_{k \in \mathbb{N}} \mu_2\left(A_k(x_1, \cdot)\right)\right) d\mu_1(x_1)$$

$$\overset{\text{cor. } 1.6.2}{=} \sum_{k \in \mathbb{N}} \int_{X_1} \mu_2\left(A_k(x_1, \cdot)\right) d\mu_1(x_1) = \sum_{k \in \mathbb{N}} (\mu_1 \otimes \mu_2)(A_k),$$

which is the sought result. Moreover the measure $\mu_1 \otimes \mu_2$ is σ-finite since if we have with $j = 1, 2$, $X_j = \cup_{k \in \mathbb{N}} X_{j,k}$ with $X_{j,k} \in \mathcal{M}_j$ and $\mu_j(X_{j,k}) < +\infty$, we get $X_1 \times X_2 = \cup_{(k,l) \in \mathbb{N} \times \mathbb{N}} (X_{1,k} \times X_{2,l})$, and thus

$$(\mu_1 \otimes \mu_2)(X_{1,k} \times X_{2,l}) = \mu_1(X_{1,k})\mu_2(X_{2,l}) < +\infty.$$

Theorem 4.2.5 (Tonelli). *Let* $(X_1, \mathcal{M}_1, \mu_1)$ *and* $(X_2, \mathcal{M}_2, \mu_2)$ *be measure spaces where each* μ_j *is a* σ-*finite positive measure. Let* $f : X_1 \times X_2 \to \overline{\mathbb{R}}_+$ *be a measurable mapping (the product* $X_1 \times X_2$ *is equipped with the* σ-*algebra* $\mathcal{M}_1 \otimes \mathcal{M}_2$). *From Proposition 4.1.3, the mappings* $X_2 \ni x_2 \mapsto f(x_1, x_2)$, $X_1 \ni x_1 \mapsto f(x_1, x_2)$ *are measurable and we may define*

$$f_1(x_1) = \int_{X_2} f(x_1, x_2) d\mu_2(x_2), \quad f_2(x_2) = \int_{X_1} f(x_1, x_2) d\mu_1(x_1).$$

Then the mappings f_j *are measurable and we have*

$$\int_{X_1} f_1(x_1) d\mu_1(x_1) = \int_{X_2} f_2(x_2) d\mu_2(x_2) = \int_{X_1 \times X_2} f(x_1, x_2) d(\mu_1 \otimes \mu_2)(x_1, x_2).$$

$$\tag{4.2.12}$$

Proof. The following notation for (4.2.12) is certainly easier to follow:

$$\int_{X_1} \left(\int_{X_2} f(x_1, x_2) d\mu_2(x_2)\right) d\mu_1(x_1) = \int_{X_2} \left(\int_{X_1} f(x_1, x_2) d\mu_1(x_1)\right) d\mu_2(x_2)$$

$$= \iint_{X_1 \times X_2} f(x_1, x_2) d(\mu_1 \otimes \mu_2)(x_1, x_2).$$

We assume first that $f = 1_Q$ with $Q \in \mathcal{M}_1 \otimes \mathcal{M}_2$: Definition 4.2.4 gives the sought result. As a consequence, we obtain as well that result for simple functions on $X_1 \times X_2$ (Definition 1.3.2). From the approximation Theorem 1.3.3, we get the existence of a sequence of simple functions $(s_k)_{k \in \mathbb{N}}$ on $X_1 \times X_2$ such that for all $(x_1, x_2) \in X_1 \times X_2$,

$$0 \leq s_k(x_1, x_2) \uparrow f(x_1, x_2).$$

We set $s_{k,1}(x_1) = \int_{X_2} s_k(x_1, x_2)d\mu_2(x_2)$, $s_{k,2}(x_2) = \int_{X_1} s_k(x_1, x_2)d\mu_1(x_1)$. Since s_k is a simple function, we have already proven that

$$\int_{X_1} s_{k,1}d\mu_1 = \int_{X_2} s_{k,2}d\mu_2 = \int_{X_1 \times X_2} s_k d(\mu_1 \otimes \mu_2). \tag{4.2.13}$$

Using now Beppo Levi's Theorem 1.6.1 on $(X_1 \times X_2, \mathcal{M}_1 \otimes \mathcal{M}_2, \mu_1 \otimes \mu_2)$, we get

$$\lim_k \int_{X_1 \times X_2} s_k d(\mu_1 \otimes \mu_2) = \int_{X_1 \times X_2} f d(\mu_1 \otimes \mu_2). \tag{4.2.14}$$

For $x_1 \in X_1$, Beppo Levi's theorem on $(X_2, \mathcal{M}_2, \mu_2)$, applied to the non-negative increasing sequence $s_k(x_1, x_2)$ gives

$$0 \leq s_{k,1}(x_1) = \int_{X_2} s_k(x_1, x_2)d\mu_2(x_2) \uparrow \int_{X_2} f(x_1, x_2)d\mu_2(x_2) = f_1(x_1).$$

Beppo Levi's theorem on $(X_1, \mathcal{M}_1, \mu_1)$, applied to the non-negative increasing sequence $s_{k,1}(x_1)$ gives then

$$\lim_k \int_{X_1} s_{k,1}d\mu_1 = \int_{X_1} f_1 d\mu_1. \tag{4.2.15}$$

We get then

$$\int_{X_1 \times X_2} f d(\mu_1 \otimes \mu_2) \overset{(4.2.14)}{=} \lim_k \int_{X_1 \times X_2} s_k d(\mu_1 \otimes \mu_2)$$
$$\overset{(4.2.13)}{=} \lim_k \int_{X_1} s_{k,1}d\mu_1 \overset{(4.2.15)}{=} \int_{X_1} f_1 d\mu_1,$$

and we prove similarly $\int_{X_1 \times X_2} f d(\mu_1 \otimes \mu_2) = \int_{X_2} f_2 d\mu_2$, concluding the proof. $\square$

Remark 4.2.6. Lemma 1.2.14 on double series with terms in $\overline{\mathbb{R}}_+$ is a very elementary version of Tonelli's theorem.

Theorem 4.2.7 (Fubini). *Let $(X_1, \mathcal{M}_1, \mu_1)$ and $(X_2, \mathcal{M}_2, \mu_2)$ be measure spaces where each μ_j is a σ-finite positive measure. Let $f : X_1 \times X_2 \to \mathbb{C}$ be a measurable mapping (the product $X_1 \times X_2$ is equipped with the σ-algebra $\mathcal{M}_1 \otimes \mathcal{M}_2$).*

(1) If $\int_{X_1} \left(\int_{X_2} |f(x_1, x_2)| d\mu_2(x_2) \right) d\mu_1(x_1) < +\infty$, then $f \in L^1(\mu_1 \otimes \mu_2)$.

(2) *If $f \in L^1(\mu_1 \otimes \mu_2)$, then $f(x_1, \cdot) \in L^1(\mu_2)$ μ_1-a.e. in x_1, $f(\cdot, x_2) \in L^1(\mu_1)$ μ_2-a.e. in x_2 and*

$$\int_{X_1} \left(\int_{X_2} f(x_1, x_2) d\mu_2(x_2) \right) d\mu_1(x_1)$$

$$= \int_{X_2} \left(\int_{X_1} f(x_1, x_2) d\mu_1(x_1) \right) d\mu_2(x_2) \qquad (4.2.16)$$

$$= \iint_{X_1 \times X_2} f(x_1, x_2) d(\mu_1 \otimes \mu_2)(x_1, x_2).$$

Proof. To obtain (1), we need only to apply Tonelli's theorem 4.2.5 to $|f|$. Let us prove (2). We assume first that f is real valued: then we have

$$f = f_+ - f_-, \quad \text{with} \quad f_\pm \geq 0, \quad f_+(x) = \max(f(x), 0), \quad f_-(x) = \max(-f(x), 0).$$

From Tonelli's theorem and the assumption of (2), we get

$$\int_{X_1} \left(\int_{X_2} f_+(x_1, x_2) d\mu_2(x_2) \right) d\mu_1(x_1) = \int_{X_2} \left(\int_{X_1} f_+(x_1, x_2) d\mu_1(x_1) \right) d\mu_2(x_2)$$

$$= \iint_{X_1 \times X_2} f_+(x_1, x_2) d(\mu_1 \otimes \mu_2)(x_1, x_2) < +\infty,$$

and the same identity holds for f_-. As a result the $\mathcal{M}_1$ measurable functions $(f_+)_1, (f_-)_1$ belong to $L^1(\mu_1)$ (we define as in Lemma 4.2.1 for $g : X_1 \times X_2 \to \overline{\mathbb{R}}_+$ measurable, $g_1(x_1) = \int_{X_2} g(x_1, x_2) d\mu_2(x_2), g_2(x_2) = \int_{X_1} g(x_1, x_2) d\mu_1(x_1)$). From Proposition 1.7.1 (4) we get

$$(f_+)_1 < +\infty, \quad (f_-)_1 < +\infty, \quad \mu_1\text{-a.e.}$$

Similarly, we prove $(f_+)_2 < +\infty, (f_-)_2 < +\infty, \mu_2$-a.e. Since we have

$$|f(x_1, x_2)| = f_+(x_1, x_2) + f_-(x_1, x_2),$$

this gives the first part of (2). Applying the identities (4.2.16) for f_+ and f_-, we find the identity of (2) by writing a linear combination of real numbers. When f is complex valued, we may consider separately the imaginary and real parts, each of them satisfying the assumptions of (2) and thus which can be given the same treatment as above $\square$

Remark 4.2.8. Let $(X_1, \mathcal{M}_1, \mu_1)$ and $(X_2, \mathcal{M}_2, \mu_2)$ be measure spaces where each μ_j is a σ-finite positive measure. Let $f_j : X_j \to \mathbb{C}, j = 1, 2$ be mappings of $L^1(\mu_j)$. We define on $X_1 \times X_2$, the tensor product of f_1 with f_2, noted $f_1 \otimes f_2$, by $(f_1 \otimes f_2)(x_1, x_2) = f_1(x_1) f_2(x_2)$. This function is measurable (Remark 4.1.2) and Theorem 4.2.7 gives right away that $f_1 \otimes f_2$ belongs also to $L^1(\mu_1 \otimes \mu_2)$ as well as the formula

$$\iint_{X_1 \times X_2} (f_1 \otimes f_2) d(\mu_1 \otimes \mu_2) = \left(\int_{X_1} f_1 d\mu_1 \right) \left(\int_{X_2} f_2 d\mu_2 \right).$$

4.3 The Lebesgue measure on $\mathbb{R}^m$ and tensor products

The Lebesgue measure on $\mathbb{R}^m$ was constructed in Section 2.4. In the present section, we are willing to compare that measure to the tensor product of Lebesgue measures on $\mathbb{R}$, so that we can reduce the computation of multiple integrals to a succession of computations of simple integrals.

Theorem 4.3.1. *Let m_1, m_2 be integers ≥ 1. We set $m = m_1 + m_2$. With λ_d standing for the Lebesgue measure on $\mathbb{R}^d$ and $\mathcal{L}_d$ for the Lebesgue σ-algebra on $\mathbb{R}^d$ (see Theorem 2.4.2), we have $\mathcal{L}_m \supset \mathcal{L}_{m_1} \otimes \mathcal{L}_{m_2}$ and λ_m coincides with $\lambda_{m_1} \otimes \lambda_{m_2}$ on $\mathcal{L}_{m_1} \otimes \mathcal{L}_{m_2}$.*

Proof. Using the notation of Definition 1.1.3, we get from (1.2.15) and Remark 4.1.4

$$\mathcal{B}_{m_1} \otimes \mathcal{B}_{m_2} = \mathcal{B}_m = \mathcal{M}(\text{compact CAR}) \subset \mathcal{M}(\text{CAR}) = \mathcal{L}_{m_1} \otimes \mathcal{L}_{m_2}. \tag{4.3.1}$$

Theorem 2.2.14 implies that for $E_j \in \mathcal{L}_{m_j}, j = 1, 2$, there exist a F_σ set A_j and a G_δ set B_j such that $A_j \subset E_j \subset B_j$, $\lambda_{m_j}(B_j \backslash A_j) = 0$. As a result, we have

$$\underbrace{A_1 \times \mathbb{R}^{m_2}}_{F_\sigma \text{ set}} \subset E_1 \times \mathbb{R}^{m_2} \subset \underbrace{B_1 \times \mathbb{R}^{m_2}}_{G_\delta \text{ set}}.$$

It is thus enough to prove that

$$\lambda_m\big((B_1 \backslash A_1) \times \mathbb{R}^{m_2}\big) = 0, \tag{4.3.2}$$

since this implies that $E_1 \times \mathbb{R}^{m_2} \in \mathcal{L}_m$ (Theorem 2.2.14) as well as $\mathbb{R}^{m_1} \times E_2 \in \mathcal{L}_m$, so that $E_1 \times E_2 \in \mathcal{L}_m$, entailing $\mathcal{L}_{m_1} \otimes \mathcal{L}_{m_2} = \mathcal{M}(\text{Rectangles}) \subset \mathcal{L}_m$. To obtain (4.3.2), we shall use Proposition 1.4.4 (2) and prove that for all $M \geq 0$,

$$\lambda_m\big((B_1 \backslash A_1) \times \{x_2 \in \mathbb{R}^{m_2}, |x_2| \leq M\}\big) = 0. \tag{4.3.3}$$

On the other hand, $\lambda_{m_1} \otimes \lambda_{m_2}$ is a positive measure defined on $\mathcal{B}_m = \mathcal{B}_{m_1} \otimes \mathcal{B}_{m_2}$, finite on the compact sets since a compact subset K of $\mathbb{R}^m$ is included in a product $\beta_1 \times \beta_2$ with $\beta_j = \{x_j \in \mathbb{R}^{m_j}, |x_j| \leq M\}$ and thus

$$(\lambda_{m_1} \otimes \lambda_{m_2})(K) \leq (\lambda_{m_1} \otimes \lambda_{m_2})(\beta_1 \times \beta_2) = \lambda_{m_1}(\beta_1)\lambda_{\mu_2}(\beta_2) < +\infty.$$

Moreover, from Theorem 2.2.14 (2) and Definition 4.2.4 we find

$$(\lambda_{m_1} \otimes \lambda_{m_2})([0, 1]^m) = \lambda_{m_1}([0, 1]^{m_1})\lambda_{m_2}([0, 1]^{m_2}) = 1.$$

Also $\lambda_{m_1} \otimes \lambda_{m_2}$ is invariant by translation since for $E \in \mathcal{B}_m$ and $t = (t_1, t_2) \in \mathbb{R}^{m_1} \times \mathbb{R}^{m_2}$, we have

$$(\lambda_{m_1} \otimes \lambda_{m_2})(E + t) = \int_{\mathbb{R}^{m_1}} \left(\int_{\mathbb{R}^{m_2}} \mathbf{1}_{E + (t_1, t_2)}(x_1, x_2) d\lambda_{m_2}(x_2) \right) d\lambda_{m_1}(x_1)$$

and by translation invariance of λ_{m_2}, we get

$$(\lambda_{m_1} \otimes \lambda_{m_2})(E+t) = \int_{\mathbb{R}^{m_1}} \left(\int_{\mathbb{R}^{m_2}} \mathbf{1}_{E+(t_1,t_2)}(x_1, x_2 + t_2) d\lambda_{m_2}(x_2) \right) d\lambda_{m_1}(x_1)$$

$$= \int_{\mathbb{R}^{m_1}} \left(\int_{\mathbb{R}^{m_2}} \mathbf{1}_{E+(t_1,0)}(x_1, x_2) d\lambda_{m_2}(x_2) \right) d\lambda_{m_1}(x_1),$$

so that using Fubini's theorem, we find

$$(\lambda_{m_1} \otimes \lambda_{m_2})(E+t) = \int_{\mathbb{R}^{m_2}} \left(\int_{\mathbb{R}^{m_1}} \mathbf{1}_{E+(t_1,0)}(x_1, x_2) d\lambda_{m_1}(x_1) \right) d\lambda_{m_2}(x_2)$$

and using translation invariance of λ_{m_1}, we get

$$(\lambda_{m_1} \otimes \lambda_{m_2})(E+t) = \int_{\mathbb{R}^{m_2}} \left(\int_{\mathbb{R}^{m_1}} \mathbf{1}_{E+(t_1,0)}(x_1 + t_1, x_2) d\lambda_{m_1}(x_1) \right) d\lambda_{m_2}(x_2)$$

$$= \int_{\mathbb{R}^{m_2}} \left(\int_{\mathbb{R}^{m_1}} \mathbf{1}_E(x_1, x_2) d\lambda_{m_1}(x_1) \right) d\lambda_{m_2}(x_2) = (\lambda_{m_1} \otimes \lambda_{m_2})(E).$$

As a result, from Theorem 2.4.2 $\lambda_{m_1} \otimes \lambda_{m_2}$ and λ_m coincide on $\mathcal{B}_m$. This implies (4.3.3) since $\lambda_{m_1}(B_1 \backslash A_1) = 0$. We have proven that

$$\mathcal{B}_m = \mathcal{B}_{m_1} \otimes \mathcal{B}_{m_2} \subset \mathcal{L}_{m_1} \otimes \mathcal{L}_{m_2} \subset \mathcal{L}_m, \tag{4.3.4}$$

$$A \in \mathcal{B}_m \implies (\lambda_{m_1} \otimes \lambda_{m_2})(A) = \lambda_m(A). \tag{4.3.5}$$

Moreover, for $Q \in \mathcal{L}_{m_1} \otimes \mathcal{L}_{m_2}$, since $Q \in \mathcal{L}_m$, there exist an F_σ set A and a G_δ set B such that $A \subset Q \subset B, \lambda_m(B \backslash A) = 0$. Since A is a Borel set, we get

$$\lambda_m(Q) = \overbrace{\lambda_m(Q \backslash A)}^{\leq \lambda_m(B \backslash A) = 0} + \lambda_m(A) = (\lambda_{m_1} \otimes \lambda_{m_2})(A).$$

Moreover as $B \backslash A$ is also a Borel set, we have

$$(\lambda_{m_1} \otimes \lambda_{m_2})(Q) = \overbrace{(\lambda_{m_1} \otimes \lambda_{m_2})(Q \backslash A)}^{\leq (\lambda_{m_1} \otimes \lambda_{m_2})(B \backslash A) = \lambda_m(B \backslash A) = 0} + (\lambda_{m_1} \otimes \lambda_{m_2})(A)$$

proving that λ_m coincides with $\lambda_{m_1} \otimes \lambda_{m_2}$ on $\mathcal{L}_{m_1} \otimes \mathcal{L}_{m_2}$, completing the proof. $\quad\square$

4.4   Notes

Sections 4.1–4.2 clearly belong to Chapter 1 and we could have logically exposed their content there. However, it was our wish to reduce as much as possible the exposition of the general theory and to delay the introduction of multiple integrals.

We have seen $\mathcal{L}_{m_1+m_2} \supset \mathcal{L}_{m_1} \otimes \mathcal{L}_{m_2}$ and that inclusion can be shown to be strict. In fact $\mathcal{L}_m$ is *complete* (Property (5) in Theorem 2.2.1), whereas $\mathcal{L}_{m_1} \otimes \mathcal{L}_{m_2}$ is not complete. However it can be proven (see Exercise 4.5.3) that

$$(\mathbb{R}^{m_1+m_2}, \mathcal{L}_{m_1+m_2}, \lambda_{m_1+m_2})$$

is the completion of the measure space $(\mathbb{R}^{m_1} \times \mathbb{R}^{m_2}, \mathcal{L}_{m_1} \otimes \mathcal{L}_{m_2}, \lambda_{m_1} \otimes \lambda_{m_2})$.

Let us review the names of mathematicians encountered in this chapter:

Guido FUBINI (1879–1943) was one of the greatest Italian mathematicians; he was expelled from Italy in 1938 by the antisemitic laws of the Mussolini regime and emigrated to the US, where the Princeton *Institute for Advanced Study* offered him a position.

Leonida TONELLI (1885–1946) was also an Italian mathematician.

4.5 Exercises

Exercise 4.5.1. *Let $\mathcal{L}$ be the Lebesgue σ-algebra on $\mathbb{R}$. Checking a set $\{a\} \times A$, where $A \subset \mathbb{R}, A \notin \mathcal{L}$, show that $\mathcal{L} \otimes \mathcal{L}$ is not complete.*

Answer. In Exercise 2.8.19, we have constructed a subset A of the real line which does not belong to the Lebesgue σ-algebra (our construction depended heavily on the Axiom of Choice). With λ_2 standing for the Lebesgue measure on $\mathbb{R}^2$, we have

$$\{a\} \times A \subset \{a\} \times \mathbb{R}, \quad \lambda_2(\{a\} \times \mathbb{R}) = \sum_{k \in \mathbb{Z}} \lambda_2(\{a\} \times [k, k+1[) = 0.$$

Nevertheless $\{a\} \times A$ does not belong to $\mathcal{L} \otimes \mathcal{L}$, otherwise using Proposition 4.1.3, we would find

$$\mathcal{L} \ni (\{a\} \times B)(a, \cdot) = \{x_2 \in \mathbb{R}, (a, x_2) \in \{a\} \times A\} = A,$$

contradicting $A \notin \mathcal{L}$.

Exercise 4.5.2. *Let $(X_j, d_j), j = 1, 2$ be two separable metric spaces. We define on $(X_1 \times X_2)^2$, $d((x_1, x_2), (y_1, y_2)) = \max(d_1(x_1, y_1), d_2(x_2, y_2))$.*

(1) *Show that d is a distance on $X = X_1 \times X_2$ such that both projections $X_1 \times X_2 \to X_j, j = 1, 2$, are continuous. Show that d defines the product topology on $X_1 \times X_2$.*

(2) *Show that (X, d) is separable.*

(3) *Show that every open set in X is a countable union of products of open balls.*

(4) *Show that the Borel σ-algebra of X equals the tensor product of the Borel σ-algebras on each X_j.*

Answer. (1) The mapping d is valued in $\mathbb{R}_+$ (see (1.2.7), (1.2.8), (1.2.9)) symmetric since d_j are symmetric, satisfying the triangle inequality (as the d_j), separated (as the d_j). The projections π_j are continuous since if

$$\lim_n d\big((x_{1,n}, x_{2,n}), (x_1, x_2)\big) = 0,$$

this implies $\lim_n d_j(x_{j,n}, x_j) = 0$. Since the product topology $\mathscr{O}_p$ on $X_1 \times X_2$ is defined as the weakest (coarsest) topology making these projections continuous, we find that $\mathscr{O}_p \subset \mathscr{O}_d$, where $\mathscr{O}_d$ is the topology defined by the distance d on $X_1 \times X_2$. On the other hand we have for $x_j \in X_j, r > 0$, with obvious notation,

$$B_d\big((x_1, x_2), r\big) = B_{d_1}(x_1, r) \times B_{d_2}(x_2, r)$$

so that the topology $\mathscr{O}_d$ on $X_1 \times X_2$ generated by the open d-balls[1], is equal to the topology generated by the products of open balls. Since the products of open balls belong to $\mathscr{O}_p$, we find that $\mathscr{O}_d$ is included in $\mathscr{O}_p$ and thus $\mathscr{O}_p = \mathscr{O}_d$.[2]

(2) Let $(x_{j,n})_{n \in \mathbb{N}}$ be a dense subset of X_j: then $D = \big\{(x_{1,m}, x_{2,n})\big\}_{(m,n) \in \mathbb{N}^2}$ is countable and dense in X.

(3) Let Ω be an open subset of X. We consider the countable family of balls

$$(\dagger) \qquad \mathcal{C}_\Omega = \{B_d(y, r)\}_{\substack{y \in D, r \in \mathbb{Q}_+^* \\ B_d(y,r) \subset \Omega}}.$$

Let $x_0 \in \Omega$: then $B_d(x_0, r_0) \subset \Omega$ with some positive $r_0 \in \mathbb{Q}$. We can find $y_0 \in D$ such that $d(x_0, y_0) < r_0/2$: this implies $x_0 \in B(y_0, r_0/2) \subset \Omega$ (since $d(y_0, z) < r_0/2 \Longrightarrow d(x_0, z) < \frac{r_0}{2} + \frac{r_0}{2} = r_0 \Longrightarrow z \in B_d(x_0, r_0) \subset \Omega$). As a result,

$$\Omega = \cup_{\substack{y \in D, r \in \mathbb{Q}_+^* \\ B_d(y,r) \subset \Omega}} B_d(y, r),$$

giving the result since $B_d(y, r)$ is a product of open balls.[3]

(4) We have, with obvious notation $\mathcal{B}_1 \otimes \mathcal{B}_2 \subset \mathcal{B}$ since $\mathcal{B}$ is a σ-algebra such that the projections are measurable (since they are continuous), thus contains $\mathcal{B}_1 \otimes \mathcal{B}_2$. Moreover we have $\mathcal{B}_1 \otimes \mathcal{B}_2 \subset \mathcal{B} = \mathcal{M}(\mathcal{C}) \subset \mathcal{B}_1 \otimes \mathcal{B}_2$ since each element of $\mathcal{C}$ is a product of balls, proving the result.

---

[1] If X is a set and $(\mathscr{O}_i)_{i \in I}$ is a family of topologies on X, then $\mathscr{O} = \cap_{i \in I} \mathscr{O}_i$ is also a topology on X. Let $\mathcal{F}$ be a family of subsets of X: since $\mathcal{P}(X)$ is a topology on X, we may define the topology on X generated by $\mathcal{F}$ as the intersection of topologies on X containing $\mathcal{F}$: this is the coarsest topology on X containing $\mathcal{F}$.

[2] Taking $d((x_1, x_2), (y_1, y_2)) = d_1(x_1, y_1) + d_2(x_2, y_2)$ does not change significantly the argument, although $\beta = B_d((x_1, x_2), r)$ is no longer a product of open balls, it is a union of products since $(z_1, z_2) \in \beta$ implies that $B_{d_1}(z_1, r/2) \times B_{d_2}(z_2, r/2) \subset \beta$.

[3] Here also, taking $d((x_1, x_2), (y_1, y_2)) = d_1(x_1, y_1) + d_2(x_2, y_2)$ does not change significantly the argument, although in that case B_d is not a product of balls. However, defining $\mathcal{C}_{X_j}, j = 1, 2$ as in ($\dagger$), we find that $B_d(y, r)$ is a union – necessarily countable – of products $B_1 \times B_2$ with $B_j \in \mathcal{C}_j$.

Exercise 4.5.3. *Let m_1, m_2 be positive integers and let us set $m = m_1 + m_2$.*

(1) *Prove that $\mathcal{B}_m = \mathcal{B}_{m_1} \otimes \mathcal{B}_{m_2}$, where $\mathcal{B}_d$ stands for the Borel σ-algebra on $\mathbb{R}^d$.*

(2) *Prove that $\mathcal{L}_{m_1} \otimes \mathcal{L}_{m_2} \subset \mathcal{L}_m$, where $\mathcal{L}_d$ stands for the Lebesgue σ-algebra on $\mathbb{R}^d$ and that the inclusion is strict.*

(3) *Prove that $(\mathbb{R}^m, \mathcal{L}_m, \lambda_m)$ is the completion of the measure space (see Exercise 2.8.13) $(\mathbb{R}^{m_1} \times \mathbb{R}^{m_2}, \mathcal{L}_{m_1} \otimes \mathcal{L}_{m_2}, \lambda_{m_1} \otimes \lambda_{m_2})$.*

Answer. (1) See Exercise 4.5.2.

(2) See Theorem 4.3.1 and Exercise 4.5.1 for the strict inclusion.

(3) From Theorem 4.3.1, we know that λ_m coincides with $\lambda_{m_1} \otimes \lambda_{m_2}$ on $\mathcal{L}_{m_1} \otimes \mathcal{L}_{m_2}$. Let $P \in \mathcal{L}_m$: there exists a F_σ set A (thus in $\mathcal{B}_m$), a G_δ set B (thus in $\mathcal{B}_m$), such that

$$A \subset P \subset B, \quad \lambda_m(B \backslash A) = 0.$$

Now $A \in \mathcal{B}_m = \mathcal{B}_{m_1} \otimes \mathcal{B}_{m_2} \subset \mathcal{L}_{m_1} \otimes \mathcal{L}_{m_2}$, we find

$$P = \underbrace{P \backslash A}_{\in \mathcal{L}_m} \cup \underbrace{A}_{\in \mathcal{L}_{m_1} \otimes \mathcal{L}_{m_2}}, \quad P \backslash A \subset \underbrace{B \backslash A}_{\in \mathcal{L}_{m_1} \otimes \mathcal{L}_{m_2}}, \quad \lambda_m(B \backslash A) = 0.$$

so that P belongs to the completion of $\mathcal{L}_{m_1} \otimes \mathcal{L}_{m_2}$ for the measure $\lambda_{m_1} \otimes \lambda_{m_2}$ which coincides with λ_m on $\mathcal{L}_{m_1} \otimes \mathcal{L}_{m_2}$. Since the measure space $(\mathbb{R}^m, \mathcal{L}_m, \lambda_m)$ is complete and contains $(\mathbb{R}^{m_1} \times \mathbb{R}^{m_2}, \mathcal{L}_{m_1} \otimes \mathcal{L}_{m_2}, \lambda_{m_1} \otimes \lambda_{m_2})$, this proves the result: in fact if $(\mathbb{R}^m, \mathcal{C}, \mu)$ is the completion of $(\mathbb{R}^{m_1} \times \mathbb{R}^{m_2}, \mathcal{L}_{m_1} \otimes \mathcal{L}_{m_2}, \lambda_{m_1} \otimes \lambda_{m_2})$, the σ-algebra $\mathcal{C}$ is generated by $\mathcal{L}_{m_1} \otimes \mathcal{L}_{m_2}$ and the subsets of its negligible sets and since $\lambda_{m_1} \otimes \lambda_{m_2}$ coincides with λ_m on $\mathcal{L}_{m_1} \otimes \mathcal{L}_{m_2}$, $\mathcal{C}$ is included in $\mathcal{L}_m$.

Exercise 4.5.4. *Let ϕ be a continuous function supported in $[0,1]$ such that $\int \phi(t)dt = 1$. We define on $\mathbb{R}^2$ the following function:*

$$f(x_1, x_2) = H(x_2)\phi(x_2 - E(x_2))\Big\{\phi(x_1 - E(x_2)) - \phi(x_1 - E(x_2) - 1)\Big\}$$

where E is the integer part (floor function, see footnote on page 16) and $H = \mathbf{1}_{\mathbb{R}_+}$ the Heaviside function.

(1) *Prove that*

$$\int\Big(\int f(x_1, x_2)dx_2\Big)dx_1 = 1, \quad \int\Big(\int f(x_1, x_2)dx_1\Big)dx_2 = 0.$$

(2) *Comment on this example.*

Answer. (1) We have $\int f(x_1, x_2)dx_1 = 0$ and

$$\int f(x_1, x_2)dx_2 = \sum_{n \geq 0} \int_n^{n+1} \phi(x_2 - n)\big(\phi(x_1 - n) - \phi(x_1 - n - 1)\big)dx_2$$

$$= \sum_{n \geq 0} \big(\phi(x_1 - n) - \phi(x_1 - n - 1)\big) = \phi(x_1),$$

and thus $\int\Big(\int f(x_1, x_2)dx_2\Big)dx_1 = 1$.

(2) The assumptions of Fubini's theorem cannot be satisfied: computing for instance

$$\int |f(x_1, x_2)| dx_1$$

$$= H(x_2)|\phi(x_2 - E(x_2))| \int |\phi(x_1 - E(x_2)) - \phi(x_1 - E(x_2) - 1)| dx_1$$

$$= H(x_2)|\phi(x_2 - E(x_2))| \underbrace{\int |\phi(t + 1) - \phi(t)| dt}_{\substack{=c_\phi > 0 \\ \text{since } \phi \text{ is not periodic}}},$$

so that

$$\int \left(\int |f(x_1, x_2)| dx_1 \right) dx_2$$

$$= c_\phi \sum_{n \geq 0} \int_n^{n+1} |\phi(x_2 - n)| dx_2 = c_\phi \sum_{n \geq 0} \int |\phi(t)| dt = +\infty.$$

Exercise 4.5.5. *Let ϕ be a non-negative smooth function supported in $(0, 1)$ such that $\int \phi(t) dt = 1$. Let ψ be a non-negative smooth 1-periodic function defined on $\mathbb{R}$ such that ψ vanishes in a neighborhood of 0. We define on $\mathbb{R}^2$ the function*

$$f(x_1, x_2) = H(x_2)\psi(x_2)\big(\phi(x_1 - x_2) - \phi(x_1 - x_2 - 1)\big),$$

where H is the Heaviside function.

(1) *Give an example of functions ϕ, ψ satisfying the above assumptions. Prove that f is a smooth function.*

(2) *Prove that $\int f(x_1, x_2) dx_1 = 0$ and calculate $\int f(x_1, x_2) dx_2$.*

(3) *Comment on this example.*

Answer. (1) The function ρ in Exercise 2.8.6 (with $m = 1$) is smooth non-negative with support $[-1, 1]$. We can take

$$\phi(t) = \frac{\rho(4t - 2)}{\int \rho(4s) ds} : \quad \phi \text{ is smooth } \geq 0 \text{ with support } [1/4, 3/4] \text{ and integral 1.}$$

We can take $\psi(t) = \sum_{n \in \mathbb{Z}} \phi(t + n)$. We claim that this function is smooth, 1-periodic and vanishes on $[-1/4, 1/4] + \mathbb{Z}$: in fact the function $t \mapsto \phi(t - n)$ is supported in $[n + \frac{1}{4}, n + \frac{3}{4}]$, so these functions have disjoint supports for different n, implying that ψ is smooth 1-periodic with support $\mathbb{Z} + [\frac{1}{4}, \frac{3}{4}]$, thus vanishing on $\mathbb{Z} + [-\frac{1}{4}, \frac{1}{4}]$. As a result, the function $x \mapsto H(x)\psi(x)$ is smooth and f is a smooth function.

(2) We have $\int f(x_1, x_2)dx_1 = 0$ and

$$\int f(x_1, x_2)dx_2 = \int_0^{+\infty} \psi(x_2) \overbrace{\left(\phi(x_1 - x_2) - \phi(x_1 - x_2 - 1)\right)}^{\substack{\text{vanishes for } x_2 \\ \text{outside } (x_1 - 2, x_1)}} dx_2$$

$$= \int \phi(x_1 - x_2) \underbrace{\left(H(x_2)\psi(x_2) - H(x_2 - 1)\psi(x_2 - 1)\right)}_{\omega(x_2)} dx_2.$$

The function ω vanishes for $x_2 > 1$ since ψ is 1-periodic and also for $x_2 < 0$; as a result,

$$\int f(x_1, x_2)dx_2 = \int_0^1 \phi(x_1 - x_2)\psi(x_2)dx_2.$$

(3) We find thus

$$I = \int \left(\int f(x_1, x_2)dx_2\right) dx_1 = \int \left(\int \phi(x_1 - x_2)\psi(x_2)\mathbf{1}_{[0,1]}(x_2)dx_2\right) dx_1,$$

and using Tonelli's theorem, this gives

$$I = \int \phi(t)dt \int_0^1 \psi(s)ds = 1.$$

This is a smooth version of the counterexample of Exercise 4.5.4. Of course computing

$$\int |f(x_1, x_2)|dx_1 = H(x_2)\psi(x_2) \int |\phi(x_1 - x_2) - \phi(x_1 - x_2 - 1)|dx_1$$

$$= H(x_2)\psi(x_2) \int |\phi(1 + t) - \phi(t)|dt,$$

and as in Exercise 4.5.4(3), $\int \left(\int |f(x_1, x_2)|dx_1\right) dx_2 = +\infty$. On the other hand,

$$\int |f(x_1, x_2)|dx_2 = \int_0^{+\infty} \psi(x_2)|\phi(x_1 - x_2) - \phi(x_1 - x_2 - 1)|dx_2,$$

and using Tonelli's theorem, we get as well

$$\int \left(\int |f(x_1, x_2)|dx_2\right) dx_1 = \iint \psi(x_2)H(x_2)|\phi(t + 1) - \phi(t)|dtdx_2 = +\infty.$$

This is a second example proving that the assumption (1) in Fubini's theorem 4.2.7 cannot be dispensed with.

Exercise 4.5.6. *We consider* $(\mathbb{R}, \mathcal{P}(\mathbb{R}), \mathfrak{h}_0)$ *where* $\mathfrak{h}_0$ *is the counting measure and* $(\mathbb{R}, \mathcal{L}_1, \lambda_1)$ *the Lebesgue measure on* $\mathbb{R}$.

  (1) *Prove that* $D = \{(x_1, x_2) \in \mathbb{R}^2, x_1 = x_2\}$ *belongs to* $\mathcal{L}_1 \otimes \mathcal{P}(\mathbb{R})$.

  (2) *Calculate* $\int 1_D(x_1, x_2) d\lambda_1(x_1)$ *and* $\int 1_D(x_1, x_2) d\mathfrak{h}_0(x_2)$.

  (3) *Comment on the previous example.*

Answer. (1) As a closed subset of $\mathbb{R}^2$, the diagonal D is a Borel set of $\mathbb{R}^2$, thus (see Exercise 4.5.3) belongs to $\mathcal{B}_2 = \mathcal{B}_1 \otimes \mathcal{B}_1 \subset \mathcal{L}_1 \otimes \mathcal{P}(\mathbb{R})$.

(2) Since $x_1 \mapsto 1_D(x_1, x_2)$ vanishes λ_1-a.e., we have $\int 1_D(x_1, x_2) d\lambda_1(x_1) = 0$, and on the other hand $\int 1_D(x_1, x_2) d\mathfrak{h}_0(x_2) = 1$.

(3) As a result, we have

$$\int \left(\int 1_D(x_1, x_2) d\lambda_1(x_1) \right) d\mathfrak{h}_0(x_2) = 0,$$

and $\int \left(\int 1_D(x_1, x_2) d\mathfrak{h}_0(x_2) \right) d\lambda_1(x_1) = +\infty$. This proves that the assumption of σ-finiteness in Tonelli's theorem is not superfluous (the counting measure is σ finite only on countable sets).

Exercise 4.5.7. *Prove* $\lim_{A \to +\infty} \int_0^A \frac{\sin x}{x} dx = \frac{\pi}{2}$, *using Fubini's theorem and the identity* $1/x = \int_0^{+\infty} e^{-tx} dt$ *for* $x > 0$.

Answer. We have for $A > 0$,

$$\int_0^A \frac{\sin x}{x} dx = \int_0^A \sin x \left(\int_0^{+\infty} e^{-tx} dt \right) dx,$$

so that using Fubini's theorem,

$$\int_0^A \frac{\sin x}{x} dx = \int_0^{+\infty} \left(\int_0^A \mathrm{Im}\{e^{x(i-t)}\} dx \right) dt = \int_0^{+\infty} \Big[\mathrm{Im}\, \frac{e^{x(i-t)}}{i-t}\Big]_{x=0}^{x=A} dt$$

$$= \int_0^{+\infty} \mathrm{Im} \left(\frac{(-t-i)}{t^2+1}\left(e^{A(i-t)} - 1\right) \right) dt$$

$$= \int_0^{+\infty} \frac{1}{t^2+1} \left(1 - e^{-tA} \cos A - t e^{-tA} \sin A\right) dt.$$

Lebesgue's dominated convergence theorem ensures that

$$\lim_{A \to +\infty} \int_0^A \frac{\sin x}{x} dx = \int_0^{+\infty} \frac{dt}{t^2+1} = \frac{\pi}{2}.$$

Exercise 4.5.8. *Let* $(X, \mathcal{M}, \mu)$ *be a measure space where* μ *is a positive measure.*

  (1) *Let* S *be defined by* (3.2.20). *Show that for* $1 \le p < +\infty$, S *is dense in* $L^p(\mu)$.

  (2) *Show that for* $1 \le p < +\infty$, $C_c^0(\mathbb{R}^n)$ *is dense in* $L^p(\mathbb{R}^n)$.

(3) *Let Ω be an open subset of $\mathbb{R}^n$. Show that for $1 \leq p < +\infty$, $C_c^0(\Omega)$ is dense in $L^p(\Omega)$.*

(4) *Let $\rho \in C_c^\infty(\mathbb{R}^n; \mathbb{R}_+)$, $\int_{\mathbb{R}^n} \rho(x)dx = 1$ (cf. Exercise 2.8.6). For $\epsilon > 0, x \in \mathbb{R}^n, u \in L^1_{\text{loc}}(\mathbb{R}^n)$, we set*

$$u_\epsilon(x) = \int_{\mathbb{R}^n} u(y)\rho\left(\frac{x-y}{\epsilon}\right)\frac{dy}{\epsilon^n}.$$

Show that it is meaningful and that $u_\epsilon \in C^\infty(\mathbb{R}^n)$.

(5) *Let $1 \leq p < +\infty$. Show that for $u \in L^p(\mathbb{R}^n)$, we have $u_\epsilon \in L^p(\mathbb{R}^n)$ and $\lim_{\epsilon \to 0_+} u_\epsilon = u$ in $L^p(\mathbb{R}^n)$.*

(6) *We replace ρ in (4) by $e^{-\pi|x|^2}$ where $|x|$ is the Euclidean norm. Show that for $u \in L^1(\mathbb{R}^n)$, u_ϵ is analytic and $\lim_{\epsilon \to 0_+} u_\epsilon = u$ in $L^1(\mathbb{R}^n)$. Assuming $u \in C_c^0(\mathbb{R}^n)$, show that this method provides a proof of the Stone–Weierstrass Theorem.*

Answer. (1) is Proposition 3.2.11,

(2), (3) are proven in Theorem 3.4.3.

(4) The function $y \mapsto u(y)\rho((x-y)/\varepsilon)$ is compactly supported in $|y| \leq \varepsilon + |x|$, so that the integrand defining u_ε is indeed in L^1_{comp} for each x. Moreover the function $x \mapsto u(y)\rho((x-y)/\varepsilon)$ is smooth and we have

$$\sup_{|x|<M}\left|u(y)\rho^{(k)}((x-y)/\varepsilon)\varepsilon^{-n-k}\right| \leq |u(y)|\|\rho^{(k)}\|_{L^\infty}\varepsilon^{-n-k}\mathbf{1}\big(|y| \leq \varepsilon + M\big),$$

which is $\in L^1(\mathbb{R}^n)$. We may apply Theorem 3.3.4 to get $u_\varepsilon \in C^\infty$ along with

$$u_\varepsilon^{(k)}(x) = \int u(y)\rho^{(k)}((x-y)/\varepsilon)\varepsilon^{-n-k}dy.$$

(5) We note $\rho_\epsilon(t) = \rho(t\epsilon^{-1})\epsilon^{-n}$ and $u_\epsilon = \rho_\epsilon * u$. This function belongs to $C^\infty(\mathbb{R}^n)$. Jensen's inequality (Theorem 3.1.3) implies for $u \in L^p(\mathbb{R}^n)$

$$\|\rho_\epsilon * u\|^p_{L^p(\mathbb{R}^n)} = \int_{\mathbb{R}^n}\left|\int_{\mathbb{R}^n}\rho_\epsilon(x-y)u(y)dy\right|^p dx$$

$$\leq \iint_{\mathbb{R}^n \times \mathbb{R}^n}\rho_\epsilon(x-y)|u(y)|^p dydx = \int_{\mathbb{R}^n}|u(y)|^p dy = \|u\|^p_{L^p(\mathbb{R}^n)}$$

entailing $u_\epsilon \in L^p$. Moreover using again Jensen's inequality, we get

$$\int_{\mathbb{R}^n}|(u*\rho_\epsilon)(x) - u(x)|^p dx = \int_{\mathbb{R}^n}\left|\int_{\mathbb{R}^n}\big(u(x-\epsilon t) - u(x)\big)\rho(t)dt\right|^p dx$$

$$\leq \int_{\mathbb{R}^n}\int_{\mathbb{R}^n}|u(x-\epsilon t) - u(x)|^p\rho(t)dtdx = \int_{\mathbb{R}^n}\|\tau_{\epsilon t}u - u\|^p_{L^p(\mathbb{R}^n)}\rho(t)dt.$$

Exercise 3.7.15 proves pointwise convergence towards 0 of $\|\tau_{\epsilon t}u - u\|_{L^p}^p\rho(t)$, which is dominated by $2^p\|u\|_{L^p}^p\rho(t) \in L^1$. We get $\lim_{\epsilon\to 0}\|u * \rho_\epsilon - u\|_{L^p(\mathbb{R}^n)} = 0$.

(6) We have

$$u_\epsilon(x) = \int_{\mathbb{R}^n} e^{-\pi|x-t|^2\epsilon^{-2}}u(t)\epsilon^{-n}dt$$

and for $u \in L^1(\mathbb{R}^n)$, we have $u_\epsilon \in L^1(\mathbb{R}^n)$ (even $\|u_\epsilon\|_{L^1(\mathbb{R}^n)} \le \|u\|_{L^1(\mathbb{R}^n)}$ using the previous proof). We may extend u_ϵ to $\mathbb{C}^n$, defining for $z = x + iy$ $(x, y \in \mathbb{R}^n)$,

$$u_\epsilon(x + iy) = \int_{\mathbb{R}^n} e^{-\pi\sum_{j=1}^n(z_j-t_j)^2\epsilon^{-2}}u(t)\epsilon^{-n}dt. \tag{4.5.1}$$

This integral converges since

$$\left|e^{-\pi\sum_{j=1}^n(z_j-t_j)^2\epsilon^{-2}}\right| = e^{-\pi|x-t|^2\epsilon^{-2}}e^{\pi\epsilon^{-2}y^2}.$$

Holomorphy of (4.5.1) on $\mathbb{C}^n$ follows from Theorem 3.3.7, inducing analyticity on $\mathbb{R}^n$. The proof of the convergence in $L^1(\mathbb{R}^n)$ of u_ϵ is proven as in the previous question. Let $\varphi \in C_c^0(\mathbb{R}^n)$ and

$$\varphi_\epsilon(x) - \varphi(x) = \int_{\mathbb{R}^n} e^{-\pi|x-t|^2\epsilon^{-2}}\big(\varphi(t) - \varphi(x)\big)\epsilon^{-n}dt$$

$$= \int_{\mathbb{R}^n} e^{-\pi|t|^2}\big(\varphi(x - \epsilon t) - \varphi(x)\big)dt.$$

We have for $A > 0$,

$$|\varphi_\epsilon(x) - \varphi(x)| \le \int_{|t|\le A}|\varphi(x - \epsilon t) - \varphi(x)|dt + 2\|\varphi\|_{L^\infty}\int_{|t|>A}e^{-\pi|t|^2}dt,$$

and thus

$$\sup_{x\in\mathbb{R}^n}|\varphi_\epsilon(x) - \varphi(x)|$$

$$\le A^n|\mathbb{B}^n|\sup_{|x_1-x_2|\le\epsilon A}|\varphi(x_1) - \varphi(x_2)| + 2\|\varphi\|_{L^\infty}\int_{|t|>A}e^{-\pi|t|^2}dt.$$

Since φ is uniformly continuous, we get for all $A > 0$,

$$\limsup_{\epsilon\to 0_+}\|\varphi_\epsilon - \varphi\|_{L^\infty} \le 2\|\varphi\|_{L^\infty}\int_{|t|>A}e^{-\pi|t|^2}dt.$$

Taking the limit when $A \to +\infty$, we find $\lim_{\epsilon\to 0_+}\|\varphi_\epsilon - \varphi\|_{L^\infty} = 0$. As a result φ is a uniform limit of a sequence of analytic functions, restrictions to $\mathbb{R}^n$ of holomorphic functions on $\mathbb{C}^n$ (entire functions). The radius of convergence of the power series defining these entire functions is infinite, so that, on every compact set, these functions are a uniform limit of polynomials.

Exercise 4.5.9. *Find an example of a monotone class $\mathcal{M}$ on a set X, such that $\emptyset, X \in \mathcal{M}$, but such that $\mathcal{M}$ is not a σ-algebra.*

Answer. Let X be an uncountable set and

$$\mathcal{M} = \{A \subset X, A \text{ countable}\}.$$

Then $\mathcal{M}$ is obviously a monotone class, but is not stable by complement, so is not a σ-algebra. Taking $\mathcal{M}' = \mathcal{M} \cup \{X\}$, we get a monotone class: let $(A_n)_{n\in\mathbb{N}}$ be an increasing sequence in $\mathcal{M}'$. If all A_n are different from X, $\cup A_n$ is countable and thus belongs to $\mathcal{M}$. If one of the $A_n = X$, then $\cup A_n = X \in \mathcal{M}'$. Let $(B_n)_{n\in\mathbb{N}}$ be a decreasing sequence in $\mathcal{M}'$. If all B_n are different from X, $\cap B_n$ is countable and thus belongs to $\mathcal{M}$. If $B_N = X$, then since the sequence is decreasing, $B_0 - \cdots - B_N - X$; either the sequence is stationary equal to X and then $\cap B_n = X$, or B_{N+1} is countable and $\cap_{n\in\mathbb{N}} B_n$ is countable, in both cases in $\mathcal{M}'$.

Exercise 4.5.10. *Let X be a set.*

(1) *Let $(\mathcal{M}_i)_{i\in I}$ be a family of monotone classes on X (see Definition 4.2.2). Prove that $\mathcal{N} = \cap_{i\in I} \mathcal{M}_i$ is a monotone class on X.*

(2) *Let $\mathcal{F}$ be an algebra on X, i.e., a non-empty subset of the powerset $\mathcal{P}(X)$ such that*

$$A \in \mathcal{F} \Longrightarrow A^c \in \mathcal{F}, \tag{4.5.2}$$
$$A, B \in \mathcal{F} \Longrightarrow A \cup B \in \mathcal{F}, A \cap B \in \mathcal{F}. \tag{4.5.3}$$

Prove that the smallest monotone class containing $\mathcal{F}$ is the smallest σ-algebra containing $\mathcal{F}$ (Monotone Class Theorem).

Answer. (1) Obvious from the definition.

(2) Since a σ-algebra is a monotone class, setting

$$m(\mathcal{F}) = \bigcap_{\substack{\mathcal{M} \text{ monotone class} \\ \text{containing } \mathcal{F}}} \mathcal{M}, \qquad s(\mathcal{F}) = \bigcap_{\substack{\mathcal{M} \ \sigma\text{-algebra} \\ \text{containing } \mathcal{F}}} \mathcal{M},$$

we find $m(\mathcal{F}) \subset s(\mathcal{F})$. To get the required equality, it is enough to prove that $m(\mathcal{F})$ is a σ-algebra: since $m(\mathcal{F})$ contains $\mathcal{F}$, this will imply $m(\mathcal{F}) \supset s(\mathcal{F})$. We know that $m(\mathcal{F})$ is not empty since $\mathcal{F}$ is assumed to be non-empty. Let $(A_n)_{n\in\mathbb{N}}$ be a sequence of $m(\mathcal{F})$; to prove that $\cup_{n\in\mathbb{N}} A_n$ belongs to $m(\mathcal{F})$, it is enough to prove that $m(\mathcal{F})$ is stable by finite union. In fact, we have

$$\cup_{n\in\mathbb{N}} A_n = \cup_{n\in\mathbb{N}} B_n, \qquad B_n = \cup_{0\leq k\leq n\in\mathbb{N}} A_k,$$

and if we know that each B_n belongs to $m(\mathcal{F})$, the monotone class property will imply the result. Inductively, it is enough to prove that $A_1, A_2 \in m(\mathcal{F})$ implies $A_1 \cup A_2 \in m(\mathcal{F})$. Let $E \in \mathcal{F}$ (which is non-empty) and let us define

$$\mathcal{N}_E = \{A \in m(\mathcal{F}), A \cap E, A^c \cap E, A \cap E^c \in m(\mathcal{F})\}.$$

Then $\mathcal{N}_E$ is a monotone class. Note first that $\mathcal{N}_E$ contains $\mathcal{F}$ and thus is non-empty. Let $(A_n)_{n\in\mathbb{N}}$ be an increasing sequence of $\mathcal{N}_E$. We have

$$\cup_n A_n \in m(\mathcal{F}), \quad (\cup_n A_n) \cap E = \cup_n \underbrace{(A_n \cap E)}_{\in m(\mathcal{F}),\ \text{increasing}} \in m(\mathcal{F}),$$

$$(\cup_n A_n)^c \cap E = \cap_n \underbrace{A_n^c \cap E}_{\in m(\mathcal{F}),\ \text{decreasing}} \in m(\mathcal{F}),$$

$$(\cup_n A_n) \cap E^c = \cup_n \underbrace{A_n \cap E^c}_{\in m(\mathcal{F}),\ \text{increasing}} \in m(\mathcal{F}).$$

Let $(B_n)_{n\in\mathbb{N}}$ be an decreasing sequence of $\mathcal{N}_E$. We have

$$\cap_n B_n \in m(\mathcal{F}), \quad (\cap_n B_n) \cap E = \cap_n \underbrace{(B_n \cap E)}_{\in m(\mathcal{F}),\ \text{decreasing}} \in m(\mathcal{F}),$$

$$(\cap_n B_n)^c \cap E = \cup_n \underbrace{(B_n^c \cap E)}_{\in m(\mathcal{F}),\ \text{increasing}} \in m(\mathcal{F}),$$

$$(\cap_n B_n) \cap E^c = \cap_n \underbrace{B_n \cap E^c}_{\in m(\mathcal{F}),\ \text{decreasing}} \in m(\mathcal{F}).$$

Since $\mathcal{N}_E$ is a monotone class containing $\mathcal{F}$, it contains $m(\mathcal{F})$ and thus is equal to $m(\mathcal{F})$. Let us now consider for $B \in m(\mathcal{F})$,

$$\mathcal{N}_B = \{A \in m(\mathcal{F}), A \cap B, A^c \cap B, A \cap B^c \in m(\mathcal{F})\}.$$

Reasoning as above $\mathcal{N}_B$ is a monotone class; moreover it contains $\mathcal{F}$ since for $E \in \mathcal{F}, B \in m(\mathcal{F}) = \mathcal{N}_E$, we have $B \cap E, B \cap E^c, B^c \cap E \in m(\mathcal{F})$. Since $\mathcal{N}_B$ is also included in $m(\mathcal{F})$, it is thus equal to $m(\mathcal{F})$. We have $X, \emptyset \in \mathcal{F}$ since $\mathcal{F}$ is non-empty and for $E \in \mathcal{F}, X = E^c \cup E \in \mathcal{F}, X^c = \emptyset$. As a result if $A \in m(\mathcal{F})$, since $X \in m(\mathcal{F})$, we have

$$A^c = A^c \cap X \in m(\mathcal{F}),$$

so that $m(\mathcal{F})$ is stable by complement. For $A, B \in m(\mathcal{F})$, we find

$$(A \cup B)^c = A^c \cap B^c \in m(\mathcal{F}) \Longrightarrow A \cup B \in m(\mathcal{F})$$

so that $m(\mathcal{F})$ is stable by finite union. As a result, from the remarks at the beginning, $m(\mathcal{F})$ is a σ-algebra.

Exercise 4.5.11.

(1) *Calculate*

$$I(a, \alpha) = \int_0^{+\infty} \frac{dx}{(x^2 + a^2)^\alpha}, \quad \textit{for } a > 0 \textit{ and } \alpha > 1/2.$$

(2) *Calculate* $J(\alpha) = \iint_{\mathbb{R}^2_+} \dfrac{dxdy}{(1+x^2+y^2)^\alpha}$ *for* $\alpha > 1$.

Answer. (1) Using Proposition 2.3.2, setting $x = a\tan t$, $I(a,\alpha) =$

$$\int_0^{\pi/2} \frac{a(1+\tan^2 t)dt}{(a^2+a^2\tan^2 t)^\alpha} = a^{1-2\alpha}\int_0^{\pi/2}(\cos t)^{2\alpha-2}dt = \frac{a^{1-2\alpha}\Gamma(\alpha-\frac{1}{2})\pi^{1/2}}{2\Gamma(\alpha)}.$$

according to Lemma 10.5.7 on the Wallis integrals.

(2) Using Fubini's theorem, we get for $\alpha > 1$,

$$J(\alpha) = \int_0^{+\infty} I(\sqrt{1+y^2},\alpha)dy = \frac{\Gamma(\alpha-\frac{1}{2})\pi^{1/2}}{2\Gamma(\alpha)}\int_0^{+\infty}(1+y^2)^{\frac{1}{2}-\alpha}dy$$

$$= \frac{\Gamma(\alpha-\frac{1}{2})\pi^{1/2}}{2\Gamma(\alpha)}I(1,\alpha-\frac{1}{2}) = \frac{\Gamma(\alpha-\frac{1}{2})}{2\Gamma(\alpha)}\frac{\Gamma(\alpha-1)\pi}{2\Gamma(\alpha-\frac{1}{2})} = \frac{\pi}{4(\alpha-1)}.$$

N.B. Using the results of the next chapter on change of variables, it is easier to calculate

$$J(\alpha) = \frac{\pi}{2}\int_0^{+\infty} r(1+r^2)^{-\alpha}dr = \frac{\pi}{4(\alpha-1)}[(1+r^2)^{1-\alpha}]_{r=+\infty}^{r=0} = \frac{\pi}{4(\alpha-1)}.$$

Exercise 4.5.12.

(1) *Calculate the volume $|\mathbb{B}^n|$ of the unit Euclidean ball in $\mathbb{R}^n$.*

(2) *Calculate the volume of the simplex*

$$\Sigma_n = \{x \in \mathbb{R}^n, \forall j, x_j \geq 0, x_1 + \cdots + x_n \leq 1\}.$$

(3) *Let $p \in [1,+\infty)$. Calculate the volume of the unit ball of $\mathbb{R}^n$ for the norm* $\|x\|_p = \left(\sum_{1\leq j\leq n}|x_j|^p\right)^{1/p}$.

Answer. (1) We consider on $\mathbb{R}^n_x \times \mathbb{R}^+_t$ the product measure $\lambda_n \otimes \lambda_1$. We define

$$I = \iint_{\{(x,t)\in\mathbb{R}^n\times\mathbb{R}_+,\,\|x\|\leq t\}} te^{-t^2}dtdx = \int_0^{+\infty} t^{1+n}e^{-t^2}dt|\mathbb{B}^n| = \frac{|\mathbb{B}^n|\Gamma(\frac{n}{2}+1)}{2},$$

and we have also

$$I = \int_{\mathbb{R}^n}\left(\int_{\|x\|}^{+\infty} te^{-t^2}dt\right)dx = \frac{1}{2}\int_{\mathbb{R}^n} e^{-\|x\|^2}dx = \frac{\pi^{n/2}}{2},$$

so that

$$|\mathbb{B}^n| = \frac{2\pi^{n/2}}{n\Gamma(n/2)}. \tag{4.5.4}$$

(2) We have

$$|\Sigma_n| = \int_{\mathbb{R}^n} H(x_1)\ldots H(x_n)H(1 - x_1 - \cdots - x_n)dx_1 \ldots dx_n.$$

We study first for $f \geq 0$ measurable and $a = (a_j)_{1\leq j\leq n} \in (0, +\infty)^n$,

$$I_n(f, a) = \int_{\mathbb{R}^n_+} f\Big(\sum_{1\leq j\leq n} x_j\Big) \prod_{1\leq j\leq n} \frac{x_j^{a_j-1}}{\Gamma(a_j)}dx,$$

and we claim that

$$I_n(f, a) = \int_0^{+\infty} \frac{t^{-1+\sum_{1\leq j\leq n} a_j}}{\Gamma(\sum_{1\leq j\leq n} a_j)} f(t)dt. \tag{4.5.5}$$

That property is true for $n = 1$, and assuming that it is true for some $n \geq 1$, we check

$$I_{n+1}(f, a) = \int_{\mathbb{R}_+} \int_{\mathbb{R}^n_+} f\Big(\sum_1^n x_j + x_{n+1}\Big)\frac{x_{n+1}^{a_{n+1}-1}}{\Gamma(a_{n+1})} \prod_{1\leq j\leq n} \frac{x_j^{a_j-1}}{\Gamma(a_j)}dxdx_{n+1}$$

$$= \int_{\mathbb{R}_+} \frac{x_{n+1}^{a_{n+1}-1}}{\Gamma(a_{n+1})} I_n(\tau_{-x_{n+1}}f, (a_j)_{1\leq j\leq n})dx_{n+1}$$

$$= \iint_{\mathbb{R}^2_+} \frac{x_{n+1}^{a_{n+1}-1}}{\Gamma(a_{n+1})} \frac{t^{-1+\sum_{1\leq j\leq n} a_j}}{\Gamma(\sum_{1\leq j\leq n} a_j)} f(t + x_{n+1})dtdx_{n+1}$$

$$= \iint_{\mathbb{R}^2} H(x_{n+1})H(s - x_{n+1})f(s)\frac{x_{n+1}^{a_{n+1}-1}}{\Gamma(a_{n+1})} \frac{(s - x_{n+1})^{-1+\sum_{1\leq j\leq n} a_j}}{\Gamma(\sum_{1\leq j\leq n} a_j)}dsdx_{n+1}$$

$$= \int_0^{+\infty} f(s)s^{-1+\sum_1^{n+1} a_j}ds\frac{B(a_{n+1}, \sum_1^n a_j)}{\Gamma(\sum_{1\leq j\leq n} a_j)\Gamma(a_{n+1})},$$

where the Beta function is given by (10.5.17). Formula (10.5.18) yields (4.5.5). Applying this to $a_j = 1$, $f(t) = H(1 - t)$, we obtain

$$|\Sigma_n| = \frac{1}{\Gamma(n)} \int_0^1 t^{n-1}dt = \frac{1}{n!}. \tag{4.5.6}$$

(3) We start over with the computations of (1), this time with

$$J_p = \iint_{\{(x,t)\in\mathbb{R}^n\times\mathbb{R}_+, \|x\|_p\leq t\}} t^{p-1}e^{-t^p}dtdx = \int_0^{+\infty} t^{p+n-1}e^{-t^p}dtV_n(p)$$

$$= \frac{V_n(p)\Gamma(\frac{n}{p} + 1)}{p},$$

and we have also

$$I = \int_{\mathbb{R}^n} \left(\int_{\|x\|_p}^{+\infty} t^{p-1} e^{-t^p} dt \right) dx = \frac{1}{p} \int_{\mathbb{R}^n} e^{-\|x\|_p^p} dx$$

$$= \frac{1}{p} \left(\int_{\mathbb{R}} e^{-|t|^p} dt \right)^n = \frac{2^n \Gamma(1/p)^n}{p^{n+1}}$$

so that

$$V_n(p) = \left(\frac{2}{p} \right)^n \frac{\Gamma(1/p)^n}{\Gamma(\frac{n}{p}+1)} = \left(\frac{2}{p} \right)^n \frac{p\Gamma(1/p)^n}{n\Gamma(n/p)}. \tag{4.5.7}$$

N.B. Note that the above formula for $p = 1$ gives the volume

$$\lambda_n \left(\{ x \in \mathbb{R}^n, \sum_1^n |x_j| \le 1 \} \right) = \frac{2^n}{n!} = 2^n \lambda_n(\Sigma_n),$$

so that we have found another way to proving (4.5.6).

Exercise 4.5.13. *We consider the following functions, defined on $\mathbb{R}^2$ by*

$$f_1(x,y) = \begin{cases} \dfrac{x^2 - y^2}{x^2 + y^2} & \text{if } (x,y) \neq 0, \\ 0 & \text{if } (x,y) = 0, \end{cases} \quad f_2(x,y) = \begin{cases} \dfrac{x - y}{(x^2 + y^2)^{3/2}} & \text{if } (x,y) \neq 0, \\ 0 & \text{if } (x,y) = 0. \end{cases}$$

Calculate for $j = 1, 2$,

$$\int_0^1 \left(\int_0^1 f_j(x,y) dy \right) dx, \qquad \int_0^1 \left(\int_0^1 f_j(x,y) dx \right) dy.$$

Comment on the result.

Answer. The function f_1 is bounded measurable since

$$f_1(x,y) = \mathbf{1}_{\mathbb{R}^2 \setminus \{(0,0)\}}(x,y) R(x,y),$$

where R is a continuous function on $\mathbb{R}^2 \setminus \{(0,0)\}$, such that $|R(x,y)| \le 1$. As a result, if Ω is an open subset of $\mathbb{R}$ which does not contain 0,

$$f_1^{-1}(\Omega) = \{(x,y) \in \mathbb{R}^2 \setminus \{(0,0)\}, R(x,y) \in \Omega\} = R^{-1}(\Omega)$$

and $R^{-1}(\Omega)$ is an open subset of $\mathbb{R}^2 \setminus \{(0,0)\}$ thus an open subset of $\mathbb{R}^2$. If Ω contains 0

$$f_1^{-1}(\Omega) = R^{-1}(\Omega) \cup \{(0,0)\}, \text{ union of an open set and a closed set, thus a Borel set.}$$

The function f_2 is also measurable (and unbounded) for the same reasons. We calculate for $y > 0$,

$$\int_0^1 \frac{x^2 - y^2}{x^2 + y^2}\, dx = \int_0^1 \left(1 - \frac{2y^2}{x^2 + y^2}\right) dx = 1 - 2y^2 \frac{1}{y} \arctan \frac{1}{y} = 1 - 2y \arctan\left(\frac{1}{y}\right).$$

We note that $y \mapsto y \arctan(\frac{1}{y})$ is continuous on $[0, 1]$ and that $\lim_{y \to 0_+} y \arctan(\frac{1}{y}) = 0$. We have

$$\int_0^1 \Big(1 - \underbrace{2y}_{u'(y)} \underbrace{\arctan(1/y)}_{v(y)}\Big) dy = 1 - \left([y^2 \arctan(1/y)]_0^1 - \int_0^1 y^2 \frac{1}{1 + y^{-2}}(-y^{-2})\, dy\right)$$

$$= 1 - \frac{\pi}{4} - \int_0^1 \frac{y^2}{1 + y^2}\, dy = 1 - \frac{\pi}{4} - 1 + \int_0^1 \frac{1}{1 + y^2}\, dy = 0.$$

The value of $\int_0^1 \left(\int_0^1 f_1(x, y)\, dy\right) dx$ is identical. In fact the function f_1 is locally integrable so that

$$I_1 = \iint_{[0,1] \times [0,1]} f_1(x, y)\, dx\, dy = \iint_{[0,1] \times [0,1]} f_1(y, x)\, dx\, dy$$

$$= - \iint_{[0,1] \times [0,1]} f_1(x, y)\, dx\, dy = -I_1,$$

(the second equality follows from the change of variables $(x, y) \mapsto (y, x)$) which implies $I_1 = 0$. The assumptions of Fubini's theorem 4.2.7 are fulfilled and the double integral I_1 is indeed the iteration of simple integrals.

It is a different story for f_2, for which we cannot argue as above although $f_2(x, y) = -f_2(y, x)$. The function f_2 is measurable, but not locally integrable near the origin since the polar coordinates change of variables gives

$$|f_2(x, y)|\, dx\, dy = |\cos\theta - \sin\theta|\, r^{-1} dr\, d\theta.$$

We calculate for $y > 0$,

$$J(y) = \int_0^1 \frac{x - y}{(x^2 + y^2)^{3/2}}\, dx = \left[(x^2 + y^2)^{-1/2}\right]_1^0 - y \int_0^1 (x^2 + y^2)^{-3/2} dx$$

$$= y^{-1} - (1 + y^2)^{-1/2} - y\left[(x^2 + y^2)^{-1/2} x y^{-2}\right]_{x=0}^{x=1}$$

$$= y^{-1} - (1 + y^2)^{-1/2} - y^{-1}(1 + y^2)^{-1/2}$$

$$= (1 + y^2)^{-1/2}\left(-1 + \frac{(1 + y^2)^{1/2} - 1}{y}\right)$$

$$= (1 + y^2)^{-1/2}\left(-1 + \frac{y}{((1 + y^2)^{1/2} + 1)}\right).$$

We may then calculate

$$\int_0^1 J(y)dy = -\int_0^1 (1+y^2)^{-1/2}dy + \int_0^1 \frac{y}{1+y^2+(1+y^2)^{1/2}}dy$$

$$= \left[\ln(y+(1+y^2)^{1/2})\right]_1^0 + \int_0^{\text{arcsinh}(1)} \frac{\sinh t}{1+\sinh^2 t+(1+\sinh^2 t)^{1/2}}\cosh t\, dt$$

$$= -\ln(1+\sqrt{2}) + \int_0^{\text{arcsinh}(1)} \frac{\sinh t}{\cosh^2 t+\cosh t}\cosh t\, dt$$

$$= -\ln(1+\sqrt{2}) + \int_0^{\text{arcsinh}(1)} \frac{\sinh t}{\cosh t+1}dt$$

$$= -\ln(1+\sqrt{2}) + \left[\ln(\cosh t+1)\right]_0^{\text{arcsinh}(1)}$$

$$= -\ln(1+\sqrt{2}) + \left[\ln(\cosh t+1)\right]_0^{\text{arcsinh}(1)}, \quad \text{and since arcsinh}(1) = \ln(1+\sqrt{2}),$$

$$= -\ln(1+\sqrt{2}) + \ln\left(\cosh(\ln(1+\sqrt{2}))+1\right) - \ln 2$$

$$= -\ln(1+\sqrt{2}) + \ln\left(\frac{1+\sqrt{2}}{2}+\frac{1}{2}\frac{1}{1+\sqrt{2}}+1\right) - \ln 2$$

$$= -\ln(1+\sqrt{2}) + \ln\left(\frac{1}{2}+\frac{\sqrt{2}}{2}+\frac{\sqrt{2}}{2}-\frac{1}{2}+1\right) - \ln 2 = -\ln 2 \neq 0.$$

If for $x > 0$, we calculate

$$K(x) = \int_0^1 \frac{x-y}{(x^2+y^2)^{3/2}}dy = -J(x)$$

we shall find

$$\int_0^1 K(x)dx = \ln 2,$$

so that both integrals in the Exercise for $j = 2$ make sense with two differing values $\ln 2$ and $-\ln 2$. This does not contradict Fubini's theorem since the assumptions of integrability on the product space are not satisfied. This simple example is a useful reminder that formal manipulations of integrals without prior checking of hypotheses could lead to errors. The iteration of simple integrals does not depend on the order of integration provided the function is integrable on the product space. Also, we can remark that the fact that both integrals make sense is not sufficient to ensure their equality.

Let us give another example, algebraically simpler than the one above. We define the measurable function

$$F_2(x,y) = \begin{cases} \frac{x-y}{\max(x^3,y^3)}, & \text{if } x \geq 1 \text{ and } y \geq 1, \\ 0, & \text{otherwise.} \end{cases}$$

For $x < 1$, $F_2(x, y) = 0$. We calculate for $x \geq 1$,

$$\int_{\mathbb{R}} F_2(x, y)dy = \int_1^x \frac{x - y}{x^3}dy + \int_x^{+\infty} \frac{x - y}{y^3}dy$$

$$= x^{-2}(x - 1) - x^{-3}\left(\frac{x^2}{2} - \frac{1}{2}\right) + x\frac{x^{-2}}{2} - x^{-1} = -x^{-2} + x^{-3}2^{-1}.$$

We have thus

$$\int_{\mathbb{R}}\left(\int_{\mathbb{R}} F_2(x, y)dy\right) dx = \int_1^{+\infty}(-x^{-2} + x^{-3}2^{-1})dx = -1 + 2^{-1}2^{-1} = -3/4.$$

The same calculation gives $\int_{\mathbb{R}}\left(\int_{\mathbb{R}} F_2(x, y)dx\right) dy = 3/4$. The above remarks on f_2 are true as well for F_2.

Exercise 4.5.14.

(1) *For $z \in \mathbb{C}\backslash\mathbb{R}_-$, we define*

$$\mathrm{Log}\, z = \oint_{[1,z]} \frac{d\xi}{\xi}.$$

Show that it makes sense and coincides with $\ln z$ for $z \in \mathbb{R}_+^*$. Show that

$$\exp(\mathrm{Log}\, z) = z \quad \text{for } z \in \mathbb{C}\backslash\mathbb{R}_-.$$

Calculate $\mathrm{Log}(\exp z)$, for z such that $\exp(z) \notin \mathbb{R}_-^*$.

(2) *Show that for $\mathrm{Re}\, z > 0$,*

$$\int_{\mathbb{R}} e^{-\pi z t^2}\, dt = \exp -(\mathrm{Log}\, z)/2 = z^{-1/2}.$$

(3) *Show that*

$$\int_{\mathbb{R}_+} \frac{\ln x}{x^2 - 1}dx = \frac{\pi^2}{4}, \qquad \int_{\mathbb{R}_+}\left(\frac{\arctan x}{x}\right)^2 dx = \pi \ln 2.$$

Answer. (1) is treated in Theorem 10.5.1.

(2) From Theorem 3.3.7 $z \mapsto \int_{\mathbb{R}} e^{-\pi z t^2}\, dt$ is a holomorphic function on $\{\mathrm{Re}\, z > 0\}$ which coincides with $\exp(-\frac{\ln z}{2})$ for $z > 0$. By analytic continuation, these two functions coincide on $\{\mathrm{Re}\, z > 0\}$.

(3) We have

$$\int_0^1 \frac{\ln x}{x^2 - 1}dx = \int_1^{+\infty} \frac{\ln(y^{-1})}{y^{-2} - 1}\frac{dy}{y^2} = \int_1^{+\infty} \frac{\ln y}{y^2 - 1}dy,$$

so that

$$I = \int_0^{+\infty} \frac{\ln x}{x^2 - 1} dx = 2 \int_1^{+\infty} \frac{\ln x}{x^2 - 1} dx = 2 \int_1^{+\infty} \frac{\ln x}{x^2} \sum_{k \geq 0} x^{-2k} dx.$$

Using Corollary 1.6.2 of Beppo Levi's theorem, we get

$$I = 2 \sum_{k \geq 1} \int_1^{+\infty} x^{-2k} \ln x \, dx = 2 \sum_{k \geq 1} \int_0^{+\infty} e^{-(2k-1)t} t \, dt$$

$$= 2 \sum_{k \geq 1} \int_0^{+\infty} e^{-s} s \, ds (2k-1)^{-2} = 2\Gamma(2) \sum_{k \geq 1} (2k-1)^{-2} = \frac{\pi^2}{4},$$

since $\Gamma(2) = 1$ and

$$\frac{\pi^2}{6} = \sum_{n \geq 1} n^{-2} = \sum_{k \geq 1} (2k-1)^{-2} + \sum_{k \geq 1} (2k)^{-2} = \sum_{k \geq 1} (2k-1)^{-2} + 2^{-2} \frac{\pi^2}{6},$$

which implies $\sum_{k \geq 1} (2k-1)^{-2} = \pi^2 \left(\frac{1}{6} - \frac{1}{24} \right) = \frac{\pi^2}{8}$.

We calculate first

$$J = \iiint_{[0,1] \times [0,1] \times \mathbb{R}_+} \frac{1}{(1 + x^2 z^2)(1 + y^2 z^2)} dx \, dy \, dz.$$

For $x, y \in \mathbb{R}_+^*$, we have

$$\int_0^A \frac{1}{(1 + x^2 z^2)(1 + y^2 z^2)} dz = (y^2 - x^2)^{-1} [y \arctan(yz) - x \arctan(xz)]_{z=0}^{z=A}$$

$$= \frac{y \arctan(Ay) - x \arctan(Ax)}{y^2 - x^2} \xrightarrow[A \to \infty]{} \frac{\pi}{2(x + y)},$$

and thus

$$J = \iint_{[0,1]^2} \frac{\pi \, dx \, dy}{2(x + y)} = \frac{\pi}{2} \int_0^1 [\ln(x + y)]_{y=0}^{y=1} dx = \frac{\pi}{2} \int_0^1 \left(\ln(x + 1) - \ln x \right) dx$$

$$= \frac{\pi}{2} [(x + 1) \ln(x + 1) - x \ln x]_0^1 = \frac{\pi}{2} 2 \ln 2 = \pi \ln 2.$$

On the other hand, we have

$$J = \iint_{[0,1] \times \mathbb{R}_+} \frac{1}{1 + x^2 z^2} \left[\frac{\arctan yz}{z} \right]_{y=0}^{y=1} dx \, dz$$

$$= \int_{\mathbb{R}_+} \left[\frac{\arctan xz}{z} \right]_{x=0}^{x=1} \left[\frac{\arctan yz}{z} \right]_{y=0}^{y=1} dz = \int_{\mathbb{R}_+} \left(\frac{\arctan z}{z} \right)^2 dz,$$

which is the sought result.

Chapter 5

Diffeomorphisms of Open Subsets of $\mathbb{R}^n$ and Integration

5.1 Differentiability

Definition 5.1.1. Let U be an open subset of $\mathbb{R}^n$, $x_0 \in U$ and let $f : U \to \mathbb{R}^m$. We shall say that f is differentiable at x_0 if there exist a linear map $A : \mathbb{R}^n \to \mathbb{R}^m$, $r_0 > 0$ and a mapping $\epsilon : B(0, r_0) \to \mathbb{R}^m$ such that for all $|h| < r_0$,

$$f(x_0 + h) = f(x_0) + Ah + \epsilon(h)|h|, \quad \lim_{h \to 0} \epsilon(h) = 0. \tag{5.1.1}$$

Here $|h|$ stands for the Euclidean norm of h, but we may choose any other norm on $\mathbb{R}^n$. We say that A is the differential of f at x_0 and we write $f'(x_0) = A$.

N.B. Note that the definition above is consistent since if for $r_0 > 0$ and for all $|h| < r_0$,

$$f(x_0 + h) = f(x_0) + A_1 h + \epsilon_1(h)|h|, \quad \lim_{h \to 0} \epsilon_1(h) = 0,$$

$$f(x_0 + h) = f(x_0) + A_2 h + \epsilon_2(h)|h|, \quad \lim_{h \to 0} \epsilon_2(h) = 0,$$

we get $(A_1 - A_2)h = (\epsilon_1(h) - \epsilon_2(h))|h|$ and thus for all $T \in \mathbb{R}^n$ such that $|T| = 1$ and for all $s \in (-r_0, r_0)$, this gives

$$(A_1 - A_2)T = \epsilon_1(sT) - \epsilon_2(sT) = \lim_{s \to 0}\big(\epsilon_1(sT) - \epsilon_2(sT)\big) = 0, \quad \text{i.e., } A_1 = A_2.$$

Remark 5.1.2. (1) We note also that $f'(x_0)$ is a $m \times n$ matrix (m rows, n columns) as a linear map from $\mathbb{R}^n$ into $\mathbb{R}^m$.

(2) If f is differentiable at a point x, then the *partial derivatives* $(\frac{\partial f}{\partial x_j}(x))_{1 \leq j \leq n}$ of f exist, i.e., for all $1 \leq j \leq n$, with $\mathbf{e}_j$ the jth vector of the canonical basis of $\mathbb{R}^n$,

$$\lim_{\substack{t \to 0 \\ t \in \mathbb{R}^*}} \frac{f(x + t\mathbf{e}_j) - f(x)}{t} = \frac{\partial f}{\partial x_j}(x).$$

In fact the differentiability of f at x implies $f(x_0+te_j) = f(x_0)+A(te_j)+\epsilon(te_j)|t|$, so that for $0 < |t| < r_0$, we get

$$\big(f(x + te_j) - f(x)\big)t^{-1} = Ae_j + \epsilon(te_j)|t|t^{-1},$$

which implies $\frac{\partial f}{\partial x_j}(x) = Ae_j = f'(x)e_j$ and thus

$$f'(x)h = f'(x)\left(\sum_{1\leq j\leq n} h_j e_j \right) = \sum_{1\leq j\leq n} h_j f'(x)e_j = \sum_{1\leq j\leq n} \frac{\partial f}{\partial x_j}(x)h_j.$$

The first-order Taylor–Young formula (5.1.1) can thus be written for

$$h = (h_1,\ldots, h_n) \in \mathbb{R}^n, \ |h| < r_0, \text{ as}$$
$$f(x + h) = f(x) + \sum_{1\leq j\leq n} \frac{\partial f}{\partial x_j}(x)h_j + \epsilon(h)|h|. \qquad (5.1.2)$$

Note that $f(x) = \big(f_1(x),\ldots, f_m(x)\big)$ belongs to $\mathbb{R}^m$ and that

$$\frac{\partial f}{\partial x_j}(x) = \begin{pmatrix} \frac{\partial f_1}{\partial x_j} \\ \vdots \\ \frac{\partial f_m}{\partial x_j} \end{pmatrix}.$$

Finally, $f'(x)$ is the $m \times n$ matrix

$$\begin{pmatrix} \dfrac{\partial f_1}{\partial x_1} & \cdots & \dfrac{\partial f_1}{\partial x_n} \\ \cdots & \dfrac{\partial f_i}{\partial x_j} & \cdots \\ \dfrac{\partial f_m}{\partial x_1} & \cdots & \dfrac{\partial f_m}{\partial x_n} \end{pmatrix}_{\substack{1\leq i\leq m \\ 1\leq j\leq n}}. \qquad (5.1.3)$$

(3) Conversely, the existence of partial derivatives at a point does not ensure differentiability (not even continuity), as shown by the following example. We set

$$f(x, y) = \begin{cases} \frac{xy}{x^2+y^2} & \text{for } (x, y) \neq (0, 0), \\ 0 & \text{if } (x, y) = (0, 0). \end{cases}$$

That function is discontinuous at 0 (for $\epsilon \neq 0$, we have $f(\epsilon, \epsilon) = 1/2$) and thus is not differentiable at 0 (Formula (5.1.1) implies continuity at x_0). However, we have for all x, y, $f(x,0) = 0$, $f(0, y) = 0$, which implies $\frac{\partial f}{\partial x}(x, 0) = 0 = \frac{\partial f}{\partial y}(0, y)$.

(4) However if the partial derivatives exist and are continuous on an open set U, then f is continuously differentiable on U, i.e., is differentiable on U with $U \ni x \mapsto f'(x)$ continuous. Let us prove the previous statement. We consider $x \in U$;

there exists $r > 0$ such that the open ball $B(x, r) \subset U$. For $h = (h_1, \ldots, h_n) \in \mathbb{R}^n$, such that $|h| < r$, we have

$$
f(x + h) - f(x) = f(x_1 + h_1, \ldots, x_n + h_n) - f(x_1, \ldots, x_n)
$$
$$
= f\left(x + \sum_{1 \le j \le n} h_j e_j\right) - f\left(x + \sum_{2 \le j \le n} h_j e_j\right)
$$
$$
+ f\left(x + \sum_{2 \le j \le n} h_j e_j\right) - f\left(x + \sum_{3 \le j \le n} h_j e_j\right)
$$
$$
\cdots
$$
$$
+ f(x + h_n e_n) - f(x),
$$

so that

$$
f(x + h) - f(x) - \sum_{1 \le j \le n} \frac{\partial f}{\partial x_j}(x) h_j
$$
$$
= \sum_{1 \le j \le n} \left\{ f\left(x + h_j e_j + \sum_{j < k \le n} h_k e_k\right) - f\left(x + \sum_{j < k \le n} h_k e_k\right) - \frac{\partial f}{\partial x_j}(x) h_j \right\}
$$
$$
= \sum_{1 \le j \le n} \left\{ \int_0^1 \frac{\partial f}{\partial x_j}\left(x + \sum_{j < k \le n} h_k e_k + \theta h_j e_j\right) d\theta h_j - \frac{\partial f}{\partial x_j}(x) h_j \right\}
$$
$$
= \sum_{1 \le j \le n} h_j \int_0^1 \left\{ \frac{\partial f}{\partial x_j}\left(x + \sum_{j < k \le n} h_k e_k + \theta h_j e_j\right) - \frac{\partial f}{\partial x_j}(x) \right\} d\theta.
$$

As a result, we have

$$
\underbrace{\left| f(x + h) - f(x) - \sum_{1 \le j \le n} \frac{\partial f}{\partial x_j}(x) h_j \right|}_{= \eta(h)}
$$
$$
\le |h| \sum_{1 \le j \le n} \sup_{\theta \in [0,1]} \left| \frac{\partial f}{\partial x_j}\left(x + \theta \sum_{j < k \le n} h_k e_k\right) - \frac{\partial f}{\partial x_j}(x) \right|,
$$

with $\lim_{h \to 0} \eta(h) = 0$, thanks to the continuity of the partial derivatives. This proves the differentiability of f at x and the continuity of $f'(x)$ follows from (5.1.3).

Proposition 5.1.3. *Let U be a convex open subset of $\mathbb{R}^n$ and let*

$$
f : U \to \mathbb{R}^m, \quad f(x) = (f_1(x), \ldots, f_m(x))
$$

be a differentiable mapping on U. Then for $x, y \in U$,

$$
\|f(y) - f(x)\|_{\mathbb{R}^m} \le \|y - x\|_{\mathbb{R}^n} \sup_{\theta \in (0,1)} \|f'(x + \theta(y - x))\|.
$$

For a $(m \times n)$ matrix A, we set $\|A\| = \sup_{\|T\|_{\mathbb{R}^n}=1} \|AT\|_{\mathbb{R}^m}$, where $\|T\|_{\mathbb{R}^d}$ is the Euclidean norm of $T \in \mathbb{R}^d$.

We prove a more general statement with the following lemma whose second property implies the proposition.

Lemma 5.1.4.

(1) *Let E be a normed real vector space, let $a < b$ be real numbers and let $\phi : [a, b] \to E$ be a continuous mapping, differentiable on (a, b) so that there exists $M \in \mathbb{R}_+$ such that for all $t \in (a, b)$, $\|\phi'(t)\| \leq M$. Then*

$$\|\phi(b) - \phi(a)\| \leq M(b - a).$$

(2) *Let E, F be normed vector spaces, let U be an open set of E, let $x_0, x_1 \in U$ such that $(x_0, x_1) = \{(1 - \theta)x_0 + \theta x_1\}_{\theta \in (0,1)} \subset U$ and let $f : U \to F$ be a continuous mapping which is differentiable on (x_0, x_1). Then*

$$\|f(x_1) - f(x_0)\| \leq \|x_1 - x_0\| \sup_{x \in (x_0, x_1)} \|f'(x)\|.$$

(3) *Let E be a normed vector space, let U be an open set of E, let $x_0, x_1 \in U$ such that $(x_0, x_1) \subset U$ and let $f : U \to \mathbb{R}$ be a continuous mapping which is differentiable on (x_0, x_1). Then there exists $x \in (x_0, x_1)$ such that*

$$f(x_1) - f(x_0) = f'(x)(x_1 - x_0).$$

Proof of the lemma. (1) We may assume by rescaling that $a = 0, b = 1$. Let $\epsilon > 0$ be given. We define

$$T_\epsilon = \{t \in [0, 1], \|\phi(t) - \phi(0)\| - Mt - \epsilon t \leq \epsilon\}.$$

By continuity of ϕ, T_ϵ is a closed subset of $[0, 1]$, contains 0 (the lhs of the inequality vanishes at 0) and thus by continuity, T_ϵ contains a neighborhood of 0. Defining $c = \sup T_\epsilon$ we have $c > 0$ and since T_ϵ is closed, $c \in T_\epsilon$. Let us assume that $c < 1$. We can find $t > c$ such that

$$\left\| \frac{\phi(t) - \phi(c)}{t - c} \right\| \leq \|\phi'(c)\| + \epsilon$$

implying

$$\begin{aligned}
\|\phi(t) - \phi(0)\| &\leq \|\phi(t) - \phi(c)\| + \|\phi(c) - \phi(0)\| \\
&\leq (t - c)\|\phi'(c)\| + \epsilon(t - c) + Mc + \epsilon(c + 1) \\
&\leq (t - c)M + \epsilon(t - c) + Mc + \epsilon(c + 1) \\
&= Mt + \epsilon t + \epsilon,
\end{aligned}$$

so that $t \in T_\epsilon$, which is impossible since $t > c = \sup T_\epsilon$. As a result $c = 1$ and thus

$$\forall \epsilon > 0, \quad \|\phi(1) - \phi(0)\| \le M + 2\epsilon,$$

implying the result (1). Property (2) follows immediately by applying (1) to $\phi(\theta) = f(x_\theta)$. Let us prove the equality (3). We consider $\phi : [0, 1] \to \mathbb{R}$ defined by $\phi(\theta) = f(x_\theta)$. The function ϕ is continuous on $[0, 1]$ and also differentiable on $(0, 1)$ with $\phi'(\theta) = f'(x_\theta)(x_1 - x_0)$. Applying the Mean Value Theorem (see, e.g., Lemma 5.10.2) to ϕ gives the result (3). $\qquad\square$

N.B. We have proven in (1), (2) an *inequality*, whereas the 1D mean value theorem provides an equality. There is no equivalent of the 1D result when the function f is valued into a space with dimension greater than 2: consider for instance the analytic mapping $[0, 2\pi] \ni l \mapsto e^{it} = f(l) \in \mathbb{C}$. We have $f(2\pi) - f(0) = 0$ and there does not exist any $c \in (0, 2\pi)$ such that $f(2\pi) - f(0) = 2\pi f'(c)$ since $f'(c) = ie^{ic}$ has modulus 1.

5.2　Linear transformations

Proposition 5.2.1. *Let T be a linear isomorphism of $\mathbb{R}^n$ and let E be a Borel set of $\mathbb{R}^n$. Then $T(E)$ is also a Borel set. For $E \in \mathcal{B}_n$, we set $\mu(E) = \lambda_n(T(E))$. Then $\mu = \lambda_n([0, 1]^n)\lambda_n$, i.e.,*

$$\lambda_n(T(E)) = \lambda_n(T([0, 1]^n))\lambda_n(E). \tag{5.2.1}$$

Proof. We note first that $T(E) = (T^{-1})^{-1}(E)$ and since T^{-1} is continuous (since linear), it is also Borel-measurable, so that $(T^{-1})^{-1}(E) \in \mathcal{B}_n$ for $E \in \mathcal{B}_n$. Moreover μ is indeed a Borel measure (i.e., a positive measure defined on $\mathcal{B}_n$ finite on compact sets): $\mu(\emptyset) = \lambda_n(T(\emptyset)) = \lambda_n(\emptyset) = 0$, and if $(E_k)_{k \in \mathbb{N}}$ is a sequence of pairwise disjoint Borel sets, the injectivity of T implies for $k \ne l$, $\emptyset = T(E_k \cap E_l) = T(E_k) \cap T(E_l)$, and we have

$$\mu\big(\cup_{k \in \mathbb{N}} E_k\big) = \lambda_n\big(T(\cup_{k \in \mathbb{N}} E_k)\big) = \lambda_n\big(\cup_{k \in \mathbb{N}} T(E_k)\big) = \sum_{k \in \mathbb{N}} \lambda_n(T(E_k)) = \sum_{k \in \mathbb{N}} \mu(E_k).$$

Moreover for K compact, we have $\mu(K) = \lambda_n(\overbrace{T(K)}^{\text{compact}}) < +\infty$. Finally, μ is invariant by translation since for $x \in \mathbb{R}^n$ and $E \in \mathcal{B}_n$,

$$\mu(E + x) = \lambda_n(T(E + x)) = \lambda_n(T(E) + Tx) = \lambda_n(T(E)) = \mu(E).$$

We have also

$$\mu([0, 1]^n) = \mu([-1/2, 1/2]^n) = \lambda_n(T([-1/2, 1/2]^n))$$

$$\ge \lambda_n\big((T^{-1})^{-1}(\overbrace{]-1/2, 1/2[^n}^{\text{open set containing } 0})\big) > 0,$$

where the last inequality follows from (1) in Theorem 2.4.2. As a result, for $E \in \mathcal{B}_n$, we have

$$\frac{\mu(E)}{\mu([0,1]^n)} = \lambda_n(E), \quad \text{i.e.,} \quad \mu(E) = \lambda_n\big(T([0,1]^n)\big)\lambda_n(E). \qquad \Box$$

N.B. According to Lemma 1.4.3, μ is defined as the (direct) image of the Lebesgue measure by T^{-1}: for $E \in \mathcal{B}_n$,

$$(T^{-1})_*(\lambda_n)(E) = \lambda_n\big((T^{-1})^{-1}(E)\big) = \lambda_n(T(E)) = \mu(E).$$

Introducing the notation $T^*(\lambda_n) = (T^{-1})_*(\lambda_n)$ for the inverse image, we have the following general framework.

Let $(Y, \mathcal{N}, \nu)$ be a measure space where ν is a positive measure. Let $f : X \to Y$ be a bijective mapping. We define the inverse image $f^*(\nu)$ (or pullback by f) of the measure ν as

$$\mu = f^*(\nu) = (f^{-1})_*(\nu), \quad \text{i.e.,} \quad \mu(A) = \nu(f(A)),$$

for $A \in \mathcal{M} = \{A \subset X, f(A) \in \mathcal{N}\}$. $\mathcal{M}$ is indeed a σ-algebra on X from Lemma 1.4.3: it is the largest σ-algebra on X such that f^{-1} is measurable. The mapping f is also measurable, since for $B \in \mathcal{N}$, $f(f^{-1}(B)) = B \in \mathcal{N}$.

When X, Y are topological spaces, $\mathcal{N}$ is the Borel σ-algebra on Y and f is an homeomorphism, $\mathcal{M}$ is the Borel σ-algebra on X: in fact $\mathcal{M}$ contains the open subsets of X since f is an open mapping, as a homeomorphism, proving that $\mathcal{M} \supset \mathcal{B}_X$. On the other hand, if $A \in \mathcal{M}$,

$$A = f^{-1}\big(\underbrace{f(A)}_{\in \mathcal{B}_Y}\big) \in \mathcal{B}_X,$$

since f is measurable as a continuous mapping.

Proposition 5.2.2. *Let T be a linear isomorphism of $\mathbb{R}^n$. Then* $\lambda_n\big(T([0,1]^n)\big) = |\det T|$.

For instance, for

$$T = \begin{pmatrix} 2 & 1 \\ 1/2 & 1 \end{pmatrix},$$

the determinant is equal to $3/2$ which is the area of the parallelogram P in Figure 5.1.

Analogously, for

$$T = \begin{pmatrix} 1 & 0 & 0 \\ 3/4 & 3/4 & 1/4 \\ 0 & 1/4 & 1/2 \end{pmatrix},$$

the determinant is $5/16$ and is equal to the volume of the parallelepiped Q in Figure 5.2.

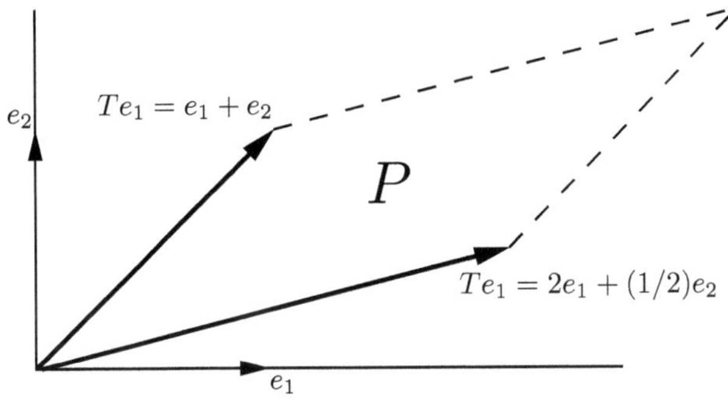

Figure 5.1: Parallelogram

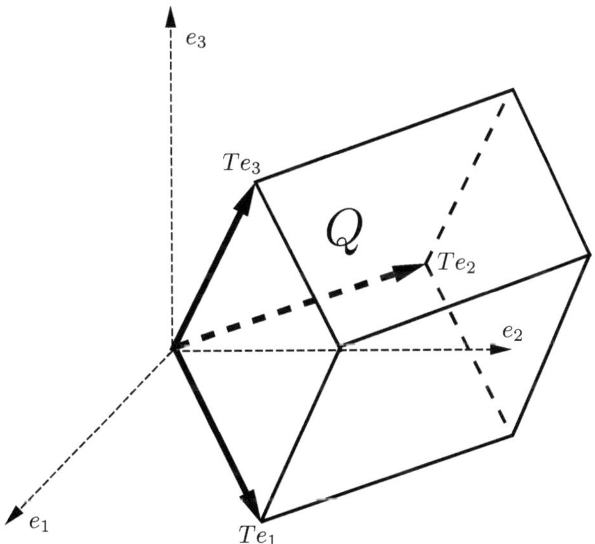

Figure 5.2: Parallelepiped

Proof of the proposition. Let us set $\Delta_T = \lambda_n\big(T([0,1]^n)\big)$. In the previous proof, we have seen that $\Delta_T > 0$ and $\lambda_n(T(E)) = \Delta_T \lambda_n(E)$, for any Borel set E. For T_1, T_2 linear isomorphisms, setting $Q_1 = [0,1]^n$, we find

$$\Delta_{T_2 T_1} = \lambda_n\big((T_2 T_1)(Q_1)\big) = \lambda_n\big(T_2\big(T_1(Q_1)\big)\big) = \Delta_{T_2} \lambda_n\big(T_1(Q_1)\big) \tag{5.2.2}$$
$$= \Delta_{T_2} \Delta_{T_1} \lambda_n(Q_1) = \Delta_{T_2} \Delta_{T_1}.$$

We have also $\Delta_{\mathrm{Id}} = 1$. We want to prove

$$\Delta_T = \lambda_n\big(T([0,1]^n)\big) = |\det T|, \tag{5.2.3}$$

for all invertible matrices T (matrix of T in the canonical basis).

CASE I. That formula holds for a diagonal matrix T: in fact if

$$T = \begin{pmatrix} a_1 & 0 & 0 \\ 0 & \ddots & 0 \\ 0 & 0 & a_n \end{pmatrix}$$

assuming all the $a_j > 0$, $T(Q_1) = \prod_{1 \le j \le n}[0, a_j]$ and Theorem 2.4.2 (1) implies

$$\lambda_n\big(T(Q_1)\big) = \prod_{1 \le j \le n} a_j = |\det T|.$$

If some a_j are negative, we have to replace in $T(Q_1)$ the interval $[0, a_j]$ by $[a_j, 0]$ so that the result is unchanged.

CASE II. The formula holds for T symmetric, i.e., whenever $T = {}^tT$: in that case T is diagonal in an orthonormal basis and there exists an invertible matrix P and a diagonal matrix D such that ${}^tPP = I, D = P^{-1}TP$. We get from (5.2.2) and CASE I,

$$\Delta_T = \Delta_{PDP^{-1}} = \Delta_P \Delta_D \Delta_{P^{-1}} = \Delta_D = |\det D| = |\det T|.$$

CASE III. The formula holds when T is an isometry, i.e., if ${}^tTT = I$ (this implies $|\det T| = 1$). In fact denoting by B_1 the closed Euclidean ball of $\mathbb{R}^n$, we have $T(B_1) = B_1$ since for $x \in B_1$, $\|Tx\| = \|x\| \le 1$. Conversely, we have $x = TT^{-1}x$ with $\|T^{-1}x\| = \|TT^{-1}x\| = \|x\| \le 1$. From Proposition 5.2.1, we find

$$\lambda_n(B_1) = \lambda_n\big(T(B_1)\big) = \Delta_T \lambda_n(B_1) \Longrightarrow \Delta_T = 1,$$

since $\lambda_n(B_1) > 0$ as B_1 contains a non-empty open set.

CASE IV. Let us tackle the general case. Let T be an invertible matrix. Then the matrix ${}^tTT$ is positive definite, i.e., symmetric with positive eigenvalues. As a consequence, there exists a matrix P such that ${}^tPP = I$, a positive definite diagonal matrix D such that

$${}^tTT = {}^tPDP \text{ (implying } (\det T)^2 = \det D). \quad \text{We define} \quad |T| = {}^tPD^{1/2}P.$$

The matrix $|T|$ is invertible as a product of invertible matrices and $T|T|^{-1}$ is an isometry since

$$\begin{aligned}
{}^t\big(T|T|^{-1}\big)T|T|^{-1} &= {}^t\big(|T|^{-1}\big){}^tTT|T|^{-1} \\
&= {}^t\big(P^{-1}D^{-1/2}({}^tP)^{-1}\big){}^tPDPP^{-1}D^{-1/2}({}^tP)^{-1} \\
&= P^{-1}D^{-1/2}({}^tP^{-1}){}^tPDD^{-1/2}({}^tP)^{-1} = P^{-1}({}^tP)^{-1} = I.
\end{aligned}$$

As a consequence, since $T = T|T|^{-1}|T|$, we find from CASES I, II, III,

$$\Delta_T = \Delta_{T|T|^{-1}}\Delta_{|T|} = \Delta_{|T|} = \Delta_{D^{1/2}} = |\det D^{1/2}| = |\det T|,$$

where the last equality follows from $(\det T)^2 = \det D = (\det D^{1/2})^2$. The proof of Proposition 5.2.2 is complete. Note that along with (5.2.1), we obtain that for $E \in \mathcal{B}_n$ and T a linear isomorphism of $\mathbb{R}^n$, $\lambda_n(T(E)) = |\det T|\lambda_n(E)$, i.e.,

$$\int_{\mathbb{R}^n} \mathbf{1}_{T(E)}(y)dy = |\det T| \int_{\mathbb{R}^n} \mathbf{1}_E(x)dx = \int_{\mathbb{R}^n} \mathbf{1}_{T(E)}(Tx)|\det T|dx. \qquad (5.2.4)$$

$\square$

Proposition 5.2.3. *Let T be a linear isomorphism of $\mathbb{R}^n$ and let $f \in L^1(\mathbb{R}^n)$. Then $f \circ T \in L^1(\mathbb{R}^n)$ and*

$$\int_{\mathbb{R}^n} f(y)dy = \int_{\mathbb{R}^n} f(Tx)|\det T|dx. \qquad (5.2.5)$$

Remark 5.2.4. We need to verify first that $f \circ T$ actually makes sense, which is easy but needs verification: the function f is defined modulo equality a.e. and it should also be the case of $f \circ T$. Let us then consider $f \in \mathcal{L}^1(\mathbb{R}^n)$, i.e., a measurable function $f : \mathbb{R}^n \to \mathbb{C}$ such that $\int_{\mathbb{R}^n} |f(x)|dx < +\infty$. Let $f_1 : \mathbb{R}^n \to \mathbb{C}$ be a.e. equal to f, i.e., $\{x \in \mathbb{R}^n, f(x) \neq f_1(x)\} = N$ is a Lebesgue set with measure 0. Since T is a homeomorphism, it is Borel-measurable and $T^{-1}(E) \in \mathcal{B}_n$ when $E \in \mathcal{B}_n$. Now since N belongs to the Lebesgue σ-algebra, thanks to Theorem 2.2.14, there exist Borel sets A, B with $A \subset N \subset B$, $\lambda_n(B \backslash A) = 0$. We find that

$$\underbrace{T^{-1}(A)}_{\in \mathcal{B}_n} \subset T^{-1}(N) \subset \underbrace{T^{-1}(B)}_{\in \mathcal{B}_n},$$

$$\lambda_n\big(T^{-1}(B) \backslash T^{-1}(A)\big) = \lambda_n\big(T^{-1}(B \backslash A)\big) \underbrace{=}_{(5.2.4)} \lambda_n(B \backslash A)|\det T|^{-1} = 0,$$

proving that $T^{-1}(N)$ belongs to the Lebesgue σ-algebra (T is proven Lebesgue-measurable). Moreover, since $\lambda_n(A) = 0$, (A is a subset of N) we find $\lambda_n(B) = \lambda_n(B \backslash A) + \lambda_n(A) = 0$, as well as

$$\lambda_n(T^{-1}(A)) = \lambda_n(T^{-1}(B)) = 0 \implies \lambda_n(T^{-1}(N)) = 0.$$

We have thus

$$\{y \in \mathbb{R}^n, f(Ty) \neq f_1(Ty)\} = \{y \in \mathbb{R}^n, Ty \in N\} = T^{-1}(N),$$

and $T^{-1}(N)$ is a Lebesgue set with measure 0, so that $f \circ T = f_1 \circ T$ a.e.

Remark 5.2.5. We go on with a trivial remark: this is indeed the absolute value of the determinant which should appear in Formula (5.2.5) and this does not contradict the habit of the reader with changes of variable in one dimension: with $f \in C_c(\mathbb{R})$, we have indeed

$$\int_{\mathbb{R}} f(y)dy = \int_{\mathbb{R}} f(-2x)2dx$$

since with the standard method

$$\int_{\mathbb{R}} f(y)dy = \int_{-\infty}^{+\infty} f(y)dy = \int_{+\infty}^{-\infty} f(-2x)(-2)dx$$

$$= \int_{-\infty}^{+\infty} f(-2x)2dx = \int_{\mathbb{R}} f(-2x)2dx.$$

Proof. Let us assume first that f is non-negative; using the approximation Theorem 1.3.3 we find a sequence $(s_k)_{k\in\mathbb{N}}$ of simple functions converging pointwise increasingly towards f. With $s_k = \sum_{1\le j\le N_k} \alpha_{j,k} \mathbf{1}_{A_{j,k}}$ (we may assume $\alpha_{j,k} > 0$), from Lemma 1.6.3 and (5.2.4), we get

$$\int_{\mathbb{R}^n} s_k(y)dy = \sum_{1\le j\le N_k} \alpha_{j,k} \int_{\mathbb{R}^n} \mathbf{1}_{A_{j,k}}(y)dy$$

$$= \sum_{1\le j\le N_k} \alpha_{j,k} \int_{\mathbb{R}^n} \mathbf{1}_{A_{j,k}}(Tx)|\det T|dx$$

$$= \int_{\mathbb{R}^n} \left(\sum_{1\le j\le N_k} \alpha_{j,k} \mathbf{1}_{A_{j,k}}(Tx) \right)|\det T|dx$$

$$= \int_{\mathbb{R}^n} s_k(Tx)|\det T|dx.$$

Beppo Levi's theorem 1.6.1 implies

$$\int_{\mathbb{R}^n} f(y)dy = \lim_{k\to+\infty} \int_{\mathbb{R}^n} s_k(y)dy = \lim_{k\to+\infty} \int_{\mathbb{R}^n} s_k(Tx)|\det T|dx$$

$$= \int_{\mathbb{R}^n} f(Tx)|\det T|dx.$$

For $f \in L^1(\mathbb{R}^n)$, the decomposition $f = (\operatorname{Re} f)_+ - (\operatorname{Re} f)_- + i(\operatorname{Im} f)_+ - i(\operatorname{Im} f)_-$ and the previous case give (5.2.5). $\qquad\square$

5.3 Change-of-variables formula

Definition 5.3.1 (C^1 diffeomorphism). Let U, V be open subsets of $\mathbb{R}^n$ and let $\kappa : U \to V$. We shall say that κ is a C^1 diffeomorphism from U onto V if it is a bijection of class C^1 as well as κ^{-1}. For each $x \in U$, the linear bijective mapping $\kappa'(x)$ is called the Jacobian matrix of κ and the determinant $\det(\kappa'(x))$ is called the Jacobian determinant. Let us recall that for

$$U \ni x = (x_1,\dots,x_n) \mapsto \kappa(x) = (\kappa_1(x),\dots,\kappa_n(x)) \in V,$$

we have

$$\kappa'(x) = \begin{pmatrix} \frac{\partial \kappa_1}{\partial x_1} & \cdots & \frac{\partial \kappa_1}{\partial x_n} \\ \cdots & \frac{\partial \kappa_i}{\partial x_j} & \cdots \\ \frac{\partial \kappa_n}{\partial x_1} & \cdots & \frac{\partial \kappa_n}{\partial x_n} \end{pmatrix}_{1 \le i,j \le n} \qquad (i \text{ row index, } j \text{ column index}). \qquad (5.3.1)$$

Moreover with $\nu = \kappa^{-1}$, since for all $x \in U$, $(\nu \circ \kappa)(x) = x$, we have

$$\nu'\big(\kappa(x)\big)\kappa'(x) = I, \quad \text{i.e.,} \quad \nu'\big(\kappa(x)\big) = \kappa'(x)^{-1}.$$

When a diffeomorphism κ is of class C^k for some $k \ge 1$ (resp. C^∞, resp. analytic) as well as κ^{-1}, we shall say that κ is a C^k-diffeomorphism (resp. C^∞-diffeomorphism, resp. analytic-diffeomorphism).

Remark 5.3.2. Let U be an open subset of $\mathbb{R}^n$, $x_0 \in U$ and let $\kappa : U \to \mathbb{R}^n$ be a C^1 mapping such that $\det \kappa'(x_0) \ne 0$. Then the *Inverse Function Theorem* ensures that there exists an open neighborhood U_0 of x_0 and an open set V_0 such that $\kappa : U_0 \to V_0$ is a C^1 diffeomorphism from U_0 onto V_0. This fundamental result of differential calculus reduces the problem of local invertibility of a C^1 mapping to a linear algebra problem, that is the invertibility of a $n \times n$ matrix (Jacobian matrix). When κ is of class C^k for some $k \ge 1$ and such that $\det \kappa'(x_0) \ne 0$, the inverse function theorem provides a local C^k-diffeomorphism.

Proposition 5.3.3. *Let $\kappa : U \to V$ be a C^1 diffeomorphism of open subsets U, V of $\mathbb{R}^n$. Then if A is a Borel subset of U, $\kappa(A)$ is a Borel subset of V. If E is a Lebesgue-measurable subset of U, then $\kappa(E)$ is a Lebesgue-measurable subset of V.*

Proof. The first assertion is obvious since $\kappa(A) = (\kappa^{-1})^{-1}(A)$ and $\nu = \kappa^{-1}$ is continuous, thus Borel-measurable (Proposition 1.2.5, Lemma 1.2.9). To check the next assertion it suffices to prove

$$A \text{ is a Borel set with null measure} \implies \nu^{-1}(A) \text{ has null measure.} \qquad (5.3.2)$$

If (5.3.2) holds, then for $E \subset A$, with A Borel set with null measure, we obtain $\nu^{-1}(E) \subset \nu^{-1}(A) = B$, where B is a Borel set with null measure. Since the Lebesgue σ-algebra is generated by the Borel σ-algebra and the subsets of Borel sets with null measure, Lemma 1.1.4 will provide the result. Property (5.3.2) follows from the next proposition. $\qquad \square$

Proposition 5.3.4. *Let U, V be open subsets of $\mathbb{R}^n$ and let $\kappa : U \to V$ be a C^1 diffeomorphism. Let A be a Borel subset of U. Then $\kappa(A)$ is a Borel subset of V and*

$$\lambda_n(\kappa(A)) = \int_A |\det \kappa'(x)| dx.$$

More generally, for $f \ge 0$ measurable on V,

$$\int_V f(y)dy = \int_U f(\kappa(x))|\det \kappa'(x)|\, dx.$$

Proof. Let P be a compact rational rectangle (product of compact intervals of $\mathbb{R}$ with rational endpoints) included in U. Let $\epsilon > 0$ be given. By uniform continuity on the compact P, there exists δ (depending on ϵ and P) such that[1]

$$\sup_{\substack{|x_1 - x_2| \leq \delta \\ x_1, x_2 \in P_0}} \|\kappa'(x_1) - \kappa'(x_2)\| + |\det \kappa'(x_1) - \det \kappa'(x_2)| \leq \epsilon.$$

We define also

$$\sup_{x \in P} \|\kappa'(x)^{-1}\| = M \ (< +\infty \text{ since } P \text{ is compact}).$$

We may write $P = \cup_{1 \leq j \leq N} Q_j$ where the Q_j are compact rational rectangles with sides $\rho \leq \delta$ such that $Q_j \cap Q_k$ is included in a hyperplane whenever $j \neq k$: since $P = \prod_{1 \leq l \leq n} I_l$ where each I_l is a compact interval of $\mathbb{R}$ with rational endpoints (a compact rational interval), we may write I_l as a finite union of compact rational intervals $I_{l,r}$ with length[2] ρ, such that for $r \neq s$, $I_{l,r} \cap I_{l,s}$ is either empty or reduced to a single point. As a result, we get

$$P = \bigcup_{\substack{1 \leq r_1 \leq N_1 \\ \cdots \\ 1 \leq r_n \leq N_n}} \underbrace{\left(\prod_{1 \leq l \leq n} I_{l,r_l} \right)}_{\substack{\text{compact rational} \\ \text{rectangle } Q}}.$$

Let a_j be the center of mass of Q_j so that $Q_j = \{x, \ |x - a_j| \leq \rho/2\}$. Let us set

$$\gamma(x) = \kappa'(a_j)^{-1} \kappa(x).$$

Using the mean value inequality (Proposition 5.1.3) and the convexity of Q_j, we get for $x \in Q_j$,

$$|\gamma(x) - \gamma(a_j)| \leq \sup_{x \in Q_j} \|\kappa'(a_j)^{-1} \kappa'(x)\| \ |x - a_j|.$$

Moreover, we have $\kappa'(a_j)^{-1} \kappa'(x) - \mathrm{Id} = \kappa'(a_j)^{-1}(\kappa'(x) - \kappa'(a_j))$ so that

$$\|\kappa'(a_j)^{-1} \kappa'(x)\| \leq 1 + M\epsilon.$$

This implies $\sup_{x \in Q_j} |\gamma(x) - \gamma(a_j)| \leq (1 + M\epsilon)\rho/2$, and thus

$$\lambda_n(\gamma(Q_j)) \leq (1 + M\epsilon)^n \rho^n = (1 + M\epsilon)^n \lambda_n(Q_j).$$

---

[1] We shall note here $|x|$ for the sup norm of $x \in \mathbb{R}^n$ and with a $d \times d$ matrix A, we define $\|A\| = \sup_{|x|=1} |Ax|$.

[2] Possible since each I_l has a rational length m_l: we must find integers $N_1, \ldots, N_n$ such that $m_1/N_1 = \cdots = m_n/N_n \leq \delta$. To do this it is enough to find an integer N_1 such that for all $k \in \{1, \ldots, n\}$, $N_1 m_k/m_1 = N_k \in \mathbb{N}$. Since m_k/m_1 are rational numbers, it suffices to take N_1 as a multiple of the product of denominators. This gives the above equality and the inequality holds for a large enough multiple.

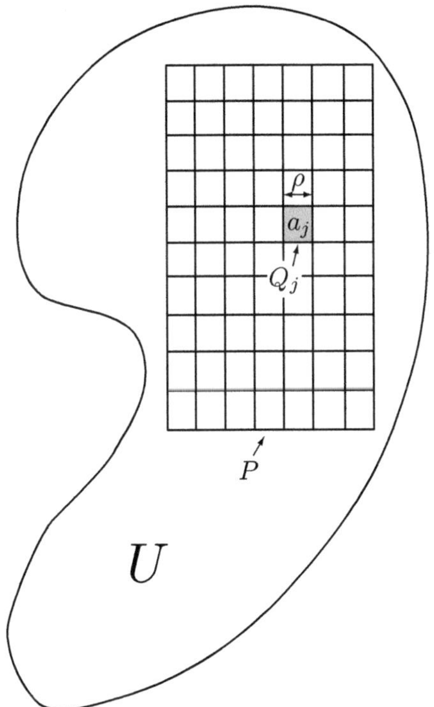

We have already proven that for a linear isomorphism T and a Borel set E,

$$\lambda_n(T(E)) = |\det T|\,\lambda_n(E).$$

This implies $\lambda_n(\gamma(Q_j)) = |\det \kappa'(a_j)|^{-1}\lambda_n(\kappa(Q_j))$ and thus

$$\lambda_n(\kappa(Q_j)) \leq |\det \kappa'(a_j)|(1 + M\epsilon)^n \lambda_n(Q_j).$$

Since for $x \in Q_j$, $|\det \kappa'(a_j)| \leq \epsilon + |\det \kappa'(x)|$, we get

$$\lambda_n(\kappa(Q_j)) \leq (\epsilon + |\det \kappa'(x)|)(1 + M\epsilon)^n \lambda_n(Q_j).$$

Integrating that inequality on Q_j, we find

$$\lambda_n(\kappa(Q_j))\lambda_n(Q_j) \leq \left(\epsilon\lambda_n(Q_j) + \int_{Q_j} |\det \kappa'(x)|dx\right)(1 + M\epsilon)^n \lambda_n(Q_j),$$

so that

$$\lambda_n(\kappa(Q_j)) \leq \left(\epsilon\lambda_n(Q_j) + \int_{Q_j} |\det \kappa'(x)|dx\right)(1 + M\epsilon)^n.$$

From $P = \cup_{1\leq j\leq N}Q_j$, we find $\kappa(P) = \cup_{1\leq j\leq N}\kappa(Q_j)$; moreover

$$\lambda_n(P) = \sum_{1\leq j\leq N} \lambda_n(Q_j)$$

since $m(Q_j \cap Q_l) = \emptyset$ if $j \neq l$. Consequently, for all $\epsilon > 0$,

$$\lambda_n(\kappa(P)) \leq \sum_{1 \leq j \leq N} \lambda_n(\kappa(Q_j)) \leq (1 + M\epsilon)^n \left(\epsilon \lambda_n(P) + \int_P |\det \kappa'(x)| dx \right).$$

Taking the infimum for $\epsilon > 0$, we obtain

$$\lambda_n(\kappa(P)) \leq \int_P |\det \kappa'(x)| dx, \tag{5.3.3}$$

for every compact rational rectangle.

Let us now consider a Borel subset A of U and Ω an open set of U containing A. From Lemma 1.2.6 we know that we may write Ω as a countable union of compact rational rectangles. Thanks to Lemma 2.4.4, *it is also possible to make these compact rational rectangles with an intersection of null measure whenever they are distinct.* Since $A \subset \Omega = \cup_{k \in \mathbb{N}} P_k \subset U$, we have

$$\lambda_n(\kappa(A)) \leq \sum_{\mathbb{N}} \lambda_n(\kappa(P_k)) \underset{(5.3.3)}{\leq} \sum_{\mathbb{N}} \int_{P_k} |\det \kappa'(x)| dx = \int_\Omega |\det \kappa'(x)| dx.$$

The measure $|\det \kappa'(x)| dx$ is outer regular (the Riesz representation theorem 2.2.14 implies that the positive Radon measure $\varphi \in C_c(U) \mapsto \int_U \varphi(x) |\det \kappa'(x)| dx$ provides a regular measure which is the measure with density $|\det \kappa'(x)|$ with respect to the Lebesgue measure), so that

$$\lambda_n(\kappa(A)) \leq \int_A |\det \kappa'(x)| dx. \tag{5.3.4}$$

In particular this implies that if A is a Borel set with null measure, then $\kappa(A)$ (which is a Borel set) has also null measure. Also, for B a Borel subset of V, with $A = \kappa^{-1}(B)$ we find

$$\int_V \mathbf{1}_B(y) dy = \lambda_d(B) \leq \int_{\kappa^{-1}(B)} |\det \kappa'(x)| dx = \int_U \mathbf{1}_B(\kappa(x)) |\det \kappa'(x)| dx.$$

Using Beppo Levi's theorem 1.6.1 and Theorem 1.3.3 (approximation by simple functions), we obtain for $f \geq 0$, Borel measurable defined on V,

$$\int_V f(y) dy \leq \int_U f(\kappa(x)) |\det \kappa'(x)| dx.$$

Switching U with V, we get

$$\int_U f(\kappa(x)) |\det \kappa'(x)| dx \leq \int_V f(\kappa(\nu(y))) |\det \kappa'(\nu(y))| |\det \nu'(y)| dy = \int_V f(y) dy,$$

so that for $f \geq 0$, Borel measurable defined on V, we obtain

$$\int_V f(y)dy = \int_U f(\kappa(x)) \, |\det \kappa'(x)|dx. \tag{5.3.5}$$

A non-negative Lebesgue-measurable function f is the pointwise limit of a sequence of simple functions coinciding a.e. with simple Borel functions so that $f = f_0$ a.e. with f_0 a Borel non-negative function. This implies that (5.3.5) holds for $f \geq 0$ Lebesgue measurable. The proof of Proposition 5.3.4 is complete. $\qquad\square$

Applying this proposition to $|f|, (\operatorname{Re} f)_\pm, (\operatorname{Im} f)_\pm$ for $f \in L^1(V)$, we obtain the following result.

Theorem 5.3.5. *Let U, V be open subsets of $\mathbb{R}^n$, let $\kappa : U \to V$ be a C^1 diffeomorphism and let $f \in L^1(V)$. Then $|\det \kappa'| f \circ \kappa$ belongs to $L^1(U)$ and*

$$\int_V f(y)dy = \int_U f(\kappa(x)) \, |\det \kappa'(x)| \, dx. \tag{5.3.6}$$

5.4 Examples, polar coordinates in $\mathbb{R}^n$

Polar coordinates in $\mathbb{R}^2$

We check first

$$\kappa :] \, 0, +\infty[\times] - \pi, \pi[\longrightarrow \mathbb{C}\backslash\mathbb{R}_- = \mathbb{R}^2\backslash(\mathbb{R}_- \times \{0\})$$
$$(r, \theta) \mapsto re^{i\theta} = (r\cos\theta, r\sin\theta)$$

$$\nu = \kappa^{-1} : \quad \mathbb{C}\backslash\mathbb{R}_- \longrightarrow \,]0, +\infty[\times] - \pi, \pi[$$
$$z \mapsto |z|, \operatorname{Im}(\operatorname{Log} z)$$

where the complex logarithm is defined on $\mathbb{C}\backslash\mathbb{R}_-$ by (10.5.1). We have in particular proven in Section 10.5 that for $z \in \mathbb{C}\backslash\mathbb{R}_-$, $\exp(\ln z) = z$ and for $|\operatorname{Im} z| < \pi$, $\operatorname{Log} e^z = z$. The Jacobian matrix $\mathcal{J}$ of κ and its Jacobian determinant J are

$$\mathcal{J} = \begin{pmatrix} \frac{\partial x}{\partial r} & \frac{\partial x}{\partial \theta} \\ \frac{\partial y}{\partial r} & \frac{\partial y}{\partial \theta} \end{pmatrix} = \begin{pmatrix} \cos\theta & -r\sin\theta \\ \sin\theta & r\cos\theta \end{pmatrix}, \quad J = \begin{vmatrix} \cos\theta & -r\sin\theta \\ \sin\theta & r\cos\theta \end{vmatrix} = r.$$

For $f \in L^1(\mathbb{R}^2)$, we have since $\mathbb{R}_- \times \{0\}$ has null Lebesgue measure in $\mathbb{R}^2$, using the diffeomorphism κ and Theorem 5.3.5,

$$\iint_{\mathbb{R}^2} f(x, y)dxdy = \iint_{\mathbb{R}_+ \times (-\pi, \pi)} f(r\cos\theta, r\sin\theta)rdrd\theta. \tag{5.4.1}$$

Spherical coordinates in $\mathbb{R}^3$

We define

$$\kappa :]0,+\infty[\times]0,\pi[\times]-\pi,\pi[\quad \longrightarrow \quad \mathbb{R}^3\backslash\{(x,y,z), x \le 0, y = 0\}$$
$$(r,\phi,\theta) \quad \mapsto \quad (r\cos\theta\sin\phi, r\sin\theta\sin\phi, r\cos\phi)$$

and we have

$$\kappa^{-1} = \nu : \mathbb{R}^3\backslash\{(x,y,z), x \le 0, y = 0\} \longrightarrow]0,+\infty[\times]0,\pi[\times]-\pi,\pi[$$
$$(x,y,z) \mapsto \left((x^2+y^2+z^2)^{1/2}, \operatorname{Im}\operatorname{Log}(z+i\sqrt{x^2+y^2}), \operatorname{Im}\operatorname{Log}(x+iy)\right)$$

which makes sense since $x+iy \notin \mathbb{R}_-$ and $z+i\sqrt{x^2+y^2} \notin \mathbb{R}_-$ (otherwise $x = y = 0$). The Jacobian matrix $\mathcal{J}$ of κ and its Jacobian determinant J are

$$\mathcal{J} = \begin{pmatrix} \frac{\partial x}{\partial r} & \frac{\partial x}{\partial \phi} & \frac{\partial x}{\partial \theta} \\ \frac{\partial y}{\partial r} & \frac{\partial y}{\partial \phi} & \frac{\partial y}{\partial \theta} \\ \frac{\partial z}{\partial r} & \frac{\partial z}{\partial \phi} & \frac{\partial z}{\partial \theta} \end{pmatrix} = \begin{pmatrix} \cos\theta\sin\phi & r\cos\theta\cos\phi & -r\sin\theta\sin\phi \\ \sin\theta\sin\phi & r\sin\theta\cos\phi & r\cos\theta\sin\phi \\ \cos\phi & -r\sin\phi & 0 \end{pmatrix},$$

$$J = \begin{vmatrix} \cos\theta\sin\phi & r\cos\theta\cos\phi & -r\sin\theta\sin\phi \\ \sin\theta\sin\phi & r\sin\theta\cos\phi & r\cos\theta\sin\phi \\ \cos\phi & -r\sin\phi & 0 \end{vmatrix}$$

$$= r^2\sin\phi \begin{vmatrix} \cos\theta\sin\phi & \cos\theta\cos\phi & -\sin\theta \\ \sin\theta\sin\phi & \sin\theta\cos\phi & \cos\theta \\ \cos\phi & -\sin\phi & 0 \end{vmatrix}$$

$$= (r^2\sin\phi)(\cos^2\phi + \sin^2\phi) = r^2\sin\phi.$$

As a result we have for $f \in L^1(\mathbb{R}^3)$,

$$\iiint_{\mathbb{R}^3} f(x,y,z)\,dxdydz$$
$$= \iiint_{\substack{r>0,|\theta|<\pi, \\ 0<\phi<\pi}} f(r\sin\phi\cos\theta, r\sin\phi\sin\theta, r\cos\phi)r^2\sin\phi\,drd\phi d\theta. \tag{5.4.2}$$

It is interesting to note that it is not necessary to go through the previous computation to obtain (5.4.2). We may skip as well the fact that κ is a diffeomorphism by simply iterating two-dimensional changes in polar coordinates as follows. We have

$$\iiint_{\mathbb{R}^3} f(x,y,z)\,dxdydz = \iiint_{\substack{z\in\mathbb{R},\rho>0 \\ |\theta|<\pi}} f(\rho\cos\theta, \rho\sin\theta, z)\rho\,dzd\rho d\theta$$

$$= \iiint_{\substack{r>0,|\theta|<\pi, \\ 0<\phi<\pi}} f(r\sin\phi\cos\theta, r\sin\phi\sin\theta, r\cos\phi)r\sin\phi\,rdrd\phi d\theta$$

$$= \iiint_{\substack{r>0,|\theta|<\pi, \\ 0<\phi<\pi}} f(r\sin\phi\cos\theta, r\sin\phi\sin\theta, r\cos\phi)r^2\sin\phi\,drd\phi d\theta,$$

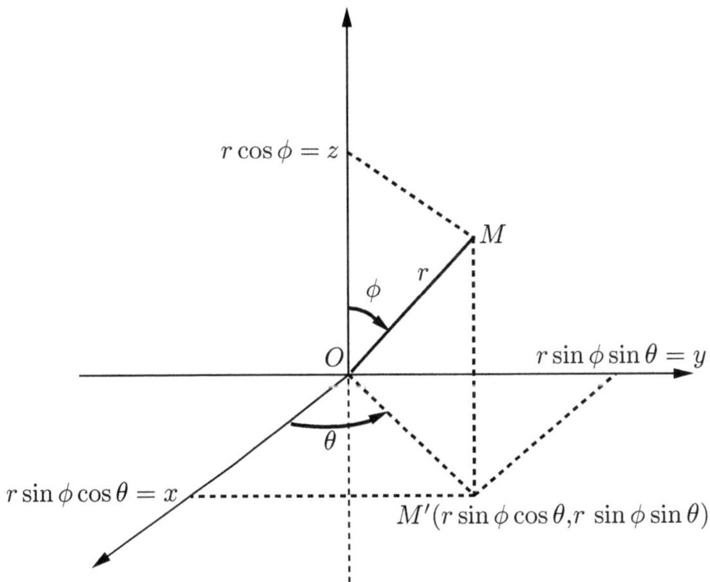

Figure 5.3: SPHERICAL COORDINATES: $r > 0, |\theta| < \pi, 0 < \phi < \pi$

where the first equality is the polar coordinates change in the plane (x, y) and the second equality comes from the polar coordinates change in the half-plane (z, ρ) $(\rho \geq 0)$.

Polar coordinates in $\mathbb{R}^n$

It is possible to build upon the two-dimensional formula to get all dimensions inductively as follows. We write, using the n-dimensional formula,

$$\iint_{\mathbb{R}^n_x \times \mathbb{R}_z} f(x, z)\,dx\,dz = \int_{\mathbb{R}^+_\rho \times \mathbb{S}^{n-1}_\omega \times \mathbb{R}_z} f(\rho\omega, z)\rho^{n-1}\,d\rho\,d\omega\,dz.$$

Then we use 2D polar coordinates in the half-plane z, ρ with

$$z = r\cos\phi, \rho = r\sin\phi, 0 < \phi < \pi,$$

to get

$$\iint_{\mathbb{R}^n \times \mathbb{R}} f(x, z)\,dx\,dz = \int_{\mathbb{S}^{n-1}_\omega \times (0,\pi)_\phi \times \mathbb{R}^+_r} f(r\omega\sin\phi, r\cos\phi)r^n(\sin\phi)^{n-1}\,d\omega\,d\phi\,dr,$$

so that

$$d_{\mathbb{S}^n}(\sigma) = d_{\mathbb{S}^n}(\omega\sin\phi \oplus \cos\phi) = (\sin\phi)^{n-1}d_{\mathbb{S}^{n-1}}(\omega)\,d\phi. \tag{5.4.3}$$

We have proven, say for $f \in C_c(\mathbb{R}^n)$,

$$\int_{\mathbb{R}^n} f(x)dx = \int_0^{+\infty} \int_{\mathbb{S}^{n-1}} f(r\omega)d_{\mathbb{S}^{n-1}}(\omega)r^{n-1}dr, \qquad (5.4.4)$$

where $d_{\mathbb{S}^{n-1}}$ is defined inductively by (5.4.3). We have seen

$$\text{2D}: \begin{cases} x_1 = r\cos\theta \\ x_2 = r\sin\theta \end{cases} \qquad\qquad \begin{array}{c} d_{\mathbb{S}^1}(\theta) = d\theta \\ |\theta| < \pi \end{array}$$

$$\text{3D}: \begin{cases} x_1 = r\cos\theta\sin\phi \\ x_2 = r\sin\theta\sin\phi \\ x_3 = r\cos\phi \end{cases} \qquad\qquad \begin{array}{c} d_{\mathbb{S}^2}(\theta,\phi) = \sin\phi\, d\phi d\theta \\ |\theta| < \pi, 0 < \phi < \pi \end{array}$$

$$\text{4D}: \begin{cases} x_1 = r\cos\theta\sin\phi_1\sin\phi_2 \\ x_2 = r\sin\theta\sin\phi_1\sin\phi_2 \\ x_3 = r\cos\phi_1\sin\phi_2 \\ x_4 = r\cos\phi_2 \end{cases} \qquad \begin{array}{c} d_{\mathbb{S}^3}(\theta,\phi_1,\phi_2) = \sin^2\phi_2\sin\phi_1\, d\phi_2 d\phi_1 d\theta \\ |\theta| < \pi, 0 < \phi_1, \phi_2 < \pi. \end{array}$$

In n dimensions, the spherical coordinates are

$$\begin{cases} x_1 = r\cos\theta\sin\phi_1\sin\phi_2\ldots\sin\phi_{n-3}\sin\phi_{n-2} \\ x_2 = r\sin\theta\sin\phi_1\sin\phi_2\ldots\sin\phi_{n-3}\sin\phi_{n-2} \\ x_3 = r\cos\phi_1\sin\phi_2\ldots\sin\phi_{n-3}\sin\phi_{n-2} \\ \qquad\qquad \ldots \\ x_{n-1} = r\cos\phi_{n-3}\sin\phi_{n-2} \\ x_n = r\cos\phi_{n-2} \end{cases}$$

with

$$d_{\mathbb{S}^{n-1}}(\theta,\phi_1,\ldots,\phi_{n-2}) = (\sin\phi_{n-2})^{n-2}(\sin\phi_{n-3})^{n-3}\ldots\sin\phi_1 d\phi_{n-2}\ldots d\phi_1 d\theta$$
$$(5.4.5)$$

$$|\theta| < \pi, \quad 0 < \phi_j < \pi, \quad 1 \leq j \leq n-2. \qquad (5.4.6)$$

An alternative way is to use the homogeneity and to define, say for a continuous function on the sphere,

$$\int_{\mathbb{S}^{n-1}} f(\sigma)d\sigma = \int_{\mathbb{R}^n} f(\frac{x}{|x|})\chi(|x|)dx \quad \text{where} \quad \int_{\mathbb{R}^+} \chi(r)r^{n-1}dr = 1. \qquad (5.4.7)$$

It is not difficult to prove that this formula does not depend on χ satisfying (5.4.7). A good choice can be $\chi(r) = e^{-r}/\Gamma(n)$. Another way would be more geometrical and simply use the fact that the sphere is a smooth hypersurface of $\mathbb{R}^n$, without resorting as above to some homogeneity property. We may define the Euclidean surface measure on $\mathbb{S}^{n-1}$, say for f continuous on $\mathbb{R}^n$,

$$\int_{\mathbb{S}^{n-1}} f(\sigma)d\sigma = \lim_{\varepsilon \to 0} \int_{\mathbb{R}^n} f(x)\rho\left(\frac{|x|-1}{\varepsilon}\right)\varepsilon^{-1}dx, \quad \rho \in C_c^\infty(\mathbb{R}), \int \rho = 1.$$

A useful computation is the $n-1$ area of $\mathbb{S}^{n-1}$, using polar coordinates and $1 = \int e^{-\pi|x|^2}dx$; we get

$$1 = |\mathbb{S}^{n-1}| \int_0^{+\infty} e^{-\pi r^2} r^{n-1}dr = |\mathbb{S}^{n-1}| \int_0^{+\infty} e^{-x}(x/\pi)^{(n-1)/2}\pi^{-1/2}\frac{1}{2}x^{-1/2}dx$$

$$= \frac{1}{2}|\mathbb{S}^{n-1}|\pi^{-n/2}\Gamma(n/2),$$

yielding

$$|\mathbb{S}^{n-1}| = \frac{2\pi^{n/2}}{\Gamma(n/2)}, \tag{5.4.8}$$

e.g.,

$$|\mathbb{S}^1| = 2\pi, \quad |\mathbb{S}^2| = \frac{2\pi^{3/2}}{\frac{1}{2}\Gamma(1/2)} = 4\pi, \quad |\mathbb{S}^3| = \frac{2\pi^2}{\Gamma(2)} = 2\pi^2. \tag{5.4.9}$$

We can check that this is consistent with Formula (4.5.4) since

$$|\mathbb{B}^n|_n = \int_0^1 r^{n-1}dr|\mathbb{S}^{n-1}|_{n-1} = \frac{2\pi^{n/2}}{n\Gamma(n/2)}.$$

We obtain in particular that the volume of a Euclidean ball with radius R, $B_n(R)$ in $\mathbb{R}^n$ is

$$\lambda_n(B_n(R)) = V(R) = \frac{2\pi^{n/2}}{n\Gamma(n/2)}R^n.$$

The reader will have noticed that, with $V(r)$ as the n-volume of the Euclidean ball with radius r and $S(r)$ the $(n-1)$-volume of the Euclidean sphere with radius r, we have

$$V'(r) = S(r),$$

which is suggested by the following picture, indicating that the shaded volume is $V(r+dr) - V(r) \sim S(r)dr$, i.e., $V'(r) = S(r)$.

Note that to integrate a function f on the sphere of center x_0 and radius R in $\mathbb{R}^n$, we get

$$\int_{|x-x_0|=R} f(\omega)d\omega = \int_{\mathbb{S}^{n-1}} f(x_0 + R\sigma)d\sigma R^{n-1}.$$

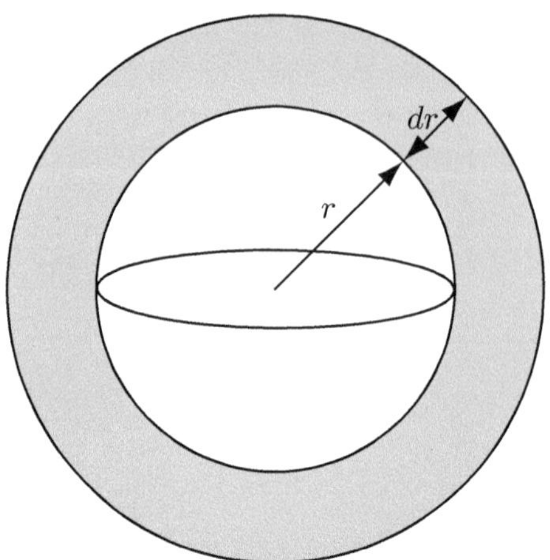

We have also for $A \in O(n)$ (the orthogonal group in n dimensions, i.e., $n \times n$ matrices with ${}^t\!AA = \mathrm{Id}$),

$$\int_{\mathbb{S}^{n-1}} f(A\omega)\,d\omega = \int_{\mathbb{S}^{n-1}} f(\omega)\,d\omega, \tag{5.4.10}$$

since $\int_{\mathbb{S}^{n-1}} f(A\omega)\,d\omega = \int_{\mathbb{R}^n} f(Ax/|x|)e^{-|x|}\,dx/\Gamma(n) = \int_{\mathbb{R}^n} f(y/|y|)e^{-|y|}\,dy/\Gamma(n)$.

5.5   Integration on a C^1 hypersurface of the Euclidean $\mathbb{R}^n$

Definition 5.5.1. Let Σ be a subset of the Euclidean $\mathbb{R}^n$ ($n \geq 2$). We shall say that Σ is a C^1 hypersurface of $\mathbb{R}^n$ if there exists a function $\rho \in C^1(\mathbb{R}^n; \mathbb{R})$ such that

$$\Sigma = \{x \in \mathbb{R}^n, \rho(x) = 0\}, \quad d\rho(x) \neq 0 \quad \text{for } x \in \Sigma.$$

A function ρ satisfying these properties will be called a defining function for Σ.

N.B. Using the implicit function theorem, it implies that Σ is locally the graph of a C^1 function of $(n-1)$ variables. For instance we may assume that $(\partial\rho/\partial x_n)(x_0) \neq 0$ at some point $x_0 \in \Sigma$ and thus we may find a neighborhood U_0 of x_0 such that $\Sigma \cap U_0$ appears as the graph $\{\mathbb{R}^{n-1} \times \mathbb{R} \ni (x', x_n) \in U_0, x_n = \alpha(x')\}$ where $\rho(x', \alpha(x')) \equiv 0$.

Let f be a compactly supported continuous function defined on $\mathbb{R}^n$. We want to define the positive Radon measure

$$f \mapsto \int_{\Sigma} f\,d\sigma$$

using the Euclidean embedding of Σ into $\mathbb{R}^n$.

Lemma 5.5.2. *Let Σ be a C^1 hypersurface of $\mathbb{R}^n$, with a defining function ρ, let $\theta \in C_c^\infty(\mathbb{R}; \mathbb{R}_+)$ such that $\int \theta(t)dt = 1$ and let $f \in C_c(\mathbb{R}^n)$. Then the following limit exists:*

$$\lim_{\epsilon \to 0_+} \int_{\mathbb{R}^n} \theta\left(\frac{\rho(x)}{\epsilon}\right) \epsilon^{-n}\|d\rho(x)\|f(x)dx.$$

That limit does not depend on the choice of the defining function ρ of Σ, nor on the choice of the function θ. This limit defines a positive Radon measure with support Σ.

Proof. If $\operatorname{supp} f \subset \Sigma^c$, then the limit above is 0: since $\operatorname{supp} f$ is compact and Σ is closed, we have $\operatorname{dist}(\operatorname{supp} f, \Sigma) > 0$, which implies that $\rho(x) \geq \epsilon_0 > 0$ on the support of f, implying that $\theta(\rho(x)/\epsilon)$ vanishes for $x \in \operatorname{supp} f$ and ϵ small enough (depending only on the support of θ and on ϵ_0).

We may thus assume that $\operatorname{supp} f \cap \Sigma \neq \emptyset$. Since $\operatorname{supp} f \cap \Sigma$ is a compact set, we can find a finite cover of it by open sets $U_1, \ldots, U_N$ such that, in each U_j, the defining function ρ appears as a coordinate. We have

$$\operatorname{supp} f \subset \cup_{1 \leq j \leq N} U_j \cup \Sigma^c$$

and a partition of unity argument (Theorem 2.1.3) shows that

$$f = f_0 + \sum_{1 \leq j \leq N} f_j, \quad \operatorname{supp} f_0 \subset \Sigma^c, \quad \operatorname{supp} f_j \subset U_j.$$

As above the contribution of f_0 is 0, and by linearity, we have only to consider the case where f is supported in a subset U_j (denoted by U). We may assume for instance that, on U_j, $\partial\rho/\partial x_n \neq 0$ and consider the local diffeomorphism

$$(x_1, \ldots, x_{n-1}, x_n) = x \mapsto \kappa(x) = (x_1, \ldots, x_{n-1}, \rho(x)), \quad \nu = \kappa^{-1}.$$

We have by the change of variable formula,

$$\begin{cases} y' = x' \\ y_n = \rho(x', x_n) \end{cases} \quad \begin{cases} x' = y' \\ x_n = \alpha(y', y_n) \end{cases}$$

with $\rho(x', \alpha(x', y_n)) \equiv y_n$,

$$\int_{\mathbb{R}^n} \theta\left(\frac{\rho(x)}{\epsilon}\right) \epsilon^{-1}\|d\rho(x)\|f(x)dx = \int_U \theta\left(\frac{\rho(x)}{\epsilon}\right) \epsilon^{-1}\|d\rho(x)\|f(x)dx$$

$$= \int_{V=\kappa(U)} \theta\left(\frac{\rho(\nu(y))}{\epsilon}\right) \epsilon^{-1}\|d\rho(\nu(y))\|f(\nu(y))|\nu'(y)|dy$$

$$= \int_{V=\kappa(U)} \theta\left(\frac{y_n}{\epsilon}\right) \epsilon^{-1}\left((\partial\rho/\partial x')^2 + (\partial\rho/\partial x_n)^2\right)^{1/2} f(y', \alpha(y', y_n)) \left|\frac{\partial\alpha}{\partial y_n}\right| dy$$

whose limit when ϵ goes to zero is

$$\int f(y', \alpha(y', 0))((\partial\rho/\partial x')^2 + (\partial\rho/\partial x_n)^2)^{1/2}\left|\frac{\partial\alpha}{\partial y_n}\right| dy'.$$

We note that $\frac{\partial\rho}{\partial x'} + \frac{\partial\rho}{\partial x_n}\frac{\partial\alpha}{\partial x'} = 0$, $\frac{\partial\rho}{\partial x_n}\frac{\partial\alpha}{\partial y_n} = 1$, so that the limit is

$$\int f(x', \alpha(x', 0))\left(\left|\frac{\partial\alpha}{\partial x'}(x', 0)\right|^2 + 1\right)^{1/2} dx'. \tag{5.5.1}$$

This proves also that the result does not depend on the choice of the function θ satisfying the required assumptions and also that this defines a positive Radon measure with support Σ. We need to verify that this Radon measure does not depend on the choice of the defining function ρ. By localization and partition of unity, we may consider a coordinate chart U near a point of Σ and two defining functions ρ_1, ρ_2 for Σ defined on U neighborhood of 0. As seen above, we may assume that $\partial\rho_1/\partial x_n \neq 0$ and

$$(x', x_n) \in \{\rho_1 = 0\} \cap \Sigma \iff x_n = \alpha_1(x', 0), \quad \alpha_1 \in C^1(U),$$

so that $\rho_2(x', \alpha_1(x', 0)) \equiv 0$ near the origin, which implies

$$\partial\rho_2/\partial x' + (\partial\rho_2/\partial x_n)(\partial\alpha_1/\partial x') = 0 \implies \partial\rho_2/\partial x_n \neq 0, \text{ at } 0,$$

otherwise $\partial\rho_2/\partial x_n = 0, \partial\rho_2/\partial x' = 0$ at 0, contradicting the assumption $d\rho_2 \neq 0$ at Σ. Now $\rho_2 = 0$ is equivalent to $x_n = \alpha_2(x', 0)$ as well as to $x_n = \alpha_1(x', 0)$, proving that $\alpha_1(x', 0) = \alpha_2(x', 0) = \alpha(x', 0)$ near the origin and (5.5.1) holds there. The proof of the lemma is complete. $\qquad\qquad\square$

Definition 5.5.3. Let Σ be a C^1 hypersurface of $\mathbb{R}^n$, with a defining function ρ. We define the simple layer on Σ as the positive Radon measure with support Σ given by

$$C_c(\mathbb{R}^n) \ni f \mapsto \int_\Sigma f d\sigma = \lim_{\epsilon \to 0+} \int_{\mathbb{R}^n} \theta\left(\frac{\rho(x)}{\epsilon}\right) \epsilon^{-1} \|d\rho(x)\| f(x) dx.$$

Definition 5.5.4. Let Ω be an open set of $\mathbb{R}^n$: Ω will be said to have a C^1 boundary if for all $x_0 \in \partial\Omega$, there exists a neighborhood U_0 of x_0 in $\mathbb{R}^n$ and a C^1 function $\rho_0 \in C^1(U_0; \mathbb{R})$ such that $d\rho_0$ does not vanish and $\Omega \cap U_0 = \{x \in U_0, \rho_0(x) < 0\}$.

Note that $\partial\Omega \cap U_0 = \{x \in U_0, \rho_0(x) = 0\}$ since the implicit function theorem shows that, if $(\partial\rho_0/\partial x_n)(x_0) \neq 0$ for some $x_0 \in \partial\Omega$, the mapping $x \mapsto (x_1, \ldots, x_{n-1}, \rho_0(x))$ is a local C^1-diffeomorphism.

Theorem 5.5.5 (Gauss–Green formula). *Let Ω be an open set of $\mathbb{R}^n$ with a C^1 boundary, X a C^1 vector field on Ω, continuous on $\bar{\Omega}$. Then we have, if X is compactly supported or Ω is bounded,*

$$\int_\Omega (\operatorname{div} X) dx = \int_{\partial\Omega} \langle X, \nu \rangle d\sigma, \tag{5.5.2}$$

where ν is the exterior unit normal and $d\sigma$ is the Euclidean measure on $\partial\Omega$.

Proof. We may assume that $\Omega = \{x \in \mathbb{R}^n, \rho(x) < 0\}$, where $\rho : \mathbb{R}^n \longrightarrow \mathbb{R}$ is C^1 and such that $d\rho \neq 0$ at $\partial\Omega$. The exterior normal to the open set Ω is defined on (a neighborhood of) $\partial\Omega$ as $\nu = \|d\rho\|^{-1}d\rho$. We can reformulate the theorem as

$$\int_\Omega \mathrm{div} X \; dx = \int \langle X, \nu \rangle \delta(\rho(x)) \|d\rho(x)\| = \lim_{\epsilon \to 0_+} \int \langle X, d\rho(x) \rangle \theta(\rho(x)/\epsilon) dx/\epsilon$$

where $\theta \in C_c(\mathbb{R})$ has integral 1. Since it is linear in X, it is enough to prove it for $a(x)\partial_{x_1}$, with $a \in C_c^1$. We have, with $\psi = 1$ on $(1, +\infty)$, $\psi = 0$ on $(-\infty, 0)$,

$$\int_\Omega \mathrm{div} X \; dx = \int_{\rho(x)<0} \frac{\partial a}{\partial x_1}(x) dx = \lim_{\epsilon \to 0_+} \int \frac{\partial a}{\partial x_1}(x)\psi(-\rho(x)/\epsilon) dx$$

$$= \lim_{\epsilon \to 0_+} \int a(x)\psi'(-\rho(x)/\epsilon)\epsilon^{-1}\frac{\partial \rho}{\partial x_1}(x) dx$$

$$= \lim_{\epsilon \to 0_+} \int \langle a(x)\partial_{x_1}, d\rho \rangle \psi'(-\rho(x)/\epsilon)\epsilon^{-1} dx$$

$$= \lim_{\epsilon \to 0_+} \int \langle X, d\rho \rangle \theta(\rho(x)/\epsilon)\epsilon^{-1} dx,$$

with $\theta(t) = \psi'(-t)$, $\int_{-\infty}^{+\infty} \theta(t)dt = \int_{-\infty}^{+\infty} \psi'(-t)dt = \int_{-\infty}^{+\infty} \psi'(t)dt = 1$. $\square$

In two dimensions, we get the Green–Riemann formula

$$\iint_\Omega \left(\frac{\partial P}{\partial x} + \frac{\partial Q}{\partial y}\right) dxdy = \int_{\partial\Omega} Pdy - Qdx, \tag{5.5.3}$$

since with $X = P\partial_x + Q\partial_y$, $\Omega \equiv \rho(x,y) < 0$, the lhs of (5.5.3) and (5.5.2) are the same, whereas the rhs of (5.5.2) is, if $\rho(x,y) = f(x) - y$ on the support of X,

$$\iint \langle X, d\rho \rangle \delta(\rho) dxdy = \lim_{\varepsilon \to 0_+} \iint (P(x,y)f'(x) - Q(x,y))\theta((f(x) - y)/\varepsilon) dxdy/\varepsilon$$

$$= \int (P(x, f(x))f'(x) - Q(x, f(x))) dx = \int_{\partial\Omega} Pdy - Qdx.$$

Corollary 5.5.6. *Let Ω be an open subset of $\mathbb{R}^n$ with a C^1 boundary, $u, v \in C^2(\bar{\Omega})$. Then*

$$\int_\Omega (\Delta u)(x)v(x)dx = \int_\Omega u(x)(\Delta v)(x)dx + \int_{\partial\Omega} \left(v\frac{\partial u}{\partial \nu} - u\frac{\partial v}{\partial \nu}\right) d\sigma, \tag{5.5.4}$$

$$\int_\Omega \nabla u \cdot \nabla v dx = -\int_\Omega u\Delta v dx + \int_{\partial\Omega} u\frac{\partial v}{\partial \nu} d\sigma, \tag{5.5.5}$$

where $\Delta = \sum_{1 \leq j \leq n} \partial_{x_j}^2$ is the Laplace operator and $\frac{\partial u}{\partial \nu} = \nabla u \cdot \nu$ where ν is the unit exterior normal.

Proof. We have $v\Delta u = \mathrm{div}\,(v\nabla u) - \nabla u \cdot \nabla v$ so that $v\Delta u - u\Delta v = \mathrm{div}(v\nabla u - u\nabla v)$ providing the first formula from Green's formula (5.5.2). The same formula written as $\nabla u \cdot \nabla v = -u\Delta v + \mathrm{div}\,(u\nabla v)$ entails the second formula. $\square$

5.6   More on Hausdorff measures on $\mathbb{R}^m$

We begin with a result on the structure of open subsets of $\mathbb{R}^m$, that could have been proven in Chapter 1. It will be useful in our study of Hausdorff measures.

Theorem 5.6.1. *Let Ω be an open subset of $\mathbb{R}^m$ and let $r > 0$ be given. There exists a countable pairwise disjoint family $\{B_n\}_{n\in\mathbb{N}}$ of open Euclidean balls with radii smaller than r such that $\overline{B_n} \subset \Omega$ and*

$$\lambda_m\big(\Omega\backslash(\cup_{n\in\mathbb{N}}B_n)\big) = 0.$$

N.B. The reader will find a less precise (but as useful and simpler to prove) statement in Exercise 5.10.12.

Proof. We have seen in Lemma 2.4.4 that for a given open set Ω, we could find a countable family of compact rational rectangles $\{Q_n\}_{n\in\mathbb{N}}$ such that for $n \neq n'$ $Q_n \cap Q_{n'}$ is included in a hyperplane parallel to the axes. Also the image of a compact rational rectangle by a dilation of a suitably chosen integer ratio is a compact rectangle with integer sides, thus a finite union of translations of $[0,1]^m$ with intersections included in a hyperplane. Performing the inverse dilation, we see that each Q_n is a finite union of *cubes* (rectangles whose sides have the same length) such that the intersection of two different cubes is included in a hyperplane. As a result, we could assume that Ω is an open cube whose sides are all smaller than r. We shall assume only that Ω has finite measure and that a Euclidean ball included in Ω has a radius automatically smaller than r.

Let $\Omega = \Omega_0$ be an open set such that $\lambda_m(\Omega_0) < +\infty$. As noted above, there exists a countable family $(C_{n,0})_{n\in\mathbb{N}}$ of compact cubes such that the family $(\mathring{C}_{n,0})_{n\in\mathbb{N}}$ is pairwise disjoint and

$$\Omega_0 = \cup_{n\in\mathbb{N}}C_{n,0}, \quad \lambda_m(\Omega_0) = \sum_{n\in\mathbb{N}} \lambda_m(C_{n,0}).$$

For each $C_{n,0}$, we consider the inscribed open Euclidean ball $B_{n,0}$ and we have

$$\lambda_m(B_{n,0}) = \alpha_m\lambda_m(C_{n,0}),$$

with a constant $\alpha_m \in (0,1)$ depending only on m (note that the $B_{n,0}$ are pairwise disjoint as subsets of $\mathring{C}_{n,0}$). Let us choose $\beta \in (1, \frac{1}{1-\alpha_m})$. We have $\lambda_m(C_{n,0}\backslash B_{n,0}) = (1 - \alpha_m)\lambda_m(C_{n,0})$, so that

$$\lambda_m(\Omega_0\backslash \cup_{\mathbb{N}} B_{n,0}) = (1 - \alpha_m)\lambda_m(\Omega_0).$$

Since $\beta > 1$, we may find a finite subset $\mathcal{N}_0$ such that

$$\lambda_m(\Omega_0\backslash \cup_{\mathcal{N}_0} B_{n,0}) \leq \beta(1 - \alpha_m)\lambda_m(\Omega_0).$$

We consider now the open set

$$\Omega_1 = \Omega_0 \backslash \big(\cup_{n \in \mathcal{N}_0} \overline{B_{n,0}}\big), \quad (B_{n,0}) \text{ pairwise disjoint open Euclidean balls}$$

$$\overline{B_{n,0}} \subset \Omega_0, \quad \lambda_m(\Omega_1) \leq \beta(1 - \alpha_m)\lambda_m(\Omega_0).$$

Let $k \geq 1$ be an integer. Let us assume that we have found some open subsets

$$\Omega_0 \supset \Omega_1 \supset \cdots \supset \Omega_k, \quad \mathcal{N}_0, \ldots, \mathcal{N}_{k-1} \text{ finite sets,}$$

$$\text{for each } 0 \leq j < k,$$

$$(B_{n,j})_{n \in \mathcal{N}_j} \text{ pairwise disjoint open Euclidean balls}, \quad \overline{B_{n,j}} \subset \Omega_j,$$

$$\Omega_{j+1} = \Omega_j \backslash \big(\cup_{n \in \mathcal{N}_j} \overline{B_{n,j}}\big), \quad \lambda_m(\Omega_{j+1}) \leq \beta(1 - \alpha_m)\lambda_m(\Omega_j), \quad 0 \leq j < k.$$

We consider the open set Ω_k (which has finite measure as a subset of Ω_0) and using what was done for Ω_0, we can find a finite set $\mathcal{N}_k$, and a pairwise disjoint set of open Euclidean balls $(B_{n,k})_{n \in \mathcal{N}_k}$ such that $\overline{B_{n,k}} \subset \Omega_k$,

$$\Omega_{k+1} = \Omega_k \backslash \big(\cup_{n \in \mathcal{N}_k} \overline{B_{n,k}}\big), \quad \lambda_m(\Omega_{k+1}) \leq \beta(1 - \alpha_m)\lambda_m(\Omega_k),$$

so that we have constructed an open set Ω_{k+1} such that the above properties are true up to $k + 1$. We can thus perform that construction for every $k \geq 1$. We find in particular inductively for $k \geq 1$,

$$\lambda_m(\Omega_k) \leq \big(\beta(1 - \alpha_m)\big)^k \lambda_m(\Omega_0).$$

We consider now $\cup_{j \geq 0} \big(\cup_{n \in \mathcal{N}_j} B_{n,j}\big)$. This is a pairwise disjoint union: in the first place $B_{n,j} \cap B_{n,j'} = \emptyset$ for $j \neq j'$, say $j < j'$, since

$$B_{n,j} \cap B_{n',j'} \subset \Omega_{j+1}^c \cap \Omega_{j'} \subset \Omega_{j+1}^c \cap \Omega_{j+1} = \emptyset.$$

Moreover for a given j the family $(B_{n,j})_{n \in \mathcal{N}_j}$ is pairwise disjoint. We have also $\overline{B_{n,j}} \subset \Omega_j \subset \Omega_0$, and for $k \geq 1$,

$$\lambda_m\big(\Omega_0 \backslash (\cup_{j \geq 0} \cup_{n \in \mathcal{N}_j} B_{n,j})\big) \leq \lambda_m\big(\Omega_0 \backslash (\cup_{0 \leq j \leq k} \cup_{n \in \mathcal{N}_j} B_{n,j})\big)$$

$$= \lambda_m\big(\underbrace{\Omega_0 \backslash (\cup_{0 \leq j \leq k} \cup_{n \in \mathcal{N}_j} \overline{B_{n,j}})}_{=\Omega_{k+1}}\big)$$

$$\leq \big(\beta(1 - \alpha_m)\big)^{k+1} \lambda_m(\Omega_0).$$

As a result, since $\beta(1 - \alpha_m) \in (0, 1)$, $\lambda_m\big(\Omega_0 \backslash (\cup_{j \geq 0} \cup_{n \in \mathcal{N}_j} B_{n,j})\big) = 0$. $\qquad \square$

Let $m \geq 1$ be an integer. We define[3]

$$v_m = \frac{1}{2^m} \frac{\pi^{m/2}}{\Gamma(1 + \frac{m}{2})} = \frac{\lambda_m(\mathbb{B}^m)}{2^m}. \tag{5.6.1}$$

For $\nu \in \mathbb{S}^{m-1}$, we shall denote by $\nu^\perp$ the hyperplane orthogonal to ν and for $y \in \mathbb{R}^m$, we shall denote by $y + \mathbb{R}\nu$ the affine line with direction ν through y.

---

[3]See Exercise 4.5.12.

Definition 5.6.2 (Steiner symmetrization). Let A be a Borel subset of $\mathbb{R}^m$. The *Steiner symmetrization* of A with respect to $\nu \in \mathbb{S}^{n-1}$ is defined as

$$\sigma_\nu(A) = \bigcup_{\substack{y \in \nu^\perp \\ |t| \le \frac{1}{2}\lambda_1(A\cap(y+\mathbb{R}\nu))}} \{y + t\nu\}.$$

Lemma 5.6.3. *Let A, ν be as above. Then the set $\sigma_\nu(A)$ is a Borel set, symmetric with respect to $\nu^\perp$ and $\lambda_m(\sigma_\nu(A)) = \lambda_m(A)$. Moreover we have*

$$\operatorname{diam}_2(\sigma_\nu(A)) \le \operatorname{diam}_2(A).$$

Proof. Note that from Lemma 1.2.9 the Borel σ-algebra on the line $y + \mathbb{R}\nu$ (a closed set of $\mathbb{R}^m$) is made with the Borel subsets of $\mathbb{R}^m$ included in that line. As a result, $A \cap (y + \mathbb{R}\nu)$ is a Borel set of the line $y + \mathbb{R}\nu$ and one can take its Lebesgue measure. The symmetry is obvious since $y + t\nu \in \sigma_\nu(A), y \in \nu^\perp$, imply $y - t\nu \in \sigma_\nu(A)$. We have also from Fubini's theorem,

$$\lambda_m(A) = \iint_{\nu^\perp \times \mathbb{R}\nu} \mathbf{1}_A(y \oplus z)\,dy\,dz$$

$$= \int_{y \in \nu^\perp} \lambda_1(A \cap (y + \mathbb{R}\nu))\,dy$$

$$= \int_{y \in \nu^\perp} \int_{|t| \le \frac{1}{2}\lambda_1(A\cap(y+\mathbb{R}\nu))} dt\,dy$$

$$= \int_{\mathbb{R}^m} \mathbf{1}\left\{x = y \oplus t\nu \in \nu^\perp \oplus \mathbb{R}\nu, |t| \le \frac{1}{2}\lambda_1(A \cap (y + \mathbb{R}\nu))\right\} dx$$

$$= \lambda_m(\sigma_\nu(A)).$$

The mapping $\mathbb{R}^m \ni x = y \oplus t\nu \mapsto (\lambda_1(A \cap (y + \mathbb{R}\nu)), t) \in \mathbb{R}^2$ is measurable since

$$\lambda_1(A \cap (y + \mathbb{R}\nu)) = \int_{\mathbb{R}\nu} \mathbf{1}_A(y \oplus z)\,dz$$

so that Proposition 4.1.3 and Theorem 1.2.7 imply that $\sigma_\nu(A)$ is a Borel set. We consider now for $j = 1, 2$, $x_j = y_j \oplus t_j\nu \in \sigma_\nu(A)$. We know that for $j = 1, 2$,

$$I_j = \{\theta \in \mathbb{R}, y_j + \theta\nu \in A\} \ne \emptyset, \quad |t_j| \le \lambda_1(I_j)/2.$$

Claim. For I_1, I_2, non-empty Borel subsets of $\mathbb{R}$,

$$\lambda_1(I_1) + \lambda_1(I_2) \le 2 \sup_{\theta_j \in I_j} |\theta_1 - \theta_2|.$$

Let us take provisionally this claim for granted. Then we get, when the diameter of A is finite,

$$\|x_1 - x_2\|^2 = \|y_1 - y_2\|^2 + (t_1 - t_2)^2 \le \|y_1 - y_2\|^2 + \frac{1}{4}\left(\lambda_1(I_1) + \lambda_1(I_2)\right)^2$$

$$\le \|y_1 - y_2\|^2 + \sup_{\theta_j \in I_j} |\theta_1 - \theta_2|^2 \le \operatorname{diam}_2(A)^2,$$

entailing $\mathrm{diam}_2(\sigma_\nu(A)) \leq \mathrm{diam}_2(A)$. We are left with the proof of the above claim. We may assume that I_j are both bounded, otherwise the rhs of the inequality to be proven is $+\infty$. We set then $a_j = \inf I_j, b_j = \sup I_j$. We may assume by symmetry that $b_2 \geq b_1$. Let us suppose first that $a_2 \geq a_1$; it is enough to prove

$$b_1 - a_1 + b_2 - a_2 \leq 2(b_2 - a_1),$$

which is equivalent to $b_2 \geq b_1 + a_1 - a_2$, which is satisfied since $b_2 \geq b_1$ and $a_1 - a_2 \leq 0$. Still with $b_2 \geq b_1$, we assume now $a_2 \leq a_1$ and we have to prove

$$2 \max(b_2 - a_1, b_1 - a_2) \geq b_1 - a_1 + b_2 - a_2.$$

When $b_2 - a_1 \geq b_1 - a_2$ it amounts to proving

$$2(b_2 - a_1) \geq b_1 - a_1 + b_2 - a_2 \iff b_2 - a_1 \geq b_1 - a_2 \text{ (hypothesis)}.$$

When $b_2 - a_1 \leq b_1 - a_2$, we have to prove

$$2(b_1 - a_2) \geq b_1 - a_1 + b_2 - a_2 \iff b_1 - a_2 \geq b_2 - a_1 \text{ (hypothesis)},$$

completing the proof of the claim. The proof of Lemma 5.6.3 is complete. $\square$

Lemma 5.6.4. *Let* $\nu, \omega \in \mathbb{S}^{m-1}$ *such that* $\omega \cdot \nu = 0$ *and let* A *be a Borel set symmetrical with respect to* $\omega^\perp$. *Then* $\sigma_\nu(A)$ *is also symmetrical with respect to* $\omega^\perp$.

Proof. We have

$$\sigma_\nu(A) = \bigcup_{\substack{y \in \nu^\perp \\ |t| \leq \frac{1}{2}\lambda_1(A \cap (y + \mathbb{R}\nu))}} \{y + t\nu\} = \bigcup_{\substack{z \in \nu^\perp \cap \omega^\perp, s \in \mathbb{R} \\ |t| \leq \frac{1}{2}\lambda_1(A \cap (z + s\omega + \mathbb{R}\nu))}} \{z + s\omega + t\nu\},$$

so that, denoting $\mathrm{sym}_{\omega^\perp}(B)$ the symmetric of B with respect to $\omega^\perp$, we find

$$\mathrm{sym}_{\omega^\perp}(\sigma_\nu(A)) = \bigcup_{\substack{z \in \nu^\perp \cap \omega^\perp, s \in \mathbb{R} \\ |t| \leq \frac{1}{2}\lambda_1(A \cap (z + s\omega + \mathbb{R}\nu))}} \{z - s\omega + t\nu\} = \bigcup_{\substack{z \in \nu^\perp \cap \omega^\perp, s \in \mathbb{R} \\ |t| \leq \frac{1}{2}\lambda_1(A \cap (z - s\omega + \mathbb{R}\nu))}} \{z + s\omega + t\nu\}.$$

Since we have

$$A \cap (z - s\omega + \mathbb{R}\nu) = \mathrm{sym}_{\omega^\perp}(A) \cap \mathrm{sym}_{\omega^\perp}(z + s\omega + \mathbb{R}\nu)$$
$$= \mathrm{sym}_{\omega^\perp}(A \cap (z + s\omega + \mathbb{R}\nu)),$$

we find

$$\mathrm{sym}_{\omega^\perp}(\sigma_\nu(A)) = \bigcup_{\substack{z \in \nu^\perp \cap \omega^\perp, s \in \mathbb{R} \\ |t| \leq \frac{1}{2}\lambda_1(A \cap (z + s\omega + \mathbb{R}\nu))}} \{z + s\omega + t\nu\} = \sigma_\nu(A),$$

proving the lemma. $\square$

Lemma 5.6.5 (Isodiametric inequality). *Let A be a Borel subset of $\mathbb{R}^m$. With v_m given in (5.6.1), we have*

$$\lambda_m(A) \le v_m(\mathrm{diam}_2(A))^m,$$

where diam_2 *stands for the Euclidean diameter of A:* $\mathrm{diam}_2(A) = \sup_{x,y \in A} \|x - y\|_2$ *where* $\|x\|_2$ *is the Euclidean norm.*

N.B. This lemma says that the Lebesgue measure of A is smaller than the Lebesgue measure of the ball with diameter $\mathrm{diam}_2(A)$. This statement is far from obvious for the Euclidean norm since it is possible to find Borel sets A which are *not* included in a ball with diameter $\mathrm{diam}_2 A$. Let us consider for instance in $\mathbb{R}^2$ the

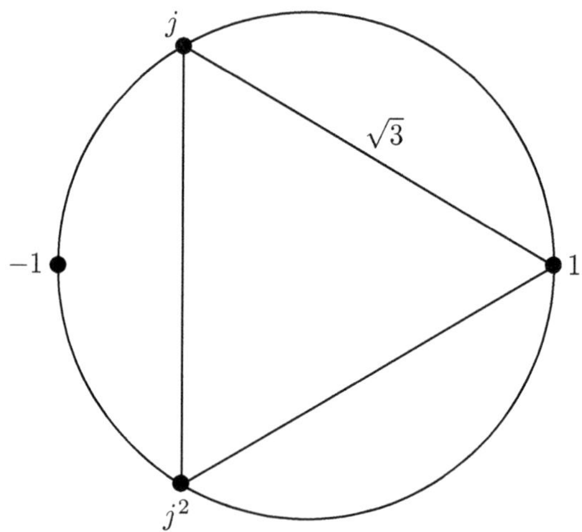

Figure 5.4: TRIANGLE WITH DIAMETER $\sqrt{3}$, CIRCUMSCRIBED CIRCLE WITH DIAMETER 2.

(equilateral) triangle T with vertices $1, j = e^{2i\pi/3}, j^2 = e^{-2i\pi/3}$. We have

$$\mathrm{diam}_2(T) = |1 - e^{2i\pi/3}| = \left|\frac{3}{2} - i\frac{\sqrt{3}}{2}\right| = \sqrt{\frac{9}{4} + \frac{3}{4}} = \sqrt{3}.$$

However the circumscribed circle of that triangle is the unit circle, thus has diameter $2 > \sqrt{3}$: it is not possible to find a circle with diameter $\mathrm{diam}_2(T)$ containing T. On the other hand, we have indeed

$$\lambda_2(T) = \frac{3\sqrt{3}}{4} \le v_2 \, \mathrm{diam}_2(T)^2 = \frac{\pi}{4}3.$$

Note also that for the d_∞ distance, it is obvious that a bounded set A is included in a cube with sides parallel to the axes equal to $\mathrm{diam}_\infty A$. Since A is bounded, $\bar{A}$ is compact with the same diameter as A, we can apply Lemma 2.6.9.

Proof of the lemma. Let $e_1, \ldots, e_m$ be the canonical basis of $\mathbb{R}^m$ and

$$A_m = (\sigma_{e_m} \circ \cdots \circ \sigma_{e_1})(A).$$

We have from Lemma 5.6.3 that $\mathrm{diam}_2(A_m) \leq \mathrm{diam}_2(A)$ and $\lambda_m(A_m) = \lambda_m(A)$. Moreover the set A_m is symmetrical with respect to all hyperplanes $e_1^\perp, \ldots, e_m^\perp$, since the symmetry of B with respect to a hyperplane $\omega^\perp$ induces the same symmetry for $\sigma_\nu(B)$ whenever $\nu \cdot \omega = 0$ (Lemma 5.6.4). As a result the set A_m is symmetric with respect to the origin: this implies that

$$A_m \subset \bar{B}\left(0, \frac{1}{2}\mathrm{diam}_2(A_m)\right) \quad \text{(Euclidean ball)}.$$

In fact, if $\|x\|_2 > \frac{1}{2}\mathrm{diam}_2(A_m)$ then x cannot belong to A_m otherwise the symmetry of A_m will imply that $-x$ belongs as well to A_m with

$$\mathrm{diam}_2(A_m) \geq d_2(x, -x) = 2\|x\|_2 > \mathrm{diam}_2(A_m),$$

which is impossible. Finally we have

$$\lambda_m(A) = \lambda_m(A_m) \leq \lambda_m\left(B\left(0, \frac{1}{2}\mathrm{diam}_2(A_m)\right)\right)$$
$$= v_m(\mathrm{diam}_2(A_m))^m \leq v_m(\mathrm{diam}_2(A))^m,$$

concluding the proof of Lemma 5.6.5.

Remark 5.6.6. The statement of Lemma 5.6.5 is true as well for A in the Lebesgue σ-algebra. In fact, thanks to Theorem 2.2.14, we can then find E, F Borel sets such that

$$E \subset A \subset F, \quad \lambda_m(F \cap E^c) = 0,$$

so that from the lemma,

$$\lambda_m(A) = \lambda_m(E) \leq v_m(\mathrm{diam}_2 E)^m \leq v_m(\mathrm{diam}_2 A)^m. \qquad \square$$

Theorem 5.6.7. *Let m be a positive integer. The Hausdorff measure $\mathfrak{h}_m$ on the metric space $(\mathbb{R}^m, d_\infty)$ (see Theorem 2.6.10) is equal to the product of the Hausdorff measure on the metric space $(\mathbb{R}^m, d_2)$ (d_2 is the Euclidean distance) by the constant v_m defined in (5.6.1). For $\varepsilon > 0$, we define for $E \subset X$,*

$$\mathfrak{h}^*_{m,\varepsilon,d_2}(E) = \inf\left\{\sum_{n \in \mathbb{N}}(\mathrm{diam}_2 U_n)^\kappa, \quad E \subset \cup_{n \in \mathbb{N}} U_n, \ U_n \ open, \ \mathrm{diam}_2 U_n \leq \varepsilon\right\},$$

where diam_2 *stands for the Euclidean diameter. We have*

$$\mathfrak{h}_m = \mathfrak{h}_{m,d_\infty} = v_m \mathfrak{h}_{m,d_2}.$$

Proof. We recall first the obvious inequalities $d_\infty \leq d_2 \leq m^{1/2}d_\infty$, and we note that this implies for E subset of $\mathbb{R}^m$,

$$\{\text{open covering } (U_n)_{n\in\mathbb{N}} \text{ of } E, \text{diam}_2(U_n) \leq \varepsilon\}$$
$$\subset \{\text{open covering } (U_n)_{n\in\mathbb{N}} \text{ of } E, \text{diam}_\infty(U_n) \leq \varepsilon\}$$
$$\subset \{\text{open covering } (U_n)_{n\in\mathbb{N}} \text{ of } E, \text{diam}_2(U_n) \leq m^{1/2}\varepsilon\},$$

so that, since $d_2/d_\infty \geq 1$ and $(d_\infty/d_2)^m \geq m^{-m/2}$, we get

$$\mathfrak{h}^*_{m,\varepsilon,d_2}(E) \geq \mathfrak{h}^*_{m,\varepsilon,d_\infty}(E) \geq m^{-m/2}\mathfrak{h}^*_{m,m^{1/2}\varepsilon,d_2}(E),$$

entailing

$$(\sharp) \qquad \mathfrak{h}^*_{m,d_2}(E) \geq \mathfrak{h}^*_{m,d_\infty}(E) \geq m^{-m/2}\mathfrak{h}^*_{m,d_2}(E).$$

Note also that the measure $v_m\mathfrak{h}_{m,d_2}$ is defined on the Borel σ-algebra $\mathcal{B}_m$, is translation invariant and is finite on compact sets (from the previous inequalities). To obtain $v_m\mathfrak{h}_{m,d_2} = \lambda_m = \mathfrak{h}_{m,d_\infty}$, we need only to prove that

$$(\flat) \qquad v_m\mathfrak{h}_{m,d_2}([0,1]^m) = 1.$$

Let $\varepsilon > 0$ be given. Thanks to Theorem 5.6.1, it is possible to find a sequence $(B_n)_{n\in\mathbb{N}}$ of pairwise disjoint open Euclidean balls with (Euclidean) diameter $\leq \varepsilon$, included in $(0,1)^m$ such that

$$[0,1]^m = \cup_{\mathbb{N}} B_n \cup Z, \quad \lambda_m(Z) = 0,$$
$$1 = \lambda_m([0,1]^m) = \sum_n \lambda_m(B_n) = \sum_n v_m \,\text{diam}_2(B_n)^m,$$

implying that (see (2.6.2))

$$v_m\mathfrak{h}^*_{m,\varepsilon,d_2}([0,1]^m) \leq v_m\mathfrak{h}^*_{m,\varepsilon,d_2}(\cup_{\mathbb{N}} B_n) + v_m\mathfrak{h}^*_{m,\varepsilon,d_2}(Z)$$
$$\leq 1 + \underbrace{v_m\mathfrak{h}^*_{m,d_2}(Z)}_{\text{inequality }(\sharp)} = 1,$$

and thus $v_m\mathfrak{h}^*_{m,d_2}([0,1]^m) \leq 1$. On the other hand, if the previous inequality were strict, for all $\epsilon > 0$, all $\delta > 0$, we could find an open covering of $[0,1]^m$ by a sequence of sets (U_n) with diameter $\leq \epsilon$ such that

$$1 = \lambda_m([0,1]^m) \leq \sum_n \lambda_m(U_n) \underbrace{\leq}_{\text{Lemma 5.6.5}} \sum_n v_m(\text{diam}_2(U_n))^m$$
$$\leq v_m\mathfrak{h}^*_{m,d_2}([0,1]^m) + \delta < 1,$$

if $\delta = (1 - v_m\mathfrak{h}^*_{m,d_2}([0,1]^m))/2$. This inequality entails $1 < 1$ and thus cannot hold. $\qquad\square$

5.7 Cantor sets

Perfect sets, Nowhere dense sets

Definition 5.7.1. Let X be a topological space.

(1) A subset A of X is said to be *perfect* if it is closed without isolated point, i.e.,

$$\bar{A} = A \quad \text{and} \quad \forall a \in A, \forall V \in \mathcal{V}_a, \quad (V \backslash \{a\}) \cap A \neq \emptyset.$$

(2) A subset A of X is said to be *nowhere dense* (or *rare*) when $\overset{\circ}{\bar{A}} = \emptyset$.

It is easy to find perfect sets (e.g., closed balls with positive radius in $\mathbb{R}^n$) or closed sets which are not perfect such as $\mathbb{Z}$ (all points are isolated) or $(-\infty, 0] \cup \{1/2\} \cup [1, +\infty)$ ($1/2$ is the only isolated point).

Theorem 5.7.2 (Cantor–Bendixson theorem). *Let (X, d) be a separable complete metric space and let F be a closed subset of X. Then F is the disjoint union $P \cup C$, where C is countable and P is perfect.*

Proof. Let $D = \{q_k\}_{k \in \mathbb{N}}$ be a countable dense subset of X. Every open set of X is a (necessarily countable) union of open balls $B(q_k, r)$ where $r \in \mathbb{Q}_+$: if Ω is an open set of X, then for $x \in \Omega$, the open ball $B(x, r) \subset \Omega$ for some positive rational r. Then there exists $q_k \in D$ such that $d(q_k, x) < r/2$, which implies that

$$x \in B(q_k, r/2) \subset B(x, r) \subset \Omega.$$

As a result the set $\{B(q, r)\}_{q \in D, r \in \mathbb{Q}_+}$ is a countable basis for the topology of X.

Let F be a closed set of X. A point $x \in F$ is said to be a *condensation point* of F if $\forall V \in \mathcal{V}_x$, $V \cap F$ is uncountable. Let P be the set of condensation points of F and $C = F \backslash P$. Considering $B(q, r) \cap F$, $q \in D, r \in \mathbb{Q}_+$, we find a countable basis $\{U_n\}_{n \in \mathbb{N}}$ for the topology of F. By definition of P, we have

$$C = \bigcup_{\substack{n \in \mathbb{N} \\ U_n \text{ countable}}} U_n :$$

in fact, if $x \in C$, there exists $n \in \mathbb{N}$ such that $x \in U_n$ countable. Conversely, if U_n is a countable open subset of F, then every point in U_n belongs to C, so that C is countable, as a countable union of countable sets. Let $x \in P$ and let V be a neighborhood of x in F. Then V is uncountable and since C is countable, V contains uncountably many points of P. Moreover P is closed in F, as the complement of C, open in F as a union of open sets. As a result, P is closed in X and

$$F = P \cup C, \quad P \text{ perfect}, \ C \text{ countable}, \ P \cap C = \emptyset. \qquad \square$$

Cantor ternary set

We want to construct a subset of the real line which is perfect and nowhere dense, i.e., closed without isolated point and with empty interior. *Cantor's ternary set* is an excellent example. We shall use the following notation: let $J = [a, b]$ be a compact interval of the real line. We shall denote by

$$J_0 = \left[a, a + \frac{b-a}{3}\right], \qquad \text{the first third of } J, \tag{5.7.1}$$

$$J_2 = \left[a + \frac{2(b-a)}{3}, b\right], \qquad \text{the last third of } J. \tag{5.7.2}$$

We start with $I = [0, 1]$ and we have

$$I_0 = \left[0, \frac{1}{3}\right] \qquad I_2 = \left[\frac{2}{3}, 1\right],$$

$$I_{00} = \left[0, \frac{1}{9}\right] \qquad I_{02} = \left[\frac{2}{9}, \frac{3}{9}\right] \qquad I_{20} = \left[\frac{6}{9}, \frac{7}{9}\right] \qquad I_{22} = \left[\frac{8}{9}, \frac{9}{9}\right]$$

$$\cdots\cdots\cdots\cdots\cdots\cdots\cdots\cdots\cdots\cdots\cdots\cdots\cdots\cdots$$

$$\text{for } \alpha = (\alpha_1, \ldots, \alpha_n) \in \{0, 2\}^n, \ x_\alpha = \sum_{1 \le j \le n} \frac{\alpha_j}{3^j}, \quad I_\alpha = [x_\alpha, x_\alpha + 3^{-n}]. \tag{5.7.3}$$

We verify inductively that for $\alpha = (\alpha_1, \ldots, \alpha_n) \in \{0, 2\}^n$,

$$I_{\alpha 0} = [x_\alpha, x_\alpha + 3^{-n-1}] = [x_{\alpha 0}, x_{\alpha 0} + 3^{-n-1}],$$

$$I_{\alpha 2} = [x_\alpha + 2 \times 3^{-n-1}, x_\alpha + 3^{-n}] = [x_{\alpha 2}, x_{\alpha 2} + 3^{-n-1}].$$

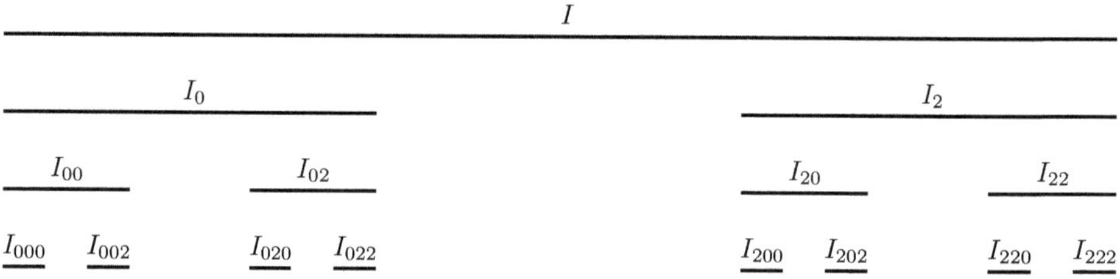

I

I_0 I_2

I_{00} I_{02} I_{20} I_{22}

I_{000} I_{002} I_{020} I_{022} I_{200} I_{202} I_{220} I_{222}

Figure 5.5: INTERVALS $I_\alpha, \alpha \in \{0, 2\}^{1,2,3}$.

For an integer $n \ge 1$, we define the compact set K_n by

$$K_n = \bigcup_{\alpha \in \{0,2\}^n} I_\alpha, \tag{5.7.4}$$

and we note that $(I_\alpha)_{\alpha \in \{0,2\}^n}$ are 2^n pairwise disjoint compact intervals with length 3^{-n} (true for $n = 1$ and if true for some $n \ge 1$, also true for $n+1$: we have for $\alpha \in \{0,2\}^n$, $I_{\alpha 0}, I_{\alpha 2}$ pairwise disjoint with length 3^{-n-1}). As a result, we have

$$\lambda_1(K_n) = 2^n \times 3^{-n}. \tag{5.7.5}$$

We note also that $K_n \supset K_{n+1}$ by construction since $I_\alpha \supset I_{\alpha 0} \cup I_{\alpha 2}$. We define then at last the Cantor ternary set K_∞ by

$$K_\infty = \cap_{n \geq 1} K_n = \bigcap_{n \geq 1} \left(\cup_{\alpha \in \{0,2\}^n} I_\alpha \right). \tag{5.7.6}$$

Lemma 5.7.3. *The Cantor ternary set K_∞ is a compact subset of $[0,1]$ with Lebesgue measure 0 which is equipotent to $\mathbb{R}$. Moreover K_∞ has no isolated points and has an empty interior. The set K_∞ is totally discontinuous, i.e., the connected component of each of its points is reduced to a singleton.*

Proof. K_∞ is a compact set as an intersection of compact sets and its Lebesgue measure must be smaller than $(2/3)^n$ for each n so is zero. As a result K_∞ cannot contain an interval with positive measure, thus has an empty interior and is totally discontinuous. Let us check the mapping

$$
\begin{array}{rcl}
\Phi : \{0,2\}^{\mathbb{N}^*} & \longrightarrow & K_\infty, \\[4pt]
\alpha & \mapsto & \sum_{1 \leq j} \frac{\alpha_j}{3^j}.
\end{array} \tag{5.7.7}
$$

Let us prove first that Φ is indeed valued in K_∞. From (5.7.4) and (5.7.3), with $(\alpha_1, \ldots, \alpha_n) \in \{0,2\}^n$, we have $x_\alpha = \sum_{1 \leq j \leq n} \frac{\alpha_j}{3^j} \in K_n$. As a result, for $\alpha \in \{0,2\}^{\mathbb{N}^*}$,

$$\sum_{1 \leq j} \frac{\alpha_j}{3^j} = \lim_n \underbrace{\sum_{1 \leq j \leq n} \frac{\alpha_j}{3^j}}_{\in K_n \subset K_m, \text{ for } n \geq m.} \in \cap_{m \geq 1} K_m = K_\infty.$$

The mapping Φ is one-to-one since for $\alpha', \alpha'' \in \{0,2\}^{\mathbb{N}^*}$ and

$$\alpha'_j = \alpha''_j \quad \text{for } 1 \leq j < N, \quad \alpha'_N < \alpha''_N,$$

we have necessarily $\alpha'_N = 0, \alpha''_N = 2$ and

$$\Phi(\alpha') = \sum_{j \geq 1} \frac{\alpha'_j}{3^j} = \sum_{1 \leq j < N} \frac{\alpha''_j}{3^j} + \sum_{j \geq N+1} \frac{\alpha'_j}{3^j} \leq \sum_{1 \leq j < N} \frac{\alpha''_j}{3^j} + 3^{-N-1} 2 \frac{1}{1 - \frac{1}{3}}$$

$$= \sum_{1 \leq j < N} \frac{\alpha''_j}{3^j} + 3^{-N} < \sum_{1 \leq j \leq N} \frac{\alpha''_j}{3^j} = \Phi(\alpha'').$$

Let us prove now that Φ is onto; let $x \in K_\infty$. Then for all $n \geq 1$, there exists $\alpha^{(n)} = (\alpha_1^{(n)}, \ldots, \alpha_n^{(n)}) \in \{0,2\}^n$ such that $x \in I_\alpha$, i.e.,

$$x_{\alpha^{(n)}} \leq x \leq x_{\alpha^{(n)}} + 3^{-n} \implies x = \lim_n x_{\alpha^{(n)}} = \lim_n \left(\sum_{1 \leq j \leq n} \frac{\alpha_j^{(n)}}{3^j} \right). \tag{5.7.8}$$

Claim. We may assume that both inequalities above are strict, otherwise the answer is clear: On the one hand, if $x = x_{\alpha(n)}$ for some $n \geq 1$, then

$$x = \Phi(\alpha_1^{(n)}, \ldots, \alpha_n^{(n)}, 0, 0, \ldots).$$

On the other hand if $x = x_{\alpha(n)} + 3^{-n}$ for some $n \geq 1$, then

$$x = \sum_{1 \leq j \leq n} \frac{\alpha_j^{(n)}}{3^j} + \underbrace{\sum_{j \geq n+1} \frac{2}{3^j}}_{3^{-n-1}2\frac{1}{1-1/3}=3^{-n}} = \Phi(\alpha_1^{(n)}, \ldots, \alpha_n^{(n)}, 2, 2, \ldots),$$

proving the claim[4].

We know also that $x \in [0,1]$ so that $x = \sum_{j \geq 1} \frac{x_j}{3^j}, \quad x_j \in \{0,1,2\}$ and

$$0 \leq x - \sum_{1 \leq j \leq n} \frac{x_j}{3^j} \leq \sum_{j > n} \frac{x_j}{3^j} \leq 3^{-n-1}2\frac{1}{1 - \frac{1}{3}} = 3^{-n},$$

so that eventually with the strict inequalities of (5.7.8),

$$3^n \underbrace{\sum_{1 \leq j \leq n} \frac{x_j}{3^j}}_{\in \mathbb{N}} \leq 3^n x \leq 3^n \sum_{1 \leq j \leq n} \frac{x_j}{3^j} + 1, \quad 3^n \underbrace{x_{\alpha(n)}}_{\in \mathbb{N}} < 3^n x < 3^n x_{\alpha(n)} + 1,$$

implying $\sum_{1 \leq j \leq n} x_j 3^{n-j} = 3^n x_{\alpha(n)} = \sum_{1 \leq j \leq n} \alpha_j^{(n)} 3^{n-j}$. The latter identity implies

$$\underbrace{x_1}_{\in \mathbb{N}} + \underbrace{\sum_{2 \leq j \leq n} x_j 3^{1-j}}_{\in [0,6 \times 3^{-2} \times \frac{3}{2}) = [0,1)} = \underbrace{\alpha_1^{(n)}}_{\in \mathbb{N}} + \underbrace{\sum_{2 \leq j \leq n} \alpha_j^{(n)} 3^{1-j}}_{\in [0,1)},$$

so that, taking the floor function of each side (see the footnote on page 16), we get $x_1 = \alpha_1^{(n)}$ and similarly $x_j = \alpha_j^{(n)}$ for $1 \leq j \leq n$, so that each x_j belongs to $\{0,2\}$, proving that x belongs to the image of Φ. We have obtained in particular the following description of the Cantor ternary set[5].

Lemma 5.7.4. $K_\infty = \{x \in [0,1], \exists (x_j)_{j \geq 1}, x_j \in \{0,2\}, x = \sum_{j \geq 1} \frac{x_j}{3^j}\}.$

The bijectivity of Φ and Section 10.1 prove that $\operatorname{card} K_\infty = \operatorname{card}\{0,2\}^{\mathbb{N}^*} = \operatorname{card}\{0,1\}^{\mathbb{N}} = \operatorname{card} \mathcal{P}(\mathbb{N}) = \operatorname{card} \mathbb{R}$. Let us finally prove that K_∞ has no isolated point. Let x be in K_∞: then for each $n \geq 1$, there exists $\alpha^{(n)} \in \{0,2\}^n$ such that

$$K_\infty \ni x_{\alpha(n)} \leq x \leq x_{\alpha(n)} + 3^{-n} \in K_\infty$$

and thus $\big([x-3^{-n}, x+3^{-n}] \setminus \{x\}\big) \cap K_\infty \neq \emptyset$, completing the proof of the lemma. $\square$

---

[4]Note that for instance 1/3, written as 0.1 in its development in base 3 can also be written as 0.02222222222222 . . .

[5]Similarly, the development of 1 in base 3 can be written as $1 = 0.222222222222\ldots$

Lemma 5.7.5. *The Hausdorff dimension (see Definition 2.6.8) of the Cantor ternary set K_∞ is*

$$\log_3 2 = \frac{1}{\log_2 3} = \frac{\ln 2}{\ln 3} \approx 0.6309.$$

Proof. We have $K_\infty \subset \cup_{\alpha \in \{0,2\}^n} I_\alpha$ with $\operatorname{diam} I_\alpha = 3^{-n}$ so that

$$\mathfrak{h}^*_{\kappa,3^{-n}}(K_\infty) \leq \sum_{\alpha \in \{0,2\}^n} (\operatorname{diam} I_\alpha)^\kappa = 2^n 3^{-n\kappa} = e^{n(\ln 2 - \kappa \ln 3)},$$

implying for $\kappa_0 = \ln 2 / \ln 3$, that

$$(\sharp) \qquad\qquad \mathfrak{h}_{\kappa_0}(K_\infty) \leq 1, \qquad \text{and for } \kappa > \kappa_0,\ \mathfrak{h}_\kappa(K_\infty) = 0.$$

The main point in the proof is to estimate $\mathfrak{h}_{\kappa_0}(K_\infty)$ from below by a positive quantity. Let $\epsilon > 0$ be given and let $(V_n)_{n \in \mathbb{N}}$ be a covering of K_∞ by open sets with diameter $\leq \epsilon$. By compactness of K_∞, we may extract a finite covering and since each V_n is a union of open intervals, we may find a finite collection $(J_l)_{1 \leq l \leq L}$ of open intervals with diameter smaller than ϵ (assumed $< 1/3$) such that

$$K_\infty \subset \cup_{1 \leq l \leq L} J_l, \quad K_\infty \cap J_l \neq \emptyset, \quad \sum_n (\operatorname{diam} V_n)^{\kappa_0} \geq \sum_{1 \leq l \leq L} (\operatorname{diam} J_l)^{\kappa_0}.$$

For each l, there exists a unique $n_l \geq 1$ such that $3^{-n_l-1} \leq \operatorname{diam} J_l < 3^{-n_l}$ and moreover J_l meets exactly one $(I_\alpha)_{\alpha \in \{0,2\}^{n_l}}$: it must meet one such interval otherwise the intersection with K_∞ would be empty and could not meet two since the distance between two such intervals is at least 3^{-n_l} by construction. We have moreover

$$\operatorname{diam} J_l \geq 3^{-n_l-1} \implies 3^{\kappa_0}(\operatorname{diam} J_l)^{\kappa_0} \geq 2^{-n_l} \implies 2^j 3^{\kappa_0}(\operatorname{diam} J_l)^{\kappa_0} \geq 2^{j-n_l}.$$

Since J_l meets only one $I_{\alpha^{(l)}}$, $\alpha^{(l)} \in \{0,2\}^{n_l}$, it meets at most 2^{j-n_l} intervals I_β for $\beta \in \{0,2\}^j, j \geq n_l$. As a consequence, we have for $j \geq \max_{1 \leq l \leq L} n_l$,

$$\begin{aligned}
2^j &= \operatorname{card}\{\text{connected component of } K_j\} \\
&\leq \sum_{1 \leq l \leq N} \operatorname{card}\{\text{connected component of } K_j \text{ meeting } J_l\} \\
&\leq \sum_{1 \leq l \leq N} 2^{j-n_l} \leq \sum_{1 \leq l \leq N} 2^j 3^{\kappa_0}(\operatorname{diam} J_l)^{\kappa_0},
\end{aligned}$$

so that $\sum_n (\operatorname{diam} V_n)^{\kappa_0} \geq \sum_{1 \leq l \leq L} (\operatorname{diam} J_l)^{\kappa_0} \geq 3^{-\kappa_0} = 1/2$ and thus

$$(\flat) \qquad\qquad \mathfrak{h}_{\kappa_0}(K_\infty) \geq 1/2,$$

implying the result from $(\flat)$, $(\sharp)$, Lemma 2.6.7 and Definition 2.6.8. $\qquad\square$

The Cantor function

With K_n defined in (5.7.4) for $n \geq 1$, we define

$$\psi_n(x) = \frac{1}{|K_n|} \int_0^x \mathbf{1}_{K_n}(t)dt, \quad |K_n| = \lambda_1(K_n) = (2/3)^n. \tag{5.7.9}$$

The function ψ_n is continuous on $\mathbb{R}$, with value 0 (resp. 1) for $x \leq 0$ (resp. $x \geq 1$) and is monotone increasing. We have with I_α defined in (5.7.3),

$$\psi_{n+1}(x) - \psi_n(x) = (3/2)^n \int_0^x \sum_{\alpha \in \{0,2\}^n} \left(\frac{3}{2}\mathbf{1}_{I_{\alpha 0}}(t) + \frac{3}{2}\mathbf{1}_{I_{\alpha 2}}(t) - \mathbf{1}_{I_\alpha}(t) \right) dt$$

$$= (3/2)^n \sum_{\alpha \in \{0,2\}^n} \underbrace{\int_0^x \left(\frac{1}{2}\mathbf{1}_{I_{\alpha 0}}(t) + \frac{1}{2}\mathbf{1}_{I_{\alpha 2}}(t) - \mathbf{1}_{I_{\alpha 1}}(t) \right) dt}_{\substack{=0 \text{ if } x \leq x_\alpha \\ \text{or } x \geq x_\alpha + 3^{-n}}}$$

$$= (3/2)^n \sum_{\alpha \in \{0,2\}^n} \mathbf{1}_{I_\alpha}(x) \int_0^x \left(\frac{1}{2}\mathbf{1}_{I_{\alpha 0}}(t) + \frac{1}{2}\mathbf{1}_{I_{\alpha 2}}(t) - \mathbf{1}_{I_{\alpha 1}}(t) \right) dt$$

$$= (3/2)^n \sum_{\alpha \in \{0,2\}^n} \mathbf{1}_{I_\alpha}(x)$$

$$\int_0^x \left(\frac{1}{2}\mathbf{1}_{[0,1]}\left(\frac{t - x_{\alpha 0}}{3^{-n-1}}\right) + \frac{1}{2}\mathbf{1}_{[0,1]}\left(\frac{t - x_{\alpha 2}}{3^{-n-1}}\right) - \frac{1}{2}\mathbf{1}_{[0,1]}\left(\frac{t - x_{\alpha 1}}{3^{-n-1}}\right) \right) dt$$

$$= (3/2)^n \frac{1}{2} \sum_{\alpha \in \{0,2\}^n} \mathbf{1}_{I_\alpha}(x) \left(\int_{-3^{n+1}x_{\alpha 0}}^{(x-x_{\alpha 0})3^{n+1}} \mathbf{1}_{[0,1]}(s)ds3^{-n-1} \right.$$

$$\left. + \int_{-3^{n+1}x_{\alpha 2}}^{(x-x_{\alpha 2})3^{n+1}} \mathbf{1}_{[0,1]}(s)ds3^{-n-1} - \int_{-3^{n+1}x_{\alpha 1}}^{(x-x_{\alpha 1})3^{n+1}} \mathbf{1}_{[0,1]}(s)ds3^{-n-1} \right).$$

As a result we have (note $\int_{\mathbb{R}} \mathbf{1}_{[0,1]}(s)ds = 1$),

$$|\psi_{n+1}(x) - \psi_n(x)| \leq \frac{3^n}{2^{n+1}} \sum_{\alpha \in \{0,2\}^n} \mathbf{1}_{I_\alpha}(x)3^{-n-1} \times 3 \leq 2^{-n-1}.$$

Consequently, the sequence (ψ_n) is converging uniformly on $\mathbb{R}$ towards a function Ψ, the so-called *Cantor function*, which is continuous monotone increasing, with value 0 (resp. 1) for $x \leq 0$ (resp. $x \geq 1$). Moreover, from the calculation above if $x \notin K_n = \cup_{\alpha \in \{0,2\}^n} I_\alpha$, we have $\psi_{n+1}(x) = \psi_n(x)$ and since $K_n \supset K_{n+1} \supset K_{n+l}$ for $l \geq 2$, we have $x \notin K_{n+l}$ for $l \geq 2$, so that $\psi_{n+2}(x) = \psi_{n+1}(x) = \psi_n(x)$ and $\psi_{n+l}(x) = \psi_n(x)$ for $l \geq 0$, proving

$$x \notin K_n \implies \Psi(x) = \psi_n(x).$$

We see also that ψ_n is piecewise affine with $\psi_n' = \mathbf{1}_{K_n}/|K_n|$, so that Ψ is constant on each connected component of the complement of K_n, and since $K_\infty^c = \cup_{n\geq 1} K_n^c$, this implies that Ψ is constant on each connected component of K_∞^c, i.e., is almost everywhere differentiable with a null derivative. Nevertheless the function Ψ is monotone increasing such that $\Psi(0) = 0, \Psi(1) = 1$.

Let us calculate the weak derivative of Ψ. We define for $\phi \in C_c^1(\mathbb{R})$,

$$\langle \Psi', \phi \rangle = -\int_{\mathbb{R}} \Psi(x)\phi'(x)dx = \lim_{h\to 0} \int_{\mathbb{R}} \Psi(x)\big(\phi(x) - \phi(x+h)\big)h^{-1}dx,$$

and thus

$$\langle \Psi', \phi \rangle = \lim_{h\to 0_+} \int_{\mathbb{R}} \big(\Psi(x) - \Psi(x-h)\big)\phi(x)h^{-1}dx.$$

Since Ψ is monotone increasing, it implies that the linear form

$$C_c^1(\mathbb{R}) \ni \phi \mapsto \langle \Psi', \phi \rangle$$

is non-negative, i.e., takes non-negative values for ϕ valued in $\mathbb{R}_+$. As a result, for $\phi \in C_c^1(\mathbb{R})$ and $\chi \in C_c^1(\mathbb{R}; [0,1])$ equal to 1 near the support of ϕ, we have

$$\langle \Psi', \phi \rangle = \langle \Psi', \underbrace{\chi\phi - \chi\|\phi\|_{L^\infty(\mathbb{R})}}_{\leq 0} \rangle + \langle \Psi', \chi \rangle \|\phi\|_{L^\infty(\mathbb{R})},$$

so that $\langle \Psi', \phi \rangle \leq \langle \Psi', \chi \rangle \|\phi\|_{L^\infty(\mathbb{R})}$, and thus $\langle \Psi', -\phi \rangle \leq \langle \Psi', \chi \rangle \|\phi\|_{L^\infty(\mathbb{R})}$, entailing

$$|\langle \Psi', \phi \rangle| \leq \langle \Psi', \chi \rangle \|\phi\|_{L^\infty(\mathbb{R})},$$

and the linear form $C_c^1(\mathbb{R}) \ni \phi \mapsto \langle \Psi', \phi \rangle$ can be extended as a positive Radon measure, i.e., a positive linear form on $C_c(\mathbb{R})$: let $\phi \in C_c(\mathbb{R})$ and let ϕ_n be a sequence in $C_c^1(\mathbb{R})$ converging to ϕ in $C_c(\mathbb{R})$ (uniform convergence on $\mathbb{R}$ with support $\phi_n \subset L$ fixed compact). Then for $\chi = 1$ near L,

$$|\langle \Psi', \phi_{n+k} \rangle - \langle \Psi', \phi_n \rangle| = \langle \Psi', \phi_{n+k} - \phi_n \rangle| \leq \langle \Psi', \chi \rangle \|\phi_{n+k} - \phi_n\|_{L^\infty},$$

so that we may define

$$\langle \Psi', \phi \rangle = \lim_n \langle \Psi', \phi_n \rangle$$

and get a positive Radon measure (the definition is independent of the approximating sequence ϕ_n). As a result, the measure μ constructed using Theorem 2.2.1 is supported in K_∞: if $\phi \in C_c^1(\mathbb{R})$ with $\operatorname{supp}\phi \subset K_\infty^c$, we find

$$\langle \Psi', \phi \rangle = -\int \phi'(x)\Psi(x)dx = 0$$

since Ψ is constant on each connected component of K_∞^c. Moreover, as a consequence of the following lemma, μ has no atoms (see Exercise 1.9.24, page 61).

Lemma 5.7.6. *Let Φ be a continuous monotone increasing function on $\mathbb{R}$. Then the distribution derivative of Φ is a Radon measure μ (the so-called Cantor measure when $\Phi = \Psi$) without atoms.*

Proof. The fact that Φ' is a positive Radon measure μ is proven above. Now let $a \in \mathbb{R}$. According to Theorem 2.2.1, for $\epsilon > 0$, we have

$$\mu(\{a\}) \leq \mu((a - \epsilon, a + \epsilon)) = \sup_{\phi \in C_c^0((a-\epsilon,a+\epsilon);[0,1])} \langle \mu, \phi \rangle \leq \langle \mu, \phi_\epsilon \rangle,$$

where ϕ_ϵ is non-negative C^1, compactly supported, equal to 1 on $(a - \epsilon, a + \epsilon)$, supported on $(a - 2\epsilon, a + 2\epsilon)$. We have

$$0 \leq \langle \mu, \phi_\epsilon \rangle = - \int \Phi(x)\phi_\epsilon'(x)dx$$

$$= - \int_{a-2\epsilon}^{a-\epsilon} \Phi(x)\phi_\epsilon'(x)dx - \int_{a+\epsilon}^{a+2\epsilon} \Phi(x)\phi_\epsilon'(x)dx$$

$$= - \int_{a-2\epsilon}^{a-\epsilon} \big(\Phi(x) - \Phi(a)\big)\phi_\epsilon'(x)dx - \int_{a+\epsilon}^{a+2\epsilon} \big(\Phi(x) - \Phi(a)\big)\phi_\epsilon'(x)dx$$

$$\leq \sup_{|x-a|\leq 2\epsilon} |\Phi(x) - \Phi(a)| \int |\phi_\epsilon'(x)|dx.$$

We may choose

$$\phi_\epsilon(x) = \theta\left(\frac{x - a}{\epsilon}\right)$$

where θ is a fixed function valued in $[0, 1]$, equal to 1 on $[-1, 1]$ and supported in $[-2, 2]$ so that we get

$$\mu(\{a\}) \leq \sup_{|x-a|\leq 2\epsilon} |\Phi(x) - \Phi(a)| \int |\theta'(t)|dt \xrightarrow[\epsilon \to 0]{} 0,$$

by continuity of Φ. $\qquad\square$

We have proven the following

Proposition 5.7.7. *The Cantor function Ψ is a continuous monotone increasing function defined on $\mathbb{R}$ by the uniform limit of the sequence $(\psi_n)_{n\geq 1}$ given by (5.7.9). That function is equal to 0 (resp. 1) on $(-\infty, 0]$ (resp. $[1, \infty)$). Its weak derivative (or distribution derivative) is a positive Radon measure without atoms whose support is the Cantor ternary set K_∞ (which has Lebesgue measure 0). The function Ψ is differentiable on the open set K_∞^c where its derivative is 0.*

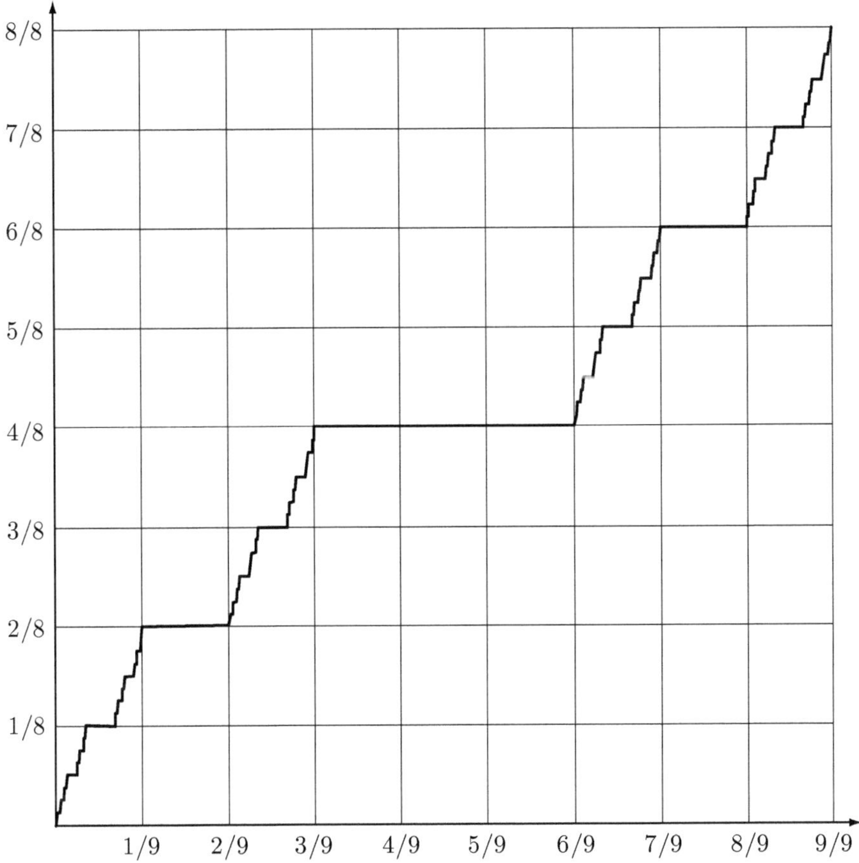

Figure 5.6: The Cantor function

Lebesgue and Borel measurability

Let us consider the function F defined by

$$[0,1] \ni x \mapsto F(x) = \Psi(x) + x \in [0,2], \tag{5.7.10}$$

where Ψ is the Cantor function defined above. F is strictly increasing continuous and thus one-to-one, with $F(0) = 0, F(1) = 2$, so that it is also onto (the continuous image of the interval $[0,1]$ is a compact interval contained in $[0,2]$ and containing $0,2$ so is $[0,2]$). Moreover F is an open mapping since the image $F(]a,b[)$ is an interval contained in $]F(a),F(b)[$ which contains $F(a+\epsilon),F(b-\epsilon)$ for all $\epsilon > 0$ small enough, thus by continuity of F, we have $F(]a,b[) =]F(a),F(b)[$.

As a result F^{-1} is continuous and F is a homeomorphism. We have also

$$F([0,1]\backslash K_\infty) = F\big((0,1)\cap K_\infty^c\big) = F\Big(\bigsqcup_{\substack{n\geq 1\\ \alpha\in\{0,2\}^n}} I_{\alpha 1}\Big) = \bigsqcup_{\substack{n\geq 1\\ \alpha\in\{0,2\}^n}} F(I_{\alpha 1}),$$

where J_1 stands for the open middle third of the interval J. As a consequence, we have

$$\lambda_1\big(F([0,1]\backslash K_\infty)\big) = \sum_{\substack{n\geq 1\\ \alpha\in\{0,2\}^n}} \lambda_1\big(F(I_{\alpha 1})\big) = \sum_{\substack{n\geq 1\\ \alpha\in\{0,2\}^n}} \lambda_1\big(I_{\alpha 1}\big) = 1,$$

since Ψ is constant on each interval $I_{\alpha 1}$ and we have

$$\lambda_1([0,2]) = \lambda_1(F([0,1])) = \lambda_1\big(F([0,1]\backslash K_\infty)\big) + \lambda_1\big(F(K_\infty)\big)$$
$$\Longrightarrow \lambda_1\big(F(K_\infty)\big) = 1. \tag{5.7.11}$$

The restriction of F to K_∞ is thus a homeomorphism from the Cantor ternary set K_∞ which has measure 0 onto $F(K_\infty)$ which has measure 1.

Lemma 5.7.8. *Let A be a Lebesgue measurable subset of $\mathbb{R}$ with positive measure. Then there exists a non-measurable set $E \subset A$.*

Proof. We may assume that $A\cap(-N_0, N_0)$ has positive measure for some $N_0 \in \mathbb{N}^*$ (otherwise $\lambda_1(A) = 0$) so that we may assume that A is bounded. As in Exercise 2.8.19, we define an equivalence relation on $\mathbb{R}$ by $x \equiv y$ meaning $x - y \in \mathbb{Q}$. We consider the quotient set of A by this equivalence relation and using the Axiom of Choice, we choose a representative in A for each class. Let E be the subset of A which is that set of representatives: for any $y \in A$, we find $x \in E, q \in \mathbb{Q}$ such that $y = x + q$. Consequently

$$A \subset \cup_{q\in\mathbb{Q},|q|\leq 2N_0}(E+q) = B \Longrightarrow 0 < \lambda_1(A) \leq \lambda_1(B) < +\infty.$$

For $q_1, q_2 \in \mathbb{Q}, q_1 \neq q_2$, we have $(E+q_1)\cap(E+q_2) = \emptyset$ since $y = x_1 + q_1 = x_2 + q_2$, for $q_j \in \mathbb{Q}, x_j \in E$ implies $x_1 = x_2$ and thus $q_1 = q_2$. Using the translation invariance of the Lebesgue measure, we get, assuming E measurable,

$$0 < \lambda_1(B) = \sum_{q\in\mathbb{Q},|q|\leq 2N_0} \lambda_1(E) \Longrightarrow \lambda_1(E) > 0 \Longrightarrow \lambda_1(B) = +\infty,$$

which is a contradiction. The set E cannot be Lebesgue measurable. $\qquad\square$

Lemma 5.7.9. *The function F defined by (5.7.10) is a homeomorphism from $[0,1]$ onto $[0,2]$ such that $\lambda_1(K_\infty) = 0$, $\lambda_1\big(F(K_\infty)\big) = 1$. The inverse homeomorphism F^{-1} is not Lebesgue measurable.*

Proof. The first part is proven in (5.7.11). Let D be a subset of $F(K_\infty)$ which does not belong to the Lebesgue σ-algebra (it is possible since the measure of $F(K_\infty)$ is positive). Then $F^{-1}(D)$ is a subset of K_∞ and thus belongs to the Lebesgue

σ-algebra since K_∞ has Lebesgue measure 0. Now

$$(F^{-1})^{-1}\big(\underbrace{F^{-1}(D)}_{\substack{\text{Lebesgue} \\ \text{measurable}}}\big) = D,$$

so that F^{-1} is continuous and is *not* Lebesgue measurable. $\qquad\square$

It is not that surprising: let $f : \mathbb{R} \to \mathbb{R}$ be Borel–Borel measurable: it means that f is measurable whenever we equip source and target[6] with the Borel σ-algebra. Of course when f is continuous, it is Borel–Borel measurable (Proposition 1.2.5). However, if we equip the target with the Lebesgue σ-algebra, there is no reason that f should be Borel–Lebesgue measurable since it may happen that the inverse image of a Lebesgue measurable set with measure 0 does not belong to the Borel σ-algebra: even if we equip both source and target with the Lebesgue σ-algebra, it does happen in the example above with $f = F^{-1}$ that the inverse image of a Lebesgue measurable set with measure 0 does not belong to the Lebesgue σ-algebra. However in Proposition 5.3.3, we have seen that if f is a C^1 diffeomorphism, it is Lebesgue–Lebesgue measurable (and of course Borel–Borel measurable).

Remark 5.7.10. Considering

$$[0,2] \xrightarrow[\text{homeomorphism}]{F^{-1}} [0,1] \xrightarrow[\text{Lebesgue–Lebesgue meas.}]{\mathbf{1}_{F^{-1}(D)}} \mathbb{R}$$

we see that the composition $(\mathbf{1}_{F^{-1}(D)} \circ F^{-1})(x) = \mathbf{1}_D(x)$ is not Lebesgue–Borel measurable since D does not belong to the Lebesgue σ-algebra. However $\mathbf{1}_{F^{-1}(D)}$ is indeed Lebesgue–Lebesgue measurable since $F^{-1}(D)$ belongs to the Lebesgue σ-algebra as a subset of the Cantor ternary set, which is a Borel set with measure 0. On the other hand, the composition

$$A \xrightarrow[\text{Lebesgue–Borel meas.}]{f} B \xrightarrow[\text{Borel–Borel meas.}]{g} C$$

is obviously Lebesgue–Borel measurable from Lemma 1.1.6.

Theorem 5.7.11. *Let $m \geq 1$ be an integer, let $\mathcal{B}_m$ be the Borel σ-algebra on $\mathbb{R}^m$ and let $\mathcal{L}_m$ be the Lebesgue σ-algebra on $\mathbb{R}^m$. Then the following cardinality results hold:*

(1) $\operatorname{card}\mathcal{B}_m = \mathfrak{c} = \operatorname{card}\mathbb{R}$,

(2) $\operatorname{card}\mathcal{L}_m = 2^{\mathfrak{c}} = \operatorname{card}\mathcal{P}(\mathbb{R})$.

Proof. The proof is given in the Exercises (with detailed answers) 5.10.7, 5.10.8, 5.10.9. $\qquad\square$

---

[6] Given two measurable spaces $(X, \mathcal{M}), (Y, \mathcal{N})$, a measurable mapping $f : X \to Y$ is said to be $\mathcal{M} - \mathcal{N}$ measurable.

A Cantor set with positive measure

Let $\theta \in (0,1]$ be given and let $(\theta_n)_{n \geq 1}$ be a sequence of positive numbers such that $\sum_{n \geq 1} 2^{n-1}\theta_n = \theta$. With $I = [0,1]$, we define

$$I_0 \cup I_2 = I \backslash I_1, \quad I_1 = \left(\frac{1-\theta_1}{2}, \frac{1+\theta_1}{2} \right), \quad I_0, I_2 \text{ compact intervals, } \max I_0 < \min I_2,$$

$$|I_1| = \theta_1, \quad |I_0| = |I_2| = (1-\theta_1)/2.$$

We define

$$I_{00} \cup I_{02} = I_0 \backslash I_{01}, \quad I_{01} = \left(m_0 - \frac{\theta_2}{2}, m_0 + \frac{\theta_2}{2} \right), \quad m_0 \text{ midpoint of } I_0,$$

$$I_{20} \cup I_{22} = I_2 \backslash I_{21}, \quad I_{21} = \left(m_2 - \frac{\theta_2}{2}, m_0 + \frac{\theta_2}{2} \right), \quad m_2 \text{ midpoint of } I_2,$$

$$\text{for } \alpha \in \{0,2\} : |I_{\alpha 1}| = \theta_2,$$

$$\text{for } \alpha \in \{0,2\}^2: |I_\alpha| = \left(\frac{1-\theta_1}{2} - \theta_2 \right) \frac{1}{2} = 2^{-2}(1 - \theta_1 - 2\theta_2).$$

Let $N \geq 1$ and assume that we have constructed 2^N compact pairwise disjoint intervals $I_\alpha, \alpha \in \{0,2\}^N$, included in $[0,1]$ with length

$$2^{-N}\left(1 - \sum_{1 \leq j \leq N} 2^{j-1}\theta_j \right)$$

and that the complement in $[0,1]$ of $\cup_{\alpha \in \{0,2\}^N} I_\alpha$ is the disjoint union of $2^N - 1$ open intervals $I_1, I_{01}, I_{21}, \ldots, I_{\beta,1}, \beta \in \{0,2\}^{N-1}$ (note that $1 + 2 + \cdots + 2^{N-1} = 2^N - 1$) with $|I_{\gamma 1}| = \theta_{j+1}$ if $\gamma \in \{0,2\}^j$. We have indeed

$$2^N 2^{-N}\left(1 - \sum_{1 \leq j \leq N} 2^{j-1}\theta_j \right) + \sum_{0 \leq j \leq N-1} 2^j \theta_{j+1} = 1.$$

We define then for each $\alpha \in \{0,2\}^N$ the open interval $I_{\alpha 1}$ as the mid-interval of I_α with length θ_{N+1}, its complement in $I_\alpha = I_{\alpha 0} \cup I_{\alpha 2}$ where $I_{\alpha 0}, I_{\alpha 2}$ are disjoint compact intervals with length

$$\frac{1}{2}(|I_\alpha| - \theta_{N+1}) = 2^{-N-1}\left(1 - \sum_{1 \leq j \leq N} 2^{j-1}\theta_j \right) - \theta_{N+1}/2$$

$$= 2^{-N-1}\left(1 - \sum_{1 \leq j \leq N+1} 2^{j-1}\theta_j \right),$$

indeed the expected result. Since the $I_\alpha, \alpha \in \{0,2\}^N$ are 2^N pairwise disjoint compact intervals, this produces 2^{N+1} pairwise disjoint compact intervals $I_\alpha, \alpha \in \{0,2\}^{N+1}$. The complement in $[0,1]$ of $\cup_{\alpha \in \{0,2\}^{N+1}} I_\alpha$ is the disjoint union of the complement of $\cup_{\alpha \in \{0,2\}^N} I_\alpha$ with the intervals $I_{\alpha 1}, \alpha \in \{0,2\}^N$: it is indeed the disjoint union of $1 + 2 + \cdots + 2^{N-1} + 2^N$ intervals $I_1, I_{01}, I_{21}, \ldots, I_{\beta,1}, \beta \in \{0,2\}^N$. We define

$$K_n^{(\theta)} = \cup_{\alpha \in \{0,2\}^n} I_\alpha, \quad K^{(\theta)} = \cap_{n \geq 1} K_n^{(\theta)}. \tag{5.7.12}$$

We note that the mapping $n \mapsto K_n^{(\theta)}$ is decreasing so that $K^{(\theta)}$ is a compact subset of $[0,1]$. We have also

$$|K_n^{(\theta)}| - 1 - \sum_{1 \leq j \leq n} 2^{j-1}\theta_j \longrightarrow |K^{(\theta)}| - 1 - 0.$$

Note that if $\theta = 1$ with the[7] choice $\theta_j = 3^{-j}$, we recover the ternary Cantor set K_∞ constructed above. When $\theta \in (0,1)$ the compact set $K^{(\theta)}$ has positive measure $1 - \theta$, but an empty interior since, with complements in $[0,1]$, we have

$$\left(K^{(\theta)}\right)^c = \cup_{n \geq 1}\left(K_n^{(\theta)}\right)^c = \cup_{n \geq 1}\left(\cup_{\alpha \in \{0,2\}^{n-1}} I_{\alpha 1}\right).$$

Let x be a point in $K^{(\theta)}$: then for each $n \geq 1$, $x \in K_n^{(\theta)} = \cup_{\alpha \in \{0,2\}^n} I_\alpha$. Thus for each $n \geq 1$ there exists $\alpha^{(n)} \in \{0,2\}^n$ such that $x \in I_{\alpha^{(n)}} = I_{\alpha^{(n)}0} \sqcup I_{\alpha^{(n)}1} \sqcup I_{\alpha^{(n)}2}$ and we can find $x_n \in I_{\alpha^{(n)}1} \subset \left(K^{(\theta)}\right)^c$ such that

$$|x - x_n| \leq |I_{\alpha^{(n)}}| = 2^{-n}\left(1 - \sum_{1 \leq j \leq n} 2^{j-1}\theta_j\right) \leq 2^{-n} \implies x = \lim_n x_n,$$

and thus x belongs to the closure of $\left(K^{(\theta)}\right)^c$; $\left(K^{(\theta)}\right)^c$ is consequently a dense open set of $[0,1]$ so that

$$K^{(\theta)} \subset \overline{(K^{(\theta)})^c} = \left(\overset{\circ}{K^{(\theta)}}\right)^c \implies \overset{\circ}{K^{(\theta)}} \subset \left(K^{(\theta)}\right)^c$$

$$\implies \overset{\circ}{K^{(\theta)}} \subset \left(K^{(\theta)}\right)^c \cap K^{(\theta)} = \emptyset.$$

As a result $K^{(\theta)}$ is a compact set of positive measure when $\theta < 1$, with empty interior (thus totally discontinuous) and also without isolated points: The proof above entails that for $x \in K^{(\theta)}$, for each $n \geq 1$ there exists $\alpha^{(n)} \in \{0,2\}^n$ such that $x \in I_{\alpha^{(n)}} = [a_n, b_n]$ where $0 < b_n - a_n = 2^{-n}\left(1 - \sum_{1 \leq j \leq n} 2^{j-1}\theta_j\right) \leq 2^{-n}$. Since the endpoints of I_α belong to K_n and also by construction to all $K_m, m \geq n$, both points a_n, b_n belong to $K^{(\theta)}$, providing a sequence $(x_n)_{n \geq 1}$ of points of $K^{(\theta)}$, distinct from x such that $x = \lim_n x_n$.

---

[7] We have indeed $\sum_{j \geq 1} 2^{j-1}3^{-j} = 3^{-1}\frac{1}{1-\frac{2}{3}} = 1$.

5.8   Category and measure

Definition 5.8.1. Let X be a topological space and $A \subset X$.

(1) The subset A is of *first category* in X when it is a countable union of rare subsets (see Definition 5.7.1). Such a subset is also said to be *meager*.

(2) The subset A of X is of *second category* in X when it is not of first category.

(3) A topological space X is a *Baire space* if for any sequence $(F_n)_{n\in\mathbb{N}}$ of closed sets with empty interiors, the union $\cup_{n\in\mathbb{N}}F_n$ is also with empty interior. Equivalently, X is a *Baire space* if for any sequence $(U_n)_{n\in\mathbb{N}}$ of dense open sets, the intersection $\cap_{n\in\mathbb{N}}U_n$ is also dense.

N.B. Note that a subset of a set of first category is also of first category: if $B \subset A$ with A of first category in a topological space X, then

$$B \subset A = \cup_\mathbb{N} A_n, \ \overset{\circ}{\overline{A_n}} = \emptyset \implies B = \cup_\mathbb{N}(B \cap A_n), \quad \overset{\circ}{\overline{B \cap A_n}} \subset \overset{\circ}{\overline{A_n}} = \emptyset.$$

The proof of the two following theorems is given in the Appendix (Theorems 10.2.39, 10.2.40).

Theorem 5.8.2 (Baire theorem). *Let (X, d) be a complete metric space and $(F_n)_{n\geq 1}$ be a sequence of closed sets with empty interiors. Then the interior of $\cup_{n\geq 1}F_n$ is also empty.*

N.B. The statement of that theorem is equivalent to saying that, in a complete metric space, given a sequence $(U_n)_{n\geq 1}$ of open dense sets the intersection $\cap_{n\geq 1}U_n$ is also dense.

Theorem 5.8.3. *Let X be a locally compact topological space (Hausdorff topological space such that each point has a compact neighborhood) and $(F_n)_{n\geq 1}$ be a sequence of closed sets with empty interiors. Then the interior of $\cup_{n\geq 1}F_n$ is also empty.*

Corollary 5.8.4. *A metric complete space, as well as a locally compact space are both Baire spaces and are both of second category in themselves, provided they are not empty. A non-empty Baire space is of second category in itself.*

Proof. Let X be a Baire space; if it were of first category in itself, it would be a countable union $\cup_\mathbb{N} A_n$ with $\overset{\circ}{\overline{A_n}} = \emptyset$, thus we would have $X = \cup_\mathbb{N}\overline{A_n}$ and by the Baire property, $X = \overset{\circ}{X}$ would be empty. $\qquad\square$

For a topological space, the category in itself is indeed a topological notion, as proven by the following lemma.

Lemma 5.8.5. *Let X, Y be topological spaces and let $\kappa : X \to Y$ be a homeomorphism. If X is of second category in itself, then Y is also of second category in itself.*

Proof. We note first that for a subset B of Y, since κ is a homeomorphism

$$\kappa^{-1}(\overline{B}) = \overline{\kappa^{-1}(B)}, \quad \kappa^{-1}(\mathring{B}) = \widering{\kappa^{-1}(B)}. \tag{5.8.1}$$

In fact, we have $\kappa^{-1}(B) \subset \kappa^{-1}(\overline{B})$ (a closed set by continuity of κ) so that $\overline{\kappa^{-1}(B)} \subset \kappa^{-1}(\overline{B})$. We have also $B = \kappa(\kappa^{-1}(B)) \subset \kappa(\overline{\kappa^{-1}(B)})$ (a closed set by continuity of κ^{-1}) so that $\overline{B} \subset \kappa(\overline{\kappa^{-1}(B)})$ and $\kappa^{-1}(\overline{B}) \subset \overline{\kappa^{-1}(B)}$, giving the first equality in (5.8.1). The second equality can be deduced by complementation, using (1.2.1). If Y were of first category, we would have $Y = \cup_{\mathbb{N}} B_n$, $\overline{B_n} = \emptyset$ and thus

$$X = \kappa^{-1}(Y) = \cup_{\mathbb{N}}\kappa^{-1}(B_n), \quad \overline{\kappa^{-1}(B_n)} = \kappa^{-1}(\overline{B_n}),$$

and $\mathrm{interior}\big(\overline{\kappa^{-1}(B_n)}\big) = \kappa^{-1}(\mathrm{interior}(\overline{B_n})) = \emptyset$, contradicting the assumption on X. $\qquad\square$

Lemma 5.8.6. *Let X be a complete metric space and let A be a subset of X such that A contains a closed set F with a non-empty interior. Then A is of second category in X.*

Proof. If A were of first category in X, so would be F, and we would have

$$F = \cup_{\mathbb{N}} B_n, \quad \overset{\circ}{\overline{B_n}} = \emptyset.$$

The complete metric space F would be a countable union of closed sets with empty interiors since

$$F = \cup_{\mathbb{N}} \underbrace{(\overline{B_n} \cap F)}_{\text{closure of } B_n \text{ in } F}, \quad \mathrm{interior}_F(\overline{B_n} \cap F) = \overset{\circ}{\overline{B_n}} \cap F = \emptyset,$$

contradicting the Baire theorem. $\qquad\square$

Note that $\mathbb{Q}$ is a meager subset of $\mathbb{R}$, thus of first category in $\mathbb{R}$, i.e., "small" in the sense of category but $\mathbb{Q}$ is dense in $\mathbb{R}$. On the other hand the notions of category and measure are unrelated: a set can be of first category (small in the sense of category) and large in the Lebesgue measure sense. Also a set can have a Lebesgue measure 0 and be of second category: the following lemma is provides some examples.

Lemma 5.8.7.

(1) *The Cantor ternary set is a compact space, and so is of second category in itself, but it is of first category in the interval $[0,1]$ with the usual topology.*

(2) *The Cantor sets with positive measure constructed in Section 5.7 are of first category in $[0,1]$.*

(3) *There exists a subset of $[0,1]$ which has Lebesgue measure 1 and which is of first category.*

(4) *There exists a subset of $[0,1]$ which has Lebesgue measure 0 and which is of second category.*

Proof. The Cantor sets are closed sets, and also with empty interior, so they are of first category in $[0, 1]$. To convince the reader that the notions of size given respectively by the Lebesgue measure and by the category are unrelated, we can also give an example of a set of first category, "small" in the sense of category, but with full Lebesgue measure in $[0, 1]$. We have seen with the construction of Cantor sets with positive measure that for any integer $k \geq 1$, we can construct a compact subset $\mathscr{C}_k$ of $[0, 1]$ such that

$$\operatorname{int}(\mathscr{C}_k) = \emptyset, \quad |\mathscr{C}_k| \geq \frac{k - 1}{k}.$$

We define then $A = \cup_{k \geq 1} \mathscr{C}_k$ and we have $|A| \geq \sup_{k \geq 1} |\mathscr{C}_k| \geq \sup_{k \geq 1}(1 - \frac{1}{k}) = 1$. Moreover, A is obviously of first category as a countable union of compact sets with empty interior.

Here is an example of a set of second category in $\mathbb{R}$, i.e., "large" in the sense of category, but with Lebesgue measure 0 (small in the sense of the Lebesgue measure). We define for $\mathbb{Q} = \{x_n\}_{n \geq 1}$,

$$A = \cap_{m \geq 1} U_m, \quad U_m = \cup_{n \geq 1}]x_n - 2^{-n-m}, x_n + 2^{-n-m}[.$$

The Lebesgue measure $|A|$ is such that

$$|A| \leq \inf_{m \geq 1} \sum_{n \geq 1} 2^{1-n-m} = \inf_{m \geq 1} 2^{-m+1} = 0.$$

If A were meager, we would have a sequence (A_k) of subsets of $\mathbb{R}$ with $\operatorname{int}(\overline{A_k}) = \emptyset$, so that

$$\mathbb{R} = A \cup A^c = \cup_k A_k \cup A^c = \cup_k \overline{A_k} \cup A^c = \cup_k \overline{A_k} \cup \cup_m U_m^c.$$

We note that $\operatorname{int}(U_m^c) = \emptyset$ since $\overline{U_m} \supset \overline{\mathbb{Q}} = \mathbb{R}$. We would have written $\mathbb{R}$ as a countable union of closed sets with empty interiors: this is not possible from the Baire theorem. $\qquad\square$

5.9 Notes

Ivar BENDIXSON (1861–1935) was a Swedish mathematician.

EUCLID (325 BC–265 BC). Euclid of Alexandria is the most prominent mathematician of antiquity, author of the fundamental treatise *The Elements*.

George GREEN (1793–1841) was an English mathematician. The Gauss–Green formula proved above appears as a particular case of *Stokes' theorem*.

Pierre-Simon LAPLACE (1749–1827) was a French mathematician. He had a considerable influence on the developments of the calculus of probabilities and celestial mechanics.

Isaac NEWTON (1642–1727) was an English physicist and mathematician. He was one of the most influential scientists of all times. His book *Philosophiæ Naturalis Principia Mathematica*, published in 1687, set up the foundations of Mechanics for more than two centuries until the scientific revolutions of Relativity and Quantum mechanics in the twentieth century.

Jakob STEINER (1796–1863) was a Swiss mathematician and geometer.

Brook TAYLOR (1685–1731) was an English mathematician. He published *Methodus incrementorum directa et inversa* in 1715, in which he introduced a version of what is now known as Taylor's formula. He took sides with Isaac Newton, creator of the *Calculus of fluxions*, in the violent controversy with Gottfried Wilhelm Leibniz (inventor of the *Infinitesimal calculus*) about priorities on the invention of Calculus. Today, both Newton and Leibniz are considered as scientific geniuses who transformed radically the mathematics and science of their times.

William Henry YOUNG (1863–1942) was an English mathematician. His name is associated to B. Taylor for the following theorems:

Theorem 5.9.1 (Taylor–Young formula). *Let $k \in \mathbb{N}$, let U be an open set of $\mathbb{R}^n$, let $f : U \to \mathbb{R}^m$ of class C^k and let $x_0 \in U$. If the function f is $k + 1$ times differentiable at x_0, there exists $\epsilon : U \to \mathbb{R}^m$ with $\lim_{x \to x_0} \epsilon(x) = 0$ such that*

$$f(x) = \sum_{0 \le j < k+1} \frac{1}{j!} f^{(j)}(x_0)(x - x_0)^j + \epsilon(x)|x - x_0|^{k+1}. \qquad (5.9.1)$$

Note that $f^{(j)}(x_0)$ is the symmetric jth linear form given by

$$\frac{f^{(j)}(x_0)}{j!} T^j = \sum_{\alpha \in \mathbb{N}^n, |\alpha|=j} \frac{(\partial_x^\alpha f)(x_0)}{\alpha!} T^\alpha, \qquad (5.9.2)$$

where for $\alpha = (\alpha_1, \ldots, \alpha_n) \in \mathbb{N}^n$,

$$|\alpha| = \sum_{1 \le l \le n} \alpha_l, \quad \partial_x^\alpha = \partial_{x_1}^{\alpha_1} \ldots \partial_{x_n}^{\alpha_n}, \quad \alpha! = \alpha_1! \ldots \alpha_n! \qquad (5.9.3)$$

Theorem 5.9.2 (Taylor–Lagrange formula). *Let $k \in \mathbb{N}$, let U be an open set of $\mathbb{R}^n$, let $f : U \to \mathbb{R}^m$ of class C^k. Let $x_0, x_1 \in U$ and assume that the function f is $k + 1$ times differentiable on $(x_0, x_1) = \{(1 - \theta)x_0 + \theta x_1\}_{\theta \in (0,1)}$. Then*

$$f(x_1) = \sum_{0 \le j \le k} \frac{1}{j!} f^{(j)}(x_0)(x_1 - x_0)^j + R_k(x_1, x_0), \qquad (5.9.4)$$

$$|R_k(x_1, x_0)| \le \frac{|x_1 - x_0|^{k+1}}{(k + 1)!} \sup_{(x_0, x_1)} \|f^{(k+1)}(x)\|, \qquad (5.9.5)$$

where $\|f^{(k+1)}\|$ stands for the norm of the multilinear form, i.e.,

$$\|f^{(l)}(y)\| = \sup_{\substack{|T_j|=1 \\ 1\le j\le l}} |f^{(l)}(y)(T_1,\ldots,T_l)|.$$

When $m = 1$, for k, U, f, x_0, x_1 as above, there exists $x \in (x_0, x_1)$ such that

$$R_k(x_1, x_0) = \frac{1}{(k+1)!} f^{(k+1)}(x)(x_1 - x_0)^{k+1}.$$

Theorem 5.9.3 (Taylor formula with integral remainder). *Let $k \in \mathbb{N}$, let U be a convex open set of $\mathbb{R}^n$, let $f : U \to \mathbb{R}^m$ of class C^{k+1}. Then for $x_1, x_0 \in U$*

$$
\begin{aligned}
f(x_1) = {}& \sum_{0\le j\le k} \frac{1}{j!} f^{(j)}(x_0)(x_1 - x_0)^j \\
& + \int_0^1 \frac{(1-\theta)^k}{k!} f^{(k+1)}\big(x_0 + \theta(x_1 - x_0)\big)\,d\theta(x_1 - x_0)^{k+1}.
\end{aligned}
\tag{5.9.6}
$$

The three theorems above are proven in Exercise 5.10.1.

Our next chapter studies the convolution and Young's inequalities for $L^p(\mathbb{R}^n)$ spaces:

$$\|u * v\|_{L^r(\mathbb{R}^n)} \le \|u\|_{L^p(\mathbb{R}^n)}\|v\|_{L^q(\mathbb{R}^n)}, \quad 1 - \frac{1}{r} = 1 - \frac{1}{p} + 1 - \frac{1}{q}, \ 1 \le p, q, r. \tag{5.9.7}$$

5.10   Exercises

Exercise 5.10.1. *Prove Theorems 5.9.1, 5.9.2, 5.9.3.*

Answer. We start with a one-dimensional lemma.

Lemma 5.10.2 (Mean Value Theorem). *Let $\varphi : [0, 1] \to \mathbb{R}$ be a continuous function which is differentiable on $(0, 1)$. Then there exists $t \in (0, 1)$ such that $\varphi(1) - \varphi(0) = \varphi'(t)$.*

Proof. The continuous function $[0, 1] \ni t \mapsto \psi(t) = \varphi(t) - \varphi(0) - t(\varphi(1) - \varphi(0))$ is such that $\psi(0) = \psi(1) = 0$. Since the image by ψ of $[0,1]$ is a compact interval $[m, M]$, either $m = M$ and ψ is constant on $[0, 1]$, so that

$$\forall t \in [0, 1], \quad \varphi(t) = \varphi(0) + t(\varphi(1) - \varphi(0)) \implies \varphi'(t) = \varphi(1) - \varphi(0),$$

or $\psi(t_0) = m < M = \psi(t_1)$. In the latter case t_0 or t_1 belong to $(0, 1)$ (we have $\psi(0) = \psi(1)$). As a result ψ has an extremum at a point $t \in (0, 1)$ and its derivative must vanish there: $0 = \psi'(t) = \varphi'(t) - \big(\varphi(1) - \varphi(0)\big)$. $\qquad\square$

(1) *Let us prove first the one-dimensional version ($m = 1$) in Theorem 5.9.2.* We introduce, following the notation of this theorem,

$$\varphi(\theta) = f(x_1) - \sum_{0 \le j \le k} \frac{f^{(j)}(x_\theta)}{j!}(x_1 - x_\theta)^j, \quad x_\theta = (1 - \theta)x_0 + \theta x_1, \ \theta \in [0, 1].$$

We note that $\varphi(1) = 0$ and we define

$$\psi(\theta) = \varphi(\theta) - \varphi(0)(1 - \theta)^{k+1}, \quad \text{so that } \psi(0) = 0 = \psi(1).$$

We may apply Lemma 5.10.2 to ψ and we get that there exists some $\theta \in (0, 1)$ with $\psi'(\theta) = 0$, i.e.,

$$0 = - \sum_{0 \le j \le k} \frac{f^{(j+1)}(x_\theta)}{j!}(x_1 - x_0)(x_1 - x_\theta)^j$$

$$+ \sum_{1 \le j \le k} \frac{f^{(j)}(x_\theta)}{j!}j(x_1 - x_\theta)^{j-1}(x_1 - x_0) + \varphi(0)(k + 1)(1 - \theta)^k,$$

implying since $x_1 - x_\theta = (1 - \theta)(x_1 - x_0)$,

$$(k + 1)\varphi(0)(1 - \theta)^k = \frac{f^{(k+1)}(x_\theta)}{k!}(x_1 - x_0)^{k+1}(1 - \theta)^k,$$

i.e.,

$$\varphi(0) = \frac{f^{(k+1)}(x_\theta)}{(k + 1)!}(x_1 - x_0)^{k+1},$$

which is the sought result.

(2) *Let us prove now the multi-dimensional inequality in Theorem 5.9.2.* Lemma 5.1.4 provides the result for $k = 0$. Let us assume that $k \ge 1$. The function f is thus assumed to be $C^k \subset C^1$. We note that the function

$$U \ni x \mapsto f'(x) \in \mathcal{L}(\mathbb{R}^n, \mathbb{R}^m) \equiv \mathbb{R}^{mn}$$

is of class C^{k-1} and k times differentiable on (x_0, x_1). We calculate

$$\frac{d}{d\theta}\big(f(x_\theta)\big) = f'(x_\theta)(x_1 - x_0)$$

$$= \sum_{0 \le j \le k-1} \frac{f'^{(j)}(x_0)}{j!}(x_\theta - x_0)^j(x_1 - x_0) + R_{k-1}(f')(x_\theta, x_0)(x_1 - x_0),$$

with (induction hypothesis)

$$(\natural) \qquad \|R_{k-1}(f')(x_\theta, x_0)\|_{\mathbb{R}^{mn}} \le \frac{\sup_{(x_0, x_\theta)} \|f'^{(k)}(x)\|_{\mathcal{M}^k_{n,mn}}}{k!}\|x_\theta - x_0\|^k_{\mathbb{R}^n},$$

where $\mathcal{M}^k_{n,mn}$ is the vector space of k multilinear forms from $(\mathbb{R}^n)^k$ to $\mathbb{R}^{mn}$. Since the function $[0,1] \ni \theta \mapsto f(x_\theta)$ is C^1 and the sum is a polynomial in θ, the function $[0,1] \ni \theta \mapsto R_{k-1}(f')(x_\theta, x_0)(x_1 - x_0)$ is also C^0 and we can integrate from 0 to 1 and get

$$f(x_1) - f(x_0) = \sum_{0 \le j \le k-1} \frac{f^{(j+1)}(x_0)}{j!} \frac{1}{j+1}(x_1 - x_0)^{j+1}$$

$$+ \int_0^1 R_{k-1}(f')(x_\theta, x_0)(x_1 - x_0)d\theta.$$

The estimate ($\natural$) implies for $\theta \in [0,1]$,

$$\|R_{k-1}(f')(x_\theta, x_0)(x_1 - x_0)\|_{\mathbb{R}^m}$$

$$\le \frac{\sup_{(x_0,x_\theta)} \|f'^{(k)}(x)\|_{\mathcal{M}^k_{n,mn}}}{k!} \|x_\theta - x_0\|_{\mathbb{R}^n}^k \|x_1 - x_0\|_{\mathbb{R}^n}$$

$$\le \frac{\theta^k}{k!} \|x_1 - x_0\|_{\mathbb{R}^n}^{k+1} \sup_{(x_0,x_1)} \|f^{(k+1)}(x)\|_{\mathcal{M}^{k+1}_{n,m}}.$$

We obtain thus $f(x_1) = \sum_{0 \le j \le k} \frac{f^{(j)}(x_0)}{j!}(x_1 - x_0)^j + R_k(f)(x_1, x_0)$, with

$$\|R_k(f)(x_1, x_0)\|_{\mathbb{R}^m} \le \|x_1 - x_0\|_{\mathbb{R}^n}^{k+1} \sup_{(x_0,x_1)} \|f^{(k+1)}(x)\|_{\mathcal{M}^{k+1}_{n,m}} \frac{1}{(k+1)!},$$

which is the sought result.

(3) *Let us prove Theorem 5.9.3.* Let $x, x + h \in U$. From the convexity of U, we may define for $\theta \in [0,1]$, $\varphi(\theta) = f(x + \theta h)$. If $k = 0$, we have $\varphi \in C^1([0,1]; \mathbb{R}^m)$,

$$\varphi(\theta) = \varphi(0) + \int_0^\theta \varphi'(s)ds = \varphi(0) + \int_0^1 \varphi'(\theta t)dt\theta.$$

If $k \ge 1$, the function φ is $C^{k+1}([0,1]; \mathbb{R}^m)$ and we may assume inductively

$$\varphi(\theta) = \sum_{0 \le j \le k-1} \frac{\varphi^{(j)}(0)}{j!}\theta^j + \int_0^1 \underbrace{\varphi^{(k)}(\theta t)}_{u(t)} \underbrace{\frac{(1-t)^{k-1}}{(k-1)!}}_{v'(t)} dt\theta^k.$$

Integrating by parts in the integral I, we get

$$I = \frac{\varphi^{(k)}(0)}{k!}\theta^k - \theta^k \int_0^1 \varphi^{(k+1)}(\theta t)\theta \frac{(1-t)^k}{k!}(-1)dt,$$

providing

$$\varphi(\theta) = \sum_{0 \le j \le k} \frac{\varphi^{(j)}(0)}{j!}\theta^j + \int_0^1 \varphi^{(k+1)}(\theta t)\frac{(1-t)^k}{k!}dt\theta^k$$

and the theorem by taking $\theta = 1$ and noting that $\varphi^{(j)}(\theta) = f^{(j)}(x + \theta h)h^j$.

(4) *Let us prove finally Theorem* 5.9.1. When $k = 0$, by definition of differentiability at x_0, we have

$$f(x_0 + h) = f(x_0) + f'(x_0)h + \epsilon(h)|h|, \quad \lim_{h \to 0} \epsilon(h) = 0.$$

For $k \geq 1$, the function $h \mapsto f'(x_0 + h)$ is C^{k-1} and k times differentiable at 0, so that inductively for $\theta \in [0, 1]$,

$$f'(x_0 + \theta h)h = \sum_{0 \leq j \leq k-1} \frac{f'^{(j)}(x_0)}{j!}(\theta h)^j h + \epsilon(\theta h)|h\theta|^k h, \quad \lim_{h \to 0} \epsilon(h) = 0.$$

Since the sum is a polynomial in θ and $\theta \mapsto f'(x_0 + \theta h)h$ is $C^{k-1} \subset C^0$, we obtain that $\theta \mapsto \epsilon(\theta h)|h\theta|^k$ is continuous and by integration with respect to $\theta \in [0, 1]$,

$$f(x_0 + h) - f(x_0) = \sum_{0 \leq j \leq k-1} \frac{f^{(j+1)}(x_0)}{j!}\frac{1}{j+1}h^{j+1} + \int_0^1 \epsilon(\theta h)|h\theta|^k h\, d\theta,$$

so that $\int_0^1 \epsilon(\theta h)|h\theta|^k h\, d\theta = |h|^{k+1}\int_0^1 \epsilon(\theta h)\theta^k\, d\theta\frac{h}{|h|}$ and

$$\left|\int_0^1 \epsilon(\theta h)\theta^k\, d\theta\frac{h}{|h|}\right| \leq \frac{1}{k+1} \sup_{\theta \in [0,1]} |\epsilon(\theta h)| = \epsilon_1(h).$$

We have indeed $\lim_{h \to 0} \epsilon_1(h) = 0$, concluding the proof.

Exercise 5.10.3. *Let E be a normed real vector space and let $f : [0, 1] \to E$ and $g : [0, 1] \to \mathbb{R}$ be continuous mappings, both differentiable on $(0, 1)$ such that for all $t \in (0, 1), \|f'(t)\| \leq g'(t)$. Prove that*

$$\|f(1) - f(0)\| \leq g(1) - g(0).$$

Answer. Let $\epsilon > 0$ be given. We define

$$T_\epsilon = \{t \in [0, 1], \|f(t) - f(0)\| - g(t) + g(0) - \epsilon t \leq \epsilon\}.$$

By continuity of f, g, T_ϵ is a closed subset of $[0, 1]$, contains 0 (the lhs of the inequality vanishes at 0) and thus by continuity, T_ϵ contains a neighborhood of 0. Defining $c = \sup T_\epsilon$ we have $c > 0$ and since T_ϵ is closed, $c \in T_\epsilon$. Let us assume that $c < 1$. We can find $t > c$ such that

$$\left\|\frac{f(t) - f(c)}{t - c}\right\| \leq \|f'(c)\| + \epsilon/2, \quad g'(c) \leq \frac{g(t) - g(c)}{t - c} + \epsilon/2,$$

implying

$$\|f(t) - f(0)\| \leq \|f(t) - f(c)\| + \|f(c) - f(0)\|$$
$$\leq (t - c)\|f'(c)\| + \frac{\epsilon(t - c)}{2} + g(c) - g(0) + \epsilon(c + 1)$$
$$\leq (t - c)g'(c) + \frac{\epsilon(t - c)}{2} + g(c) - g(0) + \epsilon(c + 1)$$
$$\leq g(t) - g(c) + \epsilon(t - c) + g(c) - g(0) + \epsilon(c + 1)$$
$$= g(t) - g(0) + \epsilon t + \epsilon,$$

so that $t \in T_\epsilon$, which is impossible since $t > c = \sup T_\epsilon$. As a result $c = 1$ and thus

$$\forall \epsilon > 0, \quad \|f(1) - f(0)\| \leq g(1) - g(0) + 2\epsilon,$$

implying the result.

Exercise 5.10.4. *Let U be an open subset of $\mathbb{R}^n$ and let $f : U \to \mathbb{R}^n$ be a C^1 injective mapping such that, for all $x \in U$, $\det f'(x) \neq 0$. Prove that $f(U)$ is an open subset of $\mathbb{R}^n$ and that f is a diffeomorphism from U onto $f(U)$.*

Answer. Let $x \in U$. Since $\det f'(x) \neq 0$, the inverse function theorem implies that there exists an open neighborhood $W(x)$ of x such that $f_{|W(x)}$ is a C^1 diffeomorphism from $W(x)$ onto $f(W(x))$. As a result,

$$f(U) = \cup_{x \in U} \underbrace{f(W(x))}_{\text{open}} \implies f(U) \text{ is open.}$$

As a consequence, $f : U \to V = f(U)$ is a C^1 bijection of open subsets of $\mathbb{R}^n$. Let Ω be an open subset of U: as above we prove that $f(\Omega)$ is an open subset of V and thus the inverse mapping is continuous and f is a homeomorphism from U onto V. The inverse function theorem implies that f^{-1} is C^1, completing the proof.

Exercise 5.10.5.

(1) *Prove that the mapping $(0, 1) \times (-\pi, \pi) \ni (r, \theta) \mapsto (r\cos\theta, r\sin\theta)$ is an analytic diffeomorphism from $(0, 1) \times (-\pi, \pi)$ onto*

$$\{z \in \mathbb{C}, |z| < 1\} \backslash (-1, 0].$$

(2) *Prove that the mapping $(0, 1) \times (-\pi, \pi] \ni (r, \theta) \mapsto (r\cos\theta, r\sin\theta)$ onto $\{z \in \mathbb{C}, 0 < |z| < 1\}$ is analytic and bijective, but is not a homeomorphism.*

Answer. (1) With $\phi(r, \theta) = (r\cos\theta, r\sin\theta)$, the mapping ϕ is analytic and bijective from $(0, +\infty) \times (-\pi, \pi)$ onto $\mathbb{C}\backslash\mathbb{R}_-$ with inverse mapping (also analytic)

$$\psi(x, y) = \left(\sqrt{x^2 + y^2}, \operatorname{Im}\big(\operatorname{Log}(x + iy)\big)\right),$$

where $\operatorname{Log} z$ is defined for $z \in \mathbb{C}\backslash\mathbb{R}_-$ by (10.5.1).

(2) Extending ϕ to $(0,1) \times (-\pi, \pi]$ keeps of course analyticity as well as bijectivity since the injective image of $(0,1) \times \{\pi\}$ is $(-1,0)$. However, it is not a homeomorphism: we have

$$\lim_{y \to 0_+} \psi\left(-\frac{1}{2} + iy\right) = (1/2, \pi), \quad \lim_{y \to 0_+} \psi\left(-\frac{1}{2} - iy\right) = (1/2, -\pi),$$

since for $0 < \phi < \pi$, $\mathrm{Log}(e^{i\phi}) = i\phi$, $\mathrm{Log}(e^{-i\phi}) = -i\phi$ and thus

$$\lim_{\phi \to \pi_-} \mathrm{Im}(\mathrm{Log}(e^{i\phi})) = \pi, \quad \lim_{\phi \to (-\pi)_+} \mathrm{Im}(\mathrm{Log}(e^{i\phi})) = -\pi.$$

Exercise 5.10.6. *Let Q be a non-degenerate real symmetric $n \times n$ matrix and let $m > 0$ be given. We define*

$$\Sigma_{Q,m} = \{x \in \mathbb{R}^n, \langle Qx, x \rangle = m\}.$$

(1) *Prove that $\Sigma_{Q,m}$ is an analytic hypersurface of $\mathbb{R}^n$.*
(2) *Assuming that the index of Q (the index is the number of negative eigenvalues) equals 0, prove that $\Sigma_{Q,m}$ is diffeomorphic to the unit Euclidean sphere of $\mathbb{R}^n$ (compact and connected for $n \geq 2$).*
(3) *Assuming that the index of Q equals 1, and $n \geq 2$ prove that $\Sigma_{Q,m}$ is diffeomorphic to the hyperboloid*

$$\left\{x \in \mathbb{R}^n, \sum_{1 \leq j \leq n-1} x_j^2 = x_n^2 + 1\right\},$$

which is non-compact, with two connected components when $n = 2$, connected if $n \geq 3$ (hyperboloid with one sheet).
(4) *Assuming that the index of Q equals 2, and $n \geq 3$ prove that $\Sigma_{Q,m}$ is diffeomorphic to the hyperboloid*

$$\left\{x \in \mathbb{R}^n, \sum_{1 \leq j \leq n-2} x_j^2 = x_{n-1}^2 + x_n^2 + 1\right\},$$

which is non-compact, has two connected component if $n = 3$ (hyperboloid with two sheets), is connected if $n \geq 4$.
(5) *We assume that $n \geq 2$. Let r be the index of Q. If $r = n-1$, prove that $\Sigma_{Q,m}$ is non-compact with two connected components. Prove that if $r < n-1$, then $\Sigma_{Q,m}$ is connected, non-compact for $r \geq 1$.*

Answer. (1) The differential of $\langle Qx, x \rangle$ is $2Qx$ and thus does not vanish at $\Sigma_{Q,m}$ since $m \neq 0$. Moreover the matrix Q can be diagonalized in an orthonormal basis, i.e.,

$$Q = PD^tP, \quad {}^tPP = \mathrm{Id}, \quad D \text{ diagonal with the eigenvalues of } Q \text{ as entries.}$$

(2) Defining $x = Py$ (a linear isomorphism), we get

$${}^tP(\Sigma_{Q,m}) = \left\{ y \in \mathbb{R}^n, \ \sum_{1 \le j \le n} \lambda_j y_j^2 = m \right\}$$

$$= \left\{ y \in \mathbb{R}^n, \ \sum_{\substack{1 \le j \le n \\ \lambda_j > 0}} \lambda_j y_j^2 = m + \sum_{\substack{1 \le j \le n \\ \lambda_j < 0}} |\lambda_j| y_j^2 \right\}$$

and thus, dividing by m the previous equations, we find the answer to questions (2), (3) and (4), except for the connectedness issues, addressed below. The arc-connectedness of the unit Euclidean sphere is obvious since if $\|x_0\| = \|x_1\| = 1$ with Euclidean norm in $\mathbb{R}^n$ ($n \ge 2$), we may consider a plane Π containing x_0, x_1: the intersection of Π with the unit sphere $\mathbb{S}^{n-1}$ is a circle (thus arc-connected).
(5) *Let us assume that $n \ge 2$ and the index $r = n - 1$.* We may thus assume that $\Sigma_{Q,m}$ is given by the equation

$$x_1^2 = 1 + \sum_{2 \le j \le n} x_j^2.$$

It has two connected components:

$$\underbrace{\{x \in \mathbb{R}^n, x_1 = \sqrt{1 + \|x'\|^2}}_{\Sigma_+} \sqcup \underbrace{\{x \in \mathbb{R}^n, x_1 = -\sqrt{1 + \|x'\|^2}}_{\Sigma_+}$$

$$\Sigma_\pm = F_\pm(\mathbb{R}^{n-1}), \quad F_\pm(x') = (\pm\sqrt{1 + \|x'\|^2}, x'), \quad \Sigma_+ \cap \Sigma_- = \emptyset.$$

Let us assume that $n \ge 3$ and the index $1 \le r \le n - 2$. The equation of $\Sigma_{Q,m}$ is

$$\|x''\|^2 = 1 + \|x'\|^2, \quad x' \in \mathbb{R}^r, x'' \in \mathbb{R}^{n-r}.$$

Then $\Sigma_{Q,m}$ is arc-connected. We consider

$$(x_0, y_0) \text{ and } (x_1, y_1) \in \mathbb{R}^{n-r} \times \mathbb{R}^r, \quad \|x_j\|_{\mathbb{R}^{n-r}}^2 = 1 + \|y_j\|_{\mathbb{R}^r}^2, \quad j = 0, 1.$$

We define for $\theta \in [0, 1]$,

$$y(\theta) = (1 - \theta)y_0 + \theta y_1, \ r(\theta) = \sqrt{1 + \|y(\theta)\|^2}, \ \xi(\theta) \in \mathbb{S}^{n-r-1}, \begin{cases} \xi(0) = x_0/\|x_0\|, \\ \xi(1) = x_1/\|x_1\|, \end{cases}$$

which is possible with a continuous ξ since $\mathbb{S}^{n-r-1}$ is arc-connected ($n - r - 1 \ge 1$). We have with $x(\theta) = r(\theta)\xi(\theta)$, $x(0) = x_0, x(1) = x_1$,

$$\|x(\theta)\|^2 = r(\theta)^2 = 1 + \|y(\theta)\|^2, \quad \text{i.e., } (x(\theta), y(\theta)) \in \Sigma_{Q,m},$$

proving the arc-connectedness of $\Sigma_{Q,m}$.

Exercise 5.10.7. *Let X be a set and let $\mathcal{E} \subset \mathcal{P}(X)$ be a family of subsets of X such that $\emptyset \in \mathcal{E}$. We want to describe $\mathcal{M}(\mathcal{E})$, the σ-algebra generated by $\mathcal{E}$ (see Definition 1.1.3). We define*

$$\mathcal{E}_c = \{E^c\}_{E \in \mathcal{E}}, \quad \mathcal{E}_\sigma = \{\cup_{\mathbb{N}} E_k\}_{E_k \in \mathcal{E}}. \tag{5.10.1}$$

Let Ω be the set of countable ordinals, as defined and studied in Propositions 10.1.35, 10.1.37 and Remark 10.1.36 (Ω is the first uncountable ordinal). We define, following Definition 10.1.42,

$$\mathcal{F}_1 = \mathcal{E} \cup \mathcal{E}_c, \tag{5.10.2}$$
for $x \in \Omega$ with an immediate predecessor y, $\quad \mathcal{F}_x = (\mathcal{F}_y)_\sigma \cup \left((\mathcal{F}_y)_\sigma\right)_c,$ $\quad$ (5.10.3)
for $x \in \Omega$ a limit ordinal, $\quad \mathcal{F}_x = \cup_{y < x} \mathcal{F}_y.$ $\quad$ (5.10.4)

(1) *Prove that $\mathcal{E} \subset \mathcal{E}_\sigma$.*
(2) *Prove that $\mathcal{F}_x \subset \mathcal{M}(\mathcal{E})$ for all $x \in \Omega$.*
(3) *Prove that $\cup_{x \in \Omega} \mathcal{F}_x = \mathcal{M}(\mathcal{E})$.*

Answer. (1) Obvious since $\emptyset \in \mathcal{E}$.
(2) Note that if $\mathcal{F}_y \subset \mathcal{M}(\mathcal{E})$ for all $y < x$, then $\mathcal{F}_x \subset \mathcal{M}(\mathcal{E})$: this is obvious for a limit ordinal and if x has an immediate predecessor y, then $\mathcal{F}_y \subset \mathcal{M}(\mathcal{E})$ implies $(\mathcal{F}_y)_\sigma \subset \mathcal{M}(\mathcal{E})$ and $\left((\mathcal{F}_y)_\sigma\right)_c \subset \mathcal{M}(\mathcal{E})$ so that in that case as well $\mathcal{F}_x \subset \mathcal{M}(\mathcal{E})$. Now since $\mathcal{F}_1 \subset \mathcal{M}(\mathcal{E})$, we may use transfinite induction (see Theorem 10.1.19) and conclude that (2) holds.
(3) It is enough to prove that $\cup_{x \in \Omega} \mathcal{F}_x$ is a σ-algebra since it contains $\mathcal{E}$ and we already know $\cup_{x \in \Omega} \mathcal{F}_x \subset \mathcal{M}(\mathcal{E})$. We note that the empty set belongs to $\cup_{x \in \Omega} \mathcal{F}_x$, which is also stable by complementation as is each $\mathcal{F}_x$: it is true for $x = 1$ and if true for all $y < x$, it is obvious for x when x is a limit ordinal and also true when x has an immediate predecessor. We may use transfinite induction to conclude. We need to prove that $\cup_{x \in \Omega} \mathcal{F}_x$ is stable by countable unions. We consider $(E_j)_{j \in \mathbb{N}}$ with $E_j \in \mathcal{F}_{x_j}$, $x_j \in \Omega$. According to Proposition 10.1.37, the countable family $\{x_j\}_{j \in \mathbb{N}}$ of countable ordinals has an upper bound $x \in \Omega$. As a consequence, for all $j \in \mathbb{N}$, $E_j \in \mathcal{F}_x$ and thus

$$\cup_{j \in \mathbb{N}} E_j \in (\mathcal{F}_x)_\sigma.$$

Since Ω has no largest element[8], x has an immediate successor $x + 1$ and $(\mathcal{F}_x)_\sigma \subset \mathcal{F}_{x+1}$, implying $\cup_{j \in \mathbb{N}} E_j \in \mathcal{F}_{x+1}$, completing the proof.

Exercise 5.10.8 (Cardinality of the Borel σ-algebra). *Let $\mathcal{B}$ be the Borel σ-algebra of $\mathbb{R}^m$.*

(1) *Prove that $\mathcal{B}$ is generated by a family of sets $\mathcal{E}$ containing the empty set and such that $\operatorname{card} \mathcal{E} = \mathfrak{c} = \operatorname{card} \mathbb{R}$.*

---

[8]Ω has no largest element otherwise we would find $x \in \Omega$ with $\Omega = (\rightarrow, x) \cup \{x\}$ and since $(\rightarrow, x)$ is countable, this would imply that Ω is countable.

(2) *Let Ω be the first uncountable ordinal. We define $\mathcal{F}_x$ for each x in Ω as in Exercise 5.10.7. Prove that $\operatorname{card}\mathcal{F}_x = \mathfrak{c}$ for each $x \in \Omega$.*

(3) *Prove that $\operatorname{card}\mathcal{B} = \mathfrak{c}$.*

Answer. (1) We consider $\mathcal{E} = \{B(x,r)\}_{x \in \mathbb{R}^m, r \geq 0}$. Thanks to Theorem 10.1.20, we have

$$\mathfrak{c} = \mathfrak{c}^m = \operatorname{card}(\mathbb{R}^m) \leq \operatorname{card}\mathcal{E} \leq \operatorname{card}(\mathbb{R}^m \times \mathbb{R}_+) = \mathfrak{c}^{m+1} = \mathfrak{c}.$$

(2) We define $P = \{x \in \Omega, \operatorname{card}\mathcal{F}_x = \mathfrak{c}\}$. We note that

$$\operatorname{card}\mathcal{E} \leq \operatorname{card}\mathcal{F}_1 \leq \operatorname{card}\mathcal{E} + \operatorname{card}(\mathcal{E}_c) = 2\operatorname{card}\mathcal{E} = \operatorname{card}\mathcal{E}$$

since $\operatorname{card}\mathcal{E} = \mathfrak{c}$ is infinite[9] and this implies that $1 \in P$. Let $x \in \Omega$; suppose that $y \in P$ for all $y < x$. Then if x has an immediate predecessor y,

$$\operatorname{card}\mathcal{F}_y = \mathfrak{c} \leq \operatorname{card}\mathcal{F}_x \leq \operatorname{card}\big((\mathcal{F}_y)_\sigma\big) + \operatorname{card}\big(((\mathcal{F}_y)_\sigma)_c\big).$$

Noting that from (10.1.5) and Section 10.1, we have

$$\mathfrak{c} = \operatorname{card}\mathcal{F}_y \leq \operatorname{card}\big((\mathcal{F}_y)_\sigma\big) \leq \operatorname{card}\big(\mathcal{F}_y^{\mathbb{N}}\big) = \mathfrak{c}^{\aleph_0} = 2^{\aleph_0^2} = 2^{\aleph_0} = \mathfrak{c},$$

we obtain $\mathfrak{c} = \operatorname{card}\mathcal{F}_x$. If x is a limit ordinal, then $\mathcal{F}_x$ is a countable union $((\to, x)$ is countable) of sets with cardinal $\mathfrak{c}$, so that

$$\mathfrak{c} \leq \operatorname{card}\mathcal{F}_x \leq \aleph_0\mathfrak{c} \leq \mathfrak{c}^2 = \mathfrak{c}.$$

In all cases $x \in P$. By a transfinite induction (see Theorem 10.1.19), we get $P = \Omega$.

(3) From Exercise 5.10.7, we know that

$$\mathcal{B} = \cup_{x \in \Omega}\mathcal{F}_x$$

and thus we can conclude $\mathfrak{c} \leq \operatorname{card}\mathcal{B} \leq \mathfrak{c}\operatorname{card}\Omega \leq \mathfrak{c}^2 = \mathfrak{c}$.

Exercise 5.10.9 (Cardinality of the Lebesgue σ-algebra). *Prove that the cardinality of the Lebesgue σ-algebra $\mathcal{L}_m$ on $\mathbb{R}^m$ is equal to $2^\mathfrak{c}$, the cardinal of $\mathcal{P}(\mathbb{R})$.*

Answer. We have obviously $\operatorname{card}\mathcal{L}_m \leq \operatorname{card}\mathcal{P}(\mathbb{R}^m) = 2^{\operatorname{card}\mathbb{R}^m} = 2^{\mathfrak{c}^m} = 2^\mathfrak{c}$. On the other hand, the ternary Cantor set K_∞ (5.7.6) (see also Lemma 5.7.3) is a Borel set with Lebesgue measure 0 and same cardinality as $\mathbb{R}$. Thus $\mathcal{P}(K_\infty) \subset \mathcal{L}_1$ and thus $\otimes_{1 \leq j \leq m}\mathcal{P}(K_\infty) \subset \mathcal{L}_m$, implying

$$2^\mathfrak{c} = 2^{m\mathfrak{c}} = (2^\mathfrak{c})^m \leq \operatorname{card}(\mathcal{L}_m) \leq 2^\mathfrak{c}$$

and the result.

---

[9]Theorem 10.1.20 proves much more: for every infinite cardinal, we have $x^2 = x$, so that $x \leq 2x \leq x^2 = x$.

Exercise 5.10.10. *Let (X, d) be a separable metric space and let $\mathcal{B}_X$ its Borel σ-algebra. Prove that $\mathcal{B}_X$ is generated by a countable family of sets $\mathcal{E}$.*

Answer. Let $D = \{a_n\}_{n \in \mathbb{N}}$ be a countable dense subset of X. Let us consider the countable family $\mathcal{E} = \{B(a_n, \epsilon)\}_{n \in \mathbb{N}, \epsilon \in \mathbb{Q}_+^*}$ of open balls. Let U be an open subset of X. Then for each $x \in U$, there exists $r_x \in \mathbb{Q}_+^*$ such that the ball $B(x, 2r_x) \subset U$. Since D is dense in X, we may find some $n_x \in \mathbb{N}$ with $d(x, a_{n_x}) < r_x$. As a consequence

$$x \in B(a_{n_x}, r_x) \subset U,$$

where the inclusion follows from the triangle inequality:

$$d(y, a_{n_x}) < r_x \implies d(y, x) \le d(y, a_{n_x}) + d(a_{n_x}, x) < r_x + r_x$$
$$\implies y \in B(x, 2r_x) \subset U.$$

We get finally that $U = \cup_{x \in U} B(a_{n_x}, r_x)$. As a result, with $\mathcal{O}$ standing for the open subsets of X,

$$\mathcal{B}_X = \mathcal{M}(\mathcal{O}) \supset \mathcal{M}(\mathcal{E}) \supset \mathcal{O} \implies \mathcal{B}_X = \mathcal{M}(\mathcal{E}).$$

Exercise 5.10.11. *Let (X, d) be a separable infinite metric space and let $\mathcal{B}_X$ its Borel σ-algebra. Prove that $\operatorname{card} \mathcal{B}_X = \mathfrak{c}$.*

Answer. In the first place, since X is not finite, it contains a subset $\{x_n\}_{n \in \mathbb{N}}$ equipotent to $\mathbb{N}$. Each subset $X_A = \{x_n\}_{n \in A}$, with $A \subset \mathbb{N}$ belongs to the Borel σ-algebra $\mathcal{B}_X$ as a countable union of singletons (which are closed sets). We have thus an injection of $\mathcal{P}(\mathbb{N})$ into $\mathcal{B}_X$, proving that

$$(*) \qquad\qquad 2^{\aleph_0} = \mathfrak{c} \le \operatorname{card} \mathcal{B}_X.$$

Let $\mathcal{E}$ be a countable family of sets generating the σ-algebra $\mathcal{B}_X$, as in Exercise 5.10.10. Let Ω be the first uncountable ordinal. We define $\mathcal{F}_\alpha$ for each α in Ω as in Exercise 5.10.7. We claim that

$$(**) \qquad\qquad \text{for each } \alpha \in \Omega, \quad \operatorname{card} \mathcal{F}_\alpha \le \mathfrak{c}.$$

We note that $\operatorname{card} \mathcal{F}_1 \le \operatorname{card} \mathcal{E} + \operatorname{card}(\mathcal{E}_c) = 2 \operatorname{card} \mathcal{E} = \operatorname{card} \mathcal{E} = \aleph_0 \le \mathfrak{c}$. Let $\alpha \in \Omega$; suppose that $\operatorname{card} \mathcal{F}_\beta \le \mathfrak{c}$ for all $\beta < \alpha$. Then if α has an immediate predecessor β,

$$\operatorname{card} \mathcal{F}_\alpha \le \operatorname{card}(\mathcal{F}_\beta)_\sigma + \operatorname{card}(((\mathcal{F}_\beta)_\sigma)_c) \le \operatorname{card}(\mathbb{R}^{\mathbb{N}}) + \operatorname{card}(\mathbb{R}^{\mathbb{N}}) = 2^{\aleph_0^2} = 2^{\aleph_0} = \mathfrak{c}.$$

If α is a limit ordinal, then $\mathcal{F}_\alpha$ is a countable union $((\to, \alpha)$ is countable$)$ of sets with cardinal $\le \mathfrak{c}$, so that

$$\operatorname{card} \mathcal{F}_\alpha \le \aleph_0 \mathfrak{c} \le \mathfrak{c}^2 = \mathfrak{c}.$$

By a transfinite induction (see Theorem 10.1.19), we get that property $(**)$ holds. From Exercise 5.10.7, we know that

$$\mathcal{B}_X = \cup_{\alpha \in \Omega} \mathcal{F}_\alpha$$

and thus we can conclude $\operatorname{card} \mathcal{B}_X \le \mathfrak{c} \operatorname{card} \Omega \le \mathfrak{c}^2 = \mathfrak{c}$. The inequality $(*)$ gives the result.

Exercise 5.10.12. *Let Ω be an open subset of $\mathbb{R}^m$. Prove that for any $\varepsilon > 0$, there exists a pairwise disjoint covering $(K_{n,\varepsilon})_{n\in\mathbb{N}}$ of Ω with $\mathrm{diam}_2(K_{n,\varepsilon}) \leq \varepsilon$.*

Answer. Let $\varepsilon > 0$ be given. For each $x \in \Omega$, there exists $r(x) \in \mathbb{Q} \cap (0, \varepsilon/2)$ such that $\bar{B}(x, r(x)) \subset \Omega$. Defining $D = \mathbb{Q}^m \cap \Omega$, for each $x \in \Omega$, we can find $a_x \in D$ such that $d(x, a_x) \leq r(x)/2$: as a consequence $x \in \bar{B}(a_x, r(x)/2) \subset \Omega$, since

$$|y - a_x|_2 \leq r(x)/2 \implies |y - x|_2 \leq r(x) \implies y \in \bar{B}(x, r(x)) \subset \Omega.$$

We have thus $\Omega = \bigcup_{\substack{a \in D_0 \subset D \\ r \in \Delta \subset \mathbb{Q} \cap (0, \varepsilon/2)}} \bar{B}(a, r)$ so that

$$\Omega = \bigcup_{n\in\mathbb{N}} B_n, \quad B_n \text{ closed ball with diameter}_2 \leq \varepsilon.$$

We define now

$$K_0 = B_0, K_1 = B_1 \backslash B_0, \ \dots, K_{n+1} = B_{n+1} \backslash (B_0 \cup \cdots \cup B_n), \ \dots.$$

We have obviously

$$\mathrm{diam}_2(K_n) \leq \mathrm{diam}_2(B_n) \leq \varepsilon, \quad \bigcup_{n\in\mathbb{N}} K_n = \bigcup_{n\in\mathbb{N}} B_n = \Omega, \tag{5.10.5}$$

$$\text{and also for } 0 \leq n_1 < n_2, \quad K_{n_1} \cap K_{n_2} \subset B_{n_1} \cap B_{n_1}^c = \emptyset. \tag{5.10.6}$$

As a consequence, $\lambda_m(\Omega) = \sum_{n\in\mathbb{N}} \lambda_m(K_{n,\varepsilon})$.

Exercise 5.10.13. *Calculate the $n-1$-dimensional area of the unit sphere $\mathbb{S}^{n-1}$ of $\mathbb{R}^n$ by using the explicit change in polar coordinates.*

Answer. We have

$$|\mathbb{B}^{n+1}|_{n+1} = \int_{x\in\mathbb{R}^{n+1}, \|x\|_2 \leq 1} dx = \frac{1}{n+1} \int_{\mathbb{S}^n} d_{\mathbb{S}^n}\sigma,$$

so that using (5.4.3), we find

$$|\mathbb{B}^{n+1}|_{n+1} = \frac{1}{n+1} \int_0^\pi (\sin\phi)^{n-1} d\phi |\mathbb{S}^{n-1}|_{n-1}.$$

The computation of the Wallis integrals in Lemma 10.5.7 gives

$$\frac{|\mathbb{S}^n|_n}{n+1} = |\mathbb{B}^{n+1}|_{n+1} = \frac{|\mathbb{S}^{n-1}|_{n-1}}{n+1} \int_0^\pi (\sin\phi)^{n-1} d\phi = \frac{|\mathbb{S}^{n-1}|_{n-1}}{n+1} \frac{\sqrt{\pi}\,\Gamma(\frac{n}{2})}{\Gamma(\frac{n+1}{2})},$$

so that $|\mathbb{S}^n|_n = |\mathbb{S}^{n-1}|_{n-1} \frac{\sqrt{\pi}\,\Gamma(\frac{n}{2})}{\Gamma(\frac{n+1}{2})}$, $|\mathbb{S}^1|_1 = 2\pi$ and thus

$$|\mathbb{S}^n|_n = 2\pi \prod_{2\leq j\leq n} \frac{\sqrt{\pi}\,\Gamma(\frac{j}{2})}{\Gamma(\frac{j+1}{2})} = 2\pi^{1+\frac{n-1}{2}} \frac{\Gamma(1)}{\Gamma(\frac{n+1}{2})},$$

recovering (5.4.8): $|\mathbb{S}^{n-1}|_{n-1} = \dfrac{2\pi^{\frac{n}{2}}}{\Gamma(\frac{n}{2})}.$

Exercise 5.10.14. *Prove that* $\operatorname{card}\{\alpha \in \mathbb{N}^d, |\alpha| = l\} = C^{d-1}_{l+d-1}.$

Answer. We start with

(♮)
$$C^{d-1}_{l+d-1} = \sum_{0 \leq j \leq l} C^{d-2}_{j+d-2}$$

which is true for $l = 0$, and since $C^{d-1}_{l+d} = C^{d-1}_{l+d-1} + C^{d-2}_{l+d-1}$ is proven by induction on l: we have

$$C^{d-1}_{l+1+d-1} = \underbrace{\sum_{0 \leq j \leq l} C^{d-2}_{j+d-2}}_{\text{induction hypothesis}} + C^{d-2}_{l+d-1} = \sum_{0 \leq j \leq l+1} C^{d-2}_{j+d-2},$$

proving (♮). Now,

$$\operatorname{card}\{\alpha \in \mathbb{N}^d, |\alpha| = l\} = \sum_{0 \leq j \leq l} \operatorname{card}\{\beta \in \mathbb{N}^{d-1}, |\beta| = j\},$$

providing the proof by induction on d of the sought formula.

Exercise 5.10.15.

(1) *We consider a norm on* $\mathbb{R}^n$, *denoted by* $\|\cdot\|$. *Find an iff condition on the real numbers* α, β *so that*

$$\int_{\mathbb{R}^n} \frac{dx}{(1+\|x\|)^\beta} < +\infty, \qquad \int_{\|x\| \leq 1} \frac{dx}{\|x\|^\alpha} < +\infty.$$

(2) *We assume that* $n \geq 2$ *and we set, with* $\|\cdot\|$ *standing for a norm on* $\mathbb{R}^{n-1}$, *and for* $\lambda > 0$,

$$C_{1,\lambda} = \{(x_1, x') \in \mathbb{R} \times \mathbb{R}^{n-1}, \|x'\| \leq \lambda|x_1|\}.$$

Give an iff condition on the real numbers α, β *so that*

$$\int_{C_{1,\lambda}} \frac{dx}{(1+|x_1|)^\beta} < +\infty, \qquad \text{for all compact } K \int_{C_{1,\lambda} \cap K} \frac{dx}{|x_1|^\alpha} < +\infty.$$

Show that this provides a proof of (1) *without using a change of variables.*

Answer. (1) The answer is $\beta > n$ and $\alpha < n$. Since all the norms on $\mathbb{R}^n$ are equivalent (see, e.g., Exercise 1.9.8), we may assume that $\|\cdot\|$ is the Euclidean norm and use polar coordinates (see Section 5.4). We need only to check the 1D integrals

$$\int_0^{+\infty} r^{n-1}(1+r)^{-\beta} dr < +\infty \iff n-1-\beta < -1, \quad \text{i.e., } \beta > n,$$

and

$$\int_0^1 r^{n-1-\alpha}\,dr < +\infty \Longleftrightarrow n-1-\alpha > -1, \quad \text{i.e., } \alpha < n.$$

(2) Let us use Fubini's theorem for these positive measurable functions. With V_{n-1} equal to the $(n-1)$-dimensional Lebesgue measure of the unit ball of $\mathbb{R}^{n-1}$ for the norm $\|\ \|$, we find

$$\int_{C_{1,\lambda}} \frac{dx}{(1+|x_1|)^\beta} = \int_{\mathbb{R}} \left(\int_{|x'|\le\lambda|x_1|} dx' \right) \frac{dx_1}{(1+|x_1|)^\beta}$$

$$= V_{n-1} \int_{\mathbb{R}} \frac{\lambda^{n-1}|x_1|^{n-1}}{(1+|x_1|)^\beta} dx_1 < +\infty$$

if and only if $\beta > n$. Similarly if the condition in (2) holds for all compact sets K, it is satisfied in particular for $\{(x_1, x') \in \mathbb{R} \times \mathbb{R}^{n-1}, |x'| \le \lambda, |x_1| \le 1\}$ and we obtain

$$\int_0^1 \frac{\lambda^{n-1}|x_1|^{n-1}}{|x_1|^\alpha} dx_1 < +\infty \Longrightarrow \alpha < n.$$

Conversely, if that condition holds and if K is a compact set, K is included in a Euclidean ball with center 0 and finite radius on which the integral is finite following the same computation. We note then that

$$\mathbb{R}^n = \cup_{1\le j\le n}\left\{x \in \mathbb{R}^n, \max_{1\le k\le n} |x_k| = |x_j|\right\}$$

so that the integral over $\mathbb{R}^n$ is a finite sum of integrals on conical sets of type

$$\left\{(x_1, x') \in \mathbb{R} \times \mathbb{R}^{n-1}, \max_{2\le k\le n} |x_k| \le |x_1|\right\}$$

for which the calculation is done.

Exercise 5.10.16. *Let n be an integer ≥ 2. For $x \in \mathbb{R}^n$, we denote by $\|x\|$ the Euclidean norm of x.*

(1) *Calculate the volume of the ellipsoid*

$$\left\{x \in \mathbb{R}^n, \sum_{1\le j\le n} \frac{x_j^2}{a_j^2} \le 1\right\}$$

  (a_j are positive parameters).

(2) *Let A be a $n \times n$ real symmetric positive definite matrix. Calculate*

$$\int_{\mathbb{R}^n} e^{-\pi\langle Ax,x\rangle}\,dx.$$

(3) *Let B be a $n \times n$ invertible real symmetric matrix. Calculate*

$$\lim_{\epsilon \to 0_+} \int_{\mathbb{R}^n} e^{-\pi\epsilon\|x\|^2} e^{-i\pi\langle Bx,x\rangle} dx.$$

(4) *Let A, B be $n \times n$ real symmetric matrices such that $A \gg 0$ (i.e., $\langle Ax, x\rangle \geq d\|x\|^2$, $d > 0$). Calculate*

$$\lim_{\epsilon \to 0_+} \int_{\mathbb{R}^n} e^{-\pi\epsilon\|x\|^2} e^{-\pi\langle(A+iB)x,x\rangle} dx.$$

Answer. (1) Performing a linear change of variables, $y_j = a_j x_j$, we get that the volume is

$$|\mathbb{B}^n| \prod_{1 \leq j \leq n} a_j,$$

where $|\mathbb{B}^n|$ is given by (4.5.4) in Exercise 4.5.12.

(2) With the change of variables $x = A^{-1/2}y$, we find $(\det A)^{-1/2}$.

(3) The matrix B can be diagonalized in an orthonormal basis:

$$D = {}^t PBP, \quad D = \begin{pmatrix} d_1 & & 0 \\ & \ddots & \\ 0 & & d_n \end{pmatrix} \text{ diagonal, } {}^t PP = \mathrm{Id}.$$

The linear change of variables $x = Py$ gives

$$\int_{\mathbb{R}^n} e^{-\epsilon\pi\|x\|^2} e^{-i\pi\langle Bx,x\rangle} dx = \int_{\mathbb{R}^n} e^{-\epsilon\pi\|y\|^2} e^{-i\pi\langle Dy,y\rangle} dy = \prod_{1 \leq j \leq n} \int_{\mathbb{R}} e^{-\pi t^2(\epsilon+id_j)} dt.$$

Using question (2) in Exercise 4.5.14, we obtain

$$\prod_{1 \leq j \leq n} (\epsilon + id_j)^{-1/2} \xrightarrow[\epsilon \to 0_+]{} \prod_{1 \leq j \leq n} |d_j|^{-1/2} e^{-i\frac{\pi}{4}\,\mathrm{sign}(d_j)} = |\det B|^{-1/2} e^{-i\frac{\pi}{4}\mathrm{signature}(B)},$$

where

$$\mathrm{signature}(B)$$
$$= \text{number of positive eigenvalues of } B - \text{number of negative eigenvalues of } B. \quad (5.10.7)$$

(4) We have

$$I(\epsilon) = \int_{\mathbb{R}^n} e^{-\pi\epsilon\|x\|^2} e^{-\pi\langle(A+iB)x,x\rangle} dx$$

$$= (\det A)^{-1/2} \int_{\mathbb{R}^n} e^{-\pi\epsilon\|A^{-1/2}y\|^2} e^{-\pi\langle(\mathrm{Id}+iA^{-1/2}BA^{-1/2})y,y\rangle} dy.$$

The real symmetric matrix $A^{-1/2}BA^{-1/2}$ can be diagonalized (eigenvalues $\lambda_j, 1 \leq j \leq n$) in an orthonormal basis and using the previous calculation we obtain

$$\lim_{\epsilon \to 0_+} I(\epsilon) = (\det A)^{-1/2} \prod_{\substack{\mu_j \text{ eigenvalues of} \\ \mathrm{Id} + iA^{-1/2}BA^{-1/2}}} \mu_j^{-1/2}.$$

The μ_j are equal to $1 + i\lambda_j$ where the λ_j are the eigenvalues of the real symmetric matrix $A^{-1/2}BA^{-1/2}$. If $(\nu_j)_{1 \leq j \leq n}$ are the (positive) eigenvalues of the positive-definite matrix A, we have

$$A = PD_A{}^tP, \quad P \in O(n), \quad D_A = \mathrm{diagonal}(\nu_1, \ldots, \nu_n),$$
$$A^{-1/2}BA^{-1/2} = QD_B{}^tQ, \quad Q \in O(n), \quad D_B = \mathrm{diagonal}(\lambda_1, \ldots, \lambda_n),$$

and thus with $A^{1/2} = PD_A^{1/2}{}^tP$, we have $A + iB = A^{1/2}Q(\mathrm{Id} + iD_B){}^tQA^{1/2}$, so that

$$\det(A + iB) = \det A \prod_{1 \leq j \leq n} (1 + i\lambda_j) = \prod_{1 \leq j \leq n} \nu_j(1 + i\lambda_j).$$

As a result $\lim_{\epsilon \to 0_+} I(\epsilon)$ is equal to a particular determination of $(\det A)^{-1/2}$ given by

$$\prod_{1 \leq j \leq n} \nu_j^{-1/2}(1 + i\lambda_j)^{-1/2} = \prod_{1 \leq j \leq n} \nu_j^{-1/2} e^{-\frac{1}{2}\mathrm{Log}(1 + i\lambda_j)}.$$

The reader may consult the section entitled *The logarithm of a nonsingular symmetric matrix* on page 463 of the Appendix for a further discussion on this topic. The following lemma may be useful for future reference.

Lemma 5.10.17. *Let A, B be $n \times n$ real symmetric matrices such that A is positive definite. Then there exists an invertible $n \times n$ real matrix R such that*

$${}^tRAR = \mathrm{Id}, \quad and \quad {}^tRBR \text{ is a diagonal matrix.}$$

Proof. There exists $P \in O(n)$ such that ${}^tPAP = D_A$ where D_A is the diagonal matrix with diagonal $(\lambda_1, \ldots, \lambda_n)$ where the λ_j are the (positive) eigenvalues of A. We may consider the real symmetric matrix $A^{-1/2}BA^{-1/2}$, where $A^{-1/2} = PD_A^{-1/2}{}^tP$: there exists $Q \in O(n)$ such that ${}^tQA^{-1/2}BA^{-1/2}Q = D_B$ where D_B is the diagonal matrix with diagonal $(\mu_1, \ldots, \mu_n)$ where the μ_j are the eigenvalues of $A^{-1/2}BA^{-1/2}$. We have thus with the invertible matrix $R = A^{-1/2}Q$,

$${}^tQA^{-1/2}AA^{-1/2}Q = {}^tRAR = \mathrm{Id}, \quad {}^tQA^{-1/2}BA^{-1/2}Q = {}^tRBR = D_B,$$

so that the quadratic forms $x \mapsto \langle Ax, x \rangle$ and $x \mapsto \langle Bx, x \rangle$ can be simultaneously diagonalized. $\qquad\square$

Exercise 5.10.18. *Using a change of variables, calculate the integrals*

$$I = \iint_{x>0,\ y>0,\ x+y<a} \frac{3y}{\sqrt{1+(x+y)^3}}\,dxdy, \qquad a>0,$$

$$J = \iint_{x>0,\ y>0} |x^4 - y^4| e^{-(x+y)^2}\,dxdy.$$

Answer. With $H = \mathbf{1}_{\mathbb{R}_+}$, we have

$$I = \iint H(x)H(y-x)H(a-y)\frac{3(y-x)}{\sqrt{1+y^3}}\,dxdy$$

$$= \iint H(x)H(y-x)H(a-y)H(y)\frac{3(y-x)}{\sqrt{1+y^3}}\,dxdy,$$

so that

$$I = \int_0^a (1+y^3)^{-1/2}3\left(\int_0^y (y-x)dx\right)dy$$

$$= \int_0^a (1+y^3)^{-1/2}3\left(y^2 - \frac{y^2}{2}\right)dy = (1+a^3)^{1/2} - 1.$$

For the second integral we set $x = u - v, y = u + v$, so that

$$J = 2\iint H(u-v)H(u+v)2|v|2|u|2(u^2+v^2)e^{-4u^2}\,dudv$$

and thus

$$J = 2^4 \int_0^{+\infty}\left(\int H(u-|v|)|v|(u^2+v^2)dv\right)ue^{-4u^2}\,du$$

$$= 2^5 \int_0^{+\infty}\left(\int_0^u v(u^2+v^2)dv\right)ue^{-4u^2}\,du$$

$$= 3\times 2^3 \int_0^{+\infty} u^5 e^{-4u^2}\,du = 3\times 2^{-4}\Gamma(3) = \frac{3}{8}.$$

Chapter 6

Convolution

6.1 The Banach algebra $L^1(\mathbb{R}^n)$

Let $u, v \in C_c(\mathbb{R}^n)$. For all $x \in \mathbb{R}^n$, the mapping $y \mapsto u(x-y)v(y)$ is continuous with compact support $\subset \operatorname{supp} v$. We may thus consider

$$(u * v)(x) = \int_{\mathbb{R}^n} u(x-y)v(y)dy. \tag{6.1.1}$$

We shall say that $u * v$ is the convolution of u with v. For a given x, the change of variables $y' = x - y$ shows that $u * v = v * u$. Theorem 3.3.1 implies readily that $u * v$ is continuous and moreover if $x \notin \operatorname{supp} u + \operatorname{supp} v$, then for all $y \in \operatorname{supp} v$, $x - y \notin \operatorname{supp} u$ (otherwise $x = x - y + y \in \operatorname{supp} u + \operatorname{supp} v$) so that for all $y \in \mathbb{R}^n, u(x-y)v(y) = 0$. As a result, $(\operatorname{supp} u + \operatorname{supp} v)^c \subset \{u * v = 0\}$ and thus $\{u * v \neq 0\} \subset \operatorname{supp} u + \operatorname{supp} v$. Since $\operatorname{supp} u + \operatorname{supp} v$ is compact (as a sum of compact sets), we have

$$\operatorname{supp}(u * v) \subset \operatorname{supp} u + \operatorname{supp} v = \{x + y\}_{\substack{x \in \operatorname{supp} u \\ y \in \operatorname{supp} v}} \tag{6.1.2}$$

and $u * v \in C_c(\mathbb{R}^n)$. Moreover convolution is associative, since for $u, v, w \in C_c(\mathbb{R}^n)$, we have

$$((u*v)*w)(x) = \int_{\mathbb{R}^n}(u*v)(x-y)w(y)dy = \iint_{\mathbb{R}^n \times \mathbb{R}^n} u(x-y-z)v(z)w(y)dydz$$

$$= \iint_{\mathbb{R}^n \times \mathbb{R}^n} u(x-z)v(z-y)w(y)dydz = \int_{\mathbb{R}^n} u(x-z)(v*w)(z)dz = (u*(v*w))(x).$$

Proposition 6.1.1. *The binary operation of $C_c(\mathbb{R}^n)$ given by $(u, v) \mapsto u * v$ is associative, commutative and distributive with respect to addition and such that*

$$\|u * v\|_{L^1(\mathbb{R}^n)} \leq \|u\|_{L^1(\mathbb{R}^n)}\|v\|_{L^1(\mathbb{R}^n)}. \tag{6.1.3}$$

Proof. The estimate is the only point to be proven. For $u, v \in C_c(\mathbb{R}^n)$, we have

$$\|u * v\|_{L^1(\mathbb{R}^n)} \leq \int_{\mathbb{R}^n} \left| \int_{\mathbb{R}^n} u(x-y)v(y)dy \right| dx \leq \iint_{\mathbb{R}^n \times \mathbb{R}^n} |u(x-y)||v(y)|dydx$$

$$= \|u\|_{L^1(\mathbb{R}^n)} \int_{\mathbb{R}^n} |v(y)|dy = \|u\|_{L^1(\mathbb{R}^n)}\|v\|_{L^1(\mathbb{R}^n)}.$$

With $u_0(x) = \exp -\pi|x|^2$, we have $\|u_0\|_{L^1(\mathbb{R}^n)} = 1$ and

$$\|u_0 * u_0\|_{L^1(\mathbb{R}^n)} = \int |(u_0 * u_0)(x)|dx = \iint e^{-\pi|x-y|^2 - \pi|y|^2} dydx = 1,$$

proving that the estimate (6.1.3) is optimal. $\square$

Proposition 6.1.2. *Let $k \in \mathbb{N}$, $\varphi \in C_c^k(\mathbb{R}^n)$ and let $u \in L^1_{\mathrm{loc}}(\mathbb{R}^n)$ (i.e., $\forall K$ compact, $u\mathbf{1}_K \in L^1(\mathbb{R}^n)$). We define*

$$(\varphi * u)(x) = \int_{\mathbb{R}^n} \varphi(x-y)u(y)dy. \tag{6.1.4}$$

The function $\varphi * u$ belongs to $C^k(\mathbb{R}^n)$ and if $u \in L^1(\mathbb{R}^n)$, then $\varphi * u$ belongs to $L^1(\mathbb{R}^n)$ and is such that $\|\varphi * u\|_{L^1(\mathbb{R}^n)} \leq \|\varphi\|_{L^1(\mathbb{R}^n)}\|u\|_{L^1(\mathbb{R}^n)}$. Moreover, we have $\mathrm{supp}(\varphi * u) \subset \mathrm{supp}\,\varphi + \mathrm{supp}\,u$, where the support of u is defined by (2.8.12).

Proof. Let $x \in \mathbb{R}^n$ be given. The function $y \mapsto u(y)\varphi(x-y)$ is supported in $x - \mathrm{supp}\,\varphi = \{x - z\}_{z \in \mathrm{supp}\,\varphi}$, a compact set (since $\mathrm{supp}\,\varphi$ is compact). Since φ is bounded, the function $y \mapsto u(y)\varphi(x-y)$ belongs to $L^1_{\mathrm{comp}}(\mathbb{R}^n)$, so that (6.1.4) makes sense. Theorem 3.3.4 shows that $\varphi * u$ belongs to $C^k(\mathbb{R}^n)$: indeed, we have

$$|\varphi^{(k)}(x-y)u(y)| \leq |u(y)|\mathbf{1}_{\mathrm{supp}\,\varphi}(x-y)\sup|\varphi^{(k)}|$$

so that for K compact, since $K - \mathrm{supp}\,\varphi = \{x - z\}_{x \in K, z \in \mathrm{supp}\,\varphi}$ is also compact, we have

$$\sup_{x \in K} |\varphi^{(k)}(x-y)u(y)| \leq |u(y)|\mathbf{1}_{K - \mathrm{supp}\,\varphi}(y)\sup|\varphi^{(k)}| \in L^1(\mathbb{R}^n_y).$$

Whenever $u \in L^1(\mathbb{R}^n)$, the inequality on L^1-norms is proven as (6.1.3).

Let us prove now the inclusion of supports. Since $\mathrm{supp}\,\varphi$ is compact and $\mathrm{supp}\,u$ is closed, the set $\mathrm{supp}\,u + \mathrm{supp}\,\varphi$ is closed: if $\lim_k(y_k + z_k) = x$, with $y_k \in \mathrm{supp}\,u$, $z_k \in \mathrm{supp}\,\varphi$, extracting a subsequence, we get $\lim_l z_{k_l} = z \in \mathrm{supp}\,\varphi$ and $\lim_l(y_{k_l} + z_{k_l}) = x$, so that the sequence y_{k_l} is converging and since $\mathrm{supp}\,u$ is closed $\mathrm{supp}\,u \ni \lim_l y_{k_l} = x - z$, proving $x \in \mathrm{supp}\,u + \mathrm{supp}\,\varphi$. We consider now the open set $V_0 = (\mathrm{supp}\,u + \mathrm{supp}\,\varphi)^c$. For all $y \in \mathbb{R}^n$, we have

$$V_0 - y \subset (\mathrm{supp}\,\varphi)^c \quad \text{or} \quad y \notin \mathrm{supp}\,u, \tag{6.1.5}$$

otherwise, we could find y_0 such that $V_0 - y_0 \cap (\operatorname{supp}\varphi) \neq \emptyset$ and $y_0 \in \operatorname{supp} u$. This would imply the existence of $x \in V_0$ such that $x - y_0 \in \operatorname{supp}\varphi$ and thus

$$V_0 \ni x = x - y_0 + y_0 \in \operatorname{supp}\varphi + \operatorname{supp} u = V_0^c,$$

which is impossible. As a result (6.1.5) implies that for $x \in V_0$, and $y \in \mathbb{R}^n$, we have $\varphi(x - y) = 0$ or $y \notin \operatorname{supp} u$. Since the domain of integration in (6.1.4) is $\operatorname{supp} u$, this implies $(\varphi * u)(x) = 0$ and $(\operatorname{supp} u + \operatorname{supp}\varphi)^c \subset \left(\operatorname{supp}(\varphi * u)\right)^c$, which is the sought result. $\qquad\square$

Proposition 6.1.3. *Let Ω be an open set of $\mathbb{R}^n$, let $u \in L^1_{\mathrm{loc}}(\Omega)$ and let V be open $\subset \Omega$. Then*

$$u\big|_V = 0 \iff \forall \varphi \in C_c(V), \int u(x)\varphi(x)dx = 0.$$

N.B. This result is important for distribution theory: a function in $L^1_{\mathrm{loc}}(\Omega)$ can be viewed as a Radon measure on Ω, i.e., a continuous linear form on $C_c(\Omega)$. For $u \in L^1_{\mathrm{loc}}(\Omega)$, we define the linear form l_u,

$$C_c(\Omega) \ni \varphi \mapsto l_u(\varphi) = \int_\Omega \varphi(x)u(x)dx,$$

which is continuous since

$$\left| \int_\Omega \varphi(x)u(x)dx \right| \leq \sup |\varphi(x)| \int_{\operatorname{supp}\varphi} |u(x)|dx.$$

This proposition proves that the mapping $u \mapsto l_u$ is injective.

Proof of the proposition. The condition is obviously necessary. Let us prove that it is sufficient. Let K be a compact set included in V and let $\chi_K \in C_c(V; [0, 1])$, $\chi_K = 1$ on K. With

$$\rho \in C_c^\infty(\mathbb{R}^n; \mathbb{R}_+), \int \rho(x)dx = 1, \ \operatorname{supp}\rho = \{\|x\| \leq 1\}, \ \epsilon > 0, \ \rho_\epsilon(\cdot) = \rho(\cdot/\epsilon)\epsilon^{-n},$$

we obtain $\quad (\rho_\epsilon * \chi_K u)(x) = \int u(y) \overbrace{\chi_K(y)\rho_\epsilon(x - y)}^{\in C_c(V)} dy = 0.$

As a consequence, we have

$$\|\chi_K u\|_{L^1(\mathbb{R}^n)} \leq \|\chi_K u - \varphi\|_{L^1(\mathbb{R}^n)} + \|\varphi - \varphi * \rho_\epsilon\|_{L^1(\mathbb{R}^n)} + \|\varphi * \rho_\epsilon - \chi_K u * \rho_\epsilon\|_{L^1(\mathbb{R}^n)}$$
$$\leq 2\|\chi_K u - \varphi\|_{L^1(\mathbb{R}^n)} + \|\varphi - \varphi * \rho_\epsilon\|_{L^1(\mathbb{R}^n)}. \tag{6.1.6}$$

Lemma 6.1.4. *Let $\varphi \in C_c^k(\mathbb{R}^n)$. Then $\varphi * \rho_\epsilon \in C_c^\infty(\mathbb{R}^n)$ and $\varphi * \rho_\epsilon \to \varphi$ in $C_c^k(\mathbb{R}^n)$ when ϵ goes to 0.*

Proof of the lemma. We have indeed $(\varphi * \rho_\epsilon)(x) = \int \varphi(x - \epsilon y)\rho(y)dy$, so that

$$|(\varphi * \rho_\epsilon)(x) - \varphi(x)| \leq \int \rho(y)|\varphi(x - \epsilon y) - \varphi(x)|dy \leq \sup_{|x_1-x_2|\leq\epsilon} |\varphi(x_1) - \varphi(x_2)|,$$

which goes to 0 with ϵ. Similar estimates hold for derivatives of order $\leq k$, and moreover we have supp $(\varphi * \rho_\epsilon) \subset \text{supp}\,\varphi + \epsilon\mathbb{B}^n \subset \text{supp}\,\varphi + \epsilon_0\mathbb{B}^n$ for $\epsilon \leq \epsilon_0$, yielding the lemma. $\qquad\square$

We go on with the proof of Proposition 6.1.3. From (6.1.6) and Lemma 6.1.4, we obtain

$$\|\chi_K u\|_{L^1(\mathbb{R}^n)} \leq 2 \inf_{\varphi \in C_c(V)} \|\chi_K u - \varphi\|_{L^1(\mathbb{R}^n)} = 0,$$

since $\chi_K u \in L^1(V)$. Thus we have $\chi_K u = 0$ for all compact sets $K \subset V$, and since $\chi_K = 1$ on K, and V is a countable union of compact sets, we find that $u = 0$ a.e. on V. $\qquad\square$

Theorem 6.1.5. *There exists a unique bilinear mapping*

$$\begin{array}{ccc} L^1(\mathbb{R}^n) \times L^1(\mathbb{R}^n) & \to & L^1(\mathbb{R}^n) \\ (u, v) & \mapsto & u * v \end{array}$$

such that if $u, v \in C_c(\mathbb{R}^n)$, $u * v$ is the convolution of u and v and

$$\|u * v\|_{L^1(\mathbb{R}^n)} \leq \|u\|_{L^1(\mathbb{R}^n)}\|v\|_{L^1(\mathbb{R}^n)}.$$

The space $L^1(\mathbb{R}^n)$ is a commutative Banach algebra[1] for addition and convolution.

Proof. Uniqueness: if $\star$ is another mapping with the same properties, $u, v \in L^1(\mathbb{R}^n)$, $\varphi, \psi \in C_c(\mathbb{R}^n)$,

$$u \star v - u * v$$
$$= (u - \varphi) \star v + \varphi \star (v - \psi) + \varphi \star \psi - (u - \varphi) * v - \varphi * (v - \psi) - \varphi * \psi,$$

using $\varphi * \psi = \varphi \star \psi$, and with $L^1(\mathbb{R}^n)$ norms,

$$\|u \star v - u * v\| \leq 2\|u - \varphi\|\|v\| + 2\|v - \psi\|\|\varphi\|.$$

The density of $C_c(\mathbb{R}^n)$ in $L^1(\mathbb{R}^n)$ and the above inequality entail $u * v = u \star v$. To prove existence, we consider sequences $(\varphi_k), (\psi_k)$ in $C_c(\mathbb{R}^n)$, converging in $L^1(\mathbb{R}^n)$: it is easily proven that $\varphi_k * \psi_k$ are Cauchy sequences since (with $L^1(\mathbb{R}^n)$ norms),

$$\|\varphi_{k+l} * \psi_{k+l} - \varphi_k * \psi_k\| \leq \|\varphi_{k+l} - \varphi_k\|\|\psi_{k+l}\| + \|\psi_{k+l} - \psi_k\|\|\varphi_k\|.$$

Moreover, using the same inequality, we prove that the limit does not depend on the choice of the sequences φ_k, ψ_κ but only on their limits. $\qquad\square$

---

[1] A complex Banach space B equipped with a multiplication $*$ which is associative, distributive with respect to the addition, such that for $\lambda \in \mathbb{C}$ and $x, y \in B$, $(\lambda x) * y = \lambda(x * y) = x * (\lambda y)$ and so that $\|x * y\| \leq \|x\|\|y\|$ is called a Banach algebra. When the multiplication is commutative the Banach algebra is said to be commutative. When the multiplication has a unit element, the Banach algebra is said to be unital.

Proposition 6.1.6. *Let $u, v \in L^1(\mathbb{R}^n)$. Then for almost all x,*

$$\int |u(x-y)v(y)|\,dy \; < +\infty.$$

Defining $h(x) = \int u(x-y)v(y)\,dy$, we have $h \in L^1(\mathbb{R}^n)$,

$$\|h\|_{L^1(\mathbb{R}^n)} \leq \|u\|_{L^1(\mathbb{R}^n)}\|v\|_{L^1(\mathbb{R}^n)} \quad \text{and } h = u * v .$$

Proof. We consider the measurable function F on $\mathbb{R}^{2n}$, given by $F(x,y) = u(x-y)v(y)$. We have

$$\int\left(\int |F(x,y)|\,dx\right)dy = \int\left(\int |u(x-y)|\,dx\right)|v(y)|\,dy$$

$$= \|u\|_{L^1(\mathbb{R}^n)}\|v\|_{L^1(\mathbb{R}^n)} < +\infty.$$

As a result, $F \in L^1(\mathbb{R}^{2n})$ and Fubini's theorem implies that

$$h(x) = \int F(x,y)\,dy$$

is an L^1 function of x. We have also proven that $\|h\|_{L^1(\mathbb{R}^n)} \leq \|u\|_{L^1(\mathbb{R}^n)}\|v\|_{L^1(\mathbb{R}^n)}$. Since for $u, v \in C_c(\mathbb{R}^n)$, we have $h = u*v$, Theorem 6.1.5 yields the conclusion. $\square$

Lemma 6.1.7. *The Banach algebra $L^1(\mathbb{R}^n)$ is not unital.*

Proof. Let us assume that $L^1(\mathbb{R}^n)$ has a unit ν. We would have for all $x \in \mathbb{R}^n$, $e^{-\pi|x|^2} = \int e^{-\pi|x-y|^2}\nu(y)\,dy$ and thus for all $\xi \in \mathbb{R}^n$,

$$(\dagger) \qquad \int e^{-\pi|x|^2}e^{-2i\pi x\cdot\xi}\,dx = \int e^{-\pi|x|^2}e^{-2i\pi x\cdot\xi}\,dx \int e^{-2i\pi y\cdot\xi}\nu(y)\,dy.$$

Claim. For $\tau \in \mathbb{R}$,

$$\int_{\mathbb{R}} e^{-\pi t^2}e^{-2i\pi t\tau}\,dt = e^{-\pi\tau^2}. \tag{6.1.7}$$

To prove this claim, we note that

$$F(\tau) = \int_{\mathbb{R}} e^{-\pi t^2}e^{-2i\pi t\tau}e^{\pi\tau^2}\,dt = \int_{\mathbb{R}} e^{-\pi(t+i\tau)^2}\,dt,$$

so that $F'(\tau) = \int_{\mathbb{R}} \frac{d}{i\,dt}\left(e^{-\pi(t+i\tau)^2}\right)dt = 0$ and $F(\tau) = F(0) = 1$, proving the Claim. Applying this to $(\dagger)$, we get $e^{-\pi|\xi|^2} = e^{-\pi|\xi|^2}\int e^{-2i\pi y\cdot\xi}\nu(y)\,dy$. Thanks to the Riemann–Lebesgue Lemma 3.4.5, $\xi \mapsto \int e^{-2i\pi y\cdot\xi}\nu(y)\,dy$ is a continuous function with limit 0 at infinity, so we cannot have $\int e^{-2i\pi y\cdot\xi}\nu(y)\,dy = 1$ which is a consequence of the previous equality. $\square$

6.2 L^p Estimates for convolution, Young's inequality

Lemma 6.2.1. *Let $(X, \mathcal{M}, \mu)$ be a measure space where μ is a σ-finite positive measure. Let $1 \le r \le \infty, 1/r + 1/r' = 1$. For $u \in L^r(\mu), w \in L^{r'}(\mu)$, the product uw belongs to $L^1(\mu)$. Moreover we have*

$$\|u\|_{L^r(\mu)} = \sup_{\|w\|_{L^{r'}(\mu)}=1} |\langle u, w\rangle|, \quad \text{with} \quad \langle u, w\rangle = \int_X u\bar{w}\,d\mu.$$

Proof. The first statement follows from Hölder's inequality (Theorem 3.1.6). Also that inequality implies for $\|w\|_{L^{r'}} = 1$ that

$$\left|\int_X u\bar{w}\,d\mu\right| \le \|u\|_{L^r(\mu)} \implies \|u\|_{L^r(\mu)} \ge \sup_{\|w\|_{L^{r'}(\mu)}=1} \left|\int_X u\bar{w}\,d\mu\right|.$$

We assume first that $1 < r < +\infty$. Taking $w = \alpha|u|^{r-1}$, with $u = \alpha|u|, |\alpha| \equiv 1$ (we define $\alpha = u/|u|$ on $\{u \ne 0\}$, $\alpha = 1$ on $\{u = 0\}$: α is easily seen to be a measurable function), we find for $u \ne 0$ in L^r,

$$\|w\|_{L^{r'}}^{r'} = \int_X |u|^{(r-1)r'=r}\,d\mu = \|u\|_{L^r}^r > 0,$$

and $\int_X u\bar{w} = \int_X u\bar{\alpha}|u|^{r-1} = \int_X |u|\alpha\bar{\alpha}|u|^{r-1} = \|u\|_{L^r}^r$. We obtain thus

$$\langle u, w/\|w\|_{L^{r'}}\rangle = \|u\|_{L^r}^{r-\frac{r}{r'}=r(1-\frac{1}{r'})=1},$$

proving the result.

We assume now $r = 1$. We take $w = \mathbf{1}_{u \ne 0}\frac{u}{|u|}$, so that we find for $u \ne 0$ in L^1,

$$\|w\|_{L^\infty} = 1, \quad \int_X u\bar{w}\,d\mu = \int |u|\,d\mu = \|u\|_{L^1}, \quad \text{proving the result in that case.}$$

We assume $r = +\infty, \mu(X) < +\infty$. Let $u \in L^\infty(\mu), u \ne 0$, and let $\epsilon > 0$: then we have

$$+\infty > \mu\Big(\underbrace{\{x \in X, |u(x)| \ge \|u\|_{L^\infty(\mu)} - \epsilon\}}_{G_\epsilon}\Big) > 0.$$

We define for $\epsilon \in (0, \|u\|_{L^\infty(\mu)})$, $w = \frac{\bar{u}\mathbf{1}_{G_\epsilon}}{|u|\mu(G_\epsilon)}$, so that $\|w\|_{L^1(\mu)} = 1$. We have also

$$\langle u, w\rangle = \int_X |u|\frac{\mathbf{1}_{G_\epsilon}}{\mu(G_\epsilon)}\,d\mu \ge \|u\|_{L^\infty(\mu)} - \epsilon,$$

so that $\sup_{\|w\|_{L^1}=1} |\langle u, w\rangle| \ge \|u\|_{L^\infty(\mu)} - \epsilon$. Since the latter is true for all $\epsilon > 0$, this gives the result.

We assume $r = +\infty$, μ σ-finite. Let $X = \cup_\mathbb{N} X_N, \mu(X_N) < +\infty$. We may assume that the sequence $(X_N)_{N \in \mathbb{N}}$ is increasing. Let $u \in L^\infty(\mu), u \neq 0$. We define for $\epsilon \in (0, \|u\|_{L^\infty(\mu)})$,

$$G_{\epsilon,N} = \{x \in X_N, |u(x)| \geq \|u\|_{L^\infty(\mu)} - \epsilon\}.$$

Since $G_\epsilon = \cup_{N \in \mathbb{N}} G_{\epsilon,N} = \{x \in X, |u(x)| \geq \|u\|_{L^\infty(\mu)} - \epsilon\}$ which has a positive measure, Proposition 1.4.4(2) implies

$$\lim_N \mu(G_{\epsilon,N}) = \mu(G_\epsilon) > 0 \implies \exists N_\epsilon, \forall N \geq N_\epsilon, \mu(G_{\epsilon,N}) > 0.$$

We define $w = \dfrac{\bar{u} 1_{G_{\epsilon,N_\epsilon}}}{|u|\mu(G_{\epsilon,N_\epsilon})}$, so that $\|w\|_{L^1(\mu)} = 1$, and we have

$$\langle u, w \rangle = \int_X |u| \frac{1_{G_{\epsilon,N_\epsilon}}}{\mu(G_{\epsilon,N_\epsilon})} d\mu \geq \|u\|_{L^\infty(\mu)} - \epsilon,$$

proving the result in that case as well. The proof of the lemma is complete. □

Theorem 6.2.2 (Young's inequality). *Let $p, q, r \in [1, +\infty]$ such that*

$$1 - \frac{1}{r} = 1 - \frac{1}{p} + 1 - \frac{1}{q}. \tag{6.2.1}$$

Then for $u, v \in C_c(\mathbb{R}^n)$, we have

$$\|u * v\|_{L^r(\mathbb{R}^n)} \leq \|u\|_{L^p(\mathbb{R}^n)} \|v\|_{L^q(\mathbb{R}^n)}. \tag{6.2.2}$$

Moreover the bilinear mapping $C_c(\mathbb{R}^n)^2 \ni (u, v) \mapsto u * v \in L^r(\mathbb{R}^n)$ can be extended to a bilinear mapping from $L^p(\mathbb{R}^n) \times L^q(\mathbb{R}^n)$ into $L^r(\mathbb{R}^n)$ satisfying (6.2.2).

Proof. (1) *We note first that if $r = 1$, then $p = q = 1$ and the inequality is already* proven as well as the unique extension property.
(2) *Moreover if $r = +\infty$, then $1/p + 1/q = 1$, the requested inequality is*

$$\|u * v\|_{L^\infty(\mathbb{R}^n)} \leq \|u\|_{L^p(\mathbb{R}^n)} \|v\|_{L^q(\mathbb{R}^n)},$$

which follows immediately from Hölder's inequality (Theorem 3.1.6). The extension property holds obviously for $1 \leq p, q < +\infty$. If $p = +\infty = r$, then $q = 1$ and

$$(u * v)(x) = \int u(x - y)v(y)dy,$$

and $(u, v) \mapsto u * v$ is a bilinear continuous mapping from $L^\infty \times L^1$ into L^∞ satisfying (6.2.2).
(3) *We may thus assume that $r \in]1, +\infty[$. If $p = +\infty$ (resp. $q = +\infty$), we have* $1 + 1/r = 1/q$ (resp. $1 + 1/r = 1/p$), so that $r = +\infty$, a case now excluded. If

$p = 1$ we have $q = r$; if $q = r = 1$, the inequality is proven. We thus may assume that $1 \leq p < +\infty$, $1 < q, r < +\infty$. Let $w \in C_c(\mathbb{R}^n)$. We consider

$$(u * v * w)(0) = \int (u * v)(y)w(-y)dy = \iint u(y - x)v(x)w(-y)dydx,$$

we define

$$t = \frac{1}{p}, \quad s = \frac{1}{q}, \quad \sigma = 1 - \frac{1}{r}, \quad u_0 = |u|^p, v_0 = |v|^q, w_0 = |w|^{1/\sigma},$$

and we find

$$(\sharp) \qquad\qquad |(u * v * w)(0)| \leq \iint u_0^t(y - x)v_0^s(x)w_0^\sigma(-y)dydx.$$

We note that

$$1 - t + 1 - s = \sigma, \text{ i.e., } 1 - t + 1 - s + 1 - \sigma = 1, \quad 1 - t, 1 - s, 1 - \sigma \geq 0.$$

Lemma 6.2.3. *Let u_0, v_0, w_0 be non-negative functions in $L^1(\mathbb{R}^n)$ with norm 1. Let $s, t, \sigma \in [0, 1]$ such that $1 - t + 1 - s + 1 - \sigma = 1$. Then*

$$\iint u_0^t(y - x)v_0^s(x)w_0^\sigma(-y)dydx \leq 1.$$

Proof of the lemma. We have for $u_0(y - x), v_0(x), w_0(-y)$ positive,

$$t \operatorname{Log} u_0(y - x) + s \operatorname{Log} v_0(x) + \sigma \operatorname{Log} w_0(-y)$$

$$= \left[(1 - t) \underbrace{\begin{pmatrix} 0 \\ 1 \\ 1 \end{pmatrix}}_{a_1} + (1 - s) \underbrace{\begin{pmatrix} 1 \\ 0 \\ 1 \end{pmatrix}}_{a_2} + (1 - \sigma) \underbrace{\begin{pmatrix} 1 \\ 1 \\ 0 \end{pmatrix}}_{a_3} \right] \cdot \underbrace{\begin{pmatrix} \operatorname{Log} u_0(y - x) \\ \operatorname{Log} v_0(x) \\ \operatorname{Log} w_0(-y) \end{pmatrix}}_{L}.$$

Consequently, we obtain, using the convexity of the exponential function,

$$u_0^t(y - x)v_0^s(x)w_0^\sigma(-y)$$
$$= \exp\left[(1 - t)(a_1 \cdot L) + (1 - s)(a_2 \cdot L) + (1 - \sigma)(a_3 \cdot L)\right]$$
$$\leq (1 - t) \exp(a_1 \cdot L) + (1 - s) \exp(a_2 \cdot L) + (1 - \sigma) \exp(a_3 \cdot L),$$

so that

$$\iint u_0^t(y - x)v_0^s(x)w_0^\sigma(-y)dydx$$

$$\leq \iint \Big\{ (1 - t)v_0(x)w_0(-y) + (1 - s)u_0(y - x)w_0(-y) \qquad\qquad (6.2.3)$$

$$+ (1 - \sigma)u_0(y - x)v_0(x) \Big\} dydx = 1,$$

concluding the proof of the lemma. $\qquad\qquad\qquad\qquad\qquad\qquad\qquad\qquad\square$

Going back to the proof of the theorem, we note that the previous lemma and ($\sharp$) imply

$$|(u * v * w)(0)| \leq \iint \Big\{ (1-t)v_0(x)w_0(-y) + (1-s)u_0(y-x)w_0(-y)$$
$$+ (1-\sigma)u_0(y-x)v_0(x) \Big\} dy dx. \tag{6.2.4}$$

We get thus with $1/r + 1/r' = 1$, $\check{w}(x) = w(-x)$, $\langle u, v \rangle = \int u\bar{w}$,

$$|\langle u * v, \check{w} \rangle| \leq (1-t)\|v\|_{L^q}^q \|w\|_{L^{r'}}^{r'} + (1-s)\|u\|_{L^p}^p \|w\|_{L^{r'}}^{r'} + (1-\sigma)\|u\|_{L^p}^q \|v\|_{L^q}^q.$$

Let us assume $\|u\|_{L^p} = \|v\|_{L^q} = \|w\|_{L^{r'}} = 1$. We have then $|\langle u * v, \check{w} \rangle| \leq 1$ so that by homogeneity,

$$|\langle u * v, w \rangle| \leq \|u\|_{L^p}\|v\|_{L^q}\|w\|_{L^{r'}}. \tag{6.2.5}$$

Since we have assumed that $r \in (1, +\infty]$, we know that $r' \in [1, +\infty)$ and $C_c(\mathbb{R}^n)$ is dense in $L^{r'}(\mathbb{R}^n)$ (Theorem 3.4.1). Inequality (6.2.5) implies for $u, v, w \in C_c(\mathbb{R}^n)$, $W \in L^{r'}(\mathbb{R}^n)$,

$$\left| \int \underbrace{(u * v)(x)}_{\substack{C_c(\mathbb{R}^n) \\ \subset L^r(\mathbb{R}^n)}} \underbrace{\overline{W(x)}}_{L^{r'}(\mathbb{R}^n)} dx \right| \leq |\langle u * v, W - w \rangle| + |\langle u * v, w \rangle|$$
$$\leq \|u * v\|_{L^r}\|W - w\|_{L^{r'}} + \|u\|_{L^p}\|v\|_{L^q}\|w\|_{L^{r'}}.$$

As a result for $u, v \in C_c(\mathbb{R}^n)$, $W \in L^{r'}(\mathbb{R}^n)$, and $\epsilon > 0$, there exists $w \in C_c(\mathbb{R}^n)$ such that $\|W - w\|_{L^{r'}} \leq \epsilon$ and thus

$$|\langle u * v, W \rangle| \leq \epsilon\|u * v\|_{L^r} + \|u\|_{L^p}\|v\|_{L^q}(\|W\|_{L^{r'}} + \epsilon),$$

which implies $|\langle u * v, W \rangle| \leq \|u\|_{L^p}\|v\|_{L^q}\|W\|_{L^{r'}}$ and from Lemma 6.2.1 this gives $\|u * v\|_{L^r} \leq \|u\|_{L^p}\|v\|_{L^q}$.

To prove that the mapping $(u, v) \mapsto u * v$ can be continuously extended from $C_c(\mathbb{R}^n)^2$ into $L^r(\mathbb{R}^n)$ to a continuous mapping from $L^p \times L^q$ into L^r, we may assume that $p, q \in [1, +\infty)$. For $(u, v) \in L^p \times L^q$ and (u_k, v_k) sequences in $C_c(\mathbb{R}^n)$ converging towards u, v respectively in L^p, L^q, we note that the sequence $(u_k * v_k)$ is a Cauchy sequence in L^r since

$$\|u_{k+l} * v_{k+l} - u_k * v_k\|_{L^r} = \|(u_{k+l} - u_k) * v_{k+l} + u_k * (v_{k+l} - v_k)\|_{L^r}$$
$$\leq \|u_{k+l} - u_k\|_{L^p}\|v_{k+l}\|_{L^q} + \|v_{k+l} - v_k\|_{L^q}\|u_k\|_{L^p},$$

and the numerical sequences $(\|v_k\|_{L^q})_k, (\|v_k\|_{L^q})_k$ are bounded. We may define $u * v$ for $(u, v) \in L^p \times L^q$ as the limit in L^r of $u_k * v_k$. That limit does not depend on the approximating sequences, thanks to the same inequality: with $\tilde{u}_k, \tilde{v}_k$ other approximating sequences, we have

$$u_k * v_k - \tilde{u}_k * \tilde{v}_k = (u_k - \tilde{u}_k) * v_k + \tilde{u}_k * (v_k - \tilde{v}_k),$$

and thus $\|u_k * v_k - \tilde{u}_k * \tilde{v}_k\|_{L^r} \leq \|u_k - \tilde{u}_k\|_{L^p}\|v_k\|_{L^q} + \|\tilde{u}_k\|_{L^p}\|v_k - \tilde{v}_k\|_{L^q}$, entailing that $\lim_k u_k * v_k = \lim_k \tilde{u}_k * \tilde{v}_k$ in L^r. $\qquad\square$

There is a more constructive approach to the definition of the convolution product between $L^p(\mathbb{R}^n)$ and $L^q(\mathbb{R}^n)$ for p, q, r satisfying (6.2.1). The case $r = +\infty$ is settled directly by Hölder's inequality. We assume in the sequel that $1 \leq r < +\infty$.

Let $u \in L^p(\mathbb{R}^n), v \in L^q(\mathbb{R}^n)$, both non-negative functions. Then the function $(x, y) \mapsto u(y - x)v(x)$ is measurable and Tonelli's theorem 4.2.5 implies that

$$(u * v)(y) = \int u(y - x)v(x)dx$$

is a measurable non-negative function of y. Moreover choosing $w(y) = \mathbf{1}_{\mathbb{B}^n}(y/k)$, inequalities (6.2.4), (6.2.5) entail that $\int_{|y| \leq k}(u * v)(y)dy$ is finite for all k. As a result the non-negative function $u * v$ is locally integrable (thus almost everywhere finite). We use now Lemma 6.2.1: for B with finite measure and $\lambda > 0$,

$$\left(\int_{B \cap \{y, (u*v)(y) \leq \lambda\}} \left((u * v)(y) \right)^r dy \right)^{1/r}$$
$$= \sup_{\substack{w \geq 0 \\ \|w\|_{L^{r'}} = 1}} \int_{B \cap \{y, (u*v)(y) \leq \lambda\}} (u * v)(y)w(y)dy,$$

and inequality (6.2.5) implies

$$\int_{B \cap \{y, (u*v)(y) \leq \lambda\}} \left((u * v)(y) \right)^r dy \leq \|u\|_{L^p(\mathbb{R}^n)}^r \|v\|_{L^q(\mathbb{R}^n)}^r,$$

which proves that for u, v non-negative respectively in $L^p(\mathbb{R}^n)$ and $L^q(\mathbb{R}^n)$ for p, q, r satisfying (6.2.1), we find that $u * v$ belongs to $L^r(\mathbb{R}^n)$ and (6.2.2) holds. Now if u, v are respectively in $L^p(\mathbb{R}^n)$ and $L^q(\mathbb{R}^n)$, we may write

$$u = (\operatorname{Re} u)_+ - (\operatorname{Re} u)_- + i(\operatorname{Im} u)_+ - i(\operatorname{Im} u)_-,$$

and define $u * v = (\operatorname{Re} u)_+ * (\operatorname{Re} v)_+ + \cdots$. The bilinearity is obvious as well as the continuity $L^p * L^q \subset L^r$. To obtain the inequality (6.2.2), we use again inequalities (6.2.4), (6.2.5). We sum-up our discussion.

Definition 6.2.4. Let $p, q, r \in [1, +\infty]$ satisfying (6.2.1). For $u \in L^p(\mathbb{R}^n)$ and $v \in L^q(\mathbb{R}^n)$, we define

$$(u * v)(y) = \int u(y - x)v(x)dx$$

which is a locally integrable function (thus a.e. finite).

Theorem 6.2.5. *Let $p, q, r \in [1, +\infty]$ satisfying (6.2.1). The mapping*

$$L^p(\mathbb{R}^n) \times L^q(\mathbb{R}^n) \ni (u, v) \mapsto u * v \in L^r(\mathbb{R}^n),$$

is continuous and (6.2.2) holds.

6.3 Weak L^p spaces

Definition 6.3.1. Let $p \in [1, +\infty)$. We define the *weak-$L^p(\mathbb{R}^n)$ space* $L^p_w(\mathbb{R}^n)$ as the set of measurable functions $u : \mathbb{R}^n \to \mathbb{C}$ such that

$$\sup_{t > 0} t^p \lambda_n \big(\{x \in \mathbb{R}^n, |u(x)| > t\}\big) = \Omega_p(u) < +\infty, \tag{6.3.1}$$

where λ_n is the Lebesgue measure on $\mathbb{R}^n$.

Remark 6.3.2. (1) We have $L^p(\mathbb{R}^n) \subset L^p_w(\mathbb{R}^n)$: let $u \in L^p(\mathbb{R}^n)$. We have for $t > 0$,

$$t^p \lambda_n(\{|u| > t\}) = \int_{|u| > t} t^p dx \leq \int_{|u| > t} |u(x)|^p dx \leq \|u\|^p_{L^p(\mathbb{R}^n)},$$

so that, with $\Omega_p(u)$ defined in (6.3.1), we have

$$\Omega_p(u) \leq \|u\|^p_{L^p(\mathbb{R}^n)}. \tag{6.3.2}$$

(2) For $x \in \mathbb{R}^n$, we define $v_p(x) = |x|^{-n/p}$ (a measurable function). For $R > 0$, we have

$$\int_{B(0,R)} v_p(x)^p dx = \int_{B(0,R)} |x|^{-n} dx \geq |\mathbb{S}^{n-1}| \int_0^R dr/r = +\infty,$$

so that v_p is not in $L^p_{\mathrm{loc}}(\mathbb{R}^n)$. On the other hand, we have for $t > 0$,

$$t^p \lambda_n\big(\{|x|^{-n/p} > t\}\big) = t^p t^{-\frac{p}{n}n} \lambda_n(\mathbb{B}^n) = \lambda_n(\mathbb{B}^n),$$

so that v_p belongs to $L^p_w(\mathbb{R}^n)$.

Lemma 6.3.3. *Let* $p \in [1, +\infty)$. *Then* $L^p_w(\mathbb{R}^n)$ *is a* $\mathbb{C}$-*vector space. For* $u, v \in L^p_w(\mathbb{R}^n), \alpha \in \mathbb{C}$, *we have*

$$\big(\Omega_p(\alpha u)\big)^{\frac{1}{p}} = |\alpha| \big(\Omega_p(u)\big)^{\frac{1}{p}}, \qquad \big(\Omega_p(u + v)\big)^{\frac{1}{p}} \leq 2^{\frac{1}{p}} \big(\Omega_p(u)^{\frac{1}{p}} + \Omega_p(v)^{\frac{1}{p}}\big).$$

Remark 6.3.4. The mapping $L^p_w(\mathbb{R}^n) \ni u \mapsto \big(\Omega_p(u)\big)^{\frac{1}{p}}$ is a *quasi-norm*: it satisfies the first two properties (separation and homogeneity) in (1.2.12), but fails to satisfy the triangle inequality, although a substitute is available with a constant $2^{1/p} > 1$. We shall see below (Lemma 6.3.5) that when $p \in (1, +\infty)$, we can find a true norm equivalent to this quasi-norm.

Proof of the lemma. Let α, β be non-zero complex numbers and let $u, v \in L^p_w$. Since for $t > 0$, $|\alpha u| \leq t/2$ and $|\beta v| \leq t/2$ imply $|\alpha u + \beta v| \leq t$, we have

$$\{|\alpha u + \beta v| > t\} \subset \{|\alpha u| > t/2\} \cup \{|\beta v| > t/2\},$$

and thus

$$t^p \lambda_n(\{|\alpha u + \beta v| > t\})$$
$$\leq (2|\alpha|)^p \left(\frac{t}{2|\alpha|}\right)^p \lambda_n(\{|\alpha u| > t/2\}) + (2|\beta|)^p \left(\frac{t}{2|\beta|}\right)^p \lambda_n(\{|\beta v| > t/2\})$$
$$\leq (2|\alpha|)^p \Omega_p(u) + (2|\beta|)^p \Omega_p(v),$$

so that $\Omega_p(\alpha u + \beta v) \leq (2|\alpha|)^p \Omega_p(u) + (2|\beta|)^p \Omega_p(v) < +\infty$, proving the vector space property. The first homogeneity equality in the lemma is obvious, let us prove the second one. We may of course assume that both quantities $\Omega_p(u), \Omega_p(v)$ are positive ($\Omega_p(u) = 0$ implies $u = 0$ a.e.). Let $\theta \in (0,1)$. Since for $t > 0$, $|u| \leq (1 - \theta)t$ and $|\beta v| \leq \theta t$ imply $|u + v| \leq t$, we have

$$\{|u + v| > t\} \subset \{|u| > t(1 - \theta)\} \cup \{|v| > t\theta\},$$

so that

$$t^p \lambda_n(\{|u + v| > t\})$$
$$\leq (1 - \theta)^{-p} t^p (1 - \theta)^p \lambda_n(\{|u| > t(1 - \theta)\}) + \theta^{-p} t^p \theta^p \lambda_n(\{|v| > t\theta\}) \quad (*)$$
$$\leq (1 - \theta)^{-p} \Omega_p(u) + \theta^{-p} \Omega_p(v).$$

We consider now the function $(0, 1) \ni \theta \mapsto (1 - \theta)^{-p} a + \theta^{-p} b = \phi_{a,b}(\theta)$, where a, b are positive parameters. We have

$$\phi'_{a,b}(\theta) = p(1 - \theta)^{-p-1} a - p\theta^{-p-1} b,$$

and the minimum of ϕ is attained at θ such that $(1 - \theta)^{-p-1} a = \theta^{-p-1} b$, i.e.,

$$\frac{\theta}{1 - \theta} = (b/a)^{\frac{1}{p+1}}, \quad \text{i.e.,} \quad \theta = \frac{(b/a)^{\frac{1}{p+1}}}{1 + (b/a)^{\frac{1}{p+1}}} = \frac{b^{\frac{1}{p+1}}}{a^{\frac{1}{p+1}} + b^{\frac{1}{p+1}}},$$

with $\phi_{a,b} = (1 - \theta)^{-p} a + \theta^{-p} b = (a^{\frac{1}{p+1}} + b^{\frac{1}{p+1}})^{p+1}$ at this point. We infer from $(*)$ that

$$\left(\Omega_p(u + v)\right)^{\frac{1}{p}} \leq \left(\Omega_p(u)^{\frac{1}{p+1}} + \Omega_p(v)^{\frac{1}{p+1}}\right)^{\frac{p+1}{p}} \leq 2^{\frac{1}{p}}\left(\Omega_p(u)^{\frac{1}{p}} + \Omega_p(v)^{\frac{1}{p}}\right),$$

where the last inequality comes from the sharp elementary[2]

$$(a^{\frac{1}{p+1}} + b^{\frac{1}{p+1}})^{\frac{p+1}{p}} \leq 2^{\frac{1}{p}}(a^{\frac{1}{p}} + b^{\frac{1}{p}}). \qquad \square$$

---

[2]We have from Hölder's inequality for a, b positive,

$$a^{\frac{1}{p+1}} + b^{\frac{1}{p+1}} \leq \left((a^{\frac{1}{p+1}})^{\frac{p+1}{p}} + (b^{\frac{1}{p+1}})^{\frac{p+1}{p}}\right)^{\frac{p}{p+1}} (1^{\frac{p+1}{1}} + 1^{\frac{p+1}{1}})^{\frac{1}{p+1}} = 2^{\frac{1}{p+1}}(a^{\frac{1}{p}} + b^{\frac{1}{p}})^{\frac{p}{p+1}}.$$

The constant $2^{\frac{1}{p+1}}$ is easily shown to be sharp by taking $a = b$.

Lemma 6.3.5. *Let $p \in (1, +\infty)$ and let p' be its conjugate exponent. For $u \in L^p_w(\mathbb{R}^n)$, we define*

$$N_p(u) = \sup_{\substack{A \ measurable \\ with \ finite \ positive \\ measure}} \lambda_n(A)^{-1/p'} \int_A |u(x)|dx. \tag{6.3.3}$$

Then N_p is a norm on $L^p_w(\mathbb{R}^n)$ which is equivalent to the quasi-norm $\Omega_p(\cdot)^{1/p}$.

Proof. Tonelli's Theorem 4.2.5 gives for a measurable subset A of $\mathbb{R}^n$,

$$\int_A |u(x)|dx = \iint \mathbf{1}_A(x) H(|u(x)| - t) H(t) dt dx, \quad \text{with } H = \mathbf{1}_{\mathbb{R}_+}.$$

As a result, for $T \geq 0$ and A measurable with finite measure, we have

$$\int_A |u(x)|dx = \int_0^{+\infty} \lambda_n\big(A \cap \{|u| > t\}\big) dt$$

$$= \int_0^T \lambda_n\big(A \cap \{|u| > t\}\big) dt + \int_T^{+\infty} \lambda_n\big(A \cap \{|u| > t\}\big) dt$$

$$\leq T\lambda_n(A) + \int_T^{+\infty} \lambda_n\big(\{|u| > t\}\big) dt.$$

$$\leq T\lambda_n(A) + \int_T^{+\infty} \Omega_p(u) t^{-p} dt = T\lambda_n(A) + \Omega_p(u)\frac{T^{1-p}}{p-1}.$$

We choose $T = \lambda_n(A)^{-1/p}\Omega_p(u)^{1/p}$ and we find

$$\int_A |u(x)|dx \leq \lambda_n(A)^{1/p'}\Omega_p(u)^{1/p} + \frac{1}{p-1}\lambda_n(A)^{-\frac{1}{p}+1}\Omega_p(u)^{1+\frac{1}{p}-1}$$

$$= \lambda_n(A)^{1/p'}\Omega_p(u)^{1/p}\frac{p}{p-1},$$

proving

$$N_p(u) \leq \frac{p}{p-1}\Omega_p(u)^{1/p}. \tag{6.3.4}$$

For $t > 0$, and X_k measurable with finite measure, we have

$$t^p\lambda_n\big(\{|u| > t\} \cap X_k\big) = t^p \int_{\{|u|>t\}\cap X_k} dx \leq t^{p-1}\int_{\{|u|>t\}\cap X_k} |u(x)|dx$$

$$\leq t^{p-1}N_p(u)\lambda_n\big(\{|u| > t\} \cap X_k\big)^{1/p'},$$

so that $t\lambda_n\big(\{|u| > t\} \cap X_k\big)^{1/p} \leq N_p(u)$. Since λ_n is σ-finite, this implies

$$\Omega_p(u) \leq N_p(u)^p. \tag{6.3.5}$$

We see now that N_p is finite ≥ 0 on L_w^p from (6.3.4). Moreover $N_p(u) = 0$ implies from (6.3.5) that $\lambda_n(\{|u| > t\}) = 0$ for all $t > 0$ and since

$$\{u \neq 0\} = \cup_{n \geq 1}\{|u| > 1/n\},$$

we find $u = 0$, a.e. Moreover, for $\alpha \in \mathbb{C}$ and $u \in L_w^p$, we have

$$N_p(\alpha u) = \sup_{\substack{A \text{ measurable} \\ \text{with finite measure} > 0}} \lambda_n(A)^{-1/p'} \int_A |\alpha u(x)|dx = |\alpha| N_p(u).$$

Eventually, for $u, v \in L_w^p$ and A measurable with finite measure, we have

$$\lambda_n(A)^{-1/p'} \int_A |u(x) + v(x)|dx$$

$$\leq \lambda_n(A)^{-1/p'} \int_A |u(x)|dx + \lambda_n(A)^{-1/p'} \int_A |v(x)|dx \leq N_p(u) + N_p(v),$$

which implies $N_p(u + v) \leq N_p(u) + N_p(v)$, proving that N_p is a norm on $L_w^p(\mathbb{R}^n)$ and concluding the proof of the lemma. $\qquad\square$

Proposition 6.3.6. *Let $p \in (1, +\infty)$. Then $L_w^p(\mathbb{R}^n)$ is a Banach space for the norm* (6.3.3).

Proof. Let us consider a Cauchy sequence $(u_k)_{k \in \mathbb{N}}$ in $L_w^p(\mathbb{R}^n)$: in particular for every measurable subset A with finite measure, we find that $(u_{k|A})_{k \in \mathbb{N}}$ is a Cauchy sequence in $L^1(A)$, thus convergent with limit v_A. Since the Lebesgue measure on $\mathbb{R}^n$ is σ-finite, we find a measurable function u such that for every A measurable with finite measure, $\lim_k \|u_k - u\|_{L^1(A)} = 0$. We check now for a measurable subset A with finite measure,

$$\lambda_n(A)^{-1/p'} \int_A |u_k(x) - u(x)|dx$$

$$\leq \lambda_n(A)^{-1/p'} \int_A |u_k(x) - u_l(x)|dx + \lambda_n(A)^{-1/p'} \int_A |u_l(x) - u(x)|dx$$

$$\leq N_p(u_k - u_l) + \lambda_n(A)^{-1/p'} \|u_l - u\|_{L^1(A)}.$$

Let $\epsilon > 0$ be given. There exists N_ϵ such that for $k, l \geq N_\epsilon$, we have $N_p(u_k - u_l) \leq \epsilon/2$. We know also that for $l \geq L_{\epsilon, A}$, we have $\lambda_n(A)^{-1/p'} \|u_l - u\|_{L^1(A)} \leq \epsilon/2$. We take $k \geq N_\epsilon$ and we choose $l = \max(N_\epsilon, L_{\epsilon, A})$: we find

$$\lambda_n(A)^{-1/p'} \int_A |u_k(x) - u(x)|dx \leq \epsilon.$$

As a result u belongs to $L_w^p(\mathbb{R}^n)$ and $N_p(u_k - u) \leq \epsilon$ for $k \geq N_\epsilon$, proving the completeness of $L_w^p(\mathbb{R}^n)$. $\qquad\square$

6.4   The Hardy–Littlewood–Sobolev inequality

We begin with a lemma, following [43].

Lemma 6.4.1. *Let $p, q, r > 1$ be real numbers such that*

$$1 - \frac{1}{p} + 1 - \frac{1}{q} = 1 - \frac{1}{r} = \frac{1}{r'}$$

and let f, g be non-negative measurable functions such that $\|f\|_{L^p(\mathbb{R}^n)} = 1 = \|g\|_{L^{r'}(\mathbb{R}^n)}$. Setting $\tau = n/q$, we define $T_\tau(f, g) = \iint f(x)|x - y|^{-\tau} g(y)dydx$ and we have

$$T_\tau(f, g) = \tau \int_{\mathbb{R}^3_+ \times \mathbb{R}^n \times \mathbb{R}^n} t_3^{-\tau-1} H(t_3 - |x - y|)$$

$$H(f(x) - t_1)H(g(y) - t_2)dt_1 dt_2 dt_3 dxdy. \quad (6.4.1)$$

Setting for $t_j \geq 0$,

$$u_1(t_1) = \int_{\mathbb{R}^n} H(f(x) - t_1)dx, \quad u_2(t_2) = \int_{\mathbb{R}^n} H(g(y) - t_2)dy, \quad u_3(t_3) = \beta_n t_3^n,$$

with $\beta_n = |\mathbb{B}^n|$ (see (4.5.4), (5.4.8)), and

$$m(t_1, t_2, t_3) = \max\big(u_1(t_1), u_2(t_2), u_3(t_3)\big),$$

we have

$$T_\tau(f, g) \leq \tau \int_{\mathbb{R}^3_+} t_3^{-\tau-1} \frac{u_1(t_1)u_2(t_2)u_3(t_3)}{m(t)} dt_1 dt_2 dt_3, \quad (6.4.2)$$

$$p \int_0^{+\infty} t_1^{p-1} u_1(t_1)dt_1 = r' \int_0^{+\infty} t_2^{r'-1} u_2(t_2)dt_2 = 1. \quad (6.4.3)$$

Proof. We have for $\tau > 0$,

$$\tau \int_0^{+\infty} t^{-\tau-1} H(t - |x|)dt = \tau \int_{|x|}^{+\infty} t^{-\tau-1}dt = [t^{-\tau}]_{t=+\infty}^{t=|x|} = |x|^{-\tau}$$

and thus

$$T_\tau(f, g) = \iint f(x)|x - y|^{-\tau} g(y)dydx$$

$$= \int_{\mathbb{R}^n \times \mathbb{R}^n \times \mathbb{R}_+} f(x)g(y)\tau t_3^{-\tau-1} H(t_3 - |x - y|)dxdydt_3$$

$$= \tau \int_{\mathbb{R}^3_+ \times \mathbb{R}^n \times \mathbb{R}^n} t_3^{-\tau-1} H(t_3 - |x - y|)H(f(x) - t_1)H(g(y) - t_2)dt_1 dt_2 dt_3 dxdy,$$

proving (6.4.1). We have thus

$$
T_\tau(f,g) \le \tau \int_{\substack{\mathbb{R}^3_+ \times \mathbb{R}^n \times \mathbb{R}^n \\ m(t)=u_3(t_3)}} t_3^{-\tau-1} H(f(x)-t_1)H(g(y)-t_2)dt_1 dt_2 dt_3 dx dy
$$

$$
+\tau \int_{\substack{\mathbb{R}^3_+ \times \mathbb{R}^n \times \mathbb{R}^n \\ m(t)=u_2(t_2)}} t_3^{-\tau-1} H(t_3-|x-y|)H(f(x)-t_1)dt_1 dt_2 dt_3 dx dy
$$

$$
+\tau \int_{\substack{\mathbb{R}^3_+ \times \mathbb{R}^n \times \mathbb{R}^n \\ m(t)=u_1(t_1)}} t_3^{-\tau-1} H(t_3-|x-y|)H(g(y)-t_2)dt_1 dt_2 dt_3 dx dy,
$$

so that

$$
T_\tau(f,g) \le \tau \int_{\mathbb{R}^3_+, m(t)=u_3(t_3)} t_3^{-\tau-1} u_1(t_1)u_2(t_2)dt
$$

$$
+\tau \int_{\mathbb{R}^3_+, m(t)=u_2(t_2)} t_3^{-\tau-1} \beta_n t_3^n u_1(t_1)dt
$$

$$
+\tau \int_{\mathbb{R}^3_+, m(t)=u_1(t_1)} t_3^{-\tau-1} \beta_n t_3^n u_2(t_2)dt
$$

$$
= \tau \int_{\mathbb{R}^3_+} t_3^{-\tau-1} \frac{u_1(t_1)u_2(t_2)u_3(t_3)}{m(t)} dt_1 dt_2 dt_3.
$$

Moreover, we have

$$
p \int_0^{+\infty} t_1^{p-1} u_1(t_1)dt_1 = \int_{\mathbb{R}^n} \int_0^{+\infty} p t_1^{p-1} H(f(x)-t_1)dx dt_1 = \int_{\mathbb{R}^n} f(x)^p dx = 1
$$

and

$$
r' \int_0^{+\infty} t_2^{r'-1} u_2(t_2)dt_2 = \int_{\mathbb{R}^n} \int_0^{+\infty} r' t_2^{r'-1} H(g(y)-t_2)dy dt_2 = \int_{\mathbb{R}^n} g(y)^{r'} dx = 1,
$$

completing the proof of the lemma. $\qquad\qquad\qquad\qquad\qquad\qquad\qquad\qquad\square$

Lemma 6.4.2. *Let $p,q,r,f,g,\tau,T_\tau,\beta_n,u_1,u_2$ as in the previous lemma. Then we have*

$$
T_\tau(f,g) \le \frac{n\beta_n^{\tau/n}}{n-\tau} \int_{\mathbb{R}^2_+} \min\left(u_1(t_1)^{1-\frac{\tau}{n}} u_2(t_2), u_1(t_1)u_2(t_2)^{1-\frac{\tau}{n}}\right)dt_1 dt_2. \qquad (6.4.4)
$$

Proof. For $t \in \mathbb{R}^3_+$, we set $V(t) = \frac{u_1(t_1)u_2(t_2)u_3(t_3)}{m(t)}$.
Let us assume that $u_1(t_1) \ge u_2(t_2)$. In that case we have

$$
\int_0^{+\infty} t_3^{-\tau-1} V(t_1,t_2,t_3)dt_3 = \int_0^{+\infty} t_3^{-\tau-1} \frac{u_1(t_1)u_2(t_2)u_3(t_3)}{\max(u_1(t_1),u_3(t_3))} dt_3
$$

$$
= u_1(t_1)u_2(t_2)\left(\int_{\mathbb{R}_+, \beta_n t_3^n \le u_1(t_1)} t_3^{-\tau-1+n} \beta_n dt_3 u_1(t_1)^{-1} + \int_{\mathbb{R}_+, \beta_n t_3^n > u_1(t_1)} t_3^{-\tau-1} dt_3 \right)
$$

$$= u_1(t_1)u_2(t_2)\beta_n\left(u_1(t_1)^{-1}\left[\frac{t_3^{n-\tau}}{n-\tau}\right]_{t_3=0}^{t_3=u_1(t_1)^{1/n}\beta_n^{-1/n}} + \beta_n^{-1}\left[\frac{t_3^{-\tau}}{\tau}\right]_{t_3=+\infty}^{t_3=u_1(t_1)^{1/n}\beta_n^{-1/n}}\right)$$

$$= u_1(t_1)u_2(t_2)\beta_n\left(u_1(t_1)^{-1+\frac{n-\tau}{n}}\frac{\beta_n^{-1+\frac{\tau}{n}}}{n-\tau} + \tau^{-1}\beta_n^{-1+\frac{\tau}{n}}u_1(t_1)^{-\tau/n}\right)$$

$$= u_1(t_1)^{1-\frac{\tau}{n}}u_2(t_2)\beta_n^{\tau/n}\frac{n}{\tau(n-\tau)}.$$

If we have instead $u_1(t_1) \le u_2(t_2)$, we find

$$\int_0^{+\infty} t_3^{-\tau-1}V(t_1,t_2,t_3)dt_3 = u_2(t_2)^{1-\frac{\tau}{n}}u_1(t_1)\beta_n^{\tau/n}\frac{n}{\tau(n-\tau)}.$$

From (6.4.2) and the previous estimates, we obtain

$$T_\tau(f,g) \le \frac{n\beta_n^{\tau/n}}{n-\tau}\int_{\mathbb{R}_+^2} \mathbf{1}\big(u_1(t_1) \ge u_2(t_2)\big)u_1(t_1)^{1-\frac{\tau}{n}}u_2(t_2)^{1-\frac{\tau}{n}}u_2(t_2)^{\frac{\tau}{n}}dt_1dt_2$$

$$+ \frac{n\beta_n^{\tau/n}}{n-\tau}\int_{\mathbb{R}_+^2} \mathbf{1}\big(u_1(t_1) \le u_2(t_2)\big)u_1(t_1)^{1-\frac{\tau}{n}}u_2(t_2)^{1-\frac{\tau}{n}}u_1(t_1)^{\frac{\tau}{n}}dt_1dt_2$$

$$= \frac{n\beta_n^{\tau/n}}{n-\tau}\int_{\mathbb{R}_+^2} u_1(t_1)^{1-\frac{\tau}{n}}u_2(t_2)^{1-\frac{\tau}{n}}\big(\min(u_1(t_1),u_2(t_2))\big)^{\tau/n}dt_1dt_2,$$

which is (6.4.4). $\qquad\square$

Lemma 6.4.3. *Let $p,q,r,f,g,\tau,T_\tau,\beta_n,u_1,u_2$ as in the previous lemmas. We define*

$$J = \int_{\mathbb{R}_+^2} \min\big(u_1(t_1)^{1-\frac{\tau}{n}}u_2(t_2),u_1(t_1)u_2(t_2)^{1-\frac{\tau}{n}}\big)dt_1dt_2. \qquad (6.4.5)$$

Then with

$$J_1 = \int_0^{+\infty}u_1(t_1)\int_0^{t_1^{p/r'}} u_2(t_2)^{1-\frac{\tau}{n}}dt_2dt_1, \quad J_2 = \int_0^{+\infty}u_2(t_2)\int_0^{t_2^{r'/p}} u_1(t_1)^{1-\frac{\tau}{n}}dt_1dt_2,$$

we have $J \le J_1 + J_2$. Moreover, we have

$$J_1 \le \frac{1}{pr'}\left(\frac{p'\tau}{n}\right)^{\tau/n}, \quad J_2 \le \frac{1}{pr'}\left(\frac{r\tau}{n}\right)^{\tau/n}.$$

Proof. We have

$$J \le \iint_{0\le t_1, 0\le t_2\le t_1^{p/r'}}(u_1(t_1)u_2(t_2))^{1-\frac{\tau}{n}}\min(u_1(t_1),u_2(t_2))^{\tau/n}dt_1dt_2$$

$$+ \iint_{0\le t_2, 0\le t_1\le t_2^{r'/p}}(u_1(t_1)u_2(t_2))^{1-\frac{\tau}{n}}\min(u_1(t_1),u_2(t_2))^{\tau/n}dt_1dt_2$$

and thus

$$J \le \int_0^{+\infty} u_1(t_1)\left(\int_0^{t_1^{p/r'}} u_2(t_2)^{1-\frac{\tau}{n}}\,dt_2\right)dt_1$$

$$+ \int_0^{+\infty} u_2(t_2)\left(\int_0^{t_2^{r'/p}} u_1(t_1)^{1-\frac{\tau}{n}}\,dt_1\right)dt_2.$$

From Hölder's inequality, since $1 - \frac{\tau}{n} = 1/q'$, we find, choosing $m = \frac{r'-1}{q'}$,

$$\int_0^{t_1^{p/r'}} u_2(t_2)^{1-\frac{\tau}{n}}\,dt_2 = \int_0^{t_1^{p/r'}} t_2^m u_2(t_2)^{1-\frac{\tau}{n}} t_2^{-m}\,dt_2$$

$$\le \left(\underbrace{\int_0^{t_1^{p/r'}} t_2^{mq'} u_2(t_2)\,dt_2}_{=1/r' \text{ from } (6.4.3)}\right)^{1/q'}\left(\int_0^{t_1^{p/r'}} t_2^{-mq}\,dt_2\right)^{1/q}.$$

We note also that

$$mq = \frac{r'-1}{q'}q < 1 \iff \frac{r'-1}{q'} < 1/q \iff r' < q' \text{ which holds since } \frac{1}{p'} + \frac{1}{q'} = \frac{1}{r'}.$$

As a result, we have

$$J_1 \le \int_0^{+\infty} u_1(t_1)\left(\frac{1}{r'}\right)^{1/q'}\left((t_1^{p/r'})^{1-mq}(1-mq)^{-1}\right)^{1/q}\,dt_1.$$

Since

$$\frac{p(1-mq)}{r'q} = \frac{p}{r'q}\left(1 - \frac{(r'-1)}{q'}q\right) = \frac{p}{r'}\left(1 - \frac{r'}{q'}\right) = p\left(\frac{1}{r'} - \frac{1}{q'}\right) = \frac{p}{p'} = p - 1,$$

we obtain, using (6.4.3),

$$J_1 \le \int_0^{+\infty} u_1(t_1)t_1^{p-1}\,dt_1\left(\frac{1}{r'}\right)^{1/q'}(1-mq)^{-1/q}$$

$$= \frac{1}{p}\left(\frac{1}{r'}\right)^{1/q'}(1-mq)^{-1/q} = \frac{1}{pr'}\left(\frac{1}{r'} - \frac{mq}{r'}\right)^{-1/q} = \frac{1}{pr'}\left(\frac{1}{r'} - \frac{q}{q'r}\right)^{-1/q}$$

$$= \frac{1}{pr'}\left(\frac{1}{r'} - \frac{(q-1)}{r}\right)^{-1/q} = \frac{1}{pr'}\left(1 - \frac{q}{r}\right)^{-1/q} = \frac{1}{pr'}\left(\frac{1}{q} - \frac{1}{r}\right)^{-1/q}q^{-1/q}$$

$$= \frac{1}{pr'}\left(\frac{1}{p'}\right)^{-1/q}q^{-1/q} = \frac{1}{pr'}\left(\frac{p'}{q}\right)^{1/q} = \frac{1}{pr'}\left(\frac{p'\tau}{n}\right)^{\tau/n}.$$

To estimate J_2 from above is analogous: we have, choosing $\mu = \frac{p-1}{q'}$,

$$\int_0^{t_2^{r'/p}} u_1(t_1)^{1-\frac{\tau}{n}}dt_1 = \int_0^{t_2^{r'/p}} t_1^{\mu} u_1(t_1)^{1-\frac{\tau}{n}} t_1^{-\mu} dt_1$$

$$\leq \left(\underbrace{\int_0^{t_2^{r'/p}} t_1^{\mu q'} u_1(t_1)dt_1}_{=1/p}\right)^{1/q'} \left(\int_0^{t_2^{r'/p}} t_1^{-\mu q}dt_1\right)^{1/q}.$$

We check $\mu q < 1$ by the same calculation, exchanging the roles of p and r': p' is replaced by r and pr' replaced by $r'p$ is unchanged. $\qquad\square$

Theorem 6.4.4 (Hardy–Littlewood–Sobolev inequality). *Let $p, q, r \in (1, +\infty)$ such that $\frac{1}{p'} + \frac{1}{q'} = \frac{1}{r'}$. There exists $C > 0$ such that, for all $F \in L^p(\mathbb{R}^n)$,*

$$\|(F * |\cdot|^{-n/q})\|_{L^r(\mathbb{R}^n)} \leq C\|F\|_{L^p(\mathbb{R}^n)}.$$

The constant C can be taken as $q'\beta_n^{1/q}\frac{1}{pr'}\left(\left(\frac{p'}{q}\right)^{1/q} + \left(\frac{r}{q}\right)^{1/q}\right).$

Proof. For $f = |F|/\|F\|_{L^p}, \|g\|_{L^{r'}} = 1$, we have proven from (6.4.4) and Lemma 6.4.3,

$$T_\tau(f,g) \leq \frac{n\beta_n^{\tau/n}}{n-\tau}\frac{1}{pr'}\left(\left(\frac{r}{q}\right)^{1/q} + \left(\frac{r}{q}\right)^{1/q}\right) = \beta_n^{1/q}q'\frac{1}{pr'}\left(\left(\frac{p'}{q}\right)^{1/q} + \left(\frac{r}{q}\right)^{1/q}\right),$$

providing the sought result. $\qquad\square$

6.5 Notes

The names of mathematicians encountered in this chapter follow.

Godfrey H. HARDY (1877–1947) was a prominent British mathematician.

John E. LITTLEWOOD (1885–1977) was a British mathematician, a pioneer of Fourier analysis in collaboration with Raymond PALEY (1907–1933).

Serguei SOBOLEV (1908–1989) was a Russian mathematician, author of several fundamental contributions to functional analysis. His name is attached to the so-called Sobolev spaces. He introduced in the 1930s a theory for weak solutions to PDE, similar to distribution theory, later developed in greater generality by the French mathematician Laurent SCHWARTZ (1915–2002).

6.6 Exercises

Exercise 6.6.1. *Let $p \in [1, +\infty]$ and let $u \in L^p(\mathbb{R}^n), v \in L^{p'}(\mathbb{R}^n)$. Prove that $u * v$ is a bounded continuous function on $\mathbb{R}^n$.*

Answer. We already know from Theorem 6.2.4 (and in fact Hölder's inequality) that $u * v$ belongs to L^∞ with $\|u * v\|_{L^\infty} \leq \|u\|_{L^p}\|v\|_{L^{p'}}$. We may assume that $1 \leq p < +\infty$ (if $p = +\infty$, then $p' = 1$ and we may use the commutativity of convolution). Moreover, we have

$$(u * v)(x + h) - (u * v)(x) = \int \big(u(x + h - y) - u(x - y)\big)v(y)dy,$$

and using the notation of Exercise 3.7.15, with $\check{u}(t) = u(-t)$, we have

$$(u * v)(x + h) - (u * v)(x) = \int (\tau_{x+h}\check{u} - \tau_x\check{u})(y)v(y)dy$$

so that $|(u * v)(x + h) - (u * v)(x)| \leq \|\tau_{x+h}\check{u} - \tau_x\check{u}\|_{L^p(\mathbb{R}^n)}\|v\|_{L^{p'}(\mathbb{R}^n)}$, and thus

$$|(u * v)(x + h) - (u * v)(x)| \leq \|\tau_h(\tau_x\check{u}) - \tau_x\check{u}\|_{L^p(\mathbb{R}^n)}\|v\|_{L^{p'}(\mathbb{R}^n)}.$$

Since $\tau_x\check{u} \in L^p(\mathbb{R}^n)$, we may apply Exercise 3.7.15 to get

$$\lim_{h \to 0} \|\tau_h(\tau_x\check{u}) - \tau_x\check{u}\|_{L^p(\mathbb{R}^n)} = 0,$$

entailing the continuity of $u * v$.

Exercise 6.6.2. *We define $E = \{(x_1, x_2) \in \mathbb{R}^2, x_1 - x_2 \notin \mathbb{Q}\}$. Show that E cannot contain a set $A_1 \times A_2$ with A_1, A_2 measurable with positive Lebesgue measure.*

Answer. Reductio ad absurdum: let us assume that $E \supset A_1 \times A_2$ with A_1, A_2 measurable with positive measure. We may assume

$$0 < \lambda_1(A_j) < +\infty, \quad \text{for } j = 1, 2,$$

and we define $\varphi(x_1) = \int_{\mathbb{R}} \mathbf{1}_{A_1}(x_1 + x_2)\mathbf{1}_{A_2}(x_2)dx_2$. The function φ is continuous, since with the notation of Exercise 3.7.15 we have

$$\varphi(x + h) - \varphi(x) = \int_{\mathbb{R}} \big[\tau_{-x-h}(\mathbf{1}_{A_1}) - \tau_{-x}(\mathbf{1}_{A_1})\big](y)\mathbf{1}_{A_2}(y)dy$$

so that since $\tau_{-x}(\mathbf{1}_{A_1}) \in L^1(\mathbb{R}^n)$, we get from Exercise 3.7.15,

$$|\varphi(x + h) - \varphi(x)| \leq \|\tau_{-h}\big(\tau_{-x}(\mathbf{1}_{A_1})\big) - \tau_{-x}(\mathbf{1}_{A_1})\|_{L^1(\mathbb{R}^n)} \xrightarrow[h \to 0]{} 0.$$

The function φ is thus continuous on $\mathbb{R}$ valued in $\mathbb{R}_+$. Moreover, we have

$$\int_{\mathbb{R}} \varphi(x_1)dx_1 = \iint_{\mathbb{R} \times \mathbb{R}} \mathbf{1}_{A_1}(x_1 + x_2)\mathbf{1}_{A_2}(x_2)dx_2dx_1 = \lambda_1(A_1)\lambda_1(A_2) \in \mathbb{R}_+^*.$$

As a consequence, there exists $x_1 \in \mathbb{R}$ such that $\varphi(x_1) > 0$; we have then $x_1 \in A_1 - A_2$, otherwise

$$\forall x_2 \in A_2, \quad x_1 + x_2 \notin A_1,$$

which implies $\mathbf{1}_{A_1}(x_1 + x_2)\mathbf{1}_{A_2}(x_2) = 0$ for all $x_2 \in \mathbb{R}$ and thus $\varphi(x_1) = 0$. As a result, we have

$$\emptyset \neq \{\varphi > 0\} \subset A_1 - A_2.$$

Moreover, we have $(A_1 - A_2) \cap \mathbb{Q} = \emptyset$, otherwise

$$\exists x_1 \in A_1, \exists x_2 \in A_2, x_1 - x_2 \in \mathbb{Q} \implies (x_1, x_2) \notin E$$

contradicting $A_1 \times A_2 \subset E$. We have proven $A_1 - A_2 \subset \mathbb{Q}^c$ and thus

$$\emptyset \neq \{\varphi > 0\} \subset \mathbb{Q}^c.$$

But the non-empty open set $\{\varphi > 0\}$ contains a non-empty open interval $]a, b[, a < b$; the density of $\mathbb{Q}$ in $\mathbb{R}$ implies $]a, b[\cap \mathbb{Q} \neq \emptyset$, which is incompatible with the above inclusion.

Exercise 6.6.3. *Let $\rho \in L^1(\mathbb{R}^n)$ with integral 1. For $\epsilon > 0$, we define $\rho_\epsilon(x) = \epsilon^{-n}\rho(x/\epsilon)$.*

(1) *Let $p \in [1, +\infty[$ and let $u \in L^p(\mathbb{R}^n)$. Show that $u * \rho_\epsilon$ converges with limit u in $L^p(\mathbb{R}^n)$ when $\epsilon \to 0_+$.*

(2) *Let us take $u = \mathbf{1}_{[0,1]}$ and $\rho(x) = e^{-\pi\|x\|^2}$. Show that $u * \rho_\epsilon$ does not converge in $L^\infty(\mathbb{R})$.*

Answer. (1) We have seen in Theorem 6.2.5 that $L^1(\mathbb{R}^n) * L^p(\mathbb{R}^n) \subset L^p(\mathbb{R}^n)$, and we have

$$\int |(u * \rho_\epsilon)(x) - u(x)|^p dx = \int \left| \int (u(x - \epsilon y) - u(x))\rho(y)dy \right|^p dx,$$

so that with the notation of Exercise 3.7.15, using Jensen's inequality (Theorem 3.1.3),

$$\|u * \rho_\epsilon - u\|^p_{L^p(\mathbb{R}^n)} \leq \int \frac{|\rho(y)|}{\|\rho\|_{L^1(\mathbb{R}^n)}} \|\tau_{\epsilon y} u - u\|^p_{L^p(\mathbb{R}^n)} dy \|\rho\|^p_{L^1(\mathbb{R}^n)}.$$

From the same Exercise 3.7.15, we find that $0 = \lim_{\epsilon \to 0} \|\tau_{\epsilon y} u - u\|_{L^p(\mathbb{R}^n)}$ and since

$$|\rho(y)| \|\tau_{\epsilon y} u - u\|_{L^p(\mathbb{R}^n)} \leq 2\|u\|_{L^p(\mathbb{R}^n)}|\rho(y)| \in L^1(\mathbb{R}^n),$$

we may apply Lebesgue's dominated convergence theorem to get the sought result. (2) (see also Exercise 4.5.8 for analogous results). From Exercise 6.6.1, the functions $u * \rho_\epsilon$ are continuous. If the sequence of continuous functions $u * \rho_\epsilon$ were converging in $L^\infty(\mathbb{R}^n)$, the convergence would be uniform and the limit would be a continuous function v. This would imply the convergence of $(u * \rho_\epsilon)_{|[-2,2]}$ towards $v_{|[-2,2]}$ in $L^1([-2,2])$. But we know from the previous question that $u * \rho_\epsilon$ converges towards u in $L^1(\mathbb{R}^n)$: this would imply that the continuous function v would satisfy

$$0 = \int_0^1 |v(x) - 1| dx + \int_{[-2,0] \cup [1,2]} |v(x)| dx = 0 \Longrightarrow v = \mathbf{1}_{[0,1]} \text{ on } [-2,2],$$

which is impossible since v is continuous. We can say a little bit more, since the expressions are quite explicit here. We have for $\epsilon > 0$,

$$(u * \rho_\epsilon)(x) = \int_{\mathbb{R}} \mathbf{1}_{[0,1]}(x - \epsilon y) e^{-\pi y^2} dy = \int_{(x-1)/\epsilon}^{x/\epsilon} e^{-\pi y^2} dy.$$

Consequently for $x \in]0,1[$, $x/\epsilon \to +\infty$ and $(x-1)/\epsilon \to -\infty$ so that

$$\lim_{\epsilon \to 0_+} (u * \rho_\epsilon)(x) = \begin{cases} 1 & \text{for } 0 < x < 1, \\ 1/2 = \int_{-\infty}^0 e^{-\pi y^2} dy = \int_0^{+\infty} e^{-\pi y^2} dy & \text{for } x = 0, 1, \\ 0 & \text{for } x \notin [0,1], \end{cases}$$

since for $x > 1$,

$$0 \le \int_{(x-1)/\epsilon}^{x/\epsilon} e^{-\pi y^2} dy \le e^{-\pi (x-1)^2 \epsilon^{-2}} \epsilon^{-1} \xrightarrow[\epsilon \to 0_+]{} 0.$$

The case $x < 0$ is dealt with analogously. The pointwise limit is actually discontinuous at 0 and 1.

Exercise 6.6.4. *Let $p_1, \ldots, p_k, q \in [1, +\infty]$ such that*

$$\sum_{1 \le j \le k} \frac{1}{p_j} = k - 1 + \frac{1}{q}.$$

Show that $\quad \|u_1 * \cdots * u_k\|_{L^q(\mathbb{R}^n)} \le \|u_1\|_{L^{p_1}(\mathbb{R}^n)} \cdots \|u_k\|_{L^{p_k}(\mathbb{R}^n)}.$

Answer. For $k = 2$, this is Young's inequality since

$$\frac{1}{p_1} + \frac{1}{p_2} = 1 + \frac{1}{q}, \quad \text{i.e.,} \quad 1 - \frac{1}{p_1} + 1 - \frac{1}{p_2} = 1 - \frac{1}{q}.$$

We have for $k \ge 2$, if $\frac{1}{p'} + \frac{1}{q'} = \frac{1}{r'}$, using induction on k,

$$\|u_1 * \cdots * u_k * u_{k+1}\|_{L^r} \le \|u_1 * \cdots * u_k\|_{L^q} \|u_{k+1}\|_{L^p} \le \|u_1\|_{L^{p_1}} \cdots \|u_k\|_{L^{p_k}} \|u_{k+1}\|_{L^p}$$

$$\frac{1}{q'} = \sum_{1 \le j \le k} \frac{1}{p_j'} \Longrightarrow \frac{1}{r'} = \left(\sum_{1 \le j \le k} \frac{1}{p_j'} \right) + \frac{1}{p'},$$

i.e.,

$$\sum_{1 \le j \le k} \frac{1}{p_j} + \frac{1}{p} = k + 1 - \frac{1}{r'} = k + \frac{1}{r}, \qquad \text{qed.}$$

Exercise 6.6.5. *Let $n \in \mathbb{N}^*$, $a \in]1, +\infty[$, $p \in]1, +\infty[$ and let $k \in C^1(\mathbb{R}^n \backslash \{0\})$ homogeneous with degree $-n/a$. We define*

$$\gamma = n\left(1 - \frac{1}{a} - \frac{1}{p}\right) \qquad \text{and we assume } \gamma \in]0, 1[.$$

(1) *Show that for $x \neq 0$, $|k(x)| \leq C_0 |x|^{-n/a}$. For $u \in L^p_{\text{comp}}(\mathbb{R}^n)$, we define $(k * u)(x) = \int k(y) u(x - y) dy$. Show that $k * u$ is meaningful and that for $R > 0$,*

$$\int_{|y| \leq R} |y|^{-n/a} |u(x - y)| dy \leq c_{n,p} \|u\|_{L^p} R^\gamma.$$

(2) *Show that for $u \in L^p_{\text{comp}}(\mathbb{R}^n)$, $k * u$ is an Hölderian function with index γ.*

Answer. (1) For $x \neq 0$, we have

$$|k(x) = k(x/|x|)|x|^{-n/a}| \leq |x|^{-n/a} \sup_{\mathbb{S}^{n-1}} |k|.$$

We have also

$$\int_{|y| \leq R} |y|^{-n/a} |u(x - y)| dy \leq \|u\|_{L^p} \left(\int_{|y| \leq R} |y|^{-np'/a} dy\right)^{1/p'}$$

$$\leq C \|u\|_{L^p} \left(\int_0^R r^{n-1-\frac{np'}{a}} dr\right)^{1/p'}$$

$$= C' \|u\|_{L^p} R^{\frac{n}{p'} - \frac{n}{a}} = C' \|u\|_{L^p} R^\gamma,$$

since $n - 1 - \frac{np'}{a} = np'(\frac{1}{p'} - \frac{1}{a}) - 1 = np'(1 - \frac{1}{p} - \frac{1}{a}) - 1 = \gamma p' - 1 > -1$. As a result,

$$(k * u)(x) = \int_{y \in \text{supp}\, u} k(x - y) u(y) dy$$

is a bounded measurable function since $u \in L^p_{\text{comp}}$ and $k \in L^{p'}_{\text{loc}}$: we have indeed

$$-\frac{np'}{a} > -n$$

since $1 - \frac{1}{p} - \frac{1}{a} > 0 \implies \frac{1}{p'} > \frac{1}{a} \implies \frac{p'}{a} < 1$. We can prove as well that $k * u$ is continuous, but the next question provides a sharper Hölderian regularity.

(2) For $u \in L^p_{\text{comp}}(\mathbb{R}^n)$, $x, h \in \mathbb{R}^n$, we have

$$(k * u)(x + h) - (k * u)(x) = \int k(y) u(x + h - y) dy - \int k(y) u(x - y) dy$$

$$= \int k(y + h) u(x - y) dy - \int k(y) u(x - y) dy,$$

so that

$$|(k * u)(x + h) - (k * u)(x)| \leq \underbrace{\int_{|y|<2|h|} |k(y + h) - k(y)|\, |u(x - y)| dy}_{I_1}$$

$$+ \underbrace{\int_{|y|\geq 2|h|} |k(y + h) - k(y)|\, |u(x - y)| dy}_{I_2}.$$

To handle I_1 we note that $|k(y)| \leq C_0|y|^{-n/a}$ and $|k(y + h)| \leq C_0|y + h|^{-n/a}$, so that, using the estimate of question (1), we get

$$I_1 \leq C_0 C'(2|h|)^\gamma \|u\|_{L^p} + C_0 C'(3|h|)^\gamma \|u\|_{L^p}.$$

We have thus, using the mean value theorem and the homogeneity of k' for I_2,

$$|(k * u)(x + h) - (k * u)(x)|$$

$$\leq 2C_0 C'(3|h|)^\gamma \|u\|_{L^p} + C''|h| \int_{|y|\geq 2|h|} \sup_{\theta\in[0,1]} |y + \theta h|^{-\frac{n}{a}-1}|u(x - y)| dy.$$

If $|y| \geq 2|h|$, we have $|y + \theta h| \geq |y| - |h| \geq \frac{1}{2}|y|$ and the factor of C'' is bounded above by

$$\omega = |h|2^{\frac{n}{a}+1}\|u\|_{L^p} \left(\int_{|y|\geq 2|h|} |y|^{-(\frac{n}{a}+1)p'} dy \right)^{1/p'}$$

$$= |h|2^{\frac{n}{a}+1}\|u\|_{L^p} \left(\int_{2|h|}^{+\infty} r^{n-1-\frac{np'}{a}-p'} dr \right)^{1/p'} |\mathbb{S}^{n-1}|^{1/p'}$$

and since

$$n - 1 - \frac{np'}{a} - p' = np'\left(\frac{1}{p'} - \frac{1}{a} \right) - 1 - p' = \gamma p' - p' - 1 = p'(\gamma - 1) - 1 < -1,$$

we get

$$\omega \leq C'''\|u\|_{L^p}|h|(|h|^{p'(\gamma-1)})^{1/p'} = C'''\|u\|_{L^p}|h|^\gamma.$$

Exercise 6.6.6. *Let n be an integer ≥ 3. For $x \in \mathbb{R}^n$, we denote by $\|x\|$ the Euclidean norm of x. Let $p \in [1, +\infty]$; a measurable function $f : \mathbb{R}^n \to \mathbb{C}$ is said to belong to L^p_{loc} when for all compact subsets K of $\mathbb{R}^n$, $\mathbf{1}_K f \in L^p(\mathbb{R}^n)$.*

(1) *We define $E(x) = \|x\|^{2-n}$ and $p_n = \frac{n}{n-2}$. Show that E belongs to $\cap_{1\leq p<p_n} L^p_{\mathrm{loc}}$, and $E \notin L^{p_n}_{\mathrm{loc}}$.*

(2) *Let $q \in]n/2, +\infty]$ and let F be a function in $L^q(\mathbb{R}^n)$ with compact support. We define*
$$C_F(x) = \int_{\mathbb{R}^n} \|x - y\|^{2-n} F(y) dy.$$

Show that C_F belongs to L^∞_{loc}.

(3) *Let φ be a function in $C_c^2(\mathbb{R}^n)$. Show that C_φ is twice differentiable.*

(4) *Let $\epsilon > 0$ be given and $\varphi \in C_c^2(\mathbb{R}^n)$. Let $\chi \in C^\infty(\mathbb{R}, [0,1])$ such that*
$$\chi(t) = \begin{cases} 0 & \text{for } t \leq 1, \\ 1 & \text{for } t \geq 2. \end{cases}$$

We set $\Delta\varphi = \displaystyle\sum_{1 \leq j \leq n} \frac{\partial^2 \varphi}{\partial x_j^2}$, $\quad I(\varphi, \epsilon) = \displaystyle\int_{\mathbb{R}^n} \|y\|^{2-n} \chi(\|y\|/\epsilon)(\Delta\varphi)(y) dy.$

Show that
$$\lim_{\epsilon \to 0_+} I(\varphi, \epsilon) = C_{\Delta\varphi}(0) \quad \text{and} \quad C_{\Delta\varphi}(0) = \alpha_n \varphi(0),$$

where α_n is a constant depending only on n (hint: calculate $\Delta(\theta(\|x\|))$ where θ is twice differentiable vanishing near 0).

(5) *Let F be as in question (2). Show that for any function φ, compactly supported and twice differentiable*
$$\int_{\mathbb{R}^n} C_F(x)(\Delta\varphi)(x) dx = \alpha_n \int_{\mathbb{R}^n} F(x)\varphi(x) dx.$$

Answer. (1) We have $\displaystyle\int_{\|x\| \leq R} \|x\|^{p(2-n)} dx = |\mathbb{S}^{n-1}| \int_0^R r^{p(2-n)+n-1} dr$ which is finite iff
$$p(2-n) + n - 1 > -1, \quad \text{i.e.,} \quad p < \frac{n}{n-2} = p_n.$$

(2) For a given $x \in \mathbb{R}^n$, the function $y \mapsto \|x-y\|^{2-n}$ belongs to L^p_{loc} for $1 \leq p < p_n$. As a result, with $K = \mathrm{supp}\, F$ (a compact set), the function
$$y \mapsto \|x - y\|^{2-n} \mathbf{1}_K(y) = G_x(y)$$

belongs to $L^p(\mathbb{R}^n)$ for $1 \leq p < p_n$. If q' is the conjugate exponent of $q > n/2$, we have $1/q < 2/n$ and
$$\frac{1}{q'} = 1 - \frac{1}{q} > 1 - \frac{2}{n} = \frac{n-2}{n} \quad \text{i.e.,} \quad q' < \frac{n}{n-2} = p_n,$$

so that the function G_x belongs to $L^{q'}(\mathbb{R}^n)$. From Hölder's inequality, the product $G_x F$ belongs to L^1 and
$$|C_F(x)| \leq \|F\|_{L^q} \|G_x\|_{L^{q'}}.$$

But we have for L compact and $x \in L$,

$$\|G_x\|_{L^{q'}}^{q'} = \int_K \|y - x\|^{q'(2-n)} dy = \int_{-x+K} \|t\|^{q'(2-n)} dt$$

$$\leq \int_{-L+K} \|t\|^{q'(2-n)} dt < +\infty,$$

since $K - L = \{a - b\}_{a \in K, b \in L}$ is compact and $q'(2 - n) > -n$ (since $q' < \frac{n}{n-2}$ from above).

(3) Indeed, with $C_\varphi(x) = \int \|y\|^{2-n} \varphi(x - y) dy$, defining $\varphi_{jk} = \frac{\partial^2 \varphi}{\partial x_j \partial x_k}$ (a C_c^0 function), we see that for a compact set M,

$$\sup_{x \in M} \|y\|^{2-n} |\varphi_{jk}(x - y)| \leq \|y\|^{2-n} \mathbf{1}_{M - \mathrm{supp}\varphi}(y) \|\varphi_{jk}\|_{L^\infty} \in L^1,$$

since if $x - y \in \mathrm{supp}\varphi$, $y = y - x + x \in M - \mathrm{supp}\varphi$ which is compact.

(4) Since $\varphi \in C_c^2$ and thus $\triangle\varphi \in C_c^0$, we have

$$\|y\|^{2-n} \chi(\|y\|/\epsilon) |(\triangle\varphi)(y)| \leq \|y\|^{2-n} |(\triangle\varphi)(y)| \in L^1.$$

Moreover for $y \neq 0$, $\lim_{\epsilon \to 0_+} \chi(\|y\|/\epsilon) = 1$, Lebesgue's dominated convergence theorem gives the first result. Moreover integrating by parts in the simple integrals in x_j (on C_c^1 functions), we get

$$I(\varphi, \epsilon) = \sum_{1 \leq j \leq n} \int \|y\|^{2-n} \chi(\|y\|/\epsilon) \varphi_{jj}(y) dy$$

$$= \sum_{1 \leq j \leq n} \int \frac{\partial^2}{\partial y_j^2} \left(\|y\|^{2-n} \chi(\|y\|/\epsilon) \right) \varphi(y) dy.$$

We note that for $x \neq 0$, $\partial_{x_j}(\|x\|) = x_j/\|x\|$ since $\partial_{x_j}(\|x\|^2) = 2x_j$, so that for θ as in the statement of question (4),

$$\triangle(\theta(\|x\|)) = \sum_{1 \leq j \leq n} \frac{\partial}{\partial x_j} \left[\theta'(\|x\|) \frac{x_j}{\|x\|} \right]$$

$$= \sum_{1 \leq j \leq n} \theta''(\|x\|) \frac{x_j^2}{\|x\|^2} + \theta'(\|x\|) \left(\frac{1}{\|x\|} - \frac{x_j}{\|x\|^2} \frac{x_j}{\|x\|} \right).$$

Denoting $r = \|x\|$, we get

$$\triangle(\theta(r)) = \theta''(r) + \frac{1}{r} n\theta'(r) - \frac{1}{r^3} r^2 \theta'(r) = \theta''(r) + \frac{n-1}{r} \theta'(r).$$

We may now calculate

$$\triangle(r^{2-n} \chi(r/\epsilon)) = (2 - n)(1 - n) r^{-n} \chi(r/\epsilon)$$

$$+ 2(2 - n) r^{1-n} \chi'(r/\epsilon) \epsilon^{-1} + r^{2-n} \chi''(r/\epsilon) \epsilon^{-2}$$

$$+ \frac{n-1}{r} \left[(2 - n) r^{1-n} \chi(r/\epsilon) + r^{2-n} \chi'(r/\epsilon) \epsilon^{-1} \right]$$

$$\overbrace{= \chi(r/\epsilon)\left[(2-n)(1-n)r^{-n} + (n-1)(2-n)r^{-n}\right]}^{=0}$$
$$+ \chi'(r/\epsilon)[2(2-n)r^{1-n}\epsilon^{-1} + (n-1)r^{1-n}\epsilon^{-1}]$$
$$+ \chi''(r/\epsilon)\epsilon^{-2}r^{2-n}$$
$$= \chi'(r/\epsilon)\epsilon^{-1}r^{1-n}(3-n) + \chi''(r/\epsilon)\epsilon^{-2}r^{2-n}.$$

We find

$$I(\varphi,\epsilon) = \int \varphi(y)\left[\chi'(\|y\|/\epsilon)\epsilon^{-1}\|y\|^{1-n}(3-n) + \chi''(\|y\|/\epsilon)\epsilon^{-2}\|y\|^{2-n}\right]dy$$

$$= \int \varphi(\epsilon y)\left[\chi'(\|y\|)\epsilon^{-1}\epsilon^{1-n}\|y\|^{1-n}(3-n) + \chi''(\|y\|)\epsilon^{-2}\epsilon^{2-n}\|y\|^{2-n}\right]\epsilon^n dy$$

$$= \int \varphi(\epsilon y)\left[\chi'(\|y\|)\|y\|^{1-n}(3-n) + \chi''(\|y\|)\|y\|^{2-n}\right]dy,$$

and since the function between the brackets is C_c^∞, we get

$$C_{\triangle\varphi}(0) = \lim_{\epsilon\to 0} I(\varphi,\epsilon) = \varphi(0)|\mathbb{S}^{n-1}| \int_0^{+\infty} [\chi'(r)(3-n) + \chi''(r)r]dr$$

$$= \varphi(0)|\mathbb{S}^{n-1}| \left((3-n) - \int_0^{+\infty} \chi'(r)dr \right)$$

$$= \varphi(0) \underbrace{|\mathbb{S}^{n-1}|(2-n)}_{\alpha_n}.$$

(5) Thanks to Fubini's theorem we have, with $\psi_y(x) = \psi(x+y)$,

$$\int C_F(x)(\triangle\varphi)(x)dx = \iint \|x-y\|^{2-n}F(y)(\triangle\varphi)(x)dxdy$$

$$= \iint \|x\|^{2-n}F(y)(\triangle\varphi)(x+y)dxdy$$

$$= \int \left[\int \|x\|^{2-n}(\triangle\varphi)(x+y)dx\right] F(y)dy = \int \left[\int \|x\|^{2-n}(\triangle\varphi)_y(x)dx\right] F(y)dy$$

$$= \int \left[\int \|x\|^{2-n}(\triangle\varphi_y)(x)dx\right] F(y)dy = \int C_{\triangle\varphi_y}(0)F(y)dy$$

$$= \int \alpha_n\varphi_y(0)F(y)dy = \int \alpha_n\varphi(y)F(y)dy.$$

We may note that with F of class C_c^2, the above equality gives $\triangle(\alpha_n^{-1}C_F) = F$ and gives a solution to the equation

$$\triangle u = F$$

for $F \in L^{\frac{n}{2}+\delta}$ with $\delta > 0$ and F with compact support.

Exercise 6.6.7. *Let $n \geq 1$ be an integer. For $x \in \mathbb{R}^n$, $\|x\|$ stands for the Euclidean norm of x. For $(t, x) \in \mathbb{R} \times \mathbb{R}^n$, we define*

$$E(t, x) = \begin{cases} (4\pi t)^{-\frac{n}{2}} \exp -\dfrac{\|x\|^2}{4t} & \text{if } \ t > 0, \\ 0 & \text{if } \ t \leq 0. \end{cases}$$

(1) *Show that for all $T \in \mathbb{R}$, E belongs to $L^1(]-\infty, T] \times \mathbb{R}^n)$.*

(2) *For $t > 0$, we define on $\mathbb{R}^n$ the function $e(t)$ by $e(t)(x) = E(t, x)$. Show that for all $t > 0$, $e(t) \in L^1(\mathbb{R}^n)$.*

(3) *Let ψ be in $L^1(\mathbb{R}^n)$. For $t > 0$, we set $u(t) = e(t) * \psi$. Show that for all $t > 0$, $u(t) \in L^1(\mathbb{R}^n)$ and*

$$\lim_{t \to 0_+} u(t) = \psi \quad \text{in } L^1(\mathbb{R}^n).$$

(4) *We assume in the sequel that $\psi \in C_c(\mathbb{R}^n)$. For $t > 0$ and $x \in \mathbb{R}^n$, we set $U(t, x) = u(t)(x)$. Show that $U \in C^\infty(\mathbb{R}_+^* \times \mathbb{R}^n)$.*

(5) *Show that for $t > 0$, $x \in \mathbb{R}^n$, we have $\dfrac{\partial U}{\partial t}(t, x) = \displaystyle\sum_{1 \leq j \leq n} \dfrac{\partial^2 U}{\partial x_j^2}(t, x)$.*

Answer. (1) The function E is positive measurable. For $T > 0$, we have

$$\int_0^T (4\pi t)^{-n/2} \int_{\mathbb{R}^n} \exp -\frac{\|x\|^2}{4t} dx dt = \int_0^T dt = T.$$

(2) For $t > 0$, the same calculation proves $\int_{\mathbb{R}^n} E(t, x) dx = 1$.

(3) The function $u(t)$ belongs to L^1 as a convolution of L^1 functions. We have for $t > 0$,

$$u(t)(x) = \int_{\mathbb{R}^n} \psi(x - y)(4\pi t)^{-n/2} \exp -\frac{\|y\|^2}{4t} dy = \int_{\mathbb{R}^n} \psi(x + z(4\pi t)^{1/2}) e^{-\pi \|z\|^2} dz.$$

Let ψ_k be a sequence of continuous functions with compact support converging towards ψ in L^1 and let us set $\epsilon = (4\pi t)^{1/2}$. We have

$$u(t) - \psi = e(t) * \psi - \psi = e(t) * (\psi - \psi_k) - (\psi - \psi_k) + e(t) * \psi_k - \psi_k$$

and thus for a parameter $M > 0$,

$$\|u(t) - \psi\|_{L^1} \leq 2\|\psi - \psi_k\|_{L^1} + \iint_{\mathbb{R}_x^n \times \{\|z\| \leq M\}} |\psi_k(x + \epsilon z) - \psi_k(x)| e^{-\pi \|z\|^2} dz dx$$

$$+ \iint_{\mathbb{R}_x^n \times \{\|z\| > M\}} |\psi_k(x + \epsilon z) - \psi_k(x)| e^{-\pi \|z\|^2} dz dx.$$

Lebesgue's dominated convergence theorem applied to the first integral for $\epsilon \to 0_+$ (pointwise convergence towards 0 follows from the continuity of ψ_k, domination is due to the uniform compact support for $0 \leq \epsilon \leq 1$) provides

$$\limsup_{t \to 0_+} \|u(t) - \psi\|_{L^1} \leq 2\|\psi - \psi_k\|_{L^1} + \int_{\{\|z\| > M\}} e^{-\pi \|z\|^2} dz 2\|\psi_k\|_{L^1}.$$

As a result, taking the limit when k goes to $+\infty$, we get

$$\limsup_{t \to 0_+} \|u(t) - \psi\|_{L^1} \le \int_{\{\|z\| > M\}} e^{-\pi \|z\|^2} dz \, 2 \|\psi\|_{L^1}$$

for all $M > 0$, implying the sought result.

(4) We write for $t > 0, x \in \mathbb{R}^n$,

$$U(t, x) = \int_{\mathbb{R}^n} \psi(y)(4\pi t)^{-n/2} \exp{-\frac{\|x - y\|^2}{4t}} dy$$

and defining $F(t, x, y) = \psi(y)(4\pi t)^{-n/2} \exp{-\frac{\|x-y\|^2}{4t}}$, we see that

(i) $\int_{\mathbb{R}^n} |F(t, x, y)| dy < \infty$,

(ii) $(t, x) \mapsto F(t, x, y)$ is C^∞ on $\mathbb{R}_+^* \times \mathbb{R}^n$,

(iii) for all compact $K \subset \mathbb{R}_+^* \times \mathbb{R}^n$, $\int_{\mathbb{R}^n} \sup_{(t,x) \in K} |\partial_t^k \partial_x^\alpha F(t, x, y)| dy < +\infty$.

The last point follows from the following identity (easily proven by induction on $k + |\alpha|$):

$$\partial_t^k \partial_x^\alpha F(t, x, y) = \psi(y) Q_{k\alpha}(t^{-1/2}, x - y) \exp{-\frac{\|x - y\|^2}{4t}}$$

where $Q_{k\alpha}$ is a polynomial. The function U is thus C^∞ on $\mathbb{R}_+^* \times \mathbb{R}^n$.

(5) We calculate then directly on $\mathbb{R}_+^* \times \mathbb{R}^n$,

$$\frac{\partial U}{\partial t} = \int_{R^u} \psi(y)(4\pi t)^{-n/2} \exp{-\frac{\|x - y\|^2}{4t}} \left[-\frac{n}{2t} + \frac{\|x - y\|^2}{4t^2} \right] dy$$

and

$$\frac{\partial U}{\partial x_j} = \int_{R^n} \psi(y)(4\pi t)^{-n/2} \exp{-\frac{\|x - y\|^2}{4t}} \left[-\frac{(x_j - y_j)}{2t} \right] dy,$$

which gives

$$\frac{\partial^2 U}{\partial x_j^2} = \int_{R^n} \psi(y)(4\pi t)^{-n/2} \exp{-\frac{\|x - y\|^2}{4t}} \left[\frac{(x_j - y_j)^2}{4t^2} - \frac{1}{2t} \right] dy$$

and then

$$\sum_{1 \le j \le n} \frac{\partial^2 U}{\partial x_j^2} = \int_{R^n} \psi(y)(4\pi t)^{-n/2} \exp{-\frac{\|x - y\|^2}{4t}} \left[\frac{\|x - y\|^2}{4t^2} - \frac{n}{2t} \right] dy = \frac{\partial U}{\partial t}.$$

Exercise 6.6.8. *Let $(X, \mathcal{M}, \mu)$ be a measure space where μ is a positive measure. Let $p \in [1, +\infty), q \in [1, +\infty]$. We define the Lorentz space $L^{p,q}(X)$ as the set of measurable functions $f : X \to \mathbb{C}$ such that*

$$\left(t^p \mu(\{x \in X, |f(x)| > t\}) \right)^{1/p} \in L^q\left(\mathbb{R}_+, \frac{dt}{t}\right).$$

We define the following quantities on $L^{p,q}(X)$:

$$\text{for } p, q \in [1, +\infty), \ \|f\|_{L^{p,q}(X)} = \left(\int_0^{+\infty} \left(t^p \mu(\{x \in X, |f(x)| > t\}) \right)^{\frac{q}{p}} \frac{dt}{t} \right)^{\frac{1}{q}}, \quad (6.6.1)$$

$$\text{for } p \in [1, +\infty), q = +\infty, \quad \|f\|_{L^{p,\infty}(X)}^p = \sup_{t>0} t^p \mu(\{x \in X, |f(x)| > t\}), \quad (6.6.2)$$

$$\text{for } p = q = +\infty, \quad \|f\|_{L^{\infty,\infty}(X)} = \|f\|_{L^{\infty}(X)}. \quad (6.6.3)$$

(1) *Show that $L^{p,p}(X) = L^p(X)$ and $L^{p,\infty}(X) = L_w^p(X)$ (see Definition 6.3.1).*

(2) *Prove that $L^{p,q}(X)$ is a vector space and that $\|f\|_{L^{p,q}(X)}$ is a quasi-norm on this vector space.*

N.B. A very complete and accessible description of $L^{p,q}$ spaces is given in the survey article [34] by the American mathematician R. HUNT (1937–2009).

Answer. (1) The second assertion follows immediately from the very definition of $L_w^p(X)$. The first assertion is obvious by definition for $p = \infty$. If $p \in [1, +\infty)$, we have for $f \in L^p(X)$,

$$\|f\|_{L^p(X)}^p = \int_X |f(x)|^p d\mu = \int_X \left(\int_{\mathbb{R}_+} pt^{p-1} H(|f(x)| - t) dt \right) d\mu,$$

and by Tonelli's theorem,

$$\|f\|_{L^p(X)}^p = p \int_0^{+\infty} t^p \mu(\{x \in X, |f(x)| > t\}) \frac{dt}{t} = p\|f\|_{L^{p,p}(X)}^p.$$

(2) The answer is obvious for $p = q = \infty$, and is already known for $q = \infty$. We may thus assume that $p, q \in [1, +\infty)$. Let us prove that $\|\cdot\|_{L^{p,q}(X)}$ is a quasi-norm: if $\|f\|_{L^{p,q}(X)} = 0$, then for all $t > 0$, $\mu(\{x \in X, |f(x)| > t\}) = 0$ so that

$$\{x \in X, f(x) \neq 0\} = \cup_{n \in \mathbb{N}^*}\{x \in X, |f(x)| > 1/n\}$$

has measure 0 and thus $f = 0$ a.e. Moreover $\|\cdot\|_{L^{p,q}(X)}$ is positively homogeneous with degree 1, and $L^{p,q}(X)$ is stable by multiplication by a complex number since with $z \in \mathbb{C}^*$,

$$t^q \left(\mu(\{x \in X, |zf(x)| > t\}) \right)^{q/p} \frac{dt}{t} = |z|^q s^q \left(\mu(\{x \in X, |f(x)| > s\}) \right)^{q/p} \frac{ds}{s}.$$

Let f, g be in $L^{p,q}(X)$. Let $\theta \in (0, 1)$. Since for $t > 0$, $|f| \leq (1-\theta)t$ and $|g| \leq \theta t$ imply $|f + g| \leq t$, we have

$$\{|f + g| > t\} \subset \{|f| > t(1 - \theta)\} \cup \{|g| > t\theta\},$$

so that

$$t^p \mu(\{|f + g| > t\}) \leq t^p \mu(\{|f| > t(1 - \theta)\}) + t^p \mu(\{|g| > t\theta\}),$$

and thus

$$\left(t^p\mu(\{|f+g|>t\})\right)^{1/p} \le \left(t^p\mu(\{|f|>t(1-\theta)\})+t^p\mu(\{|g|>t\theta\})\right)^{1/p}$$
$$\le t\mu(\{|f|>t(1-\theta)\})^{1/p}+t\mu(\{|g|>t\theta\})^{1/p}$$

where the last inequality follows from the sharp elementary[3]

$$\forall a,b\ge 0, \forall p\ge 1, \quad (a^p+b^p)^{1/p}\le a+b.$$

We obtain

$$\left(t^p\mu(\{|f+g|>t\})\right)^{1/p} \le t\mu(\{|f|>t(1-\theta)\})^{1/p}+t\mu(\{|f|>t\theta\})^{1/p}$$

and the triangle inequality in L^q and the homogeneity give

$$\|f+g\|_{L^{p,q}(X)} \le (1-\theta)^{-1}\|f\|_{L^{p,q}(X)}+\theta^{-1}\|g\|_{L^{p,q}(X)}.$$

We may assume that both $\|f\|_{L^{p,q}(X)},\|g\|_{L^{p,q}(X)}$ are positive (otherwise f or g are 0 a.e.) and choosing $\theta=\|g\|^{1/2}/(\|f\|^{1/2}+\|g\|^{1/2})$, we get

$$\|f+g\|_{L^{p,q}(X)} \le \left(\|f\|_{L^{p,q}(X)}^{1/2}+\|g\|_{L^{p,q}(X)}^{1/2}\right)^2 \le 2\left(\|f\|_{L^{p,q}(X)}+\|g\|_{L^{p,q}(X)}\right),$$

proving the result.

Exercise 6.6.9. *Let $(X,\mathcal{M},\mu)$ be a measure space where μ is a positive measure. Let $f:X\to\mathbb{C}$ be a measurable function. We define the distribution function $f_*:[0,+\infty]\to[0,+\infty]$ of f and the decreasing rearrangement function $f^*:[0,+\infty]\to[0,+\infty]$ by*

$$f_*(l)=\mu(\{x\in X,|f(x)|>l\}), \tag{6.6.4}$$
$$f^*(s)=\inf\{t\ge 0, f_*(t)\le s\}, \tag{6.6.5}$$

with the usual convention $\inf\emptyset=+\infty$.

(1) *Prove that f_* and f^* are decreasing.*
(2) *Prove that f_* is right-continuous.*
(3) *Prove that for all $t\ge 0$, $f^*(f_*(t))\le t$ and that for all $s\ge 0$, $f_*(f^*(s))\le s$.*
(4) *Prove that f^* is right-continuous.*
(5) *Prove that f and f^* have the same distribution function (with $\overline{\mathbb{R}}_+$ equipped with the Lebesgue measure).*

---

[3] For $p\ge 1$, the function $[0,1]\ni\tau\mapsto\tau^p+(1-\tau)^p=\gamma(\tau)$ is convex as a sum of convex functions and thus for $a,b\ge 0$,

$$\tau^p+(1-\tau)^p=\gamma(\tau 1+(1-\tau)0)\le\tau\gamma(1)+(1-\tau)\gamma(0)=1\implies(a^p+b^p)\le(a+b)^p.$$

This inequality is shown to be sharp by taking $b=0$.

Answer. (1) Note that

$$f_*(+\infty) = 0, \quad f_*(0) = \mu(\{x, f(x) \neq 0\}),$$
$$f^*(+\infty) = 0, \quad f^*(0) = \inf\{t \geq 0, f_*(t) = 0\}.$$

Let $t_1 < t_2$ in $\overline{\mathbb{R}}_+$: then

$$\{x \in X, |f(x)| > t_2\} \subset \{x \in X, |f(x)| > t_1\},$$

implying readily that f_* is decreasing. Let $s_1 < s_2$ in $\overline{\mathbb{R}}_+$: then

$$(f_*)^{-1}([0, s_1]) \subset (f_*)^{-1}([0, s_2]) \implies f^*(s_2) \leq f^*(s_1).$$

(2) Let (ϵ_k) be a sequence of positive numbers decreasing towards 0. We have

$$\{|f| > t\} = \cup_{k \in \mathbb{N}}\{|f| > t + \epsilon_k\},$$

so that by Proposition 1.4.4(2) (or Beppo Levi's theorem), we get $\lim_k f_*(t+\epsilon_k) = f_*(t)$.

(3) Let $t \in \overline{\mathbb{R}}_+$. If $s = f_*(t)$, we have $f^*(s) \leq t$ (otherwise $t < f^*(s)$ and from (6.6.5) we find $f_*(t) > s = f_*(t)$ which is impossible): we have proven $f^*(f_*(t)) \leq t$. Let $s \in \overline{\mathbb{R}}_+$. We have $f_*(+\infty) = 0$, and thus if $f^*(s) = +\infty$, $f_*(f^*(s)) = 0 \leq s$. We may thus assume $f^*(s) < +\infty$. Let (t_k) be a decreasing sequence converging towards $f^*(s)$ with $f_*(t_k) \leq s$. By the already proven right-continuity of f_*, we have

$$\lim_k f_*(t_k) = f_*(f^*(s)) \implies f_*(f^*(s)) \leq s.$$

(4) Let $s \in \mathbb{R}_+$ and (ϵ_k) be a positive sequence decreasing to 0. We already know that $f^*(s + \epsilon_k) \leq f^*(s)$ and let us assume that there exists t such that

$$\forall k \in \mathbb{N}, \quad f^*(s + \epsilon_k) < t < f^*(s).$$

This implies $f_*(f^*(s + \epsilon_k)) \leq f_*(t) \leq f_*(f^*(s)) \leq s$, and thus $t \geq f^*(s)$ which contradicts the assumption.

(5) We start with a lemma.

Lemma 6.6.10. *Let* $g : [0, +\infty] \to [0, +\infty]$ *be a decreasing function. Then,*

$$g_*(t) = \sup\{s \in \overline{\mathbb{R}}_+, g(s) > t\}.$$

Proof. Let s be such that $g(s) > t$. Then, since g is decreasing, $g([0, s]) \subset (t, +\infty]$, so that $s \leq g_*(t)$ and thus $\sup\{s \in \overline{\mathbb{R}}_+, g(s) > t\} \leq g_*(t)$. Conversely, let $s_\infty = \sup\{s \in \overline{\mathbb{R}}_+, g(s) > t\}$. If $s > s_\infty$, we have $g(s) \leq t$, so that

$$(s_\infty, +\infty] \subset \{s, g(s) \leq t\} \implies \{s, g(s) > t\} \subset [0, s_\infty] \implies g_*(t) \leq s_\infty,$$

concluding the proof of the lemma. $\square$

We can apply this lemma to the decreasing function f^* to get

$$(f^*)_*(t) = \sup\{s \in \overline{\mathbb{R}}_+, f^*(s) > t\}.$$

We have

(†)
$$f_*(t) \geq (f^*)_*(t),$$

otherwise, we would have $f_*(t) < (f^*)_*(t)$ and thus we could find s such that $f_*(t) < s$ with $f^*(s) > t$, which would give $f^*(s) \leq f^*(f_*(t)) \leq t$, contradicting $f^*(s) > t$. Conversely, we note that

$$(f^*)_*(t) = \int_0^{+\infty} H\big(f^*(s) - t\big)\,ds$$

and since $f^*(f_*(t)) \leq t$, we get the first inequality

(‡) $$(f^*)_*(t) \leq \int_0^{+\infty} H\Big(f^*(s) - f^*\big(f_*(t)\big)\Big)\,ds \leq \int_0^{+\infty} H\big(f^*(t) - s\big)\,ds = f^*(t),$$

where the second inequality follows from the inclusion

$$\{s, f^*(s) > f^*(f_*(t))\} \subset \{s, s < f_*(t)\},$$

due to the implication $s \geq f_*(t) \implies f^*(s) \leq f^*(f_*(t))$. The inequalities (†), (‡) give the answer.

Exercise 6.6.11. *Let $(X, \mathcal{M}, \mu)$ be a measure space where μ is a positive measure and let*

$$1 \leq p_1 \leq p \leq p_2 \leq +\infty.$$

Prove that $L^p(\mu) \subset L^{p_1}(\mu) + L^{p_2}(\mu)$.

Answer. We may of course assume that $p_1 < p < p_2$. We note then that, for $u \in L^p(\mu)$,

$$\mu(\{x \in X, |u(x)| > 1\}) = \int_{\{|u(x)|>1\}} d\mu \leq \int_{\{|u(x)|>1\}} |u(x)|^p d\mu \leq \|u\|_{L^p(\mu)}^p < +\infty.$$

We have $u = u\mathbf{1}_{\{|u|>1\}} + u\mathbf{1}_{\{|u|\leq 1\}}$ and $u\mathbf{1}_{\{|u|\leq 1\}} \in L^\infty(\mu)$. We have also with $q = p/p_1 \geq 1$, $1/q' = 1 - p_1/p$,

$$\int_X |u\mathbf{1}_{\{|u|>1\}}|^{p_1} d\mu \leq \left(\int_X |u|^{p_1 q} d\mu\right)^{1/q} \left(\int_X \mathbf{1}_{\{|u|>1\}}^{p_1 q'} d\mu\right)^{1/q'}$$

$$= \|u\|_{L^p(\mu)}^{p_1} \mu(\{|u| > 1\})^{1 - \frac{p_1}{p}} \leq \|u\|_{L^p(\mu)}^{p_1 + (1 - \frac{p_1}{p})p} = \|u\|_{L^p(\mu)}^p < +\infty,$$

so that we have proven that $u\mathbf{1}_{\{|u|>1\}} \in L^{p_1}(\mu)$. If $p_2 = +\infty$, we use $u\mathbf{1}_{\{|u|\leq 1\}} \in L^{\infty}(\mu)$ to conclude. If $p_2 < +\infty$, we estimate

$$\int_X |u\mathbf{1}_{\{|u|\leq 1\}}|^{p_2}\, d\mu = \int_X |u\mathbf{1}_{\{|u|\leq 1\}}|^{p_2-p}|u|^p d\mu \leq \int_X |u|^p d\mu = \|u\|^p_{L^p(\mu)} < +\infty.$$

Finally we have proven more precisely that for $u \in L^p(\mu)$,

$$u = \underbrace{u\mathbf{1}_{\{|u|>1\}}}_{u_1} + \underbrace{u\mathbf{1}_{\{|u|\leq 1\}}}_{u_2}, \|u_1\|_{L^{p_1}(\mu)} \leq \|u\|^{p/p_1}_{L^p(\mu)}, \|u_2\|_{L^{p_2}(\mu)} \leq \|u\|^{p/p_2}_{L^p(\mu)}. \quad (6.6.6)$$

Chapter 7

Complex Measures

7.1 Complex measures

Definition 7.1.1. Let $(X, \mathcal{M})$ be a measurable space (see Definition 1.1.1). A complex measure on $(X, \mathcal{M})$ is a mapping $\mu : \mathcal{M} \to \mathbb{C}$ such that $\mu(\emptyset) = 0$ and for any sequence $(A_k)_{k \in \mathbb{N}}$ of pairwise disjoint elements of $\mathcal{M}$,

$$\mu(\cup_{k \in \mathbb{N}} A_k) = \sum_{k \in \mathbb{N}} \mu(A_k), \tag{7.1.1}$$

i.e., the series $\sum_{k \subset \mathbb{N}} \mu(A_k)$ converges in $\mathbb{C}$ with limit $\mu(\cup_{k \in \mathbb{N}} A_k)$. A real measure is a complex measure valued in $\mathbb{R}$.

N.B. Reading Definition 1.1.1 of a positive measure, we realize the unpleasant fact that a positive measure is not always a complex measure, since for a positive measure the convergence of the series with positive terms $\sum_{k \in \mathbb{N}} \mu(A_k)$ always holds in $\overline{\mathbb{R}}_+$, but not necessarily in $\mathbb{R}$: in the first place, some $\mu(A_k)$ could be $+\infty$ and even if every $\mu(A_k)$ is non-negative finite, it could happen that the series $\sum_{k \in \mathbb{N}} \mu(A_k) = +\infty$.

We note also that the set of complex measures on $(X, \mathcal{M})$ is a $\mathbb{C}$-vector space. We could have defined easily a vector-valued measure: with $(X, \mathcal{M})$ being a measurable space and B being a Banach space, Definition 7.1.1 gives a meaning to a B-valued measure. In particular when B is finite dimensional, the notion of an $\mathbb{R}^N$-valued measure follows easily from the notion of a real-valued measure. When B is infinite dimensional, integration theory presents specific difficulties which are beyond the scope of this book.

Remark 7.1.2. Definition 7.1.1 implies the so-called commutative convergence of the series $\sum_{k \in \mathbb{N}} \mu(A_k)$, which is equivalent to its absolute convergence (see Exercise 7.7.1). So it is a consequence of the definition of a complex measure, that for $(A_k)_{k \in \mathbb{N}}$ pairwise disjoint sets in $\mathcal{M}$, $\sum_{k \in \mathbb{N}} |\mu(A_k)| < +\infty$.

The following definition provides a good set of examples of complex measures.

Definition 7.1.3. Let $(X, \mathcal{M}, \mu)$ be a measure space where μ is a positive measure and let $h \in L^1(\mu)$. The complex measure ν defined on $\mathcal{M}$ by $\nu(E) = \int_E h d\mu$ is called the measure with density h with respect to μ and we use the notation $d\nu = h d\mu$. For $f \in L^1(\nu)$, we have $fh \in L^1(\mu)$ and

$$\int_X f d\nu = \int_X f h d\mu.$$

Remark 7.1.4. We have seen in Proposition 1.6.5 that in a measure space $(X, \mathcal{M}, \mu)$ where μ is a positive measure, given a measurable function $h : X \to \overline{\mathbb{R}}_+$, we may define a new positive measure ν on $(X, \mathcal{M})$ by

$$\nu(E) = \int_E h d\mu \quad \text{and for } f : X \to \overline{\mathbb{R}}_+ \text{ measurable} \int_X f d\nu = \int_X f \cdot h d\mu.$$

Of course when h belongs to $L^1(\mu)$, we can write

$$h = (\operatorname{Re} h)_+ - (\operatorname{Re} h)_- + i(\operatorname{Im} h)_+ - i(\operatorname{Im} h)_-,$$

and we may define the complex measure

$$d\nu = h d\mu, \quad \nu(E) = \int_E h d\mu.$$

The measure ν is the complex linear combination of the finite positive measures

$$d\nu = (\operatorname{Re} h)_+ d\mu - (\operatorname{Re} h)_- d\mu + i(\operatorname{Im} h)_+ d\mu - i(\operatorname{Im} h)_- d\mu.$$

There are various extensions of this notion when h does not belong to $L^1(\mu)$, for instance if $h = h_+ - h_-, h_\pm \geq 0$ measurable and $h_- \in L^1(\mu)$: in that case the positive measures $h_\pm d\mu$, well defined by Proposition 1.6.5, are such that $h_- d\mu$ is bounded so that $h_+ d\mu - h_- d\mu$ makes sense and is a measure.

Remark 7.1.5. More generally, on a measure space $(X, \mathcal{M})$ we may consider μ_1, μ_2 two positive measures such that

$$\{E \in \mathcal{M}, \mu_1(E) = \mu_2(E) = +\infty\} = \emptyset, \tag{7.1.2}$$

so that we may define the *signed measure*

$$\mu : \mathcal{M} \to \overline{\mathbb{R}}, \quad \mu(E) = \mu_1(E) - \mu_2(E).$$

We have of course $\mu(\emptyset) = 0$ and if $(A_k)_{k \in \mathbb{N}}$ is a pairwise disjoint sequence of $\mathcal{M}$, we have

$$\mu\big(\cup_\mathbb{N} A_k\big) = \sum_\mathbb{N} \mu(A_k).$$

To verify that the sum above converges in $\overline{\mathbb{R}}$ and that equality holds, we note that with $A = \cup_{\mathbb{N}} A_k$, either $\mu_1(A) < +\infty$ or $\mu_2(A) < +\infty$. Let us assume that the latter holds: then we have

$$\mu_1(A) = \sum_{\mathbb{N}} \mu_1(A_k), \quad \text{convergence in } \overline{\mathbb{R}}_+,$$

$$\mu_2(A) = \sum_{\mathbb{N}} \mu_2(A_k), \quad \text{convergence in } \mathbb{R}_+,$$

so that $\mu_1(A) - \mu_2(A)$ makes sense, belongs to $(-\infty, +\infty]$, and is equal to $+\infty$ if $\mu_1(A) = +\infty = \mu_1(A) - \mu_2(A) = \sum_{\mathbb{N}} \mu(A_k)$, with convergence[1] in $(-\infty, +\infty]$. If $\mu_1(A) < +\infty$, the σ-additivity property is obvious. Of course if $\mu_2(A) = +\infty$ so that $\mu_1(A) < +\infty$, the discussion is similar, leading to convergence in $[-\infty, +\infty)$. In both cases, convergence and equality hold in $\overline{\mathbb{R}}$.

7.2 Total variation of a complex measure

Definition 7.2.1. Let $(X, \mathcal{M})$ be a measurable space and let λ be a complex measure on $(X, \mathcal{M})$ (Definition 7.1.1). The *total variation measure* of λ, denoted by $|\lambda|$, is defined on $\mathcal{M}$ as

$$|\lambda|(E) = \sup_{\substack{(E_k)_{k \in \mathbb{N}} \text{ pairwise disjoint} \\ \text{with union } E, \\ E_k \in \mathcal{M}}} \sum_{\mathbb{N}} |\lambda(E_k)|. \tag{7.2.1}$$

The name total variation measure is justified by the following results proving that $|\lambda|$ is actually a positive measure on $\mathcal{M}$.

Remark 7.2.2. We may use the word partition of E for the $(E_k)_{k \in \mathbb{N}}$ although according to our definition, a partition $(X_i)_{i \in I}$ of a set X is a pairwise disjoint family of subsets of X, with union X and also such that no X_i is empty. Adding empty sets in the family does not change the sum in (7.2.1).

Proposition 7.2.3. *Let $(X, \mathcal{M})$ be a measurable space and let λ be a complex measure on $(X, \mathcal{M})$. The total variation measure of λ is a positive measure on $(X, \mathcal{M})$ with a finite total variation, i.e., such that $|\lambda|(X) < +\infty$.*

---

[1] Let (a_k) be a sequence in $\overline{\mathbb{R}}_+$ such that $\sum_{\mathbb{N}} a_k = +\infty$ and let (b_k) be a sequence in $\mathbb{R}_+$ such that $\sum_{\mathbb{N}} b_k < +\infty$. Then $\lim_{n \to +\infty} \sum_{0 \le k \le n} (a_k - b_k) = +\infty$: in the first place for each k, $a_k - b_k$ makes sense in $(-\infty, +\infty]$ and

$$\sum_{0 \le k \le n} (a_k - b_k) \ge \left(\sum_{0 \le k \le n} a_k - \sum_{\mathbb{N}} b_k \right) \xrightarrow[n \to +\infty]{} +\infty, \qquad \text{qed.}$$

Proof. We have obviously $|\lambda|(\emptyset) = 0$. Let $A \in \mathcal{M}$ and let $(A_n)_{n \in \mathbb{N}}$ be a partition of A. Let us consider $(F_k)_{k \in \mathbb{N}}$ a partition of A: we have

$$\sum_{k \in \mathbb{N}} |\lambda(F_k)| = \sum_{k \in \mathbb{N}} \left| \sum_{n \in \mathbb{N}} \lambda(A_n \cap F_k) \right| \le \sum_{n \in \mathbb{N}} \sum_{k \in \mathbb{N}} |\lambda(A_n \cap F_k)| \le \sum_{n \in \mathbb{N}} |\lambda|(A_n),$$

implying

$$|\lambda|(A) \le \sum_{n \in \mathbb{N}} |\lambda|(A_n).$$

Since Formula (7.2.1) is obviously increasing[2] with E, we may assume that for all $n \in \mathbb{N}$, $|\lambda|(A_n) < +\infty$. This implies that for all n, all $\epsilon > 0$, there exists $(E_{n,k,\epsilon})_{k \in \mathbb{N}}$ partition of A_n such that

$$|\lambda|(A_n) - \epsilon 2^{-n-1} < \sum_{k \in \mathbb{N}} |\lambda(E_{n,k,\epsilon})| \le |\lambda|(A_n).$$

Since we have $\sum_{n,k \in \mathbb{N}} |\lambda(E_{n,k,\epsilon})| \le |\lambda|(A)$ we obtain

$$\forall \epsilon > 0, \quad -\epsilon + \sum_n |\lambda|(A_n) \le |\lambda|(A) \implies \sum_n |\lambda|(A_n) \le |\lambda|(A),$$

proving that $|\lambda|$ is indeed a positive measure on $(X, \mathcal{M})$.

Let us now prove that $|\lambda|$ is bounded.

Lemma 7.2.4. *Let $X_0 \in \mathcal{M}$ such that $|\lambda|(X_0) = +\infty$. Then there exists $X_1, Y_1 \in \mathcal{M}$ such that*

$$X_0 = X_1 \cup Y_1, \quad X_1 \cap Y_1 = \emptyset, \quad |\lambda(Y_1)| \ge 1, \quad |\lambda|(X_1) = +\infty.$$

Proof of the lemma. From the assumption on X_0, for any $\epsilon > 0$, we may find a partition $(A_{k,\epsilon})_{k \in \mathbb{N}}$ of X_0 such that

$$\sum_{k \in \mathbb{N}} |\lambda(A_{k,\epsilon})| > \frac{5\sqrt{2}}{\epsilon} \implies \exists N, \text{ such that } \sum_{0 \le k \le N} |\lambda(A_{k,\epsilon})| > \frac{4\sqrt{2}}{\epsilon}, \qquad (7.2.2)$$

and according to Exercise 7.7.2, we find $J \subset \{0, \ldots, N\}$ such that

$$|\lambda(\cup_{k \in J} A_{k,\epsilon})| = \left| \sum_{k \in J} \lambda(A_{k,\epsilon}) \right| > \frac{1}{\epsilon},$$

---

[2]Let $E \subset F$ in $\mathcal{M}$ and $(E_k)_{k \in \mathbb{N}}$ be a partition of E: we have

$$\sum_{k \in \mathbb{N}} |\lambda(E_k)| \le |\lambda(F \cap E^c)| + \sum_{k \in \mathbb{N}} |\lambda(E_k)| \le |\lambda|(F).$$

and this implies

$$|\lambda(\cup_{k\in\mathbb{N}\setminus J}A_{k,\epsilon})| = |\lambda(X) - \lambda(\cup_{k\in J}A_{k,\epsilon})| > \frac{1}{\epsilon} - |\lambda(X)| > \frac{1}{2\epsilon},$$

provided ϵ is chosen ab initio such that $2\epsilon|\lambda(X)| < 1$. Moreover, since $|\lambda|$ is a positive measure and X_0 is the disjoint union of $(\cup_{k\in\mathbb{N}\setminus J}A_{k,\epsilon})$, $(\cup_{k\in J}A_{k,\epsilon})$, at least one of these sets has infinite $|\lambda|$ measure and we have proven above that both have λ measure with absolute value ≥ 1, providing we choose $\epsilon = \frac{1}{2+2|\lambda(X)|}$. This completes the proof of the lemma. $\qquad\square$

Now let us show that $|\lambda|(X) = +\infty$ leads to a contradiction. Using the previous lemma, we set $X_0 = X$ and we find X_1, Y_1 disjoint subsets of X_0 such that $|\lambda|(X_1) = +\infty$ and $|\lambda(Y_1)| \geq 1$. We may thus apply the lemma again and find X_2, Y_2 disjoint subsets of X_1 such that $|\lambda|(X_2) = +\infty$ and $|\lambda(Y_2)| \geq 1$. Let us assume that we have found $(X_1, Y_1), \ldots, (X_n, Y_n)$, with

$$X_j \cap Y_j = \emptyset, \quad X_j \cup Y_j = X_{j-1}, \quad |\lambda|(X_j) = +\infty, \quad |\lambda(Y_j)| \geq 1, \ j \geq 1.$$

we may apply the lemma and find X_{n+1}, Y_{n+1} disjoint subsets of $X_n = X_{n+1}\cup Y_{n+1}$ such that $|\lambda|(X_{n+1}) = +\infty$ and $|\lambda(Y_{n+1})| \geq 1$. As a result, we have constructed a sequence $(Y_n)_{n\geq 1}$ of elements of $\mathcal{M}$, such that $|\lambda(Y_n)| \geq 1$ and for $n, m \geq 1$,

$$Y_n \cap Y_{n+m} \subset Y_n \cap X_{n+m-1} \subset Y_n \cap X_n = \emptyset,$$

so that the $(Y_n)_{n\geq 1}$ are pairwise disjoint elements of $\mathcal{M}$ such that $|\lambda(Y_n)| \geq 1$. This is incompatible with the convergence of the series $\sum_{n\geq 1}\lambda(Y_n)$. The proof of the proposition is complete. $\qquad\square$

Definition 7.2.5 (Jordan decomposition of a real measure). Let $(X, \mathcal{M})$ be a measure space and let λ be a real measure. We define

$$\lambda_+ = \frac{1}{2}(|\lambda| + \lambda), \quad \lambda_- = \frac{1}{2}(|\lambda| - \lambda). \tag{7.2.3}$$

The positive measures $\lambda_\pm$ are both bounded and satisfy

$$|\lambda| = \lambda_+ + \lambda_-, \quad \lambda = \lambda_+ - \lambda_-. \tag{7.2.4}$$

7.3 Absolute continuity, mutually singular measures

Definition 7.3.1. Let $(X, \mathcal{M})$ be a measurable space, let μ be a positive measure on $(X, \mathcal{M})$ and let λ be a complex or a positive measure on $(X, \mathcal{M})$. We shall say that λ is absolutely continuous with respect to μ, and use the notation $\lambda \ll \mu$, whenever

$$\text{for } E \in \mathcal{M}, \quad \mu(E) = 0 \implies \lambda(E) = 0. \tag{7.3.1}$$

Lemma 7.3.2. *Let $(X, \mathcal{M})$ be a measurable space, let λ be a complex or a positive measure on $(X, \mathcal{M})$ and let $C \in \mathcal{M}$. The following properties are equivalent:*

(i) *for all $E \in \mathcal{M}$, $\lambda(E) = \lambda(E \cap C)$.*

(ii) *for all $E \in \mathcal{M}$, $E \subset C^c \Longrightarrow \lambda(E) = 0$.*

Such a set C will be called a **carrier** *of λ.*

Proof. Obviously (i) implies (ii) since $\lambda(\emptyset) = 0$. Conversely, if (ii) holds, and $E \in \mathcal{M}$, we have $\lambda(E) = \lambda(E \cap C) + \underbrace{\lambda(E \cap C^c)}_{=0} = \lambda(E \cap C)$. $\square$

Definition 7.3.3. Let $(X, \mathcal{M})$ be a measurable space, and let λ_1, λ_2 be two measures on $\mathcal{M}$. These two measures will be said to be *mutually singular* whenever they are carried by disjoint sets: there exist $A_1, A_2 \in \mathcal{M}, A_1 \cap A_2 = \emptyset$ such that A_j is a carrier of $\lambda_j, j = 1, 2$. We shall then use the notation $\lambda_1 \perp \lambda_2$.

Proposition 7.3.4. *Let $(X, \mathcal{M})$ be a measurable space, and let $\mu, \lambda, \lambda_1, \lambda_2$ be measures on $\mathcal{M}$ with μ a positive measure. Then we have*

$$\lambda_1 \perp \mu \ and \ \lambda_2 \perp \mu \Longrightarrow \lambda_1 + \lambda_2 \perp \mu, \tag{7.3.2}$$

$$\lambda_1 \ll \mu \ and \ \lambda_2 \ll \mu \Longrightarrow \lambda_1 + \lambda_2 \ll \mu, \tag{7.3.3}$$

$$\lambda_1 \ll \mu \ and \ \lambda_2 \perp \mu \Longrightarrow \lambda_1 \perp \lambda_2, \tag{7.3.4}$$

$$\lambda \ll \mu \ and \ \lambda \perp \mu \ \Longrightarrow \lambda = 0. \tag{7.3.5}$$

Proof. If $\lambda_j \perp \mu, j = 1, 2$, then there exist $A_1, A_2, A \in \mathcal{M}$ such that A_j is a carrier for λ_j, B_1, B_2 are carriers for μ and $A_j \cap B_j = \emptyset, j = 1, 2$. Then $B_1 \cap B_2$ is also a carrier for μ (obvious from (i) in Lemma 7.3.2) and $A_1 \cup A_2$ is a carrier for $\lambda_1 + \lambda_2$ since $\forall E \in \mathcal{M}$,

$$(\lambda_1 + \lambda_2)(E) = (\lambda_1 + \lambda_2)(E \cap (A_1 \cup A_2)) + (\lambda_1 + \lambda_2)(E \cap A_1^c \cap A_2^c)$$
$$= (\lambda_1 + \lambda_2)(E \cap (A_1 \cup A_2)),$$

since $\lambda_j(E \cap A_1^c \cap A_2^c) = 0$. Since we have $(A_1 \cup A_2) \cap (B_1 \cap B_2) = \emptyset$, this gives the sought result.

Let us assume now $\lambda_1 \ll \mu$ and $\lambda_2 \ll \mu$ and let $E \in \mathcal{M}$ such that $\mu(E) = 0$. Then we have $\lambda_j(E) = 0$ and the result.

If $\lambda_1 \ll \mu$ and $\lambda_2 \perp \mu$, we find $A_2, B \in \mathcal{M}$ such that $A_2 \cap B = \emptyset$, μ carried by B and λ_2 carried by A_2. We have thus for $E \in \mathcal{M}$,

$$\lambda_1(E) = \lambda_1(E \cap A_2^c) + \lambda_1(\underbrace{E \cap A_2}_{\subset A_2 \subset B^c}) = \lambda_1(E \cap A_2^c),$$

since $E \cap A_2 \subset B^c \Longrightarrow \mu(E \cap A_2) = 0 \Longrightarrow \lambda_1(E \cap A_2) = 0$. As a result λ_1 is carried by A_2^c which is disjoint from A_2, a carrier of λ_2, entailing $\lambda_1 \perp \lambda_2$.

We assume now $\lambda \ll \mu$ and $\lambda \perp \mu$. Let A, B be disjoint in $\mathcal{M}$ respective carriers for λ, μ. For $E \in \mathcal{M}$, we have $\lambda(E) = \lambda(E \cap A)$ and since $\mu(E \cap A) = $

$\mu(E \cap A \cap B) = \mu(\emptyset) = 0$, the assumption $\lambda \ll \mu$ implies $\lambda(E \cap A) = 0$ and thus $\lambda(E) = 0$. The proof of the proposition is complete. $\qquad\square$

Lemma 7.3.5. *Let $(X, \mathcal{M}, \mu)$ be a measure space where μ is a positive measure. Then μ is σ-finite if and only if there exists $w \in \mathcal{L}^1(\mu)$ such that for all $x \in X$, $0 < w(x) < 1$.*

Proof. A simple modification of Exercise 3.7.9: if μ is σ-finite, take $w = f/3$ where f is given by (3.7.4) on page 161. The same exercise provides a stronger converse. $\qquad\square$

7.4 Radon–Nikodym theorem

Theorem 7.4.1 (Radon–Nikodym Theorem). *Let $(X, \mathcal{M})$ be a measurable space, let μ be a positive σ-finite measure on $(X, \mathcal{M})$ and let λ be a complex measure on $(X, \mathcal{M})$.*

(1) *There exists a unique couple (λ_a, λ_s) of complex measures on $(X, \mathcal{M})$ such that*

$$\lambda = \lambda_a + \lambda_s, \quad \lambda_a \ll \mu, \quad \lambda_s \perp \mu.$$

(2) *There exists a unique element $h \in L^1(\mu)$ such that $d\lambda_a = h d\mu$, i.e., for all $E \in \mathcal{M}$,*

$$\lambda_a(E) = \int_E h d\mu.$$

*The couple of measures (λ_a, λ_s) is called the **Lebesgue decomposition** of λ with respect to μ and h is called the **Radon–Nikodym derivative** of λ with respect to μ.*

Proof. We shall follow the proof given by John von Neumann ([66]). Let us prove the uniqueness properties: if for $\lambda_{a,j}, \lambda_{s,j}, j = 1, 2$, complex measures, we have

$$\lambda = \lambda_{a,j} + \lambda_{s,j}, \quad \lambda_{a,j} \ll \mu, \quad \lambda_{s,j} \perp \mu,$$

then, from (7.3.3), $\lambda_{a,1} - \lambda_{a,2} \ll \mu$ and since $\lambda_{a,1} - \lambda_{a,2} = \lambda_{s,1} - \lambda_{s,2} \perp \mu$ (from (7.3.2)), property (7.3.5) implies $\lambda_{a,1} - \lambda_{a,2} = 0$ and thus $\lambda_{s,1} = \lambda_{s,2}$. Moreover, if $d\lambda_a = h_j d\mu$, $h_j \in L^1(\mu)$, $j = 1, 2$, we obtain for all $E \in \mathcal{M}$,

$$\int_E (h_1 - h_2) d\mu = 0,$$

which implies $h_1 = h_2$ from Exercise 1.9.27. We shall now prove the existence part, which is the most interesting part of this theorem.

Let us assume first that λ is a bounded positive measure. Let w be a function given by Lemma 7.3.5 and let us define the bounded positive measure ϕ by

$$d\phi = d\lambda + w d\mu.$$

For $E \in \mathcal{M}$ and $f = \mathbf{1}_E$, we have

$$\int_X f d\phi = \int_X f d\lambda + \int_X w f d\mu. \tag{7.4.1}$$

As a result, this equality holds as well for simple functions (see Definition 1.3.2), and thus for a non-negative measurable function, we apply Beppo Levi's theorem 1.6.1 and Theorem 1.3.3 (to the three positive measures $d\phi, d\lambda, w d\mu$). For $f \in L^2(\phi)$, we have

$$\left| \int_X f d\lambda \right| \leq \int_X |f| d\lambda \leq \int_X |f| d\phi \leq \|f\|_{L^2(\phi)} \phi(X)^{\frac{1}{2}}$$

$$\leq \|f\|_{L^2(\phi)} \big(\lambda(X) + \|w\|_{L^1(\mu)} \big)^{\frac{1}{2}}.$$

Consequently, the mapping $L^2(\phi) \ni f \mapsto \int_X f d\lambda$ is a continuous linear form on $L^2(\phi)$: by the classical Riesz representation theorem of continuous linear forms in a Hilbert space, we know that there exists a unique $g \in L^2(\phi)$ such that

$$\forall f \in L^2(\phi), \quad \int_X f d\lambda = \langle f, g \rangle_{L^2(\phi)}. \tag{7.4.2}$$

Let $E \in \mathcal{M}$ such that $\phi(E) > 0$; for $f = \mathbf{1}_E$ in (7.4.2), we find

$$\lambda(E) = \int_E \bar{g} d\phi,$$

and since $\lambda(E)$ is real this implies in particular that $\int_E \operatorname{Im} g d\phi = 0$, for all $E \in \mathcal{M}$, so that g is real-valued ϕ-almost everywhere. Moreover, from the inequality $\lambda(E) \leq \phi(E)$, we infer that for all $E \in \mathcal{M}$ such that $\phi(E) > 0$,

$$0 \leq \frac{1}{\phi(E)} \int_E g d\phi \leq 1,$$

and from Exercise 1.9.30 and $g \in L^2(\phi) \subset L^1(\phi)$ (due to ϕ bounded measure), this implies that $g(x) \in [0,1]$, ϕ-almost everywhere, i.e., on N^c where $N \in \mathcal{M}$ with $\phi(N) = 0$. We may replace g in (7.4.2) by $\tilde{g} = g \mathbf{1}_{N^c}$ and find that $\forall x \in X, \tilde{g}(x) \in [0,1]$, so that we may rewrite (7.4.2) as

$$\forall f \in L^2(\phi), \quad \int_X f d\lambda = \int_X f \tilde{g} d\phi = \int_X f \tilde{g} d\lambda + \int_X f \tilde{g} w d\mu,$$

$$\text{i.e.,} \quad \int_X f(1 - \tilde{g}) d\lambda = \int_X f \tilde{g} w d\mu. \tag{7.4.3}$$

Let us now define the positive measures λ_a, λ_s on $(X, \mathcal{M})$ by

$$A = \{x \in X, 0 \leq \tilde{g}(x) < 1\}, \ B = \{x \in X, 0 \leq \tilde{g}(x) = 1\} \ (\text{note } A, B \in \mathcal{M}), \tag{7.4.4}$$

$$\text{for } E \in \mathcal{M}, \quad \lambda_a(E) = \lambda(A \cap E), \quad \lambda_s(E) = \lambda(B \cap E). \tag{7.4.5}$$

Taking $f = 1_B$ in (7.4.3), we obtain $\int_B w\,d\mu = 0$, and since $w(x) > 0$ for all x, this implies $1_B w = 0$, μ-a.e. and thus $\mu(B) = 0$, so that

$$\lambda_s \perp \mu, \quad (B \text{ is a carrier of } \lambda_s).$$

In (7.4.3), we may as well take $f = (1 + \tilde{g} + \cdots + \tilde{g}^N)1_E$ for $E \in \mathcal{M}$ since $\tilde{g}$ is bounded and the measure ϕ is bounded, entailing

$$\int_{E\cap A} (1 - \tilde{g}^{N+1})d\lambda = \int_E (1 - \tilde{g}^{N+1})d\lambda = \int_E \tilde{g}(1 + \tilde{g} + \cdots + \tilde{g}^N)w\,d\mu.$$

For $x \in A$, the sequence $(1 - \tilde{g}^{N+1}(x))$ converges monotonically to 1, so that, thanks to Beppo Levi's theorem, the lhs converges to $\lambda(E \cap A) = \lambda_a(E)$. The sequence $\big(\tilde{g}(x)(1 + \tilde{g}(x) + \cdots + \tilde{g}^N(x))\big)_{N\in\mathbb{N}}$ converges increasingly towards a non-negative measurable function $h(x)$, so that

$$\forall E \in \mathcal{M}, \quad \lambda_a(E) = \int_E h\,d\mu.$$

Since $\lambda_a(X) < +\infty$, we get as well that $h \in L^1(\mu)$ and $\lambda_a \ll \mu$, which concludes the proof for a λ bounded positive measure.

Let us assume now that λ is a complex measure on $(X, \mathcal{M})$. Then, according to the decomposition into real and imaginary parts and to the Jordan decomposition (Definition 7.2.5), we have

$$\lambda = \operatorname{Re}\lambda + i\operatorname{Im}\lambda = (\operatorname{Re}\lambda)_+ - (\operatorname{Re}\lambda)_- + i(\operatorname{Im}\lambda)_+ - i(\operatorname{Im}\lambda)_-$$

where $(\operatorname{Re}\lambda)_\pm, (\operatorname{Im}\lambda)_\pm$ are bounded positive measures to which we may apply the previous result. This completes the proof of Theorem 7.4.1. $\qquad\square$

Remark 7.4.2. If λ is a positive σ-finite measure (as well as μ), then using Lemma 7.3.5 we can find a measurable function v valued in $(0, 1)$ such that $vd\lambda$ is a bounded positive measure. We can use the Lebesgue decomposition of $vd\lambda$, so that for f non-negative measurable,

$$\int_X fv\,d\lambda = \int_X fh\,d\mu + \int_X f\,d\nu_s, \quad \nu_a \ll \mu, \quad 0 \le h \in L^1(\mu), \quad \nu_s \perp \mu.$$

This implies $\int_X f\,d\lambda = \int_X fv^{-1}h\,d\mu + \int_X fv^{-1}d\nu_s$, and

$$d\lambda = v^{-1}h\,d\mu + v^{-1}d\nu_s, \quad \lambda(E) = \int_E v^{-1}h\,d\mu + \int_{E\cap C} v^{-1}d\nu_s, \quad \mu(C) = 0.$$

The positive measure $v^{-1}h\,d\mu$ is absolutely continuous with respect to μ, thanks to (3) in Proposition 1.5.4 which implies as well that $v^{-1}d\nu_s$ and $d\mu$ are mutually singular. This means that the first part of the Radon–Nikodym Theorem holds for λ a positive σ-finite measure (and of course μ positive σ-finite), although the second part may not hold since the function $v^{-1}h$ need not be in $L^1(\mu)$: however, due to the explicit construction used in Lemma 7.3.5, we see that $v^{-1}h1_{X_n}$ will belong to $L^1(\mu)$ if $\cup_{n\in\mathbb{N}}X_n = X$, $\lambda(X_n) < +\infty$.

Lemma 7.4.3 (Hahn decomposition)**.** *Let $(X, \mathcal{M})$ be a measure space and let λ be a real measure. There exists a partition $\{A_+, A_-\}$ of X with $A_\pm \in \mathcal{M}$ such that*

$$\forall E \in \mathcal{M}, \quad \lambda_\pm(E) = \pm\lambda(A_\pm \cap E).$$

Proof. See Exercise 7.7.9. $\qquad\qquad\qquad\qquad\qquad\qquad\qquad\qquad\qquad\qquad\qquad\quad\;\;\square$

Theorem 7.4.4. *Let $(X, \mathcal{M})$ be a measure space. The mapping $\lambda \mapsto |\lambda|(X)$ is a norm on the vector space $\mathcal{M}(X, \mathcal{M})$ of complex measures on $(X, \mathcal{M})$. Using the notation $\|\lambda\| = |\lambda|(X)$, $\|\|\lambda\|\| = \sup_{E \in \mathcal{M}} |\lambda(E)|$, these formulas are defining norms on $\mathcal{M}(X, \mathcal{M})$ which are equivalent and make $\mathcal{M}(X, \mathcal{M})$ a Banach space.*

Proof. Note first that $\mathcal{M}(X, \mathcal{M})$ is obviously a complex vector space and Definition 7.2.1 implies that $\|\cdot\|$ is valued in $\mathbb{R}_+$, homogeneous with degree 1, separated, and satisfies the triangle inequality. The quantity $\|\|\lambda\|\| = \sup_{E \in \mathcal{M}} |\lambda(E)|$ is such that

$$\|\|\lambda\|\| \leq \sup_{E \in \mathcal{M}} |\lambda|(E) \leq \|\lambda\| < +\infty.$$

Thus $\|\|\cdot\|\|$ is a norm on $\mathcal{M}(X, \mathcal{M})$ (we have proven finiteness, which was the only non-obvious property). Moreover, if λ is a real measure, we have from the Jordan-Hahn decomposition (Definition 7.2.5, Lemma 7.4.3),

$$|\lambda|(X) = \lambda_+(X) + \lambda_-(X) = \lambda(A_+) - \lambda(A_-) \leq 2\|\|\lambda\|\|.$$

If λ is a complex measure, we have $\lambda = \operatorname{Re}\lambda + i\operatorname{Im}\lambda$ and thus

$$\|\lambda\| \leq \|\operatorname{Re}\lambda\| + \|\operatorname{Im}\lambda\| \leq 2\|\|\operatorname{Re}\lambda\|\| + 2\|\|\operatorname{Im}\lambda\|\| = \|\|\lambda + \overline{\lambda}\|\| + \|\|\lambda - \overline{\lambda}\|\| \leq 4\|\|\lambda\|\|,$$

where the definition of $\overline{\lambda}$ is simply $\overline{\lambda}(E) = \overline{\lambda(E)}$, which makes $\overline{\lambda}$ a complex measure with the same norm as λ (true for $\|\cdot\|, \|\|\cdot\|\|$).

Let us now consider a Cauchy sequence $(\mu_n)_{n \in \mathbb{N}}$ in $\mathcal{M}(X, \mathcal{M})$. For each $E \in \mathcal{M}$, the sequence of complex numbers $(\mu_n(E))_{n \in \mathbb{N}}$ is a Cauchy sequence and we may define

$$\mu(E) = \lim_n \mu_n(E). \qquad\qquad\qquad (7.4.6)$$

We have thus obviously $\mu(\emptyset) = 0$ and finite additivity. Let $(A_k)_{k \in \mathbb{N}}$ be a pairwise disjoint sequence of elements of $\mathcal{M}$. We note first that

$$|\mu(E) - \mu_n(E)| = \lim_m |\mu_m(E) - \mu_n(E)| \leq \limsup_m \|\|\mu_m - \mu_n\|\| = \epsilon_n \xrightarrow[n \to +\infty]{} 0.$$

Using the finite additivity property of μ we have

$$\mu(\cup_{k \in \mathbb{N}} A_k) = \sum_{0 \leq k \leq N} \mu(A_k) + \mu(\cup_{k > N} A_k).$$

We have also for $n, N \in \mathbb{N}$,

$$\left| \mu(\cup_{k \in \mathbb{N}} A_k) - \sum_{0 \le k \le N} \mu(A_k) \right| = \left| \mu(\cup_{k > N} A_k) \right|$$

$$\le \left| \mu(\cup_{k>N} A_k) - \mu_n(\cup_{k>N} A_k) \right| + \left| \mu_n(\cup_{k>N} A_k) \right| \le \epsilon_n + \sum_{k > N} \left| \mu_n(A_k) \right|,$$

so that for all $n \in \mathbb{N}$, $\limsup_{N \to +\infty} \left| \mu(\cup_{k \in \mathbb{N}} A_k) - \sum_{0 \le k \le N} \mu(A_k) \right| \le \epsilon_n$, and since $\lim_n \epsilon_n = 0$, this implies

$$\lim_{N \to +\infty} \left| \mu(\cup_{k \in \mathbb{N}} A_k) - \sum_{0 \le k \le N} \mu(A_k) \right| = 0,$$

proving the convergence of the series $\sum_{k \ge 0} \mu(A_k)$ towards $\mu(\cup_{k \in \mathbb{N}} A_k)$, which is the sought σ-additivity. The proof of Theorem 7.4.4 is complete. $\qquad\square$

7.5   The dual of $L^p(X, \mathcal{M}, \mu)$, $1 \le p < +\infty$

Let $(X, \mathcal{M}, \mu)$ be a measure space where μ is a positive measure. We consider the Banach spaces $L^p(\mu)$ (see Theorem 3.2.8) and we want to determine their dual spaces whenever $1 \le p < +\infty$ and the measure μ is σ-finite. For $1 \le p < +\infty$, we shall denote by p' the conjugate index such that

$$\frac{1}{p} + \frac{1}{p'} = 1$$

($p' = p/(p-1)$ if $1 < p < +\infty$ and $p' = +\infty$ if $p = 1$).

Main result

Theorem 7.5.1. *Let $(X, \mathcal{M}, \mu)$ be a measured space where μ is a σ-finite positive measure and let $1 \le p < +\infty$. Let $\xi \in (L^p(\mu))^*$, the topological dual of $L^p(\mu)$. Then there exists a unique $g \in L^{p'}(\mu)$ such that*

$$\forall f \in L^p(\mu), \quad \langle \xi, f \rangle = \int_X f g \, d\mu, \quad \|\xi\|_{(L^p(\mu))^*} = \|g\|_{L^{p'}(\mu)},$$

so that, for $1 \le p < +\infty$, $(L^p(\mu))^* = L^{p'}(\mu)$.

N.B. We may consider the sesquilinear mapping

$$\Phi : L^p(\mu) \times L^{p'}(\mu) \longrightarrow \mathbb{C}$$
$$(f, g) \mapsto \int_X f \bar{g} \, d\mu.$$

which is well defined, thanks to Hölder's inequality (Theorem 3.1.6), and satisfy

$$|\Phi(f, g)| \le \|f\|_{L^p} \|g\|_{L^{p'}}.$$

Let us check that the mapping $L^{p'}(\mu) \ni g \mapsto \Phi_g \in (L^p(\mu))^*$ given by $\Phi_g(f) = \Phi(f, g)$ is isometric, i.e.,

$$\|\Phi_g\|_{(L^p)^*} = \sup_{\|f\|_{L^p}=1} \left| \int_X f\bar{g}\,d\mu \right| = \|g\|_{L^{p'}}. \tag{7.5.1}$$

In fact the inequality $\|\Phi_g\|_{(L^p)^*} \leq \|g\|_{L^{p'}}$ follows from Hölder's inequality and for a given $0 \neq g \in L^{p'}$ and $1 < p < +\infty$ we have, with

$$f = \frac{g}{|g|}|g|^{p'/p}\mathbf{1}_{g\neq 0}\|g\|_{L^{p'}}^{-p'/p}, \quad \|f\|_{L^p}^p = \int_X |g|^{p'}\,d\mu\|g\|_{L^{p'}}^{-p'} = 1,$$

and the equality

$$\int_X f\bar{g}\,d\mu = \int_X |g|^{1+\frac{p'}{p}}\,d\mu\|g\|_{L^{p'}}^{-p'/p} = \|g\|_{L^{p'}}^{-\frac{p'}{p}+p'} = \|g\|_{L^{p'}}.$$

The same type of argument works for $p = 1$: here $p' = +\infty$ and for $0 \neq g \in L^\infty$ we choose $\epsilon > 0$ such that $\mu(\{|g| \geq \|g\|_{L^\infty} - \epsilon\}) > 0$ and we set

$$f = \frac{g}{|g|}\frac{\mathbf{1}(|g| \geq \|g\|_{L^\infty} - \epsilon)}{\mu(\{|g| \geq \|g\|_{L^\infty} - \epsilon\})}, \quad \text{so that } \|f\|_{L^1} = 1,$$

and

$$\Phi_g(f) = \int_X |g|\underbrace{\frac{\mathbf{1}(|g| \geq \|g\|_{L^\infty} - \epsilon)}{\mu(\{|g| \geq \|g\|_{L^\infty} - \epsilon\})}}_{G_\epsilon}\,d\mu = \frac{1}{\mu(G_\epsilon)}\int_{\|g\|_{L^\infty}-\epsilon\leq|g|\leq\|g\|_{L^\infty}} |g|\,d\mu$$

$$\geq \frac{1}{\mu(G_\epsilon)}(\|g\|_{L^\infty} - \epsilon)\mu(G_\epsilon) = \|g\|_{L^\infty} - \epsilon.$$

As a result $\|\Phi_g(f)\|_{(L^1)^*} = \|g\|_{L^\infty}$. As a result the mapping

$$\psi : L^{p'}(\mu) \longrightarrow (L^p(\mu))^*, \quad \psi(g) = \Phi_g$$

is injective and isometric and thus has a closed image isomorphic to $L^{p'}(\mu)$ (thanks to the Open Mapping Theorem 10.2.43). The main difficulty of the above theorem is the proof that ψ is indeed *onto* when $1 \leq p < +\infty$. We shall see some examples (see (7.5.11)) showing that for $p = \infty$, the dual space of L^∞, i.e., the bidual of L^1 is much larger than L^1 and that the mapping ψ is not onto in general in that case[3].

Proof of the theorem. Let then $1 \leq p < +\infty$ and $\xi \in (L^p(\mu))^*$. *We assume first that $\mu(X) < +\infty$.* For $E \in \mathcal{M}$, we define

$$\lambda(E) = \xi(\mathbf{1}_E). \tag{7.5.2}$$

---

[3]It is true however that ψ is an isometric one-to-one mapping, even for $p = \infty$: for $g \in L^1$, we have $\Phi_g(\frac{g}{|g|}\mathbf{1}_{\{g\neq 0\}}) = \|g\|_{L^1}$.

If A, B are measurable and disjoint, we have $\mathbf{1}_{A \cup B} = \mathbf{1}_A + \mathbf{1}_B$, which implies that λ is finitely additive. Let us consider $E = \cup_{j \in \mathbb{N}} E_j$ with $(E_j)_{j \in \mathbb{N}}$ pairwise disjoint elements of $\mathcal{M}$. With $A_k = \cup_{j \le k} E_j$, we have

$$\|\mathbf{1}_E - \mathbf{1}_{A_k}\|_{L^p}^p = \int_{E \setminus A_k} d\mu = \mu(E \setminus A_k).$$

By the Lebesgue dominated convergence theorem, we know that $\lim_k \mu(E \setminus A_k) = 0$, and since ξ is continuous on L^p, we get that $\lim_k \lambda(A_k) = \lambda(E)$, i.e.,

$$\lambda(E) = \sum_{j \in \mathbb{N}} \lambda(E_j),$$

so that λ is a complex measure. Moreover if $\mu(E) = 0$, we have $\mathbf{1}_E = 0$, μ-a.e. and $\mathbf{1}_E = 0$ in L^p implying $\lambda(E) = 0$. As a result we have $\lambda \ll \mu$. We may apply the Radon–Nikodym Theorem 7.4.1: there exists $g \in L^1(\mu)$ such that

$$\xi(\mathbf{1}_E) = \lambda(E) = \int_E g d\mu = \int_X g \mathbf{1}_E d\mu.$$

Thus, by the linearity of ξ, for any simple function f (finite linear combination of characteristic functions of measurable sets) we get

$$\xi(f) = \int_X f g d\mu, \quad \text{which is true as well for } f \in L^\infty(\mu), \tag{7.5.3}$$

since a function in $L^\infty(\mu)$ is a uniform limit of simple functions. If $p = 1$, for all $E \in \mathcal{M}$, we have

$$\left| \int_X \mathbf{1}_E g d\mu \right| = |\xi(\mathbf{1}_E)| \le \|\xi\|_{(L^1)^*} \|\mathbf{1}_E\|_{L^1} = \mu(E) \|\xi\|_{(L^1)^*},$$

and thus $|g(x)| \le \|\xi\|_{(L^1)^*}$ μ-a.e., implying

$$\|g\|_{L^\infty(\mu)} \le \|\xi\|_{(L^1)^*}. \tag{7.5.4}$$

If $1 < p < +\infty$, we consider a measurable function α such that $\alpha g = |g|$ (see Exercise 1.9.16), and we define

$$f_n = \mathbf{1}_{E_n} |g|^{p'-1} \alpha, \quad E_n = \{|g| \le n\}.$$

We have $|\alpha| = 1$ on the set $\{g \ne 0\}$ and $p(p' - 1) = p'$ so that

$$|f|_n^p = \mathbf{1}_{E_n} |g|^{p'}, \quad |f_n| \le n^{p'},$$

and applying (7.5.3) to the L^∞ function f_n, we get

$$\xi(f_n) = \int_X \mathbf{1}_{E_n} |g|^{p'-1} \alpha g d\mu = \int_{E_n} |g|^{p'} d\mu,$$

and $\left| \int_{E_n} |g|^{p'} d\mu \right| \le \|\xi\|_{(L^p)^*} \|f_n\|_{L^p} = \|\xi\|_{(L^p)^*} \left(\int_{E_n} |g|^{p'} d\mu \right)^{1/p}$ and this implies

$$\left| \int_{E_n} |g|^{p'} d\mu \right|^{1-\frac{1}{p}=\frac{1}{p'}} \le \|\xi\|_{(L^p)^*}.$$

Beppo Levi's theorem 1.6.1 then implies that $\|g\|_{L^{p'}} \le \|\xi\|_{(L^p)^*}$. Since ξ and $f \mapsto \int f g d\mu$ coincide (and are continuous) on $L^\infty(\mu)$, which is dense in $L^p(\mu)$ (see Proposition 3.2.11), they coincide on $L^p(\mu)$ and $\|\xi\|_{(L^p)^*} = \|g\|_{L^{p'}}$. The proof is complete in the case $\mu(X) < +\infty$.

Let us now assume that $\mu(X) = +\infty$. From Lemma 7.3.5, we know that there exists $w \in L^1(\mu)$ such that $\forall x \in X, 0 < w(x) < 1$. We consider now the finite measure $d\nu = w d\mu$ ($\nu(X) = \int_X w d\mu < \infty$) and the linear isometries

$$\left. \begin{array}{ccc} L^p(\nu) & \longrightarrow & L^p(\mu) \\ F & \mapsto & F w^{1/p} \end{array} \right\}, \quad \left\{ \begin{array}{ccc} L^p(\mu) & \longrightarrow & L^p(\nu) \\ f & \mapsto & f w^{-1/p} \end{array} \right., \tag{7.5.5}$$

noting that we have

$$\|F\|_{L^p(\nu)}^p = \int_X |F|^p w d\mu = \|F w^{1/p}\|_{L^p(\mu)}^p,$$

$$\|f\|_{L^p(\mu)}^p = \int_X |f|^p w^{-1} d\nu = \|f w^{-1/p}\|_{L^p(\nu)}^p.$$

As a consequence, if $\xi \in (L^p(\mu))^*$ we can define $\eta \in (L^p(\nu))^*$ by

$$\forall F \in L^p(\nu), \quad \langle \eta, F \rangle_{(L^p(\nu))^*, L^p(\nu)} = \langle \xi, w^{1/p} F \rangle_{(L^p(\mu))^*, L^p(\mu)},$$

and

$$\|\eta\|_{(L^p(\nu))^*} = \|\xi\|_{(L^p(\mu))^*}.$$

We can use the proven result on finite measures to find $G \in L^{p'}(\nu)$ such that $\|G\|_{L^{p'}(\nu)} = \|\eta\|_{(L^p(\nu))^*}$ with $\langle \eta, F \rangle_{(L^p(\nu))^*, L^p(\nu)} = \int_X F G d\nu$ so that

$$\langle \xi, f \rangle_{(L^p(\mu))^*, L^p(\mu)} = \int_X f w^{-1/p} G w d\mu = \int_X f g d\mu, \quad g = G w^{1-\frac{1}{p}},$$

and, if $p' < \infty$,

$$\|\xi\|_{(L^p(\mu))^*}^p = \|G\|_{L^{p'}(\nu)}^p = \int_X |G|^{p'} w d\mu = \int_X (|G| w^{1-\frac{1}{p}})^{p'} d\mu = \|g\|_{L^{p'}(\mu)}^p.$$

If $p = 1, p' = \infty$, we have $g = G$ and $\|\xi\|_{(L^1(\mu))^*} = \|G\|_{L^\infty(\nu)} = \|g\|_{L^\infty(\nu)}$. The proof of the theorem is complete. $\qquad\square$

The Banach spaces c_0, ℓ^p

These are spaces of sequences of complex numbers $(x_k)_{k \geq 1}$. We have

$$c_0 = \{(x_k)_{k \geq 1}, \lim_k x_k = 0\}, \quad \|(x_k)_{k \geq 1}\|_\infty = \sup_{k \geq 1} |x_k|, \tag{7.5.6}$$

$$\text{for } p \geq 1, \ \ell^p = \{(x_k)_{k \geq 1}, \sum_{k \geq 1} |x_k|^p < +\infty\}, \quad \|(x_k)_{k \geq 1}\|_p = \left(\sum_{k \geq 1} |x_k|^p\right)^{1/p}, \tag{7.5.7}$$

$$\ell^\infty = \{(x_k)_{k \geq 1}, \sup_{k \geq 1} |x_k| < +\infty\}, \quad \|(x_k)_{k \geq 1}\|_\infty = \sup_{k \geq 1} |x_k|. \tag{7.5.8}$$

These spaces are Banach spaces, and ℓ^2 is a Hilbert space (see Theorem 3.2.8). Note also that the space c_0 is a closed subspace of ℓ^∞ (Exercise 3.7.29). The spaces c_0, ℓ^p, for $1 \leq p < +\infty$ are separable since the finite sequences of complex numbers with rational real and imaginary part are dense (Exercise 3.7.29). The space ℓ^∞ is *not* separable (see Exercise 3.7.20).

Duality results

Let us prove that $c_0^* = \ell^1$. We consider the mapping

$$\begin{array}{ccc} c_0 \times \ell^1 & \longrightarrow & \mathbb{C} \\ (x, y) & \mapsto & \sum_{k \geq 1} x_k \overline{y_k} := (x, y) \end{array} \tag{7.5.9}$$

and we have $|(x, y)| \leq \|x\|_{c_0} \|y\|_{\ell^1}$. As a consequence, we have a mapping

$$\ell^1 \ni y \mapsto j(y) \in c_0^* \text{ with } j(y) \cdot x = (x, y).$$

The mapping j is linear, sends ℓ^1 into c_0^* and that inequality proves as well that j is continuous: $\|j(y)\|_{c_0^*} \leq \|y\|_{\ell^1}$. On the other hand, for a given y in ℓ^1, $N \in \mathbb{N}^*$, choosing $x_k = y_k/|y_k|$ when $y_k \neq 0$ and $k \leq N$, $x_k = 0$ otherwise, we have $x = (x_k)_{k \geq 1} \in c_0, \|x\|_{c_0} \leq 1$,

$$\|j(y)\|_{c_0^*} = \sup_{\|x\|_{c_0} \leq 1} |(x, y)| \geq \sum_{1 \leq k \leq N} |y_k|, \quad \text{for all } N \geq 1,$$

so that $\|j(y)\|_{c_0^*} = \|y\|_{\ell^1}$. As a result $j(\ell^1)$ is a closed subspace of c_0^* which is isomorphic to ℓ^1.

We need to prove that j is onto. Let us take $\xi \in c_0^*$; we define for $j \geq 1$, $e_j = (\delta_{j,k})_{k \geq 1} \ (\in c_0)$. We choose some real numbers θ_j so that $e^{i\theta_j} \xi \cdot e_j = |\xi \cdot e_j|$ and we consider $x = (e^{i\theta_1}, \dots, e^{i\theta_n}, 0, 0, 0 \dots) \in c_0, \|x\|_{c_0} = 1$, so that

$$\xi \cdot x = \sum_{1 \leq j \leq n} e^{i\theta_j} \xi \cdot e_j = \sum_{1 \leq j \leq n} |\xi \cdot e_j|.$$

As a result, we have for all $n \geq 1$, $\sum_{1 \leq j \leq n} |\xi \cdot e_j| \leq \|\xi\|_{c_0^*} \|x\|_{c_0} = \|\xi\|_{c_0^*}$, proving that $y = (\xi \cdot e_j)_{j \geq 1} \in \ell^1$. Now, we have for $x = (x_j)_{j \geq 1} \in c_0$, by the continuity of ξ,

$$\xi \cdot x = \lim_{n \to +\infty} \sum_{1 \leq j \leq n} x_j(\xi \cdot e_j) = (x, (\xi \cdot e_j)_{j \geq 1}) = (x, y),$$

proving that $\xi = j(y)$ for some $y \in \ell^1$ and the sought surjectivity.

Theorem 7.5.1 implies that $(\ell^1)^* = \ell^\infty$ (a direct proof analogous to the previous one is also possible).

Let us now prove that $(\ell^\infty)^*$, which is the bidual of ℓ^1, is (much) larger than ℓ^1. The space c_0 is a closed proper subspace of ℓ^∞, and the Hahn–Banach theorem (Theorem 10.2.38) allows us to construct $\xi_0 \in (\ell^\infty)^*$ such that

$$\xi_{0|c_0} = 0, \quad \xi_0 \cdot x_0 = 1, \quad x_0 = (1, 1, 1, \dots) \in \ell^\infty \backslash c_0. \tag{7.5.10}$$

As a consequence, the mapping $j : \ell^1 \longrightarrow (\ell^1)^{**} = (\ell^\infty)^*$, defined in Proposition 10.3.13, is not onto since there is no $y \in \ell^1$ such that $j(y) = \xi_0$: otherwise, we would have for $x \in \ell^\infty$,

$$\langle \xi_0, x \rangle_{(\ell^\infty)^*, \ell^\infty} = \langle j(y), x \rangle_{(\ell^1)^{**}, (\ell^1)^*} = \langle x, y \rangle_{(\ell^1)^*, \ell^1} = \sum_{j \geq 1} \overline{x_j} y_j,$$

and since $\langle \xi_0, e_j \rangle_{(\ell^\infty)^*, \ell^\infty} = 0$, that would imply $y_j = 0$ for all $j \geq 1$, and $\xi_0 = 0$, contradicting (7.5.10). The next proposition summarizes the situation.

Proposition 7.5.2. *We consider the spaces* c_0, ℓ^p *defined in* (7.5.6), (7.5.7), (7.5.8). *When* $1 < p < +\infty$ *we define* $p' \in]1, +\infty[$ *by the identity* $\frac{1}{p} + \frac{1}{p'} = 1$. *Then we have*

$$(\ell^1)^* = \ell^\infty, (\ell^1)^{**} \neq \ell^1, \quad \ell^1 \text{ is not reflexive}, \tag{7.5.11}$$

$$1 < p < +\infty, \quad (\ell^p)^* = \ell^{p'}, (\ell^p)^{**} = \ell^p, \quad \ell^p \text{ is reflexive}, \tag{7.5.12}$$

$$\ell^\infty \text{ is not reflexive}, \tag{7.5.13}$$

$$c_0^* = \ell^1, c_0^{**} = (\ell^1)^* = \ell^\infty \neq c_0, \quad c_0 \text{ is not reflexive}. \tag{7.5.14}$$

Proof. The first and the fourth lines are proven above, the second line follows from Theorem 7.5.1, the third line is a consequence of Proposition 10.3.16, since ℓ^1 is not reflexive. $\qquad\square$

Examples of weak convergence

Definition of weak convergence and elementary properties related to that notion are given in Section 10.3. We consider the space $L^p(\mathbb{R})$ for some $p \in [1, +\infty[$. We want to provide some examples of a sequence $(u_k)_{k \in \mathbb{N}}$ of $L^p(\mathbb{R})$ weakly converging to 0, but not strongly converging to 0. Here we assume $1 < p < +\infty$.

- A first phenomenon is *strong oscillations*: take $u_k(x) = e^{ikx}\mathbf{1}_{[0,1]}(x)$: the L^p norm of u_k is constant equal to 1 but for $v \in L^{p'}$, the sequence

$$\langle u_k, v \rangle = \int u_k(x)\bar{v}(x)dx$$

has limit zero (a consequence of the Riemann–Lebesgue Lemma 3.4.5).

- The sequence $(u_k)_{k\in\mathbb{N}}$ may also *concentrate at a point*: take

$$u_k(x) = k^{1/p}u_1(kx),$$

where u_1 has norm 1 in L^p. Here also the L^p-norm of u_k is a constant equal to 1. However for $v \in L^{p'}$,

$$\langle u_k, v \rangle = \int u_k(x)\bar{v}(x)dx = \int u_1(t)\bar{v}(t/k)dtk^{-\frac{1}{p'}},$$

with $p, p' \in]1, +\infty[$. With $\varphi, \psi \in C^0_c(\mathbb{R})$ we have with $\psi_k(x) = k^{1/p}\psi(kx)$,

$$|\langle u_k, v \rangle| \leq |\langle u_k, v - \varphi \rangle| + |\langle u_k - \psi_k, \varphi \rangle| + |\langle \psi_k, \varphi \rangle|$$
$$\leq \|u_1\|_{L^p}\|v - \varphi\|_{L^{p'}} + \|u_1 - \psi\|_{L^p}\|\varphi\|_{L^{p'}} + |\langle \psi_k, \varphi \rangle|,$$

which implies $\limsup_k |\langle u_k, v \rangle| \leq \|u_1\|_{L^p}\|v - \varphi\|_{L^{p'}} + \|u_1 - \psi\|_{L^p}\|\varphi\|_{L^{p'}}$, and this gives the weak convergence to 0 since p, p' are both in $]1, +\infty[$.

- The sequence $(u_k)_{k\in\mathbb{N}}$ may also *escape to infinity*: take $u_k(x) = u_0(x - k)$, where u_0 has norm 1 in L^p. Reasoning as above, we need only to check

$$\int \psi(x - k)\varphi(x)dx,$$

for $\varphi, \psi \in C^0_c(\mathbb{R})$: that quantity is 0 for k large enough.

7.6 Notes

Hans HAHN (1879–1934) was an Austrian mathematician. He served as an adviser for Kurt GÖDEL (1906–1978) (see our appendix on page 414).

Otto NIKODÝM (1887–1974) was a Polish mathematician.

John VON NEUMANN (1903–1957) was a Hungarian-born American mathematician, a major scientist of the twentieth century, with fundamental contributions in Quantum Mechanics, Information Theory, Functional Analysis and Set Theory.

7.7   Exercises

Exercise 7.7.1. *Let $(a_k)_{k\in\mathbb{N}}$ be a sequence of complex numbers such that for any bijection $\sigma : \mathbb{N} \to \mathbb{N}$, the series $\sum_{k\in\mathbb{N}} a_{\sigma(k)}$ converges. Then*

$$\sum_{k\in\mathbb{N}} |a_k| < +\infty,$$

i.e., the series is absolutely converging.

Answer. Let us assume that all a_k are real valued and $\sum_{k\in\mathbb{N}} |a_k| = +\infty$. Writing

$$a_k = a_k^+ - a_k^-, \quad a_k^+ = \max(a_k, 0), \ a_k^- = \max(-a_k, 0), \quad |a_k| = a_k^+ + a_k^-,$$

we have $\sum_{k\in\mathbb{N}} a_k^+ = +\infty = \sum_{k\in\mathbb{N}} a_k^-$, otherwise if one of this series converges in $\mathbb{R}$ (say $\sum_{k\in\mathbb{N}} a_k^+ < +\infty$), since $\sum_{k\in\mathbb{N}} a_k$ is convergent, this would imply that $\sum_{k\in\mathbb{N}} a_k^- < +\infty$ and the convergence of $\sum_{k\in\mathbb{N}} |a_k|$, contradicting the assumption. Let us define

$$N_+ = \{k \in \mathbb{N}, a_k \geq 0\}, \quad N_- = \{k \in \mathbb{N}, a_k < 0\}.$$

We have from the properties of divergence

$$N_+ \sqcup N_- = \mathbb{N}, \quad \operatorname{card} N_\pm = \aleph_0.$$

Let $\{m_l\}_{l\geq 1} = N_-$, $\{\nu_j\}_{j\geq 1} = N_+$ be strictly increasing sequences.

Take n_1 such that $\quad \displaystyle\sum_{1\leq j\leq n_1} a_{\nu_j} + a_{m_1} \geq 1 \quad$ (possible since $\displaystyle\sum_{k\in N_+} a_k = +\infty$).

Take $n_2 > n_1$ such that $\quad \displaystyle\sum_{1\leq j\leq n_2} a_{\nu_j} + a_{m_1} + a_{m_2} \geq 2,$

$$\cdots\cdots\cdots\cdots\cdots\cdots\cdots\cdots\cdots\cdots\cdots\cdots$$

Take $n_l > n_{l-1}$ such that $\quad \displaystyle\sum_{1\leq j\leq n_l} a_{\nu_j} + a_{m_1} + \cdots + a_{m_l} \geq l.$

Then we can find $n_{l+1} > n_l$ such that

$$\sum_{1\leq j\leq n_{l+1}} a_{\nu_j} + a_{m_1} + \cdots + a_{m_l} + a_{m_{l+1}} \geq l+1.$$

We have thus constructed a strictly increasing sequence $(n_l)_{l\geq 1}$ of integers such that $\forall l \geq 1$, $\sum_{1\leq j\leq n_l} a_{\nu_j} + a_{m_1} + \cdots + a_{m_l} \geq l$, so that

$$\lim_{l\to+\infty} \left(\sum_{1\leq j\leq n_l} a_{\nu_j} + \sum_{1\leq j\leq l} a_{m_l} \right) = +\infty.$$

This implies that we have found a bijection σ from $\mathbb{N}$ onto $\mathbb{N}$ such that $\sum_{k\in\mathbb{N}} a_{\sigma(k)}$ diverges.

If the a_k are complex valued, and if $\sum_{k\in\mathbb{N}}|a_k| = +\infty$, then we have

$$\sum_{k\in\mathbb{N}}|\operatorname{Re} a_k| = +\infty \quad\text{or}\quad \sum_{k\in\mathbb{N}}|\operatorname{Im} a_k| = +\infty.$$

In the first case, we find a bijection of $\mathbb{N}$ such that $\sum_{k\in\mathbb{N}}\operatorname{Re} a_{\sigma(k)}$ diverges which implies that $\sum_{k\in\mathbb{N}} a_{\sigma(k)}$ diverges as well (its convergence would imply the convergence of the real parts).

Exercise 7.7.2. *Let $n \geq 1$ be an integer and let $\|\cdot\|$ be a norm on $\mathbb{R}^n$. Show that there exists a positive constant c (depending only on the norm $\|\cdot\|$ and on n) such that for all $N \geq 1$, all $v_1,\ldots,v_N \in \mathbb{R}^n$, there exists $J \subset \{1,\ldots,N\}$ such that*

$$\left\|\sum_{j\in J} v_j\right\| \geq c \sum_{1\leq j\leq N} \|v_j\|.$$

Show that for the sup-norm, c can be taken as $\frac{1}{2n}$, and for the Euclidean norm as $\frac{1}{2n\sqrt{n}}$.

Answer. Since all the norms on $\mathbb{R}^n$ are equivalent (Exercise 1.9.8), we may assume that

$$\|x\| = \max_{1\leq r\leq n}|x_r|.$$

We may also assume by homogeneity that $\sum_{1\leq j\leq N}\|v_j\| = 1$. We note that

$$\mathbb{R}^n = \cup_{1\leq l\leq n}\Gamma_l, \quad\text{with } \Gamma_l = \{x \in \mathbb{R}^n, |x_l| = \max_{1\leq r\leq n}|x_r|\},$$

so that $\sum_{1\leq j\leq N}\|v_j\| = 1 \leq \sum_{1\leq l\leq n}\sum_{\substack{1\leq j\leq N\\ v_j\in\Gamma_l}}|v_{j,l}|$ implies that we can find $l \in \{1,\ldots,n\}$ with

$$\frac{1}{n} \leq \sum_{\substack{1\leq j\leq N\\ v_j\in\Gamma_l}}|v_{j,l}| = \sum_{\substack{1\leq j\leq N\\ v_j\in\Gamma_l, v_{j,l}>0}} v_{j,l} + \sum_{\substack{1\leq j\leq N\\ v_j\in\Gamma_l, v_{j,l}<0}}(-v_{j,l}).$$

Eventually, we can find $l \in \{1,\ldots,n\}$ with

$$\frac{1}{2n} \leq \sum_{\substack{1\leq j\leq N\\ v_j\in\Gamma_l, v_{j,l}>0}} v_{j,l} \quad\text{or}\quad \frac{1}{2n} \leq \sum_{\substack{1\leq j\leq N\\ v_j\in\Gamma_l, v_{j,l}<0}}(-v_{j,l}).$$

In the first case (the second case is analogous), we have

$$\frac{1}{2n} \leq \sum_{\substack{1\leq j\leq N\\ v_j\in\Gamma_l, v_{j,l}>0}} v_{j,l} \leq \left\|\sum_{\substack{1\leq j\leq N\\ v_j\in\Gamma_l, v_{j,l}>0}} v_j\right\|$$

and we can take $J = \{1 \leq j \leq N, v_j \in \Gamma_l, v_{j,l} > 0\}$ and $c = \frac{1}{2n}$. Since

$$\max_{1 \leq l \leq n} |x_l| \leq \left(\sum_{1 \leq l \leq n} |x_l|^2 \right)^{1/2} \leq \sqrt{n} \max_{1 \leq l \leq n} |x_l|,$$

we get as well the constant for the Euclidean norm.

Exercise 7.7.3. *Let $(X, \mathcal{M}, \mu)$ be a measure space where μ is a positive measure and let λ be a complex measure on $(X, \mathcal{M})$. Prove that if $\lambda \ll \mu$ then $|\lambda| \ll \mu$ (prove that the converse is also true).*

Answer. Let $E \in \mathcal{M}$ such that $\mu(E) = 0$ and $|\lambda|(E) > 0$: we can find a partition $(E_k)_{k \in \mathbb{N}}$ of E ($E_k \in \mathcal{M}$) such that

$$\sum_{k \in \mathbb{N}} |\lambda(E_k)| \geq \frac{1}{2}|\lambda|(E) > 0,$$

which is impossible since $\mu(E) = 0 \implies \forall k, \ \mu(E_k) = 0 \implies \forall k, \ \lambda(E_k) = 0$. The converse is obvious since $\mu(E) = 0 \implies |\lambda|(E) = 0$ and since $|\lambda|(E) \geq |\lambda(E)|$ we get $\lambda(E) = 0$.

Exercise 7.7.4. *Let $(X, \mathcal{M}, \mu)$ be a measure space where μ is a positive measure and let λ be a complex measure on $(X, \mathcal{M})$. Prove that $\lambda \ll \mu$ iff*

$$\forall \epsilon > 0, \exists \delta > 0, \forall E \in \mathcal{M}, \quad \mu(E) < \delta \implies |\lambda(E)| < \epsilon. \tag{7.7.1}$$

We can write this property symbolically as $\lim_{\mu(E) \to 0} \lambda(E) = 0$ uniformly with respect to $E \in \mathcal{M}$.

Answer. If (7.7.1) holds, with $E \in \mathcal{M}$ such that $\mu(E) = 0$, we obtain immediately $\lambda(E) = 0$, proving $\lambda \ll \mu$. Let us assume conversely that (7.7.1) does not hold:

$$\exists \epsilon_0 > 0, \forall k \in \mathbb{N}, \exists E_k \in \mathcal{M}, \quad \mu(E_k) < 2^{-k} \text{ and } |\lambda(E_k)| \geq \epsilon_0.$$

We define $F_j = \cup_{k, k \geq j} E_k$ so that the sequence $(F_j)_{j \geq 1}$ is decreasing and

$$\mu(F_j) \leq \sum_{k \geq j} 2^{-k} = 2^{1-j} \implies \mu(F = \cap_{j \geq 1} F_j) = 0.$$

On the other hand we have $|\lambda|(F_j) \geq |\lambda|(E_j) \geq |\lambda(E_j)| \geq \epsilon_0$ and since $|\lambda|$ is a bounded positive measure, thanks to (3) in Proposition 1.4.4, we have $|\lambda|(F_1) < +\infty$ and we get

$$0 < \epsilon_0 \leq \lim_j |\lambda|(F_j) = |\lambda|(F) \implies |\lambda| \text{ is not absolutely continuous with respect to } \mu,$$

proving that λ is not absolutely continuous with respect to μ from the previous exercise.

Exercise 7.7.5. *Let* $(X, \mathcal{M})$ *be a measurable space and let* λ *be a complex measure carried by a set* $A \in \mathcal{M}$. *Then* $|\lambda|$ *is also carried by* A.

Answer. We have $\lambda(E) = \lambda(E \cap A)$ so that for $E \in \mathcal{M}$ and for a partition $(E_k)_{k \in \mathbb{N}}$ of E, we have
$$\sum_k |\lambda(E_k)| = \sum_k |\lambda(E_k \cap A)| \le |\lambda|(E \cap A),$$
so that $|\lambda|(E) \le |\lambda|(E \cap A) \implies |\lambda|(E \cap A) = |\lambda|(E)$.

Exercise 7.7.6. *Let* $(X, \mathcal{M})$ *be a measurable space and let* λ *be a measure on* $(X, \mathcal{M})$ *valued in* $\mathbb{R}^m$ *for some* $m \in \mathbb{N}^*$.

(1) *Give a definition of* $|\lambda|$ *such that this total variation measure is a positive bounded measure on* $(X, \mathcal{M})$ *which coincides with* $|\lambda|$ *when* λ *is a complex-valued measure (see Definition 7.2.1).*

(2) *Let* $f : X \to \mathbb{R}^m$ *be in* $L^1(|\lambda|)$. *Prove that*
$$\left\| \int_X f d|\lambda| \right\| \le \int_X \|f\| d|\lambda|. \tag{7.7.2}$$

Let T *be a closed set of* $\mathbb{R}^m$ *such that*
$$\forall E \in \mathcal{M} \text{ with } |\lambda|(E) > 0, \quad \frac{1}{|\lambda|(E)} \int_E f d|\lambda| \in T. \tag{7.7.3}$$

Prove that $f(x) \in T$, $|\lambda|$*-a.e.*

(3) *Prove that there exists a measurable function* $f : X \to \mathbb{R}^m$ *such that*
$$\forall x \in X, \ \|f(x)\| = 1, (\textit{Euclidean norm}) \quad d\lambda = f d|\lambda|.$$

This identity is called the **polar decomposition** *of the vector-valued measure* λ.

Answer. (1) We use the very same definition as in Definition 7.2.1,
$$|\lambda|(E) = \sup_{\substack{(E_k)_{k \in \mathbb{N}} \text{pairwise disjoint} \\ \text{with union } E, \\ E_k \in \mathcal{M}}} \sum_{\mathbb{N}} \|\lambda(E_k)\|, \tag{7.7.4}$$

and the proof that $|\lambda|$ is a positive measure is identical to the case where λ is complex valued in Proposition 7.2.3. To check that $|\lambda|$ is bounded requires a simple modification of the proof of Lemma 7.2.4. We modify (7.2.2) as follows:
$$\sum_{k \in \mathbb{N}} |\lambda(A_{k,\epsilon})| > \frac{(2m+1)\sqrt{m}}{\epsilon} \implies \exists N, \text{ such that } \sum_{0 \le k \le N} |\lambda(A_{k,\epsilon})| > \frac{2m\sqrt{m}}{\epsilon}, \tag{7.7.5}$$

and according to Exercise 7.7.2, we find $J \subset \{0, \dots, N\}$ such that
$$\|\lambda(\cup_{k \in J} A_{k,\epsilon})\| = \left\| \sum_{k \in J} \lambda(A_{k,\epsilon}) \right\| > \frac{1}{\epsilon},$$

and the sequel of the proof does not need any modification.

(2) We have

$$\left\|\int_X f d|\lambda|\right\| = \sup_{\|\xi\|=1}\left\langle\int_X f d|\lambda|, \xi\right\rangle = \sup_{\|\xi\|=1}\int_X \langle f, \xi\rangle d|\lambda| \le \int_X \|f\| d|\lambda|.$$

For $\eta \in T^c, \exists \rho > 0$ with $\bar{B}(\eta, \rho) \subset T^c$. If we had $|\lambda|\big(f^{-1}(\bar{B}(z, \rho))\big) > 0$, this would give, with $E = f^{-1}(\bar{B}(z, \rho))$, $\frac{1}{|\lambda|(E)}\int_E f d|\lambda| \in T$. However, we have

$$\frac{1}{|\lambda|(E)}\int_{f^{-1}(\bar{B}(z,\rho))} f d|\lambda| = \frac{1}{|\lambda|(E)}\int_{f^{-1}(\bar{B}(z,\rho))} (f - \eta)d|\lambda| + \eta,$$

and since

$$\left|\frac{1}{|\lambda|(E)}\int_{f^{-1}(\bar{B}(\eta,\rho))} (f - \eta)d|\lambda|\right| \le \frac{\rho|\lambda|(E)}{|\lambda|(E)} = \rho,$$

this would imply $\|\eta - T\| \le \rho$, which contradicts $\bar{B}(\eta, \rho) \subset T^c$. Consequently, $|\lambda|\big(f^{-1}(\bar{B}(\eta, \rho))\big) = 0$. Since the open set T^c is a countable union of closed balls, this implies that $|\lambda|\big(f^{-1}(T^c)\big) = 0$.

(3) We have obviously $\lambda \ll |\lambda|$ which is a positive bounded measure, so that we may apply the Radon–Nikodym Theorem to the m components of λ and get a function $f : X \to \mathbb{R}^m$ in $L^1(|\lambda|)$ such that

$$d\lambda = f d|\lambda|.$$

We define for $\rho > 0$, $L_\rho = \{x \in X, \|f(x)\| < \rho\}$. Let $(E_k)_{k\in\mathbb{N}}$ be a partition of L_ρ: we have, using (7.7.2),

$$\sum_{k\in\mathbb{N}} \|\lambda(E_k)\| = \sum_{k\in\mathbb{N}}\left\|\int_{E_k} f d|\lambda|\right\| \le \rho \sum_{k\in\mathbb{N}}\int_{E_k} d|\lambda| = \rho|\lambda|(L_\rho)$$

so that $|\lambda|(L_\rho) \le \rho|\lambda|(L_\rho)$, which implies $|\lambda|(L_\rho) = 0$ for $\rho < 1$. As a result $\|f\| \ge 1$, $|\lambda|$ a.e. On the other hand for $|\lambda|(E) > 0$, we have

$$\left\|\frac{1}{|\lambda|(E)}\int_E f d|\lambda|\right\| = \frac{\|\lambda(E)\|}{|\lambda|(E)} \le 1$$

and (7.7.3) implies $\|f\| \le 1$, $|\lambda|$ a.e., and eventually the sought result.

Exercise 7.7.7. *Let $\kappa \in [0, 1)$ and let $\mathfrak{h}_\kappa$ be the Hausdorff measure of dimension κ on a finite interval I of the real line with a non-empty interior (see Definition 2.6.5).*

(1) *Prove that $\lambda_1 = \mathfrak{h}_1 \ll \mathfrak{h}_\kappa$ where λ_1 is the Lebesgue measure on I. Prove that $\mathfrak{h}_\kappa$ is not σ-finite on I.*

(2) *Prove that there is no $f \in L^1(\mathfrak{h}_\kappa)$ such that $d\lambda_1 = f d\mathfrak{h}_\kappa$.*

N.B. This implies that the conclusions of the Radon–Nikodym Theorem 7.4.1 do not hold in general when the σ-finiteness of μ is not satisfied, even though λ is a bounded positive measure.

Answer. (1) We consider the measurable space $(I, \mathcal{B})$ where $\mathcal{B}$ is the Borel σ-algebra on I and $\mathfrak{h}_\kappa$ as measures on that measurable space. From Lemma 2.6.7, if A is a Borel subset of the real line,

$$\mathfrak{h}_\kappa(A) < +\infty \Longrightarrow \mathfrak{h}_1(A) = 0,$$

proving in particular the absolute continuity of $\mathfrak{h}_1$ with respect to $\mathfrak{h}_\kappa$ for $\kappa \in [0, 1)$. Moreover if $\mathfrak{h}_\kappa$ were σ-finite on I, we could find a sequence $(X_n)_{n \in \mathbb{N}}$ such that $I = \cup_\mathbb{N} X_n$ and $\mathfrak{h}_\kappa(X_n) < +\infty$. From Lemma 2.6.7, this would imply $\mathfrak{h}_1(X_n) = 0$ and thus $\mathfrak{h}_1(I) = 0$, contradicting the assumption on I.

(2) If we could find a Borel function f such that $f d\mathfrak{h}_\kappa = d\lambda_1$, this would imply for $\epsilon > 0$ and $J_\epsilon = \{t \in I, f(t) \geq \epsilon\}$,

$$+\infty > \lambda_1(J_\epsilon) = \int_{J_\epsilon} f d\mathfrak{h}_\kappa \geq \epsilon \mathfrak{h}_\kappa(J_\epsilon) \Longrightarrow \mathfrak{h}_\kappa(J_\epsilon) < +\infty \Longrightarrow \lambda_1(J_\epsilon) = 0,$$

so that $\int_{J_\epsilon} f d\mathfrak{h}_\kappa = 0$, proving that $\mathfrak{h}_\kappa(J_\epsilon) = 0$. As a result $f \leq 0$, $\mathfrak{h}_\kappa$ a.e., implying $d\lambda_1 \leq 0$.

Exercise 7.7.8. *Let $(X, \mathcal{M}, \mu)$ be a measure space where μ is a positive measure and let $f \in L^1(\mu)$. Let $d\lambda = f d\mu$ be the absolutely continuous complex measure with density f with respect to μ. Prove that*

$$d|\lambda| = |f| d\mu.$$

Answer. According to (3) in Exercise 7.7.6, there exists a measurable function w of modulus 1 such that

$$w d|\lambda| = d\lambda = f d\mu \Longrightarrow d|\lambda| = \bar{w} f d\mu,$$

implying that $\bar{w} f \geq 0$, μ-a.e. Since we have also $|\bar{w} f| = |f|$, we find $\bar{w} f = |f|$, μ-a.e., proving the sought result.

Exercise 7.7.9. *Let $(X, \mathcal{M})$ be a measurable space and let λ be a real measure on $(X, \mathcal{M})$. Show that there exists a partition of X, $\{A_+, A_-\}$, elements of $\mathcal{M}$ which are carriers respectively of λ_+, λ_- (cf. Definition 7.2.5) and*

$$\lambda_\pm(E) = \pm\lambda(E \cap A_\pm).$$

Answer. We have from the polar decomposition (Exercise 7.7.6),

$$d\lambda = u d|\lambda|, \quad |u| = 1,$$

and since λ is a real measure, u is real valued $|\lambda|$ a.e. Thus modifying u on a set of measure 0, we may assume that u takes only the values ± 1. Consequently, we have

$$d\lambda = u_+ d|\lambda| - u_- d|\lambda| = \mathbf{1}_{\{u=1\}} d|\lambda| - \mathbf{1}_{\{u=-1\}} d|\lambda|,$$

so that $u_+ + u_- = 1$ and

$$d\lambda_+ = \frac{1}{2}(d|\lambda| + d\lambda) = \mathbf{1}_{\{u=1\}} d|\lambda|, \quad d\lambda_- = \frac{1}{2}(d|\lambda| - d\lambda) = \mathbf{1}_{\{u=-1\}} d|\lambda|,$$

and we can take $A_\pm = \{u = \pm 1\}$.

Exercise 7.7.10. *Let $(X, \mathcal{M})$ be a measurable space and let λ be a real measure on $(X, \mathcal{M})$ such that there exists positive bounded measures μ_1, μ_2 with $\lambda = \mu_1 - \mu_2$. Prove that $\mu_1 \geq \lambda_+$, $\mu_2 \geq \lambda_-$.*

Answer. We have from the previous exercise for $E \in \mathcal{M}$,

$$\lambda_+(E) = \lambda(A_+ \cap E) \leq \mu_1(A_+ \cap E) \leq \mu_1(E),$$
$$\lambda_-(E) = -\lambda(A_- \cap E) \leq \mu_2(A_- \cap E) \leq \mu_2(E).$$

Exercise 7.7.11. *Let μ be a positive σ-finite Borel measure on the real line (μ is a positive measure defined on the Borel σ-algebra of $\mathbb{R}$ which is finite on compact sets) and let λ_1 be the Lebesgue measure on $\mathbb{R}$.*

(1) Show that there exist three positive Borel measures $\mu_{ac}, \mu_{sp}, \mu_{sc}$ such that

$$\mu = \mu_{ac} + \mu_{sp} + \mu_{sc}, \tag{7.7.6}$$

$$\mu_{ac} \ll \lambda_1, \quad \mu_{sp} = \sum_{k \in \mathbb{N}} \alpha_k \delta_{a_k}, \quad \text{where } a_k \in \mathbb{R}, \alpha_k > 0, \tag{7.7.7}$$

$$\mu_{sc} \perp \lambda_1, \quad \forall x \in \mathbb{R}, \ \mu_{sc}(\{x\}) = 0. \tag{7.7.8}$$

(2) Prove that the above decomposition is unique. The measure μ_{ac} is called the **absolutely continuous part** *of μ, μ_{sp} the* **pure point part** *of μ and μ_{sc} the* **singular continuous part** *of μ. A measure such that for all x, $\mu_{sc}(\{x\}) = 0$ is also said to be* **diffuse***.*

(3) Give an example of a measure μ such that $\mu = \mu_{sc}$.

Answer. (1) The Radon–Nikodym Theorem 7.4.1 implies that

$$\mu = \mu_{ac} + \mu_s, \quad d\mu_{ac} = f d\lambda_1, \ 0 \leq f \in L^1(\mathbb{R}), \quad \mu_s \perp \lambda_1,$$

where μ_s is a positive measure (note that μ_s is finite on compact sets since μ is a Borel measure). The measure μ_s is carried by a measurable set C with Lebesgue measure 0. Now Exercise 1.9.24 applied to the positive Borel measure μ_s, implies that there exists a countable subset $D = \{a_k\}_{k \in \mathbb{N}}$ of $\mathbb{R}$ such that

$$\mu_s = \underbrace{\sum_{a \in D} \mu_s(\{a\}) \delta_a}_{\mu_{sp}} + \mu_{sc},$$

where μ_{sc} is a Borel measure such that for all $x \in \mathbb{R}$, $\mu_{sc}(\{x\}) = 0$. Moreover μ_{sc} is also carried by C and thus $\mu_{sc} \perp \lambda_1$ and $\mu_{sc} \perp \mu_{sp}$ since μ_{sp} is carried by the countable set D.

(2) Let us prove the uniqueness of this decomposition. If

$$\mu = \mu_{ac,j} + \mu_{sp,j} + \mu_{sc,j}, \quad j = 1, 2,$$

with the properties of (1), we find from the uniqueness part in the Radon–Nikodym Theorem, that $\mu_{ac,1} = \mu_{ac,2}$, $\mu_{sp,1} + \mu_{sc,1} = \mu_{sp,2} + \mu_{sc,2} = \nu$. For $x \in \mathbb{R}$, we have

$$\mu_{sp,1}(\{x\}) + \mu_{sc,1}(\{x\}) = \mu_{sp,2}(\{x\}) + \mu_{sc,2}(\{x\}) \implies \mu_{sp,1}(\{x\}) = \mu_{sp,2}(\{x\}),$$

proving that $\mu_{sp,1} = \mu_{sp,2}$ and thus $\mu_{sc,1} = \mu_{sc,2}$.

(3) The *Cantor measure* Ψ' defined in Proposition 5.7.7 is the derivative of the *Cantor function* Ψ and is a positive Radon measure supported in the (compact) *Cantor ternary set* K_∞ which has Lebesgue measure 0, so that $\Psi' \perp \lambda_1$. Moreover Ψ' has no atoms (is a diffuse measure), so that $\Psi' = (\Psi')_{sc}$.

Chapter 8

Basic Harmonic Analysis on $\mathbb{R}^n$

The Fourier transform of $L^1(\mathbb{R}^n)$ functions was defined in Chapter 3 with the Riemann–Lebesgue Lemma 3.4.5. We need to extend this transformation to various other situations and it turns out that L. Schwartz' point of view to define the Fourier transformation on the very large space of tempered distributions is the simplest. However, the cost of the distribution point of view is that we have to define these objects, which is not a completely elementary matter. We have chosen here to limit our presentation to the tempered distributions, topological dual of the so-called Schwartz space of rapidly decreasing functions; this space is a Fréchet space, so its topology is defined by a countable family of semi-norms and is much less difficult to understand than the space of test functions with compact support on an open set. Proving the Fourier inversion formula on the Schwartz space is a truly elementary matter, which yields almost immediately the most general case for tempered distributions, by a duality abstract nonsense argument. This chapter may also serve the reader as a motivation to explore the more difficult local theory of distributions.

8.1   Fourier transform of tempered distributions

The Fourier transformation on $\mathscr{S}(\mathbb{R}^n)$

Definition 8.1.1. Let $n \geq 1$ be an integer. The Schwartz space $\mathscr{S}(\mathbb{R}^n)$ is defined as the vector space of C^∞ functions u from $\mathbb{R}^n$ to $\mathbb{C}$ such that, for all multi-indices. $\alpha, \beta \in \mathbb{N}^n$,

$$\sup_{x \in \mathbb{R}^n} |x^\alpha \partial_x^\beta u(x)| < +\infty.$$

Here we have used the multi-index notation: for $\alpha = (\alpha_1, \dots, \alpha_n) \in \mathbb{N}^n$ we define

$$x^\alpha = x_1^{\alpha_1} \dots x_n^{\alpha_n}, \quad \partial_x^\alpha = \partial_{x_1}^{\alpha_1} \dots \partial_{x_n}^{\alpha_n}, \quad |\alpha| = \sum_{1 \leq j \leq n} \alpha_j. \tag{8.1.1}$$

A simple example of such a function is $e^{-|x|^2}$, ($|x|$ is the Euclidean norm of x) and more generally, if A is a symmetric positive definite $n \times n$ matrix, the function

$$v_A(x) = e^{-\pi \langle Ax, x \rangle} \tag{8.1.2}$$

belongs to the Schwartz class (Exercise 8.5.1). The space $\mathscr{S}(\mathbb{R}^n)$ is a Fréchet space (see Exercise 8.5.2) equipped with the countable family of semi-norms $(p_k)_{k \in \mathbb{N}}$,

$$p_k(u) = \sup_{\substack{x \in \mathbb{R}^n \\ |\alpha|, |\beta| \le k}} |x^\alpha \partial_x^\beta u(x)|. \tag{8.1.3}$$

Definition 8.1.2. For $u \in \mathscr{S}(\mathbb{R}^n)$, we define its Fourier transform $\hat{u}$ as

$$\hat{u}(\xi) = \int_{\mathbb{R}^n} e^{-2i\pi x \cdot \xi} u(x)dx. \tag{8.1.4}$$

Lemma 8.1.3. *The Fourier transform sends continuously $\mathscr{S}(\mathbb{R}^n)$ into itself.*

Proof. Just notice that

$$\xi^\alpha \partial_\xi^\beta \hat{u}(\xi) = \int e^{-2i\pi x\xi} \partial_x^\alpha(x^\beta u)(x)dx(2i\pi)^{|\beta|-|\alpha|}(-1)^{|\beta|},$$

and since $\sup_{x \in \mathbb{R}^n}(1 + |x|)^{n+1}|\partial_x^\alpha(x^\beta u)(x)| < +\infty$, we get the result. $\square$

Lemma 8.1.4. *For a symmetric positive definite $n \times n$ matrix A, we have*

$$\widehat{v_A}(\xi) = (\det A)^{-1/2} e^{-\pi \langle A^{-1}\xi, \xi \rangle}, \tag{8.1.5}$$

where v_A is given by (8.1.2).

Proof. In fact, diagonalizing the symmetric matrix A, it is enough to prove the one-dimensional version of (8.1.5), i.e., to check

$$\int e^{-2i\pi x\xi} e^{-\pi x^2} dx = \int e^{-\pi(x+i\xi)^2} dx\, e^{-\pi \xi^2} = e^{-\pi \xi^2},$$

where the second equality is obtained by taking the ξ-derivative of $\int e^{-\pi(x+i\xi)^2} dx$: we have indeed

$$\frac{d}{d\xi}\left(\int e^{-\pi(x+i\xi)^2} dx \right) = \int e^{-\pi(x+i\xi)^2}(-2i\pi)(x + i\xi)dx$$

$$= -i \int \frac{d}{dx}\left(e^{-\pi(x+i\xi)^2} \right) dx = 0.$$

For $a > 0$, we obtain $\int_{\mathbb{R}} e^{-2i\pi x\xi} e^{-\pi a x^2} dx = a^{-1/2} e^{-\pi a^{-1}\xi^2}$, which is the sought result in one dimension. If $n \ge 2$, and A is a positive definite symmetric matrix, there exists an orthogonal $n \times n$ matrix P (i.e., ${}^t PP = \text{Id}$) such that

$$D = {}^t PAP, \quad D = \text{diag}(\lambda_1, \ldots, \lambda_n), \text{ all } \lambda_j > 0.$$

As a consequence, we have, since $|\det P| = 1$,

$$\int_{\mathbb{R}^n} e^{-2i\pi x\cdot\xi} e^{-\pi\langle Ax,x\rangle}\,dx = \int_{\mathbb{R}^n} e^{-2i\pi(Py)\cdot\xi} e^{-\pi\langle APy,Py\rangle}\,dy$$

$$= \int_{\mathbb{R}^n} e^{-2i\pi y\cdot({}^tP\xi)} e^{-\pi\langle Dy,y\rangle}\,dy$$

$$(\text{with } \eta = {}^tP\xi) \;=\; \prod_{1\le j\le n}\int_{\mathbb{R}} e^{-2i\pi y_j\eta_j} e^{-\pi\lambda_j y_j^2}\,dy_j \;=\; \prod_{1\le j\le n}\lambda_j^{-1/2} e^{-\pi\lambda_j^{-1}\eta_j^2}$$

$$= (\det A)^{-1/2} e^{-\pi\langle D^{-1}\eta,\eta\rangle}$$

$$= (\det A)^{-1/2} e^{-\pi\langle {}^tPA^{-1}P\,{}^tP\xi,{}^tP\xi\rangle}$$

$$= (\det A)^{-1/2} e^{-\pi\langle A^{-1}\xi,\xi\rangle}. \qquad\qquad \square$$

Proposition 8.1.5. *The Fourier transformation is an isomorphism of the Schwartz class and for $u \in \mathscr{S}(\mathbb{R}^n)$, we have*

$$u(x) = \int e^{2i\pi x\xi}\hat{u}(\xi)\,d\xi. \tag{8.1.6}$$

Proof. Using (8.1.5) we calculate for $u \in \mathscr{S}(\mathbb{R}^n)$ and $\epsilon > 0$, dealing with absolutely converging integrals,

$$u_\epsilon(x) = \int e^{2i\pi x\xi}\hat{u}(\xi) e^{-\pi\epsilon^2|\xi|^2}\,d\xi$$

$$= \iint e^{2i\pi x\xi} e^{-\pi\epsilon^2|\xi|^2} u(y) e^{-2i\pi y\xi}\,dy\,d\xi$$

$$= \int u(y) e^{-\pi\epsilon^{-2}|x-y|^2}\epsilon^{-n}\,dy$$

$$= \int \underbrace{\left(u(x+\epsilon y) - u(x)\right)}_{\text{with absolute value}\le\epsilon|y|\,\|u'\|_{L^\infty}} e^{-\pi|y|^2}\,dy + u(x).$$

Taking the limit when ϵ goes to zero, we get the Fourier inversion formula

$$u(x) = \int e^{2i\pi x\xi}\hat{u}(\xi)\,d\xi. \tag{8.1.7}$$

We have also proven for $u \in \mathscr{S}(\mathbb{R}^n)$ and $\check{u}(x) = u(-x)$,

$$u = \check{\hat{u}}. \tag{8.1.8}$$

Since $u \mapsto \hat{u}$ and $u \mapsto \check{u}$ are continuous homomorphisms of $\mathscr{S}(\mathbb{R}^n)$, this completes the proof of the proposition. $\qquad\square$

Proposition 8.1.6. *Using the notation*

$$D_{x_j} = \frac{1}{2i\pi}\frac{\partial}{\partial x_j}, \quad D_x^\alpha = \prod_{j=1}^n D_{x_j}^{\alpha_j} \quad \text{with } \alpha = (\alpha_1,\dots,\alpha_n) \in \mathbb{N}^n, \qquad (8.1.9)$$

we have, for $u \in \mathscr{S}(\mathbb{R}^n)$,

$$\widehat{D_x^\alpha u}(\xi) = \xi^\alpha \hat{u}(\xi), \qquad (D_\xi^\alpha \hat{u})(\xi) = (-1)^{|\alpha|}\widehat{x^\alpha u(x)}(\xi). \qquad (8.1.10)$$

Proof. We have for $u \in \mathscr{S}(\mathbb{R}^n)$, $\hat{u}(\xi) = \int e^{-2i\pi x\cdot\xi}u(x)dx$ and thus

$$(D_\xi^\alpha \hat{u})(\xi) = (-1)^{|\alpha|}\int e^{-2i\pi x\cdot\xi}x^\alpha u(x)dx,$$

$$\xi^\alpha \hat{u}(\xi) = \int (-2i\pi)^{-|\alpha|}\partial_x^\alpha\left(e^{-2i\pi x\cdot\xi}\right)u(x)dx = \int e^{-2i\pi x\cdot\xi}(2i\pi)^{-|\alpha|}(\partial_x^\alpha u)(x)dx,$$

proving both formulas. $\qquad\qquad\qquad\qquad\qquad\qquad\qquad\qquad\qquad\qquad\qquad\quad\square$

N.B. The normalization factor $\frac{1}{2i\pi}$ leads to a simplification in Formula (8.1.10), but the most important aspect of these formulas is certainly that the Fourier transformation exchanges the operation of derivation with the operation of multiplication. For instance with

$$P(D) = \sum_{|\alpha|\le m} a_\alpha D_x^\alpha,$$

we have for $u \in \mathscr{S}(\mathbb{R}^n)$, $\widehat{Pu}(\xi) = \sum_{|\alpha|\le m} a_\alpha \xi^\alpha \hat{u}(\xi) = P(\xi)\hat{u}(\xi)$, and thus

$$(Pu)(x) = \int_{\mathbb{R}^n} e^{2i\pi x\cdot\xi}P(\xi)\hat{u}(\xi)d\xi. \qquad (8.1.11)$$

Proposition 8.1.7. *Let ϕ,ψ be functions in $\mathscr{S}(\mathbb{R}^n)$. Then the convolution $\phi * \psi$ as given by (6.1.1) belongs to the Schwartz space and the mapping*

$$\mathscr{S}(\mathbb{R}^n) \times \mathscr{S}(\mathbb{R}^n) \ni (\phi,\psi) \mapsto \phi * \psi \in \mathscr{S}(\mathbb{R}^n)$$

is continuous. Moreover we have

$$\widehat{\phi * \psi} = \hat{\phi}\hat{\psi}. \qquad (8.1.12)$$

Proof. The mapping $(x,y) \mapsto F(x,y) = \phi(x-y)\psi(y)$ belongs to $\mathscr{S}(\mathbb{R}^{2n})$ since x,y derivatives of the smooth function F are linear combinations of products $(\partial^\alpha\phi)(x-y)(\partial^\beta\psi)(y)$ and moreover

$$(1 + |x| + |y|)^N|(\partial^\alpha\phi)(x-y)(\partial^\beta\psi)(y)|$$
$$\le (1 + |x-y|)^N|(\partial^\alpha\phi)(x-y)|(1 + 2|y|)^N|(\partial^\beta\psi)(y)|$$
$$\le p(\phi)q(\psi),$$

where p, q are semi-norms on $\mathscr{S}(\mathbb{R}^n)$. This proves that the bilinear mapping $(\phi, \psi) \mapsto F(\phi, \psi)$ is continuous from $\mathscr{S}(\mathbb{R}^n) \times \mathscr{S}(\mathbb{R}^n)$ into $\mathscr{S}(\mathbb{R}^{2n})$. We have now directly $\partial_x^\alpha(\phi * \psi) = (\partial_x^\alpha \phi) * \psi$ and

$$(1 + |x|)^N |\partial_x^\alpha(\phi * \psi)| \leq \int |F(\partial^\alpha \phi, \psi)(x, y)|(1 + |x|)^N dy$$

$$\leq \int \underbrace{|F(\partial^\alpha \phi, \psi)(x, y)|(1 + |x|)^N (1 + |y|)^{n+1}}_{\leq p(F)} (1 + |y|)^{-n-1} dy,$$

where p is a semi-norm of F (thus bounded by a product of semi-norms of ϕ and ψ), proving the continuity property. Also we obtain from Fubini's theorem

$$(\widehat{\phi * \psi})(\xi) = \iint e^{-2i\pi(x-y)\cdot\xi} e^{-2i\pi y \cdot \xi} \phi(x - y)\psi(y) dy dx = \hat{\phi}(\xi)\hat{\psi}(\xi),$$

completing the proof of the proposition. $\qquad\square$

The Fourier transformation on $\mathscr{S}'(\mathbb{R}^n)$

Definition 8.1.8. Let n be an integer ≥ 1. We define the space $\mathscr{S}'(\mathbb{R}^n)$ as the topological dual of the Fréchet space $\mathscr{S}(\mathbb{R}^n)$: this space is called the space of *tempered distributions* on $\mathbb{R}^n$.

We note that the mapping

$$\mathscr{S}(\mathbb{R}^n) \supset \phi \mapsto \frac{\partial \phi}{\partial x_j} \subset \mathscr{S}(\mathbb{R}^n),$$

is continuous since for all $k \in \mathbb{N}$, $p_k(\partial\phi/\partial x_j) \leq p_{k+1}(\phi)$, where the semi-norms p_k are defined in (8.1.3). This property allows us to define by duality the derivative of a tempered distribution.

Definition 8.1.9. Let $u \in \mathscr{S}'(\mathbb{R}^n)$. We define $\partial u/\partial x_j$ as an element of $\mathscr{S}'(\mathbb{R}^n)$ by

$$\left\langle \frac{\partial u}{\partial x_j}, \phi \right\rangle_{\mathscr{S}', \mathscr{S}} = -\left\langle u, \frac{\partial \phi}{\partial x_j} \right\rangle_{\mathscr{S}', \mathscr{S}}. \tag{8.1.13}$$

The mapping $u \mapsto \partial u/\partial x_j$ is a well-defined endomorphism of $\mathscr{S}'(\mathbb{R}^n)$ since the estimates

$$\forall \phi \in \mathscr{S}(\mathbb{R}^n), \quad \left|\left\langle \frac{\partial u}{\partial x_j}, \phi \right\rangle\right| \leq C_u p_{k_u}\left(\frac{\partial \phi}{\partial x_j}\right) \leq C_u p_{k_u+1}(\phi),$$

ensure the continuity on $\mathscr{S}(\mathbb{R}^n)$ of the linear form $\partial u/\partial x_j$.

Definition 8.1.10. Let $u \in \mathscr{S}'(\mathbb{R}^n)$ and let P be a polynomial in n variables with complex coefficients. We define the product Pu as an element of $\mathscr{S}'(\mathbb{R}^n)$ by

$$\langle Pu, \phi \rangle_{\mathscr{S}', \mathscr{S}} = \langle u, P\phi \rangle_{\mathscr{S}', \mathscr{S}}. \tag{8.1.14}$$

The mapping $u \mapsto Pu$ is a well-defined endomorphism of $\mathscr{S}'(\mathbb{R}^n)$ since the estimates

$$\forall \phi \in \mathscr{S}(\mathbb{R}^n), \quad |\langle Pu, \phi \rangle| \le C_u p_{k_u}(P\phi) \le C_u p_{k_u + D}(\phi),$$

where D is the degree of P, ensure the continuity on $\mathscr{S}(\mathbb{R}^n)$ of the linear form Pu.

Lemma 8.1.11. *Let Ω be an open subset of $\mathbb{R}^n$, $f \in L^1_{\mathrm{loc}}(\Omega)$ such that, for all $\varphi \in C^\infty_c(\Omega)$, $\int f(x)\varphi(x)dx = 0$. Then we have $f = 0$.*

Proof. Let K be a compact subset of Ω and let $\chi \in C^\infty_c(\Omega)$ equal to 1 on a neighborhood of K as in Exercise 2.8.7. With ρ as in Exercise 6.6.3, we get that

$$\lim_{\epsilon \to 0_+} \rho_\epsilon * (\chi f) = \chi f \quad \text{in } L^1(\mathbb{R}^n).$$

We have $\big(\rho_\epsilon * (\chi f)\big)(x) = \int f(y) \underbrace{\chi(y)\rho\big((x-y)\epsilon^{-1}\big)\epsilon^{-n}}_{=\varphi_x(y)} dy$, with $\operatorname{supp} \varphi_x \subset \operatorname{supp} \chi$, $\varphi_x \in C^\infty_c(\Omega)$, and from the assumption of the lemma, we obtain $\big(\rho_\epsilon * (\chi f)\big)(x) = 0$ for all x, implying $\chi f = 0$ from the convergence result and thus $f = 0$, a.e. on K; the conclusion of the lemma follows since Ω is a countable union of compact sets. $\qquad\square$

Definition 8.1.12 (support of a distribution). For $u \in \mathscr{S}'(\mathbb{R}^n)$, we define the support of u and we denote by $\operatorname{supp} u$ the closed subset of $\mathbb{R}^n$ defined by

$$(\operatorname{supp} u)^c = \{x \in \mathbb{R}^n, \exists V \text{ open} \in \mathscr{V}_x, \quad u_{|V} = 0\}, \tag{8.1.15}$$

where $\mathscr{V}_x$ stands for the set of neighborhoods of x and $u_{|V} = 0$ means that for all $\phi \in C^\infty_c(V)$, $\langle u, \phi \rangle = 0$.

Proposition 8.1.13.

(1) *We have $\mathscr{S}'(\mathbb{R}^n) \supset \cup_{1 \le p \le +\infty} L^p(\mathbb{R}^n)$, with a continuous injection of each $L^p(\mathbb{R}^n)$ into $\mathscr{S}'(\mathbb{R}^n)$. As a consequence $\mathscr{S}'(\mathbb{R}^n)$ contains as well all the derivatives in the sense (8.1.13) of all the functions in some $L^p(\mathbb{R}^n)$.*

(2) *For $u \in C^1(\mathbb{R}^n)$ such that*

$$\big(|u(x)| + |du(x)|\big)(1 + |x|)^{-N} \in L^1(\mathbb{R}^n), \tag{8.1.16}$$

for some non-negative N, the derivative in the sense (8.1.13) coincides with the ordinary derivative.

Proof. (1) For $u \in L^p(\mathbb{R}^n)$ and $\phi \in \mathscr{S}(\mathbb{R}^n)$, we can define

$$\langle u, \phi \rangle_{\mathscr{S}', \mathscr{S}} = \int_{\mathbb{R}^n} u(x)\phi(x)dx, \tag{8.1.17}$$

which is a continuous linear form on $\mathscr{S}(\mathbb{R}^n)$:

$$|\langle u, \phi \rangle_{\mathscr{S}', \mathscr{S}}| \leq \|u\|_{L^p(\mathbb{R}^n)} \|\phi\|_{L^{p'}(\mathbb{R}^n)},$$

$$\|\phi\|_{L^{p'}(\mathbb{R}^n)} \leq \sup_{x \in \mathbb{R}^n} \left((1+|x|)^{\frac{n+1}{p'}} |\phi(x)|\right) C_{n,p} \leq C_{n,p} p_k(\phi), \text{ for } k \geq k_{n,p} = \frac{n+1}{p'},$$

with p_k given by (8.1.3) (when $p = 1$, we can take $k = 0$). We indeed have a continuous injection of $L^p(\mathbb{R}^n)$ into $\mathscr{S}'(\mathbb{R}^n)$: in the first place the mapping described by (8.1.17) is well defined and continuous from the estimate

$$|\langle u, \phi \rangle| \leq \|u\|_{L^p} C_{n,p} p_{k_{n,p}}(\phi).$$

Moreover, this mapping is linear and injective from Lemma 8.1.11.

(2) We have for $\phi \in \mathscr{S}(\mathbb{R}^n)$, $\chi_0 \in C_c^\infty(\mathbb{R}^n)$, $\chi_0 = 1$ near the origin,

$$A = \left\langle \frac{\partial u}{\partial x_j}, \phi \right\rangle_{\mathscr{S}', \mathscr{S}} = -\left\langle u \frac{\partial \phi}{\partial x_j} \right\rangle_{\mathscr{S}', \mathscr{S}} = -\int_{\mathbb{R}^n} u(x) \frac{\partial \phi}{\partial x_j}(x) dx$$

so that, using Lebesgue's dominated convergence theorem, we find

$$A = -\lim_{\epsilon \to 0_+} \int_{\mathbb{R}^n} u(x) \frac{\partial \phi}{\partial x_j}(x) \chi_0(\epsilon x) dx.$$

Performing an integration by parts on C^1 functions with compact support (see Proposition 2.3.2 (2)), we get

$$A = \lim_{\epsilon \to 0_+} \left\{ \int_{\mathbb{R}^n} (\partial_j u)(x) \phi(x) \chi_0(\epsilon x) dx + \epsilon \int_{\mathbb{R}^n} u(x) \phi(x) (\partial_j \chi_0)(\epsilon x) dx \right\},$$

with $\partial_j u$ standing for the ordinary derivative. We have also

$$\int_{\mathbb{R}^n} |u(x) \phi(x) (\partial_j \chi_0)(\epsilon x)| dx \leq \|\partial_j \chi_0\|_{L^\infty(\mathbb{R}^n)} \int |u(x)|(1+|x|)^{-N} dx \, p_N(\phi) < +\infty,$$

so that $\langle \frac{\partial u}{\partial x_j}, \phi \rangle_{\mathscr{S}', \mathscr{S}} = \lim_{\epsilon \to 0_+} \int_{\mathbb{R}^n} (\partial_j u)(x) \phi(x) \chi_0(\epsilon x) dx$. Since the lhs is a continuous linear form on $\mathscr{S}(\mathbb{R}^n)$ so is the rhs. On the other hand for $\phi \in C_c^\infty(\mathbb{R}^n)$, the rhs is $\int_{\mathbb{R}^n} (\partial_j u)(x) \phi(x) dx$. Since $C_c^\infty(\mathbb{R}^n)$ is dense in $\mathscr{S}(\mathbb{R}^n)$ (Exercise 8.5.3), we find that

$$\left\langle \frac{\partial u}{\partial x_j}, \phi \right\rangle_{\mathscr{S}', \mathscr{S}} = \int_{\mathbb{R}^n} (\partial_j u)(x) \phi(x) dx,$$

since the mapping $\phi \mapsto \int_{\mathbb{R}^n} (\partial_j u)(x) \phi(x) dx$ belongs to $\mathscr{S}'(\mathbb{R}^n)$, thanks to the assumption on du in (8.1.16). This proves that $\frac{\partial u}{\partial x_j} = \partial_j u$. $\square$

The Fourier transformation can be extended to $\mathscr{S}'(\mathbb{R}^n)$. We start with noticing that for T, ϕ in the Schwartz class we have, using Fubini's theorem,

$$\int \hat{T}(\xi) \phi(\xi) d\xi = \iint T(x) \phi(\xi) e^{-2i\pi x \cdot \xi} dx d\xi = \int T(x) \hat{\phi}(x) dx,$$

and we can use the latter formula as a definition.

Definition 8.1.14. Let T be a tempered distribution; the Fourier transform $\hat{T}$ of T is the tempered distribution defined by the formula

$$\langle \hat{T}, \varphi \rangle_{\mathscr{S}',\mathscr{S}} = \langle T, \hat{\varphi} \rangle_{\mathscr{S}',\mathscr{S}}. \tag{8.1.18}$$

The linear form $\hat{T}$ is obviously a tempered distribution since the Fourier transformation is continuous on $\mathscr{S}$. Thanks to Lemma 8.1.11, if $T \in \mathscr{S}$, the present definition of $\hat{T}$ and (8.1.4) coincide.

This definition gives that, with δ_0 standing as the Dirac mass at 0, $\langle \delta_0, \phi \rangle_{\mathscr{S}',\mathscr{S}} = \phi(0)$ (obviously a tempered distribution), we have

$$\widehat{\delta_0} = 1, \tag{8.1.19}$$

since $\langle \widehat{\delta_0}, \varphi \rangle = \langle \delta_0, \widehat{\varphi} \rangle = \widehat{\varphi}(0) = \int \varphi(x)dx = \langle 1, \varphi \rangle$.

Theorem 8.1.15. *The Fourier transformation is an isomorphism of $\mathscr{S}'(\mathbb{R}^n)$. Let T be a tempered distribution. Then we have*[1]

$$T = \hat{\check{\hat{T}}}, \quad \check{\hat{T}} = \hat{\check{T}}. \tag{8.1.20}$$

With obvious notation, we have the following extensions of (8.1.10),

$$\widehat{D_x^\alpha T}(\xi) = \xi^\alpha \hat{T}(\xi), \quad (D_\xi^\alpha \hat{T})(\xi) = (-1)^{|\alpha|} \widehat{x^\alpha T(x)}(\xi). \tag{8.1.21}$$

Proof. We have for $T \in \mathscr{S}'$,

$$\langle \hat{\check{\hat{T}}}, \varphi \rangle_{\mathscr{S}',\mathscr{S}} = \langle \check{\hat{T}}, \check{\varphi} \rangle_{\mathscr{S}',\mathscr{S}} = \langle \hat{T}, \hat{\check{\varphi}} \rangle_{\mathscr{S}',\mathscr{S}} = \langle T, \hat{\hat{\check{\varphi}}} \rangle_{\mathscr{S}',\mathscr{S}} = \langle T, \varphi \rangle_{\mathscr{S}',\mathscr{S}},$$

where the last equality is due to the fact that $\varphi \mapsto \check{\varphi}$ commutes[2] with the Fourier transform and (8.1.7) means

$$\hat{\hat{\check{\varphi}}} = \varphi,$$

a formula also proven true on $\mathscr{S}'$ by the previous line of equality. Formula (8.1.10) is true as well for $T \in \mathscr{S}'$ since, with $\varphi \in \mathscr{S}$ and $\varphi_\alpha(\xi) = \xi^\alpha \varphi(\xi)$, we have

$$\langle \widehat{D^\alpha T}, \varphi \rangle_{\mathscr{S}',\mathscr{S}} = \langle T, (-1)^{|\alpha|} D^\alpha \hat{\varphi} \rangle_{\mathscr{S}',\mathscr{S}} = \langle T, \widehat{\varphi_\alpha} \rangle_{\mathscr{S}',\mathscr{S}} = \langle \hat{T}, \varphi_\alpha \rangle_{\mathscr{S}',\mathscr{S}},$$

and the other part is proven the same way. $\qquad\qquad\qquad\qquad\qquad\qquad\square$

---

[1] We define $\check{T}$ as the distribution given by $\langle \check{T}, \varphi \rangle = \langle T, \check{\varphi} \rangle$ and if $T \in \mathscr{S}'$, $\check{T}$ is also a tempered distribution since $\varphi \mapsto \check{\varphi}$ is an involutive isomorphism of $\mathscr{S}$.

[2] If $\varphi \in \mathscr{S}$, we have $\widehat{\check{\varphi}}(\xi) = \int e^{-2i\pi x \cdot \xi} \varphi(-x)dx = \int e^{2i\pi x \cdot \xi} \varphi(x)dx = \hat{\varphi}(-\xi) = \check{\hat{\varphi}}(\xi)$.

The Fourier transformation on $L^1(\mathbb{R}^n)$

Theorem 8.1.16. *The Fourier transformation is linear continuous from $L^1(\mathbb{R}^n)$ into $L^\infty(\mathbb{R}^n)$ and for $u \in L^1(\mathbb{R}^n)$, we have*

$$\hat{u}(\xi) = \int e^{-2i\pi x \cdot \xi} u(x)dx, \quad \|\hat{u}\|_{L^\infty(\mathbb{R}^n)} \le \|u\|_{L^1(\mathbb{R}^n)}. \tag{8.1.22}$$

Proof. Formula (8.1.4) can be used to define directly the Fourier transform of a function in $L^1(\mathbb{R}^n)$ and this gives an $L^\infty(\mathbb{R}^n)$ function which coincides with the Fourier transform: for a test function $\varphi \in \mathscr{S}(\mathbb{R}^n)$, and $u \in L^1(\mathbb{R}^n)$, we have by the definition (8.1.18) above and Fubini's theorem

$$\langle \hat{u}, \varphi \rangle_{\mathscr{S}',\mathscr{S}} = \int u(x)\hat{\varphi}(x)dx = \iint u(x)\varphi(\xi)e^{-2i\pi x \cdot \zeta}dxd\xi = \int \tilde{u}(\xi)\varphi(\xi)d\xi$$

with $\tilde{u}(\xi) = \int e^{-2i\pi x \cdot \xi} u(x)dx$ which is thus the Fourier transform of u. $\qquad\square$

The Fourier transformation on $L^2(\mathbb{R}^n)$

Theorem 8.1.17 (Plancherel formula). *The Fourier transformation can be extended to a unitary operator of $L^2(\mathbb{R}^n)$, i.e., there exists a unique bounded linear operator $F : L^2(\mathbb{R}^n) \longrightarrow L^2(\mathbb{R}^n)$, such that for $u \in \mathscr{S}(\mathbb{R}^n)$, $Fu = \hat{u}$ and we have $F^*F = FF^* = \mathrm{Id}_{L^2(\mathbb{R}^n)}$. Moreover*

$$F^* = CF = FC, \quad F^2C = \mathrm{Id}_{L^2(\mathbb{R}^n)}, \tag{8.1.23}$$

where C is the involutive isomorphism of $L^2(\mathbb{R}^n)$ defined by $(Cu)(x) = u(-x)$. This gives the Plancherel formula: for $u, v \in L^2(\mathbb{R}^n)$,

$$\int_{\mathbb{R}^n} \hat{u}(\xi)\overline{\hat{v}(\xi)}d\xi = \int u(x)\overline{v(x)}dx. \tag{8.1.24}$$

Proof. For test functions $\varphi, \psi \in \mathscr{S}(\mathbb{R}^n)$, using Fubini's theorem and (8.1.7), we get[3]

$$(\hat{\psi}, \hat{\varphi})_{L^2(\mathbb{R}^n)} = \int \hat{\psi}(\xi)\overline{\hat{\varphi}(\xi)}d\xi = \iint \hat{\psi}(\xi)e^{2i\pi x \cdot \xi}\overline{\varphi(x)}dxd\xi = (\psi, \varphi)_{L^2(\mathbb{R}^n)}.$$

Next, the density of $\mathscr{S}$ in L^2 shows that there is a unique continuous extension F of the Fourier transform to L^2 and that extension is an isometric operator (i.e., satisfying for all $u \in L^2(\mathbb{R}^n)$, $\|Fu\|_{L^2} = \|u\|_{L^2}$, i.e., $F^*F = \mathrm{Id}_{L^2}$). We note that the operator C defined by $Cu = \check{u}$ is an involutive isomorphism of $L^2(\mathbb{R}^n)$ and that for $u \in \mathscr{S}(\mathbb{R}^n)$,

$$CF^2u = u = FCFu = F^2Cu.$$

---

[3] We have to pay attention to the fact that the scalar product $(u, v)_{L^2}$ in the complex Hilbert space $L^2(\mathbb{R}^n)$ is linear with respect to u and antilinear with respect to v: for $\lambda, \mu \in \mathbb{C}, (\lambda u, \mu v)_{L^2} = \lambda\bar{\mu}(u, v)_{L^2}$.

By the density of $\mathscr{S}(\mathbb{R}^n)$ in $L^2(\mathbb{R}^n)$, the bounded operators

$$CF^2,\ \mathrm{Id}_{L^2(\mathbb{R}^n)},\ FCF,\ F^2C,$$

are all equal. On the other hand for $u, \varphi \in \mathscr{S}(\mathbb{R}^n)$, we have

$$(F^*u, \varphi)_{L^2} = (u, F\varphi)_{L^2} = \int u(x)\overline{\widehat{\varphi}(x)}\,dx$$

$$= \iint u(x)\bar{\varphi}(\xi)e^{2i\pi x\cdot\xi}\,dx\,d\xi = (CFu, \varphi)_{L^2},$$

so that $F^*u = CFu$ for all $u \in \mathscr{S}$ and by continuity $F^* = CF$ as bounded operators on $L^2(\mathbb{R}^n)$, thus $FF^* = FCF = \mathrm{Id}$. The proof is complete. $\qquad\square$

Some standard examples of Fourier transform

Let us consider the Heaviside function defined on $\mathbb{R}$ by $H(x) = 1$ for $x > 0$, $H(x) = 0$ for $x \le 0$; as a bounded measurable function, it is a tempered distribution, so that we can compute its Fourier transform. With the notation of this section, we have, with δ_0 the Dirac mass at 0, $\check{H}(x) = H(-x)$,

$$\widehat{H} + \widehat{\check{H}} = \hat{1} = \delta_0, \quad \widehat{H} - \widehat{\check{H}} = \widehat{\mathrm{sign}}, \quad \frac{1}{i\pi} = \frac{1}{2i\pi}2\widehat{\delta_0}(\xi) = \widehat{D\,\mathrm{sign}}(\xi) = \xi\widehat{\mathrm{sign}}\xi.$$

We note that $\mathbb{R} \mapsto \ln|x|$ belongs to $\mathscr{S}'(\mathbb{R})$ and[4] we define the so-called principal value of $1/x$ on $\mathbb{R}$ by

$$\mathrm{pv}\left(\frac{1}{x}\right) = \frac{d}{dx}(\ln|x|), \tag{8.1.25}$$

so that,

$$\left\langle \mathrm{pv}\,\frac{1}{x}, \phi \right\rangle = -\int \phi'(x)\ln|x|\,dx = -\lim_{\epsilon\to0_+}\int_{|x|\ge\epsilon}\phi'(x)\ln|x|\,dx$$

$$= \lim_{\epsilon\to0_+}\left(\int_{|x|\ge\epsilon}\phi(x)\frac{1}{x}\,dx + \underbrace{(\phi(\epsilon) - \phi(-\epsilon))\ln\epsilon}_{\to 0}\right) \tag{8.1.26}$$

$$= \lim_{\epsilon\to0_+}\int_{|x|\ge\epsilon}\phi(x)\frac{1}{x}\,dx.$$

This entails $\xi\left(\widehat{\mathrm{sign}}\xi - \frac{1}{i\pi}pv(1/\xi)\right) = 0$ and from Exercise 8.5.4, we get

$$\widehat{\mathrm{sign}}\xi - \frac{1}{i\pi}pv(1/\xi) = c\delta_0,$$

---

[4] For $\phi \in \mathscr{S}(\mathbb{R})$, we have $\langle \ln|x|, \phi(x)\rangle_{\mathscr{S}'(\mathbb{R}),\mathscr{S}(\mathbb{R})} = \int_{\mathbb{R}}\phi(x)\ln|x|\,dx$.

with $c = 0$ since the lhs is odd[5]. We obtain

$$\widehat{\mathrm{sign}}(\xi) = \frac{1}{i\pi} pv \frac{1}{\xi}, \tag{8.1.27}$$

$$\widehat{pv\left(\frac{1}{\pi x}\right)} = -i\,\mathrm{sign}\,\xi, \tag{8.1.28}$$

$$\hat{H} = \frac{\delta_0}{2} + \frac{1}{2i\pi} pv\left(\frac{1}{\xi}\right) = \frac{1}{(x - i0)}\frac{1}{2i\pi} \quad \text{(see Exercise 8.5.6)}. \tag{8.1.29}$$

Let us consider now for $0 < \alpha < n$ the $L^1_{\mathrm{loc}}(\mathbb{R}^n)$ function $u_\alpha(x) = |x|^{\alpha-n}$ ($|x|$ is the Euclidean norm of x); since u_α is also bounded for $|x| \geq 1$, it is a tempered distribution. Let us calculate its Fourier transform v_α. Since u_α is homogeneous of degree $\alpha - n$, we get from Exercise 8.5.9 that v_α is a homogeneous distribution of degree $-\alpha$. On the other hand, if $S \in O(\mathbb{R}^n)$ (the orthogonal group), we have in the distribution sense[6] since u_α is a radial function, i.e., such that

$$v_\alpha(S\xi) = v_\alpha(\xi). \tag{8.1.30}$$

The distribution $|\xi|^\alpha v_\alpha(\xi)$ is homogeneous of degree 0 on $\mathbb{R}^n\backslash\{0\}$ and is also "radial", i.e., satisfies (8.1.30). Moreover on $\mathbb{R}^n\backslash\{0\}$, the distribution v_α is a C^1 function which coincides with[7]

$$\int e^{-2i\pi x\cdot\xi}\chi_0(x)|x|^{\alpha-n}dx + |\xi|^{-2N}\int e^{-2i\pi x\cdot\xi}|D_x|^{2N}\left(\chi_1(x)|x|^{\alpha-n}\right)dx,$$

where $\chi_0 \in C_c^\infty(\mathbb{R}^n)$ is 1 near 0 and $\chi_1 = 1 - \chi_0$, $N \in \mathbb{N}, \alpha + 1 < 2N$. As a result $|\xi|^\alpha v_\alpha(\xi) = c_\alpha$ on $\mathbb{R}^n\backslash\{0\}$ and the distribution on $\mathbb{R}^n$ (note that $\alpha < n$),

$$T = v_\alpha(\xi) - c_\alpha|\xi|^{-\alpha}$$

is supported in $\{0\}$ and homogeneous (on $\mathbb{R}^n$) with degree $-\alpha$. From the Exercises 8.5.7(1), 8.5.5 and 8.5.8, the condition $0 < \alpha < n$ gives $v_\alpha = c_\alpha|\xi|^{-\alpha}$. To find c_α, we compute

$$\int_{\mathbb{R}^n} |x|^{\alpha-n}e^{-\pi x^2}dx = \langle u_\alpha, e^{-\pi x^2}\rangle = c_\alpha\int_{\mathbb{R}^n} |\xi|^{-\alpha}e^{-\pi \xi^2}d\xi$$

which yields

$$2^{-1}\Gamma\left(\frac{\alpha}{2}\right)\pi^{-\frac{\alpha}{2}} = \int_0^{+\infty} r^{\alpha-1}e^{-\pi r^2}dr = c_\alpha\int_0^{+\infty} r^{n-\alpha-1}e^{-\pi r^2}dr$$

$$= c_\alpha 2^{-1}\Gamma\left(\frac{n-\alpha}{2}\right)\pi^{-\left(\frac{n-\alpha}{2}\right)}.$$

---

[5] A distribution T on $\mathbb{R}^n$ is said to be odd (resp. even) when $\check{T} = -T$ (resp. T).
[6] For $M \in Gl(n, \mathbb{R})$, $T \in \mathscr{S}'(\mathbb{R}^n)$, we define $\langle T(Mx), \phi(x)\rangle = \langle T(y), \phi(M^{-1}y)\rangle|\det M|^{-1}$.
[7] We have $\widehat{u_\alpha} = \widehat{\chi_0 u_\alpha} + \widehat{\chi_1 u_\alpha}$ and for ϕ supported in $\mathbb{R}^n\backslash\{0\}$ we get,
$$\langle\widehat{\chi_1 u_\alpha}, \phi\rangle = \langle\widehat{\chi_1 u_\alpha}|\xi|^{2N}, \phi(\xi)|\xi|^{-2N}\rangle = \langle\widehat{|D_x|^{2N}\chi_1 u_\alpha}, \phi(\xi)|\xi|^{-2N}\rangle.$$

We have proven the following lemma.

Lemma 8.1.18. *Let $n \in \mathbb{N}^*$ and $\alpha \in (0, n)$. The function $u_\alpha(x) = |x|^{\alpha-n}$ is $L^1_{\mathrm{loc}}(\mathbb{R}^n)$ and also a temperate distribution on $\mathbb{R}^n$. Its Fourier transform v_α is also $L^1_{\mathrm{loc}}(\mathbb{R}^n)$ and given by*

$$v_\alpha(\xi) = |\xi|^{-\alpha} \pi^{\frac{n}{2}-\alpha} \frac{\Gamma(\frac{\alpha}{2})}{\Gamma(\frac{n-\alpha}{2})}.$$

Fourier transform of Gaussian functions

Proposition 8.1.19. *Let A be a symmetric nonsingular $n \times n$ matrix with complex entries such that $\mathrm{Re}\, A \geq 0$. We define the Gaussian function v_A on $\mathbb{R}^n$ by $v_A(x) = e^{-\pi\langle Ax, x\rangle}$. The Fourier transform of v_A is*

$$\widehat{v_A}(\xi) = (\det A)^{-1/2} e^{-\pi\langle A^{-1}\xi, \xi\rangle}, \tag{8.1.31}$$

where $(\det A)^{-1/2}$ is defined according to Formula (10.5.8). In particular, when $A = -iB$ with a symmetric real nonsingular matrix B, we get

$$\mathrm{Fourier}(e^{i\pi\langle Bx, x\rangle})(\xi) = \widehat{v_{-iB}}(\xi) = |\det B|^{-1/2} e^{i\frac{\pi}{4}\,\mathrm{sign}\, B} e^{-i\pi\langle B^{-1}\xi, \xi\rangle}. \tag{8.1.32}$$

Proof. We use the notation of Section 10.5 (in the subsection *Logarithm of a nonsingular symmetric matrix*). Let us define Υ_+^* as the set of symmetric $n \times n$ complex matrices with a positive definite real part (naturally these matrices are nonsingular since $Ax = 0$ for $x \in \mathbb{C}^n$ implies $0 = \mathrm{Re}\langle Ax, \bar{x}\rangle = \langle(\mathrm{Re}\, A)x, \bar{x}\rangle$, so that $\Upsilon_+^* \subset \Upsilon_+$).

Let us assume first that $A \in \Upsilon_+^*$; then the function v_A is in the Schwartz class (and so is its Fourier transform). The set Υ_+^* is an open convex subset of $\mathbb{C}^{n(n+1)/2}$ and the function $\Upsilon_+^* \ni A \mapsto \widehat{v_A}(\xi)$ is holomorphic and given on $\Upsilon_+^* \cap \mathbb{R}^{n(n+1)/2}$ by (8.1.31). On the other hand the function

$$\Upsilon_+^* \ni A \mapsto e^{-\frac{1}{2}\,\mathrm{trace}\,\mathrm{Log}\, A} e^{-\pi\langle A^{-1}\xi, \xi\rangle},$$

is also holomorphic and coincides with the previous one on $\mathbb{R}^{n(n+1)/2}$. By analytic continuation this proves (8.1.31) for $A \in \Upsilon_+^*$.

If $A \in \Upsilon_+$ and $\varphi \in \mathscr{S}(\mathbb{R}^n)$, we have $\langle\widehat{v_A}, \varphi\rangle_{\mathscr{S}', \mathscr{S}} = \int v_A(x)\hat{\varphi}(x)dx$ so that $\Upsilon_+ \ni A \mapsto \langle\widehat{v_A}, \varphi\rangle$ is continuous and thus (note that the mapping $A \mapsto A^{-1}$ is an homeomorphism of Υ_+), using the previous result on Υ_+^*,

$$\langle\widehat{v_A}, \varphi\rangle = \lim_{\epsilon\to 0_+} \langle\widehat{v_{A+\epsilon I}}, \varphi\rangle = \lim_{\epsilon\to 0_+} \int e^{-\frac{1}{2}\,\mathrm{trace}\,\mathrm{Log}(A+\epsilon I)} e^{-\pi\langle(A+\epsilon I)^{-1}\xi, \xi\rangle} \varphi(\xi)d\xi,$$

and by continuity of Log on Υ_+ and dominated convergence,

$$\langle\widehat{v_A}, \varphi\rangle = \int e^{-\frac{1}{2}\,\mathrm{trace}\,\mathrm{Log}\, A} e^{-\pi\langle A^{-1}\xi, \xi\rangle} \varphi(\xi)d\xi,$$

which is the sought result. $\qquad\square$

Multipliers of $\mathscr{S}'(\mathbb{R}^n)$

Definition 8.1.20. The space $\mathscr{O}_M(\mathbb{R}^n)$ of multipliers of $\mathscr{S}(\mathbb{R}^n)$ is the subspace of the functions $f \in C^\infty(\mathbb{R}^n)$ such that,

$$\forall \alpha \in \mathbb{N}^n, \exists C_\alpha > 0, \exists N_\alpha \in \mathbb{N}, \forall x \in \mathbb{R}^n, \quad |(\partial_x^\alpha f)(x)| \le C_\alpha (1 + |x|)^{N_\alpha}. \quad (8.1.33)$$

It is easy to check that, for $f \in \mathscr{O}_M(\mathbb{R}^n)$, the operator $u \mapsto fu$ is continuous from $\mathscr{S}(\mathbb{R}^n)$ into itself, and by transposition from $\mathscr{S}'(\mathbb{R}^n)$ into itself: we define for $T \in \mathscr{S}'(\mathbb{R}^n)$, $f \in \mathscr{O}_M(\mathbb{R}^n)$,

$$\langle fT, \varphi \rangle_{\mathscr{S}', \mathscr{S}} = \langle T, f\varphi \rangle_{\mathscr{S}', \mathscr{S}},$$

and if p is a semi-norm of $\mathscr{S}$, the continuity on $\mathscr{S}$ of the multiplication by f implies that there exists a semi-norm q on $\mathscr{S}$ such that for all $\varphi \in \mathscr{S}$, $p(f\varphi) \le q(\varphi)$. A typical example of a function in $\mathscr{O}_M(\mathbb{R}^n)$ is $e^{iP(x)}$ where P is a real-valued polynomial: in fact the derivatives of $e^{iP(x)}$ are of type $Q(x)e^{iP(x)}$ where Q is a polynomial so that (8.1.33) holds.

Definition 8.1.21. Let T, S be tempered distributions on $\mathbb{R}^n$ such that $\hat{T}$ belongs to $\mathscr{O}_M(\mathbb{R}^n)$. We define the convolution $T * S$ by

$$\widehat{T * S} = \hat{T}\hat{S}. \quad (8.1.34)$$

Note that this definition makes sense since $\hat{T}$ is a multiplier so that $\hat{T}\hat{S}$ is indeed a tempered distribution whose inverse Fourier transform is meaningful. We have

$$\langle T * S, \phi \rangle_{\mathscr{S}'(\mathbb{R}^n), \mathscr{S}(\mathbb{R}^n)} = \langle \widehat{T * S}, \check{\hat{\phi}} \rangle_{\mathscr{S}'(\mathbb{R}^n), \mathscr{S}(\mathbb{R}^n)} = \langle \hat{S}, \hat{T}\check{\hat{\phi}} \rangle_{\mathscr{S}'(\mathbb{R}^n), \mathscr{S}(\mathbb{R}^n)}.$$

Proposition 8.1.22. *Let T be a distribution on $\mathbb{R}^n$ such that T is compactly supported. Then $\hat{T}$ is a multiplier which can be extended to an entire function on $\mathbb{C}^n$ such that if* $\operatorname{supp} T \subset \bar{B}(0, R_0)$,

$$\exists C_0, N_0 \ge 0, \forall \zeta \in \mathbb{C}^n, \quad |\hat{T}(\zeta)| \le C_0 (1 + |\zeta|)^{N_0} e^{2\pi R_0 |\operatorname{Im} \zeta|}. \quad (8.1.35)$$

In particular, for $S \in \mathscr{S}'(\mathbb{R}^n)$, we may define according to (8.1.34) the convolution $T * S$.

Proof. Let us first check the case $R_0 = 0$: then the distribution T is supported at $\{0\}$ and from Exercise 8.5.5 is a linear combination of derivatives of the Dirac mass at 0. Formulas (8.1.19), (8.1.21) imply that $\hat{T}$ is a polynomial, so that the conclusions of Proposition 8.1.22 hold in that case.

Let us assume that $R_0 > 0$ and let us consider a function χ that is equal to 1 in a neighborhood of $\operatorname{supp} T$ (this implies $\chi T = T$) and

$$\langle \hat{T}, \phi \rangle_{\mathscr{S}', \mathscr{S}} = \langle \widehat{\chi T}, \phi \rangle_{\mathscr{S}', \mathscr{S}} = \langle T, \chi \hat{\phi} \rangle_{\mathscr{S}', \mathscr{S}}. \quad (8.1.36)$$

On the other hand, defining for $\zeta \in \mathbb{C}^n$ (with $x \cdot \zeta = \sum x_j \zeta_j$ for $x \in \mathbb{R}^n$),

$$F(\zeta) = \langle T(x), \chi(x) e^{-2i\pi x \cdot \zeta} \rangle_{\mathscr{S}', \mathscr{S}}, \tag{8.1.37}$$

we see that F is an entire function (i.e., holomorphic on $\mathbb{C}^n$): calculating

$$
\begin{aligned}
F(\zeta + h) - F(\zeta) &= \langle T(x), \chi(x) e^{-2i\pi x \cdot \zeta}(e^{-2i\pi x \cdot h} - 1) \rangle \\
&= \langle T(x), \chi(x) e^{-2i\pi x \cdot \zeta}(-2i\pi x \cdot h) \rangle \\
&\quad + \left\langle T(x), \chi(x) e^{-2i\pi x \cdot \zeta} \int_0^1 (1-\theta) e^{-2i\theta \pi x \cdot h} d\theta (-2i\pi x \cdot h)^2 \right\rangle,
\end{aligned}
$$

and applying to the last term the continuity properties of the linear form T, we obtain that the complex differential of F is

$$\sum_{1 \le j \le n} \langle T(x), \chi(x) e^{-2i\pi x \cdot \zeta}(-2i\pi x_j) \rangle d\zeta_j.$$

Moreover the derivatives of (8.1.37) are

$$F^{(k)}(\zeta) = \langle T(x), \chi(x) e^{-2i\pi x \cdot \zeta}(-2i\pi x)^k \rangle_{\mathscr{S}', \mathscr{S}}. \tag{8.1.38}$$

To evaluate the semi-norms of $x \mapsto \chi(x) e^{-2i\pi x \cdot \zeta}(-2i\pi x)^k$ in the Schwartz space, we have to deal with a finite sum of products of type

$$\left| x^\gamma (\partial^\alpha \chi)(x) e^{-2i\pi x \cdot \zeta}(-2i\pi \zeta)^\beta \right| \le (1 + |\zeta|)^{|\beta|} \sup_{x \in \mathbb{R}^n} \left| x^\gamma (\partial^\alpha \chi)(x) e^{2\pi |x| |\operatorname{Im}\zeta|} \right|.$$

We may now choose a function χ_0 equal to 1 on $B(0,1)$, supported in $B(0, \frac{R_0 + 2\epsilon}{R_0 + \epsilon})$ such that $\|\partial^\beta \chi_0\|_{L^\infty} \le c(\beta) \epsilon^{-|\beta|}$ with $\epsilon = \frac{R_0}{1 + |\zeta|}$. We find with

$$\chi(x) = \chi_0(x/(R_0 + \epsilon)) \quad \text{(which is 1 on a neighborhood of } B(0, R_0)),$$

$$
\begin{aligned}
\sup_{x \in \mathbb{R}^n} \left| x^\gamma (\partial^\alpha \chi)(x) e^{2\pi |x| |\operatorname{Im}\zeta|} \right| &\le (R_0 + 2\epsilon)^{|\gamma|} \sup_{y \in \mathbb{R}^n} \left| (\partial^\alpha \chi_0)(y) e^{2\pi (R_0 + 2\epsilon)|\operatorname{Im}\zeta|} \right| \\
&\le (R_0 + 2\epsilon)^{|\gamma|} e^{2\pi (R_0 + 2\epsilon)|\operatorname{Im}\zeta|} c(\alpha) \epsilon^{-|\alpha|} \\
&= \left(R_0 + 2\frac{R_0}{1 + |\zeta|} \right)^{|\gamma|} e^{2\pi (R_0 + 2\frac{R_0}{1 + |\zeta|})|\operatorname{Im}\zeta|} c(\alpha) \left(\frac{1 + |\zeta|}{R_0} \right)^{|\alpha|} \\
&\le (3R_0)^{|\gamma|} e^{2\pi R_0 |\operatorname{Im}\zeta|} e^{4\pi R_0} c(\alpha) R_0^{-|\alpha|} (1 + |\zeta|)^{|\alpha|},
\end{aligned}
$$

yielding

$$|F^{(k)}(\zeta)| \le e^{2\pi R_0 |\operatorname{Im}\zeta|} C_k (1 + |\zeta|)^{N_k},$$

which implies that $\mathbb{R}^n \ni \xi \mapsto F(\xi)$ is indeed a multiplier. We have also

$$\langle T, \chi \hat{\phi} \rangle_{\mathscr{S}', \mathscr{S}} = \langle T(x), \chi(x) \int_{\mathbb{R}^n} \phi(\xi) e^{-2i\pi x \xi} d\xi \rangle_{\mathscr{S}', \mathscr{S}}.$$

Since the function F is entire we have for $\phi \in C_c^\infty(\mathbb{R}^n)$, using (8.1.38) and Fubini's theorem on $\ell^1(\mathbb{N}) \times L^1(\mathbb{R}^n)$,

$$\int_{\mathbb{R}^n} F(\xi)\phi(\xi)d\xi = \sum_{k\geq 0} \langle T(x), \chi(x)(-2i\pi x)^k \rangle \int_{\text{supp}\,\phi} \frac{\xi^k}{k!}\phi(\xi)d\xi. \qquad (8.1.39)$$

On the other hand, since $\hat{\phi}$ is also entire (from the discussion on F or directly from the integral formula for the Fourier transform of $\phi \in C_c^\infty(\mathbb{R}^n)$), we have

$$\langle T, \chi\hat{\phi} \rangle = \langle T(x), \chi(x) \sum_{k\geq 0} (\hat{\phi})^{(k)}(0)x^k/k! \rangle$$

$$= \langle T(x), \chi(x) \underbrace{\lim_{N\to +\infty} \sum_{0\leq k\leq N} (\hat{\phi})^{(k)}(0)x^k/k!}_{\text{convergence in } C_c^\infty(\mathbb{R}^n)} \rangle$$

$$= \lim_{N\to +\infty} \sum_{0\leq k\leq N} \langle T(x), \chi(x)x^k/k! \rangle \int_{\mathbb{R}^n} \phi(\xi)(-2i\pi\xi)^k d\xi.$$

Thanks to (8.1.39), that quantity is equal to $\int_{\mathbb{R}^n} F(\xi)\phi(\xi)d\xi$. As a result, the tempered distributions $\hat{T}$ and F coincide on $C_c^\infty(\mathbb{R}^n)$, which is dense in $\mathscr{S}(\mathbb{R}^n)$ (see Exercise 8.5.3) and so $\hat{T} = F$, concluding the proof. $\qquad\square$

8.2 The Poisson summation formula

Wave packets

We define for $x \in \mathbb{R}^n$, $(y, \eta) \in \mathbb{R}^n \times \mathbb{R}^n$,

$$\varphi_{y,\eta}(x) = 2^{n/4}e^{-\pi(x-y)^2}e^{2i\pi(x-y)\cdot\eta} = 2^{n/4}e^{-\pi(x-y-i\eta)^2}e^{-\pi\eta^2}, \qquad (8.2.1)$$

$$\text{where for } \zeta = (\zeta_1, \dots, \zeta_n) \in \mathbb{C}^n, \quad \zeta^2 = \sum_{1\leq j\leq n} \zeta_j^2. \qquad (8.2.2)$$

We note that the function $\varphi_{y,\eta}$ is in $\mathscr{S}(\mathbb{R}^n)$ and with L^2 norm 1. In fact, $\varphi_{y,\eta}$ appears as a *phase translation* of a normalized Gaussian. The following lemma introduces the *wave packets transform* as a Gabor wavelet.

Lemma 8.2.1. *Let u be a function in the Schwartz class $\mathscr{S}(\mathbb{R}^n)$. We define*

$$(Wu)(y,\eta) = (u, \varphi_{y,\eta})_{L^2(\mathbb{R}^n)} = 2^{n/4}\int u(x)e^{-\pi(x-y)^2}e^{-2i\pi(x-y)\cdot\eta}dx \qquad (8.2.3)$$

$$= 2^{n/4}\int u(x)e^{-\pi(y-i\eta-x)^2}dx\, e^{-\pi\eta^2}. \qquad (8.2.4)$$

For $u \in L^2(\mathbb{R}^n)$, the function Tu defined by

$$(Tu)(y + i\eta) = e^{\pi \eta^2} Wu(y, -\eta) = 2^{n/4} \int u(x) e^{-\pi(y + i\eta - x)^2} dx \qquad (8.2.5)$$

is an entire function. The mapping $u \mapsto Wu$ is continuous from $\mathscr{S}(\mathbb{R}^n)$ to $\mathscr{S}(\mathbb{R}^{2n})$ and isometric from $L^2(\mathbb{R}^n)$ to $L^2(\mathbb{R}^{2n})$. Moreover, we have the reconstruction formula

$$u(x) = \iint_{\mathbb{R}^n \times \mathbb{R}^n} (Wu)(y, \eta) \varphi_{y,\eta}(x) \, dy \, d\eta. \qquad (8.2.6)$$

Proof. For u in $\mathscr{S}(\mathbb{R}^n)$, we have

$$(Wu)(y, \eta) = e^{2i\pi y\eta} \widehat{\Omega}^1(\eta, y)$$

where $\widehat{\Omega}^1$ is the Fourier transform with respect to the first variable of the $\mathscr{S}(\mathbb{R}^{2n})$ function $\Omega(x, y) = u(x) e^{-\pi(x-y)^2} 2^{n/4}$. Thus the function Wu belongs to $\mathscr{S}(\mathbb{R}^{2n})$. It makes sense to compute

$$2^{-n/2}(Wu, Wu)_{L^2(\mathbb{R}^{2n})}$$

$$= \lim_{\epsilon \to 0_+} \int u(x_1) \overline{u}(x_2) e^{-\pi[(x_1-y)^2 + (x_2-y)^2 + 2i(x_1-x_2)\eta + \epsilon^2 \eta^2]} \, dy \, d\eta \, dx_1 \, dx_2. \qquad (8.2.7)$$

Now the last integral on $\mathbb{R}^{4n}$ converges absolutely and we can use Fubini's theorem. Integrating with respect to η involves the Fourier transform of a Gaussian function and we get $\epsilon^{-n} e^{-\pi \epsilon^{-2}(x_1-x_2)^2}$. Since

$$2(x_1 - y)^2 + 2(x_2 - y)^2 = (x_1 + x_2 - 2y)^2 + (x_1 - x_2)^2,$$

integrating with respect to y yields a factor $2^{-n/2}$. We are left with

$$(Wu, Wu)_{L^2(\mathbb{R}^{2n})}$$

$$= \lim_{\epsilon \to 0_+} \int u(x_1) \, \overline{u}(x_2) e^{-\pi(x_1-x_2)^2/2} \epsilon^{-n} e^{-\pi \epsilon^{-2}(x_1-x_2)^2} \, dx_1 \, dx_2. \qquad (8.2.8)$$

Changing the variables, the integral is

$$\lim_{\epsilon \to 0_+} \int u(s + \epsilon t/2) \, \overline{u}(s - \epsilon t/2) e^{-\pi \epsilon^2 t^2/2} e^{-\pi t^2} \, dt \, ds = \|u\|^2_{L^2(\mathbb{R}^n)}$$

by Lebesgue's dominated convergence theorem: the triangle inequality and the estimate $|u(x)| \leq C(1 + |x|)^{-n-1}$ imply, with $v = u/C$,

$$|v(s + \epsilon t/2) \, \overline{v}(s - \epsilon t/2)| \leq (1 + |s + \epsilon t/2|)^{-n-1}(1 + |s + \epsilon t/2|)^{-n-1}$$

$$\leq (1 + |s + \epsilon t/2| + |s - \epsilon t/2|)^{-n-1}$$

$$\leq (1 + 2|s|)^{-n-1}.$$

Eventually, this proves that for $u \in \mathscr{S}(\mathbb{R}^n)$,

$$\|Wu\|^2_{L^2(\mathbb{R}^{2n})} = \|u\|^2_{L^2(\mathbb{R}^n)}, \tag{8.2.9}$$

so that by density of $\mathscr{S}(\mathbb{R}^n)$ in $L^2(\mathbb{R}^n)$,

$$W : L^2(\mathbb{R}^n) \to L^2(\mathbb{R}^{2n}) \quad \text{with} \quad W^*W = \mathrm{id}_{L^2(\mathbb{R}^n)}. \tag{8.2.10}$$

Noticing first that $\iint Wu(y,\eta)\varphi_{y,\eta}\,dy\,d\eta$ belongs to $L^2(\mathbb{R}^n)$ (with a norm smaller than $\|Wu\|_{L^1(\mathbb{R}^{2n})}$) and applying Fubini's theorem, we get from the polarization of (8.2.9) for $u, v \in \mathscr{S}(\mathbb{R}^n)$,

$$(u, v)_{L^2(\mathbb{R}^n)} = (Wu, Wv)_{L^2(\mathbb{R}^{2n})} = \iint Wu(y,\eta)(\varphi_{y,\eta}, v)_{L^2(\mathbb{R}^n)}\,dy\,d\eta$$

$$= \left(\iint Wu(y,\eta)\varphi_{y,\eta}\,dy\,d\eta, v\right)_{L^2(\mathbb{R}^n)},$$

yielding $u = \iint Wu(y,\eta)\varphi_{y,\eta}\,dy\,d\eta$, which is the result of the lemma. $\qquad\square$

Poisson's formula

The following lemma is in fact the Poisson summation formula for Gaussian functions in one dimension.

Lemma 8.2.2. *For all complex numbers z, the following series are absolutely convergent and*

$$\sum_{m\in\mathbb{Z}} e^{-\pi(z+m)^2} = \sum_{m\in\mathbb{Z}} e^{-\pi m^2} e^{2i\pi mz}. \tag{8.2.11}$$

Proof. We set $\omega(z) = \sum_{m\in\mathbb{Z}} e^{-\pi(z+m)^2}$. The function ω is entire and 1-periodic since for all $m \in \mathbb{Z}$, $z \mapsto e^{-\pi(z+m)^2}$ is entire and for $R > 0$,

$$\sup_{|z|\leq R} |e^{-\pi(z+m)^2}| \leq \sup_{|z|\leq R} |e^{-\pi z^2}| e^{-\pi m^2} e^{2\pi|m|R} \in \ell^1(\mathbb{Z}).$$

Consequently, for $z \in \mathbb{R}$, we obtain, expanding ω in Fourier series[8],

$$\omega(z) = \sum_{k\in\mathbb{Z}} e^{2i\pi kz} \int_0^1 \omega(x)e^{-2i\pi kx}\,dx.$$

---

[8]Note that we use this expansion only for a C^∞ 1-periodic function. The proof is simple and requires us only to compute $1 + 2\,\mathrm{Re}\sum_{1\leq k\leq N} e^{2i\pi kx} = \frac{\sin \pi(2N+1)x}{\sin \pi x}$. Then one has to show that for a smooth 1-periodic function ω such that $\omega(0) = 0$,

$$\lim_{\lambda\to+\infty} \int_0^1 \frac{\sin \lambda x}{\sin \pi x}\omega(x)\,dx = 0,$$

which is obvious since for a smooth ν (here we take $\nu(x) = \omega(x)/\sin \pi x$), $|\int_0^1 \nu(x)\sin(\lambda x)\,dx| = O(\lambda^{-1})$ by integration by parts.

We also check, using Fubini's theorem on $L^1(0,1) \times \ell^1(\mathbb{Z})$,

$$\int_0^1 \omega(x)e^{-2i\pi kx}dx = \sum_{m\in\mathbb{Z}}\int_0^1 e^{-\pi(x+m)^2}e^{-2i\pi kx}dx$$

$$= \sum_{m\in\mathbb{Z}}\int_m^{m+1} e^{-\pi t^2}e^{-2i\pi kt}dt$$

$$= \int_{\mathbb{R}} e^{-\pi t^2}e^{-2i\pi kt} = e^{-\pi k^2}.$$

So the lemma is proven for real z and since both sides are entire functions, we conclude by analytic continuation. $\qquad\square$

It is now straightforward to get the nth-dimensional version of the previous lemma: for all $z \in \mathbb{C}^n$, using the notation (8.2.2), we have

$$\sum_{m\in\mathbb{Z}^n} e^{-\pi(z+m)^2} = \sum_{m\in\mathbb{Z}^n} e^{-\pi m^2}e^{2i\pi m\cdot z}. \qquad (8.2.12)$$

Theorem 8.2.3 (Poisson summation formula). *Let n be a positive integer and let u be a function in $\mathscr{S}(\mathbb{R}^n)$. Then we have*

$$\sum_{k\in\mathbb{Z}^n} u(k) = \sum_{k\in\mathbb{Z}^n} \hat{u}(k), \qquad (8.2.13)$$

where $\hat{u}$ stands for the Fourier transform of u. In other words the tempered distribution $D_0 = \sum_{k\in\mathbb{Z}^n}\delta_k$ is such that $\widehat{D_0} = D_0$.

Proof. We write, according to (8.2.6) and to Fubini's theorem

$$\sum_{k\in\mathbb{Z}^n} u(k) = \sum_{k\in\mathbb{Z}^n} \iint Wu(y,\eta)\varphi_{y,\eta}(k)dyd\eta$$

$$= \iint Wu(y,\eta) \sum_{k\in\mathbb{Z}^n} \varphi_{y,\eta}(k)dyd\eta.$$

Now, (8.2.12), (8.2.1) give

$$\sum_{k\in\mathbb{Z}^n} \varphi_{y,\eta}(k) = \sum_{k\in\mathbb{Z}^n} \widehat{\varphi}_{y,\eta}(k),$$

so that (8.2.6) and Fubini's theorem imply the result. $\qquad\square$

8.3 Periodic distributions

The Dirichlet kernel

For $N \in \mathbb{N}$, the Dirichlet kernel D_N is defined on $\mathbb{R}$ by

$$D_N(x) = \sum_{-N \le k \le N} e^{2i\pi kx}$$

$$= 1 + 2\,\mathrm{Re} \sum_{1 \le k \le N} e^{2i\pi kx} \underset{x \notin \mathbb{Z}}{=} 1 + 2\,\mathrm{Re}\left(e^{2i\pi x}\frac{e^{2i\pi Nx} - 1}{e^{2i\pi x} - 1}\right)$$

$$= 1 + 2\,\mathrm{Re}\!\left(e^{2i\pi x - i\pi x + i\pi Nx}\right)\frac{\sin(\pi Nx)}{\sin(\pi x)} = 1 + 2\cos(\pi(N+1)x)\frac{\sin(\pi Nx)}{\sin(\pi x)}$$

$$= 1 + \frac{1}{\sin(\pi x)}\left(\sin(\pi x(2N+1)) - \sin(\pi x)\right) = \frac{\sin(\pi x(2N+1))}{\sin(\pi x)},$$

and extending by continuity at $x \in \mathbb{Z}$ that 1-periodic function, we find that

$$D_N(x) = \frac{\sin(\pi x(2N+1))}{\sin(\pi x)}. \tag{8.3.1}$$

Now, for a 1-periodic $v \in C^1(\mathbb{R})$, with

$$(D_N \star u)(x) = \int_0^1 D_N(x - t)u(t)dt, \tag{8.3.2}$$

we have

$$\lim_{N \to +\infty} \int_0^1 D_N(x-t)v(t)dt = v(x) + \lim_{N \to +\infty} \int_0^1 \sin(\pi t(2N+1))\frac{(v(x - t) - v(x))}{\sin(\pi t)}dt,$$

and the function θ_x given by $\theta_x(t) = \frac{v(x-t)-v(x)}{\sin(\pi t)}$ is continuous on $[0,1]$, and from the Riemann–Lebesgue Lemma 3.4.5, we obtain

$$\lim_{N \to +\infty} \sum_{-N \le k \le N} e^{2i\pi kx}\int_0^1 e^{-2i\pi kt}v(t)dt = \lim_{N \to +\infty}\int_0^1 D_N(x - t)v(t)dt = v(x).$$

On the other hand if v is 1-periodic and C^{1+l}, the Fourier coefficient

$$c_k(v) = \int_0^1 e^{-2i\pi kt}v(t)dt$$

$$\underset{\text{for } k \ne 0}{=} \frac{1}{2i\pi k}[e^{-2i\pi kt}v(t)]_{t=1}^{t=0} + \int_0^1 \frac{1}{2i\pi k}e^{-2i\pi kt}v'(t)dt,$$

and iterating the integration by parts, we find $c_k(v) = O(k^{-1-l})$ so that for a 1-periodic C^2 function v, we have

$$\sum_{k \in \mathbb{Z}} e^{2i\pi kx}c_k(v) = v(x). \tag{8.3.3}$$

Pointwise convergence of Fourier series

Lemma 8.3.1. *Let $u : \mathbb{R} \longrightarrow \mathbb{R}$ be a 1-periodic $L^1_{\mathrm{loc}}(\mathbb{R})$ function and let $x_0 \in [0,1]$. Let us assume that there exists $w_0 \in \mathbb{R}$ such that the Dini condition is satisfied, i.e.,*

$$\int_0^{1/2} \frac{|u(x_0 + t) + u(x_0 - t) - 2w_0|}{t}\, dt < +\infty. \tag{8.3.4}$$

Then, $\lim_{N \to +\infty} \sum_{|k| \le N} c_k(u)e^{2i\pi k x_0} = w_0$ with $c_k(u) = \int_0^1 e^{-2i\pi t k} u(t)\, dt$.

Proof. Using the above calculations, we find

$$\sum_{|k| \le N} c_k(u)e^{2i\pi k x_0} = (D_N \star u)(x_0) = w_0 + \int_0^1 \frac{\sin\big(\pi t(2N+1)\big)}{\sin(\pi t)}\big(u(x_0 - t) - w_0\big)dt,$$

so that, using the periodicity of u and the fact that D_N is an even function, we get

$$(D_N \star u)(x_0) - w_0 = \int_0^{1/2} \frac{\sin\big(\pi t(2N+1)\big)}{\sin(\pi t)}\big(u(x_0 - t) + u(x_0 + t) - 2w_0\big)dt.$$

Thanks to the hypothesis (8.3.4), the function

$$t \mapsto \mathbf{1}_{[0,\frac{1}{2}]}(t)\frac{u(x_0 - t) + u(x_0 + t) - 2w_0}{\sin(\pi t)}$$

belongs to $L^1(\mathbb{R})$ and the Riemann–Lebesgue Lemma 3.4.5 gives the conclusion. $\qquad\square$

Theorem 8.3.2. *Let $u : \mathbb{R} \longrightarrow \mathbb{R}$ be a 1-periodic L^1_{loc} function.*

(1) *Let $x_0 \in [0,1], w_0 \in \mathbb{R}$. We define $\omega_{x_0,w_0}(t) = |u(x_0 + t) + u(x_0 - t) - 2w_0|$ and we assume that*
$$\int_0^{1/2} \omega_{x_0,w_0}(t)\frac{dt}{t} < +\infty. \tag{8.3.5}$$
Then the Fourier series $(D_N \star u)(x_0)$ converges with limit w_0. In particular, if (8.3.5) is satisfied with $w_0 = u(x_0)$, the Fourier series $(D_N \star u)(x_0)$ converges with limit $u(x_0)$. If u has a left and right limit at x_0 and is such that (8.3.5) is satisfied with $w_0 = \frac{1}{2}\big(u(x_0+0)+u(x_0-0)\big)$, the Fourier series $(D_N \star u)(x_0)$ converges with limit $\frac{1}{2}\big(u(x_0 - 0) + u(x_0 + 0)\big)$.

(2) *If the function u is Hölder-continuous[9], the Fourier series $(D_N \star u)(x)$ converges for all $x \in \mathbb{R}$ with limit $u(x)$.*

(3) *If u has a left and right limit at each point and a left and right derivative at each point, the Fourier series $(D_N \star u)(x)$ converges for all $x \in \mathbb{R}$ with limit $\frac{1}{2}\big(u(x - 0) + u(x + 0)\big)$.*

---

[9]Hölder-continuity of index $\theta \in{]0,1]}$ means that $\exists C > 0, \forall t, s, \ |u(t) - u(s)| \le C|t - s|^\theta$.

Proof. (1) follows from Lemma 8.3.1; to obtain (2), we note that for a Hölder continuous function of index $\theta \in]0,1]$, we have for $t \in]0, 1/2]$,

$$t^{-1}\omega_{x,u(x)}(t) \leq Ct^{\theta-1} \in L^1([0,1/2]).$$

(3) If u has a right derivative at x_0, it means that

$$u(x_0 + t) = u(x_0 + 0) + u'_r(x_0)t + t\epsilon_0(t), \quad \lim_{t \to 0_+} \epsilon_0(t) = 0.$$

As a consequence, for $t \in]0, 1/2]$, $t^{-1}|u(x_0+t) - u(x_0+0)| \leq |u'_r(x_0) + \epsilon_0(t)|$. Since $\lim_{t \to 0_+} \epsilon_0(t) = 0$, there exists $T_0 \in]0, 1/2]$ such that $|\epsilon_0(t)| \leq 1$ for $t \in [0, T_0]$. As a result, we have

$$\int_0^{1/2} t^{-1}|u(x_0 + t) - u(x_0 + 0)|dt$$

$$\leq \int_0^{T_0} (|u'_r(x_0)| + 1)dt + \int_{T_0}^{1/2} |u(x_0 + t) - u(x_0 + 0)|dt T_0^{-1} < +\infty,$$

since u is also L^1_{loc}. The integral $\int_0^{1/2} t^{-1}|u(x_0 - t) - u(x_0 - 0)|dt$ is also finite and the condition (8.3.5) holds with $w_0 = \frac{1}{2}(u(x_0 - 0) + u(x_0 + 0))$. The proof of the lemma is complete. $\qquad\square$

Periodic distributions

We consider now a distribution u on $\mathbb{R}^n$ which is periodic with periods $\mathbb{Z}^n$. Let $\chi \in C_c^\infty(\mathbb{R}^n; \mathbb{R}_+)$ such that $\chi = 1$ on $[0,1]^n$. Then the function χ_1 defined by

$$\chi_1(x) = \sum_{k \in \mathbb{Z}^n} \chi(x - k)$$

is C^∞ periodic[10] with periods $\mathbb{Z}^n$. Moreover since

$$\mathbb{R}^n \ni x \in \prod_{1 \leq j \leq n} [E(x_j), E(x_j) + 1[,$$

the bounded function χ_1 is also bounded from below and such that $1 \leq \chi_1(x)$. With $\chi_0 = \chi/\chi_1$, we have

$$\sum_{k \in \mathbb{Z}^n} \chi_0(x - k) = 1, \quad \chi_0 \in C_c^\infty(\mathbb{R}^n).$$

For $\varphi \in C_c^\infty(\mathbb{R}^n)$, we have from the periodicity of u,

$$\langle u, \varphi \rangle = \sum_{k \in \mathbb{Z}^n} \langle u(x), \varphi(x)\chi_0(x - k) \rangle = \sum_{k \in \mathbb{Z}^n} \langle u(x), \varphi(x + k)\chi_0(x) \rangle,$$

---

[10]Note that the sum is locally finite since for K compact subset of $\mathbb{R}^n$, $(K - k) \cap \text{supp } \chi_0 = \emptyset$ except for a finite subset of $k \in \mathbb{Z}^n$.

where the sums are finite. Now if $\varphi \in \mathscr{S}(\mathbb{R}^n)$, we have, since χ_0 is compactly supported (say in $|x| \le R_0$),

$$|\langle u(x), \varphi(x+k)\chi_0(x)\rangle| \le C_0 \sup_{|\alpha|\le N_0, |x|\le R_0} |\varphi^{(\alpha)}(x+k)|$$

$$\le C_0 \sup_{|\alpha|\le N_0, |x|\le R_0} |(1 + R_0 + |x+k|)^{n+1}\varphi^{(\alpha)}(x+k)|(1+|k|)^{-n-1}$$

$$\le p_0(\varphi)(1+|k|)^{-n-1},$$

where p_0 is a semi-norm of φ (independent of k). As a result u is a tempered distribution and we have for $\varphi \in \mathscr{S}(\mathbb{R}^n)$, using Poisson's summation formula,

$$\langle u, \varphi\rangle = \langle u(x), \underbrace{\sum_{k\in\mathbb{Z}^n} \varphi(x+k)\chi_0(x)}_{\psi_x(k)}\rangle = \langle u(x), \sum_{k\in\mathbb{Z}^n} \widehat{\psi_x}(k)\rangle.$$

Now we see that $\widehat{\psi_x}(k) = \int_{\mathbb{R}^n} \varphi(x+t)\chi_0(x)e^{-2i\pi kt}dt = \chi_0(x)e^{2i\pi kx}\hat{\varphi}(k)$, so that

$$\langle u, \varphi\rangle = \sum_{k\in\mathbb{Z}^n} \langle u(x), \chi_0(x)e^{2i\pi kx}\rangle \hat{\varphi}(k),$$

which means

$$u(x) = \sum_{k\in\mathbb{Z}^n} \langle u(t), \chi_0(t)e^{2i\pi kt}\rangle e^{-2i\pi kx} = \sum_{k\in\mathbb{Z}^n} \langle u(t), \chi_0(t)e^{-2i\pi kt}\rangle e^{2i\pi kx}.$$

Theorem 8.3.3. *Let u be a periodic distribution on $\mathbb{R}^n$ with periods $\mathbb{Z}^n$. Then u is a tempered distribution and if χ_0 is a $C_c^\infty(\mathbb{R}^n)$ function such that $\sum_{k\in\mathbb{Z}^n}\chi_0(x-k) = 1$, we have*

$$u = \sum_{k\in\mathbb{Z}^n} c_k(u)e^{2i\pi kx}, \tag{8.3.6}$$

$$\hat{u} = \sum_{k\in\mathbb{Z}^n} c_k(u)\delta_k, \quad \text{with} \quad c_k(u) = \langle u(t), \chi_0(t)e^{-2i\pi kt}\rangle, \tag{8.3.7}$$

and convergence in $\mathscr{S}'(\mathbb{R}^n)$. If u is in $C^m(\mathbb{R}^n)$ with $m > n$, the previous formulas hold with uniform convergence for (8.3.6) *and*

$$c_k(u) = \int_{[0,1]^n} u(t)e^{-2i\pi kt}dt. \tag{8.3.8}$$

Proof. The first statements are already proven and the calculation of $\hat{u}$ is immediate. If u belongs to L^1_{loc} we can redo the calculations above, choosing $\chi_0 = \mathbf{1}_{[0,1]^n}$, and get (8.3.6) with c_k given by (8.3.8). Moreover, if u is in C^m with $m > n$, we get by integration by parts that $c_k(u)$ is $O(|k|^{-m})$ so that the series (8.3.6) is uniformly converging. $\qquad\square$

Theorem 8.3.4. *Let u be a periodic distribution on $\mathbb{R}^n$ with periods $\mathbb{Z}^n$. If $u \in L^2_{\mathrm{loc}}$ (i.e., $u \in L^2(\mathbb{T}^n)$ with $\mathbb{T}^n = (\mathbb{R}/\mathbb{Z})^n$), then*

$$u(x) = \sum_{k \in \mathbb{Z}^n} c_k(u) e^{2i\pi kx}, \quad with \quad c_k(u) = \int_{[0,1]^n} u(t) e^{-2i\pi kt} dt, \qquad (8.3.9)$$

and convergence in $L^2(\mathbb{T}^n)$. Moreover $\|u\|^2_{L^2(\mathbb{T}^n)} = \sum_{k \in \mathbb{Z}^n} |c_k(u)|^2$. Conversely, if the coefficients $c_k(u)$ defined by (8.3.7) are in $\ell^2(\mathbb{Z}^n)$, the distribution u is $L^2(\mathbb{T}^n)$.

Proof. As said above the formula for the $c_k(u)$ follows from changing the choice of χ_0 to $\mathbf{1}_{[0,1]^n}$ in the discussion preceding Theorem 8.3.3. Formula (8.3.6) gives the convergence in $\mathscr{S}'(\mathbb{R}^n)$ to u. Now, since

$$\int_{[0,1]^n} e^{2i\pi(k-l)t} dt = \delta_{k,l},$$

we see from Theorem 8.3.3 that for $u \in C^{n+1}(\mathbb{T}^n)$,

$$\langle u, u \rangle_{L^2(\mathbb{T}^n)} = \sum_{k \in \mathbb{Z}^n} |c_k(u)|^2.$$

As a consequence the mapping $L^2(\mathbb{T}^n) \ni u \mapsto (c_k(u))_{k \in \mathbb{Z}^n} \in \ell^2(\mathbb{Z}^n)$ is isometric with a range containing the dense subset $\ell^1(\mathbb{Z}^n)$ (if $(c_k(u))_{k \in \mathbb{Z}^n} \in \ell^1(\mathbb{Z}^n)$, u is a continuous function); since the range is closed[11], the mapping is onto and is an isometric isomorphism from the open mapping theorem. $\square$

8.4 Notes

Johann DIRICHLET (1805–1859) was a German mathematician.

Maurice FRÉCHET (1878–1973) was a French mathematician.

Joseph FOURIER (1768–1830) was a French mathematician, inventor of the trigonometrical series, a versatile tool used now in many branches of Science.

Dennis GABOR (1900–1979) was a Hungarian-born British electrical engineer.

Oliver HEAVISIDE (1850–1925) was a British electrical engineer.

Michel PLANCHEREL (1885–1967) was a Swiss mathematician.

Denis POISSON (1781–1840) was a French mathematician.

Laurent SCHWARTZ (1915–2002) was a French mathematician, creator of the modern theory of distributions. In 1950 he became the first French recipient of the Fields medal.

---

[11]If $A : \mathcal{H}_1 \to \mathcal{H}_2$ is an isometric linear mapping between Hilbert spaces and (Au_k) is a converging sequence in $\mathcal{H}_2$, then by linearity and isometry, the sequence (u_k) is a Cauchy sequence in $\mathcal{H}_1$, thus converges. The continuity of A implies that if $u = \lim_k u_k$, we have

$$v = \lim_k Au_k = Au, \quad \text{proving that the range of } A \text{ is closed.}$$

8.5   Exercises

Exercise 8.5.1. *Let A be a positive definite $n \times n$ symmetric matrix. Prove that the function ψ_A defined by $\psi_A(x) = e^{-\langle Ax, x\rangle}$ belongs to $\mathscr{S}(\mathbb{R}^n)$.*

Answer. The function ψ_A is smooth and such that

$$x^\alpha(\partial_x^\beta \psi_A)(x) = P_{\alpha,\beta}(x)\psi_A(x),$$

where $P_{\alpha,\beta}$ is a polynomial (obvious induction). Since $\langle Ax, x\rangle \geq \delta\|x\|^2$ with a positive δ and $|P_{\alpha,\beta}(x)| \leq C(1 + \|x\|^2)^{d/2}$, where d is the degree of P, we obtain the boundedness of $x^\alpha(\partial_x^\beta\psi_A)(x)$, proving the sought result.

Exercise 8.5.2. *The Schwartz class of functions is defined by*

$$\mathscr{S}(\mathbb{R}^n) = \left\{ u \in C^\infty(\mathbb{R}^n), \forall \alpha, \beta \in \mathbb{N}^n, \ \sup_{x\in\mathbb{R}^n} |x^\alpha\partial_x^\beta u(x)| = p_{\alpha\beta}(u) < \infty \right\},$$

where $\alpha = (\alpha_1, \ldots, \alpha_n) \in \mathbb{N}^n, x^\alpha = x_1^{\alpha_1}\ldots x_n^{\alpha_n}, \beta \in \mathbb{N}^n, \partial_x^\beta = \partial_{x_1}^{\beta_1}\ldots\partial_{x_n}^{\beta_n}$. Show that the $p_{\alpha\beta}$ are semi-norms on $\mathscr{S}(\mathbb{R}^n)$, making this space a Fréchet space.

Answer. The $p_{\alpha\beta}$ are semi-norms, i.e., valued in $\mathbb{R}_+$ such that $p_{\alpha\beta}(\lambda u) = |\lambda|p_{\alpha\beta}(u)$ and they satisfy the triangle inequality. We consider a Cauchy sequence $(u_k)_{k\in\mathbb{N}}$. It means that for all α, β, for all $\epsilon > 0$, there exists $k_{\alpha\beta\epsilon}$ such that for all $k \geq k_{\alpha\beta\epsilon}, l \geq 0$,

$$p_{\alpha\beta}(u_{k+l} - u_k) \leq \epsilon.$$

Using the case $\alpha = \beta = 0$, we find a continuous function u with a uniform limit of u_k. Using the uniform convergence of the sequence $(\partial_x^\alpha u_k)_{k\in\mathbb{N}}$, we get that u is C^∞ and that the sequences $(\partial_x^\alpha u_k)_{k\in\mathbb{N}}$ are uniformly converging towards $\partial_x^\alpha u$. We write then

$$|x^\alpha\partial_x^\beta(u_k - u)(x)| = \lim_{l\to+\infty} |x^\alpha\partial_x^\beta(u_k - u_l)(x)|$$

$$\leq \limsup_l p_{\alpha\beta}(u_k - u_l) \leq \epsilon$$

for $k \geq k_{\alpha\beta\epsilon}$. We get $p_{\alpha\beta}(u_k - u) \leq \epsilon$ for $k \geq k_{\alpha\beta\epsilon}$, proving the convergence in $\mathscr{S}(\mathbb{R}^n)$.

Exercise 8.5.3. *Prove that $C_c^\infty(\mathbb{R}^n)$ is dense in the Schwartz class $\mathscr{S}(\mathbb{R}^n)$.*

Answer. Let $\chi_0 \in C_c^\infty(\mathbb{R}^n)$ equal to 1 on the unit ball. Let $\phi \in \mathscr{S}(\mathbb{R}^n)$ and let us define for $k \in \mathbb{N}^*$,

$$\phi_k(x) = \chi_0(x/k)\phi(x), \quad \phi_k \in C_c^\infty(\mathbb{R}^n), \quad \phi_k(x) - \phi(x) = \phi(x)\big(\chi_0(x/k) - 1\big),$$

and with the $p_{\alpha\beta}$ defined in Exercise 8.5.2, we have

$$
p_{\alpha\beta}(\phi_k - \phi) = \sup_{x\in\mathbb{R}^n} \left| x^\alpha \sum_{\substack{\beta'+\beta''=\beta \\ |\beta''|\geq 1}} \frac{\beta!}{\beta'!\beta''!} \partial_x^{\beta'}\phi(x)\partial_x^{\beta''}\chi_0(x/k)k^{-|\beta''|} \right.
$$
$$
\left. + \sup_{x\in\mathbb{R}^n, |x|\geq k} \left| x^\alpha(\partial_x^\beta\phi)(x)(\chi_0\left(\frac{x}{k}\right)-1) \right|,
$$
$$
\leq Ck^{-1}p_{\max(|\alpha|,|\beta|)}(\phi)p_{\max(|\alpha|,|\beta|)}(\chi_0) + k^{-1}\sup_{x\in\mathbb{R}^n}\||x|x^\alpha(\partial_x^\beta\phi)(x)|,
$$

with p_k defined in (8.1.3), proving the convergence towards ϕ in the Schwartz space of the sequence $(\phi_k)_{k\in\mathbb{N}}$.

Exercise 8.5.4. *Let $T \in \mathscr{S}'(\mathbb{R})$ such that $xT = 0$. Prove that $T = c\delta_0$.*

Answer. Let $\phi \in \mathscr{S}(\mathbb{R})$ and let $\chi_0 \in C_c^\infty(\mathbb{R}^n)$ such that $\chi_0(0) = 1$. We have

$$
\phi(x) = \chi_0(x)\phi(x) + (1 - \chi_0(x))\phi(x).
$$

Applying Taylor's formula with integral remainder (see, e.g., Theorem 5.9.3), we define the smooth function ψ by

$$
\psi(x) = \frac{(1 - \chi_0(x))}{x}\phi(x)
$$

and, applying Leibniz' formula, we see also that ψ belongs to $\mathscr{S}(\mathbb{R})$. As a result

$$
\langle T, \phi\rangle_{\mathscr{S}'(\mathbb{R}),\mathscr{S}(\mathbb{R})} = \langle T, \chi_0\phi\rangle = \langle T, \chi_0(\phi - \phi(0))\rangle + \phi(0)\langle T, \chi_0\rangle = \phi(0)\langle T, \chi_0\rangle,
$$

since the function $x \mapsto \chi_0(x)(\phi(x) - \phi(0))/x$ belongs to $C_c^\infty(\mathbb{R})$. As a result $T = \langle T, \chi_0\rangle\delta_0$.

Exercise 8.5.5. *Prove that a distribution with support $\{0\}$ is a linear combination of derivatives of the Dirac mass at 0, i.e.,*

$$
u = \sum_{|\alpha|\leq N} c_\alpha\delta_0^{(\alpha)},
$$

where the c_α are some constants.

Answer. Let $N_0 \in \mathbb{N}$ such that $|\langle u, \varphi\rangle| \leq Cp_{N_0}(\varphi)$, where the semi-norms p_k are given by (8.1.3). For $\varphi \in \mathscr{S}(\mathbb{R}^n)$, we have

$$
\varphi(x) = \sum_{|\alpha|\leq N_0} \frac{(\partial_x^\alpha\varphi)(0)}{\alpha!}x^\alpha + \underbrace{\int_0^1 \frac{(1-\theta)^{N_0}}{N_0!}\varphi^{(N_0+1)}(\theta x)d\theta\, x^{N_0+1}}_{\psi(x),\quad \psi\in C^\infty(\mathbb{R}^n)},
$$

and thus for $\chi_0 \in C_c^\infty(\mathbb{R}^n)$, $\chi_0 = 1$ near 0,

$$\langle u, \varphi \rangle = \langle u, \chi_0 \varphi \rangle = \sum_{|\alpha| \leq N_0} \frac{(\partial_x^\alpha \varphi)(0)}{\alpha!} \langle u, \chi_0(x) x^\alpha \rangle + \langle u, \chi_0(x) \psi(x) x^{N_0+1} \rangle. \quad (8.5.1)$$

We note that

$$|\langle u, \chi_0(x) \psi(x) x^{N_0+1} \rangle| \leq C_0 \sup_{|\alpha| \leq N_0} |\partial_x^\alpha (\chi_0(x) \psi(x) x^{N_0+1})|. \quad (8.5.2)$$

We can take $\chi_0(x) = \rho(x/\epsilon)$, where $\rho \in C_c^\infty(\mathbb{R}^n)$ is supported in the unit ball B_1, $\rho = 1$ in $\frac{1}{2} B_1$ and $\epsilon > 0$. We have then

$$\chi_0(x) \psi(x) x^{N_0+1} = \epsilon^{N_0+1} \rho\left(\frac{x}{\epsilon}\right) \psi\left(\epsilon\frac{x}{\epsilon}\right) \frac{x^{N_0+1}}{\epsilon^{N_0+1}} = \epsilon^{N_0+1} \rho_1\left(\frac{x}{\epsilon}\right),$$

with $\rho_1(t) = \rho(t)\psi(\epsilon t)t^{N_0+1}$, so that $\rho_1 \in C_c^\infty(\mathbb{R}^n)$ is supported in the unit ball B_1 and has all its derivatives bounded independently of ϵ. From (8.5.2), we get for all $\epsilon > 0$,

$$|\langle u, \chi_0(x) \psi(x) x^{N_0+1} \rangle| \leq C_0 \sup_{|\alpha| \leq N_0} \epsilon^{N_0+1-|\alpha|} \left|(\partial_t^\alpha \rho_1)\left(\frac{x}{\epsilon}\right)\right| \leq C_1 \epsilon,$$

which implies that the left-hand side of (8.5.2) is zero.

Exercise 8.5.6. *Let $u \in \mathscr{S}'(\mathbb{R}^n)$ and $\lambda \in \mathbb{C}$. The distribution u is said to be homogeneous with degree λ if for all $t > 0$, $u(t\cdot) = t^\lambda u(\cdot)$. Prove that the distribution u is homogeneous of degree λ if and only if Euler's equation is satisfied, namely*

$$\sum_{1 \leq j \leq n} x_j \partial_{x_j} u = \lambda u. \quad (8.5.3)$$

Answer. A distribution u on $\mathbb{R}^n$ is homogeneous of degree λ means:

$$\forall \varphi \in C_c^\infty(\mathbb{R}^n), \forall t > 0, \qquad \langle u(y), \varphi(y/t)t^{-n} \rangle = t^\lambda \langle u(x), \varphi(x) \rangle,$$

which is equivalent to $\forall \varphi \in C_c^\infty(\mathbb{R}^n), \forall s > 0, \langle u(y), \varphi(sy)s^{n+\lambda} \rangle = \langle u(x), \varphi(x) \rangle$, also equivalent to

$$\forall \varphi \in C_c^\infty(\mathbb{R}^n), \qquad \frac{d}{ds}\left(\langle u(y), \varphi(sy)s^{n+\lambda} \rangle\right) = 0 \quad \text{on } s > 0. \quad (8.5.4)$$

The differentiability property is easy to derive[12] and that

$$\langle u(y), \varphi(sy)s^{n+\lambda} \rangle = \langle u(x), \varphi(x) \rangle \qquad \text{at } s = 1.$$

---

[12]We have for $s > 0$,

$$\varphi((s+h)y) - \varphi(sy) = \varphi'(sy)hy + \int_0^1 (1-\theta)\varphi''((s+\theta h)y)\,d\theta h^2 y^2.$$

It is enough to prove that for σ in a neighborhood V of s, the function $y \mapsto \varphi^{(l)}(\sigma y)$ is bounded in $\mathscr{S}(\mathbb{R}^n)$. This is obvious, choosing for instance $V = (s/2, 2s)$.

As a consequence, we get that the homogeneity of degree λ of u is equivalent to

$$\forall s > 0, \quad \left\langle u(y), s^{n+\lambda-1}\big((n+\lambda)\varphi(sy) + \sum_{1 \le j \le n}(\partial_j\varphi)(sy)sy_j\big)\right\rangle = 0,$$

also equivalent to $0 = \langle u(y), (n+\lambda+\sum_{1\le j\le n}y_j\partial_j)(\varphi(sy))\rangle$ and by the definition of the differentiation of a distribution, it is equivalent to

$$(n+\lambda)u - \sum_{1 \le j \le n}\partial_j(y_j u) = 0,$$

which is (8.5.3) by Leibniz' rule.

Exercise 8.5.7.

(1) *Prove that the Dirac mass at 0 in $\mathbb{R}^n$ is homogeneous of degree $-n$.*

(2) *Prove that if T is a homogeneous distribution of degree λ, then $\partial_x^\alpha T$ is also homogeneous with degree $\lambda - |\alpha|$.*

(3) *Prove that the distribution $\mathrm{pv}(\frac{1}{x})$ is homogeneous of degree -1 as well as $1/(x \pm i0)$.*

(4) *For $\lambda \in \mathbb{C}$ with $\mathrm{Re}\,\lambda > -1$ we define the $L^1_{\mathrm{loc}}(\mathbb{R})$ functions*

$$x_+^\lambda = \begin{cases} x^\lambda & \text{if } x > 0, \\ 0 & \text{if } x \le 0, \end{cases} \qquad \chi_+^\lambda = \frac{x_+^\lambda}{\Gamma(\lambda+1)}. \tag{8.5.5}$$

Prove that the distributions χ_+^λ and x_+^λ are homogeneous of degree λ.

Answer. (1) We have for $t > 0$,

$$\langle \delta_0(tx), \varphi(x)\rangle = \langle \delta_0(y), \varphi(y/t)t^{-n}\rangle = t^{-n}\varphi(0) = t^{-n}\langle \delta_0, \varphi\rangle.$$

(2) Taking the derivative of the Euler equation (8.5.3), we get

$$\partial_{x_k}u + \sum_{1 \le j \le k}x_j\partial_{x_j}\partial_{x_k}u - \lambda\partial_{x_k}u = 0,$$

proving that $\partial_{x_k}u$ is homogeneous of degree $\lambda - 1$ and the result by iteration.

(3) It follows immediately from the definition (8.1.26) that the distribution $\mathrm{pv}(\frac{1}{x})$ is homogeneous of degree -1. The same is true for the distributions $\frac{1}{x\pm i0}$ as it is clear from

$$\frac{1}{x \pm i0} = \frac{d}{dx}\big(\mathrm{Log}(x \pm i0)\big) = \frac{d}{dx}\big(\ln|x| \pm i\pi\breve{H}(x)\big) = \mathrm{pv}\,\frac{1}{x} \mp i\pi\delta_0. \tag{8.5.6}$$

(4) The distributions χ_+^λ and x_+^λ are homogeneous of degree λ. By an analytic continuation argument, we can prove that χ_+^λ may be defined for any $\lambda \in \mathbb{C}$ and is a homogeneous distribution of degree λ which satisfies

$$\chi_+^\lambda = \left(\frac{d}{dx}\right)^k(\chi_+^{\lambda+k}), \qquad \chi_+^{-k} = \delta_0^{(k-1)}, \quad k \in \mathbb{N}^*.$$

Exercise 8.5.8. *Let $(u_j)_{1\leq j\leq m}$ be non-zero homogeneous distributions on $\mathbb{R}^n$ with distinct degrees $(\lambda_j)_{1\leq j\leq m}$ $(j\neq k$ implies $\lambda_j\neq\lambda_k)$. Prove that they are independent in the complex vector space $\mathscr{S}'(\mathbb{R}^n)$.*

Answer. We assume that $m\geq 2$ and that there exists some complex numbers $(c_j)_{1\leq j\leq m}$ such that $\sum_{1\leq j\leq m}c_j u_j=0$. Then applying the (Euler) operator

$$\mathcal{E}=\sum_{1\leq j\leq m}x_j\partial_{x_j},$$

we get for all $k\in\mathbb{N}$, $0=\sum_{1\leq j\leq m}c_j\mathcal{E}^k(u_j)=\sum_{1\leq j\leq m}c_j\lambda_j^k u_j$. We consider now the Vandermonde matrix $m\times m$

$$V_m=\begin{pmatrix}1 & 1 & \cdots & 1\\ \lambda_1 & \lambda_2 & \cdots & \lambda_m\\ \cdots\cdots & & & \\ \lambda_1^{m-1} & \lambda_2^{m-1} & \cdots & \lambda_m^{m-1}\end{pmatrix},\quad \det V_m=\prod_{1\leq j<k\leq m}(\lambda_k-\lambda_j)\neq 0.$$

We note that for $\varphi\in C_c^\infty(\mathbb{R}^n)$, and $X\in\mathbb{C}^m$ given by

$$X=\begin{pmatrix}c_1\langle u_1,\varphi\rangle\\ c_2\langle u_2,\varphi\rangle\\ \cdots\cdots\cdots\\ c_m\langle u_m,\varphi\rangle\end{pmatrix},$$

we have $V_m X=0$, so that $X=0$, i.e., $\forall j,\forall\varphi\in C_c^\infty(\mathbb{R}^n), c_j\langle u_j,\varphi\rangle=0$, i.e., $c_j u_j=0$ and since u_j is not the zero distribution, we get the sought conclusion $c_j=0$ for all j.

Exercise 8.5.9. *Let $T\in\mathscr{S}'(\mathbb{R}^n)$ be a homogeneous distribution of degree m. Prove that its Fourier transform is a homogeneous distribution of degree $-m-n$.*

Answer. We check

$$(\xi\cdot D_\xi)\hat{T}=-\xi\cdot\widehat{xT}=-(\widehat{D_x\cdot xT})=-\frac{n}{2i\pi}\hat{T}-\frac{1}{2i\pi}(\widehat{x\cdot\partial_x T})=-\frac{(n+m)}{2i\pi}\hat{T},$$

so that Euler's equation $\xi\cdot\partial_\xi\hat{T}=-(n+m)\hat{T}$ is satisfied.

Exercise 8.5.10. *Let $u\in\mathscr{S}'(\mathbb{R}^n)$ such that $\nabla u=(\partial_1 u,\ldots,\partial_n u)=0$. Prove that u is a constant.*

Answer. For all j, we have $\xi_j\hat{u}(\xi)=0$ and since a polynomial is a multiplier of $\mathscr{S}$, we have also $|\xi|^2\hat{u}(\xi)=0$, which implies that $\operatorname{supp}\hat{u}\subset\{0\}$. From Exercise 8.5.5, we find that $\hat{u}$ is a linear combination of derivatives of the Dirac mass at 0 and (8.1.19) implies along with (8.1.21) that u is a polynomial. Now a polynomial with a vanishing gradient is a constant (use Taylor's formula).

Chapter 9

Classical Inequalities

9.1 Riesz–Thorin interpolation theorem

Theorem 9.1.1 (Hadamard three-lines theorem). *Let $a < b$ be real numbers, let $\Omega = \{z \in \mathbb{C}, a < \operatorname{Re} z < b\}$ and let $f : \overline{\Omega} \to \mathbb{C}$ be a bounded continuous function holomorphic on Ω. We define for $x \in [a, b]$,*

$$M(x) = \sup_{y \in \mathbb{R}} |f(x + iy)|.$$

Then the function M is log-convex on $[a, b]$, i.e.,

$$M(x) \leq M(a)^{\frac{b-x}{b-a}} M(b)^{\frac{x-a}{b-a}}. \tag{9.1.1}$$

N.B. Exercise 3.7.2 provides some information about logarithmic convexity. We note here that this proposition implies in particular that if f vanishes identically on the vertical line $\{\operatorname{Re} z = a\}$ or on $\{\operatorname{Re} z = b\}$, then it should vanish identically on Ω. If $M(a), M(b)$ are both positive, then (9.1.1) reads

$$(\ln M)\big((1 - \theta)a + \theta b\big) \leq (1 - \theta) \ln M(a) + \theta \ln M(b),$$

which means convexity of $\ln M$ on $[a, b]$, i.e., log-convexity. Defining $\ln 0 = -\infty$, we recover the fact that if f vanishes on one vertical line, it vanishes on Ω.

Proof. We may of course assume without loss of generality that $a = 0, b = 1$: given $a < b$ real numbers, and f as in the proposition above, we may consider

$$\tilde{f}(z) = f\big((b - a)z + a\big),$$

which is defined on $\{z \in \mathbb{C}, 0 \leq \operatorname{Re} z \leq 1\}$. If we get the result for $\tilde{f}$, it will read for $\theta \in [0, 1]$

$$\sup_{\{\operatorname{Re}\zeta = a + \theta(b-a) = x\}} |f(\zeta)| = \sup_{\{\operatorname{Re} z = \theta\}} |\tilde{f}(z)|$$

$$\leq \left(\sup_{y \in \mathbb{R}} |\tilde{f}(iy)|\right)^{1-\theta} \left(\sup_{y \in \mathbb{R}} |\tilde{f}(1+iy)|\right)^{\theta}$$

$$= \left(\sup_{y \in \mathbb{R}} |f(a + (b-a)iy)|\right)^{1-\theta} \left(\sup_{y \in \mathbb{R}} |f(a + b - a + (b-a)iy)|\right)^{\theta}$$

$$= \left(\sup_{\operatorname{Re}\zeta = a} |f(\zeta)|\right)^{\frac{b-x}{b-a}} \left(\sup_{\operatorname{Re}\zeta = b} |f(\zeta)|\right)^{\frac{x-a}{b-a}},$$

which is the sought result.

We assume first that $M(0) = M(1) = 1$. We define for $\epsilon > 0$ the holomorphic function h_ϵ on $\operatorname{Re} z > -1/\epsilon$ given by

$$h_\epsilon(z) = \frac{1}{1 + \epsilon z}.$$

We note that $\forall z \in \partial\Omega$, $|f(z)h_\epsilon(z)| \leq 1$ (in fact $|f(z)| \leq 1$ there as well as $h_\epsilon(z)$) and moreover with $C = \sup_{\bar{\Omega}} |f|$, we have for $0 \leq \operatorname{Re} z \leq 1$, $|\operatorname{Im} z| \geq C/\epsilon$,

$$|f(z)h_\epsilon(z)| \leq C|1 + \epsilon z|^{-1} \leq C\epsilon^{-1}|\operatorname{Im} z|^{-1} \leq 1. \tag{9.1.2}$$

As a result, considering the rectangle $R_\epsilon = \{0 \leq \operatorname{Re} z \leq 1, |\operatorname{Im} z| \leq C/\epsilon\}$, we see that the continuous function $fh_\epsilon : R_\epsilon \to \mathbb{C}$ is bounded above by 1 on the boundary and is holomorphic in the interior. Applying the maximum principle, we obtain that

$$(\sharp) \qquad\qquad \forall z \in R_\epsilon, \quad |f(z)h_\epsilon(z)| \leq 1.$$

On the other hand if $z \in \bar{\Omega}$ with $|\operatorname{Im} z| > C/\epsilon$, we get from (9.1.2) the same inequality $(\sharp)$. Consequently, we have for all $\epsilon > 0$ and all $z \in \bar{\Omega}$, $|f(z)h_\epsilon(z)| \leq 1$, which implies the sought result $|f(z)| \leq 1$ for $z \in \bar{\Omega}$.

We assume now that $M(0), M(1)$ are both positive, and we introduce the function

$$F(z) = M(0)^{-(1-z)} M(1)^{-z} f(z) = f(z) e^{z(\ln M(0) - \ln M(1))} M(0)^{-1}. \tag{9.1.3}$$

The function F is holomorphic on $\Omega = \{0 < \operatorname{Re} z < 1\}$, is and bounded on $\bar{\Omega}$ since

$$\sup_{z \in \bar{\Omega}} |F(z)| \leq M(0)^{-1} e^{|\ln M(0) - \ln M(1)|} \sup_{\bar{\Omega}} |f|.$$

Moreover, on the vertical lines $\operatorname{Re} z = 0, 1$, $|F|$ is bounded above respectively by

$$M(0)M(0)^{-1} = 1, \qquad M(1)M(0)M(1)^{-1}M(0)^{-1} = 1,$$

so that we may apply the previous result to obtain

$$\forall z \in \bar{\Omega}, \quad |M(0)^{-(1-z)} M(1)^{-z} f(z)| \leq 1,$$

which is precisely the sought result.

We assume now that $M(0) \geq 0, M(1) \geq 0$. Let $\epsilon > 0$ be given. We introduce the function

$$F_\epsilon(z) = (M(0) + \epsilon)^{-(1-z)}(M(1) + \epsilon)^{-z} f(z). \tag{9.1.4}$$

Then, using the previous result, we obtain

$$\forall \epsilon > 0, \forall z \in \bar{\Omega}, \quad |f(z)| \leq |(M(0) + \epsilon)^{(1-z)}(M(1) + \epsilon)^z|,$$

which implies the result, letting $\epsilon \to 0_+$. The proof of the theorem is complete. $\square$

Theorem 9.1.2 (Riesz–Thorin Interpolation Theorem). *Let $(X, \mathcal{M}, \mu)$ be a measure space where μ is a σ-finite positive measure. Let $p_0, p_1, q_0, q_1 \in [1, +\infty]$ and let $T : L^{p_j}(\mu) \longrightarrow L^{q_j}(\mu), j = 0, 1,$ be a linear map such that*

$$\|Tu\|_{L^{q_j}(\mu)} \leq M_j \|u\|_{L^{p_j}(\mu)}, \qquad j = 0, 1.$$

For $\theta \in [0, 1]$ we define $\frac{1}{p_\theta} = \frac{1-\theta}{p_0} + \frac{\theta}{p_1}$, $\frac{1}{q_\theta} = \frac{1-\theta}{q_0} + \frac{\theta}{q_1}$. Then T is a bounded linear map from $L^{p_\theta}(\mu)$ into $L^{q_\theta}(\mu)$ and

$$\forall u \in L^{p_\theta}(\mu), \quad \|Tu\|_{L^{q_\theta}(\mu)} \leq M_0^{1-\theta} M_1^{\theta} \|u\|_{L^{p_\theta}(\mu)}. \tag{9.1.5}$$

Proof. We may of course assume that $\theta \in (0, 1)$.

[1] *Let us first assume that $p_\theta = +\infty$, so that $p_0 = p_1 = +\infty$.*
Let u be a function in $L^\infty(\mu)$: Tu belongs to $L^{q_0}(\mu) \cap L^{q_1}(\mu)$.

Claim. For $\theta \in (0, 1)$, we have $L^{q_0}(\mu) \cap L^{q_1}(\mu) \subset L^{q_\theta}(\mu)$. This is obvious if $q_\theta = +\infty$ (implying $q_0 = q_1 = +\infty$) and if $q_\theta < +\infty$, assuming that q_0, q_1 are both finite (and distinct), we find some $t \in (0, 1)$ such that

$$q_\theta = (1 - t)q_0 + tq_1, \quad \text{so that with } \frac{1}{r} = 1 - t,$$

$$\int_X |v|^{q_\theta} d\mu = \int_X |v|^{q_0(1-t)} |v|^{q_1 t} d\mu \tag{9.1.6}$$

$$\leq \||v|^{q_0(1-t)}\|_{L^r} \||v|^{q_1 t}\|_{L^{r'}} = \|v\|_{L^{q_0}}^{q_0(1-t)} \|v\|_{L^{q_1}}^{q_1 t}.$$

If $q_0 = +\infty, 1 \leq q_1 < +\infty$, we have $q_\theta = q_1/\theta$ and

$$\int_X |v|^{q_\theta} d\mu \leq \|v\|_{L^\infty}^{q_1(\frac{1}{\theta}-1)} \int_X |v|^{q_1} d\mu, \tag{9.1.7}$$

proving the claim in that case as well.

We find thus that $Tu \in L^{q_\theta}$ and when q_0, q_1 are both finite, applying (9.1.6),

$$\|Tu\|_{q_\theta}^{q_\theta} \leq \|Tu\|_{q_0}^{q_0(1-t)} \|Tu\|_{q_1}^{q_1 t} \leq M_0^{q_0(1-t)} M_1^{q_1 t} \|u\|_\infty^{q_\theta},$$

and since

$$\frac{tq_1}{q_\theta} = \frac{q_\theta - q_0}{q_1 - q_0} \frac{q_1}{q_\theta} = \frac{1 - \frac{q_0}{q_\theta}}{1 - \frac{q_0}{q_1}} = \frac{q_0^{-1} - q_\theta^{-1}}{q_0^{-1} - q_1^{-1}} = \theta, \quad \text{so that } \frac{(1-t)q_0}{q_\theta} = 1 - \theta,$$

proving (9.1.5). If $q_0 = +\infty, 1 \le q_1 < +\infty$, we have $q_\theta = q_1/\theta$ and applying (9.1.7)

$$\|Tu\|_{q_\theta}^{q_\theta} \le \|Tu\|_{q_0}^{q_1(\frac{1}{\theta}-1)} \|Tu\|_{q_1}^{q_1} \le M_0^{q_1(\frac{1}{\theta}-1)} M_1^{q_1} \|u\|_\infty^{q_\theta},$$

and since

$$\frac{q_\theta - q_1}{q_\theta} = 1 - \frac{q_\theta^{-1}}{q_1^{-1} - q_0^{-1}} = 1 - \theta, \quad \text{so that} \quad \frac{q_1}{q_\theta} = \theta,$$

this implies (9.1.5) in that case as well.

[2] *We assume now that* $1 \le p_\theta < +\infty$, $q_\theta > 1$. Let u be a function in S (defined in (3.2.20)), so that

$$u = \sum_{1 \le j \le m} \alpha_j e^{i\phi_j} \mathbf{1}_{A_j}, \quad \alpha_j > 0, \phi_j \in \mathbb{R}, \quad \mu(A_j) < +\infty, \tag{9.1.8}$$

where the A_j are pairwise disjoint elements of $\mathcal{M}$. Then Tu makes sense, belongs to $L^{q_\theta}(\mu)$ and since S is dense in $L^{p_\theta}(\mu)$ (Proposition 3.2.11), it is enough to prove that

$$\forall v \in L^{(q_\theta)'}, \quad \left| \int (Tu)v d\mu \right| \le M_0^{1-\theta} M_1^\theta \|u\|_{p_\theta} \|v\|_{(q_\theta)'}. \tag{9.1.9}$$

In fact, if we prove the above inequality, thanks to Lemma 6.2.1, this will imply that $\|Tu\|_{q_\theta} \le M_0^{1-\theta} M_1^\theta \|u\|_{p_\theta}$. Now since T is a linear operator, and S is dense in $L^{p_\theta}(\mu)$, there is a unique extension of T to a bounded linear operator from $L^{p_\theta}(\mu)$ into $L^{q_\theta}(\mu)$ with operator-norm bounded above by $M_0^{1-\theta} M_1^\theta$. To obtain (9.1.9), it is enough to prove that

$$\forall v \in S, \quad \left| \int (Tu)v d\mu \right| \le M_0^{1-\theta} M_1^\theta \|u\|_{p_\theta} \|v\|_{(q_\theta)'}, \tag{9.1.10}$$

since $q_\theta > 1$ (S is dense in $L^{(q_\theta)'}$). We may thus assume that

$$v = \sum_{1 \le k \le N} \beta_k e^{i\psi_k} \mathbf{1}_{B_k}, \quad \beta_k > 0, \psi_k \in \mathbb{R}, \quad \mu(B_k) < +\infty, \tag{9.1.11}$$

where the B_k are pairwise disjoint elements of $\mathcal{M}$. We define the entire functions

$$u(z) = \sum_{1 \le j \le m} \alpha_j^{a(z)/a(\theta)} e^{i\phi_j} \mathbf{1}_{A_j}, \qquad a(z) = \frac{1-z}{p_0} + \frac{z}{p_1}, \tag{9.1.12}$$

$$v(z) = \sum_{1 \le k \le N} \beta_k^{(1-b(z))/(1-b(\theta))} e^{i\psi_k} \mathbf{1}_{B_k}, \quad b(z) = \frac{1-z}{q_0} + \frac{z}{q_1}, \tag{9.1.13}$$

$$F(z) = \int_X \big(Tu(z)\big) v(z) d\mu, \tag{9.1.14}$$

and we note that $a(\theta) = 1/p(\theta), b(\theta) = 1/q(\theta) \in (0,1)$ since $\theta \in (0,1)$. The function F is bounded on $\{z \in \mathbb{C}, 0 \le \operatorname{Re} z \le 1\}$: we have to deal with a finite sum and

$$\operatorname{Re} a(z) \in [0,1], \quad \operatorname{Re}(1 - b(z)) \in [0,1].$$

Moreover, for $y \in \mathbb{R}$, we have

$$F(iy) = \int_X T\Big(\sum_{1 \le j \le m} \alpha_j^{\frac{a(iy)}{a(\theta)}} e^{i\phi_j} \mathbf{1}_{A_j} \Big)\Big(\sum_{1 \le k \le m} \mathbf{1}_{B_k} \beta_k^{\frac{(1-b(iy))}{(1-b(\theta))}} e^{i\psi_k} \Big) d\mu,$$

and thus

$$|F(iy)| \le M_0 \Big\| \sum_{1 \le j \le m} \alpha_j^{\frac{a(iy)}{a(\theta)}} e^{i\phi_j} \mathbf{1}_{A_j} \Big\|_{p_0} \Big\| \sum_{1 \le k \le m} \mathbf{1}_{B_k} \beta_k^{\frac{(1-b(iy))}{(1-b(\theta))}} e^{i\eta_k} \Big\|_{q_0'}.$$

Since the $(A_j)_{1 \le j \le m}$ (and the $(B_k)_{1 \le k \le N}$) are pairwise disjoint, we have

$$\Big\| \sum_{1 \le j \le m} \alpha_j^{\frac{a(iy)}{a(\theta)}} e^{i\phi_j} \mathbf{1}_{A_j} \Big\|_{p_0} = \Big\| \sum_{1 \le j \le m} \alpha_j^{\frac{\operatorname{Re} a(iy)}{a(\theta)}} \mathbf{1}_{A_j} \Big\|_{p_0} = \Big\| \sum_{1 \le j \le m} \alpha_j^{\frac{p(\theta)}{p_0}} \mathbf{1}_{A_j} \Big\|_{p_0}$$

$$= \Big(\int_X \Big(\sum_{1 \le j \le m} \alpha_j^{p(\theta)} \mathbf{1}_{A_j} \Big) d\mu \Big)^{1/p_0} = \Big(\int_X |u(\theta)|^{p(\theta)} d\mu \Big)^{1/p_0} = \|u(\theta)\|_{p(\theta)}^{p_\theta/p_0},$$

and

$$\Big\| \sum_{1 \le k \le N} \mathbf{1}_{B_k} \beta_k^{\frac{(1-b(iy))}{(1-b(\theta))}} e^{i\psi_k} \Big\|_{q_0'} = \Big\| \sum_{1 \le k \le N} \beta_k^{\frac{1-\operatorname{Re} b(iy)}{1-b(\theta)}} \mathbf{1}_{B_k} \Big\|_{q_0'} = \Big\| \sum_{1 \le k \le N} \beta_k^{\frac{q'(\theta)}{q_0'}} \mathbf{1}_{B_k} \Big\|_{q_0'}$$

$$= \Big(\int_X \Big(\sum_{1 \le k \le N} \beta_k^{q'(\theta)} \mathbf{1}_{B_k} \Big) d\mu \Big)^{1/q_0'} = \Big(\int_X |v(\theta)|^{q'(\theta)} d\mu \Big)^{1/q_0'} = \|v(\theta)\|_{q'(\theta)}^{q_\theta'/q_0'},$$

so that, for $y \in \mathbb{R}$, $|F(iy)| \le M_0 \|u(\theta)\|_{p(\theta)}^{p_\theta/p_0} \|v(\theta)\|_{q'(\theta)}^{q_\theta'/q_0'}$. We obtain similarly that

$$|F(1+iy)| \le M_1 \|u(\theta)\|_{p(\theta)}^{p_\theta/p_1} \|v(\theta)\|_{q'(\theta)}^{q_\theta'/q_1'}.$$

The last two inequalities and Theorem 9.1.1 imply for $\operatorname{Re} z \in [0,1]$,

$$|F(z)| \le \Big(M_0 \|u(\theta)\|_{p(\theta)}^{\frac{p_\theta}{p_0}} \|v(\theta)\|_{q'(\theta)}^{\frac{q_\theta'}{q_0'}} \Big)^{1-\operatorname{Re} z} \Big(M_1 \|u(\theta)\|_{p(\theta)}^{p_\theta/p_1} \|v(\theta)\|_{q'(\theta)}^{q_\theta'/q_1'} \Big)^{\operatorname{Re} z},$$

so that for $\operatorname{Re} z = \theta$, since

$$\frac{p_\theta}{p_0}(1-\theta) + \frac{p_\theta}{p_1}\theta = 1 = \frac{q_\theta'}{q_0'}(1-\theta) + \frac{q_\theta'}{q_1'}\theta,$$

we get

$$\left| \int (Tu)v d\mu \right| = |F(\theta)| \leq M_0^{1-\theta} M_1^{\theta} \|u\|_{p(\theta)} \|v\|_{q'(\theta)},$$

which is indeed (9.1.10), concluding the proof in this case.

[3] *We assume now that* $1 \leq p_\theta < +\infty$, $q_\theta = 1$ (*and thus* $q_0 = q_1 = 1, q_0' = q_1' = +\infty$). *It is enough to prove* (9.1.9) (*from Proposition 3.2.11*), *and to get it,* (9.1.10) *should be modified so that* S *is replaced by* S_∞ (*see Proposition 3.2.13*), *meaning that* (9.1.11) *must be modified so that* $\mu(B_k)$ *could be* $+\infty$. *We modify* (9.1.13) *and take* $v(z) = v$. *The rest of the proof is unchanged, following case* **[2]**. *The proof of Theorem 9.1.2 is complete.* $\square$

The Riesz–Thorin interpolation theorem appears as a direct consequence of Hadamard's three-lines theorem and is a typical example of a complex interpolation method based on a version of the maximum principle for holomorphic functions on unbounded domains. Of course holomorphic functions in an unbounded domain Ω, continuous in $\bar{\Omega}$, may fail to satisfy the maximum principle[1]. However, the Phragmén–Lindelöf principle asserts that a maximum principle result holds true, provided we impose some restriction on the growth of the class of functions: Hadamard's three lines theorem, in which we have assumed boundedness for the holomorphic function, is a good example of this technique. We give below some classical consequences of Theorem 9.1.2.

Theorem 9.1.3 (Generalized Young's inequality). *Let* $p, q, r \in [1, +\infty]$ *such that* (6.2.1) *holds. Let* $(X_1, \mathcal{M}_1, \mu_1)$ *and* $(X_2, \mathcal{M}_2, \mu_2)$ *be measure spaces where each* μ_j *is a* σ-*finite positive measure and let* $k : X_1 \times X_2 \to \mathbb{C}$ *be a measurable mapping* (*the product* $X_1 \times X_2$ *is equipped with the* σ-*algebra* $\mathcal{M}_1 \otimes \mathcal{M}_2$) *such that there exists* $M \geq 0$ *with*

$$\sup_{x_1 \in X_1} \left(\int_{X_2} |k(x_1, x_2)|^p d\mu_2(x_2) \right)^{1/p} \leq M, \tag{9.1.15}$$

$$\sup_{x_2 \in X_2} \left(\int_{X_1} |k(x_1, x_2)|^p d\mu_1(x_1) \right)^{1/p} \leq M. \tag{9.1.16}$$

The linear operator L *defined by*

$$(Lu_2)(x_1) = \int_{X_2} k(x_1, x_2) u_2(x_2) d\mu_2(x_2) \tag{9.1.17}$$

can be extended to a bounded linear operator from $L^q(\mu_2)$ *into* $L^r(\mu_1)$ *with operator-norm less than* M.

---

[1]The function e^z on $\Omega = \{z \in \mathbb{C}, \operatorname{Re} z > 0\}$ is unbounded on Ω although it has modulus 1 on $\partial\Omega$.

Remark 9.1.4. The first (resp. second) supremum can be replaced by an esssup (see (3.2.6)) in the μ_1 (resp. μ_2) sense. If $p = +\infty$ (which implies $q = 1, r = +\infty$), the hypothesis reads as

$$\text{esssup}_{(x_1,x_2)\in X_1 \times X_2} |k(x_1, x_2)| \leq M = M,$$

and the result in that case is trivial since

$$|(Lu_2)(x_1)| \leq M \|u_2\|_{L^1(\mu_2)} \implies \|Lu_2\|_{L^\infty(\mu_1)} \leq M\|u_2\|_{L^1(\mu_2)}.$$

We may thus assume that $1 \leq p < +\infty$. If $q = +\infty$ (which implies $p = 1, r = +\infty$), we get also trivially

$$|(Lu_2)(x_1)| \leq \int_{X_2} |k(x_1, x_2)||u_2(x_2)|d\mu_2(x_2)$$
$$\leq M\|u_2\|_{L^\infty(\mu_2)}$$
$$\implies \|Lu_2\|_{L^\infty(\mu_1)} \leq M\|u_2\|_{L^\infty(\mu_2)}.$$

We may thus assume that p and q are finite. We may define (9.1.17) for $u_2 = \mathbf{1}_{A_2}$, where $A_2 \in \mathcal{M}$, with $\mu_2(A_2) < +\infty$. Then we have

$$\int_{A_2} |k(x_1, x_2)|d\mu_2(x_2) \leq M\|\mathbf{1}_{A_2}\|_{L^{p'}(\mu_2)}$$
$$\leq M\mu_2(A_2)^{1/p'} < +\infty.$$

As a result for $u_2 \in S_q(\mu_2)$ (the space $S_p(\mu)$ is defined by (3.2.20)), we may define Lu_2 as an $L^\infty(\mu_1)$ function. Since for $1 \leq q < +\infty$, $S_q(\mu_2)$ is dense in $L^q(\mu_2)$ (Proposition 3.2.11), the statement of Theorem 9.1.3 can be rephrased as follows: the linear operator L defined from $S_q(\mu_2)$ into $L^\infty(\mu_1)$ can be uniquely extended as a bounded linear operator from $L^q(\mu_2)$ into $L^r(\mu_1)$ with operator-norm less than M.

N.B. Young's inequality (Theorem 6.2.2) is indeed a consequence of the above result, taking $k(x_1, x_2) = a(x_1 - x_2)$ with $x_j \in \mathbb{R}^n$, μ_j equal to the Lebesgue measure on $\mathbb{R}^n$, $M = \|a\|_{L^p(\mathbb{R}^n)}$.

Proof of the theorem. As noted in the above remark, we may assume that p, q are both finite. For $u_2 \in S_q(\mu_2)$ (also if $p' = +\infty$ for $u_2 \in S_{q,\infty}(\mu_2)$, where $S_{p,\infty}(\mu)$ is defined by (3.2.24)), we have

$$\|Lu_2\|_{L^\infty(\mu_1)} \leq M\|u_2\|_{L^{p'}(\mu_2)}. \tag{9.1.18}$$

This implies that L can be extended uniquely as a bounded linear operator from $L^{p'}(\mu_2)$ into $L^\infty(\mu_1)$ so that (9.1.18) holds true. Moreover, for $u_2 \in S_q(\mu_2)$, we have if $p > 1$ (thus $p' < +\infty$),

$$\|Lu_2\|_{L^p(\mu_1)} \underset{\text{Lemma 6.2.1}}{=} \sup_{\substack{\|w\|_{L^{p'}(\mu_1)}=1 \\ w\in S_{p'}(\mu_1)}} \left| \int_{X_1} (Lu_2)(x_1)w(x_1)d\mu(x_1) \right|$$

$$\leq \sup_{\substack{\|w\|_{L^{p'}(\mu_1)}=1 \\ w\in S_{p'}(\mu_1)}} \iint_{X_1\times X_2} |k(x_1,x_2)||u_2(x_2)||w(x_1)|d\mu_1(x_1)d\mu_2(x_2)$$

$$\leq M \sup_{\substack{\|w\|_{L^{p'}(\mu_1)}=1 \\ w\in S_{p'}(\mu_1)}} \|w\|_{L^{p'}(\mu_1)} \int_{X_2} |u_2(x_2)|d\mu_2(x_2) = M\|u_2\|_{L^1(\mu_2)}.$$

This implies that if $p > 1$, L can be extended uniquely as a bounded linear operator from $L^1(\mu_2)$ into $L^p(\mu_1)$ so that

$$\|Lu_2\|_{L^p(\mu_1)} \leq M\|u_2\|_{L^1(\mu_2)}. \tag{9.1.19}$$

Applying the Riesz–Thorin interpolation theorem 9.1.2 to the inequalities (9.1.18)-(9.1.19), we find that the linear operator L sends continuously $L^{\tilde{q}}(\mu_2)$ into $L^{\tilde{r}}(\mu_2)$ (with operator norm M) with

$$\frac{1}{\tilde{q}} = \frac{1-\theta}{1} + \frac{\theta}{p'}, \quad \frac{1}{\tilde{r}} = \frac{1-\theta}{p} + \frac{\theta}{\infty},$$

for all $\theta \in [0,1]$. From (6.2.1), we have $1/p' + 1/q' = 1/r'$ so that $p' \geq r'$ and $1 \leq p \leq r$: thus we may choose

$$[0,1] \ni \theta = 1 - \frac{p}{r} \implies \frac{1-\theta}{p} = \frac{1}{r}, \tilde{r} = r, \quad \frac{1-\theta}{1} + \frac{\theta}{p'} = 1 - \frac{1}{p} + \frac{1}{r} = \frac{1}{q}, \tilde{q} = q.$$

This completes the proof for $p > 1$. Note that if $p = 1$ then $r = q$ (which can be assumed finite from Remark 9.1.4), we have directly

$$\int_{X_1} \left(\int_{X_2} |k(x_1,x_2)||u_2(x_2)|d\mu_2(x_2) \right)^q d\mu_1(x_1)$$

$$\leq \int_{X_1}\left(\int_{X_2} (|k(x_1,x_2)|^{\frac{1}{q}}|u_2(x_2)|)^q d\mu_2(x_2) \right)\left(\int_{X_2} |k(x_1,x_2)|^{\frac{q'}{q'}}d\mu_2(x_2) \right)^{\frac{q}{q'}} d\mu_1(x_1)$$

$$\leq M^{q/q'} \iint_{X_1\times X_2} |k(x_1,x_2)||u_2(x_2)|^q d\mu_2(x_2)d\mu_1(x_1) \leq M^{\frac{q}{q'}+1}\|u_2\|^q_{L^q(\mu_2)},$$

so that in this case as well, we find that

$$\|Lu_2\|_{L^q(\mu_1)} \leq M\|u_2\|_{L^q(\mu_2)}. \tag{9.1.20}$$

The proof of Theorem 9.1.3 is complete. $\square$

Theorem 9.1.5 (Hausdorff–Young). *Let $n \geq 1$ be an integer. The Fourier transform F maps injectively and continuously $L^p(\mathbb{R}^n)$ into $L^{p'}(\mathbb{R}^n)$ for $1 \leq p \leq 2$ and*

$$\forall u \in L^p(\mathbb{R}^n), \quad \|\hat{u}\|_{L^{p'}(\mathbb{R}^n)} \leq \|u\|_{L^p(\mathbb{R}^n)}. \tag{9.1.21}$$

Proof. Note first that we have defined the Fourier transformation on the space of tempered distributions (see Definition 8.1.14), so that Proposition 8.1.13(1) provides a definition of the Fourier transform for any function in $L^p(\mathbb{R}^n)$ and that this transformation is injective on $\mathscr{S}'(\mathbb{R}^n)$, since it is an isomorphism (see Theorem 8.1.15). We have seen as well in Theorem 8.1.16 that the Fourier transformation on $L^1(\mathbb{R}^n)$ is given by the explicit formula (8.1.22) and satisfies the inequality

$$\forall u \in L^1(\mathbb{R}^n), \text{we have } \hat{u} \in L^\infty(\mathbb{R}^n) \text{ and } \quad \|\hat{u}\|_{L^\infty(\mathbb{R}^n)} \leq \|u\|_{L^1(\mathbb{R}^n)}.$$

Moreover, Theorem 8.1.17 shows that the Fourier transformation is a unitary transformation of $L^2(\mathbb{R}^n)$ so that

$$\forall u \in L^2(\mathbb{R}^n), \text{we have } \hat{u} \in L^2(\mathbb{R}^n) \text{ and } \quad \|\hat{u}\|_{L^2(\mathbb{R}^n)} = \|u\|_{L^2(\mathbb{R}^n)}.$$

Applying the Riesz–Thorin interpolation theorem 9.1.2 yields readily that the Fourier transformation is a bounded linear map from $L^p(\mathbb{R}^n)$ into $L^{p'}(\mathbb{R}^n)$ for $1 \leq p \leq 2$ since for θ ranging in $[0,1]$, we have

$$\frac{1}{p} = \frac{1-\theta}{1} + \frac{\theta}{2} = 1 - \frac{\theta}{2} \Longrightarrow \frac{1}{p'} = \frac{\theta}{2}. \qquad \square$$

N.B. The constant 1 in (9.1.21) is not sharp. The best constant can be found in a paper by E. Lieb [42] who proved that for $1 < p < 2$,

$$\sup_{\|u\|_{L^p(\mathbb{R}^n)}=1} \|\hat{u}\|_{L^{p'}(\mathbb{R}^n)} = \left(p^{1/p} p'^{-1/p'}\right)^{n/2}. \tag{9.1.22}$$

Remark 9.1.6. The mapping $L^1(\mathbb{R}^n) \ni u \mapsto \hat{u} \in L^\infty(\mathbb{R}^n)$ is one-to-one and **not onto**: if it were onto it would be a bijective continuous mapping from $L^1(\mathbb{R}^n)$ onto $L^\infty(\mathbb{R}^n)$ and thus, from the Open Mapping Theorem 10.2.43 (a direct consequence of Baire's theorem), it would be an isomorphism. Since

$$\hat{\hat{v}} = \check{v} \quad \text{for a tempered distribution } v,$$

the inverse isomorphism from $L^\infty(\mathbb{R}^n)$ onto $L^1(\mathbb{R}^n)$ would be the inverse Fourier transform $\check{}$ and this would imply that the Fourier transform of an $L^\infty(\mathbb{R}^n)$ function belongs to $L^1(\mathbb{R}^n)$. However the latter is not true since the Fourier transform of $\mathbf{1}_{[-1,1]}$ (a function in $L^\infty \cap L^1$) is

$$\int_{-1}^{1} e^{-2i\pi x\xi}\, dx = \left[\frac{e^{-2i\pi x\xi}}{-2i\pi\xi}\right]_{x=-1}^{x=1} = \frac{e^{2i\pi\xi} - e^{-2i\pi\xi}}{2i\pi\xi} = \frac{\sin(2\pi\xi)}{\pi\xi},$$

which does not belong to $L^1(\mathbb{R})$ (see Exercise 2.8.20).

9.2   Marcinkiewicz Interpolation Theorem

Definition 9.2.1. Let $p, q \in [1, +\infty]$. A (not necessarily linear) mapping

$$T : L^p(\mathbb{R}^n) \longrightarrow L^q_w(\mathbb{R}^n) = L^{q,\infty}(\mathbb{R}^n),$$

such that $\exists C \geq 0, \forall u \in L^p(\mathbb{R}^n), \quad \|Tu\|_{L^{q,\infty}(\mathbb{R}^n)} \leq C\|u\|_{L^p(\mathbb{R}^n)},$

where the Lorentz space $L^{q,\infty}(\mathbb{R}^n)$ is defined in Exercise 6.6.8 (see also Definition 6.3.1) is said to be of **weak-type** (p, q).

N.B. When $q = +\infty$, this means:

$$\exists C \geq 0, \forall u \in L^p(\mathbb{R}^n), \quad \|Tu\|_{L^\infty(\mathbb{R}^n)} \leq C\|u\|_{L^p(\mathbb{R}^n)}. \tag{9.2.1}$$

For $1 \leq q < +\infty$ this means: $\exists C \geq 0, \forall u \in L^p(\mathbb{R}^n), \forall t > 0,$

$$\lambda_n\left(\{x \in \mathbb{R}^n, |(Tu)(x)| > t\}\right) \leq \left(C\|u\|_{L^p(\mathbb{R}^n)}t^{-1}\right)^q, \tag{9.2.2}$$

where λ_n stands for the Lebesgue measure on $\mathbb{R}^n$.

Definition 9.2.2. A bounded mapping $T : L^p(\mathbb{R}^n) \longrightarrow L^q(\mathbb{R}^n)$, i.e., such that

$$\exists C \geq 0, \forall u \in L^p(\mathbb{R}^n), \quad \|Tu\|_{L^q(\mathbb{R}^n)} \leq C\|u\|_{L^p(\mathbb{R}^n)}, \tag{9.2.3}$$

will be said of **strong-type** (p, q).

Of course, a strong-type (p, q) mapping is also of weak-type (p, q), since the notions are identical for $q = +\infty$ and if $1 \leq q < +\infty$, this follows from Inequality (6.3.2) (and the related inclusion $L^q \subset L^q_w$).

Theorem 9.2.3 (Marcinkiewicz Interpolation Theorem). *Let $r \in (1, +\infty]$ and let $T : L^1(\mathbb{R}^n) + L^r(\mathbb{R}^n) \longrightarrow \{measurable\ functions\}$ be a mapping such that*

$$|T(u + v)| \leq |Tu| + |Tv|. \tag{9.2.4}$$

We assume that T is of weak-type $(1, 1)$ and (r, r) (see Definition 9.2.1). Then T is of strong-type (p, p) for all $p \in (1, r)$ (see (9.2.3)).

N.B. From the inclusion $L^p \subset L^1 + L^r$ (see Exercise 6.6.11), we see that T is indeed defined on L^p. This very useful theorem (see [45] for the 1939 original paper and [44] for a historical perspective) is also very remarkable by the fact that it is providing a strong-type information from a weak-type assumption.

Notation. Let $(X, \mathcal{M}, \mu)$ be a measure space where μ is a positive measure; we shall use the following notation, for a measurable function u and $t > 0$:

$$\omega(t, u) = \mu\left(\{x \in \mathbb{R}^n, |u(x)| > t\}\right). \tag{9.2.5}$$

With $\Omega_p(u)$ given by Definition 6.3.1 (see also Exercise 6.6.8 (1)), we find that $\Omega_p(u) = \sup_{t>0} t^p \omega(t, u)$. For $p \in [1, +\infty)$ and $u \in L^p(\mu)$, we have, using Fubini's theorem,

$$
\begin{aligned}
\int_0^{+\infty} p t^{p-1} \omega(t, u) dt &= \int_0^{+\infty} p t^{p-1} \left(\int_{\{x, |u(x)|>t\}} d\mu \right) dt \\
&= \iint_{\mathbb{R}_+ \times X} p t^{p-1} H(|u(x)| - t) d\mu(x) dt \\
&= \int_X \int_0^{|u(x)|} p t^{p-1} dt d\mu(x) = \int_X |u(x)|^p d\mu(x),
\end{aligned}
$$

so that

$$
\|u\|_{L^p(\mu)} = p^{1/p} \|t \omega(t, u)^{1/p}\|_{L^p(\mathbb{R}_+, \frac{dt}{t})}. \tag{9.2.6}
$$

On the other hand for $u \in L^\infty(\mu)$ we have, according to Definition 3.2.4,

$$
\|u\|_{L^\infty(\mu)} = \inf\{t > 0, \omega(t, u) = 0\}.
$$

Proof of Theorem 9.2.3. We use the above notation with $\mu = \lambda_n$, the Lebesgue measure on $\mathbb{R}^n$. **Let us assume first r $= +\infty$.** The weak type (∞, ∞) hypothesis means $\|Tu\|_{L^\infty} \leq C \|u\|_{L^\infty}$ and we may assume that $C = 1$. We write for $u \in L^1 + L^\infty$, $t > 0$,

$$
u = \underbrace{u 1_{\{|u|>t/2\}}}_{u_1} + \underbrace{u 1_{\{|u| \leq t/2\}}}_{u_2}
$$

and this gives

$$
|(Tu)(x)| \leq |(Tu_1)(x)| + |(Tu_2)(x)| \leq |(Tu_1)(x)| + \|u_2\|_{L^\infty} \leq |(Tu_1)(x)| + \frac{t}{2},
$$

so that we find the inclusion

$$
(\sharp) \qquad \{x, |(Tu)(x)| > t\} \subset \{x, |(Tu_1)(x)| > t/2\}.
$$

The weak-type $(1, 1)$ assumption reads $t \omega(t, Tv) \leq c_{11} \|v\|_{L^1}$ so that

$$
(\flat) \qquad
\begin{aligned}
\frac{t}{2} \lambda_n \left(\left\{ x, |(Tu_1)(x)| > \frac{t}{2} \right\} \right) &\leq c_{11} \|u_1\|_{L^1} \\
\implies \omega \left(\frac{t}{2}, Tu_1 \right) &\leq \frac{2c_{11}}{t} \int_{|u|>t/2} |u| dx.
\end{aligned}
$$

Applying Formula (9.2.6) to Tu, we find, using Tonelli's theorem and $1 < p < +\infty$,

$$
\|Tu\|_{L^p}^p = p \int_0^{+\infty} t^{p-1} \omega(t, Tu) dt
$$

$$\text{(from ($\sharp$))} \quad \leq p \int_0^{+\infty} t^{p-1} \omega\left(\frac{t}{2}, Tu_1\right) dt$$

$$\text{(from ($\flat$))} \quad \leq p \int_0^{+\infty} t^{p-1} \frac{2c_{11}}{t} \int_{|u|>t/2} |u| dx dt$$

$$= 2pc_{11} \iint_{\mathbb{R}_+ \times \mathbb{R}^n} t^{p-2} H(2|u(x)| - t)|u(x)| dt dx$$

$$= \frac{2pc_{11}}{p-1} \int_{\mathbb{R}^n} (2|u(x)|)^{p-1}|u(x)| dx = \frac{2^p pc_{11}}{p-1} \|u\|_{L^p}^p,$$

which gives the strong-type (p, p) for T.

We assume now $1 < r < +\infty$. Let $u \in L^p$, let $t > 0$ and let u_1, u_2 be defined as above. Since $|(Tu)(x)| \leq |(Tu_1)(x)| + |(Tu_2)(x)|$, we find

$$\{x, |(Tu)(x)| > t\} \subset \{x, |(Tu_1)(x)| > t/2\} \cup \{x, |(Tu_2)(x)| > t/2\},$$

and thus $\omega(t, Tu) \leq \omega(\frac{t}{2}, Tu_1) + \omega(\frac{t}{2}, Tu_2)$. Following (6.6.6), we see that $u_1 \in L^1, u_2 \in L^r$. The weak-type assumptions imply with fixed positive constants c_1, c_r,

$$\frac{t}{2}\omega\left(\frac{t}{2}, Tu_1\right) \leq c_1 \|u_1\|_{L^1}, \quad \left(\frac{t}{2}\right)^r \omega\left(\frac{t}{2}, Tu_2\right) \leq c_r^r \|u_2\|_{L^r}^r.$$

We obtain thus

$$(\natural) \quad \omega(t, Tu) \leq \frac{2c_1}{t} \int |u(x)| H(2|u(x)| - t) dx + \frac{2^r c_r^r}{t^r} \int_{0 < |u(x)| \leq t/2} |u(x)|^r dx.$$

Tonelli's theorem implies

$$\int_0^{+\infty} pt^{p-1}\omega(t, Tu) dt$$

$$\leq \iint_{\mathbb{R}_+ \times \mathbb{R}^n} pt^{p-1} \frac{2c_1}{t} |u(x)| H(2|u(x)| - t) dt dx$$

$$+ \iint_{\mathbb{R}_+ \times \mathbb{R}^n} pt^{p-1} \frac{2^r c_r^r}{t^r} \mathbf{1}_{\{0 < |u| \leq t/2\}} |u(x)|^r dt dx$$

$$= \frac{2pc_1}{p-1} \int |u(x)|(2|u(x)|)^{p-1} dx + 2^r c_r^r p \int_{|u(x)|>0} |u(x)|^r \underbrace{\int_{2|u(x)|}^{+\infty} t^{p-1-r} dt}_{\text{note that } p-r<0} dx$$

$$= \frac{2^p pc_1}{p-1} \int |u(x)|^p dx + 2^r c_r^r p \int |u(x)|^r \frac{(2|u(x)|)^{p-r}}{r - p} dx$$

$$= \|u\|_{L^p}^p \left(\frac{2^p pc_1}{p-1} + \frac{2^p c_r^r p}{r - p}\right),$$

so that $\|Tu\|_{L^p} \leq \|u\|_{L^p} 2p^{1/p}\left(\frac{c_1}{p-1} + \frac{c_r^r}{r-p}\right)^{1/p}$, concluding the proof. $\qquad \square$

9.3 Maximal function

Definition 9.3.1. Let f be a function in $L^1_{\mathrm{loc}}(\mathbb{R}^n)$. The maximal function of f, denoted by $\mathcal{M}_f$, is defined on $\mathbb{R}^n$ by

$$\mathcal{M}_f(x) = \sup_{t>0} \frac{1}{|B(x,t)|} \int_{B(x,t)} |f(y)|\,dy, \tag{9.3.1}$$

where $|B(x,t)|$ is the Lebesgue measure of the ball with center x and radius t.

Using the notation $\fint_A f\,d\mu = \int_A f\,d\mu / \mu(A)$, we find

$$\mathcal{M}_f(x) = \sup_{t>0} \fint_{B(x,t)} |f(y)|\,dy = \sup_{t>0} \fint_{\mathbb{B}^n} |f(x+tz)|\,dz.$$

We note also that the maximal function (of a measurable function) is measurable (see Exercise 9.8.3).

Remark 9.3.2. Let us evaluate $\mathcal{M}_{\mathbf{1}_{\mathbb{B}^n}}$. Let $x \in \mathbb{R}^n$. For $t \geq 1 + |x|$, we have

$$|y| \leq 1 \Longrightarrow |y - x| \leq 1 + |x| \Longrightarrow y \in \bar{B}(x,t).$$

We have thus for $t \geq 1 + |x|$, $t^{-n} |\mathbb{B}^n|^{-1} \int_{B(x,t)} \mathbf{1}_{\mathbb{B}^n}(y)\,dy = t^{-n}$, implying

$$\mathcal{M}_{\mathbf{1}_{\mathbb{B}^n}}(x) \geq (1 + |x|)^{-n} \Longrightarrow \mathcal{M}_{\mathbf{1}_{\mathbb{B}^n}} \notin L^1(\mathbb{R}^n),$$

proving that the mapping $f \mapsto \mathcal{M}_f$ **does not** send L^1 into itself. We shall see below that the maximal function of an $L^1(\mathbb{R}^n)$ function is nevertheless in $L^1_w(\mathbb{R}^n)$, proving that the mapping $f \mapsto \mathcal{M}_f$ is of weak-type $(1,1)$.

Theorem 9.3.3 (Hardy–Littlewood maximal inequality). *The mapping $f \mapsto \mathcal{M}_f$ is of weak-type $(1,1)$ and of strong-type (p,p) for all $p \in (1, +\infty]$ (see Definitions 9.2.1, 9.2.2).*

Proof. Since the mapping $f \mapsto \mathcal{M}_f$ is obviously of strong-type (∞, ∞) (since $\|\mathcal{M}_f\|_{L^\infty} \leq \|f\|_{L^\infty}$), according[2] to the Marcinkiewicz interpolation theorem 9.2.3, it is enough to prove the weak-type $(1,1)$ property:

$$\exists C_n, \forall f \in L^1(\mathbb{R}^n), \quad \sup_{t>0} t\big|\{x \in \mathbb{R}^n, \mathcal{M}_f(x) > t\}\big| \leq C_n \|f\|_{L^1(\mathbb{R}^n)}. \tag{9.3.2}$$

Note that from Remark 9.3.2, the Riesz–Thorin Theorem 9.1.2 cannot be used since the mapping fails to be of strong-type $(1,1)$. We start with a lemma.

---

[2]Note that the subadditivity property is fulfilled since

$$0 \leq (\mathcal{M}_{f+g})(x) = \sup_{t>0} \fint_{\mathbb{B}^n} |(f+g)(x+tz)|\,dz \leq \sup_{t>0} \fint_{\mathbb{B}^n} |f(x+tz)|\,dz + \sup_{t>0} \fint_{\mathbb{B}^n} |g(x+tz)|\,dz.$$

Lemma 9.3.4 (Wiener covering lemma). *Let E be a measurable subset of $\mathbb{R}^n$ such that $E \subset \cup_{j \in J} B_j$ where $(B_j)_{j \in J}$ is a family of open balls such that*

$$2\rho_0 = \sup_{j \in J} \operatorname{diam} B_j < +\infty.$$

Then there exists a countable subfamily $(B_j)_{j \in D}$ of pairwise disjoint balls such that

$$\lambda_n(E) \le 5^n \sum_{j \in D} \lambda_n(B_j).$$

Proof of the lemma. Let $B_{j_0} = B(x_0, r_0)$ be a ball[3] such that $\operatorname{diam} B_{j_0} = 2r_0 > \rho_0$. Next, we define

$$J_0 = J, \quad J_1 = \{j \in J_0, B_j \cap B_{j_0} = \emptyset\}.$$

If $j \notin J_1$, then $B_j \cap B_{j_0} \neq \emptyset$, so that $\exists y_0 \in B_j \cap B_{j_0}$ and

$$x \in B_j \implies |x - x_0| \le \underbrace{|x - y_0|}_{x, y_0 \in B_j} + \underbrace{|y_0 - x_0|}_{y_0 \in B(x_0, r_0)} \le 2\rho_0 + r_0 < 5r_0,$$

entailing $j \notin J_1 \implies B_j \subset B_{j_0}^*$ which is defined as a ball with the same center as B_{j_0} and a diameter equal to five times the diameter of B_{j_0}.

- For the family $(B_j)_{j \in J_0}$ of open balls with bounded diameters,

$$\exists j_0 \in J_0, \text{ with } J_1 = \{j \in J_0, B_j \cap B_{j_0} = \emptyset\}, \quad \begin{cases} j \in J_1 \implies B_j \cap B_{j_0} = \emptyset, \\ j \notin J_1 \implies B_j \subset B_{j_0}^*. \end{cases}$$

- Let us assume that we have found $J_0 \supset J_1 \supset \cdots \supset J_k, k \ge 1, j_0 \in J_0, \ldots, j_k \in J_k$ such that

(1) $\operatorname{diam} B_{j_0} > \dfrac{1}{2} \sup_{j \in J_0} \operatorname{diam} B_j, \ldots\ldots\ldots\ldots, \operatorname{diam} B_{j_k} > \dfrac{1}{2} \sup_{j \in J_k} \operatorname{diam} B_j,$

(2) $\{j \in J_0, j \notin J_1 \implies B_j \subset B_{j_0}^*\}, \qquad$ (3) $\{j \in J_1 \implies B_j \cap B_{j_0} = \emptyset\},$

$\ldots\ldots\ldots\ldots$

(2) $\{j \in J_{k-1}, j \notin J_k \implies B_j \subset B_{j_{k-1}}^*\},$ (3) $\{j \in J_k \implies B_j \cap B_{j_{k-1}} = \emptyset\}.$

We define then $J_{k+1} = \{j \in J_k, B_j \cap B_{j_k} = \emptyset\}$ and if $J_{k+1} \neq \emptyset$ we find $j_{k+1} \in J_{k+1}$ such that

$$\operatorname{diam} B_{j_{k+1}} > \frac{1}{2} \sup_{j \in J_{k+1}} \operatorname{diam} B_j,$$

---

[3] We may of course assume that E has positive measure, which implies that J is not empty and $\rho_0 > 0$.

fulfilling (1) for $k+1$ as well. Moreover (3) holds true for $k+1$ by construction and if $j \in J_k \backslash J_{k+1}$, we have $B_j \cap B_{j_k} \neq \emptyset$, so that $\exists y_k \in B_j \cap B_{j_k}$, $B_{j_k} = B(x_k, r_k)$, and

$$x \in B_j \implies |x - x_k| \leq \underbrace{|x - y_k|}_{x, y_k \in B_j} + \underbrace{|y_k - x_k|}_{y_k \in B(x_k, r_k)}$$

$$\leq \underbrace{\operatorname{diam} B_j}_{j \in J_k} + r_k < 2 \operatorname{diam} B_{j_k} + r_k = 5 r_k,$$

entailing $B_j \subset B_{j_k}^*$, proving (2) for $k+1$.

• As a result, assuming that all the J_k are non-empty, we find

$$J_0 \supsetneq J_1 \supsetneq \cdots \supsetneq J_k \supsetneq \ldots, \quad j_k \in J_k,$$

such that

$$\begin{cases} k \geq 1: & j \in J_{k-1} \backslash J_k \implies B_j \subset B_{j_{k-1}}^*, \\ k \geq 1: & j \in J_k \implies B_j \cap B_{j_{k-1}} = \emptyset. \end{cases}$$

The family $(B_{j_k})_{k \geq 0}$ is pairwise disjoint: we consider $k' \geq k'' + 1$. We have $j_{k'} \in J_{k'} \subset J_{k''+1}$ and $j_{k''} \in J_{k''}$ so that

$$\underbrace{B_{j_{k'}}}_{j_{k'} \in J_{k''+1}} \cap B_{j_{k''}} = \emptyset.$$

Claim. If $\sum_{k \geq 0} |B_{j_k}| < +\infty$ we have for all $j \in J_0$, $B_j \subset \cup_{k \geq 1} B_{j_{k-1}}^*$. The Claim is obvious if $j \in \cup_{k \geq 1}(J_{k-1} \backslash J_k)$. Otherwise we have

$$j \in \cap_{k \geq 1}(J_{k-1}^c \cup J_k), \text{ which means } j \in \cap_{k \geq 1} J_k:$$

in fact, we have $\cap_{k \geq 1}(J_{k-1}^c \cup J_k) = \cap_{k \geq 1} J_k$ since

$$\{\forall k \geq 1, \ j \in J_k \cup J_{k-1}^c\} \text{ and } \{\exists k_0 \geq 1, j \notin J_{k_0}\}$$
$$\implies j \in J_{k_0-1}^c, k_0 \geq 2, \text{ since } J_0^c = \emptyset,$$
$$\implies j \in J_{k_0-2}^c, k_0 \geq 3 \cdots \implies j \in J_1^c \implies j \in J_0^c = \emptyset,$$

which is impossible. If $j \in \cap_{k \geq 1} J_k$, we have $\forall k \geq 1$, $2 \operatorname{diam} B_{j_k} > \operatorname{diam} B_j$ and since the series $\sum_{k \geq 0} |B_{j_k}|$ converges, this implies $\lim_k \operatorname{diam} B_{j_k} = 0$, and $\operatorname{diam} B_j = 0$ so that the open ball B_j is empty. The claim is proven.

• Finally, we have either $\sum_{k \geq 0} |B_{j_k}| = +\infty$ (a case where the conclusion of the lemma is reached trivially) or $\sum_{k \geq 0} |B_{j_k}| < +\infty$ and the above claim implies that

$$E \subset \cup_{k \geq 1} B_{j_{k-1}}^*,$$

providing the sought answer.

• When $J_{k_0} = \emptyset$ for some $k_0 \geq 1$, we find that $J_0 = \cup_{1 \leq k \leq k_0}(J_{k-1} \backslash J_k)$ and we have obviously $\forall j \in J_0$, $B_j \subset \cup_{k \geq 1} B_{j_{k-1}}^*$, obtaining the conclusion as well in that case. The proof of the Wiener covering lemma is complete. $\qquad \square$

Let us go back to the proof of Theorem 9.3.3. Let $s > 0$ be given. If $x \in \mathbb{R}^n$ is such that $\mathcal{M}_f(x) > s$, we can find $t_{s,x} > 0$ such that

$$\frac{1}{|B(x, t_{s,x})|} \int_{B(x, t_{s,x})} |f(y)| dy > s \implies |B(x, t_{s,x})| \leq s^{-1} \|f\|_{L^1(\mathbb{R}^n)} < +\infty.$$

We consider the measurable set

$$E_s = \{x \in \mathbb{R}^n, \mathcal{M}_f(x) > s\} \subset \cup_{x \in E_s} B(x, t_{s,x}),$$

and we note that $t_{s,x}^n |\mathbb{B}^n| \leq s^{-1} \|f\|_{L^1(\mathbb{R}^n)}$ so that we may apply the Wiener covering Lemma 9.3.4. We find a sequence $(x_k)_{k \in \mathbb{N}}$ in $\mathbb{R}^n$ such that the balls $B(x_k, t_{s,x_k})$ are pairwise disjoint and

$$|E_s = \{x \in \mathbb{R}^n, \mathcal{M}_f(x) > s\}|$$

$$\leq 5^n \sum_{k \in \mathbb{N}} |B(x_k, t_{s,x_k})| \leq s^{-1} 5^n \sum_{k \in \mathbb{N}} \int_{B(x_k, t_{s,x_k})} |f(y)| dy \leq s^{-1} 5^n \int_{\mathbb{R}^n} |f(y)| dy,$$

proving $s|E_s| \leq 5^n \|f\|_{L^1(\mathbb{R}^n)}$ and the weak-type $(1,1)$ property. $\square$

Remark 9.3.5. Note that with the result of Exercise 9.8.2, this implies

$$\text{for } 1 < p \leq +\infty, \quad \|\mathcal{M}_f\|_{L^p(\mathbb{R}^n)} \leq \frac{p^{1+\frac{1}{p}}}{p-1} 5^{\frac{n}{p}} \|f\|_{L^p(\mathbb{R}^n)}. \tag{9.3.3}$$

A result due to E.M. Stein and J.-O. Stromberg [56] shows that the L^p to L^p norm of $\mathcal{M}$ can be chosen independently of the dimension n.

9.4   Lebesgue differentiation theorem, Lebesgue points

Theorem 9.4.1 (Lebesgue Differentiation Theorem). *Let f be a function in $L^1(\mathbb{R}^n)$. Then, there exists a Borel set L_f such that $\lambda_n(L_f^c) = 0$ such that*

$$\forall x \in L_f, \quad \lim_{r \to 0_+} \frac{1}{\lambda_n(B(x,r))} \int_{B(x,r)} |f(y) - f(x)| dy = 0. \tag{9.4.1}$$

The set L_f is called the set of Lebesgue points of f.

Remark 9.4.2. Note that this implies that for $f \in L^1(\mathbb{R}^n)$, for all $x \in L_f$, $\lim_{r \to 0} \fint_{B(x,r)} f(y) dy = f(x)$.

Proof. For $\rho > 0$ we define the measurable set

$$E_\rho = \{x \in \mathbb{R}^n, \limsup_{t \to 0_+} \underbrace{\frac{1}{|B(x,t)|} \int_{B(x,t)} |f(y) - f(x)| dy}_{\mathcal{N}_f(t,x)} > \rho\}. \tag{9.4.2}$$

Let $\phi \in C_c^0(\mathbb{R}^n)$. We have

$$\mathcal{N}_f(t,x) \le \fint_{B(x,t)} |f(y) - \phi(y)| dy + \fint_{B(x,t)} |\phi(y) - \phi(x)| dy + |\phi(x) - f(x)|$$

$$\le \mathcal{M}_{\phi-f}(x) + \fint_{B(x,t)} |\phi(y) - \phi(x)| dy + |\phi(x) - f(x)|.$$

Since ϕ is uniformly continuous, we get

$$\limsup_{t \to 0} \mathcal{N}_f(t,x) \le \mathcal{M}_{\phi-f}(x) + |f(x) - \phi(x)|.$$

As a result the set E_ρ defined by (9.4.2) is such that

$$E_\rho \subset \{x, |f(x) - \phi(x)| > \rho/2\} \cup \{x, \mathcal{M}_{\phi-f}(x) > \rho/2\},$$

and this implies $|E_\rho| \le |\{x, |f(x) - \phi(x)| > \rho/2\}| + |\{x, \mathcal{M}_{\phi-f}(x) > \rho/2\}|$. Using now Theorem 9.3.3, we obtain for any $\phi \in C_c^0(\mathbb{R}^n)$,

$$|E_\rho| \le \frac{2}{\rho} \int_{\mathbb{R}^n} |f(x) - \phi(x)| dx + C_n \frac{2}{\rho} \|f - \phi\|_{L^1(\mathbb{R}^n)} = \frac{2(1 + C_n)}{\rho} \|f - \phi\|_{L^1(\mathbb{R}^n)}.$$

The density of $C_c^0(\mathbb{R}^n)$ in $L^1(\mathbb{R}^n)$ implies that $|E_\rho| = 0$ for all $\rho > 0$ and since

$$\{x \in \mathbb{R}^n, \limsup_{t \to 0_+} \mathcal{N}_f(t,x) > 0\} = \cup_{k \ge 1} E_{1/k},$$

this gives as well that $|E_0| = 0$. We define $L_f = E_0^c$ and we have for $x \in L_f$, $\lim_{t \to 0} \mathcal{N}_f(t,x) = 0$, which is the sought result. $\qquad\square$

Theorem 9.4.3. *Let $f \in L_{\mathrm{loc}}^1(\mathbb{R})$. We define for $x \in \mathbb{R}$, $F(x) = \int_0^x f(y) dy$.*

(1) *Then the function F is continuous on $\mathbb{R}$, differentiable almost everywhere with derivative $f(x)$.*

(2) *The weak derivative of F is f.*

Proof. (1) The continuity of F is obvious since for $h \ge 0$,

$$F(x + h) - F(x) = \int_{[x,x+h]} f(y) dy,$$

and for $h \le 0$, $F(x + h) - F(x) = -\int_{[x+h,x]} f(y) dy$. Proposition 1.7.10 implies $\lim_{h \to 0}(F(x + h) - F(x)) = 0$. We consider now for $h \ne 0$,

$$\left| \frac{F(x + h) - F(x)}{h} - f(x) \right| \le \frac{1}{|h|} \int_{[x,x+h] \cup [x+h,x]} |f(y) - f(x)| dy$$

$$\le \frac{2}{2|h|} \int_{[x-|h|,x+|h|]} |f(y) - f(x)| dy.$$

Applying the previous theorem if $f \in L^1(\mathbb{R})$ (or Exercise 9.8.4 when $f \in L^1_{\text{loc}}(\mathbb{R})$), we find that F is differentiable at the Lebesgue points of f, with derivative f.

(2) We have for $\phi \in C_c^\infty(\mathbb{R}^n)$, using Fubini's theorem,

$$\langle F', \phi \rangle = -\int F(x)\phi'(x)dx$$

$$= -\int \phi'(x) \int \left(H(x)\mathbf{1}_{[0,x]}(y) - H(-x)\mathbf{1}_{[x,0]}(y) \right) f(y)dydx$$

$$= \int f(y)\left(-\int_{0 \le y \le x} \phi'(x)dx + \int_{x \le y \le 0} \phi'(x)dx \right) dy$$

$$= \int f(y)\left(H(y)\phi(y) + H(-y)\phi(y) \right) dy = \langle f, \phi \rangle,$$

proving the result. $\square$

Remark 9.4.4. *Almost everywhere differentiability is a very weak piece of information.* Almost everywhere differentiability of a function F is a very weak property that does not tell much about the function F: in the first place the trivial example of the Heaviside function shows that a bounded function can be differentiable almost everywhere in $\mathbb{R}$ with a zero derivative without being a constant. The much more elaborate example of the Cantor function defined in Proposition 5.7.7 shows that a continuous function can be differentiable almost everywhere with a null derivative without being a constant, so is not the integral of its a.e. derivative.

Remark 9.4.5. It may also happen that a continuous function is differentiable everywhere but with a derivative which is not integrable in the Lebesgue sense (see Exercise 9.8.5). Some other theories of integration are devised in such a way that a differentiable function is always the integral of its derivative. This is the case in particular of the so-called Henstock–Kurzweil integration theory [38] as well as some earlier theories due to Denjoy and Perron.

The distribution (or weak) derivative does not miss jumps and singularities as the notion of everywhere differentiability. Here the reader may consider only tempered distributions as in Chapter 8, but the statements are true as well for general distributions defined as local objects.

Lemma 9.4.6. *Let T be a distribution on $\mathbb{R}$ such that $T' = 0$. Then T is a constant.*

Proof. Let $\phi \in C_c^\infty(\mathbb{R})$ and let $\chi_0 \in C_c^\infty(\mathbb{R})$ with integral 1. Denoting $I(\phi) = \int_{\mathbb{R}} \phi(y)dy$, the function ψ defined by

$$\psi(x) = \phi(x) - I(\phi)\chi_0(x),$$

belongs to $C_c^\infty(\mathbb{R})$ and is the derivative of $\Psi(x) = \int_{-\infty}^x \psi(y)dy$. Note that Ψ is C^∞ and with compact support, since for x large enough

$$\Psi(x) = \int_{\mathbb{R}} \phi(y)dy - I(\phi)\int_{\mathbb{R}} \chi_0(y)dy = 0.$$

As a result, we find

$$\langle T, \phi \rangle = \langle T, \psi \rangle + I(\phi)\langle T, \chi_0 \rangle = \langle T, \Psi' \rangle + I(\phi)\langle T, \chi_0 \rangle = -\langle T', \Psi \rangle + I(\phi)\langle T, \chi_0 \rangle,$$

so that $T = \langle T, \chi_0 \rangle$. $\qquad\square$

Theorem 9.4.7. *Let F be a locally integrable function in $\mathbb{R}$ such that its distribution derivative F' is locally integrable. Then the function F is bounded continuous and for all $a \in \mathbb{R}$,*

$$F(x) = F(a) + \int_a^x F'(y)dy. \tag{9.4.3}$$

The function F is also a.e. differentiable with (ordinary) derivative $F'(x)$.

Proof. We define $G(x) = \int_a^x F'(y)dy$ and from Theorem 9.4.3, we find that the distribution derivative G' of G is equal to F' (and that G is continuous). Thus the distribution derivative of $F - G$ is zero, so that $F - G$ is the constant $F(a) - G(a) = F(a)$. The last statement follows from Theorem 9.4.3. $\qquad\square$

9.5 Gagliardo–Nirenberg inequality

Proposition 9.5.1. *For all $\phi \in C_c^1(\mathbb{R}^n)$, we have*

$$\|\phi\|_{L^{\frac{n}{n-1}}(\mathbb{R}^n)} \le \frac{1}{2} \prod_{1 \le j \le n} \left\| \frac{\partial \phi}{\partial x_j} \right\|_{L^1(\mathbb{R}^n)}^{1/n}. \tag{9.5.1}$$

Proof. The cases $n = 1, 2$ are very easy: for $n = 1$, we have

$$2\phi(x) = \int_{-\infty}^x \phi'(t)dt + \int_{+\infty}^x \phi'(t)dt \implies 2\|\phi\|_{L^\infty(\mathbb{R})} \le \|\phi'\|_{L^1(\mathbb{R})}.$$

For $n = 2$, we have, using the previous result,

$$|\phi(x_1, x_2)| \le \frac{1}{2} \int_{\mathbb{R}} |\partial_1 \phi(t_1, x_2)|dt_1, \quad |\phi(x_1, x_2)| \le \frac{1}{2} \int_{\mathbb{R}} |\partial_2 \phi(x_1, t_2)|dt_2,$$

so that

$$4\|\phi\|_{L^2(\mathbb{R}^2)}^2 \le \int_{\mathbb{R}^4} |\partial_1\phi(t_1, x_2)||\partial_2\phi(x_1, t_2)|dt_1 dt_2 dx_1 dx_2 = \|\partial_1\phi\|_{L^1(\mathbb{R}^2)}\|\partial_2\phi\|_{L^1(\mathbb{R}^2)}.$$

The cases $n \ge 3$ are more complicated and we need to start with a lemma.

Lemma 9.5.2. *Let $n \ge 2$ be an integer and let $\omega_1, \ldots, \omega_n$ be non-negative measurable functions on $\mathbb{R}^{n-1}$ so that ω_j is a function of $(x_k)_{1 \le k \le n, k \ne j}$. Then, we have*

$$\int_{\mathbb{R}^n} \omega_1^{\frac{1}{n-1}} \ldots \omega_n^{\frac{1}{n-1}} dx_1 \ldots dx_n \le \prod_{j=1}^n \left(\int_{\mathbb{R}^{n-1}} \omega_j d\widehat{x}_j \right)^{\frac{1}{n-1}},$$

where $d\widehat{x}_j = \prod_{\substack{1 \le k \le n \\ k \ne j}} dx_k$.

Proof of the lemma. For $n = 2$ we have indeed

$$\int_{\mathbb{R}^2} \omega_1(x_2)\omega_2(x_1)dx_1 dx_2 = \|\omega_1\|_{L^1(\mathbb{R})}\|\omega_2\|_{L^1(\mathbb{R})}.$$

Let us assume now that $n \geq 3$: we have

$$I_n = \int_{\mathbb{R}^n} \omega_1^{\frac{1}{n-1}}\ldots\omega_n^{\frac{1}{n-1}}dx_1\ldots dx_n = \int_{\mathbb{R}^{n-1}} \overbrace{\omega_1^{\frac{1}{n-1}}}^{\substack{\text{does not}\\\text{depend on } x_1}} \left(\int_{\mathbb{R}} \prod_{2\leq j\leq n} \omega_j^{\frac{1}{n-1}}dx_1\right)d\widehat{x_1},$$

and since $\frac{1}{n-1} + \frac{n-2}{n-1} = 1$, Hölder's inequality implies

$$I_n \leq \|\omega_1\|_{L^1(\mathbb{R}^{n-1})}^{\frac{1}{n-1}} \left\{\int_{\mathbb{R}^{n-1}} \left(\int_{\mathbb{R}} \prod_{2\leq j\leq n} \omega_j^{\frac{1}{n-1}}dx_1\right)^{\frac{n-1}{n-2}} d\widehat{x_1}\right\}^{\frac{n-2}{n-1}}.$$

We have, using the generalized Hölder's inequality of Exercise 3.7.31,

$$\int_{\mathbb{R}} \prod_{2\leq j\leq n} \omega_j^{\frac{1}{n-1}}dx_1 \leq \prod_{2\leq j\leq n} \left(\int_{\mathbb{R}} (\omega_j^{\frac{1}{n-1}})^{n-1}dx_1\right)^{\frac{1}{n-1}} = \left(\prod_{2\leq j\leq n} \int_{\mathbb{R}} \omega_j dx_1\right)^{\frac{1}{n-1}}.$$

This gives

$$I_n \leq \|\omega_1\|_{L^1(\mathbb{R}^{n-1})}^{\frac{1}{n-1}} \left\{\int_{\mathbb{R}^{n-1}} \prod_{2\leq j\leq n} \underbrace{\left(\int_{\mathbb{R}} \omega_j dx_1\right)^{\frac{1}{n-2}}}_{=\Omega_j^{\frac{1}{n-2}}} d\widehat{x_1}\right\}^{\frac{n-2}{n-1}},$$

with Ω_j independent of x_1, x_j (here $1 \neq j$ since $j \geq 2$). We may apply the induction hypothesis to obtain

$$I_n \leq \|\omega_1\|_{L^1(\mathbb{R}^{n-1})}^{\frac{1}{n-1}} \left\{\prod_{2\leq j\leq n} \|\Omega_j\|_{L^1(\mathbb{R}^{n-2})}^{\frac{1}{n-2}}\right\}^{\frac{n-2}{n-1}}$$

$$= \|\omega_1\|_{L^1(\mathbb{R}^{n-1})}^{\frac{1}{n-1}} \left\{\prod_{2\leq j\leq n} \|\Omega_j\|_{L^1(\mathbb{R}^{n-2})}\right\}^{\frac{1}{n-1}},$$

and since for $2 \leq j \leq n$,

$$\|\Omega_j\|_{L^1(\mathbb{R}^{n-2})} = \int_{\mathbb{R}^{n-2}} \int_{\mathbb{R}} \omega_j dx_1 \prod_{2\leq k\leq n, k\neq j} dx_k = \|\omega_j\|_{L^1(\mathbb{R}^{n-1})},$$

this proves the lemma. $\qquad\qquad\qquad\qquad\qquad\qquad\qquad\qquad\qquad\qquad\qquad\qquad\qquad\quad\square$

Let us go back to the proof of Proposition 9.5.1. We have

$$2|\phi(x)| \le \int_{\mathbb{R}} |\partial_j \phi(x_1,\dots,x_{j-1},t_j,x_{j+1},\dots,x_n)|dt_j = \omega_j(x),$$

where ω_j does not depend on x_j. This implies that

$$2^{\frac{n}{n-1}}|\phi(x)|^{\frac{n}{n-1}} \le \prod_{1\le j\le n} \omega_j(x)^{\frac{1}{n-1}},$$

and from Lemma 9.5.2, this implies

$$2^{\frac{n}{n-1}}\int |\phi(x)|^{\frac{n}{n-1}}dx \le \left(\prod_{1\le j\le n} \|\omega_j\|_{L^1(\mathbb{R}^{n-1})} \right)^{\frac{1}{n-1}} = \left(\prod_{1\le j\le n} \|\partial_j\psi\|_{L^1(\mathbb{R}^n)} \right)^{\frac{1}{n-1}},$$

which is (9.5.1), concluding the proof. $\qquad\square$

Proposition 9.5.3. *The space $W^{1,1}(\mathbb{R}^n)$ is defined as the set of functions $u \in L^1(\mathbb{R}^n)$ such that the distribution ∇u belongs as well to $L^1(\mathbb{R}^n)$. This space is a Banach space for the norm*

$$\|u\|_{W^{1,1}(\mathbb{R}^n)} = \|u\|_{L^1(\mathbb{R}^n)} + \|\nabla u\|_{L^1(\mathbb{R}^n)}.$$

Proof. Let $(u_k)_{k\in\mathbb{N}}$ be a Cauchy sequence in $W^{1,1}(\mathbb{R}^n)$. Then, we find $u, V \in L^1$ such that $\lim_k u_k = u, \lim \nabla u_k = V$ in the space $L^1(\mathbb{R}^n)$. Now for $\phi \in C_c^\infty(\mathbb{R}^n)$, we have

$$\int V\phi dx = \lim_k \int \phi \nabla u_k dx = \lim_k \langle \nabla u_k, \phi \rangle = -\lim_k \langle u_k, \nabla\phi \rangle$$

$$= -\lim_k \int u_k \nabla\phi dx = -\int u\nabla\phi dx - \langle \nabla u, \phi \rangle,$$

proving $V = \nabla u$. $\qquad\square$

Theorem 9.5.4 (Gagliardo–Nirenberg inequality)**.** *Let $u \in W^{1,1}(\mathbb{R}^n)$. Then u belongs to $L^{\frac{n}{n-1}}(\mathbb{R}^n)$ and is such that*

$$\|u\|_{L^{\frac{n}{n-1}}(\mathbb{R}^n)} \le \frac{1}{2} \prod_{1\le j\le n} \|\partial_j u\|_{L^1(\mathbb{R}^n)}^{1/n}. \tag{9.5.2}$$

Proof. Let $\rho \in C_c^\infty(\mathbb{R}^n;\mathbb{R}_+)$ such that $\int \rho(x)dx = 1$. For $\epsilon > 0$, we define $\rho_\epsilon(x) = \rho(x/\epsilon)\epsilon^{-n}$. The function $(u * \rho_\epsilon)(x) = \int u(y)\rho_\epsilon(x - y)dy$ is smooth (obvious from Theorem 3.3.4), belongs to $L^1(\mathbb{R}^n)$ (Proposition 6.1.1) and converges to u in $L^1(\mathbb{R}^n)$: for $\phi \in C_c^0(\mathbb{R}^n)$, we have

$$u * \rho_\epsilon - u = (u - \phi) * \rho_\epsilon + \phi * \rho_\epsilon - \phi + \phi - u,$$

so that with L^1 norms, using (6.1.3), for $\epsilon \leq 1$,

$$\|u * \rho_\epsilon - u\| \leq 2\|u - \phi\| + \int_K |(\phi * \rho_\epsilon)(x) - \phi(x)|dx,$$

where K is the compact set $\operatorname{supp}\phi + \operatorname{supp}\rho$. From Lemma 6.1.4, we find uniform convergence of the sequence of continuous functions $\phi * \rho_\epsilon$ and this implies

$$\forall \phi \in C_c^0(\mathbb{R}^n), \quad \limsup_{\epsilon \to 0} \|u * \rho_\epsilon - u\| \leq 2\|u - \phi\|.$$

The density of $C_c^0(\mathbb{R}^n)$ in $L^1(\mathbb{R}^n)$ entails that $\lim_\epsilon \|u * \rho_\epsilon - u\| = 0$. We have also

$$\rho_\epsilon * \nabla u = \nabla(\rho_\epsilon * u) \tag{9.5.3}$$

since for $\phi \in C_c^\infty(\mathbb{R}^n)$,

$$\int_{\mathbb{R}^n} (\rho_\epsilon * \nabla u)(x)\phi(x)dx = \iint \rho_\epsilon(x - y)(\nabla u)(y)\phi(x)dxdy$$

$$= \langle \nabla u, \check{\rho}_\epsilon * \phi \rangle = -\langle u, \check{\rho}_\epsilon * \nabla\phi \rangle = -\int (\rho_\epsilon * u)(x)\nabla\phi(x)dx = \langle \nabla(\rho_\epsilon * u), \phi \rangle,$$

proving (9.5.3).

Let us assume first that u belongs to $W^{1,1}(\mathbb{R}^n)$ and is compactly supported. We may apply (9.5.1) to the smooth compactly supported $\rho_\epsilon * u$. We note that the sequence $\partial_j(\rho_\epsilon * u) = \rho_\epsilon * \partial_j u$ converges in $L^1(\mathbb{R}^n)$ towards $\partial_j u$. Moreover the inequality (9.5.1) applied to $\rho_{\epsilon_1} * u - \rho_{\epsilon_2} * u$ implies that $\rho_\epsilon * u$ is a Cauchy sequence in $L^{n/n-1}(\mathbb{R}^n)$ thus converges with a limit v; since that sequence is converging towards u in $L^1(\mathbb{R}^n)$, and for $\phi \in C_c^0(\mathbb{R}^n)$, we have

$$\int v(x)\phi(x)dx = \lim_\epsilon \int (\rho_\epsilon * u)(x)\phi(x)dx = \int u(x)\phi(x)dx.$$

Lemma 8.1.11 implies $u = v$ which belongs to $L^{n/n-1}$. Inequality (9.5.2) holds true by taking the limits in (9.5.1).

Let us assume now that u belongs to $W^{1,1}(\mathbb{R}^n)$. Let χ be in $C_c^\infty(\mathbb{R}^n; [0, 1])$, equal to 1 on $B(0, 1)$ and supported in $B(0, 2)$. For $\epsilon > 0$ we have obviously (dominated convergence)

$$\lim_{\epsilon \to 0} \chi(\epsilon x)u(x) = u(x) \quad \text{in } L^1(\mathbb{R}^n).$$

Let us calculate for $\chi_\epsilon(x) = \chi(\epsilon x)$, $\nabla(u\chi_\epsilon) = \chi_\epsilon \nabla u + u\nabla\chi_\epsilon$. We have

$$\lim_{\epsilon \to 0} \int_{\mathbb{R}^n} |u(x)\chi'(\epsilon x)|dx\epsilon = 0 = \lim_{\epsilon \to 0} \int_{\mathbb{R}^n} |u(x)|(1 - \chi(\epsilon x))dx,$$

where the first equality is obvious (domination by $\|u\|_{L^1}\epsilon\|\chi'\|_{L^\infty}$) as well as the next one since

$$\int_{\mathbb{R}^n} |u(x)|(1 - \chi(\epsilon x))dx \leq \int_{|x| \geq 1/\epsilon} |u(x)|dx.$$

We have thus

$$\lim_{\epsilon \to 0} \chi_\epsilon u = u, \quad \lim_{\epsilon \to 0} \nabla(\chi_\epsilon u) = \nabla u \quad \text{in } L^1.$$

Since $u_\epsilon = \chi_\epsilon u$ is compactly supported in $W^{1,1}$, we may apply the previous result to get Inequality (9.5.2) for u_ϵ. That inequality implies as well that u_ϵ is a Cauchy sequence in $L^{n/n-1}$ and thus converges in that space towards a function v. Since the sequence u_ϵ converges in L^1 towards u, the same reasoning as above shows $v = u$ and the result.

Remark 9.5.5. The Gagliardo–Nirenberg inequality (9.5.2) has some interesting properties, beyond the most remarkable of being true. In the first place, this inequality has a scaling invariance: take $u \in W^{1,1}(\mathbb{R}^n)$ and $A \in Gl(n, \mathbb{R})$, and consider the function

$$u_A(x) = u(Ax)|\det A|^{\frac{n-1}{n}}, \quad \text{so that} \quad (\nabla u_A)(x) = (\nabla u)(Ax)A|\det A|^{\frac{n-1}{n}}.$$

We have

$$\|u_A\|_{L^{\frac{n}{n-1}}(\mathbb{R}^n)} = \left(\int |u(Ax)|^{\frac{n}{n-1}}|\det A|dx \right)^{\frac{n-1}{n}} = \|u\|_{L^{\frac{n}{n-1}}(\mathbb{R}^n)},$$

and

$$\|\nabla u_A\|_{L^1(\mathbb{R}^n)} = \int |(\nabla u)(Ax)A||\det A|^{\frac{n-1}{n}}dx = \int |(\nabla u)(y)A||\det A|^{-\frac{1}{n}}dx.$$

Considering $(\nabla u)(x)$ as a linear form on $\mathbb{R}^n$, and A as a linear endomorphism of $\mathbb{R}^n$, we have

$$\|(\nabla u)(x)A\| = \sup_{|T|=1} \|(\nabla u)(x)AT\|.$$

Let us assume now that $A = \alpha\Omega$, where $\alpha \in \mathbb{R}^*, \Omega \in O(n)$. We get then

$$\|(\nabla u)(x)A\| = |\alpha|\|(\nabla u)(x)\|, \quad |\det A| = |\alpha|^n,$$

so that $\|\nabla u_A\|_{L^1(\mathbb{R}^n)} = \|\nabla u\|_{L^1(\mathbb{R}^n)}$. Inequality (9.5.2) implies

$$\|u\|_{L^{\frac{n}{n-1}}(\mathbb{R}^n)} \leq \frac{1}{2n} \sum_{1\leq j\leq n} \int |(\partial_j u)(x)|dx \leq \frac{1}{2\sqrt{n}} \int \left(\sum_{1\leq j\leq n} |(\partial_j u)(x)|^2 \right)^{1/2} dx$$

$$= \frac{1}{2\sqrt{n}} \int \underbrace{\|\nabla u(x)\|}_{\substack{\text{Euclidean} \\ \text{norm on } \mathbb{R}^n}} dx = \frac{1}{2\sqrt{n}}\|\nabla u\|_{L^1(\mathbb{R}^n)}, \tag{9.5.4}$$

and the latter is invariant by affine similarities (generated by homothetic transformations $x \mapsto x_0 + \alpha x$, $\alpha \in \mathbb{R}^*$, and linear isometries $x \mapsto \Omega x$, $\Omega \in O(n)$).

On the other hand, we shall use Theorem 9.5.4 to prove the so-called Sobolev inequalities of the next section. Although these inequalities can be handled via some Fourier analysis methods, this is not the case for the Gagliardo–Nirenberg inequality above which involves the L^1-norm of the gradient (L^1 is not so friendly to Fourier analysis). It is thus an interesting reminder that a clever but elementary combinatorial argument such as Lemma 9.5.2 can find its way into proving a statement that is not accessible to Fourier analysis. □

9.6　Sobolev spaces, Sobolev injection theorems

We begin with a lemma.

Lemma 9.6.1. *Let $n \geq 1$ be an integer and let $p, q \in [1, +\infty)$ such that $\frac{1}{q} = \frac{1}{n} + \frac{1}{p}$. Then there exists a constant $C(p, n)$ such that for all $v \in C_c^1(\mathbb{R}^n)$,*

$$\|v\|_{L^p(\mathbb{R}^n)} \leq C(p, n)\|\nabla v\|_{L^q(\mathbb{R}^n)}.$$

Proof. When $n = 1$, we find that the sought estimate is true as well for $p = +\infty, q = 1$ (this is (9.5.1)) and for $1 \leq p < +\infty$, we cannot have $q \geq 1$. We may thus assume that $n \geq 2$.

Let us first suppose that $v \geq 0$. We define $u = v^{\frac{p(n-1)}{n}}$: we note that

$$\frac{1}{p} + \frac{1}{n} \leq 1 \Longrightarrow \frac{1}{p} \leq \frac{n-1}{n} \Longrightarrow \frac{p(n-1)}{n} \geq 1,$$

so that we have with ordinary differentiation, $\partial_j u = \frac{p(n-1)}{n} v^{\frac{p(n-1)}{n} - 1} \partial_j v$, and the function u is also C_c^1. On the other hand we have, using (9.5.1),

$$\|v\|_{L^p}^p = \|u\|_{L^{\frac{n}{n-1}}}^{\frac{n}{n-1}} \leq 2^{-\frac{n}{n-1}} \prod_{1 \leq j \leq n} \|\partial_j u\|_{L^1}^{\frac{1}{n-1}} \tag{9.6.1}$$

$$\leq 2^{-\frac{n}{n-1}} \left(\frac{p(n-1)}{n}\right)^{\frac{n}{n-1}} \underbrace{\left(\prod_{1 \leq j \leq n} \int |\partial_j v||v|^{p-\frac{p}{n}-1} dx\right)^{\frac{1}{n-1}}}_{\text{term } I},$$

and this implies that

$$\|v\|_{L^p}^{p(n-1)} \leq 2^{-n} \left(\frac{p(n-1)}{n}\right)^n \prod_{1 \leq j \leq n} \left(\|\partial_j v\|_{L^q} \|v^{\frac{np-p-n}{n}}\|_{L^{q'}}\right).$$

We note that $\frac{(np-p-n)}{n} = p(1 - \frac{1}{n} - \frac{1}{p}) = \frac{p}{q'}$, so that if $q > 1$ we have proven

$$\|v\|_{L^p}^{p(n-1)} \leq 2^{-n} \left(\frac{p(n-1)}{n}\right)^n \left(\prod_{1 \leq j \leq n} \|\partial_j v\|_{L^q}\right) \|v^p\|_{L^1}^{\frac{n}{q'}},$$

which gives (the result) for $v \not\equiv 0$,

$$\|v\|_{L^p}^n = \|v\|_{L^p}^{p(n-1)-\frac{np}{q'}} \leq 2^{-n}\left(\frac{p(n-1)}{n}\right)^n \prod_{1\leq j\leq n} \|\partial_j v\|_{L^q},$$

since $p(n-1) - \frac{np}{q'} = pn(1 - \frac{1}{n} - \frac{1}{q'}) = pn(\frac{1}{q} - \frac{1}{n}) = n$. If $q = 1$, we have in term I above, $p - \frac{p}{n} - 1 = p(1 - \frac{1}{n} - \frac{1}{p}) = 0$, so that (9.6.1) gives the answer in the case $q = 1$.

We drop now the non-negativity assumption on v. For $\epsilon > 0$, and $\chi \in C_c^\infty(\mathbb{R}^n; [0,1])$ equal to 1 near the support of v, we define the C_c^1 function u_ϵ by

$$u_\epsilon(x) = \left(v(x)^2 + \epsilon^2\right)^{\frac{1}{2}\frac{p(n-1)}{n}} \chi(x).$$

We have $\lim_{\epsilon\to 0} \|u_\epsilon\|_{L^{\frac{n}{n-1}}}^{\frac{n-1}{n}} = \lim_{\epsilon\to 0} \int \left(v(x)^2 + \epsilon^2\right)^{\frac{p}{2}} \chi(x)^{\frac{n}{n-1}} dx = \|v\|_{L^p}^p$, and calculating

$$\nabla u_\epsilon = (\nabla\chi)\left(v^2 + \epsilon^2\right)^{\frac{1}{2}\frac{p(n-1)}{n}} + \chi\frac{p(n-1)}{2n}\left(v^2 + \epsilon^2\right)^{\frac{p(n-1)}{2n}-1}2v\nabla v,$$

using $p(n-1)/n \geq 1$, we get that

$$\lim_{\epsilon\to 0}(\nabla u_\epsilon)(x) = \chi(x)\frac{p(n-1)}{2n}|v(x)|^{\frac{p(n-1)}{n}-2}2v(x)(\nabla v)(x),$$

so that with dominated convergence, we obtain

$$\lim_{\epsilon\to 0}\|\nabla u_\epsilon\|_{L^1} = \frac{p(n-1)}{n}\int |v|^{\frac{p(n-1)-n}{n}}|\nabla v| dx.$$

Applying Gagliardo–Nirenberg (9.5.4) to u_ϵ we find

$$\|v\|_{L^p}^{\frac{p(n-1)}{n}} = \lim_\epsilon \|u_\epsilon\|_{L^{\frac{n}{n-1}}} \leq \frac{1}{2\sqrt{n}}\lim_\epsilon \|\nabla u_\epsilon\|_{L^1} = \frac{p(n-1)}{2n^{3/2}}\int |v|^{\frac{p(n-1)-n}{n}}|\nabla v| dx.$$

If $q = 1$, we have $p(n-1) - n = pn(1 - \frac{1}{n} - \frac{1}{p}) = \frac{pn}{q'} = 0$, $p(n-1) = n$ and the previous inequality gives the answer. If $q > 1$, we have $p(n-1) - n = \frac{pn}{q'}$ and Hölder's inequality implies

$$\|v\|_{L^p}^{\frac{p(n-1)}{n}} \leq \frac{p(n-1)}{2n^{3/2}}\|v\|_{L^p}^{\frac{p}{q'}}\|\nabla v\|_{L^q}.$$

Since $\frac{p(n-1)}{n} - \frac{p}{q'} = p(1 - \frac{1}{n} - \frac{1}{q'}) = p(\frac{1}{q} - \frac{1}{n}) = 1$, this completes the proof of Lemma 9.6.1. $\qquad\square$

Proposition 9.6.2. *Let $p \in [1, +\infty]$ and $s \in \mathbb{N}$. We define the Sobolev space $W^{s,p}(\mathbb{R}^n)$ as the set of functions $u \in L^p(\mathbb{R}^n)$ such that the distribution derivatives $\partial^\alpha u$ belong to $L^p(\mathbb{R}^n)$ when the multi-index $\alpha \in \mathbb{N}^n$ is such that $|\alpha| \leq s$. This space is a Banach space for the norm*

$$\|u\|_{W^{s,p}(\mathbb{R}^n)} = \sum_{|\alpha| \leq s} \|\partial^\alpha u\|_{L^p(\mathbb{R}^n)}.$$

When $p = 2$, it is a Hilbert space with dot-product

$$(u, v)_{W^{s,2}(\mathbb{R}^n)} = \sum_{|\alpha| \leq s} (\partial^\alpha u, \partial^\alpha v)_{L^2(\mathbb{R}^n)}.$$

Proof. This set is obviously a vector space. Let $(u_k)_{k \in \mathbb{N}}$ be a Cauchy sequence in $W^{s,p}(\mathbb{R}^n)$. Then, we find $u, v_\alpha \in L^p$ such that $\lim_k u_k = u, \lim_k \partial^\alpha u_k = v_\alpha$ in the Banach space $L^p(\mathbb{R}^n)$. Now for $\phi \in C_c^\infty(\mathbb{R}^n)$, we have

$$\int v_\alpha \phi dx = \lim_k \int \phi \partial^\alpha u_k dx = \lim_k \langle \partial^\alpha u_k, \phi \rangle = (-1)^{|\alpha|} \lim_k \langle u_k, \partial^\alpha \phi \rangle$$

$$= (-1)^{|\alpha|} \lim_k \int u_k \partial^\alpha \phi dx = (-1)^{|\alpha|} \int u \partial^\alpha \phi dx = \langle \partial^\alpha u, \phi \rangle,$$

proving $v_\alpha = \partial^\alpha u$. $\qquad\square$

Lemma 9.6.3. *Let $p \in [1, +\infty)$ and $k \in \mathbb{N}$. Then $C_c^\infty(\mathbb{R}^n)$ is dense in $W^{k,p}(\mathbb{R}^n)$. More precisely, defining for $\epsilon > 0$, $\rho \in C_c^\infty(\mathbb{R}^n)$ such that $\int \rho(t)dt = 1$, $\chi \in C_c^\infty(\mathbb{R}^n)$ equal to 1 on a neighborhood of 0, $\rho_\epsilon(x) = \epsilon^{-n}\rho(x/\epsilon)$, $\chi_\epsilon(x) = \chi(\epsilon x)$ and*

$$R_\epsilon u = \rho_\epsilon * \chi_\epsilon u, \tag{9.6.2}$$

we have $\lim_{\epsilon \to 0} R_\epsilon u = u$ with convergence in $W^{k,p}(\mathbb{R}^n)$.

Proof. Let $u \in W^{k,p}(\mathbb{R}^n)$. The sequence of compactly supported functions $(\chi_\epsilon u)$ converges in $L^p(\mathbb{R}^n)$ towards u. We have also

$$R_\epsilon u - u = \rho_\epsilon * (\chi_\epsilon u - u) + \rho_\epsilon * u - u,$$

so that $\|R_\epsilon u - u\|_{L^p} \leq \|\chi_\epsilon u - u\|_{L^p} + \|\rho_\epsilon * u - u\|_{L^p}$ and the result for $k = 0$. For $|\alpha| \leq k$, we have

$$\partial^\alpha R_\epsilon u - \partial^\alpha u = \rho_\epsilon * \partial^\alpha(\chi_\epsilon u) - \partial^\alpha u = \rho_\epsilon * ([\partial^\alpha, \chi_\epsilon]u) + \rho_\epsilon * (\chi_\epsilon \partial^\alpha u) - \partial^\alpha u,$$

entailing

$$\|\partial^\alpha R_\epsilon u - \partial^\alpha u\|_{L^p} \leq \|R_\epsilon \partial^\alpha u - \partial^\alpha u\|_{L^p} + \sum_{\substack{\beta \leq \alpha \\ |\beta| \geq 1}} \frac{\alpha!}{\beta!} \epsilon^{|\beta|} \|\rho_\epsilon * ((\partial^\beta \chi)_\epsilon \partial^{\alpha-\beta} u)\|_{L^p},$$

which implies convergence in $W^{k,p}(\mathbb{R}^n)$ of $R_\epsilon u$. $\qquad\square$

Theorem 9.6.4. *Let $n \geq 2$ be an integer and let $p, q \in [1, +\infty)$ such that $\frac{1}{p} = \frac{1}{n} + \frac{1}{q}$. Then we have the continuous embedding*

$$W^{1,p}(\mathbb{R}^n) \hookrightarrow L^q(\mathbb{R}^n) = W^{0,q}(\mathbb{R}^n),$$

and there exists $C(p, n) > 0$ such that for all $u \in W^{1,p}(\mathbb{R}^n)$,

$$\|u\|_{L^q(\mathbb{R}^n)} \leq C(p, n)\|\nabla u\|_{L^p(\mathbb{R}^n)}. \tag{9.6.3}$$

Remark 9.6.5. Note that when p ranges in the interval $[1, n)$, we have $q = \frac{np}{n-p}$ ranging in $[\frac{n}{n-1}, +\infty)$. We shall use the notation

$$p^*(n) = \frac{np}{n-p} \qquad \text{for the } Sobolev \ conjugate \ exponent. \tag{9.6.4}$$

We may note here that in the limiting case $p = n, q = +\infty$, the above inclusion does not hold for $n \geq 2$ (however Remark 9.6.6 shows that it is true for $n = 1$). Let $\beta \in (\frac{1}{n}, 1)$ and $w(x) = \chi(x)(\ln |x|)^{1-\beta}/(1 - \beta)$, where $\chi \in C_c^\infty(\mathbb{R}^n)$ is equal to 1 on $B(0, 1/4)$ and is supported in $B(0, 1/2)$. We have

$$(\nabla w)(x) = (\ln |x|)^{-\beta}|x|^{-1}\frac{x}{|x|}\chi(x) + C_c^\infty(\mathbb{R}^n)$$

$$\implies \|\nabla w\|_{L^n}^n \leq C + C \int_0^{1/2} r^{n-1}r^{-n}|\ln r|^{-\beta n}dr = C + \int_2^{+\infty} \frac{dR}{R|\ln R|^{\beta n}} < +\infty,$$

since $n\beta > 1$. The function w is also in $L^n(\mathbb{R}^n)$ since

$$\|w\|_{L^n}^n \leq C_1 \int_0^{1/2} r^{n-1}|\ln r|^{(1-\beta)n}dr = C_1 \int_2^{+\infty} \frac{(\ln R)^{(1-\beta)n}dR}{R^{n+1}} < +\infty.$$

However w does not belong to L^∞ since $\beta < 1$.

Remark 9.6.6. In the case $n = 1$, we have then $p = 1, q = +\infty$ and it is indeed true that $W^{1,1}(\mathbb{R}) \hookrightarrow L^\infty(\mathbb{R})$. Let $u \in W^{1,1}(\mathbb{R})$. In the proof of Theorem 9.5.4, we have shown the density of $C_c^1(\mathbb{R})$ in $W^{1,1}(\mathbb{R})$: let (ϕ_k) be a sequence of functions of $C_c^1(\mathbb{R})$ converging in $W^{1,1}(\mathbb{R})$. We have

$$u(x) = u(x) - \phi_k(x) + \int_{-\infty}^x \phi_k'(t)dt \implies |u(x)| \leq |u(x) - \phi_k(x)| + \|\phi_k'\|_{L^1(\mathbb{R})},$$

and thus $|u(x)| \leq |u(x) - \phi_k(x)| + \|\phi_k' - u'\|_{L^1(\mathbb{R})} + \|u'\|_{L^1(\mathbb{R})}$. We may find a subsequence of (ϕ_k) converging almost everywhere to u so that we have a.e.,

$$|u(x)| \leq \|u'\|_{L^1(\mathbb{R})} \implies u \in L^\infty(\mathbb{R}), \ \|u\|_{L^\infty(\mathbb{R})} \leq \|u'\|_{L^1(\mathbb{R})}.$$

Proof of Theorem 9.6.4. Let $u \in W^{1,p}(\mathbb{R}^n)$. Then from Lemma 9.6.3, we have $\lim_\epsilon R_\epsilon u = u$ in $W^{1,p}(\mathbb{R}^n)$. Moreover from Lemma 9.6.1, we find that

$$\|R_\epsilon u\|_{L^q(\mathbb{R}^n)} \leq C(p, n)\|\nabla R_\epsilon u\|_{L^p(\mathbb{R}^n)}.$$

This inequality proves that $(R_\epsilon u)$ is a Cauchy sequence in $L^q(\mathbb{R}^n)$, thus converging towards some $v \in L^q(\mathbb{R}^n)$. Since $(R_\epsilon u)$ converges towards u in $W^{1,p}(\mathbb{R}^n)$, we find for $\phi \in C_c^\infty(\mathbb{R}^n)$,

$$\langle v, \phi \rangle = \lim_\epsilon \int (R_\epsilon u)\phi\, dx = \langle u, \phi \rangle \implies v = u, \quad u \in L^q(\mathbb{R}^n).$$

Passing to the limit with respect to ϵ in the inequality above gives (9.6.3). $\square$

Theorem 9.6.7. *Let $0 \le l < k$ be integers, and let $1 \le p < q < +\infty$ be real numbers such that*
$$\frac{k-l}{n} = \frac{1}{p} - \frac{1}{q}. \qquad \text{Then } W^{k,p}(\mathbb{R}^n) \hookrightarrow W^{l,q}(\mathbb{R}^n).$$

Proof. If $n = 1$, we should have $p = 1, q = +\infty, k = l+1$, and we have already seen that $W^{1,1}(\mathbb{R}) \hookrightarrow W^{0,\infty}(\mathbb{R})$, with

$$\|u\|_{L^\infty} \le \frac{1}{2}\|u'\|_{L^1} \text{ for } u, u' \in L^1$$

$$\implies \text{ for } l \in \mathbb{N} \text{ and } u^{(l)}, u^{(l+1)} \in L^1(\mathbb{R}), \ \|u^{(l)}\|_{L^\infty} \le \frac{1}{2}\|u^{(l+1)}\|_{L^1},$$

which implies for $l \in \mathbb{N}$, $W^{1+l,1}(\mathbb{R}) \hookrightarrow W^{l,\infty}(\mathbb{R})$. We assume now $n \ge 2$ and we note that Theorem 9.6.4 tackles the case $k = 1, l = 0$ with the estimate

$$\forall u \in W^{1,p}(\mathbb{R}^n), \quad \|u\|_{L^q(\mathbb{R}^n)} \le C(p,n)\|\nabla u\|_{L^p(\mathbb{R}^n)}, \quad \frac{1}{p} - \frac{1}{q} = \frac{1}{n}.$$

We note that this implies

$$\forall u \in W^{1+l,p}(\mathbb{R}^n), \quad \|\nabla^l u\|_{L^q(\mathbb{R}^n)} \le C(p,n)\|\nabla^{l+1} u\|_{L^p(\mathbb{R}^n)}, \quad \frac{1}{p} - \frac{1}{q} = \frac{1}{n},$$

which deals with the case $k = l+1$. Let us assume that for $k - l = \nu \ge 1$, we have proven

$$\forall u \in W^{\nu+l,p}(\mathbb{R}^n), \quad \|\nabla^l u\|_{L^q(\mathbb{R}^n)} \le C(p,n)\|\nabla^{l+\nu} u\|_{L^p(\mathbb{R}^n)}, \quad \frac{1}{p} - \frac{1}{q} = \frac{\nu}{n}.$$

This implies that for

$$\frac{1}{p_{\nu+1}} - \frac{1}{q_{\nu+1}} = \frac{\nu+1}{n}, \quad \frac{1}{p_{\nu+1}} - \frac{1}{q_{\nu+1}} - \frac{1}{n} = \frac{\nu}{n},$$

$$\forall u \in W^{\nu+l+1,p_{\nu+1}}(\mathbb{R}^n), \ \|\nabla^{l+1} u\|_{L^{q_\nu}(\mathbb{R}^n)} \le C(p_{\nu+1},n)\|\nabla^{l+1+\nu} u\|_{L^{p_{\nu+1}}(\mathbb{R}^n)},$$

with $\frac{1}{p_{\nu+1}} - \frac{1}{q_\nu} = \frac{\nu}{n}$, $q_\nu = \frac{nq_{\nu+1}}{n+q_{\nu+1}}$. But we have

$$\|\nabla^l u\|_{L^r(\mathbb{R}^n)} \le C(q_\nu,n)\|\nabla^{l+1} u\|_{L^{q_\nu}(\mathbb{R}^n)}, \quad \frac{1}{q_\nu} - \frac{1}{r} = \frac{1}{n},$$

so that $\frac{1}{r} = \frac{1}{q_{\nu+1}} + \frac{1}{n} - \frac{1}{n}$,   i.e., $r = q_{\nu+1}$. We have thus proven by induction on ν that

$$\forall u \in W^{\nu+l,p}(\mathbb{R}^n), \quad \|\nabla^l u\|_{L^q(\mathbb{R}^n)} \le C(p,n)\|\nabla^{l+\nu} u\|_{L^p(\mathbb{R}^n)}, \quad \frac{1}{p} - \frac{1}{q} = \frac{\nu}{n},$$

proving the sought result. $\qquad\square$

Remark 9.6.8. We have proven above that

$$W^{k,p}(\mathbb{R}^n) \hookrightarrow W^{l,q}(\mathbb{R}^n), \quad \text{for} \quad \frac{k-l}{n} = \frac{1}{p} - \frac{1}{q}, \quad 1 \le p < q < +\infty.$$

Note that in this formula, we have $k > l$ but $p < q$ so that the functions in $W^{k,p}$ have more derivatives but less Lebesgue regularity than the functions in $W^{l,q}$. This means that we can somehow trade some regularity in terms of derivatives (first index $k > l$) to buy some L^q regularity according to the fixed exchange rate given by $\frac{k-l}{n} = \frac{1}{p} - \frac{1}{q}$. We see also that Lebesgue regularity is a non-convertible currency which cannot buy a derivative regularity.

9.7 Notes

A more general definition of Sobolev spaces $W^{s,p}(\mathbb{R}^n)$ for $p \in (1, +\infty)$ and $s \in \mathbb{R}$ is

$$W^{s,p}(\mathbb{R}^n) = \{u \in \mathscr{S}'(\mathbb{R}^n), \langle D_x \rangle^s u \in L^p(\mathbb{R}^n)\}, \tag{9.7.1}$$

with $\widehat{\langle D_x \rangle^s u} = \langle \xi \rangle^s \hat{u}(\xi), \langle \xi \rangle = (1 + |\xi|^2)^{1/2}$, which makes sense since $\langle \xi \rangle^s$ belongs to the space $\mathscr{O}_M(\mathbb{R}^n)$ of multipliers of $\mathscr{S}'(\mathbb{R}^n)$ (see Definition 8.1.21). The general study of these spaces is not much more difficult than what we have done above for $s \in \mathbb{N}$, but a simple exposition would require some basic study of the *Fourier multiplier* $\langle \xi \rangle^s$, i.e., of the operator $\langle D_x \rangle^s$. For instance, we would have to prove L^p boundedness ($p \in (1, +\infty)$) for the operators $D_{x_j}\langle D_x \rangle^{-1}$, and here also a simplifying point of view would certainly be required to introduce elementary facts about pseudodifferential operators. We felt that a five-hundred-page book does not need a hundred more and decided to end the book here. Some information on the topic of pseudodifferential operators can be found in Chapter 18 in the third volume of Hörmander's treatise on Linear Partial Differential Operators [32] and also in the book [41] and the references therein.

The names of mathematicians encountered in this chapter follow.

Arnaud DENJOY (1884–1974) was a French mathematician.

Ralph HENSTOCK (1923–2007) was an English mathematician.

Jaroslav KURZWEIL (born 1926) is a Czech mathematician.

Emilio GAGLIARDO (1930–2008) was an Italian mathematician.

Ernst LINDELÖF (1870–1946) is a Finnish mathematician.

Józef MARCINKIEWICZ (1910–1940) was a Polish mathematician. He died probably during the Katyn killings perpetrated by the NKVD (Soviet secret police).

Louis NIRENBERG (born 1925) is a Canadian-born American mathematician.

Oskar PERRON (1880–1975) was a German mathematician.

Lars PHRAGMÉN (1863–1937) was a Swedish mathematician.

Olof THORIN (1912–2004) was a Swedish mathematician.

Norbert WIENER (1894–1964) was a prominent American scientist, one of the founders of modern harmonic analysis and computer science.

9.8   Exercises

Exercise 9.8.1. *Let $p, q, r \in [1, 2]$ such that (6.2.1) holds. Let $u \in L^p(\mathbb{R}^n), v \in L^q(\mathbb{R}^n)$. Prove that $\hat{u} \in L^{p'}(\mathbb{R}^n), v \in L^{q'}(\mathbb{R}^n)$ and that the product $\hat{u}\hat{v}$ belongs to $L^{r'}(\mathbb{R}^n)$. Show that*

$$u * v \in L^r(\mathbb{R}^n) \quad and \quad \widehat{u * v} = \hat{u}\hat{v}.$$

Answer. The fact that $u * v$ belongs to L^r follows from Young's inequality and we have $\hat{u} \in L^{p'}, \hat{v} \in L^{q'}$ from the Hausdorff–Young Theorem. This implies from Hölder's inequality that the product $\hat{u}\hat{v}$ belongs to $L^{r'}$ since

$$\int |\hat{u}|^{r'}|\hat{v}|^{r'}\,d\xi \leq \left(\int |\hat{u}|^{sr'}\,d\xi\right)^{1/s} \left(\int |\hat{v}|^{s'r'}\,d\xi\right)^{1/s'},$$

where we may choose

$$s = \frac{p'}{r'} \implies \frac{1}{s'} = 1 - \frac{r'}{p'} = r'\left(\frac{1}{r'} - \frac{1}{p'}\right) = \frac{r'}{q'} \implies r's' = q'.$$

The above argument extends when $r' = +\infty$ (which implies $p' = q' = +\infty$ so that $p = q = r = 1$ and $\hat{u}, \hat{v}$ belong to L^∞). We have thus

$$\|\widehat{u * v}\|_{L^{r'}(\mathbb{R}^n)} \leq \|\hat{u}\|_{L^{p'}(\mathbb{R}^n)}\|\hat{v}\|_{L^{q'}(\mathbb{R}^n)} \leq \|u\|_{L^p(\mathbb{R}^n)}\|v\|_{L^q(\mathbb{R}^n)}. \tag{9.8.1}$$

To get that $\widehat{u * v} = \hat{u}\hat{v}$, it is enough to prove it for u, v in the Schwartz space since then we shall obtain with $\varphi_k, \psi_k \in \mathscr{S}(\mathbb{R}^n)$ such that $\lim_k \varphi_k = u$ in L^p, $\lim_k \psi_k = v$ in L^q, thanks to (9.8.1),

$$\hat{u}\hat{v} = \underbrace{\lim_k \widehat{\varphi_k}}_{\substack{\text{limit} \\ \text{in } L^{p'}}} \underbrace{\lim_l \widehat{\psi_l}}_{\substack{\text{limit} \\ \text{in } L^{q'}}} = \underbrace{\lim_k \widehat{\varphi_k * \psi_k}}_{\substack{\text{limit} \\ \text{in } L^{r'}}} = \widehat{u * v}.$$

Formula (8.1.12) gives the result.

Exercise 9.8.2. *Show that if T satisfies the assumptions of Theorem 9.2.3 with $r = +\infty$ and*

$$t\omega(t, Tu) \le c_1 \|u\|_{L^1}, \qquad \|Tu\|_{L^\infty} \le c_\infty \|u\|_{L^\infty},$$

then for $1 < p < +\infty$, we have

$$\|Tu\|_{L^p} \le \frac{p^{1+\frac{1}{p}}}{p-1} c_1^{1/p} c_\infty^{1/p'} \|u\|_{L^p}.$$

Answer. We have only to revisit the proof of Theorem 9.2.3 while paying more attention to the choice of the various constants. We write for $u \in L^1 + L^\infty$, $t > 0$, $\alpha > c_\infty$,

$$u = \underbrace{u \mathbf{1}_{\{|u| > t/\alpha\}}}_{u_1} + \underbrace{u \mathbf{1}_{\{|u| \le t/\alpha\}}}_{u_2}, \tag{9.8.2}$$

and this gives

$$|(Tu)(x)| \le |(Tu_1)(x)| + |(Tu_2)(x)| \le |(Tu_1)(x)| + \|u_2\|_{L^\infty} \le |(Tu_1)(x)| + \frac{c_\infty t}{\alpha},$$

so that we find the inclusion

$$(\sharp) \qquad \{x, |(Tu)(x)| > t\} \subset \{x, |(Tu_1)(x)| > t(1 - c_\infty \alpha^{-1})\}.$$

The weak-type $(1, 1)$ assumption reads $t\omega(t, Tv) \le c_1 \|v\|_{L^1}$ so that

$$(\flat) \qquad \omega(t(1 - c_\infty \alpha^{-1}), Tu_1) \le \frac{c_1}{t(1 - c_\infty \alpha^{-1})} \int_{|u| > t/\alpha} |u| \, dx.$$

Applying Formula (9.2.6) to Tu, we find, using Tonelli's theorem and $1 < p < +\infty$,

$$\|Tu\|_{L^p}^p = p \int_0^{+\infty} t^{p-1} \omega(t, Tu) \, dt$$

$$(\text{from } (\sharp)) \quad \le p \int_0^{+\infty} t^{p-1} \omega(t(1 - c_\infty \alpha^{-1}), Tu_1) \, dt$$

$$(\text{from } (\flat)) \quad \le p \int_0^{+\infty} t^{p-1} \frac{c_1}{t(1 - c_\infty \alpha^{-1})} \int_{|u| > t/\alpha} |u| \, dx \, dt$$

$$= \frac{pc_1}{1 - c_\infty \alpha^{-1}} \iint_{\mathbb{R}_+ \times \mathbb{R}^n} t^{p-2} H(\alpha|u(x)| - t) |u(x)| \, dt \, dx$$

$$= \frac{pc_1}{(1 - c_\infty \alpha^{-1})(p - 1)} \int_{\mathbb{R}^n} (\alpha|u(x)|)^{p-1} |u(x)| \, dx$$

$$= \frac{\alpha^{p-1} pc_1}{(1 - c_\infty \alpha^{-1})(p - 1)} \|u\|_{L^p}^p.$$

We check now for $\alpha = \lambda c_\infty$ with $\lambda > 1$ (assuming of course $c_\infty > 0$),

$$\frac{\alpha^{p-1}pc_1}{(1 - c_\infty \alpha^{-1})(p-1)} = p'c_1 \frac{\lambda^p c_\infty^{p-1}}{\lambda - 1}.$$

We have proven that for any $\lambda > 1$,

$$\sup_{\|u\|_{L^p}=1} \|Tu\|_{L^p} \le (p'c_1)^{1/p} \frac{\lambda}{(\lambda-1)^{1/p}} c_\infty^{1/p'},$$

so that choosing $\lambda = p/(p-1)$ gives the sought answer.

Exercise 9.8.3. *Let $f : \mathbb{R}^n \to \mathbb{C}$ be an L^1_{loc} function. Prove that $\mathcal{M}_f$ is a measurable function (see Definition 9.3.1).*

Answer. For each $t > 0$ the function $\mathbb{R}^n \times \mathbb{R}^n \ni (x, z) \mapsto f(x + tz)$ is measurable (from Theorem 1.2.7) and Proposition 4.1.3 implies that

$$x \mapsto |B(x,t)|^{-1} \int_{B(x,t)} |f(y)|\,dy = |\mathbb{B}^n|^{-1} \int_{\mathbb{B}^n} |f(x+tz)|\,dz,$$

is measurable. Proposition 1.3.1 proves that

$$\widetilde{\mathcal{M}}_f(x) = \sup_{t \in \mathbb{Q}_+^*} \fint_{\mathbb{B}^n} |f(x+tz)|\,dz$$

is measurable. Let $\epsilon > 0$ be given. Let us consider $t > 0$ and $0 < s \in \mathbb{Q}$ such that $t \le s \le t(1+\epsilon)$; we have

$$\frac{1}{t^n |\mathbb{B}^n|} \int_{B(x,t)} |f(y)|\,dy \le \frac{1}{t^n |\mathbb{B}^n|} \int_{B(x,s)} |f(y)|\,dy \le (\tfrac{s}{t})^n \widetilde{\mathcal{M}}_f(x) \le (1+\epsilon)^n \widetilde{\mathcal{M}}_f(x),$$

which implies $\mathcal{M}_f(x) \le (1+\epsilon)^n \widetilde{\mathcal{M}}_f(x)$. Since $\widetilde{\mathcal{M}}_f(x) \le \mathcal{M}_f(x)$, we find that for any $\epsilon > 0$, $\mathcal{M}_f(x) \le (1+\epsilon)^n \widetilde{\mathcal{M}}_f(x) \le (1+\epsilon)^n \mathcal{M}_f(x)$, proving that $\mathcal{M}_f$ is equal to the measurable $\widetilde{\mathcal{M}}_f$ (this works in particular when $\mathcal{M}_f(x) = +\infty$).

Exercise 9.8.4. *Show that Theorem 9.4.1 holds for $f \in L^1_{\mathrm{loc}}(\Omega)$ where Ω is an open subset of $\mathbb{R}^n$.*

Answer. Using Exercise 2.8.10, we find a sequence $(K_j)_{j\ge1}$ of compact subsets of Ω such that $K_j \subset \mathring{K}_{j+1}$ and $\Omega = \cup_{j\ge1} K_j$; Exercise 2.8.7 provides a function $\varphi_j \in C_c^\infty(\mathring{K}_{j+1})$ equal to 1 on K_j. We may now consider the $L^1(\mathbb{R}^n)$ function $\varphi_j f$ and apply Theorem 9.4.1: we find a measurable set L_j such that $\lambda_n(L_j^c) = 0$ so that

$$\forall x \in L_j, \quad \lim_{r \to 0} \fint_{B(x,r)} |\varphi_j(y)f(y) - \varphi_j(x)f(x)|\,dy = 0.$$

In particular, for $j \geq 2$ and $x \in K_{j-1} \cap L_j$, we have $x \in \mathring{K}_j$ so that $B(x, r) \subset K_j$ for $r > 0$ small enough and this gives $\forall x \in K_{j-1} \cap L_j$,

$$0 = \lim_{r \to 0} \fint_{B(x,r)} |\varphi_j(y) f(y) - \varphi_j(x) f(x)| \, dy = \lim_{r \to 0} \fint_{B(x,r)} |f(y) - f(x)| \, dy.$$

As a result the conclusion holds whenever $x \in L = \cup_{j \geq 2} (K_{j-1} \cap L_j)$ which is a measurable subset of Ω. On the other hand we have

$$L^c \cap \Omega = \Omega \cap \bigcap_{j \geq 2} (K_{j-1}^c \cup L_j^c) \subset \cup_{j \geq 2} L_j^c \cup \underbrace{\left(\Omega \cap \cap_{j \geq 2} K_{j-1}^c\right)}_{=\emptyset},$$

so that $\lambda_n(L^c \cap \Omega) = 0$.

Exercise 9.8.5. *Let F be defined on $\mathbb{R}$ by $F(0) = 0$ and for $x \neq 0, F(x) = x^2 \sin(x^{-2})$.*

(1) *Prove that F is differentiable everywhere and calculate its derivative F'.*
(2) *Prove that F' is not locally integrable.*
(3) *Prove that the weak derivative of F is not a Radon measure.*

Answer. (1) Differentiability outside 0 is obvious with

$$x \neq 0, \quad F'(x) = 2x \sin(x^{-2}) - 2x^{-1} \cos(x^{-2}), \qquad F'(0) = \lim_{x \to 0} x \sin(x^{-2}) = 0.$$

We note in particular that F' is not continuous since $F'(\frac{1}{\sqrt{2k\pi}}) = -2\sqrt{2k\pi}$ for $k \in \mathbb{N}^*$.

(2) Since $2x \sin(x^{-2})$ is locally bounded, we have to prove that $x^{-1} \cos(x^{-2})$ is not locally integrable:

$$\int_0^1 |\cos(x^{-2})| x^{-1} dx = \frac{1}{2} \int_1^{+\infty} |\cos t| \frac{dt}{t} = +\infty \quad \text{(see Exercise 2.8.20)}.$$

(3) The weak derivative f of F is defined as a linear form on $C_c^\infty(\mathbb{R})$ functions (or as a tempered distribution, cf. Chapter 8 with Definition 8.1.8), with

$$\langle F', \varphi \rangle = - \int_{\mathbb{R}} F(x) \varphi'(x) dx.$$

Let us assume that φ is supported in $(0, +\infty)$: we have then

$$\langle F', \varphi \rangle = \int \left(2x \sin(x^{-2}) - 2x^{-1} \cos(x^{-2})\right) \varphi(x) dx.$$

We choose now $\varphi_k \in C_c^\infty((a_k, b_k); [0, 1])$ with $k \in \mathbb{N}^*$,

$$a_k = \left(2\pi k + \frac{\pi}{4}\right)^{-1/2}, \qquad b_k = \left(2\pi k - \frac{\pi}{4}\right)^{-1/2}$$

so that $x \in (a_k, b_k) \implies x^{-2} \in (2\pi k - \frac{\pi}{4}, 2\pi k + \frac{\pi}{4}) \implies \cos(x^{-2}) \in (2^{-1/2}, 1]$. As a result, we have

$$\int_{a_k}^{b_k} x^{-1} \cos(x^{-2}) \varphi_k(x) dx \geq 2^{-1/2} \left(2\pi k - \frac{\pi}{4}\right)^{1/2} \int_{a_k}^{b_k} \varphi_k(x) dx.$$

We may also assume that φ_k equals 1 on $\left[(2\pi k + \frac{\pi}{6})^{-1/2}, (2\pi k - \frac{\pi}{6})^{-1/2}\right]$, implying

$$\int_{a_k}^{b_k} x^{-1} \cos(x^{-2}) \varphi_k(x) dx \geq 2^{-1/2} \left(2\pi k - \frac{\pi}{4}\right)^{1/2} \frac{\pi}{3} \frac{1}{2} \left(2\pi k + \frac{\pi}{6}\right)^{-3/2} \geq c_0 k^{-1}.$$

Since the intervals (a_k, b_k) are pairwise disjoint, the function

$$\Phi_N(x) = \sum_{1 \leq k \leq N} \varphi_k(x)$$

is such that $\Phi_N \in C_c^{\infty}((0, +\infty); [0, 1])$ and

$$\langle F', \Phi_N \rangle \leq -c_0 \sum_{1 \leq k \leq N} \frac{1}{k} + \int_0^1 2x dx \xrightarrow[N \to +\infty]{} -\infty.$$

Exercise 9.8.6. *Let $\rho \in C_c(\mathbb{R}^n; \mathbb{R}_+)$ such that $\int \rho(z) dz = 1$. We define for $\epsilon > 0$, $\rho_\epsilon(x) = \epsilon^{-n} \rho(x/\epsilon)$ and the operator R_ϵ on $L^1_{\mathrm{loc}}(\mathbb{R}^n)$ by*

$$(R_\epsilon u)(x) = \int \rho_\epsilon(x - y) u(y) dy. \tag{9.8.3}$$

(1) *Let $1 \leq p < +\infty$. Prove that if u belongs to $L^p(\mathbb{R}^n)$, $\lim_{\epsilon \to 0_+} R_\epsilon u = u$, with L^p convergence. Moreover prove that for almost all x, $\lim_{\epsilon \to 0_+} (R_\epsilon u)(x) = u(x)$.*

(2) *Let $u \in L^\infty(\mathbb{R}^n)$. Prove that for almost all x, $\lim_{\epsilon \to 0_+} (R_\epsilon u)(x) = u(x)$. Prove that $\|R_\epsilon u\|_{L^\infty} \leq \|u\|_{L^\infty}$.*

Answer. (1) The proof of Theorem 3.4.3 answers the very first statement. Let us answer the two questions about a.e. convergence assuming only $u \in L^1_{\mathrm{loc}}$: we have $(R_\epsilon u)(x) - u(x) = \int \big(u(x - \epsilon z) - u(x)\big) \rho(z) dz$ so that for $N > 0$, assuming as we may that $\mathrm{supp}\,\rho \subset \mathbb{B}^n$,

$$\mathbf{1}_{\mathbb{B}^n}(x/N)|(R_\epsilon u)(x) - u(x)| \leq \mathbf{1}_{\mathbb{B}^n}(x/N)\|\rho\|_{L^\infty} \int_{z \in \mathrm{supp}\,\rho} |u(x - \epsilon z) - u(x)| dz.$$

We define $U(y) = \mathbf{1}_{(1+N)\mathbb{B}^n}(y) u(y)$ and we have for $\epsilon \leq 1$,

$$\mathbf{1}_{\mathbb{B}^n}(x/N)|(R_\epsilon u)(x) - u(x)| \leq \mathbf{1}_{\mathbb{B}^n}(x/N)\|\rho\|_{L^\infty} \int \mathbf{1}_{\mathbb{B}^n}(z)|U(x - \epsilon z) - U(x)| dz.$$

From the Lebesgue differentiation theorem applied to the $L^1(\mathbb{R}^n)$ function U, we have for almost every x, $\lim_{\epsilon \to 0} \int \mathbf{1}_{\mathbb{B}^n}(z)|U(x - \epsilon z) - U(x)| dz = 0$. For each positive integer N, we find a set L_N such that $|L_N^c| = 0$ and

$$\forall x \in N\mathbb{B}^n \cap L_N, \quad \lim_{\epsilon \to 0}(R_\epsilon u)(x) = u(x).$$

Since $\left\{\cup_{N\geq 1}\left(N\mathbb{B}^n \cap L_N\right)\right\}^c = \cap_{N\geq 1}\left((N\mathbb{B}^n)^c \cup L_N^c\right) \subset \cup_{N\geq 1}L_N^c$, which has measure 0, this completes the proof of a.e. convergence.

(2) The inequality $\|R_\epsilon u\|_{L^\infty} \leq \|u\|_{L^\infty}$ follows trivially from the assumptions on ρ.

Exercise 9.8.7. *Let $b \in L^1(\mathbb{R}^n)$, and $v \in L^\infty(\mathbb{R}^n)$. Prove that*

$$\lim_{\substack{t\to 0 \\ t\in\mathbb{R}^n}} \int |b(x)||v(x+t) - v(x)|dx = 0.$$

Answer. Let R, κ be positive constants. We define

$$A_{R,\kappa}(t) = \{x, |x| \leq R \text{ and } |v(x+t) - v(x)| > \kappa\}.$$

We have for $t \in \mathbb{R}^n$, for $|t| \leq R$,

$$\lambda_n(\{x, |x| \leq R, |x+t| > R\}) = \int_{|x|\leq R} H(|x+t| - R)dx$$

$$= R^n \int_{\mathbb{B}^n} H(|Ry+t| - R)dy = R^n \int_{\mathbb{B}^n} H(|y+tR^{-1}| - 1)dy$$

$$\leq R^n \int_{\mathbb{B}^n} H(|y| + |t|R^{-1} - 1)dy = R^n n^{-1}|\mathbb{S}^{n-1}|\left(1 - (1 - |t|R^{-1})^n\right)$$

$$\leq R^n n^{-1}|\mathbb{S}^{n-1}|n|t|R^{-1} = R^{n-1}|\mathbb{S}^{n-1}||t|.$$

We have also the estimates

$$\lambda_n(A_{R,\kappa}(t)) \leq \frac{1}{\kappa}\int_{|x|\leq R, |x+t|\leq R} |v(x+t) - v(x)|dx + \lambda_n(\{x, |x| \leq R, |x+t| > R\})$$

$$\leq \frac{1}{\kappa}\|\tau_{-t}v_R - v_R\|_{L^1} + |t|R^{n-1}|\mathbb{S}^{n-1}|,$$

with $v_R(x) = v(x)\mathbf{1}(|x| \leq R)$ which is an $L^1(\mathbb{R}^n)$ function. This implies that for all $\kappa > 0$, $R > 0$ we have $\lim_{t\to 0} \lambda_n(A_{R,\kappa}(t)) = 0$ and thus

$$\limsup_{t\to 0} \int |b(x)||v(x+t) - v(x)|dx$$

$$= \limsup_{t\to 0}\left\{\int_{|v(x+t)-v(x)|\leq \kappa} |b(x)||v(x+t) - v(x)|dx + \int_{|v(x+t)-v(x)|>\kappa} |b(x)||v(x+t) - v(x)|dx\right\}$$

$$\leq \kappa\|b\|_{L^1} + 2\|v\|_{L^\infty}\limsup_{t\to 0}\int_{A_{R,\kappa}(t)} |b(x)|dx + 2\|v\|_{L^\infty}\int_{|x|>R} |b(x)|dx$$

$$= \kappa\|b\|_{L^1} + 2\|v\|_{L^\infty}\int_{|x|>R} |b(x)|dx.$$

We infer the result from this inequality, letting $R \to +\infty$ and $\kappa \to 0_+$.

Exercise 9.8.8. *For $p \in [1,2]$, $t = 1/p$, draw the curve $[1/2,1] \mapsto t^{-t}(1-t)^{1-t} = p^{1/p}p'^{-1/p'}$, related to the best constant in the Hausdorff–Young inequality (9.1.22). $\left(p^{1/p}p'^{-1/p'}\right)^{n/2}$.*

Answer. We draw (see Figure 9.1) the graph of the function

$$[1/2,1] \ni t \mapsto t^{-t}(1-t)^{1-t},$$

with t standing for $1/p$.

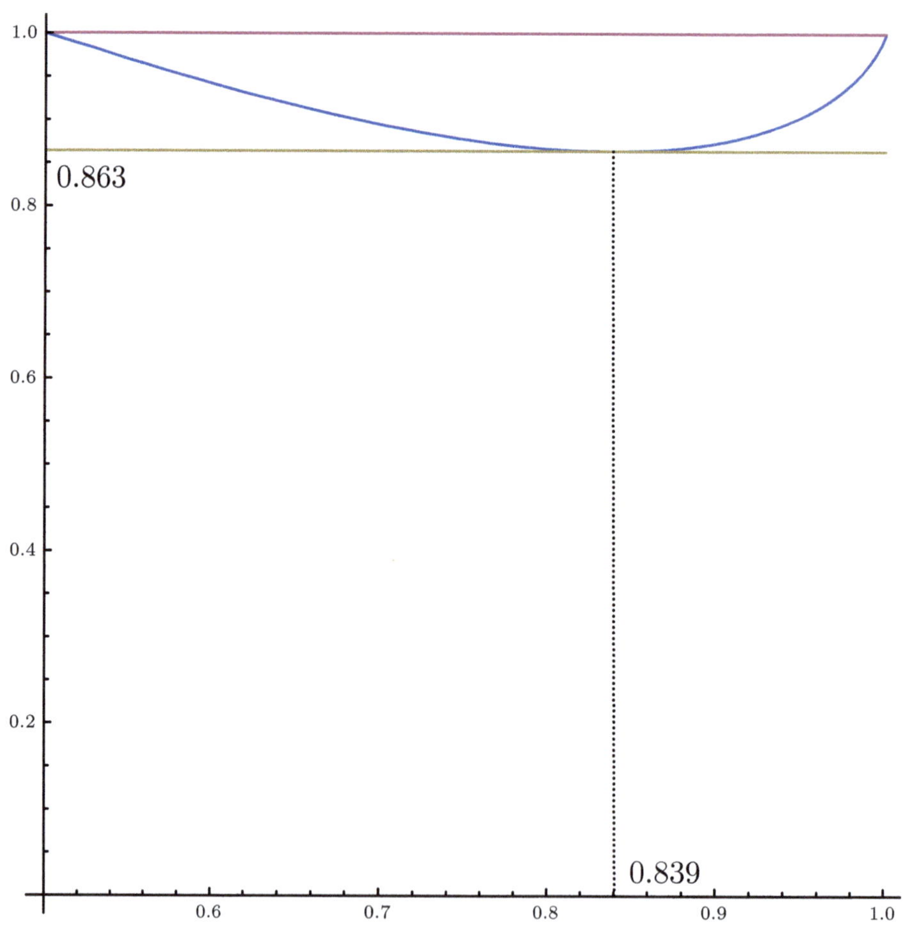

Figure 9.1: FUNCTION $t^{-t}(1-t)^{1-t}, t \in [1/2,1]$.

Chapter 10

Appendix

10.1 Set theory, cardinals, ordinals

Set theory

We shall assume that the reader is familiar with elementary set theory (e.g., definitions of union, intersection, products, of family of sets) and knows a little bit about Russell's paradox (see, e.g., Exercise 1.9.2). A simple introduction to the subject would be to solve the seven first exercises in Section 1.9. The notions of Cartesian product[1], relations, equivalence relations, partitions, quotient set, functions, images and inverse images, as well as injectivity, surjectivity, bijectivity, composition of functions shall also be assumed to be familiar to the reader.

Definition 10.1.1. Let E be a set and $\leq$ be a binary relation on E.

(1) The relation $\leq$ is said to be an *order relation* whenever it is reflexive ($x \leq x$) antisymmetric ($x \leq y, y \leq x \longrightarrow y - x$) and transitive ($x \leq y, y \leq z \Longrightarrow x \leq z$).

(2) The order relation is said to be *total* whenever for any $(x, y) \in E^2$, either $x \leq y$ or $y \leq x$.

(3) An ordered set $(E, \leq)$ is said to be *well ordered* whenever every non-empty subset of E has a smallest element, i.e.,

$$\forall A \text{ non-empty} \subset E, \exists a \in A, \quad \forall x \in A, a \leq x.$$

Note that the smallest element of a non-empty subset of E is unique, when it exists.

---

[1] The Cartesian product $\prod_{i \in I} X_i$ of a family of sets $(X_i)_{i \in I}$ is defined as the set of mappings x from I to $\cup_{i \in I} X_i$ such that, for all $i \in I$, $x(i) \in X_i$. A particular case of interest occurs when $\forall i \in I, X_i = X$; then we note $\prod_{i \in I} X_i = X^I$ which is the set of mappings from I to X. A more academic remark is concerned with the case when $I = \emptyset$: in that case, $\prod_{i \in \emptyset} X_i$ is not empty since it has a single element which is the mapping whose graph is the empty set.

Definition 10.1.2. Let $(E, \leq)$ be an ordered set.

(1) An element $a \in E$ is said to be *maximal* if $\{x \in E, x > a\} = \emptyset$.

(2) An element $a \in E$ is said to be the smallest (resp. largest) element in E if for all $x \in E$, $x \geq a$ (resp. $x \leq a$). If a smallest (resp. largest) element exists, then it is unique.

(3) Let X be a subset of E. An *upper bound* of X is an element $M \in E$ such that $X \subset (\to, M] = \{x \in E, x \leq M\}$. A *lower bound* of X is an element $m \in E$ such that $X \subset [m, \leftarrow) = \{x \in E, m \leq x\}$.

(4) Let X be a subset of E. When the set of upper bounds (resp. lower bounds) is non-empty and has a smallest element b, we call that element the *least upper bound* or *supremum* (resp. *greatest lower bound* or *infimum*).

We state below the *Axiom of Choice*, *Zorn's lemma* and *Zermelo's theorem*, three statements that can be proven to be equivalent. The Axiom of Choice plays an important rôle in measure theory, since it is a key argument to find non-measurable sets (see Exercise 2.8.19).

Axiom of choice.[2] Let I be a set and let $(A_i)_{i \in I}$ be a family of non-empty sets. Then the Cartesian product $\prod_{i \in I} A_i$ is non-empty.

Zorn's lemma. Let $(X, \leq)$ be a non-empty inductive ordered set: if Y is a totally ordered subset of X, there exists $x \in X$ which is an upper bound for Y. Then there exists a maximal element in X.

Zermelo's theorem. On any set X, one can define an order relation $\leq$ which makes $(X, \leq)$ a well-ordered set.

Obviously the set $\mathbb{N}$ of the natural integers with the usual order is indeed well ordered, and this is the basis for the familiar *induction* reasoning; considering a sequence $(\mathscr{P}_n)_{n \in \mathbb{N}}$ of statements such that $\mathscr{P}_0$ is true and $\forall n \in \mathbb{N}, \mathscr{P}_n \implies \mathscr{P}_{n+1}$ we define

$$S = \{n \in \mathbb{N}, \mathscr{P}_n \text{ is not true}\}.$$

If S is not empty, then it has a smallest element s_0 and necessarily $s_0 > 0$ since $\mathscr{P}_0$ is true; as a consequence $s_0 - 1 \in S^c$, so that $\mathscr{P}_{s_0-1}$ is true, implying that $\mathscr{P}_{s_0}$ is true, contradicting $s_0 \in S$. As a result, S should be empty and $\mathscr{P}_n$ is true for all $n \in \mathbb{N}$. In some sense, Zorn's lemma could be used in a similar way to handle a non-countable family of statements satisfying properties analogous to those of the countable family mentioned above (see Theorem 10.1.19). Of course, it is not difficult to equip a countable set X with an order relation which makes it a well-ordered set: it suffices to use the bijection with a subset of $\mathbb{N}$. However, the set $\mathbb{Q}$ of rational numbers (which is countable), with the standard order is *not* a well-ordered set; consider for instance $T = \{x \in \mathbb{Q}_+, x^2 \geq 2\}$, a set which

---

[2]This Axiom has not much to do with choosing an element in a non-empty set: the real point at stake is the case where the set I is uncountable and it is in fact in that framework that it is used to build non-measurable sets.

is bounded from below without a smallest element (exercise). This means that to construct an order relation on $\mathbb{Q}$ which makes it a well-ordered set, one has to use a different order than the classical one and, for instance, one may use an explicit bijection between $\mathbb{Q}$ and $\mathbb{N}$. The real difficulties begin when you want to construct an order relation on $\mathbb{R}$ which makes it a well-ordered set; naturally, one cannot use the standard order, e.g., since $]0, 1]$ does not have a smallest element, although it has the greatest lower bound 0. So the construction of that order relation has no relationship with the standard order on the real line and is in fact a result of set theory, dealing with order relations on $\mathscr{P}(\mathbb{N})$, the set of subsets of $\mathbb{N}$.

Cardinals

A non-empty finite set is defined as a set X such that there exists $N \in \mathbb{N}^*$ and a bijection from $\{1, \ldots, N\}$ onto X. The empty set is also finite. If $N_1, N_2 \in \mathbb{N}^*$ are such that there exists a bijection from $\{1, \ldots, N_1\}$ onto $\{1, \ldots, N_2\}$, this implies $N_1 = N_2$. We can thus define the *Cardinal* (noted $\operatorname{card} X$) of a finite set X as its number of elements and $\operatorname{card} \emptyset = 0$.

Lemma 10.1.3. *Let X be a set. The following properties are equivalent.*

(i) *X is infinite, i.e., X is not a finite set.*

(ii) *There exists a proper subset Y of X and a bijection from X onto Y.*

(iii) *There exists an injection $\phi : X \longrightarrow X$ such that $\phi(X)$ is a proper subset of X.*

Proof. Let us assume that X is finite: then if Y is a proper subset of X, its cardinal is strictly smaller than $\operatorname{card} X$, and there cannot exist a bijection from X onto Y: this proves (ii)$\Longrightarrow$(i).

Let us assume now that X is infinite: then X cannot be empty. Let $x_1 \in X$ and let us assume that for every $N \in \mathbb{N}^*$, there exists a subset $\{x_1, \ldots, x_n\} \subset X$ with N elements: this is true for $n = 1$ and assuming this for some $n \geq 1$, the set $\{x_1, \ldots, x_n\}$ must be proper in X (otherwise X would be finite) and thus there exists $x_{n+1} \in X$ such that $\operatorname{card}\{x_1, \ldots, x_n, x_{n+1}\} = n + 1$. As a result, we find a subset $N \subset X$ such that there is a bijection ϕ from $\mathbb{N}$ onto N. We consider now the mapping $\Phi : X \longrightarrow N^c \cup \phi(2\mathbb{N})$ defined by

$$\begin{cases} \Phi(x) = x, & \text{if } x \in N^c, \\ \Phi(x) = \phi(2\phi^{-1}(x)), & \text{if } x \in N. \end{cases}$$

The mapping Φ is bijective with inverse mapping Ψ,

$$\begin{cases} \Psi(x) = x, & \text{if } x \in N^c, \\ \Psi(x) = \phi(\frac{1}{2}\phi^{-1}(x)), & \text{if } x \in \phi(2\mathbb{N}). \end{cases}$$

Indeed, we have for $x \in N^c$, $(\Phi \circ \Psi)(x) = x = (\Psi \circ \Phi)(x)$. For $x \in N$, we have

$$(\Psi \circ \Phi)(x) = \Psi\big(\underbrace{\phi(2\phi^{-1}(x))}_{\in \phi(2N)}\big) = \phi\Big(\frac{1}{2}\phi^{-1}\big(\phi(2\phi^{-1}(x))\big)\Big) = x,$$

and for $x \in \phi(2N)$, $x = \phi(2n)$,

$$(\Phi \circ \Psi)(x) = \Phi\big(\phi(\frac{1}{2}\phi^{-1}(x))\big) = \Phi(\underbrace{\phi(n)}_{\in N}) = \phi\Big(2\phi^{-1}\big(\phi(n)\big)\Big) = \phi(2n) = x.$$

Now the set $Y = N^c \cup \phi(2N)$ is a proper subset of $X = N^c \cup \overbrace{\phi(2N) \cup \phi(2N+1)}^{\text{partition of } N}$ and Φ is a bijection from X onto Y: this proves (i)$\Longrightarrow$(ii). Since the equivalence between (ii) and (iii) is obvious, the proof of Lemma 10.1.3 is complete. $\square$

Remark 10.1.4. We get immediately that a subset of a finite set is finite and a superset of an infinite set is infinite.

Definition 10.1.5. Let X, Y be two sets: they are said to be equipotent whenever there exists a bijective mapping $\phi : X \longrightarrow Y$.

Remark 10.1.6. We note that a set X is equipotent to itself and for X, Y, Z sets such that X is equipotent to Y and Y is equipotent to Z, we find that X is also equipotent to Z; also X equipotent to Y is equivalent to Y equipotent to X. We refrain to say that equipotence is an equivalence relation since there is not a set of all sets. When two sets X, Y are equipotent, we shall write symbolically $\operatorname{card} X = \operatorname{card} Y$, without defining each side of the equality (note that it nevertheless consistent with the case where X is finite).

Remark 10.1.7. We have defined on page 1 the notion of countable set: we have also proven there that a countable set is either finite or equipotent to N and that a countable union of countable sets is countable. A byproduct of the proof of Lemma 10.1.3 is that every infinite set contains a set equipotent to N. We shall note $\aleph_0 = \operatorname{card} N$ (see Exercise 1.9.3).

Theorem 10.1.8. *We have*

$$\operatorname{card}(N \times N) = \operatorname{card} N = \operatorname{card} \mathbb{Q}, \quad \operatorname{card} \mathbb{R} = \operatorname{card} \mathcal{P}(N).$$

The set of real numbers is not countable.

Proof. The first equality is proven in Exercise 1.9.1, the second on page 2, the third equality and the last assertion in Exercise 1.9.5. $\square$

Theorem 10.1.9 (Schröder–Bernstein Theorem). *Let X, Y be two sets and let $f : X \longrightarrow Y$, $g : Y \longrightarrow X$ be injective mappings. Then there exists a bijective mapping from X onto Y, i.e., $\operatorname{card} X = \operatorname{card} Y$.*

Proof. We set $A_0 = X \backslash g(Y)$, and for $n \geq 0$, $A_{n+1} = g(f(A_n))$. We define for $x \in X$,

$$\Phi(x) = \begin{cases} f(x), & \text{if } x \in \cup_{n \geq 0} A_n, \\ g^{-1}(x) & \text{otherwise,} \end{cases}$$

where g^{-1} is the inverse mapping of the bijection $g : Y \to g(Y)$. Note that it is consistent since if $x \notin \cup_{n \geq 0} A_n$, then $x \in g(Y)$. The mapping Φ is one-to-one: let us assume that $\Phi(x') = \Phi(x'')$. Then if $x', x'' \in \cup_{n \geq 0} A_n$, we find $f(x') = f(x'')$ and thus from the injectivity of f, we get $x' = x''$. If $x', x'' \notin \cup_{n \geq 0} A_n$, then we find $g^{-1}(x') = g^{-1}(x'')$ and since $g : Y \to g(Y)$ is bijective, we get $x' = x''$. Let us check the case $x' \in \cup_{n \geq 0} A_n, x'' \notin \cup_{n \geq 0} A_n$: we have then

$$f(x') = g^{-1}(x'') \Longrightarrow g(f(x')) = x'' \Longrightarrow x'' \in \cup_{n \geq 0} g(f(A_n)) = \cup_{n \geq 0} A_{n+1},$$

which contradicts the assumption on x'', proving injectivity for Φ. Let us show now that Φ is onto: let $y \in Y$. If y belongs to $\cup_{n \geq 0} f(A_n) = f(\cup_{n \geq 0} A_n)$, then $y = f(x) = \Phi(x)$ for some $x \in \cup_{n \geq 0} A_n$. If $y \notin f(\cup_{n \geq 0} A_n)$, then $x = g(y) \notin \cup_{n \geq 0} A_n$: otherwise

$$y = g^{-1}(g(y)) \in \cup_{n \geq 0} g^{-1}(A_n \cap g(Y)) = \cup_{n \geq 0} g^{-1}(A_{n+1} \cap g(Y))$$
$$= \cup_{n \geq 0} g^{-1}\left(g(f(A_n)) \cap g(Y)\right) \underbrace{=}_{\substack{\text{injectivity} \\ \text{of } g}} \cup_{n \geq 0} f(A_n) = f(\cup_{n \geq 0} A_n),$$

contradicting the assumption on y. As a result, we have indeed $x = g(y) \notin \cup_{n \geq 0} A_n$ and $y = g^{-1}(x) = \Phi(x)$, which ends the proof. $\qquad\square$

Definition 10.1.10. Let X, Y be two sets. We shall say that $\operatorname{card} X \leq \operatorname{card} Y$ if X is equipotent to a subset of Y, i.e., if there exists an injection from X into Y.

Remark 10.1.11. It follows from the Schröder–Bernstein theorem that $\operatorname{card} X \leq \operatorname{card} Y$ and $\operatorname{card} Y \leq \operatorname{card} X$ imply $\operatorname{card} X = \operatorname{card} Y$. On the other hand, it is obvious that $\operatorname{card} X \leq \operatorname{card} X$ and for Z a third set,

$$\operatorname{card} X \leq \operatorname{card} Y \text{ and } \operatorname{card} Y \leq \operatorname{card} Z \Longrightarrow \operatorname{card} X \leq \operatorname{card} Z.$$

Again we shall refrain from claiming that we have found an order relation on cardinals, since in the first place we have not defined a cardinal and next, because there is not a set of all sets. We shall say that $\operatorname{card} X < \operatorname{card} Y$ whenever $\operatorname{card} X \leq \operatorname{card} Y$ and X is not equipotent to Y.

Theorem 10.1.12 (Cantor's Theorem). *Let X be a set and $\mathcal{P}(X)$ its powerset. Then* $\operatorname{card} X < \operatorname{card} \mathcal{P}(X)$.

Proof. Cf. Exercise 1.9.2. $\qquad\square$

Let $(A_i)_{i \in I}$ be a family of sets. The *disjoint union* of this family is

$$\bigsqcup_{i \in I} A_i = \bigcup_{i \in I} (A_i \times \{i\}). \tag{10.1.1}$$

We note that for $i' \neq i''$ in I, we have $(A_{i'} \times \{i'\}) \cap (A_{i''} \times \{i''\}) = \emptyset$. Let $(A_i)_{i \in I}$ be a family of sets and let $(B_i)_{i \in I}$ be a family of sets such that for each $i \in I$, A_i is equipotent to B_i. This implies obviously that $\bigsqcup_{i \in I} A_i$ is equipotent to $\bigsqcup_{i \in I} B_i$, so that we may define

$$\sum_{i \in I} \operatorname{card} A_i = \operatorname{card}\left(\bigsqcup_{i \in I} A_i\right). \tag{10.1.2}$$

Also the Cartesian product $\prod_{i \in I} A_i$ is equipotent to $\prod_{i \in I} B_i$ so that we may define as well

$$\prod_{i \in I} \operatorname{card} A_i = \operatorname{card}\left(\prod_{i \in I} A_i\right). \tag{10.1.3}$$

In particular sums and products of cardinals are commutative and associative. We have seen above $\aleph_0 + \aleph_0 = \aleph_0, \aleph_0^2 = \aleph_0$.

Let X, Y be two sets and let Y^X be the set of all mappings from X into Y: this notation is justified by the fact that a mapping ϕ from X into Y is $(\phi(x))_{x \in X}$ where each $\phi(x)$ belongs to Y. Then if X' is equipotent to X and Y' is equipotent to Y, we obtain obviously the equipotence of Y^X with $Y'^{X'}$ so that we may define

$$(\operatorname{card} Y)^{\operatorname{card} X} = \operatorname{card}(Y^X). \tag{10.1.4}$$

For instance, we have proven in Exercises 1.9.3, 1.9.5,

$$2^{\operatorname{card} X} = \operatorname{card}(\mathcal{P}(X)), \quad \mathfrak{c} = \operatorname{card} \mathbb{R} = 2^{\aleph_0}, \quad \aleph_0 < \mathfrak{c}. \tag{10.1.5}$$

Lemma 10.1.13. *Let X be a set and let $\{\omega\}$ be a singleton. Then the set X is infinite if and only if the disjoint union $X \sqcup \{\omega\}$ is equipotent to X. In other words, a cardinal number x is infinite if and only if $x = x + 1$.*

Proof. Let X be a finite set: then $\operatorname{card} X < 1 + \operatorname{card} X$. Let X be an infinite set: then X contains a set equipotent to $\mathbb{N}$, we may assume that it contains $\mathbb{N}$. We have then

$$X = \mathbb{N} \sqcup (X \backslash \mathbb{N}) \quad \text{equipotent to} \quad \underbrace{\{\omega\} \sqcup \mathbb{N}}_{\text{equipotent to } \mathbb{N}} \sqcup (X \backslash \mathbb{N}) = X \sqcup \{\omega\},$$

proving $\operatorname{card} X = 1 + \operatorname{card} X$. $\qquad\qquad\qquad\qquad\qquad\qquad\qquad\qquad\qquad\quad\square$

Remark 10.1.14. Let X, Y, Z be three sets. Then $X \times (Y \sqcup Z)$ is equipotent to $(X \times Y) \sqcup (X \times Z)$ so that with $x = \operatorname{card} X, y = \operatorname{card} Y, z = \operatorname{card} Z$,

$$x(y + z) = xy + xz.$$

Note also that $\emptyset \times X = \emptyset$, i.e., $0x = 0$.

Lemma 10.1.15. *Let X, Y, Z be three sets.*

(1) *The set $Z^{X \sqcup Y}$ is equipotent to $Z^X \times Z^Y$, so that, with $x = \operatorname{card} X, y = \operatorname{card} Y, z = \operatorname{card} Z$, $z^{x+y} = z^x z^y$.*

(2) *The set $(Z^Y)^X$ is equipotent to $Z^{Y \times X}$, i.e., $(z^y)^x = z^{yx}$.*

(3) *The set $(X \times Y)^Z$ is equipotent to the set $X^Z \times Y^Z$, i.e., $(xy)^z = x^z y^z$.*

Proof. We consider the mappings

$$\Psi: \quad \begin{array}{ccc} Z^{X \sqcup Y} & \longrightarrow & Z^X \times Z^Y \\ \phi & \mapsto & (\phi_{|X}, \phi_{|Y})' \end{array} \qquad \Gamma: \quad \begin{array}{ccc} Z^X \times Z^Y & \longrightarrow & Z^{X \sqcup Y} \\ (f, g) & \mapsto & \Gamma(f, g) \end{array}$$

where, considering X, Y as disjoint sets with union $X \sqcup Y$,

$$\text{for } x \in X, \quad \Gamma(f, g)(x) = f(x), \quad \text{for } y \in Y, \quad \Gamma(f, g)(y) = g(y).$$

We have $\Gamma \circ \Psi = \operatorname{Id}_{Z^{X \sqcup Y}}$ and $\Psi \circ \Gamma = \operatorname{Id}_{Z^X \times Z^Y}$: for

$$\phi: X \sqcup Y \to Z,$$

we have $(\Gamma \circ \Psi)(\phi) = \Gamma\big((\phi_{|X}, \phi_{|Y})\big)$ so that for $x \in X$, $(\Gamma \circ \Psi)(\phi)(x) = \phi(x)$, for $y \in Y$, $(\Gamma \circ \Psi)(\phi)(y) = \phi(y)$, i.e., $(\Gamma \circ \Psi)(\phi) = \phi$. Also for $f: X \to Z, g: Y \to Z$, we have

$$(\Psi \circ \Gamma)(f, g) = \Psi\big(\Gamma(f, g)\big) = \Psi\big(f \sqcup g\big), \quad \text{with } f \sqcup g: X \sqcup Y \to Z$$

defined by $(f \sqcup g)(x) = f(x)$ for $x \in X$, $(f \sqcup g)(y) = g(y)$ for $y \in Y$ and thus

$$(\Psi \circ \Gamma)(f, g) = \big((f \sqcup g)_{|X}, (f \sqcup g)_{|Y}\big) = (f, g),$$

proving (1).

We consider the mappings

$$\Omega: \quad \begin{array}{ccc} Z^{Y \times X} & \longrightarrow & (Z^Y)^X \\ \phi & \mapsto & (X \ni x \mapsto \phi(\cdot, x))' \end{array} \qquad \Theta: \quad \begin{array}{ccc} (Z^Y)^X & \longrightarrow & Z^{Y \times X} \\ f & \mapsto & \Theta(f), \end{array}$$

with $\Theta(f)(y, x) = f(x)(y)$. We have

$$(\Theta \circ \Omega)(\phi)(y, x) = \Omega(\phi)(x)(y) = \phi(y, x), \quad (\Omega \circ \Theta)(f)(x)(y) = \Theta(f)(y, x) = f(x)(y),$$

proving (2).

We consider the mappings

$$\Xi: \quad \begin{array}{ccc} (X \times Y)^Z & \longrightarrow & X^Z \times Y^Z \\ \phi & \mapsto & \Xi(\phi) \end{array}' \qquad \Lambda: \quad \begin{array}{ccc} X^Z \times Y^Z & \longrightarrow & (X \times Y)^Z \\ (f, g) & \mapsto & \Lambda(f, g), \end{array}$$

with $\Xi(\phi) = \big(z \mapsto \Pi_X \phi(z), z \mapsto \Pi_Y \phi(z)\big)$, $\Lambda(f, g)(z) = (f(z), g(z))$. We have

$$(\Lambda \circ \Xi)(\phi)(z) = (\Pi_X \phi(z), \Pi_Y \phi(z)) = \phi(z),$$

$$(\Xi \circ \Lambda)(f, g) = \big(z \mapsto \Pi_X \Lambda(f, g)(z), z \mapsto \Pi_Y \Lambda(f, g)(z)\big)$$
$$= (z \mapsto f(z), z \mapsto g(z)) = (f, g),$$

proving (3). $\qquad \square$

Remark 10.1.16. We note also that $\aleph_0^2 = \aleph_0$ ($\mathbb{N}^2$ is equipotent to $\mathbb{N}$) and $2\aleph_0 = \aleph_0$ and applying the previous lemma

$$\mathfrak{c} \leq \mathfrak{c} + \aleph_0 \leq 2\mathfrak{c} = 2 2^{\aleph_0} = 2^{\aleph_0+1} = 2^{\aleph_0} = \mathfrak{c} \implies \mathfrak{c} = \mathfrak{c} + \aleph_0 = 2\mathfrak{c}.$$

Moreover, we have

$$\mathfrak{c}^2 = \mathfrak{c}, \tag{10.1.6}$$

since $\mathfrak{c}^2 = 2^{\aleph_0} 2^{\aleph_0} = 2^{2\aleph_0} = 2^{\aleph_0} = \mathfrak{c}$. We note also that

$$\mathfrak{c}^{\mathfrak{c}} = (2^{\aleph_0})^{\mathfrak{c}} = 2^{\aleph_0 \mathfrak{c}} = 2^{\mathfrak{c}} > \mathfrak{c},$$

since $\mathfrak{c} \leq \aleph_0 \mathfrak{c} \leq \mathfrak{c}^2 = \mathfrak{c}$ gives $\aleph_0 \mathfrak{c} = \mathfrak{c}$. We have proven

$$\mathrm{card}(\mathbb{R}^{\mathbb{R}}) = \mathrm{card}\big(\mathcal{P}(\mathbb{R})\big) > \mathrm{card}\,\mathbb{R}. \tag{10.1.7}$$

On the other hand, considering $C(\mathbb{R};\mathbb{R})$ (set of real-valued continuous functions), we see that each $\phi \in C(\mathbb{R};\mathbb{R})$ is determined by its restriction to $\mathbb{Q}$, so that

$$\mathrm{card}\big(C(\mathbb{R};\mathbb{R})\big) \leq \mathrm{card}(\mathbb{R}^{\mathbb{Q}}) = \mathfrak{c}^{\aleph_0} = 2^{\aleph_0^2} = 2^{\aleph_0} = \mathfrak{c}.$$

On the other hand, $C(\mathbb{R};\mathbb{R})$ contains the constant functions whose cardinality is $\mathfrak{c}$. We have proven that

$$\mathrm{card}\big(C(\mathbb{R};\mathbb{R})\big) = \mathrm{card}\,\mathbb{R}. \tag{10.1.8}$$

The continuum hypothesis (CH) asserts that there is no subset of the real line which is not countable and not equipotent to $\mathbb{R}$, i.e., there is no cardinal number x such that $\aleph_0 < x < \mathfrak{c}$. Since $\mathfrak{c} = 2^{\aleph_0}$ this statement has a natural generalization. *The general continuum hypothesis* (GCH) asserts that for any non-finite cardinal $\mathfrak{a}$ there is no cardinal number x such that $\mathfrak{a} < x < 2^{\mathfrak{a}}$. The CH problem was stated in 1900 by David HILBERT (1862–1943) as the first one in his list of 23 important mathematical questions.

In 1940, Kurt GÖDEL (1906–1978) proved that (CH) cannot be disproved from the standard axioms of set theory (Zermelo–Fraenkel set theory: ZF), even adding the axiom of choice (C). In other words there is no proof of the negation of CH in ZFC. Paul COHEN (1934–2007) showed in 1963 that (CH) cannot be proven in ZFC. Both results assume that ZFC is non-contradictory.

Let us give a couple of examples of applications of Zorn's lemma 10.1 to Set Theory.

Lemma 10.1.17. *Let X, Y be two sets. Then* $\mathrm{card}\,X \leq \mathrm{card}\,Y$ *or* $\mathrm{card}\,Y \leq \mathrm{card}\,X$.

Proof. Let us consider the set $\mathscr{I} = \{(X_i, \phi_i)\}_{i \in I}$ where $\phi_i : X_i \to Y$ is injective and $X_i \subset X$. It is a non-empty set since the mapping $\phi : \emptyset \to Y$ with graph $\emptyset \times \emptyset$ is injective. We equip it with the order relation

$$(X_1, \phi_1) \leq (X_2, \phi_2) \quad \text{means} \quad X_1 \subset X_2 \text{ and } \phi_{2|X_1} = \phi_1.$$

If $\{(X_i, \phi_i)\}_{i \in J}$ is a totally ordered subset of $\mathscr{I}$, we consider $A = \cup_{i \in J} X_i$ and ϕ defined on X_i by ϕ_i: if x belongs to $X_{i'}, X_{i''}$, we have $X_{i'} \subset X_{i''}$ or $X_{i''} \subset X_{i'}$ and in both cases $\phi_{i'}(x) = \phi_{i''}(x)$, proving the consistency of the definition of ϕ, as well as its injectivity. According to Zorn's lemma, the set $\mathscr{I}$ must contain a maximal element $(\tilde{X}, \phi)$. If $\tilde{X} = X$, we have found an injection from X into Y.

If there is some $x_0 \in \tilde{X}^c$, then we claim that $\phi : \tilde{X} \to Y$ is bijective: we need only to prove that it is onto. If ϕ were not onto, we could find some $y_0 \in Y$, such that $\phi : \tilde{X} \to Y \backslash \{y_0\}$ and thus the extension of ϕ to $\tilde{X} \cup \{x_0\}$ defined by $\phi(x_0) = y_0$ would be an injection from $\tilde{X} \cup \{x_0\}$ into Y, contradicting the maximality of $(\tilde{X}, \phi)$. Thus, we have found an injection of Y into X, completing the proof of the lemma. $\qquad\square$

Lemma 10.1.18. *Let X, Y be two sets, $Y \neq \emptyset$. The inequality $\operatorname{card} X \geq \operatorname{card} Y$ is equivalent to the existence of a surjective map $p : X \to Y$.*

Proof. Let us assume that such a surjective map exists. Then the set

$$\prod_{y \in Y} p^{-1}(\{y\})$$

is the product of non-empty sets so that thanks to the Axiom of Choice 10.1, the product is non-empty: $\forall y \in Y, \exists s(y) \in X$ such that $p(s(y)) = y$. The mapping $s : Y \to X$ is injective since $p \circ s$ is injective. Conversely if $\operatorname{card} Y \leq \operatorname{card} X$, from Definition 10.1.10, we can find an injection from Y onto X, i.e., a subset Z of X which is equipotent to Y ($\psi : Z \to Y$ bijective): then we have $X = Z \sqcup (X \backslash Z)$ and we can define with $y_0 \in Y$ (assumed to be non-empty), for $x \in X$,

$$p(x) = \begin{cases} \psi(x) & \text{if } x \in Z, \\ y_0 & \text{if } x \notin Z. \end{cases}$$

Since ψ is onto, p is onto. $\qquad\square$

Theorem 10.1.19 (Principle of Transfinite Induction). *Let $(X, \leq)$ be a well-ordered set and let us assume that for each $x \in X$, $P(x)$ is a statement. We assume that for all $x \in X$,*
$$P(y) \text{ holds for all } y < x \implies P(x).$$
Then $P(x)$ is true for all $x \in X$.

Proof. Let $S = \{x \in X, P(x) \text{ does not hold}\}$. If S is not empty, it has a smallest element a. Now for all $x < a$, $P(x)$ holds true and thus $P(a)$ holds true, contradicting $a \in S$. Consequently, S is empty.

N.B. Note that the assumption implies that $P(x_0)$ holds true for the smallest element x_0 of X. $\qquad\square$

Theorem 10.1.20. *Let X be an infinite set. Then $X \times X$ is equipotent to X: for any infinite cardinal x, $x^2 = x$.*

Proof. We have seen in the proof of Lemma 10.1.3 that X contains a set X_0 equipotent to $\mathbb{N}$ and thus we can find a bijection $\psi_0 : X_0 \to X_0 \times X_0$. We consider now

$$\mathscr{F} = \{\psi : Y \to Y \times Y, \text{ bijective}, Y \text{ infinite} \subset X\}.$$

The family $\mathscr{F}$ is non-empty and ordered by $(\psi_1, Y_1) \leq (\psi_2, Y_2)$ meaning

$$Y_1 \subset Y_2, \quad \psi_{2|Y_1} = \psi_1.$$

The family $\mathscr{F}$ is inductive for that order: let $(\psi_i, Y_i)_{i \in I}$ be a totally ordered subset of $\mathscr{F}$. Setting $Y = \cup_{i \in I} Y_i$, we define for $y \in Y$, $\psi(y) = \psi_i(y)$ if $y \in Y_i$: note that this definition is consistent since if $y \in Y_i \cap Y_j$, then we have $Y_i \subset Y_j$ (or $Y_i \supset Y_j$) and the restriction of ψ_j to Y_i equals ψ_j (or the same property exchanging i with j). The mapping $\psi : Y \to Y \times Y$ is injective since for $y', y'' \in Y$, we find $i \in I$ such that $y', y'' \in Y_i$ and thus $\psi(y') = \psi(y'')$ means $\psi_i(y') = \psi_i(y'')$ implying $y' = y''$. It is also onto since for $(y', y'') \in Y \times Y$, we find $i \in I$ such that $y', y'' \in Y_i$ and thus there exists $y \in Y_i$ such that $\psi(y) = \psi_i(y) = (y', y'')$.

Applying Zorn's Lemma 10.1, we find a maximal element (ψ, Y) in $\mathscr{F}$. We have in particular $\mathfrak{a} = \mathfrak{a}^2$ with $\mathfrak{a} = \operatorname{card} Y$ and $\mathfrak{a}$ is an infinite cardinal. If $\mathfrak{a} = \operatorname{card} X$, we are done. If $\mathfrak{a} < \operatorname{card} X$, we find that $\operatorname{card}(Y^c) > \mathfrak{a}$ (otherwise $\operatorname{card} X = \operatorname{card} Y + \operatorname{card} Y^c \leq 2\mathfrak{a} \leq \mathfrak{a}^2 = \mathfrak{a}$, contradicting the assumption). As a consequence we may find a subset Z of Y^c equipotent to Y. We note that $\mathfrak{a} \leq 2\mathfrak{a}^2 = 2\mathfrak{a} \leq \mathfrak{a}^2 = \mathfrak{a}$ and we consider

$$(Y \cup Z) \times (Y \cup Z) = (Y \times Y) \cup \underbrace{(Y \times Z) \cup (Z \times Y) \cup (Z \times Z)}_{\text{with cardinal } 3\mathfrak{a}^2 = \mathfrak{a}},$$

so that, using a bijective map $\theta : Z \to (Y \times Z) \cup (Z \times Y) \cup (Z \times Z)$ we may define for $x \in Y \cup Z$,

$$\tilde{\psi}(x) = \begin{cases} \psi(x), & \text{if } x \in Y, \\ \theta(x), & \text{if } x \in Z. \end{cases}$$

The mapping $\tilde{\psi}$ is bijective from $Y \cup Z$ onto $(Y \cup Z)^2$ and extends ψ, contradicting the maximality property. The proof of Theorem 10.1.20 is complete. $\qquad\square$

Kurt Gödel proved in 1938 that the Axiom of Choice is consistent with (ZF), i.e., that, if (ZF) is consistent[3], then (ZFC) is also consistent. Paul Cohen proved in 1963 that the Axiom of choice is independent of (ZF), i.e., is not a consequence of the axioms of (ZF).

---

[3] A consistent theory is a theory that does not contain a contradiction, i.e., does not contain a proof of a statement S and a proof of its negation not S.

Ordinals

Introduction

We have seen in Definition 10.1.1(3) the notion of *well-ordered set*. Let us give a couple of examples. Of course, $\mathbb{N}$ equipped with the usual order is a well-ordered set as well as any finite ordered set. $\mathbb{Z}$ with the standard order is not a well-ordered set since it does not have a smallest element, neither is $(0, 1]$ with the order induced by $\mathbb{R}$ (no smallest element). Let us now consider

$$E = \left\{1 - \frac{1}{n}\right\}_{n \geq 1} \cup \mathbb{N}^*, \qquad (10.1.9)$$

with the order induced by the standard order on $\mathbb{Q}$. Then, although E is equipotent to $\mathbb{N}$, it is *not* isomorphic to $\mathbb{N}$ as an ordered set: before giving a proof of this, let us give a definition.

Definition 10.1.21. Let $(X, \leqslant_X), (Y, \leqslant_Y)$ be well-ordered sets. These two ordered sets are said to be isomorphic if there exists a bijective mapping $\phi : X \to Y$ that is increasing, i.e., such that $x_1 \leqslant_X x_2$ in X implies $\phi(x_1) \leqslant_Y \phi(x_2)$ in Y. Note that a mapping ϕ as above is such that ϕ^{-1} is also increasing[4]. We shall say then that

$$\mathrm{ord}\, X = \mathrm{ord}\, Y \quad (\text{the } ordinal \text{ of } X \text{ equals the ordinal of } Y), \qquad (10.1.10)$$

or that the ordered sets X, Y are *order-isomorphic*.

N.B. As for the notion of cardinal, note that we have not defined the ordinal of a well-ordered set, but only the equality between ordinals, meaning isomorphism in the natural sense for ordered sets.

Lemma 10.1.22. *Let $(A, \leqslant_A), (B, \leqslant_B)$ be two disjoint well-ordered sets. We define $X = A \cup B$ and the following relation on X:*

$$x_1 \leqslant_X x_2 \quad means \quad \begin{cases} either & x_1, x_2 \in A, \ x_1 \leqslant_A x_2 \\ or & x_1, x_2 \in B, \ x_1 \leqslant_B x_2 \\ or & (x_1, x_2) \in A \times B. \end{cases}$$

Then $(X, \leqslant_X)$ is a well-ordered set

Proof. Let us check first that $\leqslant_X$ is indeed an order relation on X: it is obviously reflexive and if $x_1 \leqslant_X x_2, x_2 \leqslant_X x_1$, either both x_1, x_2 belong to A or both belong to B and then are equal; the third case $(x_1, x_2) \in A \times B$ cannot occur since we would also have $(x_2, x_1) \in A \times B$, so that $x_2 \in A \cap B = \emptyset$. The relation is indeed antisymmetric. Let us now assume that $x_1 \leqslant_X x_2, x_2 \leqslant_X x_3$: if x_1, x_2 are both in

---

[4]Take $\phi(x_1) = y_1 \leq y_2 = \phi(x_2)$ in Y, then $x_1 \leq x_2$ otherwise $x_2 < x_1$ and $\phi(x_2) < \phi(x_1)$, contradicting the assumption.

A, then either $x_3 \in A$ and the transitivity follows from the transitivity of $\leqslant_A$ or $x_3 \in B$ and $x_1 \leqslant_x x_3$. If x_1, x_2 are both in B, then x_3 must belong to B so that $x_1 \leq x_3$. Moreover, if $x_1 \in A, x_2 \in B$, then x_3 must belong to B so that $x_1 \leqslant_x x_3$, concluding the proof of transitivity.

Let C be a non-empty subset of X: if $C \subset B$, then $\min_X C = \min_B C$. If $C \cap A \neq \emptyset$, then

$$\min_X C = \min_A (C \cap A) = c,$$

since $c \in C$ and if $x \in C$, then either $x \in B$ and $c \leqslant_x x$ or $x \in A$ and $c \leqslant_A x$ so that $c \leqslant_x x$. The proof of the lemma is complete. $\qquad\square$

Remark 10.1.23. This implies that E defined by (10.1.9) with the order induced by the order of $\mathbb{R}$ is well ordered. Also we can see that there is no bijective increasing mapping from $\mathbb{N}$ onto E. If such a mapping existed, we would have $\phi(n) = 1 - \frac{1}{n+1}$ for all $n \in \mathbb{N}$: it is true for $n = 0$ since $\phi(0)$ should be the minimum of E. Assuming that it is true up to some $N \geq 0$, we see that the minimum of $\phi(\{0, \dots, N\}^c)$ should be $\phi(N + 1)$ and also

$$\min\big(\phi(\{0, \dots, N\})\big)^c = 1 - \frac{1}{N + 2},$$

so that $\phi(N + 1) = 1 - \frac{1}{N+2}$, which was to be proven. As a result $\phi(\mathbb{N}) \cap \mathbb{N}^* = \emptyset$ and ϕ cannot be onto.

Definition 10.1.24. Let $(X, \leqslant_x)$ be a well-ordered set. A subset S of X is said to be a *segment* of X if $s \in S, x \in X, x \leqslant_x s \Longrightarrow x \in S$.

Obviously X itself, the empty set, any set

$$S_a = (\rightarrow, a) = \{x \in X, x < a\}, \quad a \in X, \qquad (10.1.11)$$

are segments of X: for the latter, $s < a, x \leq s$ imply $x < a$. Moreover if $a, b \in X$, $S_a = S_b$ implies $a = b$: otherwise $a < b$ (resp. $b < a$) and $a \in S_b = S_a$ (resp. $b \in S_a = S_b$), which is impossible.

Proposition 10.1.25. *Let $(X, \leqslant_x)$ be a well-ordered set. Any union or intersection of segments of X is again a segment of X. A segment of a segment of X is a segment of X. For each proper segment S of X (a segment $\neq X$), there exists $a \in X$ with $S = S_a$.*

Proof. We start by the proof of the third statement. If S is a proper segment of X, S^c is not empty so that we may define $a = \min S^c$. We have $S \subset S_a$: if $x \in S$ and $x \geq a$, then by the segment property, we must have $a \in S$, which is impossible since $a \in S^c$. Also we have $S_a \subset S$: if $x < a$ then $x \notin S^c$ by the minimum property of a, i.e., $x \in S$.

For the first statement, let us consider a family $(S_i)_{i \in I}$ of segments of X. If $I = \emptyset$, then $\cup_I S_i = \emptyset$ is a segment. If $I \neq \emptyset$,

$$s \in \cup_I S_i, x \leq s \Longrightarrow \exists j \in I, s \in S_j, x \leq s \Longrightarrow x \in S_j \subset \cup_I S_i.$$

Moreover to check that $\cap_I S_i$ is a segment, we may assume that $I \neq \emptyset$ (otherwise $\cap_I S_i = X$), and

$$s \in \cap_I S_i, x \leq s \Longrightarrow \forall i \in I, s \in S_i, x \leq s \Longrightarrow \forall i \in I, x \in S_i.$$

For the second statement we consider a segment Σ of a segment of X, which is either X or $(\to, a)$; the first case is trivial, and if Σ is a segment of $(\to, a)$, we find either $\Sigma = (\to, a)$ or for some $b < a$, $\Sigma = \{x \in X, x < a \text{ and } x < b\} = (\to, b)$. $\square$

Ordering of ordinals

Definition 10.1.26. Let $(X, \leqslant_X), (Y, \leqslant_Y)$ be two well-ordered sets. We shall say that

$$\operatorname{ord} X \preccurlyeq \operatorname{ord} Y$$

if X is order-isomorphic to a segment of Y. When $\operatorname{ord} X \preccurlyeq \operatorname{ord} Y$ and X is not order-isomorphic to Y, we shall write $\operatorname{ord} X \prec \operatorname{ord} Y$.

Lemma 10.1.27. *Let $(X, \leqslant_X)$ be a well-ordered set. The only order-isomorphism of X onto a segment of X is the identity of X.*

Proof. Let us assume that $\phi : X \longrightarrow (\to, a)$ is an order-isomorphism of X onto a proper segment of X ($a \in X$, see Proposition 10.1.25). We define

$$(\flat) \qquad\qquad A = \{x \in X, \phi(x) < x\},$$

and we note that $a \in A$ so that we can define $b = \min A$. We have

$$(\sharp) \qquad\qquad \phi(b) < b \Longrightarrow \phi(\phi(b)) < \phi(b) \Longrightarrow \phi(b) \in A,$$

contradicting the fact that b is the smallest element of A. We have proven that $\phi : X \to X$ is an order-isomorphism.

We want now to prove that ϕ is the identity. The set A defined in $(\flat)$ must be empty, otherwise as above its smallest element b satisfies $(\sharp)$, leading to a contradiction. As a result, we have for all $x \in X$, $x \leq \phi(x)$ and applying this result to ϕ^{-1}, we find

$$\forall x \in X, \quad x \leq \phi(x) \leq \phi^{-1}(\phi(x)) = x, \quad \text{i.e., } \phi = \operatorname{Id}. \qquad \square$$

Proposition 10.1.28. *Let $(X, \leqslant_X), (Y, \leqslant_Y)$ be well-ordered sets. Then*

$$\operatorname{ord} X \preccurlyeq \operatorname{ord} Y \text{ and } \operatorname{ord} Y \preccurlyeq \operatorname{ord} X \quad \Longrightarrow \quad \operatorname{ord} X = \operatorname{ord} Y.$$

Proof. Let $\phi : X \to T$ be an order-isomorphism of X onto a segment T of Y and let $\psi : Y \to S$ be an order-isomorphism of Y onto a segment S of X. Then

$$X \ni x \mapsto \psi(\phi(x)) \in (\psi \circ \phi)(X) = \psi(T)$$

is an order-isomorphism and $\psi(T)$ is a segment of S, thus from Proposition 10.1.25 is also a segment of X. Applying Lemma 10.1.27 shows that $\psi(T) = X$ so that $S = X$ and $\operatorname{ord} X = \operatorname{ord} Y$. $\square$

Proposition 10.1.29. *Let* $(X, \leqslant_X), (Y, \leqslant_Y)$ *be two well-ordered sets. Then either* $\operatorname{ord} X \preccurlyeq \operatorname{ord} Y$ *or* $\operatorname{ord} Y \preccurlyeq \operatorname{ord} X$.

Proof. We define

$$M = \{x \in X, \exists y \in Y, \ \operatorname{ord}(\to, x) = \operatorname{ord}(\to, y)\}.$$

We note that for each $x \in M$, there exists a unique $y \in Y$ such that $\operatorname{ord}(\to, x) = \operatorname{ord}(\to, y)$: if we have for $y_1, y_2 \in Y$ (say with $y_2 \leqslant_Y y_1$)

$$\operatorname{ord}(\to, y_1) = \operatorname{ord}(\to, y_2)$$

then $(\to, y_1)$ is order-isomorphic to its segment $(\to, y_2)$ and Lemma 10.1.27 implies $y_2 = y_1$. We have thus a mapping $\phi : M \to Y$ defined by

$$\operatorname{ord}(\to, x) = \operatorname{ord}(\to, \phi(x)).$$

Note that ϕ is injective since if $\phi(x_1) = \phi(x_2)$, say with $x_2 \leqslant_X x_1$, we find that $(\to, x_1)$ is isomorphic to its segment $(\to, x_2)$, so that Lemma 10.1.27 implies $x_2 = x_1$. Moreover ϕ is increasing since if $x_2 \leqslant_X x_1$, we must have $\phi(x_2) \leqslant_Y \phi(x_1)$, otherwise $\phi(x_1) <_Y \phi(x_2)$ with

$$\operatorname{ord}(\to, \phi(x_2)) = \operatorname{ord}(\to, x_2), \quad \operatorname{ord}(\to, x_1) = \operatorname{ord}(\to, \phi(x_1)),$$

so that $(\to, \phi(x_2))$ is isomorphic to a segment of $(\to, \phi(x_1))$ which is a proper segment of $(\to, \phi(x_2))$: this is not possible, thanks to Lemma 10.1.27. We find also that $\phi(M) = N$ is a segment of Y: let

$$t = \phi(s), s \in M, \quad \text{i.e., } \operatorname{ord}(\to, s) = \operatorname{ord}(\to, \phi(s)),$$

and let $y \leqslant_Y t = \phi(s)$. Using the isomorphism between $(\to, s)$ and $(\to, \phi(s))$, we find an isomorphism between $(\to, y)$ and $(\to, x)$ for some $x \leqslant_X s$, proving $y = \phi(x)$. This implies as well that M is a segment of X.

Suppose now that X is not isomorphic to a segment of Y: then $X \backslash M$ is not empty (otherwise, we would have an isomorphism $\phi : X \to N$ of X onto a segment of Y). If Y is not isomorphic to a segment of X, then $Y \backslash N$ is not empty (otherwise, we would have an isomorphism $\phi : M \to Y$ of a segment of X onto Y). Assuming that neither X is isomorphic to a segment of Y, nor Y is isomorphic to a segment of X, both $X \backslash M, Y \backslash N$ are non-empty. We define

$$a = \min(X \backslash M), \quad b = \min(Y \backslash N).$$

Then $(\to, a)$ is isomorphic to M and $(\to, b)$ is isomorphic to N (see Proposition 10.1.25), and since N is isomorphic to M, this implies $a \in M$, contradicting the assumption. The proof is complete. $\qquad\qquad\square$

Remark 10.1.30. Let $(X, \leqslant_X), (Y, \leqslant_Y), (Z, \leqslant_Z)$ be three well-ordered sets. Then

$$\operatorname{ord} X \preccurlyeq \operatorname{ord} Y \text{ and } \operatorname{ord} Y \preccurlyeq \operatorname{ord} Z \Longrightarrow \operatorname{ord} X \preccurlyeq \operatorname{ord} Z.$$

In fact if $\phi : X \to S$ is an isomorphism onto a segment S of Y and $\psi : Y \to T$ is an isomorphism onto a segment T of Z, we find that

$$X \ni x \mapsto \psi(\phi(x)) \in (\psi \circ \phi)(X) = \psi(S)$$

is an isomorphism onto a segment of T, which is also a segment of Z, thanks to Proposition 10.1.25.

Addition of ordinals

Let $(A, \leqslant_A), (B, \leqslant_B)$ be two well-ordered sets. We shall denote by

$$A \sqcup_+ B \tag{10.1.12}$$

the well-ordered set defined in Lemma 10.1.22 on the disjoint union $A \sqcup B$. According to the discussion on Example (10.1.9) in Remark 10.1.23, we have proven that

$$\operatorname{ord} \mathbb{N} \prec \operatorname{ord}(\mathbb{N} \sqcup_+ \mathbb{N}). \tag{10.1.13}$$

Moreover, replacing A by an order-isomorphic A' and B by an order-isomorphic B' provides $A' \sqcup_+ B'$ order-isomorphic to $A \sqcup_+ B$, so that we can give the following definition.

Definition 10.1.31. Let $(A, \leqslant_A), (B, \leqslant_B)$ be two well-ordered sets. We define the addition of ordinals,

$$\operatorname{ord} A \oplus_+ \operatorname{ord} B = \operatorname{ord}(A \sqcup_+ B).$$

Our notation emphasizes the fact that this addition is **not** commutative.

Lemma 10.1.32. *Denoting by* ω *the ordinal of* $\mathbb{N}$ *and by* k *the ordinal of a finite set with* k *elements, we have*

(1) $\omega \prec \omega \oplus_+ \omega$,
(2) $\omega = k \oplus_+ \omega$,
(3) $\omega \prec \omega \oplus_+ k$, *if* $k \geq 1$.
(4) *If* α *is an ordinal,* $\alpha \prec \alpha \oplus_+ 1$.

Proof. We prove (2): we have

$$k \oplus_+ \omega = \operatorname{ord}\left(\{1, \dots, k\} \cup \{k+1, k+2, \dots\}\right) = \operatorname{ord} \mathbb{N}^* = \operatorname{ord} \mathbb{N} = \omega.$$

Let us prove (4): let X be a well-ordered set and $\phi : X \to X \sqcup_+ \{\infty\}$ be an order-isomorphism. Let a be the (unique) element of X such that $\phi(a) = \infty$. Then for all $x \in X \backslash \{a\}$, $\phi(x) < \infty = \phi(a)$ implying $x < a$. Thus the restriction of ϕ to $X \backslash \{a\}$ is an isomorphism from $(\to, a)$ (a proper segment of X) onto X. From Lemma 10.1.27, it is impossible, proving (4). Since (4) implies (3) which implies (1), the proof of the lemma is complete. $\qquad\square$

N.B. An immediate consequence of the previous lemma is that

for every finite ordinal,	$\alpha \prec 1 \oplus_+ \alpha = \alpha \oplus_+ 1,$	(10.1.14)
for every infinite ordinal,	$\alpha = 1 \oplus_+ \alpha \prec \alpha \oplus_+ 1.$	(10.1.15)

Moreover this lemma proves as well that, given $(A, \leqslant_A), (B, \leqslant_B)$ two well-ordered sets, the well-ordered set $A \sqcup_+ B$ is order-isomorphic to A if and only if $B = \emptyset$.

Uncountable well-ordered sets

Proposition 10.1.33. *Let α be an ordinal. Then the set of all ordinals β such that $\beta \prec \alpha$ is a well-ordered set whose ordinal is α.*

In other words, let $(A, \leqslant_A)$ be a well-ordered set. The set $\mathfrak{S}_A = \{(\rightarrow, a)\}_{a \in A}$ of proper segments of A (see Proposition 10.1.25) is a well-ordered set by the inclusion relation and is order-isomorphic to A.

Proof. We consider the mapping $\phi : A \rightarrow \mathfrak{S}_A$ defined by $\phi(a) = (\rightarrow, a)$. It is obviously onto and increasing and if $\phi(a_1) = \phi(a_2)$, this implies

$$\big(\phi(a_1)\big)^c = \big(\phi(a_2)\big)^c \implies a_1 = \min\big(\phi(a_1)\big)^c = \min\big(\phi(a_2)\big)^c = a_2,$$

proving that ϕ is one-to-one and the proposition. $\qquad\square$

Theorem 10.1.34. *Any set of ordinals is well ordered. Moreover there does not exist a set of all ordinals.*

N.B. The existence of a set of all sets leads to the so-called Russell's paradox (see Exercise 1.9.2). Here as well the existence of a set of all ordinals leads to a contradiction, known as the Burali-Forti[5] paradox.

Proof. Let $\mathfrak{F} = (X_i)_{i \in I}$ be a family of well-ordered sets. From Proposition 10.1.29, we may assume that the set I is infinite. Let us assume that there is no $j \in I$ such that $\forall i \in I, \operatorname{ord} X_j \leq \operatorname{ord} X_i$, i.e.,

$$\forall j \in I, \exists i \in I, \ \operatorname{ord} X_i \prec \operatorname{ord} X_j, \quad \text{i.e., } X_i \text{ isomorphic to a proper segment of } X_j,$$

so that $\forall j \in I, \exists i \in I, \exists a_j \in X_j, \quad \operatorname{ord} X_i = \operatorname{ord}(\rightarrow, a_j)_{X_j}$. For $j_1 \in I$, there exists $j_2 \in I$ such that $\operatorname{ord} X_{i_2} \prec \operatorname{ord} X_{i_1}$ and thus we find a strictly decreasing sequence

$$\cdots \prec \operatorname{ord} X_{i_{n+1}} \prec \operatorname{ord} X_{i_n} \prec \cdots \prec \operatorname{ord} X_{i_2} \prec \operatorname{ord} X_{i_1}.$$

Thanks to Proposition 10.1.33, that sequence included in the ordinals $\prec \operatorname{ord} X_{i_1}$ should have a smallest element, which is not possible. Thus we have proven the first statement in the theorem.

Let us prove the second statement by reductio ad absurdum. Let $\mathscr{U}$ be the set of all ordinals; then it should be well ordered with an ordinal $\mathfrak{u}$ which should be the largest ordinal, contradicting (4) in Lemma 10.1.32. $\qquad\square$

---

[5]Cesare Burali-Forti (1861–1931) was an Italian mathematician. He came up in 1897 with the first discovery of a paradox in Cantor set theory.

Proposition 10.1.35. *There exists an uncountable well-ordered set Ω such that for all $x \in \Omega$, the segment $(\to, x)$ is countable. The well-ordered set Ω is unique up to an order-isomorphism. Let A be a countable well-ordered set: then $\operatorname{ord} A \prec \operatorname{ord} \Omega$.*

Proof. According to Zermelo's Theorem 10.1, the set of real numbers $\mathbb{R}$ (which is uncountable, see Theorem 10.1.8) can be well ordered (of course with an order which is not the standard one). If $\mathbb{R}$ does not have the required property, we define

$$a = \min\{x \in \mathbb{R}, (\to, x) \text{ uncountable}\}.$$

Then we take $\Omega = (\to, a)$ which is uncountable and such that for $x < a$, $(\to, x)$ is countable, proving the existence.

For the uniqueness property, let Ω_1 be a well-ordered set with the same property. If $\operatorname{ord} \Omega_1 \prec \operatorname{ord} \Omega$, then Ω_1 would be isomorphic to a proper segment of Ω, that is to a countable set, which is incompatible with the requirement that Ω_1 is uncountable.

Let A be a countable well-ordered set. Thanks to Proposition 10.1.29, A is order-isomorphic to a proper segment of Ω (since Ω is uncountable, the inequality $\operatorname{ord} \Omega \preccurlyeq \operatorname{ord} A$ is ruled out). $\qquad\square$

Remark 10.1.36. We can reformulate the previous result by saying that there exists a unique ordinal $\operatorname{ord} \Omega$, where Ω is the set of countable ordinals.

Proposition 10.1.37. *Let $\operatorname{ord} \Omega$ be as above the set of countable ordinals. Every countable subset of Ω has an upper bound.*

Proof. Let $\{x_j\}_{j\in\mathbb{N}} \subset \Omega$. The countable union of countable sets $\cup_{j\in\mathbb{N}}(\to, x_j)$ is also a countable set (see Theorem 10.1.8) and cannot be equal to Ω. Thanks to Proposition 10.1.25, it is also a (proper) segment of Ω and thus there exists $y \in \Omega$ such that

$$\cup_{j\in\mathbb{N}}(\to, x_j) = (\to, y)$$

implying that $\forall j \in \mathbb{N}, x_j \leq y$, i.e., y is indeed an upper bound for $\{x_j\}_{j\in\mathbb{N}}$. $\qquad\square$

Remark 10.1.38. Note that $\omega = \operatorname{ord} \mathbb{N}$ is the smallest infinite countable ordinal, but that, according to (4) in Lemma 10.1.32 and k finite ≥ 1,

$$\omega \prec \omega \oplus_+ 1 \prec \cdots \prec \omega \oplus_+ k \quad \text{are all countable ordinals.}$$

Moreover, it is also possible to define the (non-commutative) product of ordinals.

Definition 10.1.39. Let $(A, \leqslant_A), (B, \leqslant_B)$ be two well-ordered sets. We define the product of ordinals,

$$\operatorname{ord} B \otimes_\times \operatorname{ord} A = \operatorname{ord}(A \times B),$$

where the Cartesian product $A \times B$ is endowed with the *lexicographic* order:

$$(a_1, b_1) \leqslant_{A \times B} (a_2, b_2) \text{ means } \begin{cases} a_1 <_A a_2 \\ \text{or } a_1 = a_2, \quad b_1 \leqslant_B b_2. \end{cases}$$

Our notation emphasizes the fact that this multiplication is not commutative.

Note that this order makes $A \times B$ well ordered: let X be a non-empty subset of $A \times B$. We define

$$a_0 = \min\{a \in A, \exists b \in B, (a, b) \in X\}, \quad b_0 = \min\{b \in B, (a_0, b) \in X\},$$

and we have $(a_0, b_0) = \min X$.

Lemma 10.1.40. *With $\omega = \operatorname{ord} \mathbb{N}$, we have*

$$2 \otimes_\times \omega = \operatorname{ord}(\mathbb{N} \times \{1, 2\}) = \omega \prec \omega \otimes_\times 2 = \operatorname{ord}(\{1, 2\} \times \mathbb{N}) = \omega \oplus_+ \omega.$$

Proof. We have $\{1, 2\} \times \mathbb{N} = (\{1\} \times \mathbb{N}) \cup (\{2\} \times \mathbb{N}) \equiv \mathbb{N} \sqcup_+ \mathbb{N}$, proving the last equality. Moreover, we have $\mathbb{N} \times \{1, 2\} = (\mathbb{N} \times \{1\}) \cup (\mathbb{N} \times \{2\})$. Considering

$$\phi : \mathbb{N} \times \{1, 2\} \to \mathbb{N}, \quad \phi((n, 1)) = 2n, \quad \phi((n, 2)) = 2n + 1,$$

we see that ϕ is bijective and increasing, proving the equalities in the lhs. We have proven in Lemma 10.1.32 (1) the requested strict inequality between ordinals. $\square$

Remark 10.1.41. We can also go on with Remark 10.1.38: for k, l finite ≥ 1,

$$\omega \prec \omega \oplus_+ k \prec \omega \oplus_+ \omega = \omega \otimes_\times 2 \prec (\omega \otimes_\times 2) \oplus_+ l \prec \omega \otimes_\times 3,$$

all countable ordinals. With the powers ω^2, ω^3 (to be defined) we could find other countable ordinals.

Definition 10.1.42. Let $(X, \leqslant_X)$ be a well-ordered set.

(1) Let $a \in X$ such that $\{x \in X, x > a\} = (a, \to) \neq \emptyset$. We define the *immediate successor* of a, that we note by $a + 1$, as

$$a + 1 = \min(a, \to), \quad \text{(note that } a < a + 1\text{)}.$$

(2) Let $b \in X$ such that there exists $a \in X$ with $a + 1 = b$, i.e.,

$$b = \min(a, \to).$$

Then a is uniquely determined[6] and is called the *immediate predecessor* of b.

(3) Let $x \in X$ which has no immediate predecessor. Then x is called a *limit element* of the well-ordered set X.

---

[6]If $b = a_1 + 1 = a_2 + 1$, i.e., $\min(a_1, \to) = \min(a_2, \to)$, then

$$a_1 < a_2 \implies a_2 \in (a_1, \to) \implies b = a_1 + 1 \leq a_2 < a_2 + 1 = b, \quad \text{which is impossible.}$$

Example. Let Ω be as in Proposition 10.1.35: $\omega = \operatorname{ord} \mathbb{N}$ has no immediate predecessor, otherwise we would find a countable ordinal a such that $\omega = a + 1 = \min(a, \rightarrow)$ with $a < \omega$. If a was finite, then $a + 1$ would be also finite (impossible), and if a was not finite, a would be countable and thus such that $\omega \leq a < a + 1 = \omega$, which is impossible.

10.2 Topological matters

Filters

General properties of filters

Definition 10.2.1. Let X be a set. A subset $\mathscr{V}$ of $\mathcal{P}(X)$ such that the conditions

$$V \subset W, \ V \in \mathscr{V} \Longrightarrow W \in \mathscr{V}, \tag{10.2.1}$$

$$V_j \in \mathscr{V}, j = 1, 2 \Longrightarrow V_1 \cap V_2 \in \mathscr{V}, \tag{10.2.2}$$

$$\emptyset \notin \mathscr{V}, \quad X \in \mathscr{V}, \tag{10.2.3}$$

are fulfilled is called a *filter* on X.

Remark 10.2.2. A set X on which there exists a filter $\mathscr{V}$ is necessarily non-empty: we have $\mathcal{P}(\emptyset) = \{\emptyset\}$ and since $\emptyset \notin \mathscr{V}$, the latter is not compatible with $X \in \mathscr{V}$.

Simple examples of filters are

- On a (non-empty) topological space X, for $x \in X$,

$$\mathscr{V}_x = \{V \subset X, V \text{ neighborhood of } x\}$$

is a filter (the filter of neighborhoods of x, cf. (1.2.4), (1.2.5), (1.2.6)).
- On $\mathbb{R}^n$, $\mathscr{V}_\infty = \{V \subset \mathbb{R}^n, V^c \text{ bounded}\}$ (here bounded means included in a ball with finite radius). The first axiom is satisfied since a subset of a bounded set is bounded, the second axiom follows from $(V_1 \cap V_2)^c = V_1^c \cup V_2^c$ and the fact that a union of two bounded sets is bounded. Finally, the empty set has the unbounded complement $\mathbb{R}^n$ and the empty set, complement of $\mathbb{R}^n$, is bounded.
- On an infinite set X, $\mathscr{F}_\infty = \{V \subset X, V^c \text{ finite}\}$ is a filter (a subset of a finite set is finite, a finite union of finite sets is finite).

Definition 10.2.3. Let X be a set and $\mathscr{F}_j, j = 1, 2$ be filters on X. We shall say that $\mathscr{F}_2$ is *finer* than $\mathscr{F}_1$ when $\mathscr{F}_2 \supset \mathscr{F}_1$.

If $(\mathscr{F}_i)_{i \in I}$ is a family of filters on a set X (I non-empty), then $\mathscr{F} = \cap_{i \in I} \mathscr{F}_i$ is also a filter on X: if $V \in \mathscr{F}, V \subset W$, then W belongs to each $\mathscr{F}_i$, thus to $\mathscr{F}$. If $V', V'' \in \mathscr{F}$, then $V' \cap V''$ belongs to each $\mathscr{F}_i$, thus to $\mathscr{F}$. Moreover the empty set cannot belong to $\mathscr{F}$, since it would belong to an $\mathscr{F}_i$.

Lemma 10.2.4. *Let X be a set and $\emptyset \neq \mathscr{B} \subset \mathcal{P}(X)$ with the non-empty finite intersection property: for every finite family $B_1, \ldots, B_N$ of $\mathscr{B}$, $\cap_{1 \leq j \leq N} B_j \neq \emptyset$. Then*

$$\mathscr{F} = \{V \subset X, \exists B_1, \ldots, B_N \in \mathscr{B}, \cap_{1 \leq j \leq N} B_j \subset V\}$$

is a filter on X. It is the smallest filter on X which contains $\mathscr{B}$, called the filter generated by $\mathscr{B}$ and denoted by $\widetilde{\mathscr{B}}$.

Proof. Let $\mathscr{F} \ni V \subset W$, then $W \in \mathscr{F}$. Let $V', V'' \in \mathscr{F}$: there exists $(B_j')_{1 \leq j \leq M}$, $(B_k'')_{1 \leq k \leq N}$ in $\mathscr{B}$ such that $V' \supset \cap_{1 \leq j \leq M} B_j'$, $V'' \supset \cap_{1 \leq k \leq N} B_k''$ and thus $V' \cap V'' \supset \cap_{1 \leq j \leq M} B_j' \cap_{1 \leq k \leq N} B_k''$, proving $V' \cap V'' \in \mathscr{F}$. Finally $\emptyset \notin \mathscr{F}$ since it would imply from the definition that for $B_1, \ldots, B_N$ in $\mathscr{B}$, $\emptyset = \cap_{1 \leq j \leq N} B_j$. Also $X \in \mathscr{F}$ since there exists $B \in \mathscr{B}$ ($\mathscr{B}$ non-empty) and $B \subset X$. Moreover any filter containing $\mathscr{B}$ must contain $\mathscr{F}$. $\qquad\square$

Lemma 10.2.5. *Let $f : X \longrightarrow Y$ be a mapping and $\mathscr{F}$ be a filter on X. Then the set*

$$f(\mathscr{F}) = \{f(V)\}_{V \in \mathscr{F}}$$

has the non-empty finite intersection property and thus generates a filter on Y denoted by $\widetilde{f(\mathscr{F})}$, called the filter-image by f of the filter $\mathscr{F}$.

Proof. Note that the family $f(\mathscr{F})$ is not empty since it contains $f(X)$. Moreover, for $V_1, \ldots, V_N \in \mathscr{F}$, we have

$$\cap_{1 \leq j \leq N} f(V_j) \supset f\big(\underbrace{\cap_{1 \leq j \leq N} V_j}_{\in \mathscr{F}}\big) \neq \emptyset.$$

According to Lemma 10.2.4, $f(\mathscr{F})$ generates a filter. $\qquad\square$

Definition 10.2.6. *Let X be a set and let $\mathscr{F} = (A_i)_{i \in I}, \mathscr{G} = (B_j)_{j \in J}$ be filters on X. The filters $\mathscr{F}, \mathscr{G}$ are said to be secant if*

$$\forall (i, j) \in I \times J, \quad A_i \cap B_j \neq \emptyset.$$

Proposition 10.2.7. *Let X be a set and let $\mathscr{F}, \mathscr{G}$ be filters on X. Then the filters $\mathscr{F}, \mathscr{G}$ have a least upper bound (for the inclusion relation) if and only if they are secant.*

Proof. The condition is obviously necessary since when a filter $\mathscr{H} \supset \mathscr{F} \cup \mathscr{G}$, the intersection of two elements of $\mathscr{H}$ must be non-empty. Conversely let $\mathscr{F} = (A_i)_{i \in I}, \mathscr{G} = (B_j)_{j \in J}$ be secant filters on X. We define

$$\mathscr{H} = \{C \subset X, \exists (i, j) \in I \times J, C \supset A_i \cap B_j\}.$$

We note that $\mathscr{H}$ is a filter on X since the first property (10.2.1) is obvious, the second one (10.2.2) follows from

$$\underbrace{A_{i_1} \cap A_{i_2}}_{\in \mathscr{F}} \cap \underbrace{B_{j_1} \cap B_{j_2}}_{\in \mathscr{G}},$$

the third one (10.2.3) from the secant hypothesis. We have trivially $\mathscr{H} \supset \mathscr{F} \cup \mathscr{G}$ and if $\mathscr{K}$ is a filter on X containing $\mathscr{F} \cup \mathscr{G}$, any $A_i \cap B_j$ should belong to $\mathscr{K}$ and thus from (10.2.1), $\mathscr{H} \subset \mathscr{K}$, proving the sought result. $\qquad\square$

Definition 10.2.8. Let X be a set. An *Ultrafilter* on X is a filter $\mathscr{U}$ which is maximal for the inclusion: if a filter $\mathscr{V}$ on X contains $\mathscr{U}$, it should be equal to $\mathscr{U}$.

Proposition 10.2.9. *Let X be a set and let $\mathscr{F}_0$ be a filter on X. There exists an ultrafilter containing $\mathscr{F}_0$.*

Proof. Zornification. We consider the (non-empty) family

$$\Phi = \{\mathscr{F} \text{ filter on } X \text{ such that } \mathscr{F} \supset \mathscr{F}_0\}.$$

It is inductive since if $(\mathscr{F}_i)_{i \in I}$ is a totally ordered subset of Φ, we may consider

$$\mathscr{G} = \cup_{i \in I} \mathscr{F}_i$$

and note that it is a filter on X: let $V \in \mathscr{G}$, $W \supset V$, then $V \in \mathscr{F}_i$ for some $i \in I$, so that $W \in \mathscr{F}_i \subset \mathscr{G}$. If $V_1, V_2 \in \mathscr{G}$, since $(\mathscr{F}_i)_{i \in I}$ is totally ordered, we find $i \in I$ such that V_1, V_2 both belong to $\mathscr{F}_i$, implying that $V_1 \cap V_2 \in \mathscr{F}_i \subset \mathscr{G}$. Finally $\emptyset \notin \mathscr{G}$, otherwise it should belong to some $\mathscr{F}_i$. Applying Zorn's Lemma 10.1 yields a maximal element $\mathscr{U}$ in Φ. If $\mathscr{V}$ is a filter containing $\mathscr{U}$, it must contain $\mathscr{F}_0$, thus it belongs to Φ, thus is equal to $\mathscr{U}$ by maximality: $\mathscr{U}$ is an ultrafilter. $\qquad\square$

Lemma 10.2.10. *Let $\mathscr{U}$ be an ultrafilter on a set X. If A_1, A_2 are subsets of X such that $A_1 \cup A_2 \in \mathscr{U}$, then $A_1 \in \mathscr{U}$ or $A_2 \in \mathscr{U}$.*

Proof. Reductio ad absurdum. Let A_1, A_2 be subsets of X such that $A_1 \cup A_2 \in \mathscr{U}$, $A_1 \notin \mathscr{U}$ and $A_2 \notin \mathscr{U}$. We define

$$\mathscr{F} = \{M \subset X, A_1 \cup M \in \mathscr{U}\}.$$

This is a filter on X since if $V \supset M \in \mathscr{F}$, then $A_1 \cup V \supset A_1 \cup M \in \mathscr{U}$, implying $A_1 \cup V \in \mathscr{U}$ and $V \in \mathscr{F}$. If $V', V'' \in \mathscr{F}$, then

$$A_1 \cup (V' \cap V'') = \underbrace{(A_1 \cup V')}_{\in \mathscr{U}} \cap \underbrace{(A_1 \cap V'')}_{\in \mathscr{U}} \Longrightarrow V' \cap V'' \in \mathscr{F}.$$

Moreover $\emptyset \notin \mathscr{F}$ since $A_1 \notin \mathscr{U}$. The filter $\mathscr{F}$ contains $\mathscr{U}$ since $M \in \mathscr{U}$ implies $A_1 \cup M \in \mathscr{U}$. Finally, we see also that A_2 belongs to $\mathscr{F}$ and not to $\mathscr{U}$, contradicting the maximality of the filter $\mathscr{U}$. $\qquad\square$

Lemma 10.2.11. *Let $\mathscr{F}$ be a filter on a set X such that for any subset M of X, either $M \in \mathscr{F}$ or $M^c \in \mathscr{F}$. Then $\mathscr{F}$ is an ultrafilter.*

Proof. Let $\mathscr{G}$ be a filter containing $\mathscr{F}$. For $A \in \mathscr{G}$, we have $A^c \notin \mathscr{G}$, thus $A^c \notin \mathscr{F}$, thus $A \in \mathscr{F}$, proving the maximality of $\mathscr{F}$. $\qquad\square$

Proposition 10.2.12. *Let $f : X \to Y$ be a surjective mapping and let $\mathscr{F}$ be a filter on X. Then the filter-image by f of $\mathscr{F}$ is equal to $\{f(A)\}_{A \in \mathscr{F}}$. Moreover if $\mathscr{F}$ is an ultrafilter, so is $f(\mathscr{F}) = \{f(A)\}_{A \in \mathscr{F}}$.*

Proof. The filter-image is $\widetilde{f(\mathscr{F})}$ and is generated by $f(\mathscr{F})$: it suffices to prove that $f(\mathscr{F})$ is a filter when f is onto. If $W \supset f(A)$ with $A \in \mathscr{F}$, then

$$f^{-1}(W) \supset f^{-1}(f(A)) \supset A \Longrightarrow f^{-1}(W) \in \mathscr{F} \Longrightarrow f(f^{-1}(W)) \in f(\mathscr{F}),$$

and since f is onto[7], we have $f(f^{-1}(W)) = W$, so that $W \in f(\mathscr{F})$, proving the first property (10.2.1). Let $V_1, V_2 \in f(\mathscr{F})$: then with $A_j \in \mathscr{F}$, we have

$$V_1 \cap V_2 = f(A_1) \cap f(A_2) \supset f(\underbrace{A_1 \cap A_2}_{\in \mathscr{F}}),$$

and from the already proven first property, we get $V_1 \cap V_2 \in f(\mathscr{F})$. On the other hand, $\emptyset \notin f(\mathscr{F})$, otherwise for some $A \in \mathscr{F}$, we would have $f(A) = \emptyset$, which implies $A = \emptyset$ (impossible since $\mathscr{F}$ is a filter on X).

    If $\mathscr{F}$ is an ultrafilter on X, then $\mathscr{G} = f(\mathscr{F})$ is a filter on Y and if B is a subset of Y, either $f^{-1}(B) \supset A$ for some $A \in \mathscr{F}$ and (since f is onto),

$$B = f(f^{-1}(B)) \supset f(A) \Longrightarrow B \in \mathscr{G},$$

or $f^{-1}(B)$ does not contain any element of $\mathscr{F}$. In the latter case, since $\mathscr{F}$ is an ultrafilter (see Lemma 10.2.10) and $f^{-1}(B) \notin \mathscr{F}$,

$$X = f^{-1}(B) \cup f^{-1}(B^c) \Longrightarrow f^{-1}(B^c) \in \mathscr{F} \Longrightarrow B^c = f(f^{-1}(B^c)) \in f(\mathscr{F}).$$

As a consequence $\mathscr{G}$ is a filter on Y verifying the property of Lemma 10.2.11, and thus an ultrafilter, completing the proof. $\qquad\qquad\qquad\qquad\qquad\qquad\square$

Filters in a topological space

Definition 10.2.13. Let X be a topological space, $x \in X$ and $\mathscr{F}$ be a filter on X.

(1) The filter $\mathscr{F}$ is said to converge to x whenever it is finer than the filter $\mathscr{V}_x$ of neighborhoods of x, i.e., when $\mathscr{F} \supset \mathscr{V}_x$.

(2) The closure of the filter $\mathscr{F}$ is defined as $\cap_{A \in \mathscr{F}} \overline{A}$.

N.B. When a point x is a *limit point* of a filter $\mathscr{F}$, i.e., when $\mathscr{F}$ converges to x, then it also belongs to the closure of $\mathscr{F}$: let A be an element of $\mathscr{F}$ and let $V \in \mathscr{V}_x$. Since these sets both belong to the filter $\mathscr{F}$, we have $A \cap V \neq \emptyset$ and this[8] implies $x \in \overline{A}$.

---

[7] The inclusion $f(f^{-1}(W)) \subset W$ always holds and when f is onto and $y \in W$, there exists $x \in f^{-1}(W)$ with $y = f(x)$, so that $y \in f(f^{-1}(W))$.

[8] Applying (1.2.1) to A^c yields $(\overline{A})^c = \text{interior}(A^c)$ so that

$$x \notin \overline{A} \iff \exists V \in \mathscr{V}_x, V \subset A^c \iff \exists V \in \mathscr{V}_x, V \cap A = \emptyset.$$

Lemma 10.2.14. *Let X, Y be topological spaces, $x \in X$ and $f : X \longrightarrow Y$ be a mapping. The mapping f is continuous at x if and only if*

$$\widetilde{f(\mathscr{V}_x)} \supset \mathscr{V}_{f(x)},$$

where $\mathscr{V}_z$ stands for the filter of neighborhoods of z.

Proof. For f to be continuous at $x \in X$ means

$$\forall W \in \mathscr{V}_{f(x)}, \ \exists V \in \mathscr{V}_x \text{ such that } f(V) \subset W. \tag{10.2.4}$$

This implies that $\widetilde{f(\mathscr{V}_x)} \supset \mathscr{V}_{f(x)}$. Conversely, if the latter holds, it means

$$\forall W \in \mathscr{V}_{f(x)}, \ \exists V_1, \ldots, V_N \in \mathscr{V}_x, \ \cap_{1 \leq j \leq N} f(V_j) \subset W,$$

which implies $f\big(\cap_{1 \leq j \leq N} V_j\big) \subset W$, providing (10.2.4) since $\cap_{1 \leq j \leq N} V_j \in \mathscr{V}_x$. $\qquad\square$

Compactness and Tychonoff's Theorem

We recall first that a topological space $(X, \mathcal{O})$ is said to be a *Hausdorff space* whenever

$$\forall (x, y) \in X^2, x \neq y \implies \exists U \in \mathscr{V}_x, \exists V \in \mathscr{V}_y, \quad U \cap V = \emptyset. \tag{10.2.5}$$

Definition 10.2.15. A topological space $(X, \mathcal{O})$ is said to be compact when it is a Hausdorff space and satisfies the Borel–Lebesgue property: if $(\Omega_i)_{i \in I}$ is a family of open sets such that $X = \cup_{i \in I} \Omega_i$, there exists a finite subset J of I such that $X = \cup_{i \in J} \Omega_i$.

Remark 10.2.16. If A is a closed subset of a compact space X, then A is also compact. Using the definition in Lemma 1.2.2 of the induced topology on A, the separation property is obvious and we may assume that $A \subset \cup_{i \in I} \Omega_i$, where each Ω_i is an open subset of X. Then we have

$$X = \cup_{i \in I} \Omega_i \cup A^c$$

and since A^c is open, the compactness of X implies that $X = \cup_{i \in J} \Omega_i \cup A^c$ with a finite subset J of I. As a consequence $A \subset \cup_{i \in J} \Omega_i$, proving its compactness.

Proposition 10.2.17. *Let X be a topological space. The following properties are equivalent.*

  (i) *Any filter on X has a non-empty closure.*
  (ii) *Any ultrafilter on X is convergent.*
  (iii) *The Borel–Lebesgue property holds.*

A topological space satisfying these properties is said to be quasi-compact. A topological space is compact whenever it is a quasi-compact Hausdorff space.

Proof. (i) $\implies$ (ii). Let $\mathscr{U}$ be an ultrafilter on X: then there exists $x \in \cap_{U \in \mathscr{U}} \overline{U}$, so that $\mathscr{U}$ and $\mathscr{V}_x$ are secant (see Definition 10.2.6) and from Proposition 10.2.7, they have a least upper bound which must be $\mathscr{U}$ since it is an ultrafilter: this implies $\mathscr{U} \supset \mathscr{V}_x$ and (ii).

(ii) $\implies$ (iii). Let $(\Omega_i)_{i \in I}$ be an open covering of X and let us assume by contradiction that for all J finite subset of I, $\cup_{i \in J} \Omega_i \neq X$. Then the family

$$\mathscr{B} = \{\cap_{i \in J} \Omega_i^c\}_{J \text{ finite} \subset I}$$

has the non-empty finite intersection property: for $B_k = \cap_{i \in J_k} \Omega_i^c$, $1 \le k \le N$ and J_k finite subset of I, we have

$$\cap_{1 \le k \le N} B_k = \bigcap_{i \in \underbrace{\cup_{1 \le k \le N} J_k}_{\text{finite}}} \Omega_i^c \neq \emptyset.$$

According to Lemma 10.2.4 and to Proposition 10.2.9, there exists an ultrafilter $\mathscr{U}$ containing $\mathscr{B}$ and from the assumption (ii) there exists $x \in X$ such that $\mathscr{U} \supset \mathscr{V}_x$. The point x belongs to the closure of $\mathscr{U}$ and thus to

$$\bigcap_{i \in I} \overline{\Omega_i^c} \underbrace{=}_{\Omega_i \text{ open}} \bigcap_{i \in I} \Omega_i^c = \left(\cup_{i \in I} \Omega_i\right)^c = \emptyset,$$

which is impossible.

(iii) $\implies$ (i). Let $\mathscr{F} = (M_i)_{i \in I}$ be a filter on X with an empty closure: we have

$$\emptyset = \cap_{i \in I} \overline{M_i} \implies X = \cup_{i \in I} \underbrace{\left(\overline{M_i}\right)^c}_{\text{open}} \implies \exists J \text{ finite} \subset I, \ X = \cup_{i \in J} \left(\overline{M_i}\right)^c,$$

and thus $\cap_{i \in J} \overline{M_i} = \emptyset$ which is impossible since all M_i belong to the filter $\mathscr{F}$ which enjoys the non-empty finite intersection property. The proof of the proposition is complete. $\qquad\square$

Proposition 10.2.18. *Let X be a Hausdorff topological space.*

(1) *Let A, B be two compact disjoint subsets of X. Then there exist U, V open disjoint subsets of X such that $A \subset U$ and $B \subset V$.*

(2) *Let A be a compact subset of X. Then A is a closed subset of X.*

Proof. Since X is Hausdorff, for each $(x, y) \in A \times B$, there exists some open sets $U_x(y) \in \mathscr{V}_x, V_y(x) \in \mathscr{V}_y$ such that $U_x(y) \cap V_y(x) = \emptyset$. By the compactness of B, we have for all $x \in A$,

$$B \subset \cup_{1 \le j \le N_x} V_{y_j}(x) = W(x).$$

As a consequence, with $T(x) = \cap_{1\leq j\leq N_x} U_x(y_j)$, we have $T(x) \cap W(x) = \emptyset$, $W(x)$ open containing B and the open set $T(x) \in \mathscr{V}_x$. By the compactness of A, we have

$$A \subset \cup_{1\leq k\leq M} T(x_k).$$

We take then $U = \cup_{1\leq k\leq M} T(x_k)$, $V = \cap_{1\leq k\leq M} W(x_k)$, which are disjoint open sets containing respectively A, B, proving (1). Let A be a compact subset of X; if $a \notin A$, then A and $\{a\}$ are disjoint compact subsets and from the now proven (1), there exists an open set $V \in \mathscr{V}_a$ such that $V \cap A = \emptyset$, i.e., $V \subset A^c$, proving that A^c is open. $\qquad\square$

Proposition 10.2.19. *Let $(K_i)_{i\in I}$ be a family of compact subsets of a Hausdorff space X such that $\cap_{i\in I} K_i = \emptyset$. Then there exists a finite subset J of I such that $\cap_{i\in J} K_i = \emptyset$.*

Proof. Note that from Property (2) of Proposition 10.2.18, the K_i are closed subsets of X. For a fixed $i_0 \in I$,

$$K_{i_0} \subset \cup_{i\neq i_0, i\in I} K_i^c \Longrightarrow K_{i_0} \subset \cup_{i\in J} K_i^c, \quad J \text{ finite subset of } I.$$

As a result, $\cap_{i\in J\cup\{i_0\}} K_i = \emptyset$. $\qquad\square$

Theorem 10.2.20. *Let X, Y be topological spaces, with Y a Hausdorff space, and $f : X \longrightarrow Y$ be a continuous mapping. If X is compact, then $f(X)$ is compact.*

Proof. $f(X)$ is a Hausdorff space as a subset of a Hausdorff space. Let us assume that $f(X) \subset \cup_{i\in I} V_i$ where V_i are open subsets of Y. Then

$$X = \cup_{i\in I} \underbrace{f^{-1}(V_i)}_{\substack{\text{open} \\ \text{since } f \text{ continuous}}},$$

so that for some finite J, $X = \cup_{i\in J} f^{-1}(V_i)$, and thus $f(X) = \cup_{i\in J} f(f^{-1}(V_i)) \subset \cup_{i\in J} V_i$, proving the result. $\qquad\square$

Definition 10.2.21. Let $(X_i, \mathcal{O}_i)_{i\in I}$ be a family of topological spaces. The *product-topology* on $X = \prod_{i\in I} X_i$ is the weakest topology on X such that all canonical projections $\pi_i : X \to X_i$ are continuous.

We note that the continuity of the projections forces

$$\pi_i^{-1}(\mathcal{O}_i) = \{\pi_i^{-1}(\Omega)\}_{\Omega\in\mathcal{O}_i}$$

to belong to the product topology $\mathcal{O}$ on X. As a result $\mathcal{O}$ is the intersection of topologies containing $\cup_{i\in I} \pi_i^{-1}(\mathcal{O}_i)$, i.e., the smallest topology containing that set.

Lemma 10.2.22. *Let $(X_i, \mathcal{O}_i)_{i\in I}$ be a family of topological spaces and let $(X, \mathcal{O})$ be the product topology on $X = \prod_{i\in I} X_i$. Then*

$$\mathcal{O} = \{\cup_{\alpha\in A} \Omega_\alpha\}_{\substack{\Omega_\alpha = \prod_{i\in I} U_{i,\alpha},\ U_{i,\alpha}\in\mathcal{O}_i \\ U_{i,\alpha} = X_i\ except\ for\ a\ finite\ subset\ of\ I}}.$$

Proof. Let us call $\widetilde{\mathcal{O}}$ the set defined in the lemma. Since any product

$$\prod_{i \in I} U_{i,\alpha}, \quad U_{i,\alpha} \in \mathcal{O}_i, U_{i,\alpha} = X_i, \text{ except for a finite subset of } I,$$

belongs to $\mathcal{O}$, as a finite intersection of elements of $\cup_{i \in I} \pi_i^{-1}(\mathcal{O}_i)$, we find that

$$\cup_{i \in I} \pi_i^{-1}(\mathcal{O}_i) \subset \widetilde{\mathcal{O}} \subset \mathcal{O}. \tag{10.2.6}$$

Moreover $\widetilde{\mathcal{O}}$ is a topology on X since it is obviously stable by union and also by finite intersection: to verify this it is enough to consider

$$W = \left(\prod_{i \in I} U_i \right) \cap \left(\prod_{i \in I} V_i \right), \quad U_i, V_i \in \mathcal{O}_i, \text{ all but a finite number equal to } X_i.$$

We have indeed $W = \prod_{i \in I}(U_i \cap V_i)$ where all but a finite number of $(U_i \cap V_i)$ are equal to X_i and the others are open subsets of X_i. Since $\widetilde{\mathcal{O}}$ is proven to be a topology, the inclusions (10.2.6) imply $\widetilde{\mathcal{O}} = \mathcal{O}$. $\square$

Theorem 10.2.23 (Tychonoff). *Let $(X_i)_{i \in I}$ be a family of compact topological spaces. Then the space $X = \prod_{i \in I} X_i$ equipped with the product topology is compact.*

Proof. Let $\mathcal{U}$ be an ultrafilter on X. From Proposition 10.2.12, each $\pi_i(\mathcal{U})$ is an ultrafilter on X_i (π_i is the canonical projection from X onto X_i). By compactness of X_i, there exists $x_i \in X_i$ such that $\pi_i(\mathcal{U}) \supset \mathcal{V}_{x_i}$. Let us define $x = (x_i)_{i \in I}$ and let us prove that $\mathcal{U}$ converges to x: let $V \in \mathcal{V}_x$, so that x belongs to an open set of X contained in V. From Lemma 10.2.22, V contains a set

$$\prod_{i \in I} U_i, \quad x_i \in U_i \text{ open in } X_i, U_i = X_i, \text{ except for a finite subset } J \text{ of } I.$$

Since $U_i \in \mathcal{V}_{x_i}$, it belongs also to $\pi_i(\mathcal{U})$ and for all $i \in J$, there exists $V^{(i)} \in \mathcal{U}$ such that

$$U_i = \pi_i(V^{(i)}) \Longrightarrow \forall i \in J, \ U_i \supset \pi_i(W), W = \cap_{i \in J} V^{(i)} \text{ and } W \in \mathcal{U}.$$

Since for $i \notin J$, $U_i = X_i$, we obtain that

$$V \supset \prod_{i \in I} U_i \supset \prod_{i \in I} \pi_i(W) \supset W \Longrightarrow V \in \mathcal{U},$$

proving the convergence $\mathcal{U} \supset \mathcal{V}_x$ and quasi-compactness. To conclude, we need to prove the following result.

Lemma 10.2.24. *A product of Hausdorff spaces is also Hausdorff.*

Proof of the lemma. Let $(x_i')_{i\in I}, (x_i'')_{i\in I}$ be distinct points in X. We are thus able to find $i_0 \in I$ such that $x_{i_0}' \neq x_{i_0}''$ and consequently (since X_{i_0} is Hausdorff) we can find U_{i_0}', U_{i_0}'' disjoint open subsets of X_{i_0} with $x_{i_0}' \in U_{i_0}', x_{i_0}'' \in U_{i_0}''$. We define then

$$U' = \prod_{i\in I} V_i', \ V_{i_0}' = U_{i_0}', \text{ other } V_i' = X_i, U'' = \prod_{i\in I} V_i'', \ V_{i_0}'' = U_{i_0}'', \text{ other } V_i'' = X_i.$$

The sets U', U'' are disjoint and respective neighborhoods of $(x_i')_{i\in I}, (x_i'')_{i\in I}$. $\qquad\square$

The proof of Theorem 10.2.23 is complete. $\qquad\square$

Connectedness of topological spaces

Definition 10.2.25. A topological space is said to be connected if the only subsets of X which are both open and closed are X and $\emptyset$.

Lemma 10.2.26. *Let X be a topological space and let $(A_i)_{i\in I}$ be a family of connected subsets of X such that*

$$\forall (i', i'') \in I^2, \exists J = \{i_k\}_{1\leq k\leq N} \subset I, \ i_1 = i', i_N = i'',$$
$$\text{such that for } 1 \leq k < N, \ A_{i_k} \cap A_{i_{k+1}} \neq \emptyset.$$

Then the set $A = \cup_{i\in I} A_i$ is connected.

Proof. Using the induced topology (see Lemma 1.2.2), we assume that

$$A \subset \Omega_1 \cup \Omega_2, \quad \Omega_1 \cap \Omega_2 \cap A = \emptyset, \quad \Omega_j \text{ open subsets of } X.$$

Let us assume that $A \cap \Omega_1 \neq \emptyset$. Then there exists $x \in \Omega_1 \cap A_{i'}$ for some $i' \in I$. Since $A_{i'}$ is connected and

$$A_{i'} \subset \Omega_1 \cup \Omega_2, \ \Omega_1 \cap \Omega_2 \cap A_{i'} = \emptyset, A_{i'} \cap \Omega_1 \neq \emptyset \implies A_{i'} \cap \Omega_2 = \emptyset \implies A_{i'} \subset \Omega_1.$$

Let us now consider $i'' \in J$: applying the hypothesis, we find

$$J = \{i_k\}_{1\leq k\leq N} \subset I, \ i_1 = i', i_N = i'', 1 \leq k < N, A_{i_k} \cap A_{i_{k+1}} \neq \emptyset.$$

Assuming $A_{i_k} \subset \Omega_1$ for some $1 \leq k < N$, we have from the connectedness of $A_{i_{k+1}}$,

$$\emptyset \neq A_{i_{k+1}} \cap A_{i_k}, A_{i_{k+1}} \subset \Omega_1 \cup \Omega_2, \quad \Omega_1 \cap \Omega_2 \cap A_{i_{k+1}} = \emptyset, A_{i_{k+1}} \cap \Omega_1 \neq \emptyset,$$

and this implies $A_{i_{k+1}} \cap \Omega_2 = \emptyset$, thus $A_{i_{k+1}} \subset \Omega_1$. Since we have proven $A_{i_1} \subset \Omega_1$ this proves $A_{i''} \subset \Omega_1$ for any $i'' \in I$, entailing $A \subset \Omega_1$, proving connectedness for A. $\qquad\square$

Definition 10.2.27. Let X be a topological space. We define a binary relation on X by $x' \sim x''$ means there exists a connected subset A of X such that $x', x'' \in A$.

Remark 10.2.28. This relation is an equivalence relation: reflexivity and symmetry are obvious whereas transitivity follows from Lemma 10.2.26. The *connected components* of X are defined as the equivalence classes of that binary relation. We obtain a partition of X,

$$X = \sqcup_{i \in I} C_i, \quad \{C_i\}_{i \in I} = X/\sim \quad \text{(the quotient space)}.$$

Moreover each C_i is connected: we have $C_i = p(x_i)$, the equivalence class of a point x_i and if $x \in C_i$, then there exists A connected such that $x_i, x \in A$. Since all points of A are equivalent to x_i, this implies that

$$C_i = \bigcup_{A \text{ connected} \ni x_i} A$$

and Lemma 10.2.26 provides connectedness for C_i. Moreover if C is connected and contains $C_i = p(x_i)$, all elements of C are equivalent to x_i, so that $C = C_i$.

Theorem 10.2.29. *Let X, Y be topological spaces, let $f : X \to Y$ be a continuous mapping and let A be a connected subset of X. Then $f(A)$ is connected.*

Proof. Let us assume that V_1, V_2 are open subsets of Y such that

$$f(A) \subset V_1 \cup V_2, \quad f(A) \cap V_1 \cap V_2 = \emptyset.$$

By continuity of f, the sets $f^{-1}(V_j)$ are open in X and we have

$$A \subset f^{-1}(f(A)) \subset f^{-1}(V_1) \cup f^{-1}(V_2),$$

as well as $f^{-1}(V_1) \cap f^{-1}(V_2) \cap f^{-1}(f(A)) = \emptyset$. The connectedness of A implies $A \subset f^{-1}(V_j)$ say for $j = 1$ and thus $f(A) \subset V_1$, proving connectedness for $f(A)$. $\square$

Proposition 10.2.30. *Let X be a topological space and let A be a connected subset of X. Then the closure of A is also connected.*

Proof. We may assume that A is non-empty. Let us assume that

$$\bar{A} \subset \Omega_1 \cup \Omega_2, \quad \bar{A} \cap \Omega_1 \cap \Omega_2 = \emptyset, \ \Omega_j \text{ open}.$$

From the connectedness of A, we infer that A must be included in one Ω_j, say Ω_1. We have

$$A \subset \Omega_1 \cap A \underbrace{\subset}_{\substack{\text{from} \\ \Omega_1 \cap A \cap \Omega_2 = \emptyset}} \Omega_2^c \underbrace{\Longrightarrow}_{\Omega_2^c \text{ closed}} \bar{A} \subset \Omega_2^c \Longrightarrow \Omega_2 \cap \bar{A} = \emptyset \underbrace{\Longrightarrow}_{\bar{A} \subset \Omega_1 \cup \Omega_2} \bar{A} \subset \Omega_1,$$

proving connectedness for $\bar{A}$ as well. $\square$

Proposition 10.2.31. *The connected subsets of $\mathbb{R}$ are the intervals.*

Proof. Let C be a connected subset of the real line containing at least two distinct points $a < b$. If there exists $x \in (a, b)$ such that $x \notin C$, then

$$C \subset (-\infty, x) \cup (x, +\infty), \quad \text{a disjoint union of open sets,}$$

violating connectedness. As a result C is an interval, i.e., a subset of $\mathbb{R}$ such that

$$a, b \in C, a < b \Longrightarrow (a, b) \subset C.$$

Conversely, let I be an interval of $\mathbb{R}$ such that

$$I \subset U_1 \cup U_2, \quad U_1 \cap U_2 \cap I = \emptyset, \ U_j \text{ open.}$$

Let us assume that $I \cap U_1 \neq \emptyset$ and let $a_1 \in I \cap U_1$. If $I \cap U_2 \neq \emptyset$, we may find $a_2 \in I \cap U_2$. Since the sets $I \cap U_j, j = 1, 2$ are disjoint we have $a_1 \neq a_2$ and we may assume $a_1 < a_2$. Note that $[a_1, a_2] \subset I$ since I is an interval. We consider the set $[a_1, a_2] \cap U_1$ which is non-empty (contains a_1) and bounded above. We define

$$b = \sup\big([a_1, a_2] \cap U_1\big) \quad \text{(note that } a_1 \leq b \leq a_2, \text{ implying } b \in I\text{).}$$

The point b belongs to $I \subset U_1 \cup U_2$. If $b \in U_1$, then there exists $\epsilon > 0$ such that

$$(\sharp) \qquad\qquad\qquad [b - \epsilon, b + \epsilon] \subset U_1.$$

Moreover we have $b < a_2$ (otherwise $b = a_2$ and $b \in U_1 \cap U_2 \cap I = \emptyset$). Thus for some $\epsilon' > 0$, we have $b + \epsilon' < a_2$ and $b + \epsilon' \in U_1$, violating the supremum property defining b. As a result we have $b \in U_2$ (thus $b > a_1$) and there exists $\epsilon'' > 0$ such that

$$(\flat) \qquad\qquad\qquad [b - \epsilon'', b + \epsilon''] \subset U_2 \cap (a_1, +\infty).$$

Since $b - \epsilon''$ is not an upper bound for $[a_1, a_2] \cap U_1$, we may find

$$c \in [a_1, a_2] \cap U_1 \text{ such that } a_1 < b - \epsilon'' < c \leq b \Longrightarrow c \in U_1 \cap I \cap U_2 = \emptyset,$$

which is impossible. This proves that $I \subset U_1$ and the result. $\qquad\square$

Definition 10.2.32. A topological space X is said to be path-connected if for all $x_0, x_1 \in X$ there exists a continuous mapping $\gamma : [0, 1] \to X$ such that $\gamma(0) = x_0, \gamma(1) = x_1$.

Proposition 10.2.33. *A path-connected topological space is connected.*

Proof. Let X be a path-connected topological space. If X is non-empty, we may find $a \in X$ such that for all $x \in X$, there exists a continuous mapping $\gamma_x : [0, 1] \to X$ with $\gamma_x(0) = a, \gamma_x(1) = x$. We have thus

$$X = \cup_{x \in X} \gamma_x([0, 1]),$$

and we note that each $\gamma_x([0, 1])$ is connected (Theorem 10.2.29) and for $x_1, x_2 \in X$

$$a \in \gamma_{x_1}([0, 1]) \cap \gamma_{x_2}([0, 1]),$$

fulfilling the assumptions of Lemma 10.2.26, entailing the result. $\qquad\square$

Remark 10.2.34. The set

$$G = \left\{ \left(x, \sin \frac{1}{x} \right) \right\}_{0 < x \leq 2/\pi} \cup \left(\{0\} \times [-1, 1] \right) \tag{10.2.7}$$

is connected, not path-connected. In fact, the function

$$(0, 2/\pi] \ni x \mapsto (x, \sin(1/x))$$

is continuous so that $G_0 = \{(x, \sin \frac{1}{x})\}_{0 < x \leq 2/\pi}$ is connected (and path-connected) as the continuous image of the interval $(0, 2/\pi]$. The set G is the closure of G_0 and thus is connected (from Proposition 10.2.30). However, G is not path-connected: for a continuous mapping $\gamma : [0, 1] \to G$ such that $\gamma(t) = (x(t), y(t))$,

$$\gamma(0) = (0, 0), \quad \gamma(1) = (2/\pi, 1),$$

we may define $T = \sup\{t \in [0, 1], x(t) = 0\}$: then $0 \leq T < 1$ and $x(t) > 0$ for $t \in (T, 1]$, so that we may assume that

$$\gamma : [0, 1] \to G, x(0) = 0, y(0) \in [-1, 1], \quad x(t) > 0 \text{ for } t \in (0, 1], \gamma(1) = (2/\pi, 1).$$

By continuity of x we have

$$x((0, 1)) \supset (0, 2/\pi) \Longrightarrow \forall \epsilon \in (0, 2/\pi), \exists t_\epsilon \in (0, 1), \epsilon = x(t_\epsilon).$$

As a consequence, we have $y(t_\epsilon) = \sin(1/\epsilon)$. Since $\lim_\epsilon t_\epsilon = 0$ (otherwise there is a sequence (ϵ_k) of positive numbers with limit 0, such that, by compactness of $[0, 1]$, $\lim_k t_{\epsilon_k} = \theta > 0$ and this would imply $\lim_k x(t_{\epsilon_k}) = x(\theta) > 0$), we must have

$$y(0) = \lim_{\epsilon \to 0_+} y(t_\epsilon) = \lim_{\epsilon \to 0_+} \sin(1/\epsilon)$$

but the latter limit does not exist. So there is no such γ and G is not path-connected.

Partitions of unity in a topological space

A topological space $(X, \mathcal{O})$ is said to be *locally compact* if every point has a compact neighborhood.

Definition 10.2.35. A topological space is said to be locally compact if it is a Hausdorff space such that each point has a compact neighborhood.

Proposition 10.2.36. *In a locally compact topological space X, every point has a basis of compact neighborhoods, i.e., $\forall x \in X, \forall U \in \mathcal{V}_x, \exists L \, compact, L \in \mathcal{V}_x, L \subset U$. More generally, let K be a compact subset of a locally compact topological space and U an open set such that $K \subset U$. Then there exists an open set V with compact closure such that*

$$K \subset V \subset \overline{V} \subset U.$$

Proof. Since every point has a compact neighborhood, we can cover K with finitely many $(W_j)_{1\leq j\leq N}$ such that W_j is open with compact closure; the set $W = \cup_{1\leq j\leq N} W_j$ is also open with compact closure, since a finite union of open sets is open and the closure of a finite union is the union of the closures. If $U = X$, we can take $V = W$. Otherwise, for each $x \in U^c$, Proposition 10.2.18 shows that there exists V_x, V_x' open disjoint such that $K \subset V_x$, $\{x\} \subset V_x'$; as a result, $(U^c \cap \overline{W} \cap \overline{V_x})_{x\in U^c}$ is a family of compact sets with empty intersection: we have $V_x \cap V_x' = \emptyset$ and thus $x \notin \overline{V_x}$, so that

$$y \in \cap_{x\in U^c}\left(U^c \cap \overline{W} \cap \overline{V_x}\right) \Longrightarrow y \in U^c, y \in \overline{W} \text{ and for all } x \in U^c, \ y \in \overline{V_x}$$
$$\Longrightarrow y \in \overline{V_y} \Longrightarrow V_y \cap V_y' \neq \emptyset, \quad \text{which is not true.}$$

From Proposition 10.2.19, we can find $x_1, \dots, x_N \in U^c$ such that

$$\emptyset = \cap_{1\leq j\leq N}\left(U^c \cap \overline{W} \cap \overline{V_{x_j}}\right) \Longrightarrow \cap_{1\leq j\leq N}\left(\overline{W} \cap \overline{V_{x_j}}\right) \subset U. \tag{10.2.8}$$

We consider now the open set $V = W \cap \cap_{1\leq j\leq N} V_{x_j}$. We have by construction $K \subset V_{x_j} \cap U$ and thus $K \subset V \subset \overline{V} \subset \overline{W} \cap \cap_{1\leq j\leq N}\overline{V_{x_j}}$, which is compact and included in U from (10.2.8). $\qquad\square$

Exercise 2.8.2 contains a proof of Urysohn's Lemma, a basic element for constructing partitions of unity. For that purpose, see also Remark 2.1.4 after Theorem 2.1.3.

Hahn–Banach Theorem

We recall here the statement of the Hahn–Banach Theorem.

Definition 10.2.37. Let E be a vector space (on $\mathbb{R}$ or $\mathbb{C}$) and let $p : E \longrightarrow \mathbb{R}_+$. We shall say that p is a semi-norm on E if for $x, y \in E, \alpha$ scalar,

(1) $p(\alpha x) = |\alpha| p(x)$, (homogeneity),
(2) $p(x + y) \leq p(x) + p(y)$, (triangle inequality)[9].

Let us consider a countable family $(p_k)_{k\geq 1}$ of semi-norms on E. We shall say that the family $(p_k)_{k\geq 1}$ is separating whenever $p_k(x) = 0$ for all $k \geq 1$ implies $x = 0$.

Theorem 10.2.38 (Hahn–Banach theorem). *Let E be a vector space (on $\mathbb{R}$ or $\mathbb{C}$), let M be a subspace of E, let p be a semi-norm on E, and let ξ be a linear form on M such that*

$$\forall x \in M, \quad |\xi \cdot x| \leq p(x). \tag{10.2.9}$$

Then there exists a linear form $\widetilde{\xi}$ on E, such that

$$\widetilde{\xi}_{|M} = \xi, \quad \text{and} \quad \forall x \in E, |\widetilde{\xi} \cdot x| \leq p(x).$$

---

[9] We note that (1) implies $p(0) = 0$ but that the separation property (first in (1.2.12)) is not satisfied in general.

Baire category theorem and its consequences

René Baire (1874–1932) was a French mathematician who made a lasting landmark contribution to Functional Analysis, known today as the *Baire Category Theorem.*

Theorem 10.2.39 (Baire theorem). *Let (X, d) be a complete metric space and $(F_n)_{n \geq 1}$ be a sequence of closed sets with empty interiors. Then the interior of $\cup_{n \geq 1} F_n$ is also empty.*

N.B. The statement of that theorem is equivalent to saying that, in a complete metric space, given a sequence $(U_n)_{n \geq 1}$ of open dense sets the intersection $\cap_{n \geq 1} U_n$ is also dense. In fact, if (U_n) is a sequence of open dense sets, the sets $F_n = U_n^c$ are closed and $\operatorname{int} F_n = \emptyset \Longleftrightarrow \emptyset = \operatorname{int}(U_n^c) = (\overline{U}_n)^c \Longleftrightarrow \overline{U}_n = X$, so that

$$\operatorname{int}(\cup_{n \geq 1} F_n) = \emptyset \Longleftrightarrow \emptyset = \operatorname{int}(\cup_{n \geq 1} U_n^c) = \operatorname{int}\big((\cap_{n \geq 1} U_n)^c\big) = \left(\overline{(\cap_{n \geq 1} U_n)}\right)^c$$

which is equivalent to $\overline{(\cap_{n \geq 1} U_n)} = X$.

Proof of the theorem. Let $(U_n)_{n \geq 1}$ be a sequence of dense open sets. Let $x_0 \in X, r_0 > 0$ (we may assume that X is not empty, otherwise the theorem is trivial). Using the density of U_1, we obtain $B(x_0, r_0) \cap U_1 \neq \emptyset$ so that

$$\exists r_1 \in]0, r_0/2[, \quad B(x_0, r_0) \cap U_1 \supset B(x_1, 2r_1) \supset \tilde{B}(x_1, r_1) = \{y \in X, d(y, x_1) \leq r_1\}.$$

Let us assume that we have constructed $x_0, x_1, \ldots, x_n$ with $n \geq 1$ such that

$$B(x_k, r_k) \cap U_{k+1} \supset \tilde{B}(x_{k+1}, r_{k+1}), \quad 0 < r_{k+1} < r_k/2, \quad 0 \leq k \leq n - 1.$$

Using the density of U_{n+1}, we obtain $B(x_n, r_n) \cap U_{n+1} \neq \emptyset$ and

$$\exists r_{n+1} \in]0, r_n/2[, \quad B(x_n, r_n) \cap U_{n+1} \supset B(x_{n+1}, 2r_{n+1}) \supset \tilde{B}(x_{n+1}, r_{n+1}).$$

Since $0 < r_n \leq 2^{-n} r_0$ (induction), we have $\lim_n r_n = 0$ and $(x_n)_{n \geq 0}$ is a Cauchy sequence since for $k, l \geq n$,

$$B(x_k, r_k) \cup B(x_l, r_l) \subset B(x_n, r_n) \Longrightarrow d(x_k, x_l) < 2r_n.$$

Since the metric space X is assumed to be complete, the sequence $(x_n)_{n \geq 0}$ converges; let $x = \lim_n x_n$. We have for all $n \geq 0$, $\tilde{B}(x_{n+1}, r_{n+1}) \subset B(x_n, r_n)$ so that, for all $k \geq 1$, $\tilde{B}(x_{n+k}, r_{n+k}) \subset B(x_n, r_n)$ and thus

$$\sup_{k \geq 0} d(x_{n+k}, x_n) \leq r_n \Longrightarrow d(x, x_n) \leq r_n \Longrightarrow x \in \cap_{n \geq 1} \tilde{B}(x_n, r_n) \subset \cap_{n \geq 1} U_n$$

and $d(x, x_0) \leq r_0$. As a result, for all $x_0 \in X$, all $r_0 > 0$, the set

$$\tilde{B}(x_0, r_0) \cap \cap_{n \geq 1} U_n \neq \emptyset.$$

This implies that $U = \cap_{n \geq 1} U_n$ is dense since, for $x_0 \in X$, for any neighborhood V of x_0, there exists $r_0 > 0$ such that $V \supset B(x_0, 2r_0) \supset \tilde{B}(x_0, r_0)$, and thus $V \cap U \supset \tilde{B}(x_0, r_0) \cap U \neq \emptyset \Longrightarrow x_0 \in \overline{U}$. $\qquad\square$

Theorem 10.2.40. *Let X be a locally compact topological space (Hausdorff topological space such that each point has a compact neighborhood) and $(F_n)_{n\geq 1}$ be a sequence of closed sets with empty interiors. Then the interior of $\cup_{n\geq 1}F_n$ is also empty.*

Proof. The proof is essentially the same as for the previous theorem. Let $(U_n)_{n\geq 1}$ be a sequence of dense open sets. Let B_0 a non-empty open subset of X. Since $\overline{U}_1$ is dense, the open set $B_0 \cap U_1$ is non-empty and thus is a neighborhood of a point. Since each point in X has a basis of compact neighborhoods, $B_0 \cap U_1$ contains a compact set with non-empty interior and thus

$$B_0 \cap U_1 \supset \overline{B}_1, \quad \overline{B}_1 \text{ compact}, \ B_1 \text{ open} \neq \emptyset.$$

We get that $B_1 \cap U_2$ is a non-empty open set which contains a compact $\overline{B}_2$, B_2 open $\neq \emptyset$. Following the same procedure as in the previous proof, we may consider the compact set K defined by $K = \cap_{n\geq 1}\overline{B}_n$. The set K is non-empty, otherwise we would have $\emptyset = \cap_{1\leq n\leq N}\overline{B}_n = \overline{B}_N$ for some N, which is not possible since at each step, the set $\overline{B}_N$ is compact with non-empty interior. As a result, we have

$$\emptyset \neq K \subset \cap_{n\geq 1}U_n = U, \quad K \subset B_0,$$

and thus, for any open subset B_0 of X, the set $U \cap B_0 \neq \emptyset$, which means that $\overline{U} = X$. $\square$

Definition 10.2.41. Let X be a topological space and $A \subset X$.

- The subset A is said to be *rare* or *nowhere dense* when $\overset{\circ}{\overline{A}} = \emptyset$.
- The subset A is of *first category* when it is a countable union of rare subsets. Such a subset is also said to be *meager*.
- The subset A of X is of *second category* when it is not of first category.

A topological space X is a *Baire space* if for any sequence $(F_n)_{n\in\mathbb{N}}$ of closed sets with empty interiors, the union $\cup_{n\in\mathbb{N}}F_n$ is also with empty interior. As shown above, X is a *Baire space* if and only if, for any sequence $(U_n)_{n\in\mathbb{N}}$ of dense open sets, the intersection $\cap_{n\in\mathbb{N}}U_n$ is also dense.

The following results are classical consequences of Baire's Theorem.

Banach–Steinhaus

Theorem 10.2.42 (Banach–Steinhaus). *Let E be a Banach space, F be a normed vector space and $(L_j)_{j\in J}$ be a family of $\mathcal{L}(E, F)$ (continuous linear mappings from E to F) which is "weakly bounded", i.e., satisfies*

$$\forall u \in E, \quad \sup_{j\in J}\|L_j u\|_F < +\infty. \tag{10.2.10}$$

Then the family $(L_j)_{j\in J}$ is "strongly bounded", i.e., satisfies

$$\sup_{j\in J}\|L_j\|_{\mathcal{L}(E,F)} < +\infty. \tag{10.2.11}$$

Open mapping Theorem

Theorem 10.2.43 (Open mapping Theorem). *Let E, F be Banach spaces and let A be a bijective mapping belonging to $\mathcal{L}(E, F)$. Then A is an isomorphism, i.e.,*

$$\exists \beta, \gamma > 0, \quad \forall u \in E, \quad \beta \|u\|_E \leq \|Au\|_F \leq \gamma \|u\|_E. \tag{10.2.12}$$

10.3   Duality in Banach spaces

Definitions

All the vector spaces considered here are on the field $\mathbb{R}$ or $\mathbb{C}$, denoted by k. We recall that a Banach space is a complete normed vector space and for E, F Banach spaces, $\mathcal{L}(E, F)$ stands for the vector space of continuous linear mappings from E into F. The space $\mathcal{L}(E, F)$ is a Banach space for the norm

$$\|L\|_{\mathcal{L}(E,F)} = \sup_{\|x\|_E = 1} \|Lx\|_F. \tag{10.3.1}$$

The topological dual of E is the Banach space $E^* = \mathcal{L}(E, \mathrm{k})$ of continuous linear forms. When $\xi \in E^*, x \in E$, we shall write $\xi \cdot x$ instead of $\xi(x)$.

Theorem 10.3.1. *Let E be a Banach space and E^* its topological dual. Then*

$$\forall x \in E, \quad \|x\|_E = \sup_{\|\xi\|_{E^*} = 1} |\xi \cdot x|.$$

Proof. We have $\|\xi\|_{E^*} = \sup_{x \in E, \|x\|_E = 1} |\xi \cdot x|$. Let $0 \neq x_0 \in E$. Applying the Hahn–Banach Theorem 10.2.38 with $M = \mathrm{k}x_0$, $p(x) = \|x\|_E$, defining on M the linear form η by $\eta \cdot \lambda x_0 = \lambda \|x_0\|_E$, we have $|\eta \cdot \lambda x_0| \leq \|\lambda x_0\| = p(\lambda x_0)$ and we find a linear form ξ_0 defined on E such that

$$|\xi_0 \cdot x_0| = \|x_0\|_E, \quad \forall x \in E, \ |\xi_0 \cdot x| \leq \|x\|_E.$$

As a consequence, $\xi_0 \in E^*$ with $\|\xi_0\| = 1$. Finally we have proven

$$\|x_0\|_E = |\xi_0 \cdot x_0| \leq \sup_{\|\xi\|_{E^*} = 1} |\xi \cdot x_0| \leq \|x_0\|_E. \qquad \square$$

Weak convergence

Definition 10.3.2. Let E be a Banach space. The weak topology $\sigma(E, E^*)$ on E is the weakest topology such that for all $\xi \in E^*$ the mappings $E \ni x \mapsto \langle \xi, x \rangle_{E^*, E} \in \mathrm{k}$ are continuous.

Remark 10.3.3. Let E be a Banach space. For each $\xi \in E^*$, we define the semi-norm p_ξ on E by $p_\xi(x) = |\langle \xi, x \rangle_{E^*, E}|$; the properties of Definition 10.2.37 are obviously satisfied. Moreover the family $(p_\xi)_{\xi \in E^*}$ is separating from Theorem 10.3.1.

The neighborhoods of 0 for the weak topology on E, say $\mathcal{V}_0$, have the following basis: taking Ξ a finite subset of E^* and $r > 0$, we define

$$W_{\Xi,r} = \{x \in E, \forall \xi \in \Xi, p_\xi(x) < r\}. \tag{10.3.2}$$

Note that the $W_{\Xi,r}$ are convex and symmetric. Every neighborhood of 0 for the weak topology contains a $W_{\Xi,r}$ which is also a neighborhood of 0 for that topology. The neighborhoods $\mathcal{V}_x$ of a point x are defined as $\mathcal{V}_x = \{x + V\}_{V \in \mathcal{V}_0}$; E equipped with that topology is a Topological Vector Space. Note that the separating property of the family $(p_\xi)_{\xi \in E^*}$ is implying that the weak topology is separated (i.e., Hausdorff, see (10.2.5)): in fact $\{0\}$ is closed for the weak topology, since for $x_0 \neq 0$, from Theorem 10.3.1, there exists $\xi_0 \in E^*$ such that $\langle \xi_0, x_0 \rangle = 1$, so that

$$0 \notin x_0 + \{x \in E, p_{\xi_0}(x) < 1\}.$$

Otherwise, $1 = \langle \xi_0, x_0 \rangle = \langle \xi_0, \overbrace{x_0 + x}^{=0} \rangle - \langle \xi_0, x \rangle < 1$. Moreover, to check that the addition is continuous, we take $x_1, x_2 \in E$, W_{Ξ_0, r_0} as above a neighborhood of zero (Ξ_0 finite and $r_0 > 0$), and we try to find $W_{\Xi_j, r_j}, j = 1, 2$ such that

$$x_1 + W_{\Xi_1, r_1} + x_2 + W_{\Xi_2, r_2} \subset x_1 + x_2 + W_{\Xi_0, r_0}.$$

It is enough to take $W_{\Xi_j, r_j} = W_{\Xi_0, r_0/2}$. Checking the continuity of the multiplication by a scalar is similar: given $\lambda_0 \in \mathbf{k}$, $\mathbf{x}_0 \in \mathbf{E}$, W_{Ξ_0, r_0} as above, we want to find W_{Ξ_1, r_1} and $t_1 > 0$ such that

$$\forall t \in \mathbb{R}, \ |t| \leq t_1, \quad (\lambda_0 + \theta t)(x_0 + W_{\Xi_1, r_1}) \subset \lambda_0 x_0 + W_{\Xi_0, r_0}.$$

It is enough to require

$$t_1 W_{\Xi_1, r_1} \cup \lambda_0 W_{\Xi_1, r_1} \subset W_{\Xi_0, r_0/3}, \quad t_1 x_0 \subset W_{\Xi_0, r_0/3}.$$

This is satisfied for $\Xi_1 = \Xi_0$, $\quad |\lambda_0| r_1 < r_0/3$, $\quad t_1 r_1 < r_0/3$.

Remark 10.3.4. Let E be a Banach space; the weak topology $\sigma(E, E^*)$ on E is weaker than the norm-topology on E (also called the strong topology): this is obvious from the very definition of the weak topology since all the mappings $x \mapsto \langle \xi, x \rangle$ are continuous for the norm-topology since $p_\xi(x) = |\langle \xi, x \rangle| \leq \|\xi\|_{E^*} \|x\|_E$.

Let E be a Banach space and $x \in E$; a sequence $(x_n)_{n \in \mathbb{N}}$ in E is weakly converging to x means that

$$\forall \xi \in E^*, \quad \lim_n \langle \xi, x_n \rangle_{E^*, E} = \langle \xi, x \rangle_{E^*, E}. \quad \text{We write} \quad x_n \rightharpoonup x, \tag{10.3.3}$$

or to avoid confusion between the arrows $\rightharpoonup$ and $\to$, we may write

$$x_n \xrightarrow[\sigma(E, E^*)]{} x.$$

Proposition 10.3.5. *Let E be a Banach space and $(x_n)_{n\in\mathbb{N}}$ be a weakly converging sequence with limit x in E. Then $\|x_n\|_E$ is bounded and $\|x\|_E \leq \liminf_n \|x_n\|_E$. If $(\xi_n)_{n\in\mathbb{N}}$ is a strongly converging sequence in E^* with limit ξ, then*

$$\lim_n \langle \xi_n, x_n \rangle_{E^*,E} = \langle \xi, x \rangle_{E^*,E}.$$

Proof. We consider the sequence of linear forms on E^* given by $E^* \ni \xi \mapsto \langle \xi, x_n \rangle$. Since for all $\xi \in E^*$, the numerical sequence $\langle \xi, x_n \rangle$ is converging, we may apply the Banach–Steinhaus Theorem to get that $E^* \ni \xi \mapsto \langle \xi, x \rangle$ is continuous on E^*, i.e.,

$$\exists C > 0, \forall \xi \in E^*, \quad |\langle \xi, x \rangle| \leq C\|\xi\|_{E^*}.$$

Using Theorem 10.3.1, this implies $\|x\|_E \leq C$. The Banach–Steinhaus theorem 10.2.42 implies also that the norms of the linear forms $E^* \ni \xi \mapsto \langle \xi, x_n \rangle$ make a bounded sequence, and since that norm is $\|x_n\|_E$, we get that sequence $(\|x_n\|_E)$ is bounded. We have for $\xi \in E^*$ with $\|\xi\|_{E^*} = 1$, using again Theorem 10.3.1,

$$|\langle \xi, x \rangle| = \lim_n |\langle \xi, x_n \rangle| \leq \liminf_n \|x_n\|_E \implies \|x\|_E \leq \liminf_n \|x_n\|_E.$$

Moreover, we have for a strongly converging sequence $(\xi_n)_{n\in\mathbb{N}}$ with limit ξ in E^*,

$$|\langle \xi_n, x_n \rangle - \langle \xi, x \rangle| \leq |\langle \xi_n - \xi, x_n \rangle| + |\langle \xi, x_n - x \rangle|$$
$$\leq \underbrace{\|\xi_n - \xi\|_{E^*}}_{\to 0} \sup_n \|x_n\|_E + \underbrace{|\langle \xi, x_n - x \rangle|}_{\to 0},$$

which implies $\lim_n \langle \xi_n, x_n \rangle = \langle \xi, x \rangle$. $\qquad\qquad\square$

Remark 10.3.6. When the Banach space E is infinite dimensional, the weak topology $\sigma(E, E^*)$ is strictly weaker than the strong topology given by the norm of E. Let us prove that the unit sphere of E, $S = \{x \in E, \|x\|_E = 1\}$ is not closed in the weak topology $\sigma(E, E^*)$ if E is not finite dimensional. Let us consider $x_0 \in E$ with $\|x_0\|_E < 1$; let W_{Ξ_0, r_0} be a neighborhood of zero for the weak topology as in (10.3.2). We claim that

$$(x_0 + W_{\Xi_0, r_0}) \cap S \neq \emptyset. \tag{10.3.4}$$

This will imply that x_0 belongs to the closure of S for the $\sigma(E, E^*)$ topology. To prove (10.3.4), we consider the finite subset $\Xi_0 = \{\xi_j\}_{1 \leq j \leq N}$ of E^*; each $\ker \xi_j$ is a closed hyperplane, and since E is infinite dimensional, $\cap_{1 \leq j \leq N} \ker \xi_j$ is not reduced to $\{0\}$ (otherwise the mapping $E \ni x \mapsto L(x) = (\langle \xi_j, x \rangle)_{1 \leq j \leq N} \in \mathbb{R}^N$ would be injective and L would be an isomorphism from E onto $L(E)$, implying that E is finite dimensional). Taking now a non-zero $x_1 \in \cap_{1 \leq j \leq N} \ker \xi_j$, we see that the continuous function f on $\mathbb{R}$ given by $f(\theta) = \|x_0 + \theta x_1\|$ is such that

$$f(\mathbb{R}_+) \supset [\|x_0\|, +\infty[\implies \exists \theta \in \mathbb{R}, x_0 + \theta x_1 \in S.$$

This proves (10.3.4) since $x_0 + \theta x_1 \in x_0 + W_{\Xi_0, r_0}$ because $\langle \xi_j, x_1 \rangle = 0$ for all $j \in \{1, \ldots, N\}$.

Weak-$*$ convergence on E^*

Definition 10.3.7. Let E be a Banach space and E^* its topological dual. The weak-$*$ topology on E^*, denoted by $\sigma(E^*, E)$, is the weakest topology such that the mappings $E^* \ni \xi \mapsto \xi \cdot x \in \mathbf{k}$ are continuous for all $x \in E$. A sequence $(\xi_k)_{k \in \mathbb{N}}$ of E^* is weakly-$*$ converging means that $\forall x \in E$, the sequence $(\xi_k \cdot x)_{k \in \mathbb{N}}$ converges.

Proposition 10.3.8. *Let E be a Banach space and $(\xi_n)_{n \in \mathbb{N}}$ be a weakly-$*$ converging sequence with limit ξ in E^*. Then $\|\xi_n\|_{E^*}$ is bounded and $\|\xi\|_{E^*} \leq \liminf_n \|\xi_n\|_{E^*}$. Let $(x_n)_{n \in \mathbb{N}}$ be a strongly converging sequence in E with limit x. Then we have*

$$\lim_n \langle \xi_n, x_n \rangle_{E^*, E} = \langle \xi, x \rangle_{E^*, E}.$$

Proof. We have for $x \in E$ with $\|x\|_E = 1$,

$$|\langle \xi, x \rangle| = \lim_n |\langle \xi_n, x \rangle| \leq \liminf_n \|\xi_n\|_{E^*} \implies \|\xi\|_{E^*} \leq \liminf_n \|\xi_n\|_E.$$

From the Banach–Steinhaus Theorem 10.2.42 the sequence $(\xi_n)_{n \in \mathbb{N}}$ is bounded in the normed space E^* and we define $\sup_n \|\xi_n\|_{E^*} = M < \infty$. We have then

$$|\langle \xi_n, x_n \rangle - \langle \xi, x \rangle| \leq |\langle \xi_n, x_n - x \rangle| + |\langle \xi_n - \xi, x \rangle| \leq M\|x_n - x\|_E + |\langle \xi_n - \xi, x \rangle|,$$

and since $\lim_n \|x_n - x\|_E = 0 = \lim_n \langle \xi_n - \xi, x \rangle$, we obtain the result. $\qquad\square$

Lemma 10.3.9 (Diagonal Process). *Let $(a_{ij})_{i,j \in \mathbb{N}^*}$ be an infinite matrix of elements of a metric space A. We assume that each line is relatively compact, i.e., for all $i \in \mathbb{N}^*$, the set $\{a_{i,j}\}_{j \geq 1}$ is relatively compact. Then, there exists a strictly increasing mapping ν from $\mathbb{N}^*$ into itself such that, for all $i \in \mathbb{N}^*$, the sequence $\big(a_{i, \nu(k)}\big)_{k \in \mathbb{N}^*}$ converges.*

Proof of the lemma. We can extract a converging subsequence

$$(a_{1, n_1(k)})_{k \geq 1} \text{ from the first line } (a_{1,j})_{j \geq 1}.$$

We can extract a converging subsequence

$$(a_{2, n_1(n_2(k))})_{k \geq 1} \text{ from a subsequence of the second line } (a_{2, n_1(k)})_{j \geq 1}.$$

We can extract a converging subsequence

$$(a_{3, n_1(n_2(n_3(k)))})_{k \geq 1} \text{ from a subsequence of the third line } (a_{3, n_1(n_2(k))})_{j \geq 1}.$$

$$\cdots\cdots\cdots\cdots\cdots\cdots\cdots\cdots\cdots\cdots\cdots$$

For all $i \geq 1$, we can extract a converging subsequence

$$\big(a_{i, (n_1 \circ \cdots \circ n_i)(k)}\big)_{k \geq 1}.$$

Note that the mappings n_l are strictly increasing from $\mathbb{N}^*$ into itself and thus satisfy $\forall k \geq 1, n_l(k) \geq k$ (true for $k = 1$ and $n_l(k+1) > n_l(k) \geq k$ gives $n_l(k+1) \geq k+1$). We define

$$b_{i,k} = a_{i,\nu(k)}, \quad \text{with} \quad \nu(k) = (n_1 \circ \cdots \circ n_k)(k).$$

The mapping ν sends $\mathbb{N}^*$ into itself and is strictly increasing:

$$\nu(k+1) = (n_1 \circ \cdots \circ n_{k+1})(k+1) \underbrace{\geq}_{} (n_1 \circ \cdots \circ n_k)(k+1)$$

with $\underset{n_1 \circ \cdots \circ n_k \nearrow \text{strict}}{>} (n_1 \circ \cdots \circ n_k)(k) = \nu(k)$, since $n_{k+1}(k+1) \geq k+1$.

Moreover, the sequence $(b_{i,k})_{k, k>i}$ is a subsequence of the converging sequence

$$\big(a_{i,(n_1 \circ \cdots \circ n_i)(k)}\big)_{k \geq 1}$$

since for $k > i \geq 1$, $\nu(k) = (n_1 \circ \cdots \circ n_i)\big((n_{i+1} \circ \cdots \circ n_k)(k)\big)$ and

$$\mu_i(k+1) = (n_{i+1} \circ \cdots \circ n_{k+1})(k+1) \geq (n_{i+1} \circ \cdots \circ n_k)(k+1)$$
$$> (n_{i+1} \circ \cdots \circ n_k)(k) = \mu_i(k).$$

As a result, the sequence $(a_{i,\nu(k)})_{k \geq 1}$ is converging, which proves the lemma. $\quad\square$

Theorem 10.3.10. *Let E be a separable Banach space. The closed unit ball of E^* equipped with the weak-$*$ topology is (compact and) sequentially compact.*

Proof. Let $(\xi_j)_{j \in \mathbb{N}}$ be a sequence of E^* with $\sup_{j \in \mathbb{N}} \|\xi_j\|_{E^*} \leq 1$. Let $\{x_i\}_{i \in \mathbb{N}}$ be a countable dense part of E. For each $i \in \mathbb{N}$, we define $y_i : E^* \longrightarrow \mathbb{k}$ by $y_i(\xi) = \xi \cdot x_i$. Let us now consider the matrix with entries $(\xi_j \cdot x_i)_{i,j \in \mathbb{N}}$. For all $i \in \mathbb{N}$, we have

$$\sup_{j \in \mathbb{N}} |\xi_j \cdot x_i| \leq \|x_i\|_E$$

so that we can apply the diagonal process given by Lemma 10.3.9 and find ν strictly increasing from $\mathbb{N}$ to $\mathbb{N}$ such that $\forall i \in \mathbb{N}$, the sequence $(\xi_{\nu(k)} \cdot x_i)_{k \in \mathbb{N}}$ is converging. As a consequence, for $x \in E$,

$$|\xi_{\nu(k)} \cdot x - \xi_{\nu(l)} \cdot x|$$
$$\leq |\xi_{\nu(k)} \cdot x - \xi_{\nu(k)} \cdot x_i| + |\xi_{\nu(k)} \cdot x_i - \xi_{\nu(l)} \cdot x_i| + |\xi_{\nu(l)} \cdot x_i - \xi_{\nu(l)} \cdot x|$$
$$\leq 2\|x - x_i\|_E + |\xi_{\nu(k)} \cdot x_i - \xi_{\nu(l)} \cdot x_i|.$$

Let $\epsilon > 0$ be given and $x \in E$. Let $i \in \mathbb{N}$ such that $\|x - x_i\|_E < \epsilon/4$; since the sequence $(\xi_{\nu(k)} \cdot x_i)_{k \in \mathbb{N}}$ is converging, for $k, l \geq N_\epsilon$, $|\xi_{\nu(k)} \cdot x_i - \xi_{\nu(l)} \cdot x_i| < \epsilon/2$ and thus for $k, l \geq N_\epsilon$, $|\xi_{\nu(k)} \cdot x - \xi_{\nu(l)} \cdot x| < \epsilon$, proving the weak convergence of the sequence $(\xi_{\nu(k)})_{k \in \mathbb{N}}$. $\quad\square$

Remark 10.3.11. Let E be a Banach space and E^* its topological dual. For $x \in E, \xi \in E^*$, we define $p_x(\xi) = |\xi \cdot x|$. For each $x \in E$, p_x is (trivially) a semi-norm on E^*. The family $(p_x)_{x \in E}$ is a separating[10] (uncountable) family of semi-norms on E^*. We shall say that U is a neighborhood of 0 in the weak-$*$ topology if it contains a finite intersection of sets

$$V_{p_x, r} = \{\xi \in E^*, p_x(\xi) < r\}, \quad x \in E, r > 0.$$

The family of semi-norms $(p_x)_{x \in E}$ describes the weak-$*$ topology on E^*, also denoted by $\sigma(E^*, E)$.

Remark 10.3.12. Given a Banach space E and its topological dual E^*, we can define on E^* several weak topologies: the weak-$*$ topology $\sigma(E^*, E)$ described above, but also the weak topology on E^*, $\sigma(E^*, E^{**})$, where E^{**} is the *bidual* of E, i.e., the topological dual of the Banach space E^*. Note that the weak topology on E^* is stronger than the weak-$*$ topology, since $E \subset E^{**}$ as shown below.

Reflexivity

Proposition 10.3.13. *Let E be a Banach space. The bidual of E is defined as the (topological) dual of E^*. The mapping $E \ni x \mapsto j(x) \in E^{**}$ defined by*

$$\langle j(x), \xi \rangle_{E^{**}, E^*} = \langle \xi, x \rangle_{E^*, E}$$

is linear isometric and is an isomorphism on its image $j(E)$ which is a closed subspace of E^{**}. A Banach space is said to be reflexive when j is bijective (this implies in particular that E^{**} and E are isometrically isomorphic).

Proof. For $x \in E$, we have

$$
\begin{aligned}
\|j(x)\|_{E^{**}} &= \sup_{\|\xi\|_{E^*}=1} |\langle j(x), \xi \rangle_{E^{**}, E^*}| \\
&= \sup_{\|\xi\|_{E^*}=1} |\langle \xi, x \rangle_{E^*, E}| \underbrace{=}_{\text{thm } 10.3.1} \|x\|_E,
\end{aligned}
\tag{10.3.5}
$$

and thus j is isometric and obviously linear. The image $j(E)$ is closed: whenever a sequence $(j(x_k))_{k \geq 1}$ converges, it is also a Cauchy sequence as well as $(x_k)_{k \geq 1}$ since $\|x_k - x_l\|_E \leq \|j(x_k - x_l)\|_{E^{**}} = \|j(x_k) - j(x_l)\|_{E^{**}}$. As a result, the sequence $(x_k)_{k \geq 1}$ converges to some limit $x \in E$, and the continuity of j (consequence of the isometry property) ensures $\lim_k j(x_k) = j(x)$, proving that $j(E)$ is closed, and thus a Banach space for the norm of E^{**}. The mapping $j : E \longrightarrow j(E)$ is an isometric isomorphism of Banach spaces. $\qquad\square$

---

[10]If for some $\xi \in E^*$, we have $\forall x \in E, p_x(\xi) = 0$, it means $\forall x \in E, \xi \cdot x = 0$, i.e., $\xi = 0_{E^*}$.

Remark 10.3.14. Let E be a Banach space; then the bidual of E^* is equal to the dual of E^{**}, so that $\left(E^*\right)^{**} = \left((E^{**})\right)^*$, that we shall denote simply as E^{***}: we have by definition

$$\left(E^*\right)^{**} = \left(\left(E^*\right)^*\right)^*,$$

as well as

$$\left((E^{**})\right)^* = \left(\left(E^*\right)^*\right)^*.$$

Theorem 10.3.15 (Banach–Alaoglu). *Let E be a Banach space. The closed unit ball $\mathcal{B}$ of E^* is compact for the weak-$*$ topology.*

Proof. For each $x \in E$, the mapping $E^* \ni \xi \mapsto \xi \cdot x \in \mathbb{C}$ is continuous in the weak-$*$ topology; since $|\xi \cdot x| \leq \|\xi\|_{E^*} \|x\|_E$ we see that

$$\mathcal{B} \subset \prod_{x \in E} (\|x\|_E D_1), \quad D_1 = \{z \in \mathbb{C}, |z| \leq 1\},$$

and the product topology on $\prod_{x \in E}(\|x\|_E D_1)$ induces the weak-$*$ topology on $\mathcal{B}$. Using Tychonoff's Theorem 10.2.23, we see that the set $\mathcal{B}$ is a closed subset of a compact set and is thus compact. $\qquad\square$

Proposition 10.3.16. *Let E be a Banach space and B its closed unit ball. The following properties are equivalent.*

  (i) *E is reflexive,*
 (ii) *E^* is reflexive,*
(iii) *B is weakly compact, i.e., compact for the $\sigma(E, E^*)$ topology.*

Proof. Let us assume that (i) is satisfied. Then the mapping j defined by Proposition 10.3.13 is an isometric isomorphism from E to E^{**} and the weak-$*$ topology on E is well defined as the topology $\sigma(E = E^{**}, E^*)$, which is simply the weak topology on E. The Banach–Alaoglu theorem implies that the closed unit ball of $E^{**} = E$, which is thus B, is weak-$*$ compact, i.e., is weakly compact, proving (iii). Before going on with the proof of the proposition, we need a lemma.

Lemma 10.3.17. *Let E be a Banach space, B its closed unit ball and j be defined by Proposition 10.3.13. Then j is a homeomorphism of the topological space $\left(E, \sigma(E, E^*)\right)$ onto a dense subspace of the topological space $\left(E^{**}, \sigma(E^{**}, E^*)\right)$. The set $j(B)$ is dense for the $\sigma(E^{**}, E^*)$ topology in the closed unit ball of E^{**}.*

Proof of the lemma. The mapping $j : E \to j(E) \subset E^{**}$ is bijective and continuous whenever E is equipped with the weak topology $\sigma(E, E^*)$ and E^{**} with the weak-$*$ topology $\sigma(E^{**}, E^*)$: we consider a semi-norm q_ξ on E^{**}, $\xi \in E^*$, defined by

$$q_\xi(X) = |\langle X, \xi \rangle_{E^{**}, E^*}|.$$

We evaluate for $x \in E$, $q_\xi(j(x)) = |\langle j(x), \xi \rangle_{E^{**}, E^*}| = |\langle \xi, x \rangle_{E^*, E}| = p_\xi(x)$, where p_ξ is a semi-norm on E (for the weak topology). The previous equality proves that j is an homeomorphism from E to $j(E)$. A consequence of the isometry property of j given in Proposition 10.3.13 is that $j(B)$ is included in the closed unit ball B_{**} of E^{**}. Let $\widetilde{B}$ be the closure for $\sigma(E^{**}, E^*)$ of $j(B)$. First of all, B_{**} is $\sigma(E^{**}, E^*)$ compact from the Banach–Alaoglu theorem and thus is $\sigma(E^{**}, E^*)$ closed, so that $\widetilde{B} \subset B_{**}$. If there is some $X_0 \in B_{**} \backslash \widetilde{B}$, the Hahn–Banach theorem implies that there exists $\xi_0 \in E^*, \alpha \in \mathbb{R}, \epsilon > 0$ with

$$\forall x \in B, \quad \mathrm{Re}\langle \xi_0, x \rangle < \alpha < \alpha + \epsilon < \mathrm{Re}\langle X_0, \xi_0 \rangle.$$

Since $0 \in B$, this implies $\alpha > 0$. We may thus multiply the previous inequality by $1/\alpha$ and find $\xi_1 \in E^*, \epsilon_1 > 0$ such that

$$\forall x \in B, \quad \mathrm{Re}\langle \xi_1, x \rangle < 1 < 1 + \epsilon_1 < \mathrm{Re}\langle X_0, \xi_1 \rangle.$$

Using that B is stable by multiplication by $z \in \mathbb{C}$ with $|z| = 1$, we get $\|\xi_1\|_{E^*} \leq 1$, implying that $1 + \epsilon_1 < \mathrm{Re}\langle X_0, \xi_1 \rangle \leq \|X_0\|_{E^{**}} \leq 1$ which is impossible. The proof of the lemma is complete. $\qquad\square$

Going back to the proof of the proposition, we assume that (iii) holds. Then, using the previous lemma, we see that j is continuous from

$$(E, \sigma(E, E^*)) \text{ in } (E^{**}, \sigma(E^{**}, E^*))$$

and since B is compact for the $(E, \sigma(E, E^*))$ topology, we infer that $j(B)$ is compact. But the same lemma gives that $j(B)$ is dense for the $\sigma(E^{**}, E^*)$ topology in the closed unit ball of E^{**}, so $j(B)$ is closed and equal to the closed unit ball of E^{**}, implying that j is onto and (i).

We know now that (i) is equivalent to (iii), so that (ii) is equivalent to the compactness of the closed unit ball B_* of E^* in the topology $\sigma(E^*, E^{**})$. The Banach–Alaoglu theorem shows that B_* is compact for $\sigma(E^*, E)$ and if (i) holds, that topology is $\sigma(E^*, E^{**})$, so that (i) implies (ii).

Finally we assume that (ii) holds, i.e., E^* is reflexive. Let us first consider the norm-closed subspace $j(E)$ of E^{**}. The space E^{**} is reflexive since $E^* = E^{***}$ by (ii) and thus $E^{**} = E^{****}$. As a consequence, the unit ball of E^{**} is compact for the topology $\sigma(E^{**}, E^{***}) = \sigma(E^{**}, E^*)$ and thus the unit ball of the norm-closed subspace $j(E)$ is compact for the $\sigma(j(E), E^*) = \sigma(j(E), (j(E))^*)$ topology, which proves that $j(E)$ and thus E is reflexive. The proof of the proposition is complete. $\qquad\square$

10.4   Calculating antiderivatives

Table of classical antiderivatives

Let f be a continuous function on an open subset I of $\mathbb{R}$. We shall denote by $\int f(x)dx$ any antiderivative of f on I. The 33 most classical formulas are the following ones.

$$(1) \quad \int x^\alpha \, dx \quad = \frac{x^{\alpha+1}}{\alpha+1}, \qquad\qquad \text{for } \alpha \neq -1, \ I = (0, +\infty).$$

$$(2) \quad \int x^{-1} \, dx \quad = \ln|x|, \qquad\qquad I = \mathbb{R}^*.$$

$$(3) \quad \int e^{zx} \, dx \quad = z^{-1}e^{zx}, \qquad\qquad \text{for } z \neq 0, \ I = \mathbb{R}.$$

$$(4) \quad \int \tan x \, dx \quad = -\ln|\cos x|, \qquad\qquad I = \mathbb{R}\backslash(\tfrac{\pi}{2} + \pi\mathbb{Z}).$$

$$(5) \quad \int \cot x \, dx \quad = \ln|\sin x|, \qquad\qquad I = \mathbb{R}\backslash\pi\mathbb{Z}.$$

$$(6) \quad \int \frac{1}{\cos x} \, dx \quad = \ln\left|\tan\left(\frac{x}{2} + \frac{\pi}{4}\right)\right|, \qquad\qquad I = \mathbb{R}\backslash(\tfrac{\pi}{2} + \pi\mathbb{Z}).$$

$$(7) \quad \int \frac{1}{\sin x} \, dx \quad = \ln\left|\tan\left(\frac{x}{2}\right)\right|, \qquad\qquad I = \mathbb{R}\backslash\pi\mathbb{Z}.$$

$$(8) \quad \int \arcsin x \, dx \quad = x \arcsin x + \sqrt{1 - x^2}, \qquad\qquad I = (-1, 1).$$

$$(9) \quad \int \arccos x \, dx \quad = x \arccos x - \sqrt{1 - x^2}, \qquad\qquad I = (-1, 1).$$

$$(10) \quad \int \arctan x \, dx \quad = x \arctan x - \frac{1}{2}\ln(1 + x^2), \qquad\qquad I = \mathbb{R}.$$

$$(11) \quad \int \sin^2 x \, dx \quad = \frac{x}{2} - \frac{\sin(2x)}{4}, \qquad\qquad I = \mathbb{R}.$$

$$(12) \quad \int \cos^2 x \, dx \quad = \frac{x}{2} + \frac{\sin(2x)}{4}, \qquad\qquad I = \mathbb{R}.$$

$$(13) \quad \int \frac{1}{\cos^2 x} \, dx \quad = \tan x, \qquad\qquad I = \mathbb{R}\backslash(\tfrac{\pi}{2} + \pi\mathbb{Z}).$$

$$(14) \quad \int \frac{1}{\sin^2 x} \, dx \quad = -\cot x, \qquad\qquad I = \mathbb{R}\backslash\pi\mathbb{Z}.$$

$$(15) \quad \int \sinh x \, dx \quad = \cosh x, \qquad\qquad I = \mathbb{R}.$$

$$(16) \quad \int \cosh x \, dx \quad = \sinh x, \qquad\qquad\qquad\qquad\qquad I = \mathbb{R}.$$

$$(17) \quad \int \tanh x \, dx \quad = \ln(\cosh x), \qquad\qquad\qquad\qquad I = \mathbb{R}.$$

$$(18) \quad \int \coth x \, dx \quad = \ln|\sinh x|, \qquad\qquad\qquad\qquad I = \mathbb{R}^*.$$

$$(19) \quad \int \frac{1}{\cosh x} \, dx \quad = \arctan(\sinh x) = 2\arctan(e^x) - \frac{\pi}{2}, \qquad I = \mathbb{R}.$$

$$(20) \quad \int \frac{1}{\sinh x} \, dx \quad = \ln\left|\tanh \frac{x}{2}\right|, \qquad\qquad\qquad I = \mathbb{R}^*.$$

$$(21) \quad \int \frac{1}{\cosh^2 x} \, dx \quad = \tanh x, \qquad\qquad\qquad\qquad I = \mathbb{R}.$$

$$(22) \quad \int \frac{1}{\sinh^2 x} \, dx \quad = -\coth x, \qquad\qquad\qquad\qquad I = \mathbb{R}^*.$$

$$(23) \quad \int \tanh x \, dx \quad = \ln(\cosh x), \qquad\qquad\qquad\qquad I = \mathbb{R}.$$

$$(24) \quad \int \coth x \, dx \quad = \ln|\sinh x|, \qquad\qquad\qquad\qquad I = \mathbb{R}^*.$$

$$(25) \quad \int \frac{1}{\sqrt{x^2+1}} \, dx = \ln(x + \sqrt{x^2+1}) = \operatorname{arcsinh} x, \qquad I = \mathbb{R}.$$

$$(26) \quad \int \frac{1}{x^2+1} \, dx \quad = \arctan x, \qquad\qquad\qquad\qquad I = \mathbb{R}.$$

$$(27) \quad \int \frac{1}{1-x^2} \, dx \quad = \frac{1}{2}\ln\left|\frac{1+x}{1-x}\right| \ (= \operatorname{arctanh} x \text{ for } |x| < 1), \quad I = \mathbb{R}\setminus\{-1,1\}.$$

$$(28) \quad \int \ln x \, dx \quad = x \ln x - x, \qquad\qquad\qquad\qquad I = (0, +\infty).$$

$$(29) \quad \int \frac{1}{\sqrt{1-x^2}} \, dx = \arcsin x, \qquad\qquad\qquad\qquad I = (-1, 1).$$

$$(30) \quad \int \frac{1}{\sqrt{x^2-1}} \, dx = \ln|x + \sqrt{x^2-1}|(= \operatorname{arccosh} x \text{ for } x \geq 1), \ I = \mathbb{R}\setminus(-1,1).$$

$$(31) \quad \int \sqrt{1+x^2} \, dx = \frac{x}{2}\sqrt{1+x^2} + \frac{1}{2}\ln(x + \sqrt{x^2+1}), \qquad I = \mathbb{R}.$$

$$(32) \quad \int \sqrt{1-x^2} \, dx = \frac{x}{2}\sqrt{1-x^2} + \frac{1}{2}\arcsin x, \qquad\qquad I = (-1, 1).$$

$$(33) \quad \int \sqrt{x^2-1} \, dx = \frac{x}{2}\sqrt{x^2-1} - \frac{1}{2}\ln|x + \sqrt{x^2-1}|, \qquad I = \mathbb{R}\setminus(-1,1).$$

We have

$$\text{for } t \in \mathbb{C}, \quad \cos t = \frac{e^{it} + e^{-it}}{2}, \quad \sin t = \frac{e^{it} - e^{-it}}{2i},$$

$$\text{for } t \in \mathbb{C}\setminus\left(\frac{\pi}{2} + \pi\mathbb{Z}\right), \ \tan t = \frac{\sin t}{\cos t}. \qquad \text{For } t \in \mathbb{C}\setminus\pi\mathbb{Z}, \quad \cot t = \frac{\cos t}{\sin t},$$

as well as

$$\left[-\tfrac{\pi}{2}, \tfrac{\pi}{2}\right] \xrightarrow{\ \sin\ } [-1, 1] \xrightarrow{\ \arcsin\ } \left[-\tfrac{\pi}{2}, \tfrac{\pi}{2}\right], \quad \arcsin x = \int_0^x \frac{ds}{\sqrt{1 - s^2}},$$

$$[0, \pi] \xrightarrow{\ \cos\ } [-1, 1] \xrightarrow{\ \arccos\ } [0, \pi], \quad \arccos x = \int_x^1 \frac{ds}{\sqrt{1 - s^2}},$$

$$\left(-\tfrac{\pi}{2}, \tfrac{\pi}{2}\right) \xrightarrow{\ \tan\ } \mathbb{R} \xrightarrow{\ \arctan\ } \left(-\tfrac{\pi}{2}, \tfrac{\pi}{2}\right), \quad \arctan x = \int_0^x \frac{ds}{1 + s^2},$$

$$(0, \pi) \xrightarrow{\ \cot\ } \mathbb{R} \xrightarrow{\ \text{arccot}\ } (0, \pi), \quad \text{arccot}\, x = \int_x^{+\infty} \frac{ds}{1 + s^2}.$$

We have used

$$\text{for } t \in \mathbb{C}, \ \sinh t = \frac{e^t - e^{-t}}{2}, \quad \cosh t = \frac{e^t + e^{-t}}{2},$$

$$\text{for } t \in \mathbb{C}\setminus\left(\frac{i\pi}{2} + i\pi\mathbb{Z}\right), \ \tanh t = \frac{e^t - e^{-t}}{e^t + e^{-t}}. \qquad \text{For } t \in \mathbb{C}\setminus i\pi\mathbb{Z}, \ \coth t = \frac{e^t + e^{-t}}{e^t - e^{-t}},$$

so that

$$\mathbb{R} \xrightarrow{\ \sinh\ } \mathbb{R} \xrightarrow{\ \text{arcsinh}\ } \mathbb{R}, \quad \text{arcsinh}\, x = \ln\!\left(x + \sqrt{x^2 + 1}\right),$$

$$[0, +\infty) \xrightarrow{\ \cosh\ } [1, +\infty) \xrightarrow{\ \text{arccosh}\ } [0, +\infty), \quad \text{arccosh}\, x = \ln\!\left(x + \sqrt{x^2 - 1}\right),$$

$$\mathbb{R} \xrightarrow{\ \tanh\ } (-1, 1) \xrightarrow{\ \text{arctanh}\ } \mathbb{R}, \quad \text{arctanh}\, x = \frac{1}{2} \ln\!\left(\frac{1 + x}{1 - x}\right),$$

$$\mathbb{R}^* \xrightarrow{\ \coth\ } \mathbb{R}\setminus[-1, 1] \xrightarrow{\ \text{arccoth}\ } \mathbb{R}^*, \quad \text{arccoth}\, x = \frac{1}{2} \ln\!\left(\frac{x + 1}{x - 1}\right).$$

We have also

$$(34) \qquad \int \text{arcsinh}\, x \, dx = x\,\text{arcsinh}\, x - \sqrt{1 + x^2},$$

$$(35) \qquad \int \text{arccosh}\, x \, dx = x\,\text{arccosh}\, x - \sqrt{x^2 - 1}, \qquad \text{on } x > 1,$$

$$(36) \qquad \int \text{arctanh}\, x \, dx = x\,\text{arctanh}\, x + \frac{1}{2} \ln(1 - x^2), \quad \text{on } |x| < 1,$$

$$(37) \qquad \int \text{arccoth}\, x \, dx = x\,\text{arccoth}\, x + \frac{1}{2} \ln(x^2 - 1), \quad \text{on } |x| > 1.$$

Remark 10.4.1. With Definition (10.5.1) of the Logarithm on $\mathbb{C}\backslash\mathbb{R}_-$, and since for $t \in \mathbb{C}$, $\cos t = \cosh(it)$, $\sin t = -i\sinh(it)$,

$$\text{for } x \in [-1, 1], \quad \begin{cases} \arcsin x = -i\,\mathrm{Log}\big(ix + \sqrt{1 - x^2}\big), \\ \arccos x = -i\,\mathrm{Log}\big(x + i\sqrt{1 - x^2}\big). \end{cases} \tag{10.4.1}$$

$$\text{For } z \in \mathbb{C}\backslash \pm i[1, +\infty), \quad \arctan z = -i\,\mathrm{Log}(1 + iz) + \frac{i}{2}\mathrm{Log}(1 + z^2), \tag{10.4.2}$$

so that arctan is holomorphic on $\mathbb{C}\backslash \pm i[1, +\infty)$ with

$$\arctan'(z) = \frac{1}{1 + iz} + \frac{i}{2}\frac{2z}{1 + z^2} = \frac{1 - iz + iz}{1 + z^2} = \frac{1}{1 + z^2},$$

a meromorphic function on $\mathbb{C}$, with poles at $\pm i$ and residues $\mp i/2$.

Integrating rational fractions

Lemma 10.4.2. *Let $P(X), Q(X)$ be polynomials with complex coefficients such that Q is a normalized polynomial with degree $d \geq 1$ and P is a polynomial with degree $< d$. Let $z_1, \ldots, z_r$ be the distinct roots of Q with respective multiplicity $\mu_1, \ldots, \mu_r$. Then*

$$Q(X) = \prod_{1 \leq j \leq r} (X - z_j)^{\mu_j}, \quad d = \sum_{1 \leq j \leq r} \mu_j,$$

and the rational fraction $R = P/Q$ is

$$\frac{P(X)}{Q(X)} = \sum_{\substack{1 \leq j \leq r \\ 1 \leq m_j \leq \mu_j}} \frac{\alpha_{j,m_j}}{(X - a_j)^{m_j}}, \quad \text{with } \alpha_{j,m_j} = \frac{R_j^{(\mu_j - m_j)}(z_j)}{(\mu_j - m_j)!}$$

where the rational fraction R_j without a pole at z_j is given by

$$R_j(X) = (X - z_j)^{\mu_j} R(X).$$

Proof. We perform an induction on r, the number of distinct roots: when $r = 1$ we have a single root z_1 with multiplicity $\mu_1 = d$, so that

$$(X - z_1)^{\mu_1} \frac{P(X)}{Q(X)} = P(X) = \sum_{0 \leq k < \mu_1} \frac{P^{(k)}(z_1)}{k!}(X - z_1)^k$$

and thus

$$\frac{P(X)}{Q(X)} = \sum_{0 \leq k < \mu_1} \frac{P^{(k)}(z_1)}{k!}(X - z_1)^{\overbrace{k - \mu_1}^{-m}} = \sum_{1 \leq m \leq \mu_1} \frac{P^{(\mu_1 - m)}(z_1)}{(\mu_1 - m)!}(X - z_1)^{-m},$$

proving the result in that case with an explicit expression. Let us assume that the formula is true for some $r \geq 1$ and let us prove it when we have $r + 1$ distinct poles $z_1, \ldots, z_r, z_{r+1}$ with respective positive multiplicity $\mu_1, \ldots, \mu_r, \mu_{r+1}$ for the rational fraction P/Q. The rational fraction

$$(X - z_{r+1})^{\mu_{r+1}} \frac{P(X)}{Q(X)} = R_{r+1}(X)$$

$$= \sum_{0 \leq k < \mu_{r+1}} \frac{R_{r+1}^{(k)}(z_{r+1})}{k!} (X - z_{r+1})^k + S(X)(X - z_{r+1})^{\mu_{r+1}}$$

where the rational fraction R_{r+1} (and thus S have poles $z_1, \ldots, z_r$ with respective multiplicity $\mu_1, \ldots, \mu_r$. This yields

$$\frac{P(X)}{Q(X)} = \sum_{1 \leq m \leq \mu_{r+1}} \frac{R_{r+1}^{(\mu_1 - m)}(z_{r+1})}{(\mu_1 - m)!} (X - z_{r+1})^{-m} + S(X),$$

and we may apply the induction hypothesis to S: note that S has no polynomial part since a linear combination of rational fractions A_j/B_j with degree $B_j >$ degree A_j is a rational fraction A/B with degree $B >$ degree A. In fact we have

$$\sum_{1 \leq j \leq N} \frac{A_j}{B_j} = \frac{A_1 \prod_{2 \leq j \leq N} B_j + \cdots + A_N \prod_{1 \leq j \leq N-1} B_j}{\prod_{1 \leq j \leq N} B_j}$$

and the numerator has obviously a degree strictly smaller than the denominator since for instance

$$\mathrm{degree} \left(A_1 \prod_{2 \leq j \leq N} B_j \right) \leq \mathrm{degree}\, A_1 + \sum_{2 \leq j \leq N} \mathrm{degree}\, B_j$$

$$< \sum_{1 \leq j \leq N} \mathrm{degree}\, B_j = \mathrm{degree} \left(\prod_{1 \leq j \leq N} B_j \right).$$

We see also that for $1 \leq j \leq r$, $R = P/Q$,

$$S_j = (X - z_j)^{\mu_j} S = (X - z_j)^{\mu_j} \left(R - \sum_{1 \leq m \leq \mu_{r+1}} \frac{R_{r+1}^{(\mu_1 - m)}(z_{r+1})}{(\mu_1 - m)!} (X - z_{r+1})^{-m} \right)$$

so that, with $R_j = (X - z_j)^{\mu_j} R$, we have

$$S_j^{(l)}(z_j) = R_j^{(l)}(z_j) \quad \text{for } l < \mu_j.$$

The induction is thus provides the sought formula. $\square$

Although the above lemma is sufficient to calculate antiderivatives of any rational fraction, the next lemma may be also useful.

Lemma 10.4.3. *Let $P(X), Q(X)$ be polynomials with real coefficients such that Q is a normalized polynomial with degree $d \geq 1$ and P is a polynomial with degree $< d$. Let $a_1, \ldots, a_r$ be the distinct real roots of Q with respective multiplicity $\mu_1, \ldots, \mu_r$. Let $z_1, \bar{z}_1, \ldots, z_s, \bar{z}_s$ be the distinct non-real roots with respective multiplicity $\nu_1, \ldots, \nu_s$. Then*

$$Q(X) = \prod_{1 \leq j \leq r} (X - a_j)^{\mu_j} \prod_{1 \leq k \leq s} \left((X - \operatorname{Re} z_k)^2 + (\operatorname{Im} z_k)^2 \right)^{\nu_k},$$

$d = \sum_{1 \leq j \leq r} \mu_j + \sum_{1 \leq k \leq s} 2\nu_k$ *and the rational fraction P/Q is such that*

$$\frac{P(X)}{Q(X)} = \sum_{\substack{1 \leq j \leq r \\ 1 \leq m \leq \mu_j}} \frac{\alpha_{j,m}}{(X - a_j)^m} + \sum_{\substack{1 \leq k \leq s \\ 1 \leq n \leq \nu_k}} \frac{\beta_{k,n} X + \gamma_{k,n}}{\left((X - \operatorname{Re} z_k)^2 + (\operatorname{Im} z_k)^2 \right)^n}.$$

Proof. This follows immediately from Lemma 10.4.2 which implies

$$(\sharp) \qquad \frac{P(X)}{Q(X)} = \sum_{\substack{1 \leq j \leq r \\ 1 \leq m \leq \mu_j}} \frac{\alpha_{j,m}}{(X - a_j)^m} + \sum_{\substack{1 \leq k \leq s \\ 1 \leq n \leq \nu_k}} \left\{ \frac{\gamma_{k,n}}{(X - z_k)^n} + \frac{\overline{\gamma_{k,n}}}{(X - \bar{z}_k)^n} \right\}.$$

We have only to deal with

$$\frac{\gamma_{k,n}}{(X - z_k)^n} + \frac{\overline{\gamma_{k,n}}}{(X - \bar{z}_k)^n} = \frac{\gamma_{k,n}(X - \bar{z}_k)^n + \overline{\gamma_{k,n}}(X - z_k)^n}{\left((X - \operatorname{Re} z_k)^2 + (\operatorname{Im} z_k)^2 \right)^n}$$

$$= \frac{T(X - \operatorname{Re} z_k)}{\left((X - \operatorname{Re} z_k)^2 + (\operatorname{Im} z_k)^2 \right)^n},$$

where T is a real polynomial with degree less than n. We note that for $2p$ even integer

$$(X - \operatorname{Re} z_k)^{2p} = \left((X - \operatorname{Re} z_k)^2 + (\operatorname{Im} z_k)^2 - (\operatorname{Im} z_k)^2 \right)^p,$$

$$(X - \operatorname{Re} z_k)^{2p+1} = (X - \operatorname{Re} z_k)\left((X - \operatorname{Re} z_k)^2 + (\operatorname{Im} z_k)^2 - (\operatorname{Im} z_k)^2 \right)^p,$$

so that $T(X - \operatorname{Re} z_k)$ is a polynomial in the variable $\left((X - \operatorname{Re} z_k)^2 + (\operatorname{Im} z_k)^2 \right)$ with coefficients polynomial of degree ≤ 1, yielding the result. $\qquad\square$

Lemma 10.4.2 implies that to find an antiderivative of a rational fraction, we use the decomposition into partial fraction and we are left with finding an antiderivative of $(x - \zeta)^{-m}$ with $\zeta \in \mathbb{C}$. If $m \geq 2$, Formula (1) on page 448 gives the result. If $m = 1$ and $\zeta \in \mathbb{R}$, this is $\ln |x - \zeta|$ on $\mathbb{R} \backslash \{\zeta\}$. If $m = 1$ and $\operatorname{Im} \zeta \neq 0$, this is $\operatorname{Log}(x - \zeta)$ where the logarithm is defined by (10.5.1).

Lemma 10.4.4. *Let ζ be a complex number and let $m \geq 1$ be an integer.*

(1) *If $m \geq 2$, the meromorphic function $z \mapsto (z - \zeta)^{-m}$ has the antiderivative*

$$(z - \zeta)^{1-m}(1 - m)^{-1}.$$

(2) *With the complex logarithm defined by (10.5.1), the holomorphic function defined on $\mathbb{C} \backslash \{\zeta + \mathbb{R}_-\}$, $z \mapsto (z - \zeta)^{-1}$ has the antiderivative $\operatorname{Log}(z - \zeta)$.*

Remark 10.4.5. If our rational fraction is real, we may want to avoid altogether complex numbers and use only Lemma 10.4.3. By rescaling and translation we have only to deal with antiderivatives of x^{-m} or $(x^2+1)^{-n}x$, $(x^2+1)^{-n}$. The first case is already treated, the answer to the second case is $\frac{1}{2}\int\frac{du}{u^n}$ which is reduced to the first case. To calculate,

$$I_n(X) = \int_0^X \frac{dx}{(1+x^2)^n} = \int_0^{\arctan X} (1+\tan^2\theta)^{1-n}d\theta = \int_0^{\arctan X} (\cos\theta)^{2n-2}d\theta.$$

We have $I_1(X) = \arctan X$ and for $n \geq 1$,

$$I_{n+1}(X) = \int_0^{\arctan X} (\cos\theta)^{2n-2}(1-\sin^2\theta)d\theta$$

$$= I_n(X) + \frac{1}{2n-1}\int_0^{\arctan X} \sin\theta\frac{d}{d\theta}\big((\cos\theta)^{2n-1}\big)d\theta$$

$$= I_n(X) + \frac{\sin(\arctan X)\big(\cos(\arctan X)\big)^{2n-1}}{2n-1} - \frac{1}{2n-1}\int_0^{\arctan X} (\cos\theta)^{2n}d\theta,$$

so that the following induction relation holds:

$$\frac{2n}{2n-1}I_{n+1} = I_n + \frac{\sin(\arctan X)\big(\cos(\arctan X)\big)^{2n-1}}{2n-1}.$$

We note also that for $|\theta| < \pi/2$, $\sin\theta = \tan\theta\cos\theta = \tan\theta(1+\tan^2\theta)^{-1/2}$ so that

$$\sin(\arctan x) = \frac{x}{\sqrt{1+x^2}}, \qquad \cos(\arctan x) = \frac{1}{\sqrt{1+x^2}},$$

and

$$I_{n+1} = \frac{2n-1}{2n}I_n + \frac{1}{2n}\frac{x}{(1+x^2)^n}.$$

Antiderivatives of rational fractions of $\cos x$, $\sin x$

We want to calculate antiderivatives of $F(\cos x, \sin x)$ where F is a rational fraction. The following changes of variables will work depending on some invariance properties of the one-form $F(\cos x, \sin x)dx$.

1. $u = \sin x$, if the mapping $x \mapsto \pi - x$ leaves invariant the form $F(\cos x, \sin x)dx$. It is the case for instance of $\int \sin^4 x \cos^5 x\, dx$ since

$$\sin^4(\pi - x) \cos^5(\pi - x)d(\pi - x) = \sin^4 x \cos^5 x\, dx.$$

This can be applied to the integrals $\int \sin^k x \cos^{2l+1} x\, dx$ with k, l integers. The assumption means in fact that the function F is odd with respect to its first variable: $F(-X, Y) = -F(X, Y)$.

Lemma 10.4.6. *Let R be a rational fraction in $\mathbb{C}(X,Y)$, odd with respect to the first variable: then, there exists M_1, M_2, polynomials of two variables such that*

$$R(X,Y) = \frac{XM_1(X^2,Y)}{M_2(X^2,Y)} = XS(X^2,Y), \quad S \text{ rational fraction.}$$

Proof. We have

$$2R(X,Y) = \frac{P(X,Y)}{Q(X,Y)} - \frac{P(-X,Y)}{Q(-X,Y)}$$
$$= \frac{P(X,Y)Q(-X,Y) - P(-X,Y)Q(X,Y)}{Q(X,Y)Q(-X,Y)} = \frac{XN_1(X,Y)}{N_2(X,Y)},$$

where N_j are polynomials in $\mathbb{C}[X,Y]$, even w.r.t. X. Thus

$$2N_j(X,Y) = N_j(X,Y) + N_j(-X,Y) = M_j(X^2,Y),$$

where M_j is a polynomial. $\qquad\square$

We have thus

$$F(\cos x, \sin x)dx = \cos x\, G(\cos^2 x, \sin x)dx = G(1 - u^2, u)du.$$

2. $u = \cos x$, if the mapping $x \mapsto -x$ leaves invariant the form $F(\cos x, \sin x)dx$. It is the case of $\int \sin^5 x \cos^7 x dx$ since

$$\sin^5(-x) \cos^7(-x)d(-x) = \sin^5 x \cos^7 x dx.$$

It can be applied to $\int \sin^{2k+1} x \cos^l x dx$ with k, l integers. The assumption means in fact that the function F is odd with respect to its second variable: $F(X, -Y) = -F(X, Y)$. We have thus

$$F(\cos x, \sin x)dx = \sin x\, G(\cos x, \sin^2 x)dx = -G(u, 1 - u^2)du.$$

3. $u = \tan x$, if the mapping $x \mapsto \pi + x$ leaves invariant the form $F(\cos x, \sin x)$ dx. It is the case of $\int \sin^4 x \cos^6 x dx$ since

$$\sin^4(\pi + x) \cos^6(\pi + x)d(\pi + x) = \sin^4 x \cos^6 x dx.$$

It can be applied to $\int \sin^{2k} x \cos^{2l} x dx$ with k, l integers. The assumption means in fact that the function F is even: $F(-X, -Y) = F(X, Y)$.

Lemma 10.4.7. *Let R be an even rational fraction in $\mathbb{C}(X,Y)$: then, there exist $(M_j)_{1 \leq j \leq 4}$ polynomials of two variables such that*

$$R(X,Y) = \frac{M_1(X^2,Y^2) + XY M_2(X^2,Y^2)}{M_3(X^2,Y^2) + XY M_4(X^2,Y^2)}.$$

In particular, R is a rational fraction of X^2, Y^2, XY.

Proof. We have

$$2R(X,Y) = \frac{P(X,Y)}{Q(X,Y)} + \frac{P(-X,-Y)}{Q(-X,-Y)} \tag{10.4.3}$$

$$= \frac{P(X,Y)Q(-X,-Y) + P(-X,-Y)Q(X,Y)}{Q(X,Y)Q(-X,-Y)} = \frac{N(X,Y)}{D(X,Y)},$$

where N, D are even polynomials. Since the polynomial D in (10.4.3) is even we have

$$2D(X,Y) = \sum_{j+k \text{ even}} b_{j,k}X^jY^k(1 + (-1)^{j+k}),$$

and thus,

$$D(X,Y) = \sum_{0 \le j \le 2l} b_{j,2l-j}X^jY^{2l-j}$$

$$= \sum_{0 \le j' \le l} b_{2j',2l-2j'}X^{2j'}Y^{2l-2j'} + \sum_{0 \le j'' \le l-1} b_{2j''+1,2l-2j''-1}X^{2j''+1}Y^{2l-2j''-1}$$

$$= N_1(X^2,Y^2) + XYN_2(X^2,Y^2), \quad N_j \text{ polynomials.}$$

We found eventually some polynomials $(N_j)_{1 \le j \le 4}$ such that

$$2R(X,Y) = \frac{N_3(X^2,Y^2) + XYN_4(X^2,Y^2)}{N_1(X^2,Y^2) + XYN_2(X^2,Y^2)}. \qquad \square$$

We have thus

$$F(\cos x, \sin x)dx = G(\cos^2 x, \sin^2 x, \sin x \cos x)dx$$

$$= G(\cos^2 x, \sin^2 x, \sin x \cos x)dx$$

$$= G\left(\frac{1}{1+u^2}, \frac{u^2}{1+u^2}, \frac{u}{1+u^2}\right)\frac{du}{1+u^2}.$$

4. As a last remedy, we can use the change $u = \tan \frac{x}{2}$ which will provide a rational fraction in u.

This method extends *ne varietur* to rational fractions of $\sinh x, \cosh x$.

Abelian integrals

Let us give a couple of examples of the so-called *Abelian integrals*,

$$\int R\big(x, \varphi(x)\big)\,dx, \tag{10.4.4}$$

where R is a rational fraction.

The function φ in 10.4.4 is the square-root of a second-degree polynomial

For instance, we want to calculate $\int R(x, \sqrt{x^2+1})dx$. We set $x = \sinh t$ and we get $\int R(\sinh t, \cosh t) \cosh t\, dt$, which is a rational function of $\sinh, \cosh$, tackled above. To deal with $\int R(x, \sqrt{x^2-1})dx$, we set $x = \cosh t$ to obtain

$$\int R(\cosh t, \sinh t) \sinh t\, dt,$$

also a rational function of $\sinh, \cosh$. For $\int R(x, \sqrt{1-x^2})dx$, we set $x = \sin t$ to get $\int R(\sin t, \cos t) \cos t\, dt$, a rational function of $\sin, \cos$. The discussion above allows us to determine

$$\int R(x, \sqrt{ax^2+b+c})dx, \qquad \text{for } R \text{ a rational fraction.}$$

The function φ in 10.4.4 is $\left(\frac{ax+b}{cx+d}\right)^{1/m}$, $m \in \mathbb{N}^*$

We set $u = (ax+b/cx+d)^{\frac{1}{m}}$ so that $x = \rho(u^m)$ where ρ is a rational fraction and

$$\int R(x, \varphi(x))dx = \int R(\rho(u^m), u)\rho'(u^m)mu^{m-1}du,$$

also the antiderivative of a rational fraction.

The function φ in 10.4.4 enjoys a parametric unicursal representation

The assumption means that we can find rational fractions p, q of one variable such that $t \mapsto (p(t), q(t))$ is onto on the graph of φ. We set then $x = p(t)$ and we are reduced to the computation of

$$\int R(p(t), q(t))p'(t)dt, \qquad \text{again the antiderivative of a rational fraction.}$$

Let us give a specific example. We want to compute for $X > 0$

$$F(X) = \int_0^X R(x, x^{1/2} + x^{1/3})dx, \qquad \text{where } R \text{ is a rational fraction.}$$

We note that the mapping $t \mapsto (t^6, t^3 + t^2)$ provides a unicursal representation of φ. We set $x = t^6$ to obtain

$$F(X) = \int_0^{X^{1/6}} R(t^6, t^3 + t^2)6t^5 dt,$$

which is the antiderivative of a rational fraction.

Some Fourier integrals

We have seen a couple of explicit computations of Fourier transforms in (8.1.18), in Chapter 8, Section **Some standard examples of Fourier transform** on page 352 as well as in Proposition 8.1.19 and Theorem 8.2.3.

The computation of the antiderivative

$$\int e^{zt} P(t)\,dt,$$

where $z \in \mathbb{C}$ and P is a polynomial (of one variable) is also a computation of a Fourier (–Laplace) transform. If $\operatorname{Re} z < 0$, we have $\int_{-\infty}^{x} e^{zt}dt = z^{-1}e^{zx}$ and for $k \in \mathbb{N}$,

$$\int_{-\infty}^{x} e^{zt} t^k \, dt = \left(\frac{d}{dz}\right)^k \left(\int_{-\infty}^{x} e^{zt} dt\right) = \left(\frac{d}{dz}\right)^k (z^{-1}e^{zx})$$

$$= e^{zx}e^{-zx}\left(\frac{d}{dz}\right)^k (e^{zx} z^{-1}) = e^{zx}\left(e^{-zx}\frac{d}{dz}e^{zx}\right)^k (z^{-1}) = e^{zx}\left(\frac{d}{dz}+x\right)^k (z^{-1})$$

$$= e^{zx} \sum_{0\le l\le k} C_k^l x^l z^{-1-(k-l)}(-1)^{k-l}(k-l)! = z^{-1}e^{zx} \sum_{0\le l\le k} \frac{x^l}{l!} k! z^{-k+l}(-1)^{k-l},$$

so that for $P(t) = \sum_{0\le k\le m} a_k t^k$,

$$\int_{-\infty}^{x} P(t)e^{zt} dt = z^{-1}e^{zx} Q_P(x, z^{-1}),$$

$$Q_P(x, z^{-1}) = \sum_{0\le l\le m} \frac{x^l}{l!} \sum_{l\le k\le m} a_k k!(-1)^{k-l} z^{-k+l}.$$

We have thus for $\operatorname{Re} z > 0$,

$$\int_{0}^{x} P(t)e^{zt} dt = z^{-1}e^{zx} Q_P(x, z^{-1}) - z^{-1}Q_P(0, z^{-1}),$$

and by analytic continuation, this formula holds as well for $z \ne 0$. Note that the limit of the rhs when z goes to 0 is indeed $\int_{0}^{x} P(t)dt$: by linearity it suffices to verify this for the monomial $P(t) = t^k$. We need to check for $z \ne 0$,

$$N(x, z) = e^{zx}z^{-1} \sum_{0\le l\le k} \frac{x^l}{l!} k!(-1)^{k-l} z^{-k+l} - z^{-1}k!(-1)^k z^{-k}.$$

We have

$$N(x, z) = k!(-1)^k z^{-k-1}e^{zx}\left(\sum_{0\le l\le k} \frac{(-zx)^l}{l!} - e^{-zx}\right)$$

$$= -k!(-1)^k z^{-k-1}e^{zx} \int_{0}^{1} \frac{(1-\theta)^k}{k!} e^{-\theta zx} d\theta (-zx)^{k+1}$$

so that $N(x, z) = e^{zx} x^{k+1} \int_0^1 (1 - \theta)^k e^{-\theta zx} d\theta$, which has the expected limit $\frac{x^{k+1}}{k+1}$ when $z \to 0$.

Lemma 10.4.8. *Let $n \in \mathbb{N}^*$ and $\mathbb{R}^n \ni x \mapsto u(x) = \exp -2\pi|x|$, where $|x|$ stands for the Euclidean norm of x. The function u belongs to $L^1(\mathbb{R}^n)$ and its Fourier transform is*

$$\hat{u}(\xi) = \pi^{-(\frac{n+1}{2})} \Gamma\left(\frac{n+1}{2}\right) \left(1 + |\xi|^2\right)^{-(\frac{n+1}{2})}. \tag{10.4.5}$$

Proof. We note first that in one dimension

$$\int_{\mathbb{R}} e^{-2i\pi x\xi} e^{-2\pi|x|} dx = 2 \operatorname{Re} \int_0^{+\infty} e^{-2\pi x(1+i\xi)} dx = \frac{1}{\pi(1+\xi^2)},$$

corroborating the above formula in 1D. We want to take advantage of this to write $e^{-2\pi|x|}$ as a superposition of Gaussian functions; doing this will be very helpful since it is easy to calculate the Fourier transform of Gaussian functions (this quite natural idea seems to be used only in the wonderful textbook by Robert Strichartz [62] and we follow his method). For $t \in \mathbb{R}_+$, we have

$$e^{-2\pi t} = \int_{\mathbb{R}} e^{2i\pi t\tau} \frac{d\tau}{\pi(1+\tau^2)} = \iint_{\mathbb{R}^2} e^{2i\pi t\tau} e^{-s\pi(1+\tau^2)} H(s) ds d\tau$$

$$= \int_{\mathbb{R}_+} e^{-\pi s} s^{-1/2} e^{-\frac{\pi}{s} t^2} ds,$$

so that for $x \in \mathbb{R}^n$, $e^{-2\pi|x|} = \int_{\mathbb{R}_+} e^{-\pi s} s^{-1/2} e^{-\frac{\pi}{s}|x|^2} ds$ and thus

$$\hat{u}(\xi) = \iint_{\mathbb{R}^n \times \mathbb{R}_+} e^{-2i\pi x\xi} e^{-\pi s} s^{-1/2} e^{-\frac{\pi}{s}|x|^2} dx ds = \int_{\mathbb{R}_+} e^{-\pi s} s^{-1/2} e^{-\pi s|\xi|^2} s^{n/2} ds,$$

so that

$$\hat{u}(\xi) = \int_0^{+\infty} e^{-s} s^{(n-1)/2} \left(\pi(1 + |\xi|^2)\right)^{-(n+1)/2} ds,$$

which is the sought result. $\qquad\square$

Gaussian integrals

In Proposition 8.1.19, we have computed the Fourier transform of Gaussian functions, a typical case when the calculation of an integral does not follow from the knowledge of an antiderivative. However our definition of the Fourier transform of e^{ix^2} relied on a duality argument, and we want to connect this result with a more elementary approach. According to Formula (8.1.31), for $w_a(x) = e^{i\pi a x^2}$ we have for $a \in \mathbb{R}^*$,

$$\widehat{w_a}(\xi) = |a|^{-1/2} e^{i\frac{\pi}{4} \operatorname{sign} a} e^{-i\pi a^{-1}\xi^2}.$$

Let $\phi \in \mathscr{S}(\mathbb{R}^n)$: we have $\int w_a(x)\hat{\phi}(x)dx = \int |a|^{-1/2}e^{i\frac{\pi}{4}\,\mathrm{sign}\,a}e^{-i\pi a^{-1}\xi^2}\phi(\xi)d\xi$, and in particular for $\epsilon > 0$,

$$\int w_a(x)e^{-\pi\epsilon x^2}dx = |a|^{-1/2}e^{i\frac{\pi}{4}\,\mathrm{sign}\,a}\epsilon^{-1/2}\int e^{-i\pi a^{-1}\xi^2}e^{-\pi\epsilon^{-1}\xi^2}$$

$$= |a|^{-1/2}e^{i\frac{\pi}{4}\,\mathrm{sign}\,a}\epsilon^{-1/2}(\epsilon^{-1}+ia^{-1})^{-1/2} \xrightarrow[\epsilon\to 0_+]{} |a|^{-1/2}e^{i\frac{\pi}{4}\,\mathrm{sign}\,a},$$

proving that, for $a \in \mathbb{R}^*$,

$$\lim_{\epsilon\to 0_+}\int e^{i\pi a x^2}e^{-\pi\epsilon x^2}dx = |a|^{-1/2}e^{i\frac{\pi}{4}\,\mathrm{sign}\,a}. \tag{10.4.6}$$

For $\lambda > 0, a \in \mathbb{R}^*$, we have

$$\sigma(a,\lambda) = \int_0^\lambda e^{i\pi a x^2}dx = \frac{1}{2}\int_{-\lambda}^\lambda e^{i\pi a x^2}dx = \frac{1}{2}\lim_{\epsilon\to 0_+}\int_{-\lambda}^\lambda e^{i\pi(a+i\epsilon)x^2}dx.$$

We have also

$$2\int_\lambda^{+\infty} e^{i\pi(a+i\epsilon)x^2}dx = \int_{\lambda^2}^{+\infty} e^{i\pi(a+i\epsilon)t}t^{-1/2}dt$$

$$= \left[\frac{e^{i\pi(a+i\epsilon)t}}{i\pi(a+i\epsilon)}t^{-1/2}\right]_{t=\lambda^2}^{t=+\infty} + \frac{1}{2}\int_{\lambda^2}^{+\infty}\frac{e^{i\pi(a+i\epsilon)t}}{i\pi(a+i\epsilon)}t^{-3/2}dt,$$

so that for $\lambda \geq 1, \epsilon > 0$,

$$\left|\int_\lambda^{+\infty} e^{i\pi(a+i\epsilon)x^2}dx\right| \leq \frac{1}{2}\lambda^{-1}\pi^{-1}|a|^{-1} + \frac{2\lambda^{-1}}{4\pi|a|} = \frac{1}{\pi|a|\lambda}.$$

We have thus

$$\int_{-\lambda}^\lambda e^{i\pi(a+i\epsilon)x^2}dx - |a|^{-1/2}e^{i\frac{\pi}{4}\,\mathrm{sign}\,a}$$

$$= \int_{\mathbb{R}} e^{i\pi(a+i\epsilon)x^2}dx - |a|^{-1/2}e^{i\frac{\pi}{4}\,\mathrm{sign}\,a} - \int_{|x|>\lambda} e^{i\pi(a+i\epsilon)x^2}dx,$$

and

$$\left|\int_{-\lambda}^\lambda e^{i\pi(a+i\epsilon)x^2}dx - |a|^{-1/2}e^{i\frac{\pi}{4}\,\mathrm{sign}\,a}\right|$$

$$\leq \left|\int_{\mathbb{R}} e^{i\pi(a+i\epsilon)x^2}dx - |a|^{-1/2}e^{i\frac{\pi}{4}\,\mathrm{sign}\,a}\right| + \frac{2}{\pi|a|\lambda}$$

so that taking the limit when $\epsilon \to 0_+$ gives from (10.4.6),

$$\left|\int_{-\lambda}^\lambda e^{i\pi a x^2}dx - |a|^{-1/2}e^{i\frac{\pi}{4}\,\mathrm{sign}\,a}\right| \leq \frac{2}{\pi|a|\lambda},$$

entailing

$$\lim_{\lambda \to +\infty} \int_0^\lambda e^{i\pi a x^2}\,dx = \frac{1}{2}|a|^{-1/2}e^{i\frac{\pi}{4}\,\text{sign}\,a}. \tag{10.4.7}$$

This gives in particular the classical Fresnel integrals[11]

$$\int_{\mathbb{R}} \cos(x^2)\,dx = \sqrt{\frac{\pi}{2}} = \int_{\mathbb{R}} \sin(x^2)\,dx. \tag{10.4.8}$$

Another classical calculation (introduced in Exercise 2.8.20) yields

$$\int_0^{+\infty} \frac{\sin x}{x}\,dx = \frac{\pi}{2}. \tag{10.4.9}$$

We integrate the holomorphic function (on $\mathbb{C}^*$) e^{iz}/z on the path

$$[\epsilon, R] \cup \text{upper half-circle}(0, R) \ (\text{counterclockwise})$$
$$\cup\, [-R, -\epsilon] \cup \text{upper half-circle}(0, \epsilon) \ (\text{clockwise}),$$

we get

$$0 = 2i\int_\epsilon^R \frac{\sin x}{x}\,dx + \int_0^\pi \frac{e^{i\,\mathrm{Re}^{i\theta}}}{\mathrm{Re}^{i\theta}}i\,\mathrm{Re}^{i\theta}\,d\theta - \int_0^\pi \frac{e^{i\epsilon e^{i\theta}}}{\epsilon e^{i\theta}}i\epsilon e^{i\theta}\,d\theta.$$

The third integral has limit $i\pi$ when ϵ goes to 0. The absolute value of the second integral is bounded above by $\int_0^\pi e^{-R\sin\theta}\,d\theta$ which goes to 0 when R goes to $+\infty$ (thanks to the Lebesgue dominated convergence Theorem, but a simpler argument is also available here).

10.5   Some special functions

The complex logarithm

Logarithm on $\mathbb{C}\backslash\mathbb{R}_-$

The set $\mathbb{C}\backslash\mathbb{R}_-$ is star-shaped with respect to 1, so that we can define the principal determination of the logarithm for $z \in \mathbb{C}\backslash\mathbb{R}_-$ by the formula

$$\mathrm{Log}\,z = \oint_{[1,z]} \frac{d\zeta}{\zeta} = \int_0^1 \frac{(z-1)dt}{(1-t)+tz}. \tag{10.5.1}$$

Thanks to Theorem 3.3.7, the function Log is holomorphic on $\mathbb{C}\backslash\mathbb{R}_-$ and we have $\mathrm{Log}\,z = \ln z$ for $z \in \mathbb{R}_+^*$ and by analytic continuation

$$e^{\mathrm{Log}\,z} = z = e^{\mathrm{Re\,Log}\,z}e^{i\,\mathrm{Im\,Log}\,z}, \qquad \begin{cases} |z| & = e^{\mathrm{Re\,Log}\,z}, \\ \mathrm{Arg}\,z & = \mathrm{Im\,Log}\,z, \end{cases}$$

---

[11] Of course in the sense $\lim_{\lambda, \mu \to +\infty} \int_{-\mu}^\lambda e^{ix^2}\,dx$.

for $z \in \mathbb{C}\backslash\mathbb{R}_-$. For $z = re^{i\theta}, |\theta| < \pi$, we have for $r > 0$,

$$\mathrm{Log}(re^{i\theta}) = \oint_{[1,re^{i\theta}]} \frac{d\zeta}{\zeta} = \ln r + \int_0^\theta \frac{ire^{it}}{re^{it}}\,dt = \ln r + i\theta, \quad \mathrm{Im}\,\mathrm{Log}\,z = \theta.$$

We get also by analytic continuation, that $\mathrm{Log}\,e^z = z$ for $|\mathrm{Im}\,z| < \pi$. Note also that for $|z| < 1$, we have from Theorem 3.3.7,

$$\mathrm{Log}(1+z) = z\int_0^1 \frac{dt}{1+tz} = \sum_{k\geq 0} z(-1)^k \frac{z^k}{k+1} = \sum_{l\geq 1}(-1)^{l+1}\frac{z^l}{l}. \qquad (10.5.2)$$

Note that we have also for $|z| = 1, z \neq -1$,

$$\mathrm{Log}(1+z) = z\int_0^1 \frac{dt}{1+tz} = z\int_0^1 \lim_N \left(\sum_{0\leq k\leq N}(-1)^k t^k z^k\right)dt.$$

Since with $z = e^{i\theta}, |\theta| < \pi, t \in [0,1]$,

$$\left|\sum_{0\leq k\leq N}(-1)^k t^k z^k = \frac{1+(-1)^N(tz)^{1+N}}{1+tz}\right| \leq \frac{2}{|1+tz|} = \frac{2}{\sqrt{1+2t\cos\theta + t^2}}$$

$$\leq \frac{21\{\cos\theta \geq 0\}}{\sqrt{1+t^2}} + \frac{21\{-1 < \cos\theta \leq 0\}}{\sqrt{1-\cos^2\theta}} \in L^1([0,1]_t),$$

so that Lebesgue's dominated convergence implies

$$\mathrm{Log}(1+z) = z\lim_N \sum_{0\leq k\leq N}(-1)^k \frac{z^k}{k+1},$$

implying that (10.5.2) holds as well for $|z| = 1, z \neq -1$. We consider the following open subset of $\mathbb{C}$:

$$\{z \in \mathbb{C}, \exp z \notin \mathbb{R}_-^*\} = \{z \in \mathbb{C}, \mathrm{Im}\,z \neq \pi(2\pi)\}$$
$$= \cup_{k\in\mathbb{Z}} \underbrace{\{z \in \mathbb{C}, (2k-1)\pi < \mathrm{Im}\,z < (2k+1)\pi\}}_{\omega_k}.$$

Let $k \in \mathbb{Z}$. On the open set ω_k, the function $z \mapsto \mathrm{Log}(\exp z) - z$ is holomorphic with a null derivative. As a result for $z \in \omega_k$,

$$\mathrm{Log}(\exp z) - z = \mathrm{Log}\big(\exp(2ik\pi)\big) - 2ik\pi = \ln(1) - 2ik\pi = -2ik\pi,$$

i.e., $\mathrm{Log}(\exp z) = z - 2ik\pi$.

We sum-up these results as follows.

Theorem 10.5.1. *For $z \in \mathbb{C}\backslash\mathbb{R}_-$, we define $\operatorname{Log} z$ by (10.5.1). This is a holomorphic function on $\mathbb{C}\backslash\mathbb{R}_-$, with derivative $1/z$, and Log coincides with $\ln$ on $\mathbb{R}_+^*$.*

For $z \in \mathbb{C}\backslash\mathbb{R}_-$, $e^{\operatorname{Log} z} = z = re^{i\theta}$,

$$r = |z| = e^{\operatorname{Re}\operatorname{Log} z}, \quad \theta = \operatorname{Arg} z = \operatorname{Im}\operatorname{Log} z \in (-\pi, \pi). \quad (10.5.3)$$

For $k \in \mathbb{Z}, z \in \mathbb{C}$, $(2k-1)\pi < \operatorname{Im} z < (2k+1)\pi$, $\operatorname{Log}(e^z) = z - 2ik\pi$. (10.5.4)

$$\textit{For } z \in \mathbb{C}\backslash\{-1\}, |z| \leq 1, \quad \operatorname{Log}(1+z) = \sum_{l \geq 1}(-1)^{l+1}\frac{z^l}{l}. \quad (10.5.5)$$

Logarithm of a nonsingular symmetric matrix

Let Υ_+ be the set of symmetric nonsingular $n \times n$ matrices with complex entries and non-negative real part. The set Υ_+ is star-shaped with respect to the Id: for $A \in \Upsilon_+$, the segment $[1, A] = \big((1-t)\operatorname{Id} + tA\big)_{t \in [0,1]}$ is obviously made with symmetric matrices with non-negative real part which are invertible, since for $0 \leq t < 1$, $\operatorname{Re}\big((1-t)\operatorname{Id} + tA\big) \geq (1-t)\operatorname{Id} > 0$ and for $t = 1$, A is assumed to be invertible[12]. We can now define for $A \in \Upsilon_+$,

$$\operatorname{Log} A = \int_0^1 (A - I)\big(I + t(A - I)\big)^{-1}dt. \quad (10.5.6)$$

We note that A commutes with $(I + sA)$ (and thus with $\operatorname{Log} A$), so that, for $\theta > 0$,

$$\frac{d}{d\theta}\operatorname{Log}(A + \theta I)$$

$$= \int_0^1 \big(I + t(A + \theta I - I)\big)^{-1}dt - \int_0^1 (A + \theta I - I)t\big(I + t(A + \theta I - I)\big)^{-2}dt,$$

and since

$$\frac{d}{dt}\Big\{\big(I + t(A + \theta I - I)\big)^{-1}\Big\} = -\big(I + t(A + \theta I - I)\big)^{-2}(A + \theta I - I),$$

we obtain by integration by parts $\frac{d}{d\theta}\operatorname{Log}(A + \theta I) = (A + \theta I)^{-1}$. As a result, we find that for $\theta > 0, A \in \Upsilon_+$, since all the matrices involved are commuting,

$$\frac{d}{d\theta}\Big((A + \theta I)^{-1}e^{\operatorname{Log}(A+\theta I)}\Big) = 0,$$

---

[12]If A is a $n \times n$ symmetric matrix with complex entries such that $\operatorname{Re} A$ is positive definite, then A is invertible: if $AX = 0$, then,

$$0 = \langle AX, \bar{X}\rangle = \langle A\operatorname{Re} X, \operatorname{Re} X\rangle + \langle A\operatorname{Im} X, \operatorname{Im} X\rangle + \overbrace{\langle A\operatorname{Re} X, -i\operatorname{Im} X\rangle + \langle Ai\operatorname{Im} X, \operatorname{Re} X\rangle}^{=0 \text{ since } A \text{ symmetric}}$$

and taking the real part give $\langle \operatorname{Re} A\operatorname{Re} X, \operatorname{Re} X\rangle + \langle \operatorname{Re} A\operatorname{Im} X, \operatorname{Im} X\rangle = 0$, implying $X = 0$ from the positive-definiteness of $\operatorname{Re} A$.

so that, using the limit $\theta \to +\infty$, we get[13] that

$$\forall A \in \Upsilon_+, \forall \theta > 0, \ e^{\mathrm{Log}(A+\theta I)} = (A + \theta I),$$

and by continuity

$$\forall A \in \Upsilon_+, \quad e^{\mathrm{Log}\, A} = A, \quad \text{which implies} \quad \det A = e^{\mathrm{trace}\,\mathrm{Log}\, A}. \tag{10.5.7}$$

Using (10.5.7), we can define for $A \in \Upsilon_+$,

$$(\det A)^{-1/2} = e^{-\frac{1}{2}\,\mathrm{trace}\,\mathrm{Log}\, A} = |\det A|^{-1/2} e^{-\frac{i}{2}\,\mathrm{Im}(\mathrm{trace}\,\mathrm{Log}\, A)}. \tag{10.5.8}$$

- When A is a positive definite matrix, $\mathrm{Log}\, A$ is real valued and $(\det A)^{-1/2} = |\det A|^{-1/2}$.
- When $A = -iB$ where B is a real nonsingular symmetric matrix, we note that $B = PD^tP$ with $P \in O(n)$ and D diagonal. We see directly on the formulas (10.5.6), (10.5.1) that

$$\mathrm{Log}\, A = \mathrm{Log}(-iB) = P(\mathrm{Log}(-iD))^tP, \quad \mathrm{trace}\,\mathrm{Log}\, A = \mathrm{trace}\,\mathrm{Log}(-iD),$$

and thus, with (μ_j) the (real) eigenvalues of B, we have $\mathrm{Im}\,(\mathrm{trace}\,\mathrm{Log}\, A) = \mathrm{Im}\sum_{1 \le j \le n} \mathrm{Log}(-i\mu_j)$, where the last Log is given by (10.5.1). Finally we get,

$$\mathrm{Im}\,(\mathrm{trace}\,\mathrm{Log}\, A) = -\frac{\pi}{2} \sum_{1 \le j \le n} \mathrm{sign}\,\mu_j = -\frac{\pi}{2}\,\mathrm{sign}\, B,$$

where $\mathrm{sign}\, B$ is the signature of B. As a result, we have when $A = -iB$, B real symmetric nonsingular matrix

$$(\det A)^{-1/2} = |\det B|^{-1/2} e^{i\frac{\pi}{4}\,\mathrm{sign}\, B}. \tag{10.5.9}$$

---

[13] We have $e^{\mathrm{Log}(A+\theta)} = (A + \theta)B_A$ and with $\tau = \theta - 1$,

$$e^{\mathrm{Log}(A+\theta)} e^{-\ln \theta} = e^{C_\theta}, \quad C_\theta = A \int_0^1 (1 + tA + t\tau)^{-1}(1 + t\tau)^{-1} dt.$$

For $t, \tau \in \mathbb{R}_+$, the matrix $1 + tA + t\tau$ is invertible (see the footnote on page 463) and we have

$\mathrm{Re}\langle(1 + tA + t\tau)X, X\rangle \ge (1 + t\tau)\|X\|^2$, so that this implies $\|(1 + tA + t\tau)X\| \ge (1 + t\tau)\|X\|$

and thus $\|(1 + tA + t\tau)^{-1}\| \le (1 + t\tau)^{-1}$. We get

$$\|C_\theta\| \le \|A\| \int_0^1 (1 + t\tau)^{-2} dt = \frac{\|A\|}{1 + \tau} \implies \lim_{\theta \to +\infty} C_\theta = 0$$

$$\implies B_A = \lim_{\theta \to +\infty} (A + \theta)B_A e^{-\ln \theta} = \lim_{\theta \to +\infty} e^{\mathrm{Log}(A+\theta)} e^{-\ln \theta} = \lim_{\theta \to +\infty} e^{C_\theta} = I.$$

The Γ function

For $z \in \mathbb{C}$ with a positive real part, we define

$$\Gamma(z) = \int_0^{+\infty} t^{z-1} e^{-t} dt. \tag{10.5.10}$$

Theorem 3.3.7 implies that Γ is a holomorphic function on the half-plane $\{\operatorname{Re} z > 0\}$, and for z there, an integration by parts yields

$$\Gamma(z+1) = \int_0^{+\infty} t^z e^{-t} dt = [t^z e^{-t}]_{+\infty}^0 + \int_0^{+\infty} z t^{z-1} e^{-t} dt = z\Gamma(z).$$

We get immediately that

$$\text{for } n \in \mathbb{N}, \ \Gamma(n+1) = n! \quad \text{and } \Gamma(1/2) = \sqrt{\pi}. \tag{10.5.11}$$

The latter equality follows from (8.1.31) since

$$\Gamma(1/2) = \int_0^{+\infty} s^{-1} e^{-s^2} 2s\, ds = \int_{\mathbb{R}} e^{-s^2} ds = \sqrt{\pi}.$$

For $\operatorname{Re} z > -1, z \neq 0$, we define $\Gamma(z) = \frac{\Gamma(z+1)}{z}$: it coincides with the previous definition if $\operatorname{Re} z > 0$ from the previous identity. Let $k \geq 1$ be an integer: we may define for $\operatorname{Re} z > -k, z \notin \{-k+1, \ldots, 0\}$,

$$\Gamma(z) = \frac{\Gamma(z+k)}{z(z+1)\ldots(z+k-1)}. \tag{10.5.12}$$

The Γ function appears as a meromorphic function on $\mathbb{C}$ with simple poles at $-\mathbb{N}$ such that

$$\operatorname{Res}(\Gamma, -k) = \frac{(-1)^k}{k!}, \tag{10.5.13}$$

and the following *functional equation* holds:

$$\forall z \notin (-\mathbb{N}), \quad \Gamma(z+1) = z\Gamma(z). \tag{10.5.14}$$

Theorem 3.3.7 implies for $\operatorname{Re} z > 0$,

$$\Gamma'(z) = \int_0^{+\infty} t^{z-1} e^{-t} \ln t\, dt, \quad \Gamma''(z) = \int_0^{+\infty} t^{z-1} e^{-t} (\ln t)^2 dt. \tag{10.5.15}$$

Lemma 10.5.2 (Gauss' formula). *For $z \in \mathbb{C}\backslash(-\mathbb{N})$, we have:*

$$\Gamma(z) = \lim_n \frac{n!\, n^z}{\prod_{0 \leq j \leq n}(z+j)}. \tag{10.5.16}$$

Proof. We assume first that $\operatorname{Re} z > 0$. Lebesgue's dominated convergence theorem induces for $\operatorname{Re} z > 0$ that

$$\int_0^n t^z \left(1 - \frac{t}{n}\right)^n \frac{dt}{t} \xrightarrow[n \to +\infty]{} \Gamma(z) :$$

we have indeed pointwise convergence of

$$\mathbf{1}_{[0,n]}(t) t^{z-1} \left(1 - \frac{t}{n}\right)^n$$

towards $\mathbf{1}_{\mathbb{R}_+}(t) t^{z-1} e^{-t}$ and domination

$$\left| \mathbf{1}_{[0,n]}(t) t^{z-1} \left(1 - \frac{t}{n}\right)^n \right| \le \mathbf{1}_{\mathbb{R}_+}(t) t^{\operatorname{Re} z - 1} e^{-t} \in L^1(\mathbb{R}),$$

since for $x \in [0,1)$, $\ln(1-x) \le -x$ implies $\mathbf{1}_{[0,n]}(t)\left(1 - \frac{t}{n}\right)^n \le e^{-n\frac{t}{n}} = e^{-t}$. We check now

$$(\flat) \qquad \int_0^n t^z \left(1 - \frac{t}{n}\right)^n \frac{dt}{t} = \int_0^1 s^{z-1} n^z (1-s)^n ds = n^z B(z, n+1),$$

where the so-called Beta-functionis defined for a, b complex numbers with $\operatorname{Re} a > 0, \operatorname{Re} b > 0$ by

$$B(a,b) = \int_0^1 t^{a-1}(1-t)^{b-1} dt. \tag{10.5.17}$$

The holomorphy of the Beta function on this domain of $\mathbb{C}^2$ ($\operatorname{Re} a > 0, \operatorname{Re} b > 0$) follows from Theorem 3.3.7. Moreover, we have with $x_+ = xH(x), H = \mathbf{1}_{\mathbb{R}_+}$,

$$x_+^{a-1} * x_+^{b-1} = \int_{\mathbb{R}} H(t) t^{a-1} H(x-t)(x-t)^{b-1} dt$$

$$= H(x) x^{a+b-1} \int_0^1 s^{a-1}(1-s)^{b-1} ds = x_+^{a+b-1} B(a,b),$$

so that multiplying both sides by e^{-x}, we find

$$\text{for } \operatorname{Re} a > 0, \operatorname{Re} b > 0, \quad \Gamma(a)\Gamma(b) = \Gamma(a+b)B(a,b). \tag{10.5.18}$$

On the other hand, we prove directly by induction on n that for $\operatorname{Re} z > 0, n \in \mathbb{N}$,

$$B(z, n+1) = n! \prod_{0 \le j \le n} (z+j)^{-1}.$$

It is true for $n = 0$ since $B(z, 1) = \int_0^1 t^{z-1} dt = 1/z$ and we have

$$B(z, n+2) = \int_0^1 t^{z-1}(1-t)^{n+1} dt$$

$$= \left[z^{-1} t^z (1-t)^{n+1} \right]_0^1 - \int_0^1 z^{-1} t^z (n+1)(1-t)^n dt(-1)$$

$$= (n+1)z^{-1} B(z+1, n+1) \underset{\substack{\text{induction} \\ \text{hypothesis}}}{=} (n+1)z^{-1} n! \prod_{0 \le j \le n} (z+1+j)^{-1}$$

$$= (n+1)! \prod_{0 \le k \le n+1} (z+k)^{-1}, \qquad\qquad \text{qed.}$$

Applying this to ($\flat$), we get

$$\underbrace{n^z B(z, n+1)}_{\text{with limit } \Gamma(z)} = n! n^z \prod_{0 \le j \le n} (z+j)^{-1},$$

proving the result of the lemma for $\operatorname{Re} z > 0$. The result for $z \in (-\mathbb{N})^c$ follows from (10.5.12): if $\operatorname{Re} z > -k$, we have

$$\Gamma(z) = \frac{\Gamma(z+k)}{\prod_{0 \le l < k}(z+l)} = \lim_n \frac{n! n^{z+k}}{\prod_{0 \le l < k}(z+l) \prod_{0 \le j \le n}(z+j+k)}$$

$$= \lim_n \frac{n! n^z}{\prod_{0 \le q \le n}(z+q)} n^k \prod_{n-k < j \le n} (z+j+k)^{-1},$$

and since $n^k \prod_{n-k < j \le n}(z+j+k)^{-1} = \prod_{1 \le r \le k} \frac{n}{z+n+r}$, we have

$$\lim_n n^k \prod_{n-k < j \le n} (z+j+k)^{-1} = 1,$$

entailing the result. The proof of the lemma is complete. $\qquad\square$

Lemma 10.5.3 (Weierstrass Formula). *The function $1/\Gamma$ is entire with simple zeroes located at $(-\mathbb{N})$ and we have the strictly convergent infinite product*

$$\Gamma(z)^{-1} = z e^{\gamma z} \prod_{1 \le j \le +\infty} \left(1 + \frac{z}{j} \right) e^{-z/j}. \qquad (10.5.19)$$

Proof. Starting from Lemma 10.5.2, we find for $z \in \mathbb{C} \backslash (-\mathbb{N})$,

$$\Gamma(z) = z^{-1} \lim_n e^{z(\ln n - \sum_{1 \le j \le n} \frac{1}{j})} \prod_{1 \le j \le n} j(z+j)^{-1} e^{z/j}.$$

From Exercise 2.8.20, we know that $\lim_n \left(\sum_{1 \le j \le n} \frac{1}{j} - \ln n \right) = \gamma$, the Euler–Mascheroni constant, so that

$$\Gamma(z) = z^{-1} e^{-\gamma z} \prod_{1 \le j \le +\infty} \left(1 + \frac{z}{j} \right)^{-1} e^{z/j}.$$

The convergence of the infinite product follows from the previous formula, but we can also see directly that, with the complex logarithm and $j > |z|$,

$$\mathrm{Log}\left(\left(1 + \frac{z}{j} \right)^{-1} e^{z/j} \right) = -\frac{z}{j} + \frac{z}{j} + O\left(\frac{z^2}{j^2} \right) = O(j^{-2}).$$

As a result, the Γ function vanishes nowhere and $1/\Gamma$ is an entire function whose zeroes are simple and located at $(-\mathbb{N})$:

$$\Gamma(z)^{-1} = z e^{\gamma z} \prod_{1 \le j \le +\infty} \left(1 + \frac{z}{j} \right) e^{-z/j}. \qquad \square$$

Lemma 10.5.4 (Log-convexity of the Γ function). *The Γ function is positive on $\mathbb{R}_+^*$ and is also log-convex.*

Proof. The Γ function never vanishes and is also non-negative on $(0, +\infty)$, thus is positive there. Moreover, Cauchy–Schwarz inequality and (10.5.15) imply for $x > 0$

$$\Gamma'(x)^2 = \langle t^{x/2}, t^{x/2} \ln t \rangle^2_{L^2(\mathbb{R}_+, e^{-t} dt/t)}$$
$$< \| t^{x/2} \|^2_{L^2(\mathbb{R}_+, e^{-t} dt/t)} \| t^{x/2} \ln t \|^2_{L^2(\mathbb{R}_+, e^{-t} dt/t)} = \Gamma(x)\Gamma''(x),$$

so that

$$\frac{d^2}{dx^2} (\ln \Gamma) = \frac{d}{dx}\left(\frac{\Gamma'}{\Gamma} \right) = \frac{\Gamma''\Gamma - \Gamma'^2}{\Gamma^2} > 0. \qquad \square$$

Note that the minimum of the Gamma function on the positive half-line is

$$0.8856031944108886 \cdots = \Gamma(1.461632144845406\ldots).$$

Lemma 10.5.5. *Let G be a positive function defined on $(0, +\infty)$ such that $G(1) = 1$, G is log-convex and satisfies $G(x+1) = xG(x)$ for all $x > 0$. Then $G = \Gamma$.*

Proof. For $x > 0, n \in \mathbb{N}^*$, we have with $g = \ln G$, $g(n) = (n-1)!$ and

$$g(x+n) - g(x) = \sum_{0 \le j < n} \left(g(x+j+1) - g(x+j) \right) = \sum_{0 \le j < n} \ln(x+j),$$

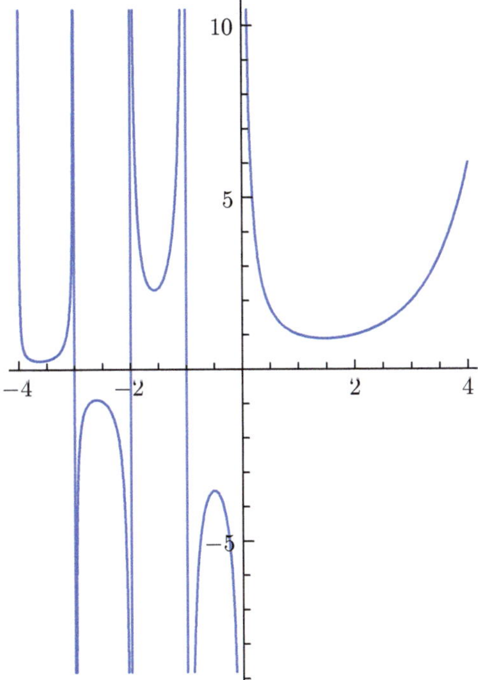

Figure 10.1: GAMMA FUNCTION ON THE REAL LINE

so that $g(x+n) - g(x) - g(n) = \ln x + \sum_{1 \le j \le n-1} \ln\left(\frac{x+j}{j}\right)$ and

$$g(x) + \ln x - x \ln n + \sum_{1 \le j \le n-1} \ln\left(\frac{x+j}{j}\right) = g(x+n) - g(n) - x \ln n. \quad (10.5.20)$$

Let $k \in \mathbb{N}^*$ with $k > x$: we have $n-1 < n < x+n < k+n$ and from the convexity of g, for $n \ge 2$,

$$\frac{g(n) - g(n-1)}{1} \le \frac{g(x+n) - g(n)}{x} \le \frac{g(n+k) - g(n)}{k} = \frac{\sum_{0 \le r < k} \ln(n+r)}{k},$$

so that

$$\ln\left(1 - \frac{1}{n}\right) \le \frac{g(x+n) - g(n) - x \ln n}{x} \le \frac{\sum_{0 \le r < k} \ln(1 + \frac{r}{n})}{k},$$

and thus $\lim_n \left(g(x+n) - g(n) - x \ln n\right)/x = 0$, which implies, thanks to (10.5.20),

$$g(x) = -\ln x + \lim_n \left(x \ln n + \sum_{1 \le j \le n-1} \ln\left(\frac{j}{x+j}\right)\right) = \ln \Gamma(x),$$

where the last equality follows from Gauss' formula (10.5.16). $\qquad \square$

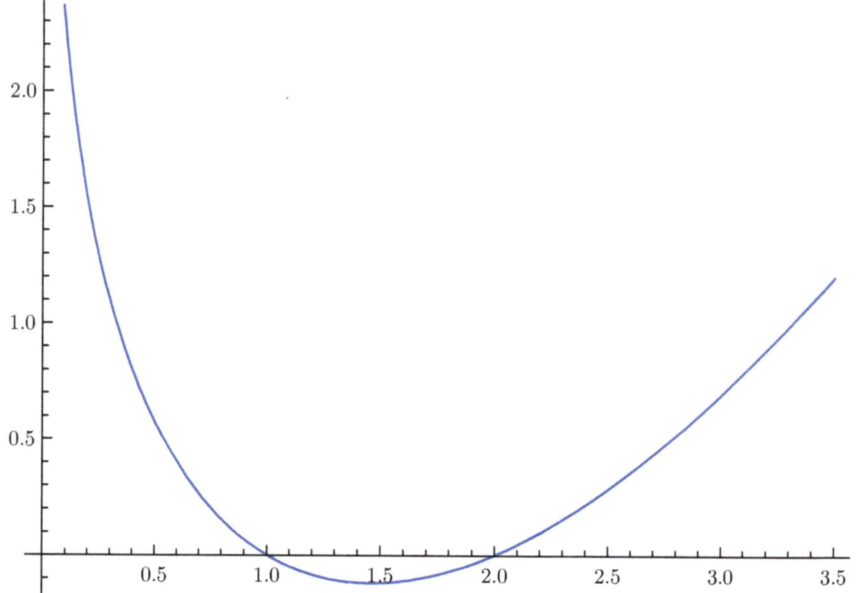

Figure 10.2: LOGARITHM OF THE GAMMA FUNCTION ON $(0, +\infty)$.

Wallis integrals

Lemma 10.5.6. *Let $q \in \mathbb{N}$. We have*

$$W_q = \int_0^{\pi/2} (\sin \theta)^q d\theta = \frac{\sqrt{\pi}\,\Gamma(\frac{q+1}{2})}{q\Gamma(\frac{q}{2})}, \quad \textit{i.e., for } p \in \mathbb{N}, \begin{cases} W_{2p} = \dfrac{\pi(2p)!}{(p!)^2 2^{2p+1}}, \\ W_{2p+1} = \dfrac{(p!)^2 2^{2p}}{(2p+1)!}. \end{cases}$$
$$(10.5.21)$$

This lemma follows from the next one.

Lemma 10.5.7. *Let $z \in \mathbb{C}$ such that $\mathrm{Re}\, z > -1$. Then*

$$\int_0^{\pi/2} (\sin \theta)^z d\theta = \frac{\sqrt{\pi}\,\Gamma(\frac{z+1}{2})}{2\Gamma(\frac{z+2}{2})}.$$

Proof. We have, with $t = \sin^2 \theta$,

$$2 \int_0^{\pi/2} (\sin \theta)^z d\theta = 2 \int_0^1 t^{z/2} (2 \sin \theta \cos \theta)^{-1} dt$$

$$= \int_0^1 t^{\frac{z-1}{2}} (1-t)^{-1/2} dt = B(\frac{z+1}{2}, 1/2) = \frac{\Gamma(\frac{z+1}{2})\Gamma(1/2)}{\Gamma(\frac{z+2}{2})},$$

where the last equality follows from (10.5.18). $\qquad\qquad\qquad\qquad\qquad\square$

Laplace equation in spherical coordinates

Lemma 10.5.8. *We have*

$$|x|^2 \Delta_{\mathbb{R}^d} = (r\partial_r)^2 + (d-2)r\partial_r + \Delta_{\mathbb{S}^{d-1}}, \tag{10.5.22}$$

where $\Delta_{\mathbb{S}^{d-1}}$ is the Laplace–Beltrami operator on the sphere $\mathbb{S}^{d-1}$.

Proof. In two dimensions, using the complex logarithm defined for $z \notin \mathbb{R}_-$ by (10.5.1) and polar coordinates

$$\begin{cases} x_1 = r\cos\theta \\ x_2 = r\sin\theta \end{cases}, \quad r > 0, |\theta| < \pi, \qquad \begin{cases} r = (x_1^2 + x_2^2)^{1/2} \\ \theta = \operatorname{Im}\operatorname{Log}(x_1 + ix_2) \end{cases}, \quad x_1 + ix_2 \notin \mathbb{R}_-,$$

we get

$$\frac{\partial}{\partial x_1} = \frac{\partial r}{\partial x_1}\frac{\partial}{\partial r} + \frac{\partial \theta}{\partial x_1}\frac{\partial}{\partial \theta} = \cos\theta\frac{\partial}{\partial r} - \frac{\sin\theta}{r}\frac{\partial}{\partial \theta},$$

$$\frac{\partial}{\partial x_2} = \frac{\partial r}{\partial x_2}\frac{\partial}{\partial r} + \frac{\partial \theta}{\partial x_2}\frac{\partial}{\partial \theta} = \sin\theta\frac{\partial}{\partial r} + \frac{\cos\theta}{r}\frac{\partial}{\partial \theta},$$

and a simple direct computation yields the two-dimensional result

$$r^2 \Delta_{\mathbb{R}^2} = (r\partial_r)^2 + \partial_\theta^2. \tag{10.5.23}$$

More generally, we get

$$\mathbb{S}^{d-1} \ni \sigma = \omega\sin\phi \oplus e_d\cos\phi, \quad \omega \in \mathbb{S}^{d-2}, \quad e_d = (0,\ldots,0,1), \quad 0 < \phi < \pi.$$

We consider the half-plane $x_d = r\cos\phi$, $\rho = r\sin\phi$, $0 < \phi < \pi$, and the two-dimensional (already proven) formula

$$r^2(\partial_{x_d}^2 + \partial_\rho^2) = (r\partial_r)^2 + \partial_\phi^2.$$

We have inductively for $d \geq 3$, $\rho^2 \Delta_{\mathbb{R}^{d-1}} = (\rho\partial_\rho)^2 + (d-3)\rho\partial_\rho + \Delta_{\mathbb{S}^{d-2}}$ and thus

$$r^2\partial_{x_d}^2 + r^2\partial_\rho^2 + r^2\Delta_{\mathbb{R}^{d-1}} = (r\partial_r)^2 + \partial_\phi^2 + r^2\rho^{-2}(\rho\partial_\rho)^2 + (d-3)r^2\rho^{-2}\rho\partial_\rho + r^2\rho^{-2}\Delta_{\mathbb{S}^{d-2}},$$

that is $r^2\Delta_{\mathbb{R}^d} = (r\partial_r)^2 + \partial_\phi^2 + (d-2)r^2\rho^{-1}\partial_\rho + \frac{\Delta_{\mathbb{S}^{d-2}}}{\sin^2\phi}$. Since

$$\frac{\partial}{\partial\rho} = \frac{\partial r}{\partial\rho}\frac{\partial}{\partial r} + \frac{\partial\phi}{\partial\rho}\frac{\partial}{\partial\phi} = \rho r^{-1}\partial_r + x_d r^{-2}\partial_\phi,$$

we get indeed

$$r^2\Delta_{\mathbb{R}^d} = (r\partial_r)^2 + (d-2)r\partial_r + \partial_\phi^2 + \frac{(d-2)}{\tan\phi}\partial_\phi + \frac{\Delta_{\mathbb{S}^{d-2}}}{\sin^2\phi}, \tag{10.5.24}$$

$$\Delta_{\mathbb{S}^{d-1}} = \partial_\phi^2 + \frac{(d-2)}{\tan\phi}\partial_\phi + \frac{\Delta_{\mathbb{S}^{d-2}}}{\sin^2\phi}. \tag{10.5.25}$$

$\square$

More calculations on the Laplace operator

In three dimensions, using the spherical coordinates

$$\begin{cases} x_1 = r\cos\theta\sin\phi \\ x_2 = r\sin\theta\sin\phi \\ x_3 = r\cos\phi \end{cases} \quad r > 0,\ 0 < \phi < \pi \text{ is the colatitude, } |\theta| < \pi \text{ is the longitude,}$$

we have

$$r^2\Delta_{\mathbb{R}^3} = (r\partial_r)^2 + r\partial_r + \partial_\phi^2 + \frac{1}{\sin^2\phi}\partial_\theta^2 + \frac{1}{\tan\phi}\partial_\phi, \tag{10.5.26}$$

which is also

$$r^2\Delta_{\mathbb{R}^3} = (r\partial_r)^2 + r\partial_r + \frac{1}{\sin^2\phi}\left((\sin\phi\,\partial_\phi)^2 + \partial_\theta^2\right).$$

In four dimensions, the spherical coordinates are

$$\begin{cases} x_1 = r\cos\theta\sin\phi_1\sin\phi_2 \\ x_2 = r\sin\theta\sin\phi_1\sin\phi_2 \\ x_3 = r\cos\phi_1\sin\phi_2 \\ x_4 = r\cos\phi_2 \end{cases} \quad 0 < \phi_1, \phi_2 < \pi, |\theta| \le \pi$$

and

$$r^2\Delta_{\mathbb{R}^4} = (r\partial_r)^2 + 2r\partial_r + \partial_{\phi_2}^2 + \frac{1}{\sin^2\phi_2}\left(\partial_{\phi_1}^2 + \frac{1}{\sin^2\phi_1}\partial_\theta^2 + \frac{1}{\tan\phi_1}\partial_{\phi_1}\right) + \frac{2}{\tan\phi_2}\partial_{\phi_2},$$

i.e.,

$$r^2\Delta_{\mathbb{R}^4} = (r\partial_r)^2 + 2r\partial_r + \partial_{\phi_2}^2 + \frac{\partial_{\phi_1}^2}{\sin^2\phi_2} + \frac{\partial_\theta^2}{\sin^2\phi_2\sin^2\phi_1} + \frac{\partial_{\phi_1}}{\sin^2\phi_2\tan\phi_1} + \frac{2\partial_{\phi_2}}{\tan\phi_2}. \tag{10.5.27}$$

In d dimensions, the spherical coordinates are

$$\begin{cases} x_1 = r\cos\theta\sin\phi_1\sin\phi_2\ldots\sin\phi_{d-3}\sin\phi_{d-2} \\ x_2 = r\sin\theta\sin\phi_1\sin\phi_2\ldots\sin\phi_{d-3}\sin\phi_{d-2} \\ x_3 = r\cos\phi_1\sin\phi_2\ldots\sin\phi_{d-3}\sin\phi_{d-2} \\ \ldots \\ x_{d-1} = r\cos\phi_{d-3}\sin\phi_{d-2} \\ x_d = r\cos\phi_{d-2} \end{cases} \quad 0 < \phi_j < \pi,\quad |\theta| < \pi.$$

We have

$$r^2 \Delta_{\mathbb{R}^d} = (r\partial_r)^2 + (d-2)r\partial_r$$

$$+ \partial^2_{\phi_{d-2}} + \frac{\partial^2_{\phi_{d-3}}}{\sin^2 \phi_{d-2}} + \cdots + \frac{\partial^2_{\phi_{d-j}}}{\sin^2 \phi_{d-2} \ldots \sin^2 \phi_{d-j+1}}$$

$$+ \cdots + \frac{\partial^2_\theta}{\sin^2 \phi_{d-2} \ldots \sin^2 \phi_1} + \frac{(d-2)}{\tan \phi_{d-2}} \partial_{\phi_{d-2}} + \frac{(d-3)}{\sin^2 \phi_{d-2} \tan \phi_{d-3}} \partial_{\phi_{d-3}}$$

$$+ \cdots + \frac{(d-j)\partial_{\phi_{d-j}}}{\sin^2 \phi_{d-2} \ldots \sin^2 \phi_{d-j+1} \tan \phi_{d-j}} + \cdots + \frac{\partial_{\phi_1}}{\sin^2 \phi_{d-2} \ldots \sin^2 \phi_2 \tan \phi_1}.$$

In other words, we have

$$\Delta_{\mathbb{S}^{d-1}} = \sum_{2 \le j \le d-1} \frac{\partial^2_{\phi_{d-j}}}{\sin^2 \phi_{d-2} \ldots \sin^2 \phi_{d-j+1}} + \frac{(d-j)\partial_{\phi_{d-j}}}{\sin^2 \phi_{d-2} \ldots \sin^2 \phi_{d-j+1} \tan \phi_{d-j}}$$

$$+ \frac{\partial^2_\theta}{\sin^2 \phi_{d-2} \ldots \sin^2 \phi_1}$$

so that, inductively, we verify

$$\Delta_{\mathbb{S}^d} = \sum_{2 \le j \le d} \frac{\partial^2_{\phi_{d+1-j}}}{\sin^2 \phi_{d-1} \ldots \sin^2 \phi_{d-j+2}}$$

$$+ \frac{(d+1-j)\partial_{\phi_{d+1-j}}}{\sin^2 \phi_{d-1} \ldots \sin^2 \phi_{d-j+2} \tan \phi_{d+1-j}} + \frac{\partial^2_\theta}{\sin^2 \phi_{d-1} \ldots \sin^2 \phi_1}$$

and indeed

$$\Delta_{\mathbb{S}^d} = \partial^2_{\phi_{d-1}} + \frac{d-1}{\tan \phi_{d-1}} \partial_{\phi_{d-1}} + \frac{1}{\sin^2 \phi_{d-1}} \Delta_{\mathbb{S}^{d-1}}.$$

Laplace–Beltrami operator

Let $(\mathcal{M}, g)$ be a Riemannian manifold of dimension n. We use the usual notation in a coordinate chart:

$$g = (g_{jk})_{1 \le j,k \le n},$$

is a symmetric positive definite matrix, with inverse matrix $g^{-1} = (g^{jk})_{1 \le j,k \le n}$,

$$ds^2 = \sum_{1 \le j,k \le n} g_{jk}(x)|dx^j||dx^k|, \quad |g| = \det g.$$

The Laplace–Beltrami operator is defined in a coordinate chart as

$$\Delta_g = |g|^{-1/2} \partial_j |g|^{1/2} g^{jk} \partial_k.$$

Note that for $u, v \in C_c^2(\mathcal{M})$, we have the selfadjointness property

$$\langle \Delta_g u, v \rangle = \langle u, \Delta_g v \rangle.$$

In fact, in a coordinate chart, we have

$$\langle \Delta_g u, v \rangle = \int \left(|g|^{-1/2} \partial_j |g|^{1/2} g^{jk} \partial_k u \right) \bar{v} |g|^{1/2} dx = - \int |g|^{1/2} g^{jk} \partial_k u \bar{\partial}_j v dx$$

$$= \int u \overline{|g|^{-1/2} \partial_k (|g|^{1/2} \underbrace{g^{jk}}_{=g^{kj}} \partial_j v)} |g|^{1/2} dx = \langle u, \Delta_g v \rangle. \qquad (10.5.28)$$

The Laplace–Beltrami operator on $\mathbb{S}^2$, with parameters θ, ϕ, $|\theta| < \pi, 0 < \phi < \pi$, is defined with

$$g = \begin{pmatrix} \sin^2 \phi & 0 \\ 0 & 1 \end{pmatrix}$$

and we recover the formula

$$\Delta_{\mathbb{S}^2} = (\sin \phi)^{-1} \left(\partial_\theta (\sin \phi)^{1-2} \partial_\theta + \partial_\phi (\sin \phi) \partial_\phi \right) = (\sin \phi)^{-2} \partial_\theta^2 + \partial_\phi^2 + \frac{1}{\tan \phi} \partial_\phi.$$

Looking at the Laplace–Beltrami operator on $\mathbb{S}^{d+1}$, we look at

$$\mathbb{S}^d \times (0, \pi) \ni (\omega, \phi) \mapsto \omega \sin \phi \oplus e_{d+1} \cos \phi \in \mathbb{S}^{d+1}$$

and we note that

$$g_{\mathbb{S}^d} = \begin{pmatrix} \sin^2 \phi \, g_{\mathbb{S}^{d-1}} & 0 \\ 0 & 1 \end{pmatrix}$$

so that

$$\Delta_{\mathbb{S}^d} = (\sin \phi)^{-d+1} \left((\sin \phi)^{d-1-2} \Delta_{\mathbb{S}^{d-1}} + \partial_\phi (\sin \phi)^{d-1} \partial_\phi \right)$$

$$= \partial_\phi^2 + (d-1)(\sin \phi)^{-d+1+d-2} \cos \phi \partial_\phi + (\sin \phi)^{-2} \Delta_{\mathbb{S}^{d-1}}$$

$$= \partial_\phi^2 + \frac{d-1}{\tan \phi} \partial_\phi + (\sin \phi)^{-2} \Delta_{\mathbb{S}^{d-1}}.$$

10.6   Classical volumes and areas

We have calculated in (4.5.4) the volume of the unit ball $\mathbb{B}^n$ of $\mathbb{R}^n$ as well as the $n-1$-dimensional "area" of the unit sphere $\mathbb{S}^{n-1}$ with Formula (5.4.8).

Cones in $\mathbb{R}^m$

We consider a measurable set $B \subset \mathbb{R}^{m-1}$ and a point $V = (0, h) \in \mathbb{R}^{m-1} \times \mathbb{R}$, $h > 0$. The cone of $\mathbb{R}^m$ with base B and vertex V is defined as

$$\Gamma(V, B) = \{X = (x, x_m) \in \mathbb{R}^{m-1} \times \mathbb{R}, \exists \lambda \geq 1, V + \lambda(X - V) \in B \times \{0\}\}.$$

This gives $\lambda x \in B$, $h + \lambda(x_m - h) = 0$, i.e., $\lambda = \frac{h}{h - x_m}$. The volume of $\Gamma(V, B)$ is

$$|\Gamma(V, B)|_m = \iint_{\frac{h}{h-x_m} x \in B, 0 \le x_m \le h} dx\, dx_m = \int_0^h |B|_{m-1} \left(\frac{h - x_m}{h}\right)^{m-1} dx_m$$

$$= |B|_{m-1} h^{-m+1} (m)^{-1} h^m,$$

that is

$$|\Gamma(V, B)|_m = \frac{|B|_{m-1} h}{m} = \frac{\text{base} \times \text{height}}{m}. \tag{10.6.1}$$

For a triangle in $\mathbb{R}^2$ ($m = 2$) or a cone in $\mathbb{R}^3$ ($m = 3$), we recover the classical formulas. Note that the cone $\Gamma(V, B)$ is the union of segments with endpoints V, $M \in B$:

$$X = (1 - \theta)V + \theta M, \quad M \in B \times \{0\}, \quad \theta \in [0, 1],$$

means that with $\lambda = \frac{1}{\theta}$,

$$V + \lambda(X - V) = V + \theta^{-1}((1 - \theta)V + \theta M - V) = M.$$

The converse follows from the fact that $B \ni M = V + \lambda(X - V)$ for some $\lambda \ge 1$ implies $X = \lambda^{-1} M + (1 - \lambda^{-1})V$.

Platonic polyhedra

Two-dimensional polygons

Before investigating the five 3-dimensional Platonic polyhedra, let us take a look at the simple two-dimensional situation. A regular polygon with k sides ($k \ge 3$) and circumscribed radius R has the area

$$A_k = \underbrace{k}_{\sharp \text{ sides}} \frac{1}{2} \underbrace{R}_{\text{base}} \underbrace{R \sin(\frac{2\pi}{k})}_{\text{height}}.$$

Note that this quantity goes to πR^2 when $k \to +\infty$. The length s of the side is $s = R|e^{2i\pi/k} - 1| = 2R\sin(\pi/k)$, so that we may define $A_k(s)$, the area of a regular polygon with k sides of length s as

$$A_k(s) = \frac{ks^2}{4\tan(\pi/k)}. \tag{10.6.2}$$

Also the perimeter $p_k = 2kR\sin(\pi/k)$ (a quantity going to $2\pi R$ when k goes to $+\infty$) and the apothem (distance from the center to a side) is

$$a_k = R|1 + e^{2i\pi/k}|\frac{1}{2} = R\cos(\pi/k).$$

We note that

$$A_k(s) = \frac{p_k a_k}{2} = \frac{2kR\sin(\pi/k)R\cos(\pi/k)}{2} = \frac{kR^2\sin(2\pi/k)}{2}. \tag{10.6.3}$$

Three-dimensional regular polyhedrons

- There are only five of them:

   Tetrahedron: 4 faces (equilateral triangles), 6 edges, 4 vertices.

   Cube: 6 faces (squares), 12 edges, 8 vertices.

   Octahedron: 8 faces (equilateral triangles), 12 edges, 6 vertices.

   Dodecahedron: 12 faces (regular pentagons), 30 edges, 20 vertices.

   Isosahedron: 20 faces (equilateral triangles), 30 edges, 12 vertices.

We want to compute their areas and their volumes, choosing as a parameter the length s of the edges. Denoting by $S_{N,k}(s)$ the area of the regular polyhedron with N faces, whose faces are regular $2D$ polygons with k sides of length s, we have

$$S_{N,k}(s) = N A_k(s). \qquad (10.6.4)$$

The apothem $a_{N,k}(s)$ is defined as the distance from the center to a face: we have, with $V_{N,k}(s)$ the volume of the regular polyhedron with N faces whose faces are regular $2D$ polygons with k sides of length s,

$$V_{N,k}(s) = N\frac{A_k(s)a_{N,k}(s)}{3} = \frac{a_{N,k}(s)S_{N,k}(s)}{3}. \qquad (10.6.5)$$

Since $S_{N,k}(s)$ is easy to determine with (10.6.4), the heart of the matter to find the volume is to determine the apothem. Note that the apothem is the radius of the inscribed sphere ($R_{N,k}(s)$ will stand for the radius of the circumscribed sphere).

- Cube, Octahedron, Tetrahedron with edge s.

   Area of the cube: $S_{6,4}(s) = 6s^2$,   Volume of the cube: $V_{6,4}(s) = s^3$.

   Area of the octahedron: $S_{8,3}(s) = 8A_3(s) = 8\dfrac{3s^2}{4\sqrt{3}} = 2\sqrt{3}s^2$,

   apothem of the octahedron (computed below), $a_{8,3}(s) = s/\sqrt{6}$,

   Volume of the octahedron: $V_{8,3}(s) = \dfrac{a_{8,3}(s)2\sqrt{3}s^2}{3} = s^3\dfrac{2}{\sqrt{3}\sqrt{6}} = s^3\dfrac{\sqrt{2}}{3}$.

We have indeed, calculating the center of a face,

$$a_{8,3}(s) = \frac{R}{3}\|(0,0,1) + (1,0,0) + (0,1,0)\| = R/\sqrt{3}, \quad 2R^2 = s^2,$$

where the last equality follows from the Pythagorean Theorem.

   Area of the tetrahedron: $S_{4,3}(s) = 4A_3(s) = \dfrac{4 \times 3s^2}{4\sqrt{3}} = s^2\sqrt{3}$,

   Volume of the tetrahedron: $V_{4,3}(s) = \dfrac{A_3(s)h}{3} = \dfrac{3s^3\sqrt{2}}{4\sqrt{3} \times 3\sqrt{3}} = \dfrac{s^3}{6\sqrt{2}}$,

with $h^2 + r^2 = s^2$ where $r = s/\sqrt{3}$ is the radius of the circumscribed cycle of the equilateral triangle with side s.

• Icosahedron, Dodecahedron. We start with the icosahedron. With coordinates in $\mathbb{C}\times\mathbb{R}$, the North pole is $V_0 = (0, R)$. Five vertices are issued from V_0 with endpoints $W_j = (re^{2i\pi j/5}, R - h)$, $j = 0, \ldots, 4$, where r is the radius of the circumscribed circle to the regular pentagon with sides s. We have

$$s^2 = r^2 + h^2, \quad r = \frac{s}{2\sin\pi/5}, \quad h^2 = s^2\left(1 - \frac{1}{4\sin^2(\pi/5)}\right).$$

The center of the face $V_0 W_0 W_{-1}$ is

$$\frac{1}{3}(2r\cos(\pi/5), R + 2(R - h)) \implies a^2 = \frac{1}{9}\left(4r^2\cos^2(\pi/5) + (3R - 2h)^2\right),$$

so that the apothem a of the icosahedron satisfies

$$a^2 = \frac{1}{9}\left(4s^2\frac{\cos^2(\pi/5)}{4\sin^2(\pi/5)} + 9R^2 + 4s^2\left(1 - \frac{1}{4\sin^2(\pi/5)}\right) - 12Rs\left(1 - \frac{1}{4\sin^2(\pi/5)}\right)^{\frac{1}{2}}\right).$$

We have also $R^2 = \|W_j\|^2 = r^2 + (R - h)^2$, so that $s^2 = r^2 + h^2 = 2Rh$ and

$$R = s\frac{\sin(\pi/5)}{\sqrt{4\sin^2(\pi/5) - 1}}.$$

We obtain

$$\begin{aligned}
a^2 &= s^2\frac{1}{9}\left(\frac{\cos^2(\pi/5)}{\sin^2(\pi/5)} + 9\frac{\sin^2(\pi/5)}{4\sin^2(\pi/5) - 1} + 4\left(1 - \frac{1}{4\sin^2(\pi/5)}\right) - 6\right) \\
&= s^2\frac{1}{9}\left(\frac{1}{\tan^2(\pi/5)} + 9\frac{\sin^2(\pi/5)}{4\sin^2(\pi/5) - 1} - \frac{1}{\sin^2(\pi/5)} - 2\right) \\
&= s^2\frac{1}{9}\left(9\frac{\sin^2(\pi/5)}{4\sin^2(\pi/5) - 1} - 3\right) = s^2\frac{1}{9}\left(\frac{3 - 3\sin^2(\pi/5)}{4\sin^2(\pi/5) - 1}\right) \\
&= s^2\frac{1}{3}\left(\frac{\cos^2(\pi/5)}{4\sin^2(\pi/5) - 1}\right).
\end{aligned}$$

Area of the icosahedron: $S_{20,3}(s) = 20A_3(s) = \dfrac{20 \times 3s^2}{4\sqrt{3}} = \dfrac{15s^2}{\sqrt{3}} = s^2 5\sqrt{3}$,

Volume of the icosahedron: $V_{20,3}(s) = \dfrac{a_{20,3}(s)S_{20,3}(s)}{3} = \dfrac{15s^3}{9}\dfrac{\cos(\pi/5)}{\sqrt{4\sin^2(\pi/5) - 1}}$,

so that[14]

$$V_{20,3}(s) = s^3\frac{5}{3}\frac{1}{\sqrt{3\tan^2(\pi/5) - 1}} = s^3\frac{5(3 + \sqrt{5})}{12}.$$

---

[14]We shall use that

$$\tan\pi/5 = \sqrt{5 - 2\sqrt{5}}, \quad \sin\pi/5 = \frac{\sqrt{2}\sqrt{5 - \sqrt{5}}}{4}, \quad \cos\pi/5 = \frac{1 + \sqrt{5}}{4}, \quad 3\tan^2(\pi/5) - 1 = (3 - \sqrt{5})^2.$$

Let us tackle finally the dodecahedron. This polyhedron is dual to the icosahedron: taking the five centers of the faces $V_0 W_j W_{j+1}, 0 \le j \le 4$, we get the top horizontal face of the dodecahedron so that the apothem of that dodecahedron is

$$\frac{1}{3}(R + 2(R - h)) = R - \frac{2h}{3} = s\frac{\sin(\pi/5)}{\sqrt{4\sin^2(\pi/5) - 1}} - \frac{2s}{3}\left(1 - \frac{1}{4\sin^2(\pi/5)}\right)^{1/2}$$

$$= \frac{s\sin(\pi/5)}{\sqrt{4\sin^2(\pi/5) - 1}} - \frac{s}{3\sin\pi/5}\sqrt{4\sin^2(\pi/5) - 1}.$$

However the length of the side of this dodecahedron is not s but

$$s' = \left\|\frac{1}{3}(2r\cos\pi/5, 3R - 2h) - \frac{1}{3}(e^{2i\pi/5}2r\cos\pi/5, 3R - 2h)\right\|$$

$$= \frac{2r\cos\pi/5 \times 2\sin\pi/5}{3} = \frac{s}{2\sin\pi/5}\frac{2\cos\pi/5 \times 2\sin\pi/5}{3} = s\frac{2\cos\pi/5}{3}.$$

As a result, we have

$$s^{-1}a_{12,5}(s) = \frac{\frac{3}{2\cos\pi/5}\sin(\pi/5)}{\sqrt{4\sin^2(\pi/5) - 1}} - \frac{\frac{3}{2\cos\pi/5}}{3\sin\pi/5}\sqrt{4\sin^2(\pi/5) - 1}$$

$$= \frac{\frac{3}{2}\tan(\pi/5)}{\sqrt{4\sin^2(\pi/5) - 1}} - \frac{\frac{3}{2}}{3\sin\pi/5}\sqrt{4\tan^2(\pi/5) - \cos^{-2}(\pi/5)}$$

$$= \frac{3}{2}\frac{\tan(\pi/5)}{\cos\pi/5\sqrt{3\tan^2(\pi/5) - 1}} - \frac{\sqrt{3\tan^2(\pi/5) - 1}}{2\sin\pi/5}$$

$$= \frac{\frac{3}{2}\tan(\pi/5)2\sin\pi/5 - (3\tan^2(\pi/5) - 1)\cos\pi/5}{2\sin\pi/5\cos\pi/5\sqrt{3\tan^2(\pi/5) - 1}}$$

$$= \frac{1}{2\sin\pi/5\sqrt{3\tan^2(\pi/5) - 1}}.$$

Area of the dodecahedron: $S_{12,5}(s) = 12A_5(s) = \dfrac{15s^2}{\tan\pi/5} = s^2 3\sqrt{5(5 + 2\sqrt{5})}$,

Volume of the dodecahedron: $V_{12,5}(s) = \dfrac{a_{12,5}(s)S_{12,5}(s)}{3}$, so that

$$V_{12,5}(s) = s^3\frac{15\cos\pi/5}{6\sin^2\pi/5\sqrt{3\tan^2(\pi/5) - 1}} = s^3\frac{15 + 7\sqrt{5}}{4}.$$

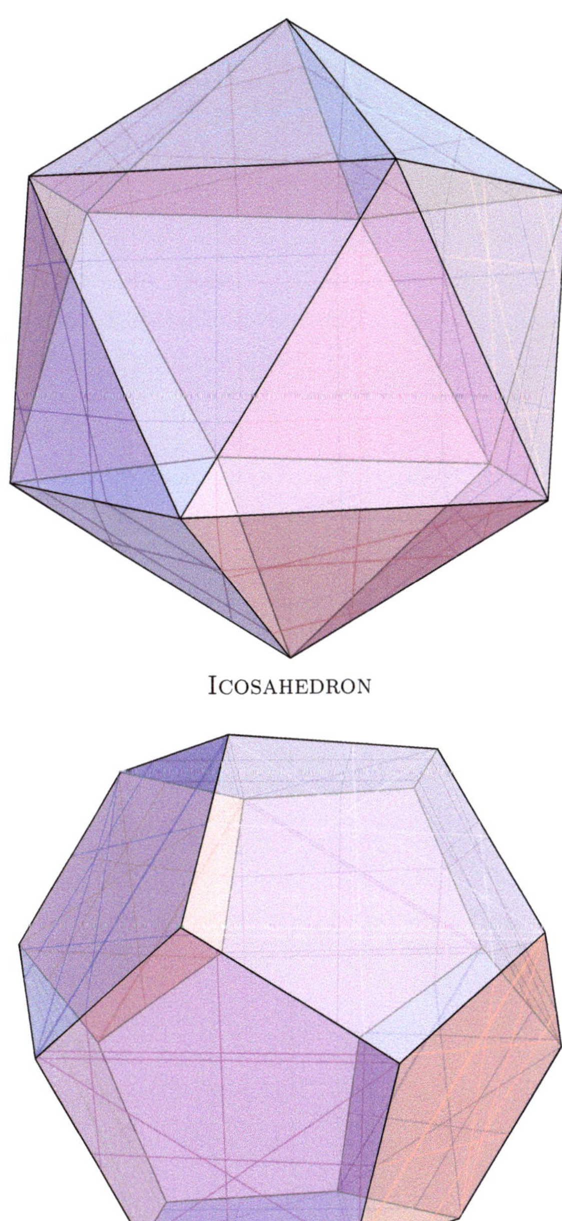

ICOSAHEDRON

DODECAHEDRON

Bibliography

[1] Luigi Ambrosio, *Corso introduttivo alla teoria geometrica della misura ed alle superfici minime*, Appunti dei Corsi Tenuti da Docenti della Scuola. [Notes of Courses Given by Teachers at the School], Scuola Normale Superiore, Pisa, 1997. MR 1736268 (2000k:49001)

[2] Luigi Ambrosio, Nicola Fusco, and Diego Pallara, *Functions of bounded variation and free discontinuity problems*, Oxford Mathematical Monographs, The Clarendon Press Oxford University Press, New York, 2000. MR 1857292 (2003a:49002)

[3] Tom M. Apostol, *Mathematical analysis*, second ed., Addison-Wesley Publishing Co., Reading, Mass.-London-Don Mills, Ont., 1974. MR 0344384 (49 #9123)

[4] ———, *Introduction to analytic number theory*, Springer-Verlag, New York, 1976, Undergraduate Texts in Mathematics. MR 0434929 (55 #7892)

[5] Hajer Bahouri, Jean-Yves Chemin, and Raphaël Danchin, *Fourier analysis and nonlinear partial differential equations*, Grundlehren der Mathematischen Wissenschaften [Fundamental Principles of Mathematical Sciences], vol. 343, Springer, Heidelberg, 2011. MR 2768550 (2011m:35004)

[6] Robert G. Bartle, *The elements of integration and Lebesgue measure*, Wiley Classics Library, John Wiley & Sons Inc., New York, 1995, Containing a corrected reprint of the 1966 original [it The elements of integration, Wiley, New York; MR0200398 (34 #293)], A Wiley-Interscience Publication. MR 1312157 (95k:28001)

[7] V.I. Bogachev, *Measure theory. Vol. I, II*, Springer-Verlag, Berlin, 2007. MR 2267655 (2008g:28002)

[8] Jean-Michel Bony, Gustave Choquet, and Gilles Lebeau, *Le centenaire de l'intégrale de Lebesgue*, C. R. Acad. Sci. Paris Sér. I Math. **332** (2001), no. 2, 85–90. MR 1813761 (2002a:28001)

[9] Nicolas Bourbaki, *Éléments de mathématique. Topologie générale. Chapitres 1 à 4*, Hermann, Paris, 1971. MR MR0358652 (50 #11111)

[10] ———, *Éléments de mathématique*, Hermann, Paris, 1976, Fonctions d'une variable réelle, Théorie élémentaire, Nouvelle édition. MR MR0580296 (58 #28327)

[11] ______, *Functions of a real variable*, Elements of Mathematics (Berlin), Springer-Verlag, Berlin, 2004, Elementary theory, Translated from the 1976 French original [MR0580296] by Philip Spain. MR 2013000

[12] ______, *Integration. I. Chapters* 1–6, Elements of Mathematics (Berlin), Springer-Verlag, Berlin, 2004, Translated from the 1959, 1965 and 1967 French originals by Sterling K. Berberian. MR 2018901 (2004i:28001)

[13] ______, *Integration. II. Chapters* 7–9, Elements of Mathematics (Berlin), Springer-Verlag, Berlin, 2004, Translated from the 1963 and 1969 French originals by Sterling K. Berberian. MR 2098271 (2005f:28001)

[14] ______, *Theory of sets*, Elements of Mathematics (Berlin), Springer-Verlag, Berlin, 2004, Reprint of the 1968 English translation [Hermann, Paris; MR0237342]. MR 2102219

[15] Haïm Brézis, *Analyse fonctionnelle*, Masson, 1983.

[16] Haïm Brézis and Elliott Lieb, *A relation between pointwise convergence of functions and convergence of functionals*, Proc. Amer. Math. Soc. **88** (1983), no. 3, 486–490. MR 699419 (84e:28003)

[17] Marc Briane and Gilles Pagès, *Le calcul intégral, licence de mathématiques*, Vuibert, Paris, France, 2000.

[18] Henri Buchwalter, *Théorie de l'intégration*, Ellipses, Paris, France, 1991.

[19] Donald L. Cohn, *Measure theory*, Birkhäuser Boston Inc., Boston, MA, 1993, Reprint of the 1980 original. MR 1454121 (98b:28001)

[20] Louis Comtet, *Analyse combinatoire. Tomes I, II*, Collection SUP: "Le Mathématicien", 4, vol. 5, Presses Universitaires de France, Paris, 1970. MR 0262087 (41 #6697)

[21] ______, *Advanced combinatorics*, enlarged ed., D. Reidel Publishing Co., Dordrecht, 1974, The art of finite and infinite expansions. MR 0460128 (57 #124)

[22] Jerôme Depauw, *Une construction de l'espace L^1 de Lebesgue*, Gaz. Math. (2011), no. 127, 5–13. MR 2791385

[23] Gerald B. Folland, *Real analysis*, second ed., Pure and Applied Mathematics (New York), John Wiley & Sons Inc., New York, 1999, Modern techniques and their applications, A Wiley-Interscience Publication. MR 1681462 (2000c:00001)

[24] Bernard R. Gelbaum, *Problems in real and complex analysis*, Problem Books in Mathematics, Springer-Verlag, New York, 1992. MR 1215048 (94d:00003)

[25] Bernard R. Gelbaum and John M.H. Olmsted, *Counterexamples in analysis*, The Mathesis Series, Holden-Day Inc., San Francisco, Calif., 1964. MR 0169961 (30 #204)

[26] ______, *Theorems and counterexamples in mathematics*, Problem Books in Mathematics, Springer-Verlag, New York, 1990. MR 1066872 (92b:00003)

[27] Casper Goffman and George Pedrick, *First course in functional analysis*, Prentice-Hall Inc., Englewood Cliffs, N.J., 1965. MR 0184064 (32 #1540)

[28] Paul R. Halmos, *Measure Theory*, D. Van Nostrand Company, Inc., New York, N. Y., 1950. MR 0033869 (11,504d)

[29] G.H. Hardy, *A mathematician's apology*, Canto, Cambridge University Press, Cambridge, 1992, With a foreword by C. P. Snow, Reprint of the 1967 edition. MR 1148590 (92j:01070)

[30] G.H. Hardy, J.E. Littlewood, and G. Pólya, *Inequalities*, Cambridge Mathematical Library, Cambridge University Press, Cambridge, 1988, Reprint of the 1952 edition. MR 944909 (89d:26016)

[31] Lars Hörmander, *The analysis of linear partial differential operators. I*, second ed., Grundlehren der Mathematischen Wissenschaften [Fundamental Principles of Mathematical Sciences], vol. 256, Springer-Verlag, Berlin, 1990, Distribution theory and Fourier analysis. MR 1065993 (91m:35001a)

[32] ______, *The analysis of linear partial differential operators. III*, Classics in Mathematics, Springer, Berlin, 2007, Pseudo-differential operators, Reprint of the 1994 edition. MR 2304165 (2007k:35006)

[33] ______, *Notions of convexity*, Modern Birkhäuser Classics, Birkhäuser Boston Inc., Boston, MA, 2007, Reprint of the 1994 edition. MR 2311920

[34] Richard A. Hunt, *On $L(p, q)$ spaces*, Enseignement Math. (2) **12** (1966), 249–276. MR 0223874 (36 #6921)

[35] Jean-Pierre Kahane, *Naissance et postérité de l'intégrale de Lebesgue*, Gaz. Math. (2001), no. 89, 5–20. MR 1858948 (2002g:01013)

[36] Sung Soo Kim, *Notes: A Characterization of the Set of Points of Continuity of a Real Function*, Amer. Math. Monthly **106** (1999), no. 3, 258–259. MR 1543429

[37] Anthony W. Knapp, *Basic real analysis*, Cornerstones, Birkhäuser Boston Inc., Boston, MA, 2005, Along with a companion volume ıt Advanced real analysis. MR 2155259 (2006c:26002)

[38] Jaroslav Kurzweil, *Henstock–Kurzweil integration: its relation to topological vector spaces*, Series in Real Analysis, vol. 7, World Scientific Publishing Co. Inc., River Edge, NJ, 2000. MR 1763305 (2001f:26008)

[39] Nicolas Lerner, *Intégration*, `http://www.math.jussieu.fr/~lerner/int.nte.html`, 2000.

[40] ______, *Lecture notes on real analysis*, `http://people.math.jussieu.fr/~lerner/realanalysis.lerner.pdf`, 2008.

[41] ______, *Metrics on the phase space and non-selfadjoint pseudo-differential operators*, Pseudo-Differential Operators. Theory and Applications, vol. 3, Birkhäuser Verlag, Basel, 2010. MR 2599384 (2011b:35002)

[42] Elliott H. Lieb, *Gaussian kernels have only Gaussian maximizers*, Invent. Math. **102** (1990), no. 1, 179–208. MR 1069246 (91i:42014)

[43] Elliott H. Lieb and Michael Loss, *Analysis*, second ed., Graduate Studies in Mathematics, vol. 14, American Mathematical Society, Providence, RI, 2001. MR 1817225 (2001i:00001)

[44] Lech Maligranda, *Marcinkiewicz interpolation theorem and Marcinkiewicz spaces*, Wiad. Mat. **48** (2012), no. 2, 157–171. MR 2986190

[45] Joseph Marcinkiewicz, *Sur l'interpolation d'opérations.*, C. R. Acad. Sci., Paris **208** (1939), 1272–1273 (French).

[46] Michel Métivier, *Probabilités*, École Polytechnique, Palaiseau, 1987, Dix leçons d'introduction. [Ten introductory lessons]. MR 971027 (89m:60001)

[47] Inder K. Rana, *An introduction to measure and integration*, Narosa Publishing House, London, 1997. MR 1645918 (99m:28002)

[48] Michael Reed and Barry Simon, *Methods of modern mathematical physics. I*, second ed., Academic Press Inc. [Harcourt Brace Jovanovich Publishers], New York, 1980, Functional analysis. MR MR751959 (85e:46002)

[49] Walter Rudin, *Real and complex analysis*, McGraw-hill, 1966.

[50] ———, *Functional analysis*, McGraw-hill, 1973.

[51] Laurent Schwartz, *Analyse I*, Hermann, Collection enseignement des sciences ♯42, Paris, 1998.

[52] ———, *Analyse II*, Hermann, Collection enseignement des sciences, ♯43, Paris, 1998.

[53] ———, *Analyse III*, Hermann, Collection enseignement des sciences, ♯44, Paris, 1998.

[54] ———, *Analyse IV*, Hermann, Collection enseignement des sciences, ♯45, Paris, 1998.

[55] Barry Simon, *Convexity*, Cambridge Tracts in Mathematics, vol. 187, Cambridge University Press, Cambridge, 2011, An analytic viewpoint. MR 2814377 (2012d:46002)

[56] E.M. Stein and J.-O. Strömberg, *Behavior of maximal functions in $\mathbf{R}^n$ for large n*, Ark. Mat. **21** (1983), no. 2, 259–269. MR 727348 (86a:42027)

[57] Elias M. Stein, *Harmonic analysis: real-variable methods, orthogonality, and oscillatory integrals*, Princeton Mathematical Series, vol. 43, Princeton University Press, Princeton, NJ, 1993, with the assistance of Timothy S. Murphy, Monographs in Harmonic Analysis, III. MR 1232192 (95c:42002)

[58] Elias M. Stein and Rami Shakarchi, *Complex analysis*, Princeton Lectures in Analysis, II, Princeton University Press, Princeton, NJ, 2003. MR 1976398 (2004d:30002)

[59] ———, *Fourier analysis*, Princeton Lectures in Analysis, vol. 1, Princeton University Press, Princeton, NJ, 2003, An introduction. MR 1970295 (2004a:42001)

[60] ———, *Real analysis*, Princeton Lectures in Analysis, III, Princeton University Press, Princeton, NJ, 2005, Measure theory, integration, and Hilbert spaces. MR 2129625 (2005k:28024)

[61] ———, *Functional analysis*, Princeton Lectures in Analysis, vol. 4, Princeton University Press, Princeton, NJ, 2011, Introduction to further topics in analysis. MR 2827930 (2012g:46001)

[62] Robert S. Strichartz, *A guide to distribution theory and Fourier transforms*, World Scientific Publishing Co. Inc., River Edge, NJ, 2003, Reprint of the 1994 original [CRC, Boca Raton; MR1276724 (95f:42001)]. MR 2000535

[63] Terence Tao, *An introduction to measure theory*, Graduate Studies in Mathematics, vol. 126, American Mathematical Society, Providence, RI, 2011. MR 2827917

[64] Michael E. Taylor, *Measure theory and integration*, Graduate Studies in Mathematics, vol. 76, American Mathematical Society, Providence, RI, 2006. MR 2245472 (2007g:28002)

[65] François Treves, *Topological vector spaces, distributions and kernels*, Academic Press, 1967.

[66] John von Neumann, *Functional Operators. I. Measures and Integrals*, Annals of Mathematics Studies, no. 21, Princeton University Press, Princeton, N. J., 1950. MR 0032011 (11,240f)

[67] Alan J. Weir, *Lebesgue integration and measure*, Cambridge University Press, London, 1973, Integration and measure, Vol. 1. MR 0480918 (58 #1067a)

[68] ———, *General integration and measure*, Cambridge University Press, London, 1974, Integration and measure, Vol. 2. MR 0480919 (58 #1067b)

Index